Lecture Notes in Mathematics

Edited by A. Dold, F. Takens and B. Teissier

Editorial Policy
for the publication of monographs

1. Lecture Notes aim to report new developments in all areas of mathematics – quickly, informally and at a high level. Monograph manuscripts should be reasonably self-contained and rounded off. Thus they may, and often will, present not only results of the author but also related work by other people. They may be based on specialized lecture courses. Furthermore, the manuscripts should provide sufficient motivation, examples and applications. This clearly distinguishes Lecture Notes from journal articles or technical reports which normally are very concise. Articles intended for a journal but too long to be accepted by most journals, usually do not have this "lecture notes" character. For similar reasons it is unusual for doctoral theses to be accepted for the Lecture Notes series.

2. Manuscripts should be submitted (preferably in duplicate) either to one of the series editors or to Springer-Verlag, Heidelberg. In general, manuscripts will be sent out to 2 external referees for evaluation. If a decision cannot yet be reached on the basis of the first 2 reports, further referees may be contacted: the author will be informed of this. A final decision to publish can be made only on the basis of the complete manuscript, however a refereeing process leading to a preliminary decision can be based on a pre-final or incomplete manuscript. The strict minimum amount of material that will be considered should include a detailed outline describing the planned contents of each chapter, a bibliography and several sample chapters.
Authors should be aware that incomplete or insufficiently close to final manuscripts almost always result in longer refereeing times and nevertheless unclear referees' recommendations, making further refereeing of a final draft necessary.
Authors should also be aware that parallel submission of their manuscript to another publisher while under consideration for LNM will in general lead to immediate rejection.

3. Manuscripts should in general be submitted in English.
Final manuscripts should contain at least 100 pages of mathematical text and should include
- a table of contents;
- an informative introduction, with adequate motivation and perhaps some historical remarks: it should be accessible to a reader not intimately familiar with the topic treated;
- a subject index: as a rule this is genuinely helpful for the reader.

Lecture Notes in Mathematics 1737

Editors:
A. Dold, Heidelberg
F. Takens, Groningen
B. Teissier, Paris

Springer
Berlin
Heidelberg
New York
Barcelona
Hong Kong
London
Milan
Paris
Singapore
Tokyo

Seiichiro Wakabayashi

Classical Microlocal Analysis in the Space of Hyperfunctions

Springer

Author

Seiichiro Wakabayashi
Institute of Mathematics
University of Tsukuba
Tsukuba-shi, Ibaraki 305-8571, Japan

E-mail: wkbysh@math.tsukuba.ac.jp

Cataloging-in-Publication Data applied for

Die Deutsche Bibliothek - CIP-Einheitsaufnahme

Wakabayashi, Seiichiro:
Classical microloca analysis in the space of hyperfunctions /
Seiichiro Wakabayashi. - Berlin ; Heidelberg ; New York ; Barcelona ;
Hong Kong ; London ; Milan ; Paris ; Singapore ; Tokyo : Springer,
2000
(Lecture notes in mathematics ; 1737)
ISBN 3-540-67603-1

Mathematics Subject Classification (2000): 35-02, 35S05, 35S30, 35A27, 35A20, 35A07, 35H10, 35A21

ISSN 0075-8434
ISBN 3-540-67603-1 Springer-Verlag Berlin Heidelberg New York

Typesetting: Camera-ready TEX output by the author
Printed on acid-free paper SPIN: 10724347 41/3143/du 543210

Preface

Many author have studied the theory of hyperfunctions from the viewpoint of "Algebraic Analysis," which is not necessarily accessible to us, studying partial differential equations (P.D.E.) in the framework of distributions. The treatment there is considably different from ours. Although we think that it is natural to work in the space of hyperfunctions for the purpose of studying P.D.E. with analytic coefficients, we do not think that "Algebraic Analysis" is indispensable for this purpose. We want to apply various methods in the framework of distributions to the studies on P.D.E. with analytic coefficients. In so doing the major difficulty is not to be able to use the "cut-off" technique. For there is obviously no non-trivial real analytic function with compact support. We shall use here "cut-off" operators (pseudodifferential operators) instead of "cut-off" functions, which map real analytic functions and hyperfunctions to real analytic functions and hyperfunctions, respectively.

In this lecture notes we attempt to establish "Classical Microlocal Analysis" in the space of hyperfunctions (or in a rather wider class of functions) which makes it possible to apply the methods in the C^∞-distribution category to the studies on P.D.E. in the hyperfunction category. Here "Classical Microlocal Analysis" means that it does not use "Algebraic Analysis" and that it is very similar to microlocal analysis in the C^∞-distribution category. Our main tool is, in some sense, integration by parts, which is equivalent to the fundamental theorem of the infinitesimal calculus. In our direction there are two books. One of them is Hörmander's book [Hr5] which gives a short introduction to the theory of hyperfunctions. The other is Treves' book [Tr2]. Treves developed in [Tr2] the theory of analytic pseudodifferential operators in the framework of distributions, which had been studied by Boutet de Monvel and Kree [BK]. On the basis of the methods in these two books, we shall establish "Classical Microlocal Analysis" in the space of hyperfunctions.

Some parts of this lecture notes are simple generalizations of the results obtained in joint work with Prof. Kajitani, and I would like to thank him for many useful discussions.

Contents

Introduction

Let $\mathcal{A}$ be the space of entire analytic functions on $\boldsymbol{C}^n$. An analytic functional is a continuous linear functional on $\mathcal{A}$ with usual topology. We say that an analytic functional u is carried by a compact subset K of $\boldsymbol{C}^n$, *i.e.*, $u \in \mathcal{A}'(K)$, if for any neighborhood ω of K in $\boldsymbol{C}^n$ there is $C_\omega \geq 0$ such that

$$|u(\varphi)| \leq C_\omega \sup_{z \in \omega} |\varphi(z)| \quad \text{for } \varphi \in \mathcal{A}.$$

We denote

$$\mathcal{A}' := \bigcup_{K \Subset \boldsymbol{R}^n} \mathcal{A}'(K).$$

The space $\mathcal{A}'$ of analytic functionals carried by $\boldsymbol{R}^n$ is very similar to the space $\mathcal{D}'$ of distributions, particularly to $\mathcal{E}'$. One can identify $\mathcal{E}'$ with a subspace of $\mathcal{A}'$, and anlytic functionals have compact "supports." For $u \in \mathcal{A}'$ we can define supp u by supp $u = \bigcap\{K;\ u \in \mathcal{A}'(K)\}$, which is called the support of u. The concept of "support" is relating to restriction mappings and sheaves. $\mathcal{A}'$ defines the sheaf $\mathcal{B}$ of hyperfunctions while $\mathcal{E}'$ does the sheaf $\mathcal{D}'$ of distributions. In order to study partial differential equations (P.D.E.) in the space of hyperfunctions it is usually sufficient to consider problems in $\mathcal{A}'$ (or $\mathcal{F}_0$ defined below). For $u \in \mathcal{A}'$ we can define the Fourier transform $\hat{u}(\xi)$ of u by

$$\hat{u}(\xi) \equiv \mathcal{F}[u](\xi) := u(e^{-iz\cdot\xi}).$$

Therefore, we can formally define pseudodifferential operators $p(x, D)$ with appropriate symbols as

$$p(x,D)u := (2\pi)^{-n} \int e^{ix\cdot\xi} p(x,\xi)\hat{u}(\xi)\, d\xi. \tag{0.1}$$

However, $p(x, D)u$ does not always belong to $\mathcal{A}'$ if $p(x, \xi)$ is not a polynomial of ξ. In microlocal analysis pseudodifferential operators play essential roles. So we need the corresponding spaces to the Schwartz spaces $\mathcal{S}$

and $\mathcal{S}'$. For $\varepsilon \in \boldsymbol{R}$ we put

$$\widehat{\mathcal{S}}_\varepsilon := \{v(\xi) \in C^\infty;\ e^{\varepsilon\langle\xi\rangle}v(\xi) \in \mathcal{S}\}.$$

We introduce the topology to $\widehat{\mathcal{S}}_\varepsilon$ in a standard way. Then the dual space $\widehat{\mathcal{S}}_\varepsilon{}'$ of $\widehat{\mathcal{S}}_\varepsilon$ is given by

$$\widehat{\mathcal{S}}_\varepsilon{}' = \{v(\xi) \in \mathcal{D}';\ e^{-\varepsilon\langle\xi\rangle}v(\xi) \in \mathcal{S}'\}.$$

If $\varepsilon \geq 0$, then $\widehat{\mathcal{S}}_\varepsilon$ is a dense subset of $\mathcal{S}$ and we can define $\mathcal{S}_\varepsilon := \mathcal{F}^{-1}[\widehat{\mathcal{S}}_\varepsilon]$ ($\subset \mathcal{S}$). By duality we can define the transposed operators ${}^t\mathcal{F}$, ${}^t\mathcal{F}^{-1}$: $\mathcal{S}_\varepsilon{}' \overset{\sim}{\to} \widehat{\mathcal{S}}_\varepsilon{}'$ for $\varepsilon \geq 0$. We rewrite ${}^t\mathcal{F} = \mathcal{F}$ since ${}^t\mathcal{F} = \mathcal{F}$ on $\mathcal{S}'$. Noting that $\widehat{\mathcal{S}}_\varepsilon{}' \supset \widehat{\mathcal{S}}_{-\varepsilon}$ for $\varepsilon \geq 0$, we define

$$\mathcal{S}_{-\varepsilon} := \mathcal{F}^{-1}[\widehat{\mathcal{S}}_{-\varepsilon}] \quad \text{for } \varepsilon \geq 0.$$

Moreover, $\mathcal{S}_{-\varepsilon}'$ is defined as the dual space of $\mathcal{S}_{-\varepsilon}$ when $\varepsilon \geq 0$. We define $\mathcal{E}_0 := \bigcap_{\varepsilon>0} \mathcal{S}_{-\varepsilon}$ and $\mathcal{F}_0 := \bigcap_{\varepsilon>0} \mathcal{S}_\varepsilon{}'$. From estimates of the Fourier transforms of analytic functoinals we see that

$$\mathcal{E}' \subset \mathcal{A}' \subset \mathcal{E}_0 \subset \mathcal{F}_0.$$

Let Γ be an open conic subset of $\boldsymbol{R}^n \times (\boldsymbol{R}^n \setminus \{0\})$. We say that $p(x,\xi) \in C^\infty(\Gamma)$ is an analytic symbol in Γ if $p(x,\xi)$ satisfies the estimates

$$\left|\partial_\xi^\alpha D_x^\beta p(x,\xi)\right| \leq CA^{|\alpha|+|\beta|}|\alpha|!|\beta|!\langle\xi\rangle^{m-|\alpha|}$$

for $(x,\xi) \in \Gamma$ with $|\xi| \geq R$ and $\alpha, \beta \in \boldsymbol{Z}_+^n$. If $p(x,\xi)$ is an analytic symbol in $\boldsymbol{R}^n \times (\boldsymbol{R}^n \setminus \{0\})$, $p(x,D)$ defined by (0.1) maps $\mathcal{E}_0$ and $\mathcal{F}_0$ to $\mathcal{E}_0$ and $\mathcal{F}_0$, respectively. However, we can not define $p(x,D)$ as an operator on $\mathcal{E}_0$ and $\mathcal{F}_0$ by (0.1) if $p(x,\xi)$ is an analytic symbol in Γ and Γ does not coincide with $\boldsymbol{R}^n \times (\boldsymbol{R}^n \setminus \{0\})$. We do not want to abandon (0.1) as the definition of $p(x,D)$. So we introduce some symbol classes which contain symbols with compact supports. We say that a symbol $a(\xi,y,\eta) \in C^\infty(\boldsymbol{R}^n \times \boldsymbol{R}^n \times \boldsymbol{R}^n)$ belongs to $S^{m_1,m_2,\delta_1,\delta_2}(R,A)$ if $a(\xi,y,\eta)$ satisfies

$$\begin{aligned}\left|\partial_\xi^{\alpha+\tilde\alpha} D_y^{\beta^1+\beta^2+\tilde\beta} \partial_\eta^{\gamma+\tilde\gamma} a(\xi,y,\eta)\right| &\leq C_{|\tilde\alpha|+|\tilde\beta|+|\tilde\gamma|} \\ &\times (A/R)^{|\alpha|+|\beta^1|+|\beta^2|+|\gamma|}\langle\xi\rangle^{m_1-|\tilde\alpha|+|\beta^1|}\langle\eta\rangle^{m_2-|\tilde\gamma|+|\beta^2|} \\ &\times \exp[\delta_1\langle\xi\rangle + \delta_2\langle\eta\rangle]\end{aligned}$$

if $\langle\xi\rangle \geq R(|\alpha| + |\beta^1|)$ and $\langle\eta\rangle \geq R(|\beta^2| + |\gamma|)$. For $a(\xi,y,\eta) \in S^{m_1,m_2,\delta_1,\delta_2}(R,A)$ we define the pseudodifferential operator $a(D_x,y,D_y)$ by

$$a(D_x,y,D_y)u := (2\pi)^{-n}\mathcal{F}_\xi^{-1}\Big[\int\Big(\int e^{iy\cdot(\eta-\xi)}a(\xi,y,\eta)\hat u(\eta)\,d\eta\Big)\,dy\Big](x)$$

for $u \in \mathcal{S}_\infty \equiv \bigcap_\varepsilon \mathcal{S}_\varepsilon$. Then we have

$$a(D_x, y, D_y) : \begin{cases} \mathcal{S}_{2\varepsilon_+ + \delta_2} \to \mathcal{S}_{\varepsilon-\delta_1}, \\ \mathcal{S}_{-\varepsilon+\delta_2} \to \mathcal{S}_{-2\varepsilon_+ - \delta_1}, \\ \mathcal{S}'_{\varepsilon-\delta_1} \to \mathcal{S}'_{2\varepsilon_+ + \delta_2}, \\ \mathcal{S}'_{-2\varepsilon_+ - \delta_1} \to \mathcal{S}'_{-\varepsilon+\delta_2} \end{cases}$$

if $R \geq 2enA$ and $\varepsilon \leq 1/R$ (see Theorem 2.3. 3 below). In particular, $a(D_x, y, D_y)$ maps continuously $\mathcal{F}_0$ to $\mathcal{F}_0$ if $\delta_1 = \delta_2 = 0$ and $R \geq 2enA$. Therefore, we get "cut-off" operators, although R must be chosen to be large at each step of the calculation. We do not fix one "cut-off" symbol and consider a family of "cut-off" symbols depending on R. This is a disadvantage in comparison with usual calculus in the ditribution category. However, we can overcome this disadvantage in most cases. Using "cut-off" operators we can define pseudodifferential operators and Fourier integral operators acting on the spaces (or the sheaves) of hyperfunctions and microfunctions. Since we must deal with operators with non-analytic symbols, the proof of the product formulas of pseudodifferential operators (and Fourier integral operators) becomes longer than usual one. This is another disadvantage of our methods. However, as a consequence, we obtain the same symbol calculus as usual one.

For $u \in \mathcal{F}_0$ the analytic wave front set $WF_A(u)$ ($\subset T^*\boldsymbol{R}^n \setminus 0$) of u is defined as follows: $(x^0, \xi^0) \in T^*\boldsymbol{R}^n \setminus 0$ does not belong to $WF_A(u)$ if there are a conic neighborhood Γ of ξ^0, $R_0 > 0$ and $\{g^R(\xi)\}_{R \geq R_0} \subset C^\infty(\boldsymbol{R}^n)$ such that $g^R(\xi) = 1$ in $\Gamma \cap \{\langle\xi\rangle \geq R\}$,

$$\left|\partial_\xi^{\alpha+\tilde{\alpha}} g^R(\xi)\right| \leq C_{|\tilde{\alpha}|}(C/R)^{|\alpha|}\langle\xi\rangle^{-|\tilde{\alpha}|}$$

if $\langle\xi\rangle \geq R|\alpha|$, and $g^R(D)u$ is analytic at x^0 for $R \geq R_0$. The precise definition that $g^R(D)u$ is analytic at x^0 will be given in Definition 1.2. 8 . Our definition of $WF_A(u)$, of course, coincides with usual definitions. Our definition of $WF_A(u)$ is very similar to the definition of the wave front set of distributions. Therefore, we can study P.D.E. in the hyperfunction category in the almost same way as in the distribution category. Our aim here is to provide microlocal analysis in the space of hyperfunctions in the same way as for distributions. As applications we shall consider microlocal uniqueness and local solvability in the last two chapters. These are still basic problems in the theory of linear partial differential operators. It is well-known in the framework of C^∞ and distributions that Carleman type estimates play an essential role in microlocal versions of the Holmgren uniqueness theorem. This is also true in the framework of analytic functions and hyperfunctions. General criteria on microlocal uniqueness will be given in Chapter 4. Microlocal uniqueness yields results not merely on propagation of analytic singularities but on analytic hypoellipticity. We can also apply the same arguments to the studies on local solvability in the framework of hyperfunctions as in the framework of distributions. We shall prove in Chapter 5 that ${}^t p(x, D)$ (resp. $p(x, D)$) satisfies energy estimates if $p(x, D)$ is locally solvable (resp. analytic hypoelliptic). We shall also show that a little strengthen estimates guarantee local solvability. So the problems on microlocal uniqueness, analytic hypoellipticity and local solvability will be reduced to the problems to derive energy estimates (or *a priori* estimates), which was carried out in the framework of C^∞ and distributions by us.

We should remark that Sjöstrand studied P.D.E. in the framework of analytic functions and distributions in [Sj], using the FBI transformation. It may be possible to deal with hyperfunctions by his methods. Using *a priori* estimates he got many remarkable results. However, we think that his theory is different from usual microlocal analysis in the distribution category, although it is new and powerful. So we will establish microlocal analysis in the space of hyperfunctions which is very similar to microlocal analysis in the framework of C^∞ and distributions.

Chapter 1

Hyperfunctions

In this chapter we shall introduce the function spaces $\mathcal{S}_\varepsilon$ and $\mathcal{S}_\varepsilon{}'$ corresponding to the Schwartz spaces $\mathcal{S}$ and $\mathcal{S}'$, respectively. These spaces play a key role in our calculus. The spaces $\mathcal{S}_{-\varepsilon}$ and $\mathcal{S}_\varepsilon{}'$ ($\varepsilon > 0$) include the space $\mathcal{A}'$ of analytic functionals. We shall define the supports and the restrictions of functions belonging to these spaces. Hyperfunctions (in a bounded open subset of $\boldsymbol{R}^n$) will be defined as residue classes of analytic functionals after the manner of Hörmander's book [Hr5] in Section 1.4. We shall prove that the presheaf $\mathcal{B}$ of hyperfunctions is a flabby sheaf. We shall also prove flabbiness of the quotient sheaf $\mathcal{B}/\mathcal{A}$ of $\mathcal{B}$ by the sheaf $\mathcal{A}$ of real analytic functions in Section 1.5.

1.1 Function spaces

Let $\varepsilon \in \boldsymbol{R}$, and denote $\langle\xi\rangle = (1+|\xi|^2)^{1/2}$, where $(\xi_1,\cdots,\xi_n) \in \boldsymbol{R}^n$ and $|\xi| = (\sum_{j=1}^n |\xi_j|^2)^{1/2}$. We define

$$\widehat{\mathcal{S}}_\varepsilon := \{v(\xi) \in C^\infty(\boldsymbol{R}^n);\ e^{\varepsilon\langle\xi\rangle}v(\xi) \in \mathcal{S}\}.$$

Here $\mathcal{S}$ denotes the Schwartz space. We introduce a family of seminorms on $\widehat{\mathcal{S}}_\varepsilon$ as follows:

$$|v|_{\widehat{\mathcal{S}}_\varepsilon,\ell} := |e^{\varepsilon\langle\xi\rangle}v(\xi)|_{\mathcal{S},\ell}$$
$$\Big(= \sup_{k+|\alpha|\le\ell}\ \sup_{\xi\in R^n} \langle\xi\rangle^k |D_\xi^\alpha e^{\varepsilon\langle\xi\rangle}v(\xi)|\Big) \quad (\ \ell = 0,1,2,\cdots),$$

where $\alpha = (\alpha_1,\cdots,\alpha_n) \in \boldsymbol{Z}_+^n$ ($= (\boldsymbol{N}\cup\{0\})^n$), $|\alpha| = \sum_{j=1}^n \alpha_j$, $D_\xi = (D_{\xi_1},\cdots,D_{\xi_n}) = -i\partial_\xi = -i(\partial/\partial\xi_1,\cdots,\partial/\partial\xi_n)$ and $D_\xi^\alpha = D_{\xi_1}^{\alpha_1}\cdots D_{\xi_n}^{\alpha_n}$.

Since $\mathcal{D}$ ($= C_0^\infty(\boldsymbol{R}^n)$) is dense in $\widehat{\mathcal{S}}_\varepsilon$, the dual space $\widehat{\mathcal{S}}_\varepsilon'$ of $\widehat{\mathcal{S}}_\varepsilon$ can be identified with $\{e^{\varepsilon\langle\xi\rangle}v(\xi) \in \mathcal{D}';\ v \in \mathcal{S}'\}$. For $\varepsilon \geq 0$ we define

$$\mathcal{S}_\varepsilon := \mathcal{F}^{-1}[\widehat{\mathcal{S}}_\varepsilon]\ (= \mathcal{F}[\widehat{\mathcal{S}}_\varepsilon] = \{u \in \mathcal{S};\ e^{\varepsilon\langle\xi\rangle}\hat{u}(\xi) \in \mathcal{S}\}),$$

where $\mathcal{F}$ and $\mathcal{F}^{-1}$ denote the Fourier transformation and the inverse Fourier transformation on $\mathcal{S}$ (or $\mathcal{S}'$), respectively, and $\hat{u}(\xi) = \mathcal{F}[u](\xi)$. We introduce the topology in $\mathcal{S}_\varepsilon$ so that $\mathcal{F} : \widehat{\mathcal{S}}_\varepsilon \longrightarrow \mathcal{S}_\varepsilon$ is homeomorphic. Denote by $\mathcal{S}_\varepsilon'$ the dual space of $\mathcal{S}_\varepsilon$ for $\varepsilon \geq 0$. Since $\mathcal{S}_\varepsilon$ is dense in $\mathcal{S}$ for $\varepsilon \geq 0$, we can regard $\mathcal{S}'$ as a subspace of $\mathcal{S}_\varepsilon'$. Then we can define the transposed operators ${}^t\mathcal{F}$ and ${}^t\mathcal{F}^{-1}$ of $\mathcal{F}$ and $\mathcal{F}^{-1}$, which map $\mathcal{S}_\varepsilon'$ and $\widehat{\mathcal{S}}_\varepsilon'$ onto $\widehat{\mathcal{S}}_\varepsilon'$ and $\mathcal{S}_\varepsilon'$, respectively. Since $\widehat{\mathcal{S}}_{-\varepsilon} \subset \widehat{\mathcal{S}}_\varepsilon'$ ($\subset \mathcal{D}'$) for $\varepsilon \geq 0$, we can define $\mathcal{S}_{-\varepsilon} := {}^t\mathcal{F}^{-1}[\widehat{\mathcal{S}}_{-\varepsilon}]$ for $\varepsilon \geq 0$, and introduce the topology so that ${}^t\mathcal{F}^{-1} : \widehat{\mathcal{S}}_{-\varepsilon} \rightarrow \mathcal{S}_{-\varepsilon}$ is homeomorphic. $\mathcal{S}_{-\varepsilon}'$ denotes the dual space of $\mathcal{S}_{-\varepsilon}$ for $\varepsilon \geq 0$. Then we have $\mathcal{S}_{-\varepsilon}' = \mathcal{F}[\widehat{\mathcal{S}}_{-\varepsilon}'] \subset \mathcal{S}' \subset \mathcal{S}_\varepsilon'$ for $\varepsilon \geq 0$ and $\mathcal{F} = {}^t\mathcal{F}$ on $\mathcal{S}'$. So we write ${}^t\mathcal{F}$ as $\mathcal{F}$. Note that $\mathcal{S}_\varepsilon$ is a Fréchet space with the topology determined by the seminorms $|u|_{\mathcal{S}_\varepsilon,\ell} := |\mathcal{F}[u]|_{\widehat{\mathcal{S}}_\varepsilon,\ell}$ ($\ell \in \boldsymbol{Z}_+$). We denote by $\mathcal{A}$ the space of entire analytic functions in $\boldsymbol{C}^n$. Let K be a compact subset of $\boldsymbol{C}^n$, and denote by $\mathcal{A}'(K)$ the space of analytic functionals carried by K, *i.e.*, $u \in \mathcal{A}'(K)$ if and only if

(i) $u : \mathcal{A} \ni \varphi \mapsto u(\varphi) \in \boldsymbol{C}$ is a linear functional, and

(ii) for any neighborhood ω of K (in $\boldsymbol{C}^n$) there is $C_\omega \geq 0$ such that

$$|u(\varphi)| \leq C_\omega \sup_{z\in\omega} |\varphi(z)| \quad \text{for } \varphi \in \mathcal{A}.$$

Put

$$\widehat{K}_{C^n} := \{z \in \boldsymbol{C}^n;\ |P(z)| \leq \sup_K |P| \text{ for any polynomial } P\}.$$

Let Ω ($\subset \boldsymbol{C}^n$) be a domain of holomorphy. We call Ω a Runge domain if every function in $\mathcal{A}(\Omega)$ can be approximated locally uniformly in Ω by polynomials, where $\mathcal{A}(\Omega)$ denotes the space of analytic functions in Ω. It is known that Ω is a Runge domain if and only if $\widehat{K}_{C^n} \cap \Omega \subset\subset \Omega$ for any $K \subset\subset \Omega$, and that K has a fundamental system of neighborhoods consisting of Runge domains if K is polynomially convex, *i.e.*, $K = \widehat{K}_{C^n}$ (see, *e.g.*, [Hr8]). Here $A \subset\subset B$ means that the closure $\overline{A}$ of A is compact and $\overline{A} \subset \mathrm{int}(B)$, where $A, B \subset \boldsymbol{R}^n$ (or $\boldsymbol{C}^n$) and $\mathrm{int}(B)$ denotes the interior of B. If K is polynomially convex, $u \in \mathcal{A}'(K)$ and φ is analytic in a neighborhood of K, then we can define $u(\varphi)$, approximating φ by entire functions.

Lemma 1.1. 1 *Let K be a compact subset of $\boldsymbol{R}^n$. Then, for $\varepsilon \geq 0$ the set*

$$\widehat{K}_\varepsilon := \{z \in \boldsymbol{C}^n;\ |\mathrm{Re}\ z - x| + |\mathrm{Im}\ z| \leq \varepsilon \quad \text{for some } x \in K\} \tag{1.1}$$

is polynomially convex. In particular, $u(\varphi)$ can be defined for $u \in \mathcal{A}'(K)$ if φ is analytic in a neighborhood of K.

Remark A direct proof of the second part of the lemma is given in Proposition 9.1.2 of [Hr5].

Proof Let $z^0 \notin \widehat{K}_\varepsilon$. This implies that $|\mathrm{Re}\ z^0 - x| + |\mathrm{Im}\ z^0| > \varepsilon$ for any $x \in K$. First assume that $\mathrm{Re}\ z^0 \in K$. Then, putting $f(z) = \exp[-(z - \mathrm{Re}\ z^0)^2]$, we have

$$\sup_{z \in \widehat{K}_\varepsilon} |f(z)| \leq \exp[\varepsilon^2] < |f(z^0)|, \tag{1.2}$$

where $z^2\,(= z \cdot z) = \sum_{j=1}^n z_j^2$ for $z = (z_1, \cdots, z_n) \in \boldsymbol{C}^n$. Since $f(z)$ is entire analytic in $\boldsymbol{C}^n$ and, therefore, can be approximated uniformly in any compact subsets of $\boldsymbol{C}^n$ by polynomials, (1.2) gives $z^0 \notin (\widehat{K}_\varepsilon)^\wedge_{\boldsymbol{C}^n}$. Next assume that $\mathrm{Re}\ z^0 \notin K$. Choose $x^0 \in K$ so that $|\mathrm{Re}\ z^0 - x^0| = \mathrm{dis}(\{\mathrm{Re}\ z^0\}, K)$ ($\equiv \inf_{x \in K} |\mathrm{Re}\ z^0 - x|$), and put

$$\begin{aligned} f(z) &= \exp[-(z - \mathrm{Re}\ z^0 - |\mathrm{Im}\ z^0| u^0)^2], \\ u^0 &= |\mathrm{Re}\ z^0 - x^0|^{-1}(\mathrm{Re}\ z^0 - x^0). \end{aligned}$$

Then we have

$$|f(z^0)| = 1 \qquad \text{and} \qquad |f(z)| = \exp[h(z)], \tag{1.3}$$

where $h(z) = |\mathrm{Im}\ z|^2 - |\mathrm{Re}\ z - \mathrm{Re}\ z^0 - |\mathrm{Im}\ z^0| u^0|^2$. Let $z^1 \in \widehat{K}_\varepsilon$, and choose $x^1 \in K$ so that $|\mathrm{Re}\ z^1 - x^1| = \mathrm{dis}(\{\mathrm{Re}\ z^1\}, K)$ ($\leq \varepsilon$). Noting that $A \equiv |\mathrm{Re}\ z^0 - x^0| + |\mathrm{Im}\ z^0| > |\mathrm{Re}\ z^1 - x^1|$, we have

$$\begin{aligned} h(z^1) &\leq (\varepsilon - |\mathrm{Re}\ z^1 - x^1|)^2 - (A - |\mathrm{Re}\ z^1 - x^1|)^2 \\ &= (\varepsilon - A)(\varepsilon + A - 2|\mathrm{Re}\ z^1 - x^1|). \end{aligned}$$

Since $A > \varepsilon$, we have $h(z^1) < 0$. This, with (1.3), proves the lemma. □

Denote $\mathcal{A}'$ ($= \mathcal{A}'(\boldsymbol{R}^n)$) $:= \bigcup_{K \Subset \boldsymbol{R}^n} \mathcal{A}'(K)$, $\mathcal{A}'(\boldsymbol{C}^n) := \bigcup_{K \Subset \boldsymbol{C}^n} \mathcal{A}'(K)$, $\mathcal{S}_\infty := \bigcap_{\varepsilon \in \boldsymbol{R}} \mathcal{S}_\varepsilon$, $\mathcal{E}_\varepsilon := \bigcap_{\delta > \varepsilon} \mathcal{S}_{-\delta}$ and $\mathcal{F}_\varepsilon := \bigcap_{\delta > \varepsilon} \mathcal{S}_\delta'$. We introduce the projective topologies to $\mathcal{S}_\infty$, $\mathcal{E}_\varepsilon$ and $\mathcal{F}_\varepsilon$. We note that $\mathcal{F}^{-1}[C_0^\infty(\boldsymbol{R}^n)] \subset \mathcal{S}_\infty$ and that $\mathcal{S}_\infty$ is dense in $\mathcal{S}_\varepsilon$ and $\mathcal{S}_\varepsilon'$ for $\varepsilon \in \boldsymbol{R}$. Let K be a compact subset of C^n and $u \in \mathcal{A}'(K)$. We can define the Fourier transform $\hat{u}(\xi)$ of u by

$$\hat{u}(\xi)\ (= \mathcal{F}[u](\xi)) = u_z(e^{-iz\cdot\xi}),$$

where $z \cdot \xi = \sum_{j=1}^n z_j\xi_j$ for $z = (z_1, \cdots, z_n) \in C^n$ and $\xi = (\xi_1, \cdots, \xi_n) \in R^n$. It is obvious that $\hat{u}(\xi)$ can be continued analytically to an entire analytic function in C^n, and that $\hat{u}(\xi) \in \widehat{\mathcal{S}}_{-\delta}$ if $K \subset \{z \in C^n;\ |\mathrm{Im}\ z| \le \varepsilon\}$ and $\delta > \varepsilon$. Since $u(P) = (P(-D_\xi)\hat{u})(0)$ for every polynomial P, the Fourier transformation $\mathcal{F}$ is injective on $\mathcal{A}'(C^n)$. So We can regard $\mathcal{A}'(K)$ as a subspace of $\mathcal{E}_\varepsilon$ ($\subset \mathcal{F}_\varepsilon$) if $K \subset \{z \in C^n;\ |\mathrm{Im}\ z| \le \varepsilon\}$.

Lemma 1.1. 2 *Let $\varepsilon \ge 0$, and let K be a compact subset of C^n such that $K \subset \{z \in C^n;\ |\mathrm{Im}\ z| \le \varepsilon\}$. If $u \in \mathcal{A}'(K)$, $\delta > \varepsilon$ and $\varphi \in \mathcal{S}_\delta$, then $\langle u, \varphi\rangle = u(\varphi)$, where $\langle\cdot,\cdot\rangle$ denotes the duality of $\mathcal{S}_\delta'$ and $\mathcal{S}_\delta$.*

Remark If $\varphi \in \mathcal{S}_\delta$, then φ can be continued analytically to an analytic function in $\{z \in C^n;\ |\mathrm{Im}\ z| < \delta\}$ (see Lemma 1.1. 3 below). Moreover, the polynomially convex hull $\widehat{K}_{C^n}$ of K is included in the convex hull $\mathrm{ch}[K]$ ($\subset \{z \in C^n;\ |\mathrm{Im}\ z| \le \varepsilon\}$). So we can define $u(\varphi)$ for $u \in \mathcal{A}'(K)$ and $\varphi \in \mathcal{S}_\delta$ with $\delta > \varepsilon$.

Proof Let $\delta > \varepsilon$, $u \in \mathcal{A}'(K)$ and $\varphi \in \mathcal{S}_\delta$. Since $u \in \mathcal{S}_{-\delta} \subset \mathcal{S}_\delta'$, we have

$$\begin{aligned}\langle u, \varphi\rangle &= \langle u_z(e^{-iz\cdot\xi}), \mathcal{F}^{-1}[\varphi](\xi)\rangle \\ &= \int_{R^n} u_z(e^{-iz\cdot\xi})\mathcal{F}^{-1}[\varphi](\xi)\,d\xi = \int_{R^n} u_z(e^{-iz\cdot\xi}\mathcal{F}^{-1}[\varphi](\xi))\,d\xi.\end{aligned}$$

Let ω be a neighborhood of K such that $\omega \subset\subset \{z \in C^n;\ |\mathrm{Im}\ z| \le (\varepsilon + \delta)/2\}$. Then it is easily seen that

$$\sup_{z\in\omega}\left|\int_{R^n\setminus\Omega_R} e^{-iz\cdot\xi}\mathcal{F}^{-1}[\varphi](\xi)\,d\xi\right| \longrightarrow 0 \quad \text{as } R \to \infty,$$

where $\Omega_R = \{\xi \in R^n;\ |\xi_j| \le R\ (\ 1 \le j \le n)\}$. Therefore, we have

$$\lim_{R\to\infty} u_z\left(\int_{\Omega_R} e^{-iz\cdot\xi}\mathcal{F}^{-1}[\varphi](\xi)\,d\xi\right) = u(\varphi).$$

Since the Riemann sum of the integral $\int_{\Omega_R} e^{-iz\cdot\xi}\mathcal{F}^{-1}[\varphi](\xi)\,d\xi$ uniformly converges to the integral in ω, we have

$$\int_{R^n} u_z(e^{-iz\cdot\xi})\mathcal{F}^{-1}[\varphi](\xi)\,d\xi = u(\varphi).$$

This proves the lemma. □

Let $\varepsilon \in \boldsymbol{R}$ and $u \in \mathcal{S}'_\varepsilon$. We can define

$$D^\alpha u\,(= D_1^{\alpha_1}\cdots D_n^{\alpha_n}u) := \mathcal{F}^{-1}[\xi^\alpha \hat{u}(\xi)](x) \in \mathcal{S}'_\varepsilon,$$

where $\alpha = (\alpha_1,\cdots,\alpha_n) \in \boldsymbol{Z}_+^n(\,= (\boldsymbol{N}\cup\{0\})^n)$, $x = (x_1,\cdots,x_n) \in \boldsymbol{R}^n$, $D = (D_1,\cdots,D_n) = -i\partial = -i(\partial/\partial x_1,\cdots,\partial/\partial x_n)$. Following [Hr5], we define

$$\begin{aligned}\mathcal{H}(u)(x,x_{n+1}) \;&:=\; (\mathrm{sgn}\ x_{n+1})\exp[-|x_{n+1}|\langle D\rangle]u(x)/2\\ &\qquad \Big(= (\mathrm{sgn}\ x_{n+1})\mathcal{F}_\xi^{-1}[\exp[-|x_{n+1}|\langle\xi\rangle]\hat{u}(\xi)](x)/2\Big) \in \mathcal{S}'_\varepsilon\end{aligned}$$

for $x_{n+1} \in \boldsymbol{R}\setminus\{0\}$. Put

$$\begin{aligned}E_0(x,x_{n+1}) \;&:=\; \mathcal{F}^{-1}_{(\xi,\xi_{n+1})}[(1+|\xi|^2+\xi_{n+1}^2)^{-1}](x,x_{n+1}),\\ P_0(x,x_{n+1}) \;&:=\; -(\partial/\partial x_{n+1})E_0(x,x_{n+1}).\end{aligned}$$

Then we have

$$E_0(x,x_{n+1}) = (2\pi)^{-(n+1)/2}r^{-(n-1)/2}K_{(n-1)/2}(r),$$

where $r = (x^2+x_{n+1}^2)^{1/2}$ and $K_\lambda(r)$ is a modified Bessel function of the second kind. It is known that

$$\begin{aligned}K_\lambda(z) \;&=\; \pi i\, e^{i\pi\lambda/2}H_\lambda^{(1)}(e^{i\pi/2}z)\\ &=\; \int_0^\infty e^{-z\cosh t}\cosh(\lambda t)\,dt \qquad (\,|\arg z| < \pi/2\,),\\ K_\lambda(z) \;&=\; \Gamma(\lambda)(z/2)^{-\lambda}/2 + o(|z|^{-\lambda}) \quad \text{as } |z|\to 0 \text{ for } \lambda > 0,\\ K_0(z) \;&=\; -\log z + o(\log|z|) \qquad \text{as } |z| \to 0,\\ K_\lambda(z) \;&=\; \left(\frac{\pi}{2z}\right)^{1/2} e^{-z}\Big[\sum_{k=0}^{m-1}\frac{(-1)^k}{k!}(\frac{1}{2}-\lambda)_k(\frac{1}{2}+\lambda)_k(2z)^{-k}\\ &\qquad\qquad + O(|z|^{-m})\Big]\\ &\qquad \text{as } |z|\to\infty \text{ if } \delta > 0 \text{ and } |\arg z| \le 3\pi/2-\delta,\end{aligned}$$

where $(\lambda)_k = \Gamma(\lambda+k)/\Gamma(\lambda)$, and that $K_\lambda(z)$ is analytic in $\boldsymbol{C}\setminus\{z\in\boldsymbol{R};\ z\le 0\}$ (see, *e.g.*, [Ol]). Moreover, we have

$$\begin{aligned}\mathcal{F}_x[P_0](\xi,x_{n+1}) \;&=\; (\mathrm{sgn}\ x_{n+1})\exp[-|x_{n+1}|\langle\xi\rangle]/2 \quad \text{if } x_{n+1}\neq 0,\\ \mathcal{H}(u)(x,x_{n+1}) \;&=\; P_0(x,x_{n+1}) \underset{x}{*} u \ (\,= \langle u(y), P_0(x-y,x_{n+1})\rangle_y)\\ &\qquad \text{if } \varepsilon \ge 0,\ |x_{n+1}| > \varepsilon \text{ and } u \in \mathcal{S}'_\varepsilon,\end{aligned}$$

since $P_0(x-y,x_{n+1}) \in \mathcal{S}_\delta(\boldsymbol{R}_y^n)$ for $\delta \ge 0$ and $(x,x_{n+1}) \in \boldsymbol{R}^{n+1}$ with $|x_{n+1}| > \delta$.

Lemma 1.1.3 *Let $u \in \mathcal{S}'_\varepsilon$, and put $U(x, x_{n+1}) = \mathcal{H}(u)(x, x_{n+1})$. Then we have the following:* (i) $U(x, x_{n+1})|_{x_{n+1}>0} \in C^\infty([0,\infty); \mathcal{S}'_\varepsilon)$, $(1 - \Delta_{x,x_{n+1}})\, U(x, x_{n+1}) = 0$ *for* $x_{n+1} \neq 0$, $U(x, +0) = u(x)/2$ *in* $\mathcal{S}'_\varepsilon$ *and* $U(x, x_{n+1}) = -U(x, -x_{n+1})$ *in* $\mathcal{S}'_\varepsilon$ *for* $x_{n+1} \neq 0$, *where* $\Delta_{x,x_{n+1}} = -\sum_{j=1}^{n+1} D_j^2$ *and* $D_{n+1} = -i\partial/\partial x_{n+1}$. (ii) $U(x, x_{n+1})$ *can be regarded as a function in* $C^\infty(\boldsymbol{R}^n \times (\boldsymbol{R} \setminus [-\varepsilon_+, \varepsilon_+]))$, *where* $\varepsilon_+ = \max\{\varepsilon, 0\}$. *Moreover, there is* $\ell \in \boldsymbol{Z}_+$ *such that*

$$|D_x^\alpha D_{n+1}^j U(x, x_{n+1})| \leq C_{\alpha,j,\delta}(1 + |x| + |x_{n+1}|)^\ell \exp[-|x_{n+1}|]$$

if $\delta > 0$ *and* $|x_{n+1}| \geq \varepsilon_+ + \delta$, *and*

$$(1 - \Delta_{x,x_{n+1}})U(x, x_{n+1}) = 0 \quad in\ \boldsymbol{R}^n \times (\boldsymbol{R} \setminus [-\varepsilon_+, \varepsilon_+]). \tag{1.4}$$

(iii) *If* $\varepsilon < 0$, *then* $u(x)$ *can be continued analytically to* $\{z \in \boldsymbol{C}^n;\ |\mathrm{Im}\ z| < -\varepsilon\}$.

Proof Since

$$\mathcal{F}_x[U(x, x_{n+1})](\xi) = (\mathrm{sgn}\ x_{n+1}) \exp[-|x_{n+1}|\langle\xi\rangle]\hat{u}(\xi)/2,$$

the assertion (i) is obvious. Note that $e^{-\varepsilon\langle\xi\rangle}\hat{u}(\xi) \in \mathcal{S}'$. Therefore, there are $\ell \in \boldsymbol{Z}_+$ and $C > 0$ such that

$$|\langle e^{-\varepsilon\langle\xi\rangle}\hat{u}(\xi), \varphi(\xi)\rangle| \leq C|\varphi|_{\mathcal{S},\ell}$$

for $\varphi \in \mathcal{S}$. This gives

$$\begin{aligned}
&|D_x^\alpha D_{n+1}^j U(x, x_{n+1})| \\
&= |\langle \exp[-\varepsilon\langle\xi\rangle]\hat{u}(\xi), \xi^\alpha D_{n+1}^j \exp[-(|x_{n+1}| - \varepsilon)\langle\xi\rangle]e^{ix\cdot\xi}/(2(2\pi)^n)\rangle_\xi| \\
&\leq C_{\alpha,j,\delta,\varepsilon}(1 + |x| + |x_{n+1}|)^\ell \exp[-|x_{n+1}|]
\end{aligned}$$

if $|x_{n+1}| \geq \varepsilon_+ + \delta\ (> \varepsilon_+)$. In fact, by Fubini's theorem we can prove that if $\delta > 0$ and $\exp[\delta\langle\xi\rangle]\hat{v}(\xi) \in \mathcal{S}'$, $v(x)$ is a function and

$$v(x) = (2\pi)^{-n}\langle \exp[\delta\langle\xi\rangle]\hat{v}(\xi), \exp[ix\cdot\xi - \delta\langle\xi\rangle]\rangle_\xi.$$

(1.4) is obvious. If $\varepsilon < 0$, then we have

$$u(x) = (2\pi)^{-n}\langle \exp[-\varepsilon\langle\xi\rangle]\hat{u}(\xi), \exp[ix\cdot\xi + \varepsilon\langle\xi\rangle]\rangle_\xi.$$

This proves the assertion (iii). □

Corollary 1.1. 4 *Let $u \in \mathcal{S}'_\varepsilon$, and put $U(x, x_{n+1}) = \mathcal{H}(u)(x, x_{n+1})$. Then*

$$\langle u, \varphi \rangle = 2 \lim_{t \to +0} \langle U(x,t), \varphi(x) \rangle_x \quad \text{for } \varphi \in \mathcal{S}_\infty. \tag{1.5}$$

Moreover, if K is a compact subset of $\boldsymbol{R}^n$ and $u \in \mathcal{A}'(K)$, then

$$u(\varphi) \, (= \langle u, \varphi \rangle) = 2 \lim_{t \to +0} \int_V U(x,t) \varphi(x) \, dx \quad \text{for } \varphi \in \mathcal{S}_\infty, \tag{1.6}$$

where V is a neighborhood of K in $\boldsymbol{R}^n$.

Remark It follows from Lemma 1.1. 3 that $\mathcal{H}(u)(x,t)$ is real analytic in x for $t > 0$ if $u \in \mathcal{F}_0$. Thus,

$$\langle u, \varphi \rangle = 2 \lim_{t \to +0} \int_{\boldsymbol{R}^n} \mathcal{H}(u)(x,t) \varphi(x) \, dx \quad \text{for } u \in \mathcal{F}_0 \text{ and } \varphi \in \mathcal{S}_\infty.$$

Proof By Lemma 1.1. 3 (1.5) is obvious. Assume that K is a compact subset of $\boldsymbol{R}^n$, V is a neighborhood of K in $\boldsymbol{R}^n$ and $u \in \mathcal{A}'(K)$. Then $U(x, x_{n+1})$ ($= \langle u(y), P_0(x - y, x_{n+1}) \rangle_y$) can be continued analytically to $\boldsymbol{R}^{n+1} \setminus K \times \{0\}$ and, therefore, $U(x, 0) = 0$ for $x \notin K$. Let ω be a complex neighborhood of K such that $\operatorname{dis}(\omega, \boldsymbol{R}^n \setminus V) > 0$. There are $C_\omega > 0$ and $C_{\omega,V} > 0$ such that

$$|U(x,t)| \le C_\omega \sup_{z \in \omega} |P_0(x - z, t)| \le C_{\omega,V} e^{-|x|}$$

if $x \in \boldsymbol{R}^n \setminus V$ and $0 < t \le 1$. By Lebesgue's convergence theorem we have

$$\lim_{t \to +0} \int_{\boldsymbol{R}^n \setminus V} U(x,t) \varphi(x) \, dx = 0 \quad \text{for } \varphi \in \mathcal{S}_\infty,$$

which gives (1.6). □

Let $a(x) \in C^\infty(\boldsymbol{R}^n)$ satisfy

$$|D^\alpha a(x)| \le C A^{|\alpha|} |\alpha|! \langle x \rangle^k \quad \text{for } x \in \boldsymbol{R}^n, \tag{1.7}$$

where $C > 0$, $A > 0$ and $k \in \boldsymbol{R}$. Then we have the following

Lemma 1.1. 5 *Multiplication by $a(x)$ is well-defined and continuous on $\mathcal{S}_\varepsilon$, i.e., the mapping $\mathcal{S}_\varepsilon \ni u \mapsto au \in \mathcal{S}_\varepsilon$ is continuous, if $|\varepsilon| < (\sqrt{n} A)^{-1}$.*

Proof Let $u \in \mathcal{S}_\infty$. Then $au \in \mathcal{S}$ is well-defined and

$$\begin{aligned}\mathcal{F}[au](\xi) &= (2\pi)^{-n} \int e^{-ix\cdot(\xi-\eta)} \langle A'x\rangle^{-2M} a(x) \langle A'D_\eta\rangle^{2M} \hat{u}(\eta)\, d\eta dx \\ &= (2\pi)^{-n} \int \mathcal{F}[\langle A'x\rangle^{-2M} a(x)](\xi-\eta) \langle A'D_\eta\rangle^{2M} \hat{u}(\eta)\, d\eta,\end{aligned}$$

where $\langle A'D_\eta\rangle^2 = 1 - \sum_{j=1}^n A'^2 (\partial/\partial\eta_j)^2$, $M = \max\{[(n+k)/2]+1, 0\}$, $A' = A/(4(1+\sqrt{2}))$ and $[c]$ denotes the largest integer $\le c$ for $c \in \boldsymbol{R}$. Note that

$$|D^\alpha \langle A'x\rangle^{-2M}| \le C_M (A/2)^{|\alpha|} |\alpha|! \langle A'x\rangle^{-2M-|\alpha|}$$

(see Lemma 2.1. 1 below). Therefore, we have

$$|D^\alpha ((\langle A'x\rangle^{-2M} a(x))| \le 2CC_M A^{|\alpha|} |\alpha|! \langle A'x\rangle^{-2M} \langle x\rangle^k.$$

Put $L = |\xi|^{-2} \sum_{j=1}^n \xi_j D_{x_j}$ for $\xi \in \boldsymbol{R}^n \setminus \{0\}$. Then we have

$$|\mathcal{F}[\langle A'x\rangle^{2M} a(x)](\xi)| \le \int |L^\ell \{\langle A'x\rangle^{-2M} a(x)\}|\, dx \le C_{k,A} (\sqrt{n} A/|\xi|)^\ell \ell!$$

if "$\xi \in \boldsymbol{R}^n \setminus \{0\}$ and $\ell \in \boldsymbol{Z}_+$" or "$\xi = 0$ and $\ell = 0$." It follows from Lemma 2.1. 1 that

$$|\mathcal{F}[\langle A'x\rangle^{2M} a(x)](\xi)| \le C'_{k,A} \langle\xi\rangle^{1/2} \exp[-(\sqrt{n}A)^{-1}\langle\xi\rangle]$$

for $\xi \in \boldsymbol{R}^n$. This yields

$$\begin{aligned}&|\langle\xi\rangle^\ell D_\xi^\alpha \mathcal{F}[au](\xi)| \\ &= (2\pi)^{-n} \langle\xi\rangle^\ell |\mathcal{F}[\langle A'x\rangle^{-2M} a(x)] * (D_\xi^\alpha \langle A'D_\xi\rangle^{2M} \hat{u})(\xi)| \\ &\le C_{k,A,\ell,|\alpha|,\varepsilon} \exp[-\varepsilon\langle\xi\rangle] |u|_{\mathcal{S}_\varepsilon, 2M+|\alpha|+\ell+n+1}\end{aligned}$$

and, therefore,

$$|au|_{\mathcal{S}_\varepsilon,\ell} \le C_{k,A,\ell,\varepsilon} |u|_{\mathcal{S}_\varepsilon, 2M+\ell+n+1},$$

if $|\varepsilon| < (\sqrt{n}A)^{-1}$. Since $\mathcal{S}_\infty$ is dense in $\mathcal{S}_\varepsilon$, this proves the lemma. □

If $a \in C^\infty(\boldsymbol{R}^n)$ satisfies the estimate (1.7) and $|\varepsilon| < (\sqrt{n}A)^{-1}$, then we can define au for $u \in \mathcal{S}'_\varepsilon$ by

$$\langle au, \varphi\rangle := \langle u, a\varphi\rangle \qquad \text{for } \varphi \in \mathcal{S}_\varepsilon.$$

We note that $au \in \mathcal{A}'(K)$ can be defined by $(au)(\varphi) = u(a\varphi)$ ($\varphi \in \mathcal{A}$), if $u \in \mathcal{A}'(K)$ and a is analytic in a neighborhood of $\widehat{K_{C^n}}$, where K is a compact subset of $\boldsymbol{C}^n$.

1.2 Supports

Definition 1.2. 1 Let $\varepsilon \geq 0$. For $u \in \mathcal{S}'_\varepsilon$ we define

$$\begin{aligned}\text{supp } u \;:=\; & \bigcap\{K;\ K \textit{ is a closed subset of } \boldsymbol{R}^n \textit{ and there exists a real} \\ & \textit{analytic function } \widetilde{U}(x, x_{n+1}) \textit{ in } \boldsymbol{R}^{n+1} \setminus K \times [-\varepsilon, \varepsilon] \\ & \textit{such that } \widetilde{U}(x, x_{n+1}) = \mathcal{H}(u)(x, x_{n+1}) \textit{ for } |x_{n+1}| > \varepsilon\}.\end{aligned}$$

Remark (i) The definition of supp u does not depend on the choice of ε satisfying $u \in \mathcal{S}'_\varepsilon$. (ii) For $u, v \in \mathcal{S}'_\varepsilon$ we have supp $(u \pm v) \subset$ supp $u \cup$ supp v, and supp $u \cap X =$ supp $v \cap X$ if supp $(u - v) \cap X = \emptyset$, where $X \subset \boldsymbol{R}^n$. (iii) If supp $u \subset K_\lambda$ ($\lambda \in \Lambda$), then supp $u \subset \bigcap_{\lambda \in \Lambda} K_\lambda$. (iv) For $u \in \mathcal{F}_\varepsilon$ there is a real analytic function $\widetilde{U}(x, x_{n+1})$ in $\boldsymbol{R}^{n+1} \setminus$ supp $u \times [-\varepsilon, \varepsilon]$ such that $\widetilde{U}(x, x_{n+1}) = \mathcal{H}(u)(x, x_{n+1})$ for $|x_{n+1}| > \varepsilon$. (v) Let $u \in \mathcal{F}_0$ and $x^0 \in \boldsymbol{R}^n$. Then $\mathcal{H}(u)(x, x_{n+1})$ can be extended to a C^2-function near $(x^0, 0) \in \boldsymbol{R}^{n+1}$ if and only if $x^0 \notin$ supp u.

Lemma 1.2. 2 (i) *Let X be an open subset of $\boldsymbol{R}^n$, and assume that $W(x, x_{n+1}) \in C^\infty(X \times (\boldsymbol{R} \setminus \{0\}))$ satisfies $(1 - \Delta_{x, x_{n+1}})W(x, x_{n+1}) = 0$, $W(x, x_{n+1}) \to 0$ in $\mathcal{D}'(X)$ as $x_{n+1} \to 0$, and $W(x, -x_{n+1}) = -W(x, x_{n+1})$ for $x \in X$ and $x_{n+1} > 0$. Then $W(x, x_{n+1})$ can be extended to a real analytic function in $X \times \boldsymbol{R}$.* (ii) *If $u \in \mathcal{S}'$,* supp u *coincides with the distribution support of u, which is the support of u as a distribution.*

Proof (i) By assumptions we can regard $W(x, x_{n+1})$ as a function in $C(\boldsymbol{R}; \mathcal{D}'(X))$ ($\subset \mathcal{D}(X \times \boldsymbol{R})$). Since $(\partial^2/\partial x_{n+1}^2)W(x, x_{n+1}) = (1 - \Delta_x)W(x, x_{n+1})$ for $x \in X$ and $x_{n+1} \neq 0$, it follows that $(\partial^2/\partial x_{n+1}^2)W(x, \pm x_{n+1}) \in C([+0, \infty); \mathcal{D}'(X))$, $(\partial^2 W/\partial x_{n+1}^2)(x, \pm 0) = 0$ in $\mathcal{D}'(X)$ and that

$$\begin{aligned}&(\partial W/\partial x_{n+1})(x, \pm\delta) \\ &\quad = (\partial W/\partial x_{n+1})(x, \pm 1) - \int_{\pm\delta}^{\pm 1} (\partial^2 W/\partial x_{n+1}^2)(x, x_{n+1})\, dx_{n+1}\end{aligned}$$

in $\mathcal{D}'(X)$ for $\delta > 0$. This yields $(\partial W/\partial x_{n+1})(x, \pm x_{n+1}) \in C([+0, \infty); \mathcal{D}'(X))$. On the other hand, we have $(\partial W/\partial x_{n+1})(x, +0) = (\partial W/\partial x_{n+1})(x, -0)$. Therefore, the mean value theorem implies that $W(x, x_{n+1}) \in C^1(\boldsymbol{R}; \mathcal{D}'(X))$. Similarly, we have $W \in C^2(\boldsymbol{R}; \mathcal{D}'(X))$. This gives $(1 - \Delta_{x, x_{n+1}})W = 0$ in $\mathcal{D}'(X \times \boldsymbol{R})$. That $(1 - \Delta_{x, x_{n+1}})$ is analytic hypoelliptic also follows from the fact that the fundamental solution

$E_0(x, x_{n+1})$ of $(1 - \Delta_{x,x_{n+1}})$ is real analytic for $(x, x_{n+1}) \neq (0,0)$, although it is well-known. So $W(x, x_{n+1})$ is real analytic in $X \times \boldsymbol{R}$. We remark that we shall prove analytic hypoellipticity of general elliptic operators in Theorem 2.8. 1 (ii) Let $(x^0, 0) \in \boldsymbol{R}^{n+1}$, and assume that there are a closed subset K of $\boldsymbol{R}^n$ and a real analytic function $\widetilde{U}(x, x_{n+1})$ in $\boldsymbol{R}^{n+1} \setminus K \times \{0\}$ satisfying $x^0 \notin K$ and $\mathcal{H}(u)(x, x_{n+1}) = \widetilde{U}(x, x_{n+1})$ for $x_{n+1} \neq 0$. $\widetilde{U}(x, x_{n+1})$ is an odd function with respect to x_{n+1}. Therefore, $\widetilde{U}(x, 0) = 0$ for $x \notin K$. Let $\varphi \in C_0^\infty(\boldsymbol{R}^n \setminus K)$. Then $\varphi(x)\widetilde{U}(x, x_{n+1}) \to 0$ in $\mathcal{S}'$ as $x_{n+1} \to +0$. On the other hand, it follows from Lemma 1.1. 3 that $\varphi(x)\mathcal{H}(u)(x, x_{n+1})$ $(= \varphi(x)\widetilde{U}(x, x_{n+1})\ ) \to \varphi(x)u(x)/2$ in $\mathcal{S}'$ as $x_{n+1} \to +0$. So we have $\varphi(x)u(x) = 0$ in $\mathcal{S}'$, which implies x^0 does not belong to the distribution support of u. Next assume that an open subset X of $\boldsymbol{R}^n$ does not meet the distribution support of u, If $\varphi \in C_0^\infty(X)$, then, by Lemma 1.1. 3 , $\varphi(x)\mathcal{H}(u)(x, x_{n+1}) \to 0$ in $\mathcal{S}'$ as $x_{n+1} \to \pm 0$, *i.e.*, $\mathcal{H}(u)(x, x_{n+1}) \to 0$ in $\mathcal{D}'(X)$ as $x_{n+1} \to \pm 0$. From Lemma 1.1. 3 and the assertion (i) we can see that $\mathcal{H}(u)(x, x_{n+1})$ can be extended to a real analytic function in $X \times \boldsymbol{R}$. and that supp $u \cap X = \emptyset$. This completes the proof. □

Lemma 1.2. 3 *For any $\varphi \in \mathcal{A}$ there exists a unique $\Phi \in C^\infty(\boldsymbol{R}^{n+1})$ such that*

$$(1 - \Delta_{x,x_{n+1}})\Phi = 0, \quad \Phi|_{x_{n+1}=0} = 0, \quad (\partial/\partial x_{n+1})\Phi|_{x_{n+1}=0} = \varphi. \qquad (1.8)$$

Moreover, Φ can be continued analytically to $\boldsymbol{C}^{n+1}$ and satisfies the following estimates; for any $R > 1$, $\delta > 0$, $\alpha \in \boldsymbol{Z}_+^n$ and $j \in \boldsymbol{Z}_+$ there is $C_{R,\delta,\alpha,j} > 0$ satisfying

$$|D_x^\alpha D_{x_{n+1}}^j \Phi(x, x_{n+1})| \\ \leq C_{R,\delta,\alpha,j}(t^{n+1} + t^{-j})e^{|x_{n+1}|} \sup_{|y|<Rt+\delta, |w|<\delta} |\varphi(x + w + iy)|$$

if $(x, x_{n+1}) \in \boldsymbol{R}^{n+1}$ and $|x_{n+1}| \leq t$.

Proof If $\Phi \in C^\infty(\boldsymbol{R}^{n+1})$ satisfies (1.8), then it follows from analytic hypoellipticity of $(1 - \Delta_{x,x_{n+1}})$ that Φ is real analytic in $\boldsymbol{R}^{n+1}$. On the other hand, by the Cauchy-Kowalevsky theorem (1.8) has a unique entire analytic solution which is an analytic continuation of Φ. In particular, (1.8) has a unique solution $\Phi \in C^\infty(\boldsymbol{R}^{n+1})$. Write $z = x+iy$ for $x, y \in \boldsymbol{R}^n$, and let $\Phi(z, x_{n+1})$ be the analytic continuation. Since $(\partial/\partial \overline{z}_j)\Phi(z, t)$ $(= (\partial/\partial x_j + i\partial/\partial y_j)\Phi(z, t)/2) = 0$ $(1 \leq j \leq n)$, $(\partial^2/\partial z_j^2)\Phi(z, t) =$

$-(\partial^2/\partial y_j^2)\Phi(z,t)$ and $u = \Phi(z,t)$ satisfies

$$\begin{cases} (\partial^2/\partial t^2)u(x,y,t) - \sum_{j=1}^n (\partial^2/\partial y_j^2)u(x,y,t) - u(x,y,t) = 0, \\ u(x,y,0) = 0, \quad (\partial u/\partial t)(x,y,0) = \varphi(x+iy). \end{cases} \tag{1.9}$$

Conversely, if u satisfies (1.9), then $(\partial/\partial x_j + i\partial/\partial y_j)u(x,y,t) = 0$ ($1 \le j \le n$) and $u(x,y,t) = \Phi(z,t)$. For a more general treatment we refer to [Wk4]. Regarding x as a parameters, (1.9) is a simple hyperbolic Cauchy problem with propagation speed 1. Let $r > 0$ and $u_0 \in C^\infty(\boldsymbol{R}^{2n})$ satisfying supp $u_0 \subset \{(x,y) \in \boldsymbol{R}^{2n};\ |y| \le r\}$. Using the Fourier transformation, we can show that

$$u(x,y,t) = \mathcal{F}_\xi^{-1}\left[\left(\sin t\sqrt{|\xi|^2-1}/\sqrt{|\xi|^2-1}\right)\tilde{u}_0(x,\xi)\right](y)$$

satisfies (1.9) with $\varphi(x+iy)$ replaced by $u_0(x,y)$, where $\tilde{u}_0(x,\xi) = \mathcal{F}_y[u_0(x,y)](\xi)$ and $\sqrt{-a} = i\sqrt{a}$ for $a > 0$. Moreover, we have

$$\begin{aligned} |D_x^\alpha D_t^j u(x,y,t)| &\le (2\pi)^{-n}\int \left||\xi|^2-1\right|^{(j-1)/2}\left|\exp[it\sqrt{|\xi|^2-1}]\right. \\ &\qquad \left. +(-1)^{j-1}\exp[-it\sqrt{|\xi|^2-1}]\right| |D_x^\alpha \tilde{u}_0(x,\xi)|/2\,d\xi \\ &\le C_j r^n(1+|t|)e^{|t|} \sup_{w\in R^n, |\beta|\le n+j} |D_x^\alpha D_w^\beta u_0(x,w)| \end{aligned} \tag{1.10}$$

for $\alpha \in \boldsymbol{Z}_+^n$ and $j \in \boldsymbol{Z}_+$. Let $R > 1$ and $T > 0$, and choose $\chi_R(s) \in C_0^\infty(\boldsymbol{R})$ so that $\chi_R(s) = 1$ ($|s| \le 1$) and supp $\chi_R \subset \{s \in \boldsymbol{R}^n;\ |s| \le R\}$. If $u_0(x,y) = \chi_R(|y|/T)\varphi(x+iy)$, then, by finite propagation property, we have $u(x,y,t) = \Phi(x+iy,t)$ for $y \in \boldsymbol{R}^n$ with $|y| \le T - |t|$. This, together with (1.10), gives

$$\begin{aligned} |D_x^\alpha D_t^j \Phi(x,t)| \;&\le\; C_{R,j}T^n(1+|t|)e^{|t|}(1+T^{-1})^{n+j} \\ &\quad\times \sup_{|y|<RT, |\beta|\le n+j} |D_x^\alpha D_y^\beta \varphi(x+iy)| \end{aligned}$$

for $|t| \le T$. Applying Cauchy's estimate (or Cauchy's integral formula), we have

$$|D_x^\alpha D_t^j \Phi(x,t)| \le C_{R,\delta,\alpha,j}(T^{n+1}+T^{-j})e^{|t|} \sup_{|y|<Rt+\delta, |w|<\delta} |\varphi(x+w+iy)|$$

if $\delta > 0$, $x \in \boldsymbol{R}^n$, $t \in \boldsymbol{R}$ and $|t| \le T$. □

We shall also need the following lemma concerning analytic continuation of solutions of $(1 - \Delta_{x,x_{n+1}})u = 0$.

Lemma 1.2.4 *Let X be an open subset of $\boldsymbol{R}^n$ and $R > 0$.* (i) *Assume that $U(x, x_{n+1})$ is a smooth function defined in a neighborhood of $X \times \{0\}$ in $\boldsymbol{R}^{n+1}$ and satisfies $(1 - \Delta_{x,x_{n+1}})U(x, x_{n+1}) = 0$ there. Moreover, if $U(x, 0)$ and $(\partial U/\partial x_{n+1})(x, 0)$ can be continued analytically to $\{z \in \boldsymbol{C}^n;$ $\mathrm{Re}\, z \in X$ and $|\mathrm{Im}\, z| < R\}$, then $U(x, x_{n+1})$ can be continued analytically to $\{(z, x_{n+1}) \in \boldsymbol{C}^n \times \boldsymbol{R};\ \mathrm{Re}\, z \in X,\ |\mathrm{Im}\, z| + |x_{n+1}| < R\}$.* (ii) *If $u \in \mathcal{D}'(X_R)$ satisfies $(1 - \Delta)u = 0$ (in X_R), then u can be regarded as an analytic function in $\{z \in \boldsymbol{C}^n;\ |\mathrm{Re}\, z - x| + |\mathrm{Im}\, z| < R$ for some $x \in X\}$, where $X_R = \{x \in \boldsymbol{R}^n;\ |x - y| < R$ for some $y \in X\}$.*

Proof (i) From analytic hypoellipticity it follows that $U(x, x_{n+1})$ is real analytic in a neighborhood of $X \times \{0\}$ in $\boldsymbol{R}^{n+1}$. By assumptions we can regard $U(x, 0)$ and $(\partial U/\partial x_{n+1})(x, 0)$ as analytic functions in $\{z \in \boldsymbol{C}^n;$ $\mathrm{Re}\, z \in X$ and $|\mathrm{Im}\, z| < R\}$. Let us consider the Cauchy problem

$$\begin{cases} (\partial^2/\partial t^2 - \Delta_y - 1)V(x, y, t) = 0, \\ V(x, y, 0) = U(x + iy, 0), \\ (\partial V/\partial t)(x, y, 0) = (\partial U/\partial x_{n+1})(x + iy, 0). \end{cases} \tag{1.11}$$

This is a simple hyperbolic Cauchy problem with propagation speed 1, regarding x as a parameter. So there is a unique solution $V(x, y, t)$ of (1.11) which is real analytic in $\{(x, y, t) \in X \times \boldsymbol{R}^n \times \boldsymbol{R};\ |y| + |t| < R\}$. Note that $U(x, t)$ is analytic and satisfies $((\partial^2/\partial t^2) - \Delta_y - 1)U(x + iy, t) = 0$ if $(x + iy, t)$ belongs to a neighborhood of $X \times \{0\}$ in $\boldsymbol{C}^n \times \boldsymbol{R}$. Therefore, $V(x, y, t) = U(x + iy, t)$ in a neighborhood of $X \times \{0\}$ in $\boldsymbol{C}^n \times \boldsymbol{R}$, which proves the assertion (i). (ii) By assumption u is real analytic in X_R. Let us consider the Cauchy problem

$$\begin{cases} (\partial^2/\partial t^2 - \sum_{j=2}^n \partial^2/\partial x_j^2 + 1)v(x, t) = 0, \\ v(x, 0) = u(x), \quad (\partial v/\partial t)(x, 0) = i(\partial u/\partial x_1)(x). \end{cases} \tag{1.12}$$

Then we have a unique solution $v(x, t)$ of (1.12) which is real analytic in $\{(x, t) \in X_R \times \boldsymbol{R};\ |x - y| + |t| < R$ for some $y \in X\}$. Moreover, we have $v(x, t) = u(x + ite_1)$ if (x, t) belongs to a neighborhood of $X_R \times \{0\}$, where $e_1 = (1, 0, \cdots, 0) \in \boldsymbol{R}^n$. Δ is rotation invariant and, therefore, we can construct real analytic function $v(x, y)$, defined in $\{(x, y) \in X_R \times \boldsymbol{R}^n;$ $|x - \tilde{x}| + |y| < R$ for some $\tilde{x} \in X\}$, such that $v_T(x, t) \equiv v(Tx, tTe_1)$ satisfies

$$\begin{cases} (\partial^2/\partial t^2 - \sum_{j=2}^n \partial^2/\partial x_j^2 + 1)v_T(x, t) = 0, \\ v_T(x, 0) = u(Tx), \quad (\partial v_T/\partial t)(x, 0) = i(\partial/\partial x_1)u(Tx) \end{cases}$$

in $\{(T^{-1}x,t) \in \boldsymbol{R}^n \times \boldsymbol{R};\ |x-\tilde{x}|+|t| < R$ for some $\tilde{x} \in X\}$ for each orthoganal matrix T of order n, where $e_1 = {}^t(1,0,\cdots,0) \in \boldsymbol{R}^n$ and x is regarded as a column vector. In fact, if T and S are orthogonal matrices of order n and satisfy $Te_1 = Se_1$, then we have $v_T(x,t) = v_S(S^{-1}Tx,t)$. It is easily seen that $v(x,y) = u(z)$ in a complex neighborhood of X_R, where $z = x+iy$. This proves the assertion (ii). □

We need the following lemma (see, *e.g.*, Theorem 7.3.2 and Lemma 7.3.7 in [Hr5]).

Lemma 1.2. 5 *Let $P(D)$ be a differential operator with constant coefficients, and let $\nu \in \mathcal{E}'(\boldsymbol{R}^n)$ satisfy the following: $\langle \nu, f(x)e^{ix\cdot\zeta}\rangle = 0$ if $f(x)$ is a polynomial and $P(-D)(f(x)e^{ix\cdot\zeta}) = 0$, where $\zeta \in \boldsymbol{C}^n$. Then there is a unique solution $u \in \mathcal{E}'(\boldsymbol{R}^n)$ of $P(D)u = \nu$. Here $\mathcal{E}'(\boldsymbol{R}^n)$ denotes the space of distributions in $\boldsymbol{R}^n$ with compact supports.*

Proposition 1.2. 6 *Let K be a compact subset of $\boldsymbol{R}^n$ and $\varepsilon \geq 0$. We put $K^\varepsilon := \{z \in \boldsymbol{C}^n;\ \mathrm{Re}\, z \in K$ and $|\mathrm{Im}\, z| \leq \varepsilon\}$. Then we have the following:* (i) *If $u \in \mathcal{A}'(K^\varepsilon)$, then $u \in \mathcal{E}_\varepsilon$ and* supp $u \subset K_\varepsilon := \{x \in \boldsymbol{R}^n;\ |x-y| \leq \varepsilon$ *for some* $y \in K\}$. *Moreover,*

$$|\tilde{U}(x,x_{n+1})| \leq C(|x|^2+x_{n+1}^2)^{-(n+2)/4}\exp[-(|x|^2+x_{n+1}^2)^{1/2}]$$

if $|x|+|x_{n+1}| \gg 1$, where $\tilde{U}(x,x_{n+1})$ is a real analytic function in $\boldsymbol{R}^{n+1}\setminus K_\varepsilon \times [-\varepsilon,\varepsilon]$ satisfying $\tilde{U}(x,x_{n+1}) = \mathcal{H}(u)(x,x_{n+1})$ for $|x_{n+1}| > \varepsilon$. (ii) *If $u \in \mathcal{A}'(\boldsymbol{C}^n) \cap \mathcal{F}_\varepsilon$ and* supp $u \subset K$, *then $u \in \mathcal{A}'(K^\varepsilon)$ and*

$$u(\partial\Phi/\partial x_{n+1}|_{x_{n+1}=0}) = \int \tilde{U}(x,x_{n+1})(1-\Delta_{x,x_{n+1}})(\chi\Phi)\,dx\,dx_{n+1}$$

for $\Phi \in C^\infty(\boldsymbol{R}^{n+1})$ with $(1-\Delta_{x,x_{n+1}})\Phi = 0$, where $\chi \in C_0^\infty(\boldsymbol{R}^{n+1})$ satisfies $\chi = 1$ near $K \times [-\varepsilon,\varepsilon]$ and $\tilde{U}(x,x_{n+1})$ is a real analytic function in $\boldsymbol{R}^{n+1}\setminus K\times[-\varepsilon,\varepsilon]$ satisfying $\tilde{U}(x,x_{n+1}) = \mathcal{H}(u)(x,x_{n+1})$ for $|x_{n+1}| > \varepsilon$.

Remark (i) We can prove that $u = 0$ if $u \in \mathcal{A}'(\boldsymbol{C}^n)$ and supp $u = \emptyset$. (ii) If $\Phi \in C^\infty(\boldsymbol{R}^{n+1})$ satisfies $(1-\Delta_{x,x_{n+1}})\Phi = 0$ in $\boldsymbol{R}^{n+1}$, then, by Lemma 1.2. 4 , $\Phi(x,x_{n+1})$ can be regarded as an entire analytic function in $\boldsymbol{C}^{n+1}$.

Proof Following [Hr5], we shall prove the proposition. (i) Assume that $u \in \mathcal{A}'(K^\varepsilon)$. Then

$$U(x,x_{n+1}) \equiv \mathcal{H}(u)(x,x_{n+1}) = u_y(P_0(x-y,x_{n+1})) \quad \text{for } |x_{n+1}| > \varepsilon.$$

Since Re $(x-y)^2 > 0$ for $x \in \boldsymbol{R}^n \setminus K_\varepsilon$ and $y \in \widehat{K_\varepsilon}$, $P(x-y,x_{n+1})$ is analytic in a complex neighborhood of $(\boldsymbol{R}^n \setminus K_{\varepsilon'}) \times \widehat{K}_\varepsilon \times \boldsymbol{R}$ with respect to (x, y, x_{n+1}) for $\varepsilon' > \varepsilon$, where $\widehat{K_\varepsilon}$ is defined as in (1.1). Therefore, $U(x, x_{n+1})$ can be extended to a real analytic function defined in $(\boldsymbol{R}^n \setminus K_\varepsilon) \times \boldsymbol{R}$ and

$$\tilde{U}(x, x_{n+1}) = u_y(P_0(x-y, x_{n+1})) \quad \text{for } (x, x_{n+1}) \in (\boldsymbol{R}^n \setminus K_\varepsilon) \times \boldsymbol{R}.$$

It follows that there are $R > 0$ and a compact complex neighborhood ω of $\widehat{K_\varepsilon}$ such that $P_0(x-y, x_{n+1})$ is analytic in a neighborhood of ω with respect to y if $|x| + |x_{n+1}| \geq R$. This yields, with some constant C,

$$|\tilde{U}(x, x_{n+1})| \leq C \sup_{y \in \omega} |P_0(x-y, x_{n+1})| \quad \text{if } |x| + |x_{n+1}| \geq R.$$

On the other hand, we have

$$|P(x-y, x_{n+1})| \leq C_{\omega,R}(|x-y|^2 + x_{n+1}^2)^{-(n+2)/4} \exp[-(|x|^2 + x_{n+1}^2)^{1/2}]$$

if $y \in \omega$ and $|x| + |x_{n+1}| \geq R$. This proves the assertion (i). (ii) Let $\chi \in C_0^\infty(\boldsymbol{R}^{n+1})$ be a function satisfying $\chi = 1$ near $K \times [-\varepsilon, \varepsilon]$, and let $\Phi \in C^\infty(\boldsymbol{R}^{n+1})$ satisfy $(1 - \Delta_{x,x_{n+1}})\Phi = 0$. Then the integral

$$\int \tilde{U}(1 - \Delta_{x,x_{n+1}})(\chi\Phi)\, dx dx_{n+1}$$

is well-defined and does not depend on the choice of χ. In fact, we can choose a compact neighborhood K' of K in $\boldsymbol{R}^n$ and $\delta > 0$ so that the boundary K' is smooth and $\chi = 1$ near $K' \times [-\varepsilon - \delta, \varepsilon + \delta]$. Then Green's formula gives

$$\begin{aligned} &\int \tilde{U}(1 - \Delta_{x,x_{n+1}})(\chi\Phi)\, dx dx_{n+1} \\ &= \int_{\partial K' \times [-\varepsilon-\delta, \varepsilon+\delta]} (\tilde{U}(\partial\Phi/\partial \boldsymbol{n}) - (\partial\tilde{U}/\partial \boldsymbol{n})\Phi)\, dS dx_{n+1} \\ &\quad + \left\{ \int_{K' \times \{\varepsilon+\delta\}} - \int_{K' \times \{-\varepsilon-\delta\}} \right\} (\tilde{U}(\partial\Phi/\partial x_{n+1}) - (\partial\tilde{U}/\partial x_{n+1})\Phi)\, dx, \end{aligned}$$

where $\boldsymbol{n}$ is the outward unit normal of $\partial K'$ and dS is the surface element of $\partial K'$. Define

$$u_1(\varphi) = \int \tilde{U}(1 - \Delta_{x,x_{n+1}})(\chi\Phi)\, dx dx_{n+1}$$

for $\varphi \in \mathcal{A}$, where Φ is a unique solution of (1.8). Then It follows from Lemma 1.2. 3 that

$$|u_1(\varphi)| \leq C_{R,\delta} \sup_{x \in K_\delta, |y| < R(\varepsilon+\delta)} |\varphi(x + iy)|$$

for any $R > 1$, where $\delta > 0$. This yields $u_1 \in \mathcal{A}'(K^\varepsilon)$. For a fixed $y \in \boldsymbol{R}^n$ $P_0(x-y, x_{n+1})$ is a distribution of (x, x_{n+1}) and we have

$$\begin{aligned}&\int_{R^{n+1}} P_0(x-y, x_{n+1})(1-\Delta_{x,x_{n+1}})(\chi\Phi)\,dx dx_{n+1}\\&= \langle (1-\Delta_{x,x_{n+1}})P_0(x-y,x_{n+1}), \chi(x,x_{n+1})\Phi(x,x_{n+1})\rangle_{x,x_{n+1}}\\&= (\partial/\partial x_{n+1})(\chi\Phi)|_{x=y,\,x_{n+1}=0}.\end{aligned} \tag{1.13}$$

Assume that $\chi \in C_0^\infty(\boldsymbol{R}^{n+1})$ satisfies $\chi = 1$ near $K_\varepsilon \times [-\varepsilon, \varepsilon]$. Then the left-hand side of (1.13) is well-defined and analytic in y when y belongs to a sufficiently small complex neighborhood of $\widehat{K_\varepsilon}$. By Lemma 1.2. 4 we may assume that Φ is entire analytic. Therefore, it follows from Lemma 1.1. 1 that u_1 operates on $(\partial\Phi/\partial x_{n+1})(x,0)$ and

$$\begin{aligned}&u_1((\partial\Phi/\partial x_{n+1})(x,0))\\&= u_{1y}\left(\int P_0(x-y,x_{n+1})(1-\Delta_{x,x_{n+1}})(\chi\Phi)\,dx dx_{n+1}\right).\end{aligned}$$

Approximating the above integral by its Riemann sum, we can show that

$$u_1((\partial\Phi/\partial x_{n+1})(x,0)) = \int U_1(x,x_{n+1})(1-\Delta_{x,x_{n+1}})(\chi\Phi)\,dx dx_{n+1},$$

where $U_1(x,x_{n+1}) = \mathcal{H}(u_1)(x,x_{n+1})$ for $|x_{n+1}| > \varepsilon$ and $U_1(x,x_{n+1})$ is a real analytic function in $\boldsymbol{R}^{n+1} \setminus K_\varepsilon \times [-\varepsilon,\varepsilon]$. Here we have used the assertion (i). We may assume that χ is an even function with respect to x_{n+1}. Then we have

$$\int H(1-\Delta_{x,x_{n+1}})(\chi\Phi)\,dx dx_{n+1} = 0,$$

where $H = U_1 - \widetilde{U}$. In fact, putting $\Phi_1(x,x_{n+1}) = (\Phi(x,x_{n+1}) - \Phi(x,-x_{n+1}))/2$, we have

$$\begin{aligned}&\int \widetilde{U}(1-\Delta_{x,x_{n+1}})(\chi\Phi)\,dx dx_{n+1} = \int \widetilde{U}(1-\Delta_{x,x_{n+1}})(\chi\Phi_1)\,dx dx_{n+1},\\&(1-\Delta_{x,x_{n+1}})\Phi_1 = 0,\\&\Phi_1(x,0) = 0, \quad (\partial\Phi_1/\partial x_{n+1})(x,0) = (\partial\Phi/\partial x_{n+1})(x,0),\\&u_1((\partial\Phi_1/\partial x_{n+1})(x,0)) = \int \widetilde{U}(1-\Delta_{x,x_{n+1}})(\chi\Phi_1)\,dx dx_{n+1},\end{aligned}$$

since $\widetilde{U}$ is an odd function with respect to x_{n+1}. Choose $\chi_1 \in C_0^\infty(\boldsymbol{R}^{n+1})$ so that $\chi = 1$ in supp χ_1 and $\chi_1 = 1$ near $K_\varepsilon \times [-\varepsilon,\varepsilon]$, and put $H_1 =$

$(1-\chi_1)H \in C^\infty(\boldsymbol{R}^{n+1})$. Then we have $\nu \equiv (1-\Delta_{x,x_{n+1}})H_1 \in C_0^\infty(\boldsymbol{R}^{n+1})$ and

$$\int((1-\Delta_{x,x_{n+1}})H_1)\Phi\, dx dx_{n+1} = 0.$$

From Lemma 1.2.5 and hypoellipticity of $(1-\Delta_{x,x_{n+1}})$ it follows that there is $f \in C_0^\infty(\boldsymbol{R}^{n+1})$ satisfying $(1-\Delta_{x,x_{n+1}})f = \nu$. Since $(1-\Delta_{x,x_{n+1}})(H_1 - f) = 0$, $H_1 - f$ is analytic and $H = H_1 - f$ outside $K_\varepsilon \times [-\varepsilon, \varepsilon]$. By the assertion (i) we have $H_1 - f \in \mathcal{S}'(\boldsymbol{R}^{n+1})$. This implies that $H_1 - f \equiv 0$, *i.e.*, $U_1 = \mathcal{H}(u)$ for $|x_{n+1}| > \varepsilon$ and $u_1 = u$, which proves the assertion (ii). □

Lemma 1.2.7 *Let X be an open subset of $\boldsymbol{R}^n$ and $\varepsilon \geq 0$. Assume that $u \in \mathcal{S}'_\varepsilon$ and $v \in \mathcal{S}'$, and represent $\mathcal{H}(u)(x, x_{n+1})$ or its analytic continuation by $U(x, x_{n+1})$.* (i) supp $(u - v) \cap X = \emptyset$ *if and only if $U(x, x_{n+1})$ can be continued analytically from $\boldsymbol{R}^n \times (\boldsymbol{R} \setminus [-\varepsilon, \varepsilon])$ to $X \times (\boldsymbol{R} \setminus \{0\})$ and $U(x, x_{n+1}) \to v(x)/2$ in $\mathcal{D}'(X)$ as $x_{n+1} \downarrow 0$.* (ii) *If $v \in C^\infty(X)$ and* supp $(u-v) \cap X = \emptyset$, *then $U(x, x_{n+1})$ can be regarded as a function in $C^\infty(X \times [0, \infty))$ and in $C^\infty(X \times (-\infty, 0])$, and $U(x, \pm 0) = \pm v(x)/2$ for $x \in X$. If v is real analytic in X, then $U(x, x_{n+1})$ can be continued analytically to a neighborhood of $X \times [0, \infty)$ and one of $X \times (-\infty, 0]$.* (iii) *Assume that X is bounded and that* supp $u \cap X_\delta = \emptyset$, *where $\delta > 0$ and X_δ denotes the δ-neighborhood of X. Then, $U(x, x_{n+1})$ can be continued analytically to the set $\{(z, z_{n+1}) \in \boldsymbol{C}^{n+1}; |\text{Re } z - x| + |(\text{Im } z, \text{Im } z_{n+1})| < \delta$ for some $x \in X\}$. In particular, for any $\rho > 0$ there is $C > 0$ such that*

$$|D_x^\alpha U(x, x_{n+1})| \leq C(2\sqrt{n}/\delta)^{|\alpha|}|\alpha|! \tag{1.14}$$

if $x \in X$ and $-\rho < x_{n+1} < \rho$.

Proof $V(x, x_{n+1}) \equiv \mathcal{H}(v)(x, x_{n+1})$ can be regarded as a real analytic function in $\boldsymbol{R}^n \times (\boldsymbol{R} \setminus \{0\})$. (i) First assume that supp $(u - v) \cap X = \emptyset$. Then $U(x, x_{n+1}) - V(x, x_{n+1})$ can be extended to a real analytic function in $X \times \boldsymbol{R}$, which is an odd function with repect to x_{n+1}. So $U(x, x_{n+1})$ is real analytic in $X \times (\boldsymbol{R} \setminus \{0\})$ and $U(x, x_{n+1}) \to v(x)/2$ in $\mathcal{D}'(X)$ as $x_{n+1} \downarrow 0$, since $V(x, x_{n+1}) \to v(x)/2$ in $\mathcal{S}'$ as $x_{n+1} \downarrow 0$. Next assume that $U(x, x_{n+1})$ is real analytic in $X \times (\boldsymbol{R} \setminus \{0\})$ and that $U(x, x_{n+1}) \to v(x)/2$ in $\mathcal{D}'(X)$ as $x_{n+1} \downarrow 0$. Put $W(x, x_{n+1}) = U(x, x_{n+1}) - V(x, x_{n+1})$. Then we have $W(x, x_{n+1}) \to 0$ in $\mathcal{D}'(X)$ as $x_{n+1} \downarrow 0$. Moreover, we have $(1 - \Delta_{x,x_{n+1}})W(x, x_{n+1}) = 0$ in $X \times (\boldsymbol{R} \setminus \{0\})$, $W(x, -x_{n+1}) = -W(x, x_{n+1})$ for $x \in X$ and $x_{n+1} > 0$. By Lemma 1.2.2 $W(x, x_{n+1})$ can be extended to a real analytic in $X \times \boldsymbol{R}^n$, which gives supp $(u - v) \cap X = \emptyset$. (ii)

Assume that $v \in C^\infty(X)$ and $\mathrm{supp}\,(u-v) \cap X = \emptyset$. Let φ and ψ be functions in $C_0^\infty(X)$ such that $\psi(x) = 1$ in $\mathrm{supp}\,\varphi$. Then, for $x_{n+1} > 0$

$$\begin{aligned}
&\varphi(x) D_x^\alpha D_{x_{n+1}}^j V(x, x_{n+1}) \\
&= (2\pi)^{-n}(\varphi(x)/2) \int \exp[ix\cdot\xi - x_{n+1}\langle\xi\rangle]\xi^\alpha (i\langle\xi\rangle)^j \mathcal{F}[\psi v](\xi)\, d\xi \\
&\quad + \langle v(y), F(x, x_{n+1}, y)\rangle_y \equiv V_{\alpha,j}^1(x, x_{n+1}) + V_{\alpha,j}^2(x, x_{n+1}),
\end{aligned}$$

where

$$\begin{aligned}
F(x, x_{n+1}, y) &= (2\pi)^{-n}(\varphi(x)/2)(1-\psi(y)) \\
&\quad \times \int \exp[i(x-y)\cdot\xi - x_{n+1}\langle\xi\rangle]\xi^\alpha (i\langle\xi\rangle)^j\, d\xi.
\end{aligned}$$

Since $\psi v \in C_0^\infty(X) \subset \mathcal{S}$, we have $V_{\alpha,j}^1(x, x_{n+1}) \in C^\infty(\boldsymbol{R}^n \times [0,\infty))$. Putting

$$L = -|x-y|^{-2} \sum_{k=1}^n (x_k - y_k) D_{\xi_k},$$

we have

$$\begin{aligned}
&\langle y\rangle^\nu D_y^\beta F(x, x_{n+1}, y) \\
&= (2\pi)^{-n}(\varphi(x)/2) \sum_{\beta' \le \beta} \binom{\beta}{\beta'} \langle y\rangle^\nu D_y^{\beta-\beta'}(1-\psi(y)) \\
&\quad \times \int e^{i(x-y)\cdot\xi} L^\ell \left\{\xi^\alpha (-\xi)^{\beta'} (i\langle\xi\rangle)^j \exp[-x_{n+1}\langle\xi\rangle]\right\} d\xi,
\end{aligned}$$

where $\nu \in \boldsymbol{Z}_+$, $\beta = (\beta_1, \cdots, \beta_n) \in \boldsymbol{Z}_+^n$, $\beta! = \beta_1! \cdots \beta_n!$, $\binom{\beta}{\beta'} = \beta!/(\beta'!(\beta-\beta')!)$ for $\beta' \in \boldsymbol{Z}_+^n$ with $\beta' \le \beta$ and $\ell \in \boldsymbol{Z}_+$. Since $|x_{n+1}^N \exp[-x_{n+1}\langle\xi\rangle]| \le N^N e^{-N}\langle\xi\rangle^{-N}$ for $x_{n+1} \ge 0$ and $N \in \boldsymbol{N}$, taking $\ell = \max\{\nu, |\alpha| + |\beta| + j + n + 1\}$, we have

$$|\langle y\rangle^\nu D_y^\beta F(x, x_{n+1}, y)| \le C_{\alpha,j,\nu,\beta}(\varphi, \psi)$$

for $x, y \in \boldsymbol{R}^n$ and $x_{n+1} \ge 0$, *i.e.*, $\{F(x, x_{n+1}, y)\}_{x \in R^n, x_{n+1} \ge 0}$ is a bounded subset of $\mathcal{S}(\boldsymbol{R}_y^n)$. We can also see that $\{(\partial/\partial x_k)F(x, x_{n+1}, y)\}_{x \in R^n, x_{n+1} \ge 0}$ $(1 \le k \le n+1)$ are bounded subsets of $\mathcal{S}(\boldsymbol{R}_y^n)$. Therefore, $V_{\alpha,j}^2(x, x_{n+1}) \in C^0(\boldsymbol{R}^n \times [0,\infty))$, which gives $V(x, x_{n+1}) \in C^\infty(X \times [0,\infty))$. By assumption we have $U(x, x_{n+1}) \in C^\infty(X \times [0,\infty))$. Similarly, we have $U(x, x_{n+1}) \in C^\infty(X \times (-\infty, 0])$. It is obvious that $U(x, \pm 0) = \pm v(x)/2$ for $x \in X$. Moreover, $V(x, x_{n+1})$ satisfies

$$\begin{cases} (1 - \Delta_{x, x_{n+1}})V(x, x_{n+1}) = 0 & \text{in } X \times [0,\infty), \\ V(x, +0) = v(x)/2, \quad (\partial V/\partial x_{n+1})(x, +0) = -\langle D\rangle v(x)/2 & \text{in } X \end{cases} \tag{1.15}$$

It is well-known that $\langle D\rangle v$ is analytic in X if v ($\in \mathcal{S}'$) is analytic in X. We shall prove this fact (analytic pseudolocality) for general analytic pseudodifferential operators in Theorem 2.6. 5 . Now assume that v is analytic in X. Apply the Cauchy-Kowalevsky theorem to (1.15), we can see that $V(x, x_{n+1})$ can be extended to a real analytic function in a neighborhood of $X \times [0,\infty)$. This proves the assertion (ii). (iii) By assumption $U(x, x_{n+1})$ is real analytic in $X_\delta \times \boldsymbol{R}$ and satisfies $(1 - \Delta_{x,x_{n+1}})U(x, x_{n+1}) = 0$. So, from Lemma 1.2. 4 it follows that $U(x, x_{n+1})$ can be continued analytically to $\{(z, z_{n+1}) \in \boldsymbol{C}^{n+1};$ $|\mathrm{Re}\ z - x| + |(\mathrm{Im}\ z, \mathrm{Im}\ z_{n+1})| < \delta$ for some $x \in X\}$. (1.14) easily follows from Cauchy's estimates. □

Definition 1.2. 8 (i) Let $\varepsilon \geq 0$, $u \in \mathcal{F}_\varepsilon$ and $x^0 \in \boldsymbol{R}^n$. We say that u is analytic at x^0 if $\mathcal{H}(u)(x, x_{n+1})$ can be continued analytically from $\boldsymbol{R}^n \times (\varepsilon, \infty)$ to a neighborhood of $\{x^0\} \times [0, \varepsilon]$ in $\boldsymbol{R}^{n+1}$. (ii) For $u \in \mathcal{F}_0$ we define

$$\text{sing supp } u := \{x \in \boldsymbol{R}^n;\ u \text{ is not analytic at } x\}.$$

We give some remarks on supp u. One can not expect that supp $u \subset K$ if $u_N \to u$ in $\mathcal{S}'_\varepsilon$ and supp $u_N \subset K$. In fact, let $u \in \mathcal{A}'(\boldsymbol{C}^n)$, and put $a_\alpha = u(z^\alpha)$ for $\alpha \in \boldsymbol{Z}^n_+$. Then there are $C > 0$ and $A > 0$ such that $|a_\alpha| \leq CA^{|\alpha|}$ for $\alpha \in \boldsymbol{Z}^n_+$. Let $\varepsilon > \sqrt{n}A$. Putting

$$u_N = \sum_{|\alpha|\leq N} a_\alpha(-\partial_x)^\alpha \delta/\alpha! \qquad \text{for } N \in \boldsymbol{Z}_+,$$

we have

$$\hat{u}_N(\xi) = \sum_{|\alpha|\leq N} a_\alpha(-i\xi)^\alpha/\alpha!,$$

and, for $M > N$,

$$\left| e^{-\varepsilon\langle\xi\rangle} \sum_{N<|\alpha|\leq M} a_\alpha(-i\xi)^\alpha/\alpha! \right| \leq C \sum_{j=N+1}^{M} (\sqrt{n}Aj/(\varepsilon e))^j/j!,$$

since $|\xi_1| + \cdots + |\xi_n| \leq \sqrt{n}|\xi|$ and $|\xi|^j e^{-\varepsilon\langle\xi\rangle} \leq (j/(\varepsilon e))^j$ for $j \in \boldsymbol{N}$. Here δ ($\in \mathcal{D}'(\boldsymbol{R}^n)$) denotes Dirac's delta function on $\boldsymbol{R}^n$. Therefore, $\{u_N\}$ is convergent in $\mathcal{S}'_\varepsilon$. For $f \in \mathcal{S}_\infty \subset \mathcal{A}$, $\sum_\alpha z^\alpha(\partial_x^\alpha f)(0)/\alpha!$ converges locally uniformly to $f(z)$ in $\boldsymbol{C}^n$. So we have

$$u(f)\ (= \langle u, f\rangle) = \sum_\alpha \langle a_\alpha(-\partial_x)^\alpha\delta/\alpha!, f\rangle$$

for $f \in \mathcal{S}_\infty$. This yields $u_N \to u$ in $\mathcal{S}'_\varepsilon$ as $N \to \infty$. It is obvious that supp $u_N \subset \{0\}$.

1.3 Localization

Let us begin with a lemma on partion of unity (see Corollary 1.4.11 in [Hr5]).

Lemma 1.3. 1 *Let F_0 and F_1 be closed subsets of $\boldsymbol{R}^n$. Then there is $\phi \in C^\infty(\boldsymbol{R}^n \setminus (F_0 \cap F_1))$ such that $\phi = 0$ near $F_0 \setminus F_1$, $\phi = 1$ near $F_1 \setminus F_0$. Moreover,* supp ϕ *is bounded if F_1 is compact.*

We shall also need an existence theorem for elliptic differential operators with constant coefficients (see, *e.g.*, Theorem 4.4.6 in [Hr5]).

Proposition 1.3. 2 *Let $P(D)$ be an elliptic differential operator with constant coefficients, and let X be an open subset of $\boldsymbol{R}^n$. Then, for any $f \in \mathcal{D}'(X)$ there exists $u \in \mathcal{D}'(X)$ such that $P(D)u = f$ in X.*

Theorem 1.3. 3 *Let K be a compact subset of $\boldsymbol{R}^n$, $\varepsilon \geq 0$ and $u \in \mathcal{F}_\varepsilon$. Then there is $v \in \mathcal{A}'(K^\varepsilon)$ such that* supp $v \subset K$ *and* supp $(u - v) \subset \overline{\boldsymbol{R}^n \setminus K}$, *where K^ε is defined as in* Proposition 1.2. 6 . *Moreover, if $v_1 \in \mathcal{A}'(K^\varepsilon)$ satisfies* supp $v_1 \subset K$ *and* supp $(u - v_1) \subset \overline{\boldsymbol{R}^n \setminus K}$, *then* supp $(v - v_1) \subset \partial K$.

Remark By Theorem 1.3. 3 one can define the restriction map from $\mathcal{F}_\varepsilon$ to $\mathcal{A}'(K^\varepsilon)/\{u \in \mathcal{A}'(K^\varepsilon);\ \mathrm{supp}\ u \subset \partial K\}$.

Proof Following [Hr5], we shall prove the theorem. From Lemma 1.3. 1 it follows that there is $\phi \in C^\infty(\boldsymbol{R}^{n+1} \setminus \partial K \times [-\varepsilon, \varepsilon])$ satisfying $\phi(x, x_{n+1}) = \phi(x, -x_{n+1})$ and

$$\phi(x, x_{n+1}) = \begin{cases} 0 & \text{if } |(x, x_{n+1})| \gg 1, \\ 1 & \text{near } (K \setminus \partial K) \times [-\varepsilon, \varepsilon], \\ 0 & \text{near } (\boldsymbol{R}^n \setminus K) \times [-\varepsilon, \varepsilon]. \end{cases}$$

Put $U(x, x_{n+1}) = \mathcal{H}(u)(x, x_{n+1})$ for $|x_{n+1}| > \varepsilon$. Then ϕU can be regarded as a function in $C^\infty(\boldsymbol{R}^{n+1} \setminus K \times [-\varepsilon, \varepsilon])$. Moreover, $(1 - \Delta_{x,x_{n+1}})(\phi U)$ can be regarded as a function in $C^\infty(\boldsymbol{R}^{n+1} \setminus \partial K \times [-\varepsilon, \varepsilon])$ and satisfies

$$(1 - \Delta_{x,x_{n+1}})(\phi U) = 0 \quad \text{near } (\boldsymbol{R}^n \setminus \partial K) \times [-\varepsilon, \varepsilon],$$

noting that $(1 - \Delta_{x,x_{n+1}})U = 0$ for $|x_{n+1}| > \varepsilon$. Since $1 - \Delta_{x,x_{n+1}}$ is elliptic, it follows from Proposition 1.3. 2 that there is $f \in C^\infty(\boldsymbol{R}^{n+1} \setminus \partial K \times [-\varepsilon, \varepsilon])$ such that $f(x, x_{n+1}) = -f(x, -x_{n+1})$ and

$$(1 - \Delta_{x,x_{n+1}})f = (1 - \Delta_{x,x_{n+1}})(\phi U) \quad \text{in } \boldsymbol{R}^{n+1} \setminus \partial K \times [-\varepsilon, \varepsilon].$$

Put $\widetilde{V} = \phi U - f \in C^\infty(\boldsymbol{R}^{n+1} \setminus K \times [-\varepsilon, \varepsilon])$. Then we have

$$(1 - \Delta_{x,x_{n+1}})\widetilde{V} = 0 \quad \text{in } \boldsymbol{R}^{n+1} \setminus K \times [-\varepsilon, \varepsilon].$$

Regarding $(1 - \phi)U$ as a function in $C^\infty(\boldsymbol{R}^{n+1} \setminus (\overline{\boldsymbol{R}^n \setminus K}) \times [-\varepsilon, \varepsilon])$, we have $U - \widetilde{V} \in C^\infty(\boldsymbol{R}^{n+1} \setminus (\overline{\boldsymbol{R}^n \setminus K}) \times [-\varepsilon, \varepsilon])$ and

$$(1 - \Delta_{x,x_{n+1}})(U - \widetilde{V}) = 0 \quad \text{in } \boldsymbol{R}^{n+1} \setminus (\overline{\boldsymbol{R}^n \setminus K}) \times [-\varepsilon, \varepsilon].$$

Define $v : \mathcal{A} \ni \varphi \mapsto v(\varphi) \in \boldsymbol{C}$ by

$$v(\varphi) = \int \widetilde{V}(1 - \Delta_{x,x_{n+1}})(\chi\Phi)\, dx dx_{n+1},$$

where χ is a function in $C_0^\infty(\boldsymbol{R}^{n+1})$ satisfying $\chi = 1$ near $K \times [-\varepsilon, \varepsilon]$ and Φ is a unique solution of (1.8). Then we can prove $v \in \mathcal{A}'(K^\varepsilon)$ by the same argument as in the proof of Proposition 1.2. 6 . Put $V = \mathcal{H}(v)(x, x_{n+1})$ for $|x_{n+1}| > \varepsilon$ and $H = \widetilde{V} - V$. Applying the same argument as in the proof of Proposition 1.2. 6 , we can regard H as a function in $C^\infty(\boldsymbol{R}^{n+1})$, which satisfies $(1 - \Delta_{x,x_{n+1}})H = 0$. Since $U - V = U - \widetilde{V} + H$ for $|x_{n+1}| > \varepsilon$, $U - V$ can be continued analytically to a function defined in $\boldsymbol{R}^{n+1} \setminus ((\overline{\boldsymbol{R}^n \setminus K}) \times [-\varepsilon, \varepsilon])$. This gives supp $(u - v) \subset \overline{\boldsymbol{R}^n \setminus K}$. $V = \widetilde{V} - H$ can be continued analytically to $\boldsymbol{R}^{n+1} \setminus K \times [-\varepsilon, \varepsilon]$ and, therefore, supp $v \subset K$. The second part of the theorem is obvious, since $V - \mathcal{H}(v_1) = (U - \mathcal{H}(v_1)) - (U - V)$. □

We denote $\mathcal{A}'_\varepsilon(K) := \{u \in \mathcal{A}'(\boldsymbol{C}^n) \cap \mathcal{F}_\varepsilon;\ \text{supp } u \subset K\}$ for a closed subset K of $\boldsymbol{R}^n$ and $\varepsilon \in \boldsymbol{R}$.

Theorem 1.3. 4 *Let K_1 and K_2 be compact subsets of $\boldsymbol{R}^n$ and $\varepsilon \geq 0$. If $u \in \mathcal{F}_\varepsilon$ and* supp $u \subset K_1 \cup K_2$, *then there are $u_1 \in \mathcal{A}'_\varepsilon(K_1)$ and $u_2 \in \mathcal{F}_\varepsilon$ such that* supp $u_2 \subset K_2$ *and $u = u_1 + u_2$.*

Proof Assume that $u \in \mathcal{F}_\varepsilon$ and supp $u \subset K_1 \cup K_2$, and put $U(x, x_{n+1}) = \mathcal{H}(u)(x, x_{n+1})$ for $|x_{n+1}| > \varepsilon$. Then U can be continued analytically to $\boldsymbol{R}^{n+1} \setminus (K_1 \cup K_2) \times [-\varepsilon, \varepsilon]$ and satisfies $(1 - \Delta_{x,x_{n+1}})U(x, x_{n+1}) = 0$ there. By Lemma 1.3. 1 we can choose $\phi \in C^\infty(\boldsymbol{R}^{n+1} \setminus (K_1 \cap K_2) \times [-\varepsilon, \varepsilon])$ so that $\phi(x, x_{n+1}) = \phi(x, -x_{n+1})$ and

$$\phi(x, x_{n+1}) = \begin{cases} 0 & \text{if } |(x, x_{n+1})| \gg 1, \\ 0 & \text{near } (K_2 \setminus K_1) \times [-\varepsilon, \varepsilon], \\ 1 & \text{near } (K_1 \setminus K_2) \times [-\varepsilon, \varepsilon]. \end{cases}$$

ϕU (resp. $(1 - \phi)U$) can be regarded as a function in $C^\infty(\boldsymbol{R}^{n+1} \setminus K_1 \times [-\varepsilon, \varepsilon])$ (resp. in $C^\infty(\boldsymbol{R}^{n+1} \setminus K_2 \times [-\varepsilon, \varepsilon])$). Moreover, $(1 - \Delta_{x,x_{n+1}})(\phi U)$

and $(1-\Delta_{x,x_{n+1}})((1-\phi)U)$ can be regarded as functions in $C^\infty(\boldsymbol{R}^{n+1}\setminus (K_1\cap K_2)\times[-\varepsilon,\varepsilon])$ and satisfies

$$(1-\Delta_{x,x_{n+1}})(\phi U)=0, \qquad (1-\Delta_{x,x_{n+1}})((1-\phi)U)=0$$

near $\{(K_1\cup K_2)\setminus(K_1\cap K_2)\}\times[-\varepsilon,\varepsilon]$. From Proposition 1.3. 2 it follows that there is $v\in C^\infty(\boldsymbol{R}^{n+1}\setminus(K_1\cap K_2)\times[-\varepsilon,\varepsilon])$ such that $v(x,x_{n+1})=-v(x,-x_{n+1})$ and

$$(1-\Delta_{x,x_{n+1}})v=(1-\Delta_{x,x_{n+1}})(\phi U) \quad \text{in } \boldsymbol{R}^{n+1}\setminus(K_1\cap K_2)\times[-\varepsilon,\varepsilon].$$

Put

$$\widetilde{U}_1=\phi U-v, \qquad \widetilde{U}_2=(1-\phi)U+v,$$

where we regard $\widetilde{U}_j$ as a function in $C^\infty(\boldsymbol{R}^{n+1}\setminus K_j\times[-\varepsilon,\varepsilon])$ ($j=1,2$). Then,

$$(1-\Delta_{x,x_{n+1}})\widetilde{U}_j=0 \quad \text{in } \boldsymbol{R}^{n+1}\setminus K_j\times[-\varepsilon,\varepsilon] \ (j=1,2).$$

Define $u_1:\mathcal{A}\ni\varphi\mapsto u_1(\varphi)\in\boldsymbol{C}$ by

$$u_1(\varphi)=\int\widetilde{U}_1(1-\Delta_{x,x_{n+1}})(\chi\Phi)\,dx dx_{n+1},$$

where χ is a function in $C_0^\infty(\boldsymbol{R}^{n+1})$ satisfying $\chi=1$ near $K_1\times[-\varepsilon,\varepsilon]$ and Φ is a unique solution of (1.8). By the same argument as in the proof of Proposition 1.2. 6 , we have $u_1\in\mathcal{A}'(K_1^\varepsilon)$. Moreover, applying the same argument as in the proof of Proposition 1.2. 6 , we can regard $H\equiv\widetilde{U}_1-\mathcal{H}(u_1)$ as a function in $C^\infty(\boldsymbol{R}^{n+1})$, which satisfies $(1-\Delta_{x,x_{n+1}})H=0$. This implies that supp $u_1\subset K_1$ and supp $(u-u_1)\subset K_2$, since $U-\mathcal{H}(u_1)=U-\widetilde{U}_1+H=\widetilde{U}_2+H$ for $|x_{n+1}|>\varepsilon$ and $\widetilde{U}_2+H\in C^\infty(\boldsymbol{R}^{n+1}\setminus K_2\times[-\varepsilon,\varepsilon])$. □

Corollary 1.3. 5 *Let X_1, X_2 and X be bounded open subsets of $\boldsymbol{R}^n$ such that $X=X_1\cup X_2$, and let $\varepsilon\geq 0$. Assume that $u_j\in\mathcal{F}_\varepsilon$ ($j=1,2$) satisfy* supp $u_j\subset\overline{X}_j$ ($j=1,2$) *and* supp $(u_1-u_2)\cap(X_1\cap X_2)=\emptyset$. *Then there is $u\in\mathcal{A}'_\varepsilon(\overline{X})$ such that* supp $(u-u_j)\cap X_j=\emptyset$ ($j=1,2$).

Proof Note that

$$\operatorname{supp}(u_1-u_2)\subset(\overline{X}_1\setminus X_2)\cup(\overline{X}_2\setminus X_1).$$

From Theorems 1.3. 3 and 1.3. 4 it follows that there are $\tilde{u}_j\in\mathcal{A}'_\varepsilon(\overline{X}_j)$ ($j=1,2$), $v_1\in\mathcal{A}'_\varepsilon(\overline{X}_1\setminus X_2)$ and $v_2\in\mathcal{A}'_\varepsilon(\overline{X}_2\setminus X_1)$ such that supp $(u_j-\tilde{u}_j)=\emptyset$ ($j=1,2$) and $\tilde{u}_1-\tilde{u}_2=v_2-v_1$. Put $u=\tilde{u}_1-v_2$ ($=\tilde{u}_2-v_1$). Then we have $u\in\mathcal{A}'_\varepsilon(\overline{X})$ and supp $(u-u_j)\cap X_j=\emptyset$ ($j=1,2$). □

Concerning singular supports we have the same results.

Theorem 1.3. 6 *Let K_1, K_2 and K be compact subsets of $\boldsymbol{R}^n$ such that $K_1 \cup K_2 \subset K$. Then, for any $u \in \mathcal{F}_0$ with* sing supp $u \subset K_1 \cup K_2$ *there are $u_1 \in \mathcal{A}'(K)$ and $u_2 \in \mathcal{F}_0$ such that $u = u_1 + u_2$ and* sing supp $u_j \subset K_j \cup \partial K$ *($j = 1, 2$). Moreover, $\mathcal{H}(u_1)$ can be continued analytically from $\boldsymbol{R}^n \times (0, \infty)$ to a neighborhood of $(K_2 \setminus K_1) \times \boldsymbol{R} \setminus \partial K \times \{0\}$.*

Remark In the above theorem one can replace K_1 with $K_1 \setminus \mathrm{int}(K_2)$. Then one can improve the result on analytic continuation of $\mathcal{H}(u_1)$.

Proof Assume that $u \in \mathcal{F}_0$ and sing supp $u \subset K_1 \cup K_2$, and put $U(x, x_{n+1}) = \mathcal{H}(u)(x, x_{n+1})$ for $x_{n+1} \neq 0$. Then there is an open neighborhood Ω_0 of $\boldsymbol{R}^n \times [0, \infty) \setminus (K_1 \cup K_2) \times \{0\}$ such that $U(x, x_{n+1})$ can be continued analytically to Ω_0. We may assume that $\Omega_0 \cap (K_1 \cap K_2) \times \{0\} = \emptyset$. By Lemma 1.3. 1 with $F_0 = K_2 \times \boldsymbol{R}$ and $F_1 = K_1 \times \{0\}$, we can choose $\phi_1 \in C^\infty(\boldsymbol{R}^{n+1} \setminus (K_1 \cap K_2) \times \{0\})$ so that

$$\phi_1(x, x_{n+1}) = \begin{cases} 0 & \text{if } |(x, x_{n+1})| \gg 1, \\ 0 & \text{near } K_2 \times \boldsymbol{R} \setminus K_1 \times \{0\}, \\ 1 & \text{near } (K_1 \setminus K_2) \times \{0\}. \end{cases}$$

Then $\phi_1 U$ (resp. $(1 - \phi_1)U$) can be regarded as a function in $C^\infty(\Omega_0 \cup \Omega_1)$ (resp. $C^\infty(\Omega_0 \cup \Omega_2)$), where $\Omega_1 = \boldsymbol{R}^{n+1} \setminus ((K_1 \cap K_2) \times \{0\} \cup \mathrm{supp}\, \phi_1)$ and $\Omega_2 = \boldsymbol{R}^{n+1} \setminus ((K_1 \cap K_2) \times \{0\} \cup \mathrm{supp}\, (1 - \phi_1))$. Moreover, $(1 - \Delta_{x, x_{n+1}})(\phi_1 U)$ can be regarded as a function in $C^\infty(\Omega_0 \cup \Omega_1 \cup \Omega_2)$ and satisfies

$$(1 - \Delta_{x, x_{n+1}})(\phi_1 U) = 0 \qquad \text{in } \Omega_1 \cup \Omega_2.$$

From Proposition 1.3. 2 there is $v_1 \in C^\infty(\Omega_0 \cup \Omega_1 \cup \Omega_2)$ such that

$$(1 - \Delta_{x, x_{n+1}})v_1 = (1 - \Delta_{x, x_{n+1}})(\phi_1 U) \quad \text{in } \Omega_0 \cup \Omega_1 \cup \Omega_2.$$

Put $V_1 = \phi_1 U - v_1 \in C^\infty(\Omega_0 \cup \Omega_1)$ and $V_2 = (1 - \phi_1)U + v_1 \in C^\infty(\Omega_0 \cup \Omega_2)$. Then we have

$$(1 - \Delta_{x, x_{n+1}})V_j = 0 \qquad \text{in } \Omega_0 \cup \Omega_j \ (\ j = 1, 2).$$

Note that $\Omega_0 \cup \Omega_1$ is a neighborhood of $(\boldsymbol{R}^n \times [0, \infty) \setminus K_1 \times \{0\}) \cup (K_2 \setminus K_1) \times \boldsymbol{R}$ and that $\Omega_0 \cup \Omega_2$ is a neighborhood of $\boldsymbol{R}^n \times [0, \infty) \setminus K_2 \times \{0\}$. We put

$$V(x, x_{n+1}) = \begin{cases} V_1(x, x_{n+1}) & (\ x_{n+1} > 0), \\ -V_1(x, -x_{n+1}) & (\ x_{n+1} < 0). \end{cases}$$

Let $\phi_2 \in C^\infty(\boldsymbol{R}^{n+1} \setminus \partial K \times \{0\})$ be a function such that $\phi_2(x, x_{n+1}) = \phi_2(x, -x_{n+1})$ and

$$\phi_2(x, x_{n+1}) = \begin{cases} 0 & \text{if } |(x, x_{n+1})| \gg 1, \\ 1 & \text{near } (K \setminus \partial K) \times \{0\}, \\ 0 & \text{near } (\boldsymbol{R}^n \setminus K) \times \{0\}. \end{cases}$$

Then $\phi_2 V$ can be regarded as a function in $C^\infty(\boldsymbol{R}^{n+1} \setminus K \times \{0\})$. Moreover, $(1 - \Delta_{x,x_{n+1}})(\phi_2 V)$ can be regarded as a function in $C^\infty(\boldsymbol{R}^{n+1} \setminus \partial K \times \{0\})$. Similarly, there is $v_2 \in C^\infty(\boldsymbol{R}^{n+1} \setminus \partial K \times \{0\})$ such that $v_2(x, x_{n+1}) = -v_2(x, -x_{n+1})$ and

$$(1 - \Delta_{x,x_{n+1}})v_2 = (1 - \Delta_{x,x_{n+1}})(\phi_2 V) \quad \text{in } \boldsymbol{R}^{n+1} \setminus \partial K \times \{0\}.$$

Applying the same argument as in the proof of Proposition 1.2. 6 , we can define $u_1 \in \mathcal{A}'(K)$ by

$$u_1(\varphi) = \int \widetilde{U}_1 (1 - \Delta_{x,x_{n+1}})(\chi\Phi)\, dx dx_{n+1} \qquad \text{for } \varphi \in \mathcal{A},$$

where $\widetilde{U}_1 = \phi_2 V - v_2 \in C^\infty(\boldsymbol{R}^{n+1} \setminus K \times \{0\})$, χ is a function in $C_0^\infty(\boldsymbol{R}^{n+1})$ satisfying $\chi = 1$ near $K \times \{0\}$ and Φ is a unique solution of (1.8). Moreover, $H \equiv \widetilde{U}_1 - \mathcal{H}(u_1)$ can be regarded as a function in $C^\infty(\boldsymbol{R}^{n+1})$ and satisfies $(1 - \Delta_{x,x_{n+1}})H = 0$ in $\boldsymbol{R}^{n+1}$. Since $\mathcal{H}(u_1) = \phi_2 V - v_2 - H = \phi_2 V_1 - v_2 - H$ for $x_{n+1} > 0$, $\mathcal{H}(u_1)$ can be continued analytically from $\boldsymbol{R}^n \times (0, \infty)$ to a neighborhood of $(\boldsymbol{R}^n \times [0, \infty) \setminus (\partial K \cup K_1) \times \{0\}) \cup ((K_2 \setminus K_1) \times \boldsymbol{R} \setminus \partial K \times \{0\})$, which gives sing supp $u_1 \subset \partial K \cup K_1$. On the other hand,

$$U - \mathcal{H}(u_1) = U - \phi_2 V_1 + v_2 + H = (1 - \phi_2)U + \phi_2 V_2 + v_2 + H \quad \text{for } x_{n+1} > 0$$

and $U - \mathcal{H}(u_1)$ can be continued analytically from $\boldsymbol{R}^n \times (0, \infty)$ to a neighborhood of $\boldsymbol{R}^n \times [0, \infty) \setminus (\partial K \cup K_2) \times \{0\}$, since $1 - \phi_2 = 0$ in a neighborhood of $(K \setminus \partial K) \times \{0\}$ in $\boldsymbol{R}^{n+1} \setminus \partial K \times \{0\}$. This implies that sing supp $(u - u_1) \subset \partial K \cup K_2$. □

Corollary 1.3. 7 *Let X_1, X_2 and X be bounded open subsets of $\boldsymbol{R}^n$ such that $X_j \subset X$ ($j = 1, 2$). Assume that $u_j \in \mathcal{F}_0$ and* sing supp $u_j \subset \overline{X}_j$ *($j = 1, 2$), and* sing supp $(u_1 - u_2) \cap (X_1 \cap X_2) = \emptyset$*, Then there is $u \in \mathcal{F}_0$ such that* sing supp $(u - u_j) \cap X_j = \emptyset$ *($j = 1, 2$) and* sing supp $u \subset \overline{X}_1 \cup \overline{X}_2 \cup \partial X$. *Moreover, $\mathcal{H}(u - u_1)$ can be continued analytically from $\boldsymbol{R}^n \times (0, \infty)$ to a neighborhood of $\{\overline{X}_1 \setminus (X_2 \cup (\partial X_1 \cap \partial X_2))\} \times \boldsymbol{R} \setminus \partial X \times \{0\}$.*

Proof It is obvious that

$$\text{sing supp } (u_1 - u_2) \subset (\overline{X}_2 \setminus X_1) \cup (\overline{X}_1 \setminus X_2).$$

From Theorem 1.3.6 there are $v_1 \subset \mathcal{A}'(\overline{X})$ and $v_2 \in \mathcal{F}_0$ such that $u_1 - u_2 = v_1 - v_2$, sing supp $v_1 \subset (\overline{X}_2 \setminus X_1) \cup \partial X$ and sing supp $v_2 \subset (\overline{X}_1 \setminus X_2) \cup \partial X$. Moreover, $\mathcal{H}(v_1)$ can be continued analytically from $\boldsymbol{R}^n \times (0, \infty)$ to a neighborhood of $\{\overline{X}_1 \setminus (X_2 \cup (\partial X_1 \cap \partial X_2))\} \times \boldsymbol{R} \setminus \partial X \times \{0\}$. Therefore, putting $u = u_1 - v_1$ $(= u_2 - v_2)$, we have $u \in \mathcal{F}_0$ and sing supp $u \subset \overline{X}_1 \cup \overline{X}_2 \cup \partial X$. Moreover, we have sing supp $(u - u_j) \cap X_j = \emptyset$ $(j = 1, 2)$, since $u - u_j = v_j$ $(j = 1, 2)$. □

1.4 Hyperfunctions

Following [Hr5], we shall define the space of hyperfunctions and study some properties of hyperfunctions.

Definition 1.4.1 Let $\varepsilon \geq 0$ and X be a bounded open subset of $\boldsymbol{R}^n$. (i) We define

$$\mathcal{B}_\varepsilon(X) := \mathcal{A}'_\varepsilon(\overline{X}) / \mathcal{A}'_\varepsilon(\partial X).$$

We also write $\mathcal{B}(X) = \mathcal{B}_0(X)$, which is called the space of hyperfunctions in X. (ii) For an open subset Y of X and $u^\circ \in \mathcal{B}_\varepsilon(X)$ the restriction $u^\circ|_Y \in \mathcal{B}_\varepsilon(Y)$ of u° to Y is defined by the residue class $[v]$ of $v \in \mathcal{A}'_\varepsilon(\overline{Y})$ which satisfies supp $(u - v) \subset \overline{X} \setminus Y$, where the residue class of $u \in \mathcal{A}'_\varepsilon(\overline{X})$ is u° in $\mathcal{B}_\varepsilon(X)$ (see Theorem 1.3.3). (iii) For $u^\circ \in \mathcal{B}_\varepsilon(X)$ we define supp $u^\circ :=$ supp $u \cap X$, where the residue class of $u \in \mathcal{A}'_\varepsilon(\overline{X})$ is u° in $\mathcal{B}_\varepsilon(X)$. (iv) For $u^\circ \in \mathcal{B}(X)$ we define sing supp $u^\circ :=$ sing supp $u \cap X$, where the residue class of $u \in \mathcal{A}'(\overline{X})$ is u° in $\mathcal{B}(X)$. (v) It follows from the remark of Theorem 1.3.3 that each u in $\mathcal{F}_\varepsilon$ uniquely determines $v \in \mathcal{B}_\varepsilon(X)$ such that supp $(u - v_1) \cap X = \emptyset$ if the residue class in $\mathcal{B}(X)$ of $v_1 \in \mathcal{A}'_\varepsilon(\overline{X})$ is v. We also call v the restriction of u to X, and denote $u|_X = v$.

Remark (i) It is obvious that supp u° and sing supp u° are well-defined. (ii) If $u \in \mathcal{B}_\varepsilon(X)$ and $Y \subset X$ is open, then supp $u|_Y =$ supp $u \cap Y$. Moreover, sing supp $u|_Y =$ sing supp $u \cap Y$ if $u \in \mathcal{B}(X)$. (iii) $\mathcal{B}_\varepsilon(X)$ can be also defined by $\mathcal{A}'_\varepsilon(\boldsymbol{R}^n)/\mathcal{A}'_\varepsilon(\boldsymbol{R}^n \setminus X)$. (iv) For $u \in \mathcal{A}'_\varepsilon(\overline{X})$ $u|_X$ $(\in \mathcal{B}_\varepsilon(X))$ is the residue class of u.

We shall define the sheaf $\mathcal{B}_\varepsilon$ and prove that the sheaf $\mathcal{B}_\varepsilon$ is a flabby sheaf (see Definition 1.4.6 and Theorem 1.4.8 below). In doing so, we need the following propositions.

Proposition 1.4. 2 *Let $P(D)$ be an elliptic differential operator with constant coefficients, and let X and Y be open subsets of $\boldsymbol{R}^n$ such that $Y \subset X$, and $K = \emptyset$ if $X \setminus Y = F \cup K$, $F \cap K = \emptyset$, F is closed in X and K is compact. If $u \in C^\infty(Y)$ satisfies $P(D)u = 0$ in Y, then there is a sequence $\{u_j\} \subset C^\infty(X)$ such that $P(D)u_j = 0$ in X and $u_j|_Y \to u$ in $C^\infty(Y)$, i.e., $D^\alpha(u_j|_Y) \to u$ uniformly in every compact subset of Y for any $\alpha \in \boldsymbol{Z}_+^n$, where $u_j|_Y$ denotes the restriction of u_j to Y.*

The above proposition is an extension of the Runge approximation theorem (see, *e.g.*, Theorem 4.4.5 in [Hr5]).

Proposition 1.4. 3 *Let K_0 and K be compact subsets of $\boldsymbol{R}^n$ with $K_0 \subset K$, and let $\varepsilon \geq 0$ and $\{\varepsilon_j\}$ be a sequence in $\boldsymbol{R}$ such that $\varepsilon_j \downarrow \varepsilon$. Assume that* (i) *$u_j \in \mathcal{F}_{\varepsilon_j}$ ($j \in \boldsymbol{N}$),* (ii) *for any neighborhood $\mathcal{U}$ of K there is $j_0 \in \boldsymbol{N}$ such that* supp *$u_j \subset \mathcal{U}$ for $j \geq j_0$, and that* (iii) *for any neighborhood $\mathcal{U}_0$ of K_0 there is $j_1 \in \boldsymbol{N}$ such that* supp *$(u_j - u_k) \subset \mathcal{U}_0$ for $j,k \geq j_1$. Then there is $u \in \mathcal{A}'_\varepsilon(K)$ ($\subset \mathcal{F}_\varepsilon$) such that for any neighborhood $\mathcal{U}_0$ of K_0 there is $j_2 \in \boldsymbol{N}$ satisfying* supp *$(u - u_j) \subset \mathcal{U}_0$ for $j \geq j_2$. Moreover, if v has the same properties as u, then* supp *$(u - v) \subset K_0$.*

Proof Following [Hr5], we shall prove the proposition. We may assume without loss of generality that (i)$'$ $u_j \in \mathcal{F}_{\varepsilon_j}$ and supp $u_j \subset K_{\varepsilon_j - \varepsilon}$ ($j = 1, 2, \cdots$), where $K_\delta = \{x \in \boldsymbol{R}^n;\ |x - y| \leq \delta$ for some $y \in K\}$, and that (ii)$'$ supp $(u_j - u_k) \subset (K_0)_{\varepsilon_j - \varepsilon}$ if $k \geq j$, modifying $\{\varepsilon_j\}$ and omitting the first several terms from $\{u_j\}$ if necessary. Let $U_j(x, x_{n+1})$ be an real analytic continuation of $\mathcal{H}(u_j)(x, x_{n+1})$ ($j \in \boldsymbol{N}$). Then, by the assumption (i)$'$, we have $U_j \in C^\infty(\boldsymbol{R}^{n+1} \setminus K_{\varepsilon_j - \varepsilon} \times [-\varepsilon_j, \varepsilon_j])$ and

$$(1 - \Delta_{x, x_{n+1}})U_j = 0 \qquad \text{in } \boldsymbol{R}^{n+1} \setminus K_{\varepsilon_j - \varepsilon} \times [-\varepsilon_j, \varepsilon_j].$$

By the assumption (ii)$'$ $U_j - U_k$ can be regarded as a function in $C^\infty(\boldsymbol{R}^{n+1} \setminus (K_0)_{\varepsilon_j - \varepsilon} \times [-\varepsilon_j, \varepsilon_j])$ and satisfies $(1 - \Delta_{x, x_{n+1}})(U_j - U_k) = 0$ there if $k \geq j$. Now we can apply Proposition 1.4. 2 with $P(D) = (1 - \Delta_{x, x_{n+1}})$, $X = \boldsymbol{R}^{n+1} \setminus K_0 \times [-\varepsilon, \varepsilon]$ and $Y = \boldsymbol{R}^{n+1} \setminus (K_0)_{\varepsilon_j - \varepsilon} \times [-\varepsilon_j, \varepsilon_j]$. In fact, assume that there are a nonvoid colsed subset F_1 of $\boldsymbol{R}^{n+1} \setminus K_0 \times [-\varepsilon, \varepsilon]$ and a nonvoid compact subset F_2 of $\boldsymbol{R}^{n+1}$ satisfying $F_1 \cap F_2 = \emptyset$ and $F_1 \cup F_2 = (K_0)_{\varepsilon_j - \varepsilon} \times [-\varepsilon_j, \varepsilon_j] \setminus K_0 \times [-\varepsilon, \varepsilon]$. Then $F_1 \cup K_0 \times [-\varepsilon, \varepsilon]$ is colsed in $\boldsymbol{R}^{n+1}$ and, therefore, compact. By assumptions there are $(x^0, x^0_{n+1}) \in F_2$ and $(y^0, y^0_{n+1}) \in K_0 \times [-\varepsilon, \varepsilon]$ such that $|x^0 - y^0| \leq \varepsilon_j - \varepsilon$ and $|x^0_{n+1} - y^0_{n+1}| \leq \varepsilon_j - \varepsilon$. Put $Q = \{(x, x_{n+1}) \in \boldsymbol{R}^{n+1};\ |x - y^0| \leq \varepsilon_j - \varepsilon$ and $|x_{n+1} - y^0_{n+1}| \leq \varepsilon_j - \varepsilon\}$ ($\subset (K_0)_{\varepsilon_j - \varepsilon} \times [-\varepsilon_j, \varepsilon_j]$). Then Q is connected and $Q = \{Q \cap (F_1 \cup K_0 \times [-\varepsilon, \varepsilon])\} \cup (Q \cap F_2)$. This is a contradiction, since $(y^0, y^0_{n+1}) \in Q \cap (F_1 \cup K_0 \times [-\varepsilon, \varepsilon])$ and $(x^0, x^0_{n+1}) \in Q \cap F_2$. Put, for $j \in \boldsymbol{N}$,

$$\begin{aligned} M_j &= \{(x, x_{n+1}) \in \boldsymbol{R}^{n+1};\ |(x, x_{n+1})| \le j \text{ and } |x-y| + |x_{n+1} - y_{n+1}| \\ &\qquad \ge 3(\varepsilon_j - \varepsilon) \text{ for any } (y, y_{n+1}) \in K_0 \times [-\varepsilon, \varepsilon]\}, \end{aligned}$$

which is a compact subset of $\boldsymbol{R}^{n+1} \setminus (K_0)_{\varepsilon_j - \varepsilon} \times [-\varepsilon_j, \varepsilon_j]$. Moreover, $M_j \uparrow (\boldsymbol{R}^{n+1} \setminus K_0 \times [-\varepsilon, \varepsilon])$ as $j \to \infty$. From Proposition 1.4. 2 there is a sequence $\{V_j\} \subset C^\infty(\boldsymbol{R}^{n+1} \setminus K_0 \times [-\varepsilon, \varepsilon])$ such that $V_j(x, x_{n+1}) = -V_j(x, -x_{n+1})$, $(1 - \Delta_{x, x_{n+1}})V_j = 0$ in $\boldsymbol{R}^{n+1} \setminus K_0 \times [-\varepsilon, \varepsilon]$ and $|U_{j+1} - U_j - V_j| \le 2^{-j}$ in M_j. Since

$$\begin{aligned} &U_{j+1} - V_1 - \cdots - V_j \\ &= U_k - V_1 - \cdots - V_{k-1} + \sum_{\nu=k}^{j} (U_{\nu+1} - U_\nu - V_\nu) \quad (j \ge k \ge 2), \end{aligned}$$

the sequence $\{U_{j+1} - V_1 - \cdots - V_j\}_{j \ge k}$ converges locally uniformly in $\boldsymbol{R}^{n+1} \setminus K_{\varepsilon_k - \varepsilon} \times [-\varepsilon_k, \varepsilon_k]$. Put $U = \lim_{j \to \infty}(U_{j+1} - V_1 - \cdots - V_j)$. Then

$$(1 - \Delta_{x, x_{n+1}})U = 0 \qquad \text{in } \mathcal{D}'(\boldsymbol{R}^{n+1} \setminus K \times [-\varepsilon, \varepsilon])$$

and $U \in C^\infty(\boldsymbol{R}^{n+1} \setminus K \times [-\varepsilon, \varepsilon])$. Applying the same argument as in the proof of Proposition 1.2. 6 , we can define $u \in \mathcal{A}'(K^\varepsilon)$ by

$$u(\varphi) = \int U(1 - \Delta_{x, x_{n+1}})(\chi \Phi)\, dx dx_{n+1} \quad \text{for } \varphi \in \mathcal{A},$$

where χ is a fuction in $C_0^\infty(\boldsymbol{R}^{n+1})$ satisfying $\chi = 1$ near $K \times [-\varepsilon, \varepsilon]$ and Φ is a unique solution of (1.8). Moreover, $U - \mathcal{H}(u)$ can be continued analytically to $\boldsymbol{R}^{n+1}$ and we have supp $u \subset K$. Since the infinite series

$$-V_1 - \cdots - V_{j-1} + \sum_{k=j}^{\infty} (U_{k+1} - U_k - V_k) \ (= U - U_j)$$

converges locally uniformly in $\boldsymbol{R}^{n+1} \setminus (K_0)_{\varepsilon_j - \varepsilon} \times [-\varepsilon_j, \varepsilon_j]$ for $j \ge 2$, we have supp $(u - u_j) \subset (K_0)_{\varepsilon_j - \varepsilon}$ for $j \ge 2$. The second part of the proposition is obvious. □

Proposition 1.4. 4 *Let X be a bounded open subset of $\boldsymbol{R}^n$, and let $\{X_\lambda\}_{\lambda \in \Lambda}$ be a family of open sets in X such that $X = \cup_{\lambda \in \Lambda} X_\lambda$. Assume that $\varepsilon \ge 0$ and $u_\lambda \in \mathcal{B}_\varepsilon(X_\lambda)$ ($\lambda \in \Lambda$) satisfy*

$$u_\lambda|_{X_\lambda \cap X_\mu} = u_\mu|_{X_\lambda \cap X_\mu} \quad \textit{for every } \lambda, \mu \in \Lambda. \tag{1.16}$$

Then there is a unique $u \in \mathcal{B}_\varepsilon(X)$ such that $u|_{X_\lambda} = u_\lambda$ for every $\lambda \in \Lambda$.

Proof Uniqueness of $u \in \mathcal{B}_\varepsilon(X)$ is obvious. In order to prove the proposition it suffices to show the following: If $v_\lambda \in \mathcal{F}_\varepsilon$, supp $v_\lambda \subset \overline{X}_\lambda$ and supp $(v_\lambda - v_\mu) \cap (X_\lambda \cap X_\mu) = \emptyset$ for any $\lambda, \mu \in \Lambda$, then there is $v \in \mathcal{F}_\varepsilon$ satisfying

$$\text{supp } v \subset \overline{X} \text{ and supp } (v - v_\lambda) \cap X_\lambda = \emptyset \text{ for any } \lambda \in \Lambda. \tag{1.17}$$

If $\{X_\lambda\}_{\lambda \in \Lambda} = \{X_1, X_2\}$, then the above assertion follows from Corollary 1.3. 5 . Next assume that $\Lambda = \boldsymbol{N}$. Then there are $w_j \in \mathcal{F}_\varepsilon$ ($j \in \boldsymbol{N}$) such that

$$\text{supp } w_j \subset \bigcup_{k=1}^{j} \overline{X}_k \text{ and supp } (w_j - v_k) \cap X_k = \emptyset \text{ for } 1 \le k \le j.$$

This yields

$$\text{supp } (w_j - w_k) \subset \bigcap_{\ell=1}^{k} (\overline{X} \setminus X_\ell) \ \Big(= \overline{X} \setminus (\bigcup_{\ell=1}^{k} X_\ell) \Big)$$

for $j \ge k \ge 1$. If $\mathcal{U}_0$ is an open neighborhood of ∂X, then $\overline{X} \setminus \mathcal{U}_0$ is a compact subset of X and there is $k \in \boldsymbol{N}$ such that $\overline{X} \setminus (\bigcup_{\ell=1}^{k} X_\ell) \subset \mathcal{U}_0$. Applying Proposition 1.4. 3 with $K = \overline{X}$ and $K_0 = \partial X$, we can show that there is $v \in \mathcal{F}_\varepsilon$ such that supp $v \subset \overline{X}$, and for any neighborhood $\mathcal{V}_0$ of ∂X there is $j_0 \in \boldsymbol{N}$ satisfying supp $(v - w_j) \subset \mathcal{V}_0$ for $j \ge j_0$. Let $k \in \boldsymbol{N}$ and $x^0 \in X_k$. Then there is $j_0 \in \boldsymbol{N}$ such that $x^0 \notin$ supp $(v - w_j)$ for $j \ge j_0$. On the other hand, supp $(w_j - v_k) \cap X_k = \emptyset$ for $j \ge k$. This shows that supp $(v - v_k) \cap X_k = \emptyset$. Now we assume that Λ is uncountable. Since X has the Linderöf property, there is $\{\lambda_j\}_{j \in \boldsymbol{N}} \subset \Lambda$ such that $X = \bigcup_{j=1}^{\infty} X_{\lambda_j}$. So there is $v \in \mathcal{F}_\varepsilon$ satisfying supp $v \subset \overline{X}$ and supp $(v - v_{\lambda_j}) \cap X_{\lambda_j} = \emptyset$ ($j \in \boldsymbol{N}$). Then, we have (1.17). In fact, by the same argument as the above, for a fixed $\lambda \in \Lambda$ there is $\tilde{v} \in \mathcal{F}_\varepsilon$ satisfying supp $\tilde{v} \subset \overline{X}$, supp $(\tilde{v} - v_{\lambda_j}) \cap X_{\lambda_j} = \emptyset$ ($j \in \boldsymbol{N}$) and supp $(\tilde{v} - v_\lambda) \cap X_\lambda = \emptyset$. It is obvious that supp $(v - \tilde{v}) \subset \partial X$. This gives (1.17). □

Definition 1.4. 5 Let X be an open subset of $\boldsymbol{R}^n$ and $\varepsilon \ge 0$, and let $\{X_\lambda\}_{\lambda \in \Lambda}$ be a family of bounded open subsets of X such that $X = \bigcup_{\lambda \in \Lambda} X_\lambda$. We define $\mathcal{B}_\varepsilon(X, \{X_\lambda\}_{\lambda \in \Lambda})$ as the collection of $\{u_\lambda\}_{\lambda \in \Lambda}$ satisfying $u_\lambda \in \mathcal{B}_\varepsilon(X_\lambda)$ and (1.16). We identify an element $\{u_\lambda\}_{\lambda \in \Lambda} \in \mathcal{B}_\varepsilon(X, \{X_\lambda\}_{\lambda \in \Lambda})$ with $\{v_\mu\}_{\mu \in M} \in \mathcal{B}_\varepsilon(X, \{\widetilde{X}_\mu\}_{\mu \in M})$ if

$$u_\lambda|_{X_\lambda \cap \widetilde{X}_\mu} = v_\mu|_{X_\lambda \cap \widetilde{X}_\mu} \quad \text{for any } \lambda \in \Lambda \text{ and } \mu \in M.$$

Then, with this identification, we define $\mathcal{B}_\varepsilon(X)$ as $\mathcal{B}_\varepsilon(X, \{X_\lambda\}_{\lambda \in \Lambda})$. We also write $\mathcal{B}(X) = \mathcal{B}_0(X)$.

Remark (i) If X is a bounded open subset of $\boldsymbol{R}^n$, then it follows from Proposition 1.4.4 that $\mathcal{B}_\varepsilon(X)$ can be identified with $\mathcal{A}'_\varepsilon(\overline{X})/\mathcal{A}'_\varepsilon(\partial X)$. So, in the above definition each element $\{v_\mu\}_{\mu\in M} \in \mathcal{B}_\varepsilon(X, \{\widetilde{X}_\mu\}_{\mu\in M})$ determines uniquely an element $\{u_\lambda\}_{\lambda\in\Lambda} \in \mathcal{B}_\varepsilon(X, \{X_\lambda\}_{\lambda\in\Lambda})$. Moreover, each element of $\mathcal{F}_\varepsilon$ determines uniquely an element of $\mathcal{B}_\varepsilon(\boldsymbol{R}^n)$. (ii) Operations on $\mathcal{B}_\varepsilon(X)$ can be naturally defined by those on $\mathcal{B}_\varepsilon(X_\lambda)$. In particular, for an open subset Y of X and $u = \{u_\lambda\} \in \mathcal{B}_\varepsilon(X)$ we define $u|_Y = \{u_\lambda|_{X_\lambda\cap Y}\} \in \mathcal{B}_\varepsilon(Y)$ and $\operatorname{supp} u = \bigcup_\lambda \operatorname{supp} u_\lambda$.

Let us define several operations in $\mathcal{A}'(\boldsymbol{R}^n)$. We have already defined differentiation and multiplication:

$$(D_j u)(\varphi) = -u(D_j\varphi), \quad (au)(\varphi) = u(a\varphi)$$

if K is a compact subset of $\boldsymbol{R}^n$, $u \in \mathcal{A}'(K)$, $\varphi \in \mathcal{A}$ and a is analytic near K. It is obvious that $\operatorname{supp} D_j u \subset \operatorname{supp} u$ and $\operatorname{supp} au \subset \operatorname{supp} u$. The tensor product $u \otimes v$ of $u \in \mathcal{A}'(\boldsymbol{R}^n)$ and $v \in \mathcal{A}'(\boldsymbol{R}^m)$ is defined by

$$(u \otimes v)(\varphi) = u_z(v_w(\varphi(z,w))) \ \ (= v_w(u_z(\varphi(z,w))))$$

for every polynomial φ in $\boldsymbol{C}^{n+m}$. Then $u\otimes v$ can be regarded as an element of $\mathcal{A}'(\boldsymbol{R}^{n+m})$. Moreover, we have $\mathcal{F}[u \otimes v] = \hat{u}(\xi)\hat{v}(\eta)$ and $\operatorname{supp} u \otimes v = \operatorname{supp} u \times \operatorname{supp} v$. Let K be a compact subset of $\boldsymbol{R}^n$, and let f be a real analytic mapping of an open subset ω of $\boldsymbol{R}^n$ on an open neighborhood $\mathcal{U}$ of K. We assume that f is a diffeomorphism. Then we can define the pull-back $f^*u \in \mathcal{A}'(f^{-1}(K))$ of $u \in \mathcal{A}'(K)$ by

$$(f^*u)(\varphi) = u((\varphi \circ h)|\det h'|) \quad \text{for } \varphi \in \mathcal{A},$$

where $h = f^{-1} : \mathcal{U} \to \omega$ and h' denotes the differential of h (the Jacobian matrix of h). The above operations on $\mathcal{A}'(\boldsymbol{R}^n)$ can be easily extended to those on $\mathcal{B}(X)$. This enables us to define $\mathcal{B}(X)$ for a real analytic manifold X in the standard way. Let X be a real analytic manifold, and let $\mathcal{K}$ be an atlas for X, *i.e.*, let $\mathcal{K}$ be a family of homeomorphisms κ of open subsets X_κ of X onto open subsets $\widetilde{X}_\kappa$ of $\boldsymbol{R}^n$ such that $X = \bigcup_{\kappa\in\mathcal{K}} X_\kappa$ and the mapping

$$\kappa' \circ \kappa^{-1} : \ \kappa(X_\kappa \cap X_{\kappa'}) \to \kappa'(X_\kappa \cap X_{\kappa'})$$

is real analytic for every $\kappa, \kappa' \in \mathcal{K}$. Then we define $\mathcal{B}(X)$ as the collection of $\{u_\kappa\}_{\kappa\in\mathcal{K}}$ satisfying $u_\kappa \in \mathcal{B}(\widetilde{X}_\kappa)$ and

$$(\kappa' \circ \kappa^{-1})^*(u_{\kappa'}|_{\kappa'(X_\kappa\cap X_{\kappa'})}) = u_\kappa|_{\kappa(X_\kappa\cap X_{\kappa'})}$$

for $\kappa, \kappa' \in \mathcal{K}$.

Definition 1.4. 6 Let X be a topological space. (i) We say that $\mathcal{F}$ is a presheaf of vector spaces (over $\boldsymbol{C}$) on X (or, simply, a presheaf on X) if the following conditions are satisfied;

(1) to every open subset U of X there is associated a vector space $\mathcal{F}(U)$ over $\boldsymbol{C}$,

(2) to every pair of open subsets U and V of X with $U \supset V$ there is associated a linear mapping $\rho_V^U : \mathcal{F}(U) \to \mathcal{F}(V)$ satisfying (a) $\rho_U^U = id$ (=identity) and (b) $\rho_W^V \circ \rho_V^U = \rho_W^U$ for open subsets U, V and W of X with $U \supset V \supset W$.

$\mathcal{F}(U)$ is called the space of sections on U and the mapping ρ_V^U are called the restriction maps. We also write $f|_V = \rho_V^U(f)$ for $f \in \mathcal{F}(U)$. (ii) A presheaf $\mathcal{F}$ of vector spaces on X is called a sheaf of vector spaces on X (or, simply, a sheaf on X) if the following conditions are satisfied;

(1) if $f \in \mathcal{F}(U)$ satisfies $f|_{U_\lambda} = 0$ for any $\lambda \in \Lambda$, then $f = 0$,

(2) if $f_\lambda \in \mathcal{F}(U_\lambda)$ ($\lambda \in \Lambda$) satisfy $f_\lambda|_{U_\lambda \cap U_\mu} = f_\mu|_{U_\lambda \cap U_\mu}$ for any $\lambda, \mu \in \Lambda$, then there is $f \in \mathcal{F}(U)$ such that $f|_{U_\lambda} = f_\lambda$ for any $\lambda \in \Lambda$,

where U is an open subset of X and $\{U_\lambda\}_{\lambda \in \Lambda}$ is a family of open subsets of X and satisfies $U = \cup_{\lambda \in \Lambda} U_\lambda$. (iii) A sheaf $\mathcal{F}$ on X is said to be flabby if for every open subset U of X the restriction map $\rho_U^X : \mathcal{F}(X) \to \mathcal{F}(U)$ is surjective.

Let X be an open subset of $\boldsymbol{R}^n$. For every open subset U of X we define $\mathcal{A}_X(U)$ as the collection of all real analytic fuctions in U. Then we can see that $\mathcal{A}_X$ is a sheaf on X. Similarly, we define the presheaf $\mathcal{B}_{\varepsilon,X}$ by associating $\mathcal{B}_\varepsilon(U)$ to every open subset U of X, where $\varepsilon \geq 0$. We also write $\mathcal{B}_\varepsilon = \mathcal{B}_{\varepsilon,R^n}$, $\mathcal{B}_X = \mathcal{B}_{0,X}$ and so on. It follows from the definition of $\mathcal{B}_\varepsilon(X)$ that $\mathcal{B}_{\varepsilon,X}$ is a sheaf on X.

Lemma 1.4. 7 *Let $\mathcal{F}$ be a sheaf on a topological space X, and let $\{U_\lambda\}_{\lambda \in \Lambda}$ be an open covering of X. If $\mathcal{F}|_{U_\lambda}$ is a flabby sheaf for $\lambda \in \Lambda$, then $\mathcal{F}$ is a flabby sheaf. Here $\mathcal{F}|_{U_\lambda}$ denotes the sheaf on U_λ defined by associating $\mathcal{F}(U)$ to every open subset U of U_λ.*

This lemma can be proved by applying Zorn's lemma (see, *e.g.*, Lemma 4.2.2 in [Kn]).

Theorem 1.4. 8 *Let X be an open subset of $\boldsymbol{R}^n$ and $\varepsilon \geq 0$. Then $\mathcal{B}_{\varepsilon,X}$ is a flabby sheaf on X.*

Remark The above theorem is still valid when X is a real analytic manifold and $\varepsilon = 0$.

Proof By Lemma 1.4. 7 it is sufficient to prove the theorem when X is bounded. Assume that X is a bounded open subset of $\boldsymbol{R}^n$, and let Y be an open subset of X and $u \in \mathcal{B}_\varepsilon(Y)$. From the remark of Definition 1.4. 5 (or Proposition 1.4. 4) it follows that there is $w \in \mathcal{A}'_\varepsilon(\overline{Y})$ such that $w|_Y = u \in \mathcal{B}_\varepsilon(Y)$. Then, putting $v = w|_X \in \mathcal{B}_\varepsilon(X)$, we have $v|_Y = u$. This implies that $\mathcal{B}_{\varepsilon,X}$ is flabby. □

1.5 Further applications of the Runge approximation theorem

Let X be a topological space, and let $\mathcal{F}$ be a presheaf on X. For a fixed $x \in X$ we introduce the following equivalence relation $\sim$ in $\bigcup_{U \ni x} \mathcal{F}(U)$: $s_1 \sim s_2$ if $s_j \in \mathcal{F}(U_j)$ ($j = 1, 2$), the U_j are open subsets of X containing x and there is an open subset V of $U_1 \cap U_2$ containing x such that $s_1|_V = s_2|_V$. We define the stalk $\mathcal{F}_x$ of $\mathcal{F}$ at x by

$$\mathcal{F}_x = \bigcup_{U \ni x} \mathcal{F}(U)/\sim, \quad i.e., \quad \mathcal{F}_x = \varinjlim_{U \ni x} \mathcal{F}(U).$$

Then we have the natural mapping $\rho_x^U : \mathcal{F}(U) \to \mathcal{F}_x$. An element of $\mathcal{F}_x$ is called a germ of sections of $\mathcal{F}$ at x. We define $\overline{\mathcal{F}}(U)$ for open subsets U of X as the collection of mappings $s : U \to \bigcup_{x \in U} \mathcal{F}_x$ satisfying the following; for each $x \in U$ there are an open neighborhood V of x in U and $t \in \mathcal{F}(V)$ such that $s(y) = \rho_y^V t$ for $y \in V$. Then, defining $\overline{\mathcal{F}}$ by associating $\overline{\mathcal{F}}(U)$ to every open subset U of X, we have a sheaf $\overline{\mathcal{F}}$ which is called the sheaf associated with the presheaf $\mathcal{F}$ (see, *e.g.*, [Kn]).

Definition 1.5. 1 (i) Let $\mathcal{F}$ and $\mathcal{G}$ be presheaves (resp. sheaves) on X. A family $h = \{h_U\}$ of linear mappings $h_U : \mathcal{F}(U) \to \mathcal{G}(U)$ is said to be a presheaf (resp. sheaf) homomorphism if $h_V(s|_V) = h_U(s)|_V$ for every pair of open subsets U and V of X with $V \subset U$ and $s \in \mathcal{F}(U)$. (ii) Let $\mathcal{F}$ and $\mathcal{G}$ be sheaves on X such that $\mathcal{F}(U) \subset \mathcal{G}(U)$ for any open subsets U of X. We say that $\mathcal{F}$ is a subsheaf of $\mathcal{G}$ if $\iota = \{\iota_U\}$ is a sheaf homomorphism, where $\iota_U : \mathcal{F}(U) \to \mathcal{G}(U)$ is the inclusion map. (iii) Let $\mathcal{G}$ be a sheaf on X, and let $\mathcal{F}$ be a subsheaf of $\mathcal{G}$. Then, associating $\mathcal{G}(U)/\mathcal{F}(U)$ to every open subset U of X we can define a presheaf on X. The sheaf associated with this presheaf is called the quotient sheaf of $\mathcal{G}$ by $\mathcal{F}$, and is denoted by $\mathcal{G}/\mathcal{F}$.

Let X be an open subset of $\boldsymbol{R}^n$. Assume that $u \in \mathcal{B}(X)$ and sing supp $u = \emptyset$. Then u can be identified with an element in $\mathcal{A}(X)$, where $\mathcal{A}(X)$ denotes the space of all real analytic functions in X. In fact, if $X = \bigcup_{\lambda \in \Lambda} X_\lambda$, $\{X_\lambda\}_{\lambda \in \Lambda}$ is a family of bounded open subsets of X and $u = \{u_\lambda\}_{\lambda \in \Lambda} \in \mathcal{B}(X)$, where $u_\lambda \in \mathcal{B}(X_\lambda)$ ($\lambda \in \Lambda$), then there are $v_\lambda \in \mathcal{A}'(\overline{X}_\lambda)$ ($\lambda \in \Lambda$) such that $v_\lambda|_{X_\lambda} = u_\lambda$ ($\lambda \in \Lambda$). Since sing supp $u = \emptyset$, we have sing supp $v_\lambda \cap X_\lambda = \emptyset$ ($\lambda \in \Lambda$) and $w_\lambda \equiv \lim_{t \to +0} 2\mathcal{H}(v_\lambda)(x,t)|_{X_\lambda} \in \mathcal{A}(X_\lambda)$ ($\lambda \in \Lambda$). We can easily see that $\{w_\lambda\}_{\lambda \in \Lambda}$ determines an element w in $\mathcal{A}(X)$. Then we identify u with w. Therefore, $\mathcal{A}(X)$ can be identified with $\{u \in \mathcal{B}(X);\ \text{sing supp}\ u = \emptyset\}$. Now we define $\mathcal{F}(U) = \mathcal{B}(U)/\mathcal{A}(U)$ for every open subset U of X and the presheaf $\mathcal{F}$ on X.

Theorem 1.5. 2 *The presheaf $\mathcal{F}$ on X defined above is a flabby sheaf. In particular, $\mathcal{F}$ can be identified with the quotient sheaf $\mathcal{B}_X/\mathcal{A}_X$.*

Proof Since $\mathcal{B}_X$ is flabby, it suffices to show that $\mathcal{F}$ is a sheaf. This can be reduced to the first problem of Cousin. However, applying the same argument as in the proofs of Propositions 1.4. 3 and 1.4. 4 , we shall give a direct proof here. Let U be an open subset of X, and let $\{U_\lambda\}_{\lambda \in \Lambda}$ be a family of open subsets of U such that $U = \bigcup_{\lambda \in \Lambda} U_\lambda$. Assume that $u_\lambda \in \mathcal{B}(U_\lambda)$ ($\lambda \in \Lambda$) satisfy $u_\lambda|_{U_\lambda \cap U_\mu} - u_\mu|_{U_\lambda \cap U_\mu} \in \mathcal{A}(U_\lambda \cap U_\mu)$ for $\lambda, \mu \in \Lambda$, *i.e.*, sing supp $(u_\lambda|_{U_\lambda \cap U_\mu} - u_\mu|_{U_\lambda \cap U_\mu}) = \emptyset$ for $\lambda, \mu \in \Lambda$. Since U has the Linderöf property and is σ compact and paracompact, we can assume without loss of generality that $\Lambda = \boldsymbol{N}$, $U_j \subset\subset U$ ($j \in \boldsymbol{N}$) and $\{U_j\}_{j \in N}$ is locally finite. Choose $v_j \in \mathcal{A}'(\overline{U}_j)$ ($j \in \boldsymbol{N}$) so that $v_j|_{U_j} = u_j \in \mathcal{B}(U_j)$ for $j \in \boldsymbol{N}$. Then, sing supp $(v_j - v_k) \cap (U_j \cap U_k) = \emptyset$ for any $j, k \in \boldsymbol{N}$. Put $w_1 = v_1$ ($\in \mathcal{A}'(\overline{U}_1)$). By Corollary 1.3. 7 there is $w_2 \in \mathcal{F}_0$ such that sing supp $w_2 \subset \overline{U}_1 \cup \overline{U}_2$, sing supp $(w_2 - v_j) \cap U_j = \emptyset$ ($j = 1,2$) and $\mathcal{H}(w_2 - v_1)(x, x_{n+1})$ can be continued analytically from $\boldsymbol{R}^n \times (0,\infty)$ to a neighborhood of $\{\overline{U}_1 \setminus (U_2 \cup (\partial U_1 \cap \partial U_2))\} \times \boldsymbol{R} \setminus \partial(U_1 \cup U_2) \times \{0\}$ ($\supset (U_1 \setminus \overline{U}_2) \times \boldsymbol{R}$). Repeating this construction we can choose $\{w_j\}_{j \in N} \subset \mathcal{F}_0$ so that sing supp $w_j \subset \bigcup_{k=1}^j \overline{U}_k$,

$$\text{sing supp } (w_{j+1} - w_j) \cap (\bigcup_{k=1}^{j} U_k) = \emptyset,$$

$$\text{sing supp } (w_{j+1} - v_{j+1}) \cap U_{j+1} = \emptyset,$$

and $W_{j+1}(x, x_{n+1}) - W_j(x, x_{n+1})$ can be continued analytically from $\boldsymbol{R}^n \times (0, \infty)$ to $(\bigcup_{k=1}^j U_k \setminus \overline{U}_{j+1}) \times \boldsymbol{R}$ ($j \in \boldsymbol{N}$), where $W_j(x, x_{n+1}) = \mathcal{H}(w_j)(x, x_{n+1})$ for $x_{n+1} \neq 0$. So we have $W_{j+1} - W_j \in C^\infty(\boldsymbol{R}^n \times (0,\infty) \cup (\bigcup_{k=1}^j U_k \setminus \overline{U}_{j+1}) \times \boldsymbol{R})$ and $(1 - \Delta_{x,x_{n+1}})(W_{j+1} - W_j) = 0$ for $j \in \boldsymbol{N}$. Let K_j ($j \in \boldsymbol{N}$)

be a compact subset of $\bigcup_{k=1}^{j} U_k \setminus \overline{U}_{j+1}$ such that $K_1 \Subset K_2 \Subset \cdots$ and $U = \bigcup_{j=1}^{\infty} \operatorname{int}(K_j)$. For example, if $K_j = \{x \in \bigcup_{k=1}^{j} U_k \setminus \overline{\bigcup_{\ell=j+1}^{\infty} U_\ell};\ \operatorname{dis}(\{x\}, \boldsymbol{R}^n \setminus (\bigcup_{k=1}^{j} U_k \setminus \overline{\bigcup_{\ell=j+1}^{\infty} U_\ell})) \geq 1/j\}$ ($j \in \boldsymbol{N}$), then $\{K_j\}$ satisfies the above conditions by virtue of paracompactness. We put, for $j \in \boldsymbol{N}$,

$$\begin{aligned} M_j &= \{(x, x_{n+1}) \in \boldsymbol{R}^{n+1};\ (x, x_{n+1}) \in K_j \times [-1,1] \text{ or } ``|(x, x_{n+1})| \leq j \\ &\qquad \text{and } x_{n+1} \geq 1/j"\} \\ &\qquad \Big(\Subset \boldsymbol{R}^n \times (0,\infty) \cup \big(\bigcup_{k=1}^{j} U_k \setminus \overline{U}_{j+1}\big) \times (-2, 0] \Big). \end{aligned}$$

Then, $M_1 \subset M_2 \subset \cdots$ and $\bigcup_{j=1}^{\infty} \operatorname{int}(M_j) = \boldsymbol{R}^n \times (0,\infty) \bigcup U \times (-1,1)$. Now we can apply Proposition 1.4.2 with $X = \boldsymbol{R}^{n+1}$ and $Y = \boldsymbol{R}^n \times (0,\infty) \cup (\bigcup_{k=1}^{j} U_k \setminus \overline{U}_{j+1}) \times (-2, 0]$. In fact, if F_1 and F_2 are closed subsets of $\boldsymbol{R}^{n+1}$ and satisfy $F_1 \cap F_2 = \emptyset$, $(x, x_{n+1}) \in F_2$ and

$$\Big\{\boldsymbol{R}^n \setminus \big(\bigcup_{k=1}^{j} U_k \setminus \overline{U}_{j+1}\big)\Big\} \times (-\infty, 0] \cup \big(\bigcup_{k=1}^{j} U_k \setminus \overline{U}_{j+1}\big) \times (-\infty, -2] = F_1 \cup F_2,$$

then $\{x\} \times (-\infty, -2] \subset F_2$ and, therefore, F_2 is not bounded since $\{x\} \times (-\infty, 0]$ and $\{x\} \times (-\infty, -2]$ are connected. From Proposition 1.4.2 it follows that there is a sequence $\{G_j\}$ such that $G_j \in C^\infty(\boldsymbol{R}^{n+1})$, $(1 - \Delta_{x,x_{n+1}})G_j = 0$ in $\boldsymbol{R}^{n+1}$ and $|W_{j+1} - W_j - G_j| \leq 2^{-j}$ in M_j for $j \in \boldsymbol{N}$. Then the sequence $\{W_j - G_1 - \cdots - G_{j-1}\}_{j \in N}$ converges locally uniformly in $\boldsymbol{R}^n \times (0,\infty)$. Putting

$$W = \lim_{j \to \infty} (W_j - G_1 - \cdots - G_{j-1}),$$

we have $(1 - \Delta_{x,x_{n+1}})W = 0$ in $\mathcal{D}'(\boldsymbol{R}^n \times (0,\infty))$ and $W \in C^\infty(\boldsymbol{R}^n \times (0,\infty))$. Since $W_j - V_j$ can be continued analytically to a neighborhood of $U_j \times \{0\}$ and

$$W - W_j = -G_1 - \cdots - G_{j-1} + \sum_{k=j}^{\infty} (W_{k+1} - W_k - G_k),$$

$W - V_j$ can be continued analytically to a neighborhood of $U_j \times \{0\}$ for $j \in \boldsymbol{N}$, where $V_j(x, x_{n+1}) = \mathcal{H}(v_j)(x, x_{n+1})$ for $x_{n+1} \neq 0$. In fact, for any $j \in \boldsymbol{N}$ there is $\ell \in \boldsymbol{N}$ such that $U_j \times [-1,1] \subset M_\ell$. Since $\sum_{k=\ell}^{\infty} (W_{k+1} - W_k - G_k)$ uniformly converges in M_ℓ and $\sum_{k=j}^{\ell-1} (W_{k+1} - W_k - G_k)$ can be continued analytically to a neighborhood of $U_j \times \{0\}$, $W - W_j$ can be continued analytically to a neighborhood of $U_j \times \{0\}$. Define

$$W(x, x_{n+1}) = -W(x, -x_{n+1}) \quad \text{for } x_{n+1} < 0.$$

Let K be a compact subset of $\boldsymbol{R}^n$, and choose $\phi_K \in C^\infty(\boldsymbol{R}^{n+1} \setminus \partial K \times \{0\})$ such that $\phi_K(x, x_{n+1}) = \phi_K(x, -x_{n+1})$ and

$$\phi_K(x, x_{n+1}) = \begin{cases} 0 & \text{if } |(x, x_{n+1})| \gg 1, \\ 1 & \text{near } (K \setminus \partial K) \times \{0\}, \\ 0 & \text{near } (\boldsymbol{R}^n \setminus K) \times \{0\}. \end{cases}$$

Let us apply the same argument as in the proof of Theorem 1.3. 3. Then there is $f_K \in C^\infty(\boldsymbol{R}^{n+1} \setminus \partial K \times \{0\})$ satisfying $f_K(x, x_{n+1}) = -f_K(x, -x_{n+1})$ and

$$(1 - \Delta_{x,x_{n+1}}) f_K = (1 - \Delta_{x,x_{n+1}})(\phi_K W) \quad \text{in } \boldsymbol{R}^{n+1} \setminus \partial K \times \{0\},$$

and $\widetilde{W}_K \equiv \phi_K W - f_K \in C^\infty(\boldsymbol{R}^{n+1} \setminus K \times \{0\})$ satisfies $(1 - \Delta_{x,x_{n+1}})\widetilde{W}_K = 0$ in $\boldsymbol{R}^{n+1} \setminus K \times \{0\}$, where $\phi_K W$ and $(1 - \Delta_{x,x_{n+1}})(\phi_K W)$ are regarded as functions in $C^\infty(\boldsymbol{R}^{n+1} \setminus K \times \{0\})$ and $C^\infty(\boldsymbol{R}^{n+1} \setminus \partial K \times \{0\})$, respectively. Define $w_K : \mathcal{A} \ni \varphi \mapsto w_K(\varphi) \in \boldsymbol{C}$ by

$$w_K(\varphi) = \int \widetilde{W}_K (1 - \Delta_{x,x_{n+1}})(\chi \Phi)\, dx dx_{n+1},$$

where χ is a function in $C_0^\infty(\boldsymbol{R}^{n+1})$ satisfying $\chi = 1$ near $K \times \{0\}$ and Φ is a unique solution of (1.8). Then $w_K \in \mathcal{A}'(K)$ and $H_K \equiv \widetilde{W}_K - \mathcal{H}(w_K)$ can be regarded as a function in $C^\infty(\boldsymbol{R}^{n+1})$ and satisfies $(1 - \Delta_{x,x_{n+1}})H_K = 0$. Let Ω_j ($j = 1, 2$) be bounded open subsets of $\boldsymbol{R}^n$, and put $K_j = \overline{\Omega}_j$ ($j = 1, 2$). Since

$$\mathcal{H}(w_{K_1}) - \mathcal{H}(w_{K_2}) = (\phi_{K_1} - \phi_{K_2})W - f_{K_1} + f_{K_2} - H_{K_1} + H_{K_2}$$

can be regarded as a function in $C^\infty((\Omega_1 \cap \Omega_2) \times \boldsymbol{R})$, we have supp $(w_{K_1} - w_{K_2}) \cap (\Omega_1 \cap \Omega_2) = \emptyset$. This implies that W uniquely determines an element w in $\mathcal{B}(\boldsymbol{R}^n)$, *i.e.*, $w = \{w|_{\overline{B}_j}\}_{j \in N}$, where $B_j = \{x \in \boldsymbol{R}^n;\ |x| < j\}$ and $w|_{\overline{B}_j}$ denotes the residue class of $w_{\overline{B}_j}$ in $\mathcal{B}(B_j)$. Put $u = w|_U \in \mathcal{B}(U)$. Then we have $u|_{U_j} - u_j \in \mathcal{A}(U_j)$ for $j \in \boldsymbol{N}$. In fact,

$$\mathcal{H}(w_{\overline{U}_j}) - \mathcal{H}(v_j) = \phi_{\overline{U}_j} W - f_{\overline{U}_j} - H_{\overline{U}_j} - V_j$$

and $W - V_j$ can be continued analytically to a neighborhood of $U_j \times \{0\}$. Thus sing supp $(w_{\overline{U}_j} - v_j) \cap U_j = \emptyset$. Therefore, $\mathcal{F}$ is a sheaf. □

We note that Theorem 1.5. 2 is still valid for a real analytic manifold X. Let X be an open subset of $\boldsymbol{R}^n$, and define

$$\begin{aligned} \mathcal{H}(X) \ := \ & \{U(x, x_{n+1}) \in C^\infty(\boldsymbol{R}^n \times (\boldsymbol{R} \setminus \{0\}) \cup X \times \{0\}); \\ & (1 - \Delta_{x,x_{n+1}})U = 0 \text{ and } U(x, x_{n+1}) = -U(x, -x_{n+1})\}. \end{aligned}$$

Then we can also define $\mathcal{B}(X) = \mathcal{H}(\emptyset)/\mathcal{H}(X)$. Let $U(x, x_{n+1}) \in \mathcal{H}(\emptyset)$. Repeating the same argument as in the proof of Theorem 1.5.2 we can show that $U(x, x_{n+1})$ uniquely determines an element $u \equiv TU \in \mathcal{B}(\boldsymbol{R}^n)$ such that $U(x, x_{n+1}) - \mathcal{H}(v)(x, x_{n+1})$ can be continued analytically to $\Omega \times \boldsymbol{R}$ if Ω is a bounded open subset of $\boldsymbol{R}^n$ and the residue class $v|_\Omega$ of v ($\in \mathcal{A}'(\overline{\Omega})$) is $u|_\Omega \in \mathcal{B}(\Omega)$. It is easy to see that U can be regarded as an element of $\mathcal{H}(X)$ if and only if $(TU)|_X = 0$ in $\mathcal{B}(X)$. Therefore, we can define the linear mapping $T : \mathcal{H}(\emptyset)/\mathcal{H}(X) \to \mathcal{B}(X)$, which is injective. Next we shall prove that T is surjective, using Proposition 1.4.2. Put

$$X_j = \{x \in X;\ |x| < j \text{ and } \mathrm{dis}(\{x\}, \boldsymbol{R}^n \setminus X) > 1/j\} \quad \text{for } j \in \boldsymbol{N}.$$

Then we have $X_1 \Subset X_2 \Subset X_3 \Subset \cdots$ and $X = \bigcup_{j=1}^\infty X_j$. Let $\{u_j\}_{j\in N}$ be an element of $\mathcal{B}(X)$, where $u_j \in \mathcal{B}(X_j)$ ($j \in \boldsymbol{N}$) satisfy $u_k|_{X_j} = u_j$ for $k \ge j$ (≥ 1). Choose $v_j \in \mathcal{A}'(\overline{X}_j)$ ($j \in \boldsymbol{N}$) so that $v_j|_{X_j} = u_j$ for $j \in \boldsymbol{N}$, and put $U_j(x, x_{n+1}) = \mathcal{H}(v_j)(x, x_{n+1})$ for $x_{n+1} \ne 0$ and $j \in \boldsymbol{N}$. Note that $U_k - U_j \in \mathcal{H}(X_j \cup (\boldsymbol{R}^n \setminus \overline{X}_k)) \subset \mathcal{H}(X_j)$ if $k \ge j$. Let $j \in \boldsymbol{N}$ be fixed. If F_1 is a nonvoid closed subset of $X \setminus X_j$, F_2 is a compact subset of $\boldsymbol{R}^n$, $X \setminus X_j = F_1 \cup F_2$ and $F_1 \cap F_2 = \emptyset$, then $F_2 = \emptyset$. In fact, assume that $F_2 \ne \emptyset$. If $x^0 \in \partial F_2$ ($\subset F_2 \subset X \setminus X_j$) and $x^0 \notin \partial X_j$, then there is an open ball B in $X \setminus \overline{X}_j$ such that $x^0 \in B$ and $B \cap F_1 \ne \emptyset$. This leads to a contradiction, since B is connected. Therefore, we have $\partial F_2 \subset \partial X_j$. Now assume that $x^0 \in \partial F_2$ and $|x^0| = j$, and put $\lambda_0 = \sup\{\lambda;\ \lambda \ge 1 \text{ and } \lambda x^0 \in F_2\}$ ($< \infty$). It is obvious that $\lambda_0 x^0 \in \partial F_2 \subset X \setminus X_j$ and $\lambda x^0 \in X \setminus X_j$ if $\lambda > \lambda_0$ and $\lambda - \lambda_0 \ll 1$. Since F_1 is closed in $X \setminus X_j$, this yields $\lambda_0 x^0 \in F_1$, which is a contradiction. So, if $x^0 \in \partial F_2$ then $\mathrm{dis}(\{x^0\}, \boldsymbol{R}^n \setminus X) = 1/j$. Assume that $x^0 \in \partial F_2$ and $\mathrm{dis}(\{x^0\}, \boldsymbol{R}^n \setminus X) = 1/j$. Then there is $y \in \boldsymbol{R}^n \setminus X$ such that $|x^0 - y| = 1/j$. Putting $\lambda_1 = \sup\{\lambda;\ \lambda \ge 0 \text{ and } x^0 + \lambda(y - x^0) \in F_2\}$ (< 1), we have $x^0 + \lambda_1(y - x^0) \in \partial F_2 \subset X \setminus X_j$ and $x^0 + \lambda_1(y - x^0) \in F_1$, which is a contradiction. Thus we have $F_2 = \emptyset$. Therefore, it follows from Proposition 1.4.2 that there are $V_j \in C^\infty(\boldsymbol{R}^n \times (\boldsymbol{R} \setminus \{0\}) \cup X \times \{0\})$ ($j \in \boldsymbol{N}$) such that $V_j(x, x_{n+1}) = -V_j(x, -x_{n+1})$,

$$|U_{j+1} - U_j - V_j| \le 2^{-j} \text{ on } M_j \text{ and } (1 - \Delta_{x,x_{n+1}})V_j = 0$$

for $j \in \boldsymbol{N}$, where

$$\begin{aligned} M_j \;=\; & \{(x, x_{n+1}) \in \boldsymbol{R}^{n+1};\ |(x, x_{n+1})| \le j \text{ and ``} x \in \overline{X}_{j-1} \text{ or} \\ & |x_{n+1}| \ge 1/j\text{''}\} \ (\Subset \boldsymbol{R}^n \times (\boldsymbol{R} \setminus \{0\}) \cup X_j \times \{0\}) \ (j \in \boldsymbol{N}) \end{aligned}$$

and $X_0 = \emptyset$. Then,

$$U \;\equiv\; U_k - V_1 - \cdots - V_{k-1} + \sum_{j=k}^{\infty} (U_{j+1} - U_j - V_j)$$

$$\Big(= \lim_{N\to\infty} (U_{N+1} - \sum_{j=1}^{N} V_j)\Big)$$

converges locally uniformly in $\boldsymbol{R}^n \times (\boldsymbol{R} \setminus \{0\})$ and satisfies $(1-\Delta_{x,x_{n+1}})U = 0$ in $\mathcal{D}'(\boldsymbol{R}^n \times (\boldsymbol{R} \setminus \{0\}))$. So we have $U \in \mathcal{H}(\emptyset)$. Moreover, we have

$$U - U_k = -V_1 - \cdots - V_{k-1} + \sum_{j=k}^{\infty} (U_{j+1} - U_j - V_j) \in \mathcal{H}(X_k),$$

and $(TU)|_{X_k} = u_k$. This proves that T is surjective.

Chapter 2

Basic calculus of Fourier integral operators and pseudodifferential operators

We shall introduce several symbol classes, which contain non-analytic symbols, and define pseudodifferential operators and Fourier integral operators. Product formulas of Fourier integral operators will be given in Sections 2.4 and 2.5, which are indispensable to our treatment. We shall prove pseudolocal properties of pseudodifferential operators in Section 2.6. As a simple application of the product formulas (and the symbol calculus) we shall construct parametrices of elliptic pseudodifferential operators in Section 2.8.

2.1 Preliminary lemmas

In this section we shall give a series of lemmas which are used frequently without quoting.

Lemma 2.1. 1 (i) $c_1 e^{-N} N^{N+1/2} \le N! \le c_2 e^{-N} N^{N+1/2}$ *for* $N \in N$, *where* c_1 *and* c_2 *are positive constants.* (ii) *For* $t \ge 1$

$$\inf_{N \in Z_+} N! t^{-N} \le c_2 \inf_{N \in N} N^{N+1/2} (et)^{-N} \le e c_2 t^{1/2} e^{-t}.$$

(iii) *For* $r \in \boldsymbol{N}$, $\alpha \in \boldsymbol{Z}_+^n$ *and* $\gamma \in \boldsymbol{Z}_+^r$ *with* $|\alpha| = |\gamma|$

$$\sum_{\alpha^1+\cdots\alpha^r=\alpha,\,(|\alpha^1|,\cdots,|\alpha^r|)=\gamma} \frac{\alpha!}{\alpha^1!\cdots\alpha^r!} = \frac{|\alpha|!}{\gamma!} \le r^{|\alpha|}.$$

In particular, $|\alpha|! \le n^{|\alpha|}\alpha!$ *for* $\alpha \in \boldsymbol{Z}_+^n$. (iv) *For* $\xi, \eta \in \boldsymbol{R}^n$

$$\langle \xi + \eta \rangle \le \sqrt{2}\langle \xi \rangle \langle \eta \rangle, \quad \langle \xi \rangle - |\eta| \le \langle \xi + \eta \rangle \le \langle \xi \rangle + |\eta|.$$

(v) *For* $s \ge 0$

$$|\partial_\xi^\alpha |\xi|^{1-s}| \le (1+\sqrt{2})^{|\alpha|}(|\alpha| + [s])!/[s]!|\xi|^{1-s-|\alpha|} \quad (\, \xi \neq 0),$$
$$|\partial_\xi^\alpha \langle \xi \rangle^{1-s}| \le (1+\sqrt{2})^{|\alpha|}(|\alpha| + [s])!/[s]!\langle \xi \rangle^{1-s-|\alpha|}.$$

(vi) *For* $N \in \boldsymbol{Z}_+$ *and* $\xi = (\xi_1, \cdots, \xi_n) \in \boldsymbol{R}^n$

$$\Big|\sum_{|\alpha|=N} \binom{N}{\alpha} \xi^\alpha\Big| = |(\xi_1 + \cdots + \xi_n)^N| \le (\sqrt{n}|\xi|)^N,$$

where $\binom{N}{\alpha} = N!/\alpha!$. (vii) $\alpha, \beta \in \boldsymbol{Z}_+^n$, $s \ge 0$ *and* $\xi \in \boldsymbol{R}^n \setminus \{0\}$

$$|\partial_\xi^\alpha(\xi^\beta|\xi|^{1-s})| \le \begin{cases} 2^{[s]+|\alpha|-|\beta|}(1+\sqrt{2})^{|\alpha|+|\beta|}|\alpha|!|\xi|^{1-s-|\alpha|+|\beta|} \\ \qquad \text{if } |\beta| \le |\alpha|, \\ (1+2\sqrt{2})^{-1}2^{[s]+|\alpha|+|\beta|+1}(1+\sqrt{2})^{|\alpha|+1}|\alpha|! \\ \times |\xi|^{1-s-|\alpha|+|\beta|} \quad \text{if } |\beta| > |\alpha|. \end{cases}$$

In particular,

$$|\partial_\xi^\alpha(\xi^\beta|\xi|^{1-s})| \le 2^{[s]+|\beta|+1}(2+2\sqrt{2})^{|\alpha|}|\alpha|!|\xi|^{1-s-|\alpha|+|\beta|}.$$

(viii) *If* $j \in \boldsymbol{Z}_+$, $\rho > 0$ *and* $A(\rho) \ge \rho^{-1}(1+\sqrt{1+2\rho^2})$, *then*

$$\begin{aligned} \left|\partial_t^j e^{-t^2}\right| &\le A(\rho)^j \sqrt{j!} e^{-t^2} \sum_{k=0}^{j} (\rho|t|)^k/\sqrt{k!} \\ &\le 2A(\rho)^j \sqrt{j!} e^{(2\rho^2-1)t^2} \qquad (2.1) \end{aligned}$$

for $t \in \boldsymbol{R}$. *Moreover,*

$$\left|\partial_\xi^\alpha e^{-\nu|\xi|^2}\right| \le e^{1/4}2^{n+|\alpha|}(\sqrt{2}+\sqrt{3})^{|\alpha|}|\alpha|!\langle \xi \rangle^{-|\alpha|}e^{-\nu|\xi|^2/4} \qquad (2.2)$$

for $\alpha \in \boldsymbol{Z}_+^n$, $0 \le \nu \le 1$ *and* $\xi \in \boldsymbol{R}^n$.

Proof The assertions (i)–(iv) and (vi) are obvious. The assertion (v) can be proved by induction on $|\alpha|$, using

$$\partial_\xi^\alpha\{|\xi|^2\partial_{\xi_j}|\xi|^{1-s}\} = (1-s)(\xi_j\partial_\xi^\alpha|\xi|^{1-s} + \alpha_j\partial_\xi^{\alpha-e_j}|\xi|^{1-s}),$$

$$\begin{aligned}|\partial_\xi^{\alpha+e_j}|\xi|^{1-s}| \le |\xi|^{-2}\Big[\sum_{k=1}^n \{2\alpha_k|\xi_k||\partial_\xi^{\alpha-e_k+e_j}|\xi|^{1-s}| \\ +\alpha_k(\alpha_k-1)|\partial_\xi^{\alpha-2e_k+e_j}|\xi|^{1-s}|\} \\ +|1-s||\xi_j||\partial_\xi^\alpha|\xi|^{1-s}| + \alpha_j|1-s||\partial_\xi^{\alpha-e_j}|\xi|^{1-s}|\Big],\end{aligned}$$

where $e_k = (\delta_{kj})_{j\to 1,2,\cdots,n} \in \boldsymbol{Z}_+^n$ and δ_{kj} denotes Kronecker's delta. The assertion (vii) easily follows from the assertion (v). The inequality (2.1) can be also proved by induction, noting that $\rho A(\rho)^2 - 2A(\rho) - 2\rho \ge 0$ and

$$(\rho|t|)^k/\sqrt{k!} = 2^{-k}((4\rho^2t^2)^k/k!)^{1/2} \le 2^{-k}e^{2\rho^2t^2}.$$

The inequality (2.2) easily follows from (2.1) with $\rho = 1/2$, using

$$\nu^{j/2}j!^{-1/2} \le \nu^{j/2}(\langle\sqrt{\nu}\xi\rangle/\sqrt{2})^{-j}e^{\langle\sqrt{\nu}\xi\rangle^2/4} \le 2^{j/2}e^{1/4}\langle\xi\rangle^{-j}e^{\nu|\xi|^2/4}$$

for $0 \le \nu \le 1$ and $j \in \boldsymbol{Z}_+$. □

Lemma 2.1. 2 *Let F be a closed subset of $\boldsymbol{R}^n$. Then there are $\chi_{k,\ell}^\varepsilon(x) \in C^\infty(\boldsymbol{R}^n)$ ($k,\ell \in \boldsymbol{Z}_+$ and $\varepsilon > 0$) such that $0 \le \chi_{k,\ell}^\varepsilon(x) \le 1$, $\chi_{k,\ell}^\varepsilon(x) = 1$ in F,* supp *$\chi_{k,\ell}^\varepsilon \subset \overline{F}_\varepsilon$ and*

$$\left|D^{\alpha^1+\alpha^2+\tilde\alpha}\chi_{k,\ell}^\varepsilon(x)\right| \le C_{|\tilde\alpha|}(C_*k/\varepsilon)^{|\alpha^1|}(C_*\ell/\varepsilon)^{|\alpha^2|}\varepsilon^{-|\tilde\alpha|}$$

if $|\alpha^1| \le k$ and $|\alpha^2| \le \ell$, where F_ε denotes ε-neighborhood of F and the $C_{|\tilde\alpha|}$ and C_* do not depend on k, ℓ and ε, and $0^0 = 1$.

Proof Let $\chi^\varepsilon(x)$ be the defining function of $F_{\varepsilon/2}$ and choose $\rho(x) \in C_0^\infty(\boldsymbol{R}^n)$ so that $\rho(x) \ge 0$, supp $\rho \subset \{x \in \boldsymbol{R}^n;\ |x| \le 1\}$ and $\int \rho(x)\,dx = 1$. We put, for $\varepsilon > 0$, $\rho_\varepsilon(x) = \varepsilon^{-n}\rho(\varepsilon^{-1}x)$ and

$$\chi_{k,\ell}^\varepsilon(x) = \chi^\varepsilon * \rho_{\varepsilon/6} * \overbrace{\rho_{\varepsilon/(6k)} * \cdots * \rho_{\varepsilon/(6k)}}^{k} * \overbrace{\rho_{\varepsilon/(6\ell)} * \cdots * \rho_{\varepsilon/(6\ell)}}^{\ell}$$

for $k,\ell \in \boldsymbol{Z}_+$, where $*$ denotes the convolution. Then $\{\chi_{k,\ell}^\varepsilon\}$ satisfies the assertion of the lemma with $C_* = 6\sup_{|\alpha|=1}\int|D^\alpha\rho(x)|\,dx$ and $C_j = 6^j\sup_{|\alpha|=j}\int|D^\alpha\rho(x)|\,dx$. □

Lemma 2.1. 3 *Let Γ_j ($j = 1, 2$) be open cones in $\boldsymbol{R}^n \setminus \{0\}$ such that $\overline{\Gamma}_1 \setminus \{0\} \subset \mathrm{int}(\Gamma_2)$. Then there are $g^R(\xi)(\equiv g^R_{\Gamma_1,\Gamma_2}(\xi)) \in C^\infty(\boldsymbol{R}^n)$ ($R \geq 2$) such that $0 \leq g^R(\xi) \leq 1$, $g^R(\xi) = 0$ for $\xi \notin \Gamma_2$, $g^R(\xi) = 1$ for $\xi \in \Gamma_1$ with $\langle \xi \rangle \geq 2R$, and*

$$\left| \partial_\xi^{\alpha+\tilde{\alpha}} g^R(\xi) \right| \leq C_{\Gamma_1,\Gamma_2,|\tilde{\alpha}|} \left(C_{\Gamma_1,\Gamma_2}/R \right)^{|\alpha|} \langle \xi \rangle^{-|\tilde{\alpha}|} \quad \textit{if } \langle \xi \rangle \geq R|\alpha|,$$

where the $C_{\Gamma_1,\Gamma_2,|\tilde{\alpha}|}$ and C_{Γ_1,Γ_2} do not depend on R.

Proof Following [Tr2], we shall give the proof. By Lemma 2.1. 2 we can choose $\phi_\ell(t) \in C^\infty(\boldsymbol{R})$ ($\ell \in \boldsymbol{Z}_+$) so that $0 \leq \phi_\ell(t) \leq 1$, $\phi_\ell(t) = 0$ for $t > 1$, $\phi_\ell(t) = 1$ for $t < 0$ and

$$|\partial_t^{j+k} \phi_\ell(t)| \leq C_k (C_* \ell)^j \quad \text{if } j \leq \ell,$$

where the C_k and C_* do not depend on ℓ. Put, for $N \in \boldsymbol{N}$ and $R > 0$,

$$h_N^R(t) := \phi_{4^N}(4^{-N} R^{-1} t - 1) - \phi_{4^{N-1}}(4^{1-N} R^{-1} t - 1). \tag{2.3}$$

Then we have

$$\sum_{N=1}^\infty h_N^R(t) = 1 - \phi_1(R^{-1}t - 1),$$

$$\sum_{N=1}^\infty h_N^R(t) = \begin{cases} 0 & (t \leq R), \\ 1 & (t \geq 2R), \end{cases}$$

$$4^{N-1} \leq t/R \leq 2 \cdot 4^N \quad \text{if } t \in \mathrm{supp}\, h_N^R, \tag{2.4}$$

$$\mathrm{supp}\, h_N^R \cap \mathrm{supp}\, h_{N'}^R = \emptyset \quad \text{if } N' \geq N + 2, \tag{2.5}$$

$$|\partial_t^{j+k} h_N^R(t)| \leq 2C_k (2/t)^k (C_*/R)^j \quad \text{if } j \leq t/(2R). \tag{2.6}$$

In fact, $t/(2R) \leq 4^N$ if $t \in \mathrm{supp}\, \phi_{4^N}(4^{-N}R^{-1}t - 1)$. Put $S_j = \{\theta = (\theta_1, \cdots, \theta_n) \in S^{n-1};\ |\theta_j| > (4n)^{-1/2}\}$ and $\widetilde{S}_j = \{\theta \in S^{n-1};\ |\theta_j| \geq 1/\sqrt{n}\}$, where $S^{n-1} = \{\theta \in \boldsymbol{R}^n;\ |\theta| = 1\}$. Then, $S^{n-1} = \cup_{j=1}^n \widetilde{S}_j$. By Lemma 2.1. 2 we can choose

$$\psi_{j,N}(\theta) \equiv \psi_{j,N}(\theta_1, \cdots, \theta_{j-1}, \theta_{j+1}, \cdots, \theta_n) \in C^\infty(S^{n-1})$$

($N \in \boldsymbol{N}$, $1 \leq j \leq n$) so that $0 \leq \psi_{j,N}(\theta) \leq 1$, $\mathrm{supp}\, \psi_{j,N} \subset S_j \cap \Gamma_2$, $\psi_{j,N}(\theta) = 1$ for $\theta \in \widetilde{S}_j \cap \Gamma_1$ and

$$\left| \partial_\theta^{\alpha+\tilde{\alpha}} \psi_{j,N}(\theta) \right| \leq C_{|\tilde{\alpha}|} (4^{N-1} C)^{|\alpha|} \quad \text{if } |\alpha| \leq 4^N,$$

where the $C_{|\tilde{\alpha}|}$ and C depend on Γ_1 and Γ_2 and do not depend on N. Putting $\psi_N(\theta) = 1 - \prod_{j=1}^n (1 - \psi_{j,N}(\theta))$, we have $\psi_N(\theta) \in C^\infty(S^{n-1})$, $0 \le \psi_N(\theta) \le 1$, supp $\psi_N \subset S^{n-1} \cap \Gamma_2$, $\psi_N(\theta) = 1$ for $\theta \in S^{n-1} \cap \Gamma_1$ and

$$\left|\partial_\theta^{\alpha+\tilde{\alpha}} \psi_N(\theta)\right| \le C'_{|\tilde{\alpha}|}(4^{N-1}C')^{|\alpha|} \quad \text{if } |\alpha| \le 4^N.$$

By induction on $|\alpha| + |\tilde{\alpha}|$ we can prove that

$$\left|\partial_\xi^{\alpha+\tilde{\alpha}}(\partial_\theta^{\beta+\tilde{\beta}}\psi_N)(\xi/|\xi|)\right| \le C_{|\tilde{\alpha}|,|\tilde{\beta}|}(4^{N-1}C_1/|\xi|)^{|\alpha|}(4^{N-1}C')^{|\beta|}|\xi|^{-|\tilde{\alpha}|} \tag{2.7}$$

if $\xi \in \boldsymbol{R}^n \setminus \{0\}$ and $|\alpha| + |\beta| \le 4^N$, where the $C_{|\tilde{\alpha}|,|\tilde{\beta}|}$ and C_1 do not depend on N. We can also prove by induction that

$$\left|\partial_\xi^{\alpha+\tilde{\alpha}}(\partial_t^{k+\tilde{k}} h_N^R)(|\xi|)\right| \le C_{|\tilde{\alpha}|,\tilde{k}}(C'_*/R)^{|\alpha|}(C_*/R)^k|\xi|^{-|\tilde{\alpha}|-\tilde{k}} \tag{2.8}$$

if $|\alpha| + k \le |\xi|/(2R)$, where the $C_{|\tilde{\alpha}|,\tilde{k}}$ and C'_* do not depend on N and R. (2.7) and (2.8) can be proved by the same induction argument, and we shall frequently use this argument. So we will give the proof of (2.8). By (2.6) (2.8) is valid with $C_{0,\tilde{k}} \ge 2^{k+1}C_{\tilde{k}}$ if $|\tilde{\alpha}| + |\tilde{\alpha}| = 0$, where the C_k are the constants in (2.6). Assume that (2.8) is valid for $|\alpha| + |\tilde{\alpha}| \le \ell$, where $\ell \in \boldsymbol{Z}_+$. Choose $\alpha, \tilde{\alpha}, e, \tilde{e} \in \boldsymbol{Z}_+^n$ so that $|\alpha| + |\tilde{\alpha}| = \ell$, $|e| + |\tilde{e}| = 1$ and $|\alpha| + |e| + k \le |\xi|/(2R)$. Then, by Lemma 2.1. 1 we have

$$\begin{aligned}
&\left|\partial_\xi^{\alpha+\tilde{\alpha}+e+\tilde{e}}(\partial_t^{k+\tilde{k}} h_N^R)(|\xi|)\right| \\
&\le C_{|\tilde{\alpha}|+|\tilde{e}|,\tilde{k}}(C'_*/R)^{|\alpha|+|e|}(C_*/R)^k|\xi|^{-|\tilde{\alpha}|-|\tilde{e}|-\tilde{k}} I^{k,\tilde{k}}_{\alpha,\tilde{\alpha},e,\tilde{e}}, \\
&I^{k,\tilde{k}}_{\alpha,\tilde{\alpha},e,\tilde{e}} = (2(1+\sqrt{2})C_*/C'_*)^{|e|}\sum_{\mu=0}^{|\tilde{\alpha}|}\frac{|\tilde{\alpha}|!}{\mu!}(1+\delta_{|\tilde{e}|,1}(|\tilde{\alpha}|-\mu)) \\
&\quad \times (2(1+\sqrt{2}))^{|\tilde{\alpha}|-\mu+|\tilde{e}|}(C_{\mu,\tilde{k}+|\tilde{e}|}/C_{|\tilde{\alpha}|+|\tilde{e}|,\tilde{k}}) \\
&\quad \times \sum_{\nu=0}^{|\alpha|}\binom{|\alpha|}{\nu}((1+\sqrt{2})/C'_*)^{|\alpha|-\nu}(|\alpha|-\nu+|e|)!(|\alpha|+|e|)^{-|\alpha|+\nu}.
\end{aligned}$$

Note that

$$\begin{aligned}
&\binom{|\alpha|}{\nu}(|\alpha|-\nu+|e|)!(|\alpha|+|e|)^{-|\alpha|+\nu} \le 1 + \delta_{|e|,1}(|\alpha|-\nu), \\
&\sum_{\nu=0}^{\infty}(\nu+1)\rho^\nu = (1-\rho)^{-2} \quad \text{if } |\rho| < 1.
\end{aligned}$$

Then we have $I^{k,\tilde{k}}_{\alpha,\tilde{\alpha},e,\tilde{e}} \le 1$ if

$$\begin{aligned} C'_* &\ge \max\{2(1+\sqrt{2}), 16(1+\sqrt{2})C_*\}, \\ C_{|\tilde{\alpha}|,\tilde{k}} &\ge \sum_{\mu<|\tilde{\alpha}|} (2(1+\sqrt{2}))^{|\tilde{\alpha}|-\mu}|\tilde{\alpha}|!C_{\mu,\tilde{k}}/\mu!, \\ C_{|\tilde{\alpha}|+1,\tilde{k}} &\ge 2\sum_{\mu=0}^{|\tilde{\alpha}|}(2(1+\sqrt{2}))^{|\tilde{\alpha}|-\mu+1}(|\tilde{\alpha}|-\mu+1)|\tilde{\alpha}|!C_{\mu,\tilde{k}+1}/\mu!. \end{aligned}$$

This proves (2.8). Put

$$\tilde{g}^R(\xi) = \sum_{N=1}^{\infty} \psi_N(\xi/|\xi|)h^R_N(|\xi|).$$

Then we have $0 \le \tilde{g}^R(\xi) \le 1$, $\tilde{g}^R(\xi) = 1$ for $\xi \in \Gamma_1$, with $|\xi| \ge 2R$, $\tilde{g}^R(\xi) = 0$ if $\xi \notin \Gamma_2$ or $|\xi| \le R$, and

$$\left|\partial_\xi^{\alpha+\tilde{\alpha}}\tilde{g}^R(\xi)\right| \le C_{|\tilde{\alpha}|}(C_2/R)^{|\alpha|}|\xi|^{-|\tilde{\alpha}|} \quad \text{if } |\alpha| \le |\xi|/(2R),$$

where the $C_{|\tilde{\alpha}|}$ and C_2 do not depend on R. If we put $g^R(\xi) = \tilde{g}^{R'}(\xi)$ with $R' = \sqrt{R^2-1}/2$, $g^R(\xi)$ has the properties of the lemma. □

Let $\varphi(\theta, x, \xi) \in C^\infty(\Omega)$ satisfy

$$\begin{aligned} &|\partial_\theta^\gamma \partial_\xi^\alpha D_x^\beta \varphi(\theta,x,\xi)| \le C_0(\theta,x,\xi)\Theta_0(\theta,x,\xi)^{|\gamma|}A_0(\theta,x,\xi)^{|\alpha|} \\ &\quad \times B_0(\theta,x,\xi)^{|\beta|}|\alpha|!|\beta|!|\gamma|!\langle\xi\rangle^{1-|\alpha|} \quad \text{in } \Omega \end{aligned}$$

for $|\alpha|+|\beta| \ge 1$, and

$$|\nabla_x\varphi(\theta,x,\xi)|\langle\xi\rangle^{-1} + |\nabla_\xi\varphi(\theta,x,\xi)| \ne 0 \quad \text{in } \Omega,$$

where $n_j \in \boldsymbol{Z}_+$ ($j = 1,2,3$), Ω is an open subset of $\boldsymbol{R}^{n_1} \times \boldsymbol{R}^{n_2} \times \boldsymbol{R}^{n_3}$, $C_0(\theta,x,\xi)$, $\Theta_0(\theta,x,\xi)$, $A_0(\theta,x,\xi)$ and $B_0(\theta,x,\xi)$ are functions defined in Ω. Put

$$\lambda(\theta,x,\xi) = |\nabla_x\varphi(\theta,x,\xi)|^2 + \langle\xi\rangle^2|\nabla_\xi\varphi(\theta,x,\xi)|^2.$$

Then we have the following

Lemma 2.1.4 *Under the above hypotheses, we have*

$$\begin{aligned} &\left|\partial_\theta^\gamma\partial_\xi^\alpha D_x^\beta \lambda(\theta,x,\xi)^{-1}\right| \le \lambda(\theta,x,\xi)^{-1}(10\Theta_0(\theta,x,\xi))^{|\gamma|} \\ &\quad \times(10A_0(\theta,x,\xi))^{|\alpha|}(10B_0(\theta,x,\xi))^{|\beta|}(\mu(\theta,x,\xi)/\lambda(\theta,x,\xi))^{|\alpha|+|\beta|+|\gamma|} \\ &\quad \times|\alpha|!|\beta|!|\gamma|!\langle\xi\rangle^{-|\alpha|} \quad \textit{in } \Omega, \end{aligned} \tag{2.9}$$

where

$$\mu(\theta,x,\xi) = 3C_0(\theta,x,\xi)^2(n_2B_0(\theta,x,\xi)^2 + n_3(A_0(\theta,x,\xi)+1/2)^2)\langle\xi\rangle^2.$$

Proof Note that

$$\sum_{k=0}^{j}(k+1)(j+1-k) = (j+1)(j+2)(j+3)/6 \le 5\cdot 2^{j-1} < 3\cdot 2^j$$

for $0 \le k \le j$. Then we have

$$\begin{aligned}\left|\partial_\theta^\gamma \partial_\xi^\alpha D_x^\beta \lambda(\theta,x,\xi)\right| &\le \mu(\theta,x,\xi)(2\Theta_0)^{|\gamma|}(2A_0)^{|\alpha|}(2B_0)^{|\beta|}\\ &\quad \times|\alpha|!|\beta|!|\gamma|!\langle\xi\rangle^{-|\alpha|} \quad \text{in } \Omega,\end{aligned}$$

where we abbreviate $\Theta_0(\theta,x,\xi)$, $\cdots$ as Θ_0, $\cdots$. Let us prove (2.9) by induction. It is obvious that (2.9) is valid if $|\alpha|+|\beta|+|\gamma| = 0$. Assume that (2.9) is valid if $|\alpha|+|\beta|+|\gamma| \le \ell$, where $\ell \in \boldsymbol{Z}_+$. Let $\gamma, e^1 \in \boldsymbol{Z}_+^{n_1}$, $\beta, e^2 \in \boldsymbol{Z}_+^{n_2}$ and $\alpha, e^3 \in \boldsymbol{Z}_+^{n_3}$ satisfy $|\alpha|+|\beta|+|\gamma| = \ell$ and $|e^1|+|e^2|+|e^3| = 1$. Noting that

$$\lambda(\theta,x,\xi)\partial_\theta^{e^1}\partial_\xi^{e^3}D_x^{e^2}\lambda(\theta,x,\xi)^{-1} = -\lambda(\theta,x,\xi)^{-1}\partial_\theta^{e^1}\partial_\xi^{e^3}D_x^{e^2}\lambda(\theta,x,\xi),$$

we have

$$\begin{aligned}&\left|\partial_\theta^{\gamma+e^1}\partial_\xi^{\alpha+e^3}D_x^{\beta+e^2}\lambda(\theta,x,\xi)^{-1}\right|\\ &\le \lambda(\theta,x,\xi)^{-1}(10\Theta_0)^{|\gamma|+|e^1|}(10A_0)^{|\alpha|+|e^3|}(10B_0)^{|\beta|+|e^2|}\\ &\quad\times(\mu(\theta,x,\xi)/\lambda(\theta,x,\xi))^{|\alpha|+|\beta|+|\gamma|+1}(|\alpha|+|e^3|)!\\ &\quad\times(|\beta|+|e^2|)!(|\gamma|+|e^1|)!\langle\xi\rangle^{-|\alpha|-|e^3|}I_{\alpha,\beta,\gamma,e^1,e^2,e^3},\\ &I_{\alpha,\beta,\gamma,e^1,e^2,e^3} = \sum_{\substack{0\le i\le|\gamma|,\,0\le j\le|\alpha|\\ 0\le k\le|\beta|,\,i+j+k>0}} 5^{-i-j-k}(\lambda/\mu)^{i+j+k-1}\\ &\qquad\qquad + \sum_{i=0}^{|\gamma|}\sum_{j=0}^{|\alpha|}\sum_{k=0}^{|\beta|}5^{-i-j-k-1}(\lambda/\mu)^{i+j+k}.\end{aligned}$$

Noting that $\lambda(\theta,x,\xi) \le C_0^2(n_2B_0^2+n_3A_0^2)\langle\xi\rangle^2 \le \mu(\theta,x,\xi)/3$, we have $I_{\alpha,\beta,\gamma,e^1,e^2,e^3} \le 1$. This proves the lemma. □

Assume that $f(\theta,x,\xi) \in C^\infty(\Omega)$ satisfy

$$\begin{aligned}&\left|\partial_\theta^{\gamma+\tilde\gamma}\partial_\xi^{\alpha+\tilde\alpha}D_x^{\beta+\tilde\beta}f(\theta,x,\xi)\right| \le C_{|\tilde\alpha|,|\tilde\beta|,|\tilde\gamma|}(\theta,x,\xi;f)\Theta(\theta,x,\xi;f)^{|\gamma|}\\ &\quad\times A(\theta,x,\xi;f)^{|\alpha|}B(\theta,x,\xi;f)^{|\beta|}\langle\xi\rangle^{-|\alpha|-|\tilde\alpha|} \quad \text{in } \Omega\\ &\quad \text{if } |\gamma| \le N_0(\theta,x,\xi;f),\ |\alpha| \le N_1(\theta,x,\xi;f) \text{ and } |\beta| \le N_2(\theta,x,\xi;f),\end{aligned}$$

where the $C_{|\tilde\alpha|,|\tilde\beta|,|\tilde\gamma|}(\theta,x,\xi;f)$, $\Theta(\theta,x,\xi;f)$, $A(\theta,x,\xi;f)$, $B(\theta,x,\xi;f)$ and $N_j(\theta,x,\xi;f)$ ($j=0,1,2$) are functions defined in Ω. We also denote by

$C_{a,b,\cdots}(A,B,\cdots)$ a constant depending on quantities a, b, $\cdots$ and A, B, $\cdots$. Let L be a differential operator defined by

$$\begin{aligned} Lf(\theta,x,\xi) &= \sum_{j=1}^{n_2} D_{x_j}\{\lambda(\theta,x,\xi)^{-1}\overline{(\partial\varphi/\partial x_j)(\theta,x,\xi)}f(\theta,x,\xi)\} \\ &\quad + \sum_{j=1}^{n_3} D_{\xi_j}\{\langle\xi\rangle^2\lambda(\theta,x,\xi)^{-1}\overline{(\partial\varphi/\partial\xi_j)(\theta,x,\xi)}f(\theta,x,\xi)\}. \end{aligned}$$

Then we have the following

Lemma 2.1. 5 *Under the above hypotheses, we have*

$$\begin{aligned} &\left|\partial_\theta^{\gamma+\tilde\gamma}\partial_\xi^{\alpha+\tilde\alpha}D_x^{\beta+\tilde\beta}L^{k+\tilde k}f(\theta,x,\xi)\right| \\ &\leq C_{|\tilde\alpha|,|\tilde\beta|,|\tilde\gamma|,\tilde k}(\theta,x,\xi;f,\varphi)\Theta_1(\theta,x,\xi;f,\varphi)^{|\gamma|} \\ &\quad \times A_1(\theta,x,\xi;f,\varphi)^{|\alpha|}B_1(\theta,x,\xi;f,\varphi)^{|\beta|}\Gamma(\theta,x,\xi;f,\varphi)^k \\ &\quad \times\Big(C_0'(\theta,x,\xi)\langle\xi\rangle^2/\lambda(\theta,x,\xi)\Big)^{\tilde k}\langle\xi\rangle^{-|\alpha|-|\tilde\alpha|-k-\tilde k} \quad \text{in } \Omega \end{aligned} \tag{2.10}$$

if $|\gamma|\leq N_0(\theta,x,\xi;f)$, $|\alpha|+k\leq N_1(\theta,x,\xi;f)$ *and* $|\beta|+k\leq N_2(\theta,x,\xi;f)$. *Here,*

$$\Theta_1(\theta,x,\xi;f,\varphi)=\max\{\Theta,4\sigma(\theta,x,\xi)N_0\Theta_0\}, \tag{2.11}$$
$$A_1(\theta,x,\xi;f,\varphi)=\max\{A,4\sigma(\theta,x,\xi)N_1A_0\}, \tag{2.12}$$
$$B_1(\theta,x,\xi;f,\varphi)=\max\{B,4\sigma(\theta,x,\xi)N_2B_0\}, \tag{2.13}$$
$$\begin{aligned} &\Gamma(\theta,x,\xi;f,\varphi) \\ &=2^4C_0\langle\xi\rangle^2\lambda^{-1}\Big\{2n_2B_0B_1+n_3(\sqrt{2A_0}+1/\sqrt{2A_0})^2A_1\Big\}, \end{aligned}\tag{2.14}$$

where we abbreviate $\Theta(\theta,x,\xi;f,\varphi)$, $\cdots$ *as* Θ, $\cdots$, *and*

$$\begin{aligned} &C_0'(\theta,x,\xi)=C_0(2n_2B_0+n_3(\sqrt{2A_0}+1/\sqrt{2A_0})^2), \\ &\sigma(\theta,x,\xi)=30C_0^2\langle\xi\rangle^2\lambda^{-1}\Big(n_2B_0^2+n_3(A_0+1/2)^2\Big). \end{aligned}$$

Proof We shall prove the lemma by induction on $k+\tilde k$. If $k+\tilde k=0$, $C_{|\tilde\alpha|,|\tilde\beta|,|\tilde\gamma|,0}\geq C_{|\tilde\alpha|,|\tilde\beta|,|\tilde\gamma|}$, $\Theta_1\geq\Theta$, $A_1\geq A$ and $B_1\geq B$, then (2.10) is valid. Now assume that (2.10) is valid if $k+\tilde k\leq\ell$, where $\ell\in Z_+$. Let $k,\tilde k,\nu,\tilde\nu\in Z_+$ satisfy $k+\tilde k=\ell$ and $\nu+\tilde\nu=1$. Then, for $|\gamma|\leq N_0$, $|\alpha|+k+\nu\leq N_1$ and $|\beta|+k+\nu\leq N_2$ we have

$$\left|\partial_\theta^{\gamma+\tilde\gamma}\partial_\xi^{\alpha+\tilde\alpha}D_x^{\beta+\tilde\beta}L^{k+\tilde k+\nu+\tilde\nu}f(\theta,x,\xi)\right|$$

$$
\begin{aligned}
&= \Big|\partial_\theta^{\gamma+\tilde\gamma}\partial_\xi^{\alpha+\tilde\alpha}D_x^{\beta+\tilde\beta}\Big\{\sum_{j=1}^{n_2} D_{x_j}(\lambda^{-1}\overline{\partial_{x_j}\varphi}L^{k+\tilde k}f) \\
&\qquad + \sum_{j=1}^{n_3} D_{\xi_j}(\langle\xi\rangle^2\lambda^{-1}\overline{\partial_{\xi_j}\varphi}L^{k+\tilde k}f)\Big\}\Big| \\
&\le C_{|\tilde\alpha|,|\tilde\beta|,|\tilde\gamma|,\tilde k+\tilde\nu}\Theta_1^{|\gamma|}A_1^{|\alpha|}B_1^{|\beta|}\Gamma^{k+\nu}\left(C_0'\langle\xi\rangle^2/\lambda\right)^{\tilde k+\tilde\nu} \\
&\qquad \times\langle\xi\rangle^{-|\alpha|-|\tilde\alpha|-k-\tilde k-1} I^{k,\tilde k,\nu,\tilde\nu}_{\alpha,\beta,\gamma,\tilde\alpha,\tilde\beta,\tilde\gamma},
\end{aligned}
$$

where

$$
\begin{aligned}
I^{k,\tilde k,\nu,\tilde\nu}_{\alpha,\beta,\gamma,\tilde\alpha,\tilde\beta,\tilde\gamma} &= (2n_2B_0B_1C_0\langle\xi\rangle^2/(\lambda\Gamma))^\nu \sum_{j_0=0}^{|\tilde\gamma|}\sum_{j_1=0}^{|\tilde\alpha|}\sum_{j_2=0}^{|\tilde\beta|+\tilde\nu}\frac{|\tilde\gamma|!|\tilde\alpha|!(|\tilde\beta|+\tilde\nu)!}{j_0!j_1!j_2!} \\
&\times\left(C_{j_1,j_2,j_0,\tilde k}/C_{|\tilde\alpha|,|\tilde\beta|,|\tilde\gamma|,\tilde k+\tilde\nu}\right)(2\sigma\Theta_0)^{|\tilde\gamma|-j_0}(2\sigma A_0)^{|\tilde\alpha|-j_1}(2\sigma B_0)^{|\tilde\beta|+\tilde\nu-j_2} \\
&\times\sum_{k_0=0}^{|\gamma|}\sum_{k_1=0}^{|\alpha|}\sum_{k_2=0}^{|\beta|+\nu}\frac{|\gamma|!|\alpha|!(|\beta|+\nu)!}{k_0!k_1!k_2!}N_0^{-|\gamma|+k_0}N_1^{-|\alpha|+k_1}N_2^{-|\beta|-\nu+k_2} \\
&\times(2\sigma N_0\Theta_0/\Theta_1)^{|\gamma|-k_0}(2\sigma N_1A_0/A_1)^{|\alpha|-k_1}(2\sigma N_2B_0/B_1)^{|\beta|+\nu-k_2} \\
&+(n_3(\sqrt{2A_0}+1/\sqrt{2A_0})^2A_1C_0\langle\xi\rangle^2/(\lambda\Gamma))^\nu \\
&\times\sum_{j_0=0}^{|\tilde\gamma|}\sum_{j_1=0}^{|\tilde\alpha|+\tilde\nu}\sum_{j_2=0}^{|\tilde\beta|}\frac{|\tilde\gamma|!(|\tilde\alpha|+\tilde\nu)!|\tilde\beta|!}{j_0!j_1!j_2!}\left(C_{j_1,j_2,j_0,\tilde k}/C_{|\tilde\alpha|,|\tilde\beta|,|\tilde\gamma|,\tilde k+\tilde\nu}\right) \\
&\times(2\sigma\Theta_0)^{|\tilde\gamma|-j_0}(2\sigma A_0)^{|\tilde\alpha|+\tilde\nu-j_1}(2\sigma B_0)^{|\tilde\beta|-j_2} \\
&\times\sum_{k_0=0}^{|\gamma|}\sum_{k_1=0}^{|\alpha|+\nu}\sum_{k_2=0}^{|\beta|}\frac{|\gamma|!(|\alpha|+\nu)!|\beta|!}{k_0!k_1!k_2!}N_0^{-|\gamma|+k_0}N_1^{-|\alpha|-\nu+k_1}N_2^{-|\beta|+k_2} \\
&\times(2\sigma N_0\Theta_0/\Theta_1)^{|\gamma|-k_0}(2\sigma N_1A_0/A_1)^{|\alpha|+\nu-k_1}(2\sigma N_2B_0/B_1)^{|\beta|-k_2}.
\end{aligned}
$$

Here we have used the estimates

$$
\begin{aligned}
&\left|\partial_\theta^\gamma\partial_\xi^\alpha D_x^\beta(\lambda^{-1}\overline{\partial_{x_j}\varphi})\right| \le 2C_0B_0\lambda^{-1}(\sigma\Theta_0)^{|\gamma|}(\sigma A_0)^{|\alpha|} \\
&\qquad\times(\sigma B_0)^{|\beta|}|\alpha|!|\beta|!|\gamma|!\langle\xi\rangle^{1-|\alpha|} \quad \text{in } \Omega, \\
&\left|\partial_\theta^\gamma\partial_\xi^\alpha D_x^\beta(\langle\xi\rangle^2\lambda^{-1}\overline{\partial_{\xi_j}\varphi})\right| \le C_0(\sqrt{2A_0}+1/\sqrt{2A_0})^2\lambda^{-1} \\
&\qquad\times(\sigma\Theta_0)^{|\gamma|}(\sigma A_0)^{|\alpha|}(\sigma B_0)^{|\beta|}|\alpha|!|\beta|!|\gamma|!\langle\xi\rangle^{2-|\alpha|} \quad \text{in } \Omega.
\end{aligned}
$$

It is easy to see that $I^{k,\tilde k,\nu,\tilde\nu}_{\alpha,\beta,\gamma,\tilde\alpha,\tilde\beta,\tilde\gamma} \le 1$ if we choose Θ_1, A_1, B_1 and Γ as (2.11)–(2.14), and

$$
2C_{|\tilde\alpha|,|\tilde\beta|,|\tilde\gamma|,\tilde k} \ge J_{|\tilde\alpha|,|\tilde\beta|,|\tilde\gamma|},
$$

$$C_{|\tilde{\alpha}|,|\tilde{\beta}|,|\tilde{\gamma}|,\tilde{k}+1} \geq 8J_{|\tilde{\alpha}|,|\tilde{\beta}|+1,|\tilde{\gamma}|} + 8J_{|\tilde{\alpha}|+1,|\tilde{\beta}|,|\tilde{\gamma}|},$$

where

$$J_{k_1,k_2,k_0} = \sum_{j_0=0}^{k_0} \sum_{j_1=0}^{k_1} \sum_{j_2=0}^{k_2} \frac{k_0!k_1!k_2!}{j_0!j_1!j_2!} \times C_{j_1,j_2,j_0,\tilde{k}} (2\sigma\Theta_0)^{k_0-j_0} (2\sigma A_0)^{k_1-j_1} (2\sigma B_0)^{k_2-j_2}.$$

This proves the lemma. □

Let $F(\theta, x, \xi) \in C^\infty(\Omega)$ satisfy

$$\left|\partial_\theta^{\gamma+\tilde{\gamma}} \partial_\xi^{\alpha+\tilde{\alpha}} D_x^{\beta+\tilde{\beta}} F(\theta, x, \xi)\right| \leq C_\gamma(\theta, x, \xi; F) C_{|\tilde{\alpha}|,|\tilde{\beta}|,|\tilde{\gamma}|}(\theta, x, \xi; F) \times A(\theta, x, \xi; F)^{|\alpha|} \boldsymbol{B}(\theta, x, \xi; F)^\beta \quad \text{in } \Omega$$

if $|\alpha| \leq N_1(\theta, x, \xi; F)$ and $\beta \leq \boldsymbol{N}_2(\theta, x, \xi; F)$, where $\boldsymbol{B}(\cdot) = (B_1(\cdot), \cdots, B_{n_2}(\cdot))$, $\boldsymbol{N}_2(\cdot) = (N_{21}(\cdot), \cdots, N_{2n_2}(\cdot))$ and $\Omega = \Omega_1 \times \Omega_2 \times \Omega_3$ is an open subset of $\boldsymbol{R}^{n_1} \times \boldsymbol{R}^{n_2} \times \boldsymbol{R}^{n_3}$. Moreover, let $\Xi(x)$ be a C^∞ map from Ω_2 to Ω_3 such that

$$|D^\beta \Xi(x)| \leq C_0(x, \Xi) \boldsymbol{B}_0(x, \Xi)^\beta \beta! \quad \text{in } \Omega_2$$

for $\beta \in \boldsymbol{Z}_+^{n_2}$ with $|\beta| \geq 1$, where $\boldsymbol{B}_0(\cdot) = (B_{01}(\cdot), \cdots, B_{0n_2}(\cdot))$. Then we have the following

Lemma 2.1.6 *Under the above hypotheses, we have*

$$\begin{aligned}&\left|D_x^{\beta+\tilde{\beta}} \left(\partial_\theta^{\gamma+\tilde{\gamma}} \partial_\xi^{\alpha+\tilde{\alpha}} D_x^{\beta'+\tilde{\beta}'} F\right)(\theta, x, \Xi(x))\right| \\ &\leq C_\gamma(\theta, x, \Xi(x); F) C_{|\tilde{\beta}|,|\tilde{\alpha}|,|\tilde{\beta}'|,|\tilde{\gamma}|}(\theta, x; \Xi, F) A(\theta, x, \Xi(x); F)^{|\alpha|} \\ &\quad \times \boldsymbol{B}(\theta, x, \Xi(x); F)^{\beta'} \boldsymbol{B}_1(\theta, x; \Xi, F)^\beta \end{aligned} \tag{2.15}$$

if $(\theta, x) \in \Omega_1 \times \Omega_2$, $|\alpha| + |\beta| \leq N_1(\theta, x, \Xi(x); F)$ and $\beta + \beta' \leq \boldsymbol{N}_2(\theta, x, \Xi(x); F)$, where $\boldsymbol{B}_1(\cdot) = (B_{11}(\cdot), \cdots, B_{1n_2}(\cdot))$ and

$$B_{1j}(\theta, x; \Xi, F) \geq \max\{4N_{2j} B_{0j}, 2B_j, 2^{n_2+4}\sqrt{n_3} C_0 A B_{0j}\} \tag{2.16}$$

$(1 \leq j \leq n_2)$. Here we abbreviate $N_{2j}(\theta, x, \Xi(x); F)$, $\cdots$ as N_{2j}, $\cdots$.

Remark If Ω_3 is an open subset of $\boldsymbol{C}^{n_3}$ and $F(\theta, x, \xi)$ analytic in ξ, then (2.15) is still valid.

Proof If $|\beta| + |\tilde{\beta}| = 0$ and $C_{0,|\tilde{\alpha}|,|\tilde{\beta}'|,|\tilde{\gamma}|}(\theta, x, \Xi; F) \geq C_{|\tilde{\alpha}|,|\tilde{\beta}'|,|\tilde{\gamma}|}(\theta, x, \Xi(x); F)$, then (2.15) is valid. Assume that (2.15) is valid if $|\beta| + |\tilde{\beta}| \leq \ell$, where

$\ell \in \boldsymbol{Z}_+$. Let $\beta, \tilde{\beta}, e, \tilde{e} \in \boldsymbol{Z}_+^{n_2}$ satisfy $|\beta| + |\tilde{\beta}| = \ell$ and $|e| + |\tilde{e}| = 1$. Then, for $|\alpha| + |\beta| + |e| \le N_1$ and $\beta + \beta' + e \le \boldsymbol{N}_2$ we have

$$\begin{aligned}
&\left| D_x^{\beta+\tilde{\beta}+e+\tilde{e}} \left(\partial_\theta^{\gamma+\tilde{\gamma}} \partial_\xi^{\alpha+\tilde{\alpha}} D_x^{\beta'+\tilde{\beta}'} F \right)(\theta, x, \Xi(x)) \right| \\
&\le C_\gamma C_{|\tilde{\beta}|+|\tilde{e}|,|\tilde{\alpha}|,|\tilde{\beta}'|,|\tilde{\gamma}|} A^{|\alpha|} \boldsymbol{B}^{\beta'} \boldsymbol{B}_1^{\beta+e} I^{\beta,\tilde{\beta},e,\tilde{e}}_{\alpha,\beta',\tilde{\alpha},\tilde{\beta}',\tilde{\gamma}}, \\
&I^{\beta,\tilde{\beta},e,\tilde{e}}_{\alpha,\beta',\tilde{\alpha},\tilde{\beta}',\tilde{\gamma}} = C_{|\tilde{\beta}|,|\tilde{\alpha}|,|\tilde{\beta}'|+|\tilde{e}|,|\tilde{\gamma}|} \boldsymbol{B}^e / (C_{|\tilde{\beta}|+|\tilde{e}|,|\tilde{\alpha}|,|\tilde{\beta}'|,|\tilde{\gamma}|} \boldsymbol{B}_1^e) \\
&\quad + \sqrt{n_3} \sum_{\tilde{\beta}^1 \le \tilde{\beta}} \binom{\tilde{\beta}}{\tilde{\beta}^1} (\tilde{\beta} - \tilde{\beta}^1 + \tilde{e})! C_{|\tilde{\beta}^1|,|\tilde{\alpha}|+|\tilde{e}|,|\tilde{\beta}'|,|\tilde{\gamma}|} (2\boldsymbol{B}_0)^{\tilde{\beta}-\tilde{\beta}^1+\tilde{e}} \\
&\quad \times C^{-1}_{|\tilde{\beta}|+|\tilde{e}|,|\tilde{\alpha}|,|\tilde{\beta}'|,|\tilde{\gamma}|} \sum_{\beta^1 \le \beta} \binom{\beta}{\beta^1} C_0 (2A\boldsymbol{B}_0)^e / \boldsymbol{B}_1^e \\
&\quad \times (\beta - \beta^1 + e)! (\beta^{\beta-\beta^1})^{-1} (2\boldsymbol{N}_2)^{\beta-\beta^1} \boldsymbol{B}_0^{\beta-\beta^1} / \boldsymbol{B}_1^{\beta-\beta^1}.
\end{aligned}$$

Here we have abbreviated $N_1(\theta, x, \Xi(x); F)$, $\cdots$ as N_1, $\cdots$. If (2.16) is satisfied and

$$\begin{aligned}
C_{|\tilde{\beta}|+1,|\tilde{\alpha}|,|\tilde{\beta}'|,|\tilde{\gamma}|} &\ge C_{|\tilde{\beta}|,|\tilde{\alpha}|,|\tilde{\beta}'|+1,|\tilde{\gamma}|} \\
&\quad + 2^{n_2} \sqrt{n_3} C_0 \sum_{\tilde{\beta}^1 \le \tilde{\beta}} \binom{\tilde{\beta}}{\tilde{\beta}^1} (\tilde{\beta} - \tilde{\beta}^1 + \tilde{e})! C_{|\tilde{\beta}^1|,|\tilde{\alpha}|+1,|\tilde{\beta}'|,|\tilde{\gamma}|} \\
&\qquad \times (2\boldsymbol{B}_0)^{\tilde{\beta}-\tilde{\beta}^1+\tilde{e}}, \\
C_{|\tilde{\beta}|,|\tilde{\alpha}|,|\tilde{\beta}'|,|\tilde{\gamma}|} &\ge \sum_{\tilde{\beta}^1 < \tilde{\beta}} \binom{\tilde{\beta}}{\tilde{\beta}^1} (\tilde{\beta} - \tilde{\beta}^1)! C_{|\tilde{\beta}^1|,|\tilde{\alpha}|,|\tilde{\beta}'|,|\tilde{\gamma}|} (2\boldsymbol{B}_0)^{\tilde{\beta}-\tilde{\beta}^1},
\end{aligned}$$

then $I^{\beta,\tilde{\beta},e,\tilde{e}}_{\alpha,\beta',\tilde{\alpha},\tilde{\beta}',\tilde{\gamma}} \le 1$. This proves the lemma. □

Lemma 2.1. 7 *Assume that $\Lambda(x,\xi) \in C^\infty(\Omega)$ satisfies*

$$\begin{aligned}
\left| \Lambda^{(\alpha)}_{(\beta)}(x,\xi) \right| &\le \varepsilon(x,\xi) A_0(x,\xi)^{|\alpha|} B_0(x,\xi)^{|\beta|} \langle\xi\rangle^{1-|\alpha|} \quad \textit{if } |\alpha| + |\beta| = 1, \\
\left| \Lambda^{(\alpha)}_{(\beta)}(x,\xi) \right| &\le C_0(x,\xi) A_0(x,\xi)^{|\alpha|} B_0(x,\xi)^{|\beta|} |\alpha|! |\beta|! \langle\xi\rangle^{1-|\alpha|} \\
&\qquad \textit{if } |\alpha| + |\beta| \ge 1
\end{aligned}$$

in Ω, where Ω is an open subset of $\boldsymbol{R}^n \times \boldsymbol{R}^{n'}$, $\varepsilon(x,\xi)$, $C_0(x,\xi)$, $A_0(x,\xi)$ and $B_0(x,\xi)$ are function defined in Ω and $\Lambda^{(\alpha)}_{(\beta)}(x,\xi) = \partial_\xi^\alpha D_x^\beta \Lambda(x,\xi)$. Define

$$\omega^\alpha_\beta(\Lambda; x, \xi) := e^{-\Lambda(x,\xi)} (e^{\Lambda(x,\xi)})^{(\alpha)}_{(\beta)}.$$

Then

$$\left|\omega^{\alpha(\gamma)}_{\beta(\delta)}(\Lambda;x,\xi)\right| \le A_1(x,\xi)^{|\gamma|}A_2(x,\xi)^{|\alpha|}B_1(x,\xi)^{|\delta|}B_2(x,\xi)^{|\beta|}$$
$$\times(|\alpha|+|\beta|+|\gamma|+|\delta|)!\langle\xi\rangle^{-|\alpha|-|\gamma|}\sum_{k=0}^{|\alpha|+|\beta|}\rho(x,\xi)^k\langle\xi\rangle^k/k! \quad (2.17)$$

in Ω *if* $\rho(x,\xi)>0$, $X(x,\xi)\ge 2$ *and*

$$A_1(x,\xi)\ge X(x,\xi)A_0(x,\xi),\quad B_1(x,\xi)\ge X(x,\xi)B_0(x,\xi),$$
$$A_2(x,\xi)\ge 3d(x,\xi)A_0(x,\xi),\quad B_2(x,\xi)\ge 3d(x,\xi)B_0(x,\xi),$$
$$d(x,\xi)=\max\{X(x,\xi),\varepsilon(x,\xi)/\rho(x,\xi),16C_0(x,\xi)/(\rho(x,\xi)X(x,\xi))\}.$$

Similarly, we have

$$\left|\partial_\xi^\alpha e^{a\langle\xi\rangle}\right| \le (A(a,\rho)(1+\sqrt{2}))^{|\alpha|}|\alpha|!\langle\xi\rangle^{-|\alpha|}e^{a\langle\xi\rangle}\sum_{k=0}^{|\alpha|}(\rho\langle\xi\rangle)^k/k!$$

if $\rho>0$, *where* $a\in\boldsymbol{R}$ *and* $A(a,\rho)\ge\min\{2+|a|/\rho,\max\{2,2|a|/\rho+1\}\}$.

Proof One can prove (2.17) by induction on $|\alpha|+|\beta|$, noting that

$$\omega^{\alpha+e(\gamma)}_{\beta+e'(\delta)}(\Lambda;x,\xi)=\omega^{\alpha(\gamma+e)}_{\beta(\delta+e')}(\Lambda;x,\xi)+\left\{\Lambda^{(e)}_{(e')}(x,\xi)\omega^\alpha_\beta(\Lambda;x,\xi)\right\}^{(\gamma)}_{(\delta)}$$

if $|e|+|e'|=1$, and

$$\sum_{\substack{0\le j\le|\gamma|,0\le k\le|\delta|\\ j+k\ge 1}}(1+\delta_{|e|,1}j)(1+\delta_{|e'|,1}k)(A_0/A_1)^j(B_0/B_1)^k\le 16/X$$

if $X\ge 2$, $A_1\ge XA_0$ and $B_1\ge XB_0$. □

2.2 Symbol classes

We shall begin with the definition of symbol classes.

Definition 2.2.1 (i) Let $m,\delta\in\boldsymbol{R}$, $R\ge 1$, $A\ge 0$ and $B\ge 0$, and let $a(x,\xi)\in C^\infty(\boldsymbol{R}^n\times\boldsymbol{R}^{n'})$, where $n'\in\boldsymbol{N}$ and $\xi=(\xi_1,\cdots,\xi_{n'})\in\boldsymbol{R}^{n'}$. We say that $a(x,\xi)\in S^{m,\delta}(\boldsymbol{R}^n\times\boldsymbol{R}^{n'};R,A,B)$ if $a(x,\xi)$ satisfies

$$\left|a^{(\alpha+\tilde\alpha)}_{(\beta+\tilde\beta)}(x,\xi)\right|\le C_{|\tilde\alpha|+|\tilde\beta|}(A/R)^{|\alpha|}(B/R)^{|\beta|}\langle\xi\rangle^{m+|\beta|-|\tilde\alpha|}e^{\delta\langle\xi\rangle}$$

for $\langle\xi\rangle \ge R(|\alpha|+|\beta|)$, where $0^0 = 1$. We also write $S^{m,\delta}(R,A,B) \equiv S^{m,\delta}(\boldsymbol{R}^n \times \boldsymbol{R}^n; R,A,B)$, $S^m(R,A,B) \equiv S^{m,0}(R,A,B)$ and $S^{m,\delta}(R,A) \equiv S^{m,\delta}(R,A,A)$. We define

$$S^+(\boldsymbol{R}^n \times \boldsymbol{R}^{n'}; R,A,B) := \bigcap_{\delta>0} S^{m,\delta}(\boldsymbol{R}^n \times \boldsymbol{R}^{n'}; R,A,B). \tag{2.18}$$

Here the right-hand side of (2.18) does not depend on the choice of $m \in \boldsymbol{R}$. (ii) Let $m_j, \delta_j \in \boldsymbol{R}$ ($j = 1,2$), $R \ge 1$, $A_j \ge 0$ ($j = 1,2$) and $B \ge 0$, and let $a(\xi,y,\eta) \in C^\infty(\boldsymbol{R}^{n'} \times \boldsymbol{R}^n \times \boldsymbol{R}^{n''})$. We say that $a(\xi,y,\eta) \in S^{m_1,m_2,\delta_1,\delta_2}(\boldsymbol{R}^{n'} \times \boldsymbol{R}^n \times \boldsymbol{R}^{n''}; R, A_1, B, A_2)$ if $a(\xi,y,\eta)$ satisfies

$$\left|\partial_\xi^{\alpha+\tilde\alpha} D_y^{\beta^1+\beta^2+\tilde\beta} \partial_\eta^{\gamma+\tilde\gamma} a(\xi,y,\eta)\right| \le C_{|\tilde\alpha|+|\tilde\beta|+|\tilde\gamma|}(A_1/R)^{|\alpha|}(B/R)^{|\beta^1|+|\beta^2|}$$
$$\times (A_2/R)^{|\gamma|}\langle\xi\rangle^{m_1+|\beta^1|-|\tilde\alpha|}\langle\eta\rangle^{m_2+|\beta^2|-|\tilde\gamma|}\exp[\delta_1\langle\xi\rangle + \delta_2\langle\eta\rangle]$$

for $\langle\xi\rangle \ge R(|\alpha|+|\beta^1|)$ and $\langle\eta\rangle \ge R(|\gamma|+|\beta^2|)$. Similarly, we define

$$S^+(\boldsymbol{R}^{n'} \times \boldsymbol{R}^n \times \boldsymbol{R}^{n''}; R, A_1, B, A_2)$$
$$:= \bigcap_{\delta>0} S^{0,0,\delta,\delta}(\boldsymbol{R}^{n'} \times \boldsymbol{R}^n \times \boldsymbol{R}^{n''}; R, A_1, B, A_2).$$

We also write $S^{m_1,m_2,\delta_1,\delta_2}(\boldsymbol{R}^n \times \boldsymbol{R}^n \times \boldsymbol{R}^n; R, A_1, B, A_2) \equiv S^{m_1,m_2,\delta_1,\delta_2}(R, A_1, B, A_2)$, $S^{m_1,m_2}(R, A_1, B, A_2) \equiv S^{m_1,m_2,0,0}(R, A_1, B, A_2)$, $S^{m_1,m_2,\delta_1,\delta_2}(R, A) \equiv S^{m_1,m_2,\delta_1,\delta_2}(R,A,A,A)$ and so on. (iii) Let $m, \delta \in \boldsymbol{R}$, $R \ge 0$, $A > 0$ and $B > 0$, and let Γ be an open conic subset of $\boldsymbol{R}^n \times (\boldsymbol{R}^{n'} \setminus \{0\})$. For $a(x,\xi) \in C^\infty(\Gamma)$ we say that $a(x,\xi) \in PS^{m,\delta}(\Gamma; R, A, B)$ if $a(x,\xi)$ satisfies

$$\left|a_{(\beta)}^{(\alpha+\tilde\alpha)}(x,\xi)\right| \le C_{|\tilde\alpha|} A^{|\alpha|} B^{|\beta|} |\alpha|!|\beta|!\langle\xi\rangle^{m-|\alpha|-|\tilde\alpha|} e^{\delta\langle\xi\rangle}$$

for $(x,\xi) \in \Gamma$ with $|\xi| \ge C$ and $\langle\xi\rangle \ge R|\alpha|$, where $C > 0$. Symbols in $PS^{m,\delta}(\Gamma; R, A, B)$ are called pseudo-analytic symbols. We define

$$PS^+(\Gamma; R, A, B) := \bigcap_{\delta>0} PS^{m,\delta}(\Gamma; R, A, B).$$

Here the right-hand side does not depend on the choice of $m \in \boldsymbol{R}$. We also define

$$AS^{m,\delta}(\Gamma; A, B) := PS^{m,\delta}(\Gamma; 0, A, B),$$

which is a symbol class of analytic symbols. Similarly, we define $AS^+(\Gamma; A, B) := PS^+(\Gamma; 0, A, B)$ and use the same abbreviations.

Remark (i) Pseudo-analytic symbols were defined by Treves [Tr2]. (ii) It is easy to see that

$$PS^{m,\delta}(\boldsymbol{R}^n \times \boldsymbol{R}^{n'}; R, A, B) \subset S^{m,\delta}(\boldsymbol{R}^n \times \boldsymbol{R}^{n'}; R_1, A, 2B)$$

if $R_1 \geq \max\{R, 1\}$. Moreover, $a(y,\eta) \in S^{0,m,0,\delta}(\boldsymbol{R}^{n'} \times \boldsymbol{R}^n \times \boldsymbol{R}^{n'}; R_1, 0, 3B, A)$ if $a(x,\xi) \in PS^{m,\delta}(\boldsymbol{R}^n \times \boldsymbol{R}^{n'}; R, A, B)$ and $R_1 \geq \max\{R, 1\}$. (iii) The functions $g^R_{\Gamma_1,\Gamma_2}(\xi)$ ($R \geq 2$) in Lemma 2.1. 3 belong to $S^0(R, C_{\Gamma_1,\Gamma_2}, 0)$. (iv) We have

$$\begin{aligned} &S^{m_1,\delta_1}(R, A_1, B_1) \cdot S^{m_2,\delta_2}(R, A_2, B_2) \\ &\qquad \subset S^{m_1+m_2,\delta_1+\delta_2}(R, A_1 + A_2, B_1 + B_2), \\ &PS^{m_1,\delta_1}(\Gamma; R, A_1, B_1) \cdot PS^{m_2,\delta_2}(\Gamma; R, A_2, B_2) \\ &\qquad \subset PS^{m_1+m_2,\delta_1+\delta_2}(\Gamma; R, A, B) \end{aligned}$$

if $A \geq \max\{A_1, A_2\}$, $A > \min\{A_1, A_2\}$, $B \geq \max\{B_1, B_2\}$, and $B > \min\{B_1, B_2\}$, and so on.

Lemma 2.2. 2 *Let $R \geq 0$, and let Γ be an open conic subset of $\boldsymbol{R}^n \times (\boldsymbol{R}^{n'} \setminus \{0\})$. Assume that $a(x,\xi) \in C^\infty(\Gamma)$ satisfies*

$$\left|a^{(\alpha)}_{(\beta)}(x,\xi)\right| \leq CA^{|\alpha|+|\beta|}|\alpha|!|\beta|!\langle\xi\rangle^{-|\alpha|} \text{ if } (x,\xi) \in \Gamma \text{ and } \langle\xi\rangle \geq R(|\alpha| + |\beta|).$$

Then there are a sequence $\{j_k\}_{k\in N} \subset \boldsymbol{N}$ and $b(x,\xi) \in C^\infty(\Gamma)$ such that $j_k \uparrow \infty$ as $k \to \infty$ and for every compact subset K of Γ and every α and β $j_k^{|\alpha|} a^{(\alpha)}_{(\beta)}(x, j_k\xi)$ converges to $b^{(\alpha)}_{(\beta)}(x,\xi)$ uniformly in K as $k \to \infty$. Moreover, we have

$$\left|b^{(\alpha)}_{(\beta)}(x,\xi)\right| \leq CA^{|\alpha|+|\beta|}|\alpha|!|\beta|!|\xi|^{-|\alpha|} \quad \text{if } (x,\xi) \in \Gamma.$$

In particular, $b(x,\xi)$ is real analytic in Γ.

Remark We want to use "cut-off" technique in studying partial differential operators in the space of hyperfunctions and to apply the same argument as in the framework of distributions. However, the above lemma shows that any symbols for "cutting-off" are not pseudo-analytic ones. So we must treat the symbol classes $S^{m,\delta}(R, A, B)$.

Proof Put $a_j(x,\xi) = a(x, j\xi)$ ($j \in \boldsymbol{N}$). Then we have $a^{(\alpha)}_{j(\beta)}(x,\xi) = j^{|\alpha|}a^{(\alpha)}_{(\beta)}(x, j\xi)$ and

$$\left|a^{(\alpha)}_{j(\beta)}(x,\xi)\right| \leq CA^{|\alpha|+|\beta|}|\alpha|!|\beta|!|\xi|^{-|\alpha|} \quad \text{for } (x,\xi) \in \Gamma \text{ and } j \in \boldsymbol{N}.$$

Therefore, the Ascoli-Arzelà theorem and a standard argument prove the lemma. □

Proposition 2.2. 3 *Let Γ and $\Gamma_j \equiv X_j \times \gamma_j$ ($1 \le j \le N$) be open conic subsets of $\boldsymbol{R}^n \times (\boldsymbol{R}^{n''} \setminus \{0\})$ such that $\Gamma \cap \{(x,\eta) \in \boldsymbol{R}^n \times \boldsymbol{R}^{n''};\ |\eta| = 1\}$ is bounded and $\Gamma \Subset \cup_{j=1}^N \Gamma_j$. Here $\Gamma^1 \Subset \Gamma^2$ for conic subsets Γ^1 and Γ^2 of $\boldsymbol{R}^n \times (\boldsymbol{R}^{n''} \setminus \{0\})$ means that $\overline{\Gamma^1} \setminus \{0\} \subset \mathrm{int}(\Gamma^2)$ and $\Gamma^1 \cap \{|\eta| = 1\}$ is bounded. Then there are symbols $\Phi_j^R(\xi, y, \eta) \in S^{0,0}(\boldsymbol{R}^{n'} \times \boldsymbol{R}^n \times \boldsymbol{R}^{n''}; R, NC_*, C(\Gamma, \{\Gamma_j\}), C(\Gamma, \{\Gamma_j\}))$ ($1 \le j \le N$, $R \ge 4$) such that $0 \le \Phi_j^R(\xi, y, \eta) \le 1$, $\mathrm{supp}\, \Phi_j^R \subset \boldsymbol{R}^{n'} \times \Gamma_j$ and*

$$\sum_{j=1}^{N} \Phi_j^R(\xi, y, \eta) = 1 \quad \text{for } (\xi, y, \eta) \in \boldsymbol{R}^{n'} \times \Gamma \text{ with } \langle \eta \rangle \ge R,$$

where C_* and $C(\Gamma, \{\Gamma_j\})$ do not depend on R and C_* depends only on n'.

Remark The proposition is still valid if $n' = 0$ and $S^{0,0}(\cdots)$ is replaced with $S^0(\boldsymbol{R}^n \times \boldsymbol{R}^{n''}; R, C(\Gamma, \{\Gamma_j\}))$.

Proof Choose open conic subsets $\Gamma_j^0 \equiv X_j^0 \times \gamma_j^0$ of $\boldsymbol{R}^n \times (\boldsymbol{R}^{n''} \setminus \{0\})$ ($1 \le j \le N$) so that $\Gamma_j^0 \Subset \Gamma_j$ ($1 \le j \le N$) and $\Gamma \Subset \cup_{j=1}^N \Gamma_j^0$. By Lemma 2.1. 3 there are $g_j^R(\eta) \in C^\infty(\boldsymbol{R}^{n''})$ ($1 \le j \le N$, $R \ge 2$) such that $0 \le g_j^R(\eta) \le 1$, $\mathrm{supp}\, g_j^R \subset \gamma_j$, $g_j^R(\eta) = 1$ for $\eta \in \gamma_j^0$ with $\langle \eta \rangle \ge 2R$ and

$$\left|\partial_\eta^{\alpha+\tilde{\alpha}} g_j^R(\eta)\right| \le C_{j,|\tilde{\alpha}|}(C_j/R)^{|\alpha|}\langle \eta \rangle^{-|\tilde{\alpha}|} \quad \text{if } \langle \eta \rangle \ge R|\alpha|$$

($1 \le j \le N$), where the $C_{j,|\tilde{\alpha}|}$ and C_j depend on γ_j and γ_j^0 and do not depend on R. It follows from Lemma 2.1. 2 that there are $\{\chi_{j,k,\ell}(x)\}_{k,\ell \in Z_+} \subset C_0^\infty(X_j)$ ($1 \le j \le N$) such that $0 \le \chi_{j,k,\ell}(x) \le 1$, $\chi_{j,k,\ell}(x) = 1$ for $x \in X_j^0$ and

$$\begin{aligned}\left|D_x^{\beta^1+\beta^2+\tilde{\beta}} \chi_{j,k,\ell}(x)\right| &\le C'_{j,|\tilde{\beta}|}(4^k C_j')^{|\beta^1|}(4^\ell C_j')^{|\beta^2|} \\ &\quad \text{if } |\beta^1| \le 2 \cdot 4^k \text{ and } |\beta^2| \le 2 \cdot 4^\ell,\end{aligned} \tag{2.19}$$

where the $C'_{j,|\tilde{\beta}|}$ and the C_j' do not depend on k and ℓ. Let $\{h_\ell^R(t)\}_{\ell \in N}$ be the family of functions defined by (2.3), and put $h_0^R(t) = \phi_1(R^{-1}t - 1)$, where $\phi_1(t)$ is a function as in the proof of Lemma 2.1. 3 . Applying the same argument as for (2.8), we can show that

$$\left|\partial_\xi^{\alpha+\tilde{\alpha}} h_\ell^R(\langle \xi \rangle)\right| \le C_{|\tilde{\alpha}|}(C_*'/R)^{|\alpha|}\langle \xi \rangle^{-|\tilde{\alpha}|} \quad \text{if } |\alpha| \le \langle \xi \rangle/(2R), \tag{2.20}$$

where the $C_{|\tilde{\alpha}|}$ and C_*' do not depend on ℓ and R. Now, put

$$\tilde{\Phi}_j^R(\xi, y, \eta) = g_j^R(\eta) \sum_{k=0}^{\infty} \sum_{\ell=1}^{\infty} \chi_{j,k,\ell}(y) h_k^R(\langle \xi \rangle) h_\ell^R(\langle \eta \rangle) \quad (1 \le j \le N).$$

Then, by (2.4), (2.5), (2.19) and (2.20) we have $0 \le \widetilde{\Phi}_j^R(\xi, y, \eta) \le 1$, $\operatorname{supp} \widetilde{\Phi}_j^R \subset \boldsymbol{R}^{n'} \times \Gamma_j$, $\widetilde{\Phi}_j^R(\xi, y, \eta) = 1$ for $(y, \eta) \in \Gamma_j^0$ with $\langle \eta \rangle \ge 2R$ and $\widetilde{\Phi}_j^R(\xi, y, \eta) \in S^{0,0}(\boldsymbol{R}^{n'} \times \boldsymbol{R}^n \times \boldsymbol{R}^{n''}; 2R, 2C'_*, 8C'_j, 2C'_* + 2C_j)$. If we put

$$\begin{aligned} \Phi_1^R(\xi, y, \eta) &= \widetilde{\Phi}_1^{R/2}(\xi, y, \eta), \\ \Phi_{j+1}^R(\xi, y, \eta) &= \widetilde{\Phi}_{j+1}^{R/2}(\xi, y, \eta) \prod_{k=1}^{j} \left(1 - \widetilde{\Phi}_k^{R/2}(\xi, y, \eta)\right) \quad (1 \le j \le N-1), \end{aligned}$$

then $\{\Phi_j^R(\xi, y, \eta)\}_{1 \le j \le N}$ has the properties of proposition, modifying C_* if necessary. □

Let $\{\phi_j(t)\}$ be as in the proof of Lemma 2.1. 3 , and put, for $R > 0$ and $\xi \in \boldsymbol{R}^n$, $\phi_0^R(\xi) = 1$ and

$$\phi_j^R(\xi) := 1 - \phi_{2j}((Rj)^{-1}\langle \xi \rangle - 2) \quad (j \ge 1).$$

Then we have $\phi_j^R \in C^\infty(\boldsymbol{R}^n)$, $0 \le \phi_j^R(\xi) \le 1$ and

$$\begin{aligned} &\phi_j^R(\xi) = \begin{cases} 0 & \text{if } \langle \xi \rangle \le 2Rj, \\ 1 & \text{if } \langle \xi \rangle \ge 3Rj, \end{cases} \\ &\left| \partial_\xi^{\alpha+\beta} \phi_j^R(\xi) \right| \le \widehat{C}_{|\beta|} (\widehat{C}/R)^{|\alpha|} \langle \xi \rangle^{-|\beta|} \quad \text{if } |\alpha| \le 2j, \end{aligned}$$

where the $\widehat{C}_{|\beta|}$ and $\widehat{C}$ do not depend on j and R. In fact, by induction on $|\alpha| + |\beta|$ we can prove that

$$\left| \partial_\xi^{\alpha+\beta} \left(\partial_t^{k+\ell} \phi_{2j} \right) ((Rj)^{-1} \langle \xi \rangle - 2) \right| \le C_{|\beta|,\ell} (\widehat{C}/R)^{|\alpha|} (C_*/R)^k \langle \xi \rangle^{-|\beta|+k}$$

if $j \ge 1$ and $|\alpha| + k \le 2j$, where the $C_{|\beta|,\ell}$ and C_* do not depend on j and R. We also put, for $j \in \boldsymbol{N}$ and $R > 0$,

$$\psi_j^R(\xi) := \phi_{j-1}^R(\xi) - \phi_j^R(\xi).$$

Then we have $\psi_j^R \in C_0^\infty(\boldsymbol{R}^n)$, $0 \le \psi_j^R(\xi) \le 1$ and

$$\begin{aligned} &\operatorname{supp} \psi_j^R \subset \{\xi \in \boldsymbol{R}^n;\ 2R(j-1) \le \langle \xi \rangle \le 3Rj\}, \\ &\operatorname{supp} \psi_j^R \cap \operatorname{supp} \psi_k^R = \emptyset \quad \text{if } k > 3j/2 + 1, \\ &\left| \partial_\xi^{\alpha+\beta} \psi_j^R(\xi) \right| \le 2\widehat{C}_{|\beta|} (\widehat{C}/R)^{|\alpha|} \langle \xi \rangle^{-|\beta|} \quad \text{if } |\alpha| \le j. \qquad (2.21) \end{aligned}$$

We shall often use $\{\phi_j^R\}$ and $\{\psi_j^R\}$. The following lemma was given in [Tr2].

Lemma 2.2. 4 *Let $m, \delta \in \boldsymbol{R}$, $R \geq 1$, $A \geq 0$ and $B \geq 0$.* (i) *Assume that $a_j(x,\xi) \in C^\infty(\boldsymbol{R}^n \times \boldsymbol{R}^{n'})$ ($j \in \boldsymbol{Z}_+$) satisfy*

$$\left|a_{j(\beta+\tilde{\beta})}^{(\alpha+\tilde{\alpha})}(x,\xi)\right| \leq C_{|\tilde{\alpha}|+|\tilde{\beta}|}(C_0/R)^j (A/R)^{|\alpha|}(B/R)^{|\beta|}\langle\xi\rangle^{m+|\beta|-|\tilde{\alpha}|}e^{\delta\langle\xi\rangle}$$

if $\langle\xi\rangle \geq R(|\alpha|+|\beta|+j)$, where $C_0 \geq 0$. Put

$$a(x,\xi) = \sum_{j=0}^{\infty} \phi_j^R(\xi) a_j(x,\xi), \tag{2.22}$$

where the $\phi_j^R(\xi)$ are the above symbols with $\boldsymbol{R}^n$ replaced by $\boldsymbol{R}^{n'}$. Then $a(x,\xi)$ belongs to $S^{m,\delta}(2R, 2A+2\widehat{C}, 2B)$ if $C_0 < R$. (ii) *Let Γ be an open conic subset of $\boldsymbol{R}^n \times (\boldsymbol{R}^{n'} \setminus \{0\})$, and assume that $a_j(x,\xi) \in C^\infty(\Gamma)$ ($j \in \boldsymbol{Z}_+$) satisfy*

$$\left|a_{j(\beta)}^{(\alpha+\tilde{\alpha})}(x,\xi)\right| \leq C_{|\tilde{\alpha}|} C_0^j A^{|\alpha|} B^{|\beta|} j!|\alpha|!|\beta|!\langle\xi\rangle^{m-|\alpha|-|\tilde{\alpha}|-j}e^{\delta\langle\xi\rangle}$$

for $(x,\xi) \in \Gamma$ with $|\xi| \geq 1$ and $\langle\xi\rangle \geq R(|\alpha|+j)$, where $C_0 \geq 0$. Then $a(x,\xi)$, which is defined by (2.22), belongs to $PS^{m,\delta}(\Gamma; 2R, A', B)$ if $C_0 < 2R$, $A' \geq \max\{A, 3\widehat{C}\}$ and $A' > \min\{A, 3\widehat{C}\}$.

Proof The assertion (i) easily follows from the fact that $|\alpha| \leq 2j$ if $\langle\xi\rangle \leq 3Rj$ and $2R(|\alpha|+|\beta|) \leq \langle\xi\rangle$, and $R(|\alpha|+|\beta|+j) \leq \langle\xi\rangle$ if $2Rj \leq \langle\xi\rangle$ and $2R(|\alpha|+|\beta|) \leq \langle\xi\rangle$. A simple calculation yields

$$\left|a_{(\beta)}^{(\alpha+\tilde{\alpha})}(x,\xi)\right| \leq B^{|\beta|}|\alpha|!|\beta|!\langle\xi\rangle^{m-|\alpha|-|\tilde{\alpha}|}e^{\delta\langle\xi\rangle}$$
$$\times \sum_{j=0}^{\infty}(C_0/(2R))^j \Big\{\sum_{\tilde{\alpha}' \leq \tilde{\alpha}} \binom{\tilde{\alpha}}{\tilde{\alpha}'} \widehat{C}_{|\tilde{\alpha}|-|\tilde{\alpha}'|} C_{|\tilde{\alpha}'|}\Big\} \sum_{\mu=0}^{|\alpha|} A^\mu (3\widehat{C})^{|\alpha|-\mu}$$

if $(x,\xi) \in \Gamma$, $|\xi| \geq 1$ and $2R|\alpha| \leq \langle\xi\rangle$. Here we have also used the fact that $j!j^k/(j^j k!) \leq 1$ for $j \in \boldsymbol{N}$ and $k \in \boldsymbol{Z}_+$, $\langle\xi\rangle \geq 2Rj$ if $\phi_j^R(\xi) \neq 0$, and $\langle\xi\rangle \leq 3Rj$ if $d\phi_j^R(\xi) \neq 0$. This proves the assertion (ii). □

2.3 Definition of Fourier integral operators

First we shall define several classes of phase functions.

Definition 2.3. 1 Let $n, n' \in \boldsymbol{N}$, and let Ω be an open conic subset of $\boldsymbol{R}^n \times (\boldsymbol{R}^{n'} \setminus \{0\})$. We say that $S(y,\xi) \in C^\infty(\Omega)$ belongs to $\mathcal{P}(\Omega; A, B, c_0, c_1, c_2, c_3)$, where $A, B \geq 0$, $c_0, c_2, c_3 \geq 0$ and $c_1 \in \boldsymbol{R}$, if $S(y,\xi)$ satisfies the following:

($\mathcal{P}$–0) There is $C \equiv C(S) > 0$ such that

$$\left|S^{(\alpha)}_{(\beta)}(y,\xi)\right| \le CA^{|\alpha|}B^{|\beta|}|\alpha|!|\beta|!\langle\xi\rangle^{1-|\alpha|}(1+\langle y\rangle^{1-|\beta|}) \quad \text{in } \Omega_1, \qquad (2.23)$$

where $\Omega_1 = \{(y,\xi) \in \Omega;\ |\xi| \ge 1\}$.

($\mathcal{P}$–1) $1 + |\nabla_\xi S(y,\xi)|^2 \ge c_0\langle y\rangle^2$ in Ω_1, where $\nabla_\xi S = (\partial_{\xi_1} S, \cdots, \partial_{\xi_n} S)$.

($\mathcal{P}$–2) There is $C_1 \equiv C_1(S) \in \boldsymbol{R}$ such that

$$\operatorname{Im} S(y,\xi) \ge -c_1|\xi| - C_1 \quad \text{in } \Omega_1.$$

($\mathcal{P}$–3) $|\nabla_y S(y,\xi)| \le c_2\langle\xi\rangle$ in Ω_1.

($\mathcal{P}$–4) $|\nabla_y S(y,\xi)| \ge c_3|\xi|$ in Ω_1.

We define $\mathcal{P}(\Omega) := \bigcup_{A,B,c_0,c_1,c_2,c_3} \mathcal{P}(\Omega; A, B, c_0, c_1, c_2, c_3)$.

Lemma 2.3. 2 *Let Ω be an open conic subset of $\boldsymbol{R}^n \times (\boldsymbol{R}^{n'} \setminus \{0\})$, and let $S(y,\xi) \in \mathcal{P}(\Omega; A, B, c_0, c_1, c_2, c_3)$ be positively homogeneous of degree 1 (in ξ). Let $(y^0,\xi^0) \in \Omega$, and put*

$$\widetilde{S}(y,\xi) = S(y,\xi) - \xi \cdot \nabla_\xi S(y^0,\xi^0) - S_0(y-y^0)\cdot\xi,$$

where $S_0 = (\partial^2 S/\partial\xi\partial y)(y^0,\xi^0)(\equiv ((\partial^2 S/\partial\xi_j\partial y_k)(y^0,\xi^0))_{j\downarrow 1,\cdots,n',k\rightarrow 1,\cdots,n})$. Here we regard y and ξ as column vectors. Then, for any $\varepsilon > 0$ there is a conic neighborhood $\widetilde{\Omega}$ of (y^0,ξ^0) in Ω such that

$$\langle\xi\rangle^{-1}|\nabla_y\widetilde{S}(y,\xi)| + |\nabla_\xi\widetilde{S}(y,\xi)| + |(\partial^2\widetilde{S}/\partial\xi\partial y)(y,\xi)| < \varepsilon$$

for $(y,\xi) \in \widetilde{\Omega}$ with $|\xi| \ge 1$, where $|T|$ denotes the matrix norm of a matrix T.

Proof We may assume that $|\xi^0| = 1$. Euler's identity gives

$$\begin{aligned} S(y,\xi) - \widetilde{S}(y,\xi) = |\xi|\Big(&S(y^0,\xi^0) + (y-y^0)\cdot\nabla_y S(y^0,\xi^0) \\ &+(\xi/|\xi| - \xi^0)\cdot\nabla_\xi S(y^0,\xi^0) + S_0(y-y_0)\cdot(\xi/|\xi|-\xi^0)\Big). \end{aligned}$$

Therefore, Taylor's formula gives

$$\begin{aligned} \widetilde{S}(y,\xi) = \frac{|\xi|}{2}\Big\{&(y-y^0)\cdot\frac{\partial^2 S}{\partial y\partial y}(y^0,\xi^0)(y-y^0) \\ &+(\xi/|\xi|-\xi^0)\frac{\partial^2 S}{\partial\xi\partial\xi}(y^0,\xi^0)(\xi/|\xi|-\xi^0) + O(|y-y^0|^3 + |\xi/|\xi|-\xi^0|^3)\Big\} \end{aligned}$$

as $|y-y^0|+|\xi/|\xi|-\xi^0|\to 0$, and proves the lemma. □

Let $n, n', n'' \in \boldsymbol{Z}_+$, and let Ω and Ω' be open conic subsets of $\boldsymbol{R}^n \times (\boldsymbol{R}^{n'}\setminus\{0\})$ and $\boldsymbol{R}^n \times (\boldsymbol{R}^{n''}\setminus\{0\})$, respectively. Let $S(y,\xi) \in \mathcal{P}(\Omega; A_0, B_0, c_0(S), c_1(S), c_2(S), c_3(S))$ and $T(y,\eta) \in \mathcal{P}(\Omega'; A_0, B_0, c_0(T), c_1(T), c_2(T), c_3(T))$, where $A_0, B_0 \ge 0$, $c_0(S), c_2(S), c_0(T), c_3(T) \ge 0$, $c_3(S), c_2(T) > 0$ and $c_1(S), c_1(T) \in \boldsymbol{R}$. We assume that a symbol $p(\xi, y, \eta) \in C^\infty(\boldsymbol{R}^{n'} \times \boldsymbol{R}^n \times \boldsymbol{R}^{n''})$ satisfies

$$\begin{aligned}\left|\partial_\xi^\alpha D_y^\beta \partial_\eta^\gamma p(\xi,y,\eta)\right| \;&\le\; C_{|\alpha|+|\gamma|}(p)(B/R)^{|\beta|}\langle\xi\rangle^{m_1+|\beta|}\langle\eta\rangle^{m_2} \\ &\quad \times \exp[\delta_1\langle\xi\rangle+\delta_2\langle\eta\rangle] \quad \text{if } \langle\xi\rangle \ge R|\beta|, \end{aligned} \tag{2.24}$$

where $R \ge 1$, $B \ge 0$, $m_1, m_2, \delta_1, \delta_2 \in \boldsymbol{R}$ and $C_k(p) \uparrow$ as $k \uparrow$. Moreover, we assume that

$$\begin{aligned}\operatorname{supp} p \;\subset\; &\{(\xi,y,\eta) \in \boldsymbol{R}^{n'} \times \boldsymbol{R}^n \times \boldsymbol{R}^{n''}; \\ &\quad (y,\xi)\in\Omega,\ (y,\eta)\in\Omega',\ |\xi|\ge 1 \text{ and } |\eta|\ge 1\}. \end{aligned}\tag{2.25}$$

Then we can define a Fourier integral operator $p_{S,T}(D_x, y, D_z)$ by

$$\begin{aligned} &p_{S,T}(D_x,y,D_z)u \\ &:= (2\pi)^{-n''}\mathcal{F}_\xi^{-1}\Big[\int\Big(\int e^{iS(y,\xi)+iT(y,\eta)}p(\xi,y,\eta)\hat{u}(\eta)\,d\eta\Big)\,dy\Big](x)\end{aligned}$$

for $u \in \mathcal{S}_\infty(\boldsymbol{R}^{n''})$. Put $q(\eta,y,\xi)=p(\xi,y,\eta)$ and define

$$\begin{aligned} &{}^r p_{S,T}(D_x,y,D_z)v := q_{T,S}(D_z,y,D_x)v \\ &= (2\pi)^{-n'}\mathcal{F}_\eta^{-1}\Big[\int\Big(\int e^{iT(y,\eta)+iS(y,\xi)}q(\eta,y,\xi)\hat{v}(\xi)\,d\xi\Big)\,dy\Big](z)\end{aligned}$$

for $v \in \mathcal{S}_\infty(\boldsymbol{R}^{n'})$. The above definitions are significant and we have the following

Theorem 2.3. 3 (i) *Assume that* $c_0(T) > 0$. *Then* $p_{S,T}(D_x, y, D_z)$ *and* ${}^r p_{S,T}(D_x, y, D_z)$ *can be extended to continuous linear operators from* $\mathcal{S}_{\varepsilon_2}(\boldsymbol{R}^{n''})$ *to* $\mathcal{S}_{\varepsilon_1}(\boldsymbol{R}^{n'})$ *and from* $\mathcal{S}_{\varepsilon_1}'(\boldsymbol{R}^{n'})$ *to* $\mathcal{S}_{\varepsilon_2}'(\boldsymbol{R}^{n''})$, *respectively, if* $\nu > 1$ *and*

$$\begin{cases} \varepsilon_2 = \delta_2 + c_1(T) + \nu c_2(T)(\delta_1 + c_1(S) + \varepsilon_1)_+/c_3(S), \\ \delta_1 + c_1(S) + \varepsilon_1 \le 1/R, \\ R \ge R_1(S,T,\nu)B,\ R \ge R_2(S,T,\nu), \end{cases} \tag{2.26}$$

where $R_j(S,T,\nu)$ $(j=1,2)$ *are constants determined by* S, T *and* ν, *and* $a_\pm = \max\{\pm a, 0\}$ *for* $a \in \boldsymbol{R}$. (ii) *Assume that* $c_0(S) > 0$.

Then $p_{S,T}(D_x, y, D_z)$ and ${}^r p_{S,T}(D_x, y, D_z)$ can be extended to continuous linear operators from $\mathcal{S}_{-\varepsilon_2}{}'(\boldsymbol{R}^{n''})$ to $\mathcal{S}_{-\varepsilon_1}{}'(\boldsymbol{R}^{n'})$ and from $\mathcal{S}_{-\varepsilon_1}(\boldsymbol{R}^{n'})$ to $\mathcal{S}_{-\varepsilon_2}(\boldsymbol{R}^{n''})$, respectively, if $\nu > 1$ and (2.26) is valid with $R_j(S,T,\nu)$ replaced by $R'_j(S,T,\nu)$ ($j = 1, 2$).

Remark (i) If $p_j(\xi, y, \eta) \in C^\infty(\boldsymbol{R}^{n'} \times \boldsymbol{R}^n \times \boldsymbol{R}^{n''})$ ($j \in \boldsymbol{N}$) satisfy the same conditions with $C_k(p)$ replaced by $C_k(p_j)$ as for $p(\xi, y, \eta)$, and $C_k(p_j) \to 0$ as $j \to \infty$ for each $k \in \boldsymbol{Z}_+$, then we have, for example,

$$p_{j\,S,T}(D_x, y, D_z)u \longrightarrow 0 \quad \text{in } \mathcal{S}_{\varepsilon_1}(\boldsymbol{R}^{n'}) \text{ as } j \to \infty$$

for $u \in \mathcal{S}_{\varepsilon_2}(\boldsymbol{R}^{n''})$ under the condition (2.26). (ii) Note that $p(\xi, y, \eta)$ also satisfies the estimates (2.24) with B and R replaced by BR'/R and R' ($\geq R$), respectively. So the condition (2.26) can be replaced by the condition

$$\begin{cases} \varepsilon_2 = \delta_2 + c_1(T) + \nu c_2(T)(\delta_1 + c_1(S) + \varepsilon_1)_+/c_3(S), \\ \delta_1 + c_1(S) + \varepsilon_1 \leq \min\{1/R, 1/R_2(S,T,\nu)\}, \ R \geq R_1(S,T,\nu)B. \end{cases}$$

(iii) Assume that $n = n' = n''$, $S(y, \xi) = -y \cdot \xi$ and $T(y, \eta) = y \cdot \eta$. Then the theorem is still valid without assuming (2.25). Moreover, we can take $R_1(S,T,\nu) = en\nu/(\nu - 1)$ and $R_2(S,T,\nu) = 0$.

Proof First assume that $c_0(T) > 0$. Define a differential operator L by

$${}^t L = (1 + |\nabla_\eta T(y,\eta)|^2)^{-1} \Big(\sum_{j=1}^{n''} (\overline{\partial_{\eta_j} T(y,\eta)}) D_{\eta_j} + 1 \Big), \tag{2.27}$$

where ${}^t L$ denotes the transposed operator of L. Applying the same argument as in the proof of Lemma 2.1. 5 , we have

$$\begin{aligned} \Big| \partial_\xi^\alpha D_y^\beta \partial_\eta^\gamma L^M (p(\xi, y, \eta)\hat{u}(\eta)) \Big| &\leq C_{|\alpha|+|\gamma|, M}(T) C_{|\alpha|+|\gamma|+M}(p) \\ &\quad \times (B_1/R)^{|\beta|} \langle \xi \rangle^{m_1 + |\beta|} \langle \eta \rangle^{m_2} \exp[\delta_1 \langle \xi \rangle + \delta_2 \langle \eta \rangle] \langle y \rangle^{-M} \\ &\quad \times \sup_{|\delta| \leq M + |\gamma|} |\partial_\eta^\delta \hat{u}(\eta)| \end{aligned} \tag{2.28}$$

if $R|\beta| \leq \langle \xi \rangle$, where $B_1 = \max\{B, 80(1 + 3n'' C(T)^2 A_0^2) B_0/c_0(T)\}$, $C(T)$ denotes the constant C in (2.23) for T and the $C_j(p)$ denote the constants in (2.24). Let $u \in \mathcal{S}_\infty(\boldsymbol{R}^{n''})$. Then $p_{S,T}(D_x, y, D_z)u$ is well-defined and

$$\begin{aligned} & p_{S,T}(D_x, y, D_z)u \\ &= (2\pi)^{-n''} \mathcal{F}_\xi^{-1} \Big[\int \Big(\int e^{iS(y,\xi) + iT(y,\eta)} L^M (p(\xi, y, \eta)\hat{u}(\eta))\, d\eta \Big) dy \Big](x), \end{aligned}$$

where $M \geq n+1$. Moreover, for $k \in \boldsymbol{Z}_+$, $\alpha \in \boldsymbol{Z}_+^{n'}$ and $\varepsilon_1 \in \boldsymbol{R}$ we have

$$\begin{aligned} I_{\varepsilon_1,k,\alpha} &\equiv \langle\xi\rangle^k D_\xi^\alpha \Big\{ e^{\varepsilon_1\langle\xi\rangle} \mathcal{F}[p_{S,T}(D_x,y,D_z)u](\xi) \Big\} \\ &= (2\pi)^{-n''} \sum_{\alpha' \leq \alpha} \binom{\alpha}{\alpha'} \int e^{iS(y,\xi)+iT(y,\eta)} \langle\xi\rangle^k e^{\varepsilon_1\langle\xi\rangle} t_{\varepsilon_1,\alpha'}(y,\xi) \\ &\qquad \times D_\xi^{\alpha-\alpha'} L^M(p(\xi,y,\eta)\hat{u}(\eta))\, d\eta dy, \end{aligned}$$

taking $M = n + |\alpha| + 1$, where

$$t_{\varepsilon,\alpha}(y,\xi) = e^{-iS(y,\xi)-\varepsilon\langle\xi\rangle} D_\xi^\alpha e^{iS(y,\xi)+\varepsilon\langle\xi\rangle}.$$

It follows from Lemma 2.1.7 (or its proof) that

$$\Big| t_{\varepsilon_1,\alpha}{}_{(\beta)}^{(\gamma)}(y,\xi) \Big| \leq C_{\varepsilon_1,|\alpha|,|\gamma|}(S)(2B_0)^{|\beta|}|\beta|!\langle\xi\rangle^{-|\gamma|}\langle y\rangle^{|\alpha|} \qquad \text{in } \Omega \cap \{|\xi| \geq 1\}.$$

Thus, by (2.28) we have

$$\begin{aligned} &\Big| D_y^\beta \Big\{ t_{\varepsilon_1,\alpha'}(y,\xi) D_\xi^{\alpha-\alpha'} L^M(p(\xi,y,\eta)\hat{u}(\eta)) \Big\} \Big| \\ &\leq C_{\varepsilon_1,|\alpha|,M}(S,T) C_{|\alpha|+M}(p)((2B_0+B_1)/R)^{|\beta|}\langle\xi\rangle^{m_1+|\beta|} \\ &\quad \times \langle\eta\rangle^{m_2} \exp[\delta_1\langle\xi\rangle + \delta_2\langle\eta\rangle]\langle y\rangle^{|\alpha|-M} \sup_{|\delta|\leq M} |\partial_\eta^\delta \hat{u}(\eta)| \end{aligned} \tag{2.29}$$

if $R|\beta| \leq \langle\xi\rangle$ and $\alpha' \leq \alpha$, where B_1 is the constant in (2.28). Let K be a differential operator defined by

$${}^tK = |\nabla_y S(y,\xi) + \nabla_y T(y,\eta)|^{-2} \sum_{j=1}^n \Big(\overline{\partial_{y_j} S(y,\xi)} + \overline{\partial_{y_j} T(y,\eta)} \Big) D_{y_j}$$

for $(\xi,y,\eta) \in \boldsymbol{R}^{n'} \times \boldsymbol{R}^n \times \boldsymbol{R}^{n''}$ satisfying $(y,\xi) \in \Omega$, $(y,\eta) \in \Omega'$ and $|\nabla_y S(y,\xi) + \nabla_y T(y,\eta)| \neq 0$. Then it follows from Lemma 2.1.5 and (2.29) that

$$\begin{aligned} &\Big| K^j \Big\{ t_{\varepsilon_1,\alpha'}(y,\xi) D_\xi^{\alpha-\alpha'} L^M(p(\xi,y,\eta)\hat{u}(\eta)) \Big\} \Big| \\ &\leq C_{\varepsilon_1,|\alpha|,M}(S,T) C_{|\alpha|+M}(p)\langle\xi\rangle^{m_1}\langle\eta\rangle^{m_2} \exp[\delta_1\langle\xi\rangle + \delta_2\langle\eta\rangle] \\ &\quad \times \langle y\rangle^{|\alpha|-M}(\Gamma(|\xi|,S,T,B,\varepsilon)/R)^j \sup_{|\delta|\leq M} |\partial_\eta^\delta \hat{u}(\eta)| \end{aligned} \tag{2.30}$$

if $\alpha' \leq \alpha$, $0 < \varepsilon < 1$, $(1-\varepsilon)c_3(S)|\xi| \geq c_2(T)\langle\eta\rangle$ and $j \leq \langle\xi\rangle/R$, where

$$\begin{aligned} \Gamma(|\xi|,S,T,B,\varepsilon) &= 2^6 n c_3(S)^{-2}\varepsilon^{-2}B_0(1+1/|\xi|)^4 C(S,T,\varepsilon) \\ &\quad \times \max\{2B_0+B_1, 2^5 \cdot 15 n c_3(S)^{-2}\varepsilon^{-2}B_0^3 C(S,T,\varepsilon)^2\}, \quad (2.31) \\ C(S,T,\varepsilon) &= C(S) + (1-\varepsilon)C(T)c_3(S)/c_2(T) \end{aligned}$$

and $C(S)$ denotes the constant C in (2.23). Fix $\varepsilon > 0$ so that $\varepsilon < 1$, and represent

$$\begin{aligned} I_{\varepsilon_1,k,\alpha} = (2\pi)^{-n''} \sum_{\alpha' \le \alpha} \binom{\alpha}{\alpha'} \Big[\Big\{ \int_{c_2(T)\langle\eta\rangle \ge (1-\varepsilon)c_3(S)|\xi|} & e^{iS(y,\xi)+iT(y,\eta)} \\ \times \langle\xi\rangle^k e^{\varepsilon_1\langle\xi\rangle} t_{\varepsilon_1,\alpha'}(y,\xi) D_\xi^{\alpha-\alpha'} L^M(p(\xi,y,\eta)\hat{u}(\eta))\, d\eta dy \\ + \int_{c_2(T)\langle\eta\rangle \le (1-\varepsilon)c_3(S)|\xi|} & e^{iS(y,\xi)+iT(y,\eta)} \langle\xi\rangle^k e^{\varepsilon_1\langle\xi\rangle} \\ \times K^j \Big\{ t_{\varepsilon_1,\alpha'}(y,\xi) D_\xi^{\alpha-\alpha'} L^M(p(\xi,y,\eta)\hat{u}(\eta)) \Big\}\, d\eta dy \Big] \end{aligned} \tag{2.32}$$

for $j \le \langle\xi\rangle/R \le j+1$, where $j \in \boldsymbol{Z}_+$. (2.29)–(2.32) give

$$\begin{aligned} \Big| I_{\varepsilon_1,k,\alpha} \Big| \le\ & C_{\varepsilon_1,\varepsilon_2,|\alpha|,k,\varepsilon}(S,T) C_{2|\alpha|+n+1}(p) \\ & \times \exp[\{c_1(S) + \varepsilon_1 + \delta_1 - (1-\varepsilon)c_3(S)(\varepsilon_2 - \delta_2 - c_1(T))/c_2(T)\}\langle\xi\rangle] \\ & \times \Big| e^{\varepsilon_2\langle\eta\rangle}\hat{u}(\eta) \Big|_{S,|\alpha|+(k+m_1)_+ +(m_2)_+ +n+n''+2} \\ & + C_{\varepsilon_1,\varepsilon_2,|\alpha|,k,\varepsilon,R}(S,T) C_{2|\alpha|+n+1}(p)(j+1)^{k+m_1+(m_2)_+ +n''} \\ & \times (\Gamma'(S,T,B,\varepsilon)/R)^j \exp[(c_1(S) + \varepsilon_1 + \delta_1 - 1/R)\langle\xi\rangle] \\ & \times \Big| e^{\varepsilon_2\langle\eta\rangle}\hat{u}(\eta) \Big|_{S,|\alpha|+n+1} \end{aligned}$$

if $j \in \boldsymbol{Z}_+$, $j \le \langle\xi\rangle/R \le j+1$ and $\varepsilon_2 \ge \delta_2 + c_1(T)$, where

$$\Gamma'(S,T,B,\varepsilon) = e\Gamma(|\xi|,S,T,B,\varepsilon)(1+1/|\xi|)^{-4}.$$

Therefore, we have

$$\begin{aligned} \Big| p_{S,T}(D_x,y,D_z)u \Big|_{S_{\varepsilon_1},\ell} \le\ & C_{\varepsilon_1,\varepsilon_2,\ell,\varepsilon,R}(S,T) C_{2\ell+n+1}(p) \\ & \times |u|_{S_{\varepsilon_2},\ell+(m_1)_+ +(m_2)_+ +n+n''+2} \end{aligned}$$

if

$$\begin{cases} \varepsilon_2 \ge \delta_2 + c_1(T) + (1-\varepsilon)^{-1}c_2(T)(\delta_1 + c_1(S) + \varepsilon_1)_+/c_3(S), \\ c_1(S) + \varepsilon_1 + \delta_1 \le 1/R, \quad R \ge 2\Gamma'(S,T,B,\varepsilon). \end{cases} \tag{2.33}$$

This proves the first part of the assertion (i). Next assume that $c_0(S) > 0$. Put $q(\eta,y,\xi) = p(\xi,y,\eta)$. Recall that ${}^r p_{S,T}(D_x,y,D_z) = q_{T,S}(D_z,y,D_x)$. Applying the same argument as before, we have

$$\Big| D_y^\beta \Big\{ \tilde{t}_{\varepsilon_2,\gamma'}(y,\eta) D_\eta^{\gamma-\gamma'} \tilde{L}^M(q(\eta,y,\xi)\hat{u}(\xi)) \Big\} \Big|$$

$$\leq C_{\varepsilon_2,|\gamma|,M}(S,T)C_{|\gamma|+M}(p)((2B_0+\widetilde{B}_1)/R)^{|\beta|}\langle\xi\rangle^{m_1+|\beta|} \times\langle\eta\rangle^{m_2}\exp[\delta_1\langle\xi\rangle+\delta_2\langle\eta\rangle]\langle y\rangle^{|\gamma|-M}\sup_{|\delta|\leq M}|\partial_\xi^\delta\hat{u}(\xi)| \tag{2.34}$$

if $R|\beta|\leq\langle\xi\rangle$ and $\gamma'\leq\gamma$,

$$\left|K^j\left\{\tilde{t}_{\varepsilon_2,\gamma'}(y,\eta)D_\eta^{\gamma-\gamma'}\widetilde{L}^M(q(\eta,y,\xi)\hat{u}(\xi))\right\}\right| \leq C_{\varepsilon_2,|\gamma|,M}(S,T)C_{|\gamma|+M}(p)\langle\xi\rangle^{m_1}\langle\eta\rangle^{m_2}\exp[\delta_1\langle\xi\rangle+\delta_2\langle\eta\rangle] \times\langle y\rangle^{|\gamma|-M}(\widetilde{\Gamma}(|\xi|,S,T,B,\varepsilon)/R)^j\sup_{|\delta|\leq M}|\partial_\xi^\delta\hat{u}(\xi)| \tag{2.35}$$

if $j\in\boldsymbol{Z}_+$, $\gamma'\leq\gamma$, $0<\varepsilon<1$, $(1-\varepsilon)c_3(S)|\xi|\geq c_2(T)\langle\eta\rangle$ and $j\leq\langle\xi\rangle/R$

for $u\in\mathcal{S}_\infty(\boldsymbol{R}^{n'})$, where

$$\tilde{t}_{\varepsilon_2,\gamma'}(y,\eta)=e^{-iT(y,\eta)+\varepsilon_2\langle\eta\rangle}D_\eta^{\gamma'}e^{iT(y,\eta)-\varepsilon_2\langle\eta\rangle},$$

$${}^t\widetilde{L}=(1+|\nabla_\xi S(y,\xi)|^2)^{-1}\Big(\sum_{j=1}^{n'}(\overline{\partial_{\xi_j}S(y,\xi)})D_{\xi_j}+1\Big),$$

$$\widetilde{B}_1=\max\{B,80(1+3n'C(S)^2A_0^2)B_0/c_0(S)\}$$

and $\widetilde{\Gamma}(|\xi|,S,T,B,\varepsilon)$ is defined by (2.31) with B_1 replaced by $\widetilde{B}_1$. Represent

$$\begin{aligned}
&\langle\eta\rangle^k D_\eta^\gamma\Big\{e^{-\varepsilon_2\langle\eta\rangle}\mathcal{F}[{}^r p_{S,T}(D_x,y,D_z)u](\eta)\Big\}\\
&=(2\pi)^{-n'}\sum_{\gamma'\leq\gamma}\binom{\gamma}{\gamma'}\Big[\int_{\Omega_\eta}e^{iS(y,\xi)+iT(y,\eta)}\langle\eta\rangle^k e^{-\varepsilon_2\langle\eta\rangle}\\
&\quad\times\tilde{t}_{\varepsilon_2,\gamma'}(y,\eta)D_\eta^{\gamma-\gamma'}\widetilde{L}^M(q(\eta,y,\xi)\hat{u}(\xi))\,d\xi dy\\
&\quad+\int_{c_2(S)\langle\xi\rangle\leq c_3(T)|\eta|/2}e^{iS(y,\xi)+iT(y,\eta)}\langle\eta\rangle^k e^{-\varepsilon_2\langle\eta\rangle}\\
&\quad\times K^{k+(m_2)_+}\Big\{\tilde{t}_{\varepsilon_2,\gamma'}(y,\eta)D_\eta^{\gamma-\gamma'}\widetilde{L}^M(q(\eta,y,\xi)\hat{u}(\xi))\Big\}\,d\xi dy\\
&\quad+\sum_{j=0}^{\infty}\int_{\Omega_{R,\eta,j}}e^{iS(y,\xi)+iT(y,\eta)}\langle\eta\rangle^k e^{-\varepsilon_2\langle\eta\rangle}\\
&\quad\times K^j\Big\{\tilde{t}_{\varepsilon_2,\gamma'}(y,\eta)D_\eta^{\gamma-\gamma'}\widetilde{L}^M(q(\eta,y,\xi)\hat{u}(\xi))\Big\}\,d\xi dy\Big],
\end{aligned}$$

where $M=|\gamma|+n+1$, $\Omega_\eta=\{(y,\xi);\ c_2(S)\langle\xi\rangle\geq c_3(T)|\eta|/2$ and $(1-\varepsilon)c_3(S)|\xi|\leq c_2(T)\langle\eta\rangle\}$, $\Omega_{R,\eta,j}=\{(y,\xi);\ (1-\varepsilon)c_3(S)|\xi|\geq c_2(T)\langle\eta\rangle$ and $Rj\leq\langle\xi\rangle\leq R(j+1)\}$ and $0<\varepsilon<1$. From (2.34) and (2.35) we have

$$\left|{}^r p_{S,T}(D_x,y,D_z)u\right|_{\mathcal{S}_{-\varepsilon_2},\ell}\leq C_{\varepsilon_1,\varepsilon_2,\ell,\varepsilon,R}(S,T)C_{2\ell+n+1}(p)$$

$$\times|u|_{\mathcal{S}_{-\varepsilon_1},\ell+(m_1)_++(m_2)_++n+n'+2}$$

if (2.33) with Γ' replaced by $\widetilde{\Gamma}'$ is valid, where

$$\widetilde{\Gamma}' = e\widetilde{\Gamma}(|\xi|, S, T, B, \varepsilon)(1+1/|\xi|)^{-4}.$$

Since

$$\begin{aligned}
\langle {}^r p_{S,T}(D_x, y, D_z)\varphi, \psi\rangle &= \langle \varphi, r_{S',T'}(D_x, y, D_z)\psi\rangle, \\
\langle p_{S,T}(D_x, y, D_z)\psi, \varphi\rangle &= \langle \psi, {}^r r_{S',T'}(D_x, y, D_z)\varphi\rangle
\end{aligned}$$

for $\varphi \in \mathcal{S}_\infty(\boldsymbol{R}^{n'})$ and $\psi \in \mathcal{S}_\infty(\boldsymbol{R}^{n''})$, where $S'(y,\xi) = S(y,-\xi)$, $T'(y,\eta) = T(y,-\eta)$ and $r(\xi,y,\eta) = p(-\xi,y,-\eta)$, the remaining part of the theorem is obvious. □

We write

$$\begin{aligned}
p_T(D_x, y, D_z) &= p_{S,T}(D_x, y, D_z) \quad \text{if } n' = n \text{ and } S(y,\xi) = -y\cdot\xi, \\
p'_S(D_x, y, D_y) &= p_{S,T}(D_x, y, D_z) \quad \text{if } n'' = n \text{ and } T(y,\eta) = y\cdot\eta, \\
p(D_x, y, D_y) &= p_T(D_x, y, D_z) \quad \text{if } n'' = n \text{ and } T(y,\eta) = y\cdot\eta.
\end{aligned}$$

Moreover, we write

$$\begin{aligned}
& p_T(x, D_z) = p_T(D_x, y, D_z) \text{ and } p(x, D) = p(D_x, y, D_y) \\
& \qquad\qquad \text{if } p(\xi, y, \eta) \equiv p(y, \eta), \\
& p'_S(D_x, y) = p'_S(D_x, y, D_y) \text{ and } p(D_x, y) = p(D_x, y, D_y) \\
& \qquad\qquad \text{if } p(\xi, y, \eta) \equiv p(\xi, y).
\end{aligned}$$

The transposed operator ${}^t p_{S,T}(D_x, y, D_z)$ of $p_{S,T}(D_x, y, D_z)$ is defined by

$$\langle {}^t p_{S,T}(D_x, y, D_z)u, \varphi\rangle = \langle u, p_{S,T}(D_x, y, D_z)\varphi\rangle$$

for $u \in \mathcal{S}_\infty(\boldsymbol{R}^{n'})$ and $\varphi \in \mathcal{S}_\infty(\boldsymbol{R}^{n''})$. Then it is obvious that

$${}^t p_{S,T}(D_x, y, D_z) = {}^r r_{S',T'}(D_x, y, D_z) = s_{T',S'}(D_z, y, D_x),$$

where $S'(y,\xi) = S(y,-\xi)$, $T'(y,\eta) = T(y,-\eta)$, $r(\xi,y,\eta) = p(-\xi,y,-\eta)$ and $s(\eta,y,\xi) = p(-\xi,y,-\eta)$. By continuity ${}^t p_{S,T}(D_x, y, D_z)$ can be defined in some $\mathcal{S}'_\varepsilon$. From Theorem 2.3. 3 it follows that $p(D_x, y, D_y)$ maps continuously $\bigcup_{\varepsilon>0}\mathcal{S}_\varepsilon$, $\bigcup_{\varepsilon>0}\mathcal{S}'_{-\varepsilon}$, $\mathcal{E}_0$ and $\mathcal{F}_0$ to $\bigcup_{\varepsilon>0}\mathcal{S}_\varepsilon$, $\bigcup_{\varepsilon>0}\mathcal{S}'_{-\varepsilon}$, $\mathcal{E}_0$ and $\mathcal{F}_0$, respectively, if $p(\xi,y,\eta) \in S^+(R,A)$ and $R \geq 2enA$.

2.4 Product formula of Fourier integral operators I

Let Ω, Ω' and Ω'' be open conic subsets of $\boldsymbol{R}^n \times (\boldsymbol{R}^{n'} \setminus \{0\})$, $\boldsymbol{R}^n \times (\boldsymbol{R}^{n''} \setminus \{0\})$ and $\boldsymbol{R}^n \times (\boldsymbol{R}^n \setminus \{0\})$, respectively, and let $S(y,\xi) \in \mathcal{P}(\Omega)$ and $T(y,\eta) \in \mathcal{P}(\Omega')$. We denote by $C(S)$, $A(S)$, $\cdots$ the constants C, A, $\cdots$ in ($\mathcal{P}$–0)–($\mathcal{P}$–4) for S. We also use the same notations $C(T)$,A(T), $\cdots$. We assume that $c_j(S), c_j(T) > 0$ ($j = 0, 3$). For simplicity we assume that $A_0 \equiv A(S) = B(S) = A(T) = B(T)$. Moreover, we assume that there are an $n'' \times n$ real matrix T_0 and $c(T) \geq 0$ such that $\tilde{T}(y,\eta) \equiv T(y,\eta) - T_0 y \cdot \eta$ satisfies the following:

($\mathcal{P}$–5) $\tilde{T}(y,\eta)$ satisfies ($\mathcal{P}$–0) with some constant $C \equiv C(\tilde{T})$ and $A = B = A_0$.

($\mathcal{P}$–6) $|\nabla_y \tilde{T}(y,\eta)| \leq c(T)\langle\eta\rangle$ and $|(\partial^2 \tilde{T}/\partial\eta\partial y)(y,\eta)| \leq c(T)$ in $\Omega' \cap \{|\eta| \geq 1\}$.

Let $U(y,\zeta) \in C^\infty(\Omega'')$ be a symbol satisfying ($\mathcal{P}$–0), ($\mathcal{P}$–3) and ($\mathcal{P}$–4) with $A = B = A_0$ and the following conditions;

($\mathcal{P}$–1)$'$ there is $c_0(U) > 0$ such that

$$|\nabla_\zeta U(y,\zeta) - \nabla_\zeta U(w,\zeta)| \geq c_0(U)|y - w| \tag{2.36}$$

for $(y,\zeta), (w,\zeta) \in \Omega'' \cap \{|\zeta| \geq 1\}$,

($\mathcal{P}$–2)$'$ there are $c_1^\pm(U)$ and $C_1^\pm(U)$ such that

$$\pm \mathrm{Im}\, U(y,\zeta) \geq -c_1^\pm(U)|\zeta| - C_1^\pm(U) \quad \text{in } \Omega'' \cap \{|\zeta| \geq 1\},$$

($\mathcal{P}$–5)$'$ $U(y,\zeta) = U_0 y \cdot \zeta + \tilde{U}(y,\zeta)$, where U_0 is an $n \times n$ real constant nonsingular matrix, and $\tilde{U}(y,\zeta)$ satisfies ($\mathcal{P}$–0) with some constant $C \equiv C(\tilde{U})$ and $A = B = A_0$,

($\mathcal{P}$–6) $|\nabla_y \tilde{U}(y,\zeta)| \leq c(U)\langle\zeta\rangle$ and $|(\partial^2 \tilde{U}/\partial\zeta\partial y)(y,\zeta)| \leq c(U)$ in $\Omega'' \cap \{|\zeta| \geq 1\}$.

Moreover, we assume that

$$c(T) \leq c_3(T)/4, \quad c(U) \leq c_3(U)/4, \tag{2.37}$$

where $c_2(U)$ and $c_3(U)$ are the constants in ($\mathcal{P}$–3) and ($\mathcal{P}$–4) for U, respectively. Let $p(\xi, w, \zeta, y, \eta) \in C^\infty(\boldsymbol{R}^{n'} \times \boldsymbol{R}^n \times \boldsymbol{R}^n \times \boldsymbol{R}^n \times \boldsymbol{R}^{n''})$ be a symbol satisfying the following:

(P–1) There are open conic subsets Ω_1' and Ω_1'' of Ω' and Ω'', respectively, and $\varepsilon_0 > 0$ such that

$$\{(y,\eta) \in \boldsymbol{R}^n \times \boldsymbol{R}^{n''};\ |w-y| \le \varepsilon_0 \text{ for some } (w,\eta) \in \Omega_1'\} \subset \Omega',$$
$$\{(y,\zeta) \in \boldsymbol{R}^n \times \boldsymbol{R}^{n};\ |w-y| \le \varepsilon_0 \text{ for some } (w,\zeta) \in \Omega_1''\} \subset \Omega''$$

and

$$\begin{aligned} \operatorname{supp} p \ \subset\ & \{(\xi,w,\zeta,y,\eta) \in \boldsymbol{R}^{n'} \times \Omega_1'' \times \Omega_1';\ (w,\xi) \in \Omega, \\ & (y,\zeta) \in \Omega'',\ |\xi| \ge 1,\ |\zeta| \ge 1 \text{ and } |\eta| \ge 1\}. \end{aligned}$$

(P–2) The estimates

$$\begin{aligned} &\left|\partial_\xi^{\alpha+\tilde\alpha} D_w^{\beta^1+\beta^2+\tilde\beta} \partial_\zeta^{\gamma+\tilde\gamma} D_y^{\lambda+\tilde\lambda} \partial_\eta^{\rho+\tilde\rho} p(\xi,w,\zeta,y,\eta)\right| \\ &\le C_{|\tilde\alpha|+|\tilde\beta|+|\tilde\gamma|+|\tilde\lambda|+|\tilde\rho|} (A_1/R_0)^{|\alpha|+|\beta^1|+|\beta^2|+|\gamma|+|\lambda|+|\rho|} \langle\xi\rangle^{m_1-|\tilde\alpha|+|\beta^1|} \\ &\quad \times \langle\zeta\rangle^{m_2-|\tilde\gamma|+|\beta^2|} \langle\eta\rangle^{m_3-|\tilde\rho|+|\lambda|} \exp[\delta_1\langle\xi\rangle + \delta_2\langle\zeta\rangle + \delta_3\langle\eta\rangle] \end{aligned}$$

hold if $\langle\xi\rangle \ge R_0(|\alpha|+|\beta^1|)$, $\langle\zeta\rangle \ge R_0(|\gamma|+|\beta^2|)$ and $\langle\eta\rangle \ge R_0(|\rho|+|\lambda|)$, where $m_j, \delta_j \in \boldsymbol{R}$ ($j=1,2,3$) and $R_0 \ge 1$.

(P–3) The estimates

$$\begin{aligned} &\left|\partial_\xi^{\alpha+\tilde\alpha} D_w^{\beta^1+\beta^2+\tilde\beta} \partial_\zeta^{\gamma+\tilde\gamma} D_y^{\lambda^1+\lambda^2+\tilde\lambda} \partial_\eta^{\rho+\tilde\rho} p(\xi,w,\zeta,y,\eta)\right| \\ &\le C_{|\tilde\alpha|+|\tilde\beta|+|\tilde\gamma|+|\tilde\lambda|+|\tilde\rho|} (A_1/R_0)^{|\alpha|+|\beta^1|+|\beta^2|+|\gamma|+|\lambda^1|+|\lambda^2|+|\rho|} \langle\xi\rangle^{m_1-|\tilde\alpha|+|\beta^1|} \\ &\quad \times \langle\zeta\rangle^{m_2-|\tilde\gamma|+|\beta^2|+|\lambda^1|} \langle\eta\rangle^{m_3-|\tilde\rho|+|\lambda^2|} \exp[\delta_1\langle\xi\rangle + \delta_2\langle\zeta\rangle + \delta_3\langle\eta\rangle] \end{aligned}$$

hold if $|w-y| \le \varepsilon_0$, $\langle\xi\rangle \ge R_0(|\alpha|+|\beta^1|)$, $\langle\zeta\rangle \ge R_0(|\gamma|+|\beta^2|+|\lambda^1|)$ and $\langle\eta\rangle \ge R_0(|\rho|+|\lambda^2|)$.

For $u \in \mathcal{S}_\infty(\boldsymbol{R}^{n''})$ we can define

$$\begin{aligned} &p_{S,-U,U,T}(D_x,w,D_w,y,D_z)u(x) \\ &= \mathcal{F}_\xi^{-1}\Big[\lim_{\nu\downarrow 0}(2\pi)^{-n-n''} \int\Big(\int\Big(\int\Big(\int e^{-\nu|\zeta|^2} e^{iS(w,\xi)-iU(w,\zeta)+iU(y,\zeta)+iT(y,\eta)} \\ &\quad \times p(\xi,w,\zeta,y,\eta)\hat u(\eta)\,d\eta\Big)\,dy\Big)\,d\zeta\Big)\,dw\Big](x). \end{aligned}$$

Lemma 2.4. 1 *Under the above hypotheses* $p_{S,-U,U,T}(D_x,w,D_w,y,D_z)$ *can be extended to a continuous linear operator from* $\mathcal{S}_\varepsilon(\boldsymbol{R}^{n''})$ *to* $\mathcal{S}_{-\delta}(\boldsymbol{R}^{n'})$ *if* $\varepsilon \ge \varepsilon(T,U,\delta_2,\delta_3)$, $\delta > \delta_1 + c_1(S)$, $R_0 \ge R(T,U,A_1,\varepsilon_0)$ *and* $\delta_2 + c_1^+(U) + c_1^-(U) \le 1/(3R_0)$.

Remark If $\delta_2 + c_1^+(U) + c_1^-(U) \le 0$, then we can prove the lemma in the same way without assuming (P–3).

Proof From Lemma 2.1. 3 (or its proof) we can choose $g^R(\zeta,\eta) \in C^\infty(\boldsymbol{R}^n \times \boldsymbol{R}^{n''})$ ($R \ge 2$) so that $0 \le g^R(\zeta,\eta) \le 1$ and

$$g^R(\zeta,\eta) = \begin{cases} 1 & \text{if } c_3(U)|\zeta| \le 2c_2(T)\langle\eta\rangle, \\ 0 & \text{if } c_3(U)|\zeta| \ge 3c_2(T)\langle\eta\rangle, \end{cases}$$

$$\left|\partial_\zeta^{\gamma+\tilde\gamma}\partial_\eta^{\rho+\tilde\rho} g^R(\zeta,\eta)\right| \le C_{|\tilde\gamma|+|\tilde\rho|}(\widehat{C}(T,U)/R)^{|\gamma|+|\rho|}\langle\zeta\rangle^{-|\tilde\gamma|-|\tilde\rho|} \quad (2.38)$$
$$\text{if } \langle\zeta\rangle \ge R|\gamma| \text{ and } \langle\eta\rangle \ge R|\rho|,$$

where $\widehat{C}(T,U)$ does not depend on R. In fact, we take $\xi = (t,\eta,\zeta) \in \boldsymbol{R}^{n+n''+1}$, $\Gamma_1 = \{(t,\eta,\zeta) \in \boldsymbol{R}^{n+n''+1} \setminus \{0\};\ c_3(U)|\zeta| < 2c_2(T)|(t,\eta)|\}$ and $\Gamma_2 = \{(t,\eta,\zeta) \in \boldsymbol{R}^{n+n''+1} \setminus \{0\};\ c_3(U)|\zeta| < 3c_2(T)|(t,\eta)|\}$ in Lemma 2.1. 3 . Then we can take

$$g^R(\zeta,\eta) = \sum_{N=0}^{\infty} \psi_N((1,\eta,\zeta)/|(1,\eta,\zeta)|)h_N^{R'}(|(1,\eta,\zeta)|),$$

where $\{\psi_N,\ h_N^R\}_{N=1,2,\dots}$ is as in the proof of Lemma 2.1. 3 , $\psi_0(\xi) = \psi_1(\xi)$, $h_0^R(t) = \phi_1(R^{-1}t-1)$ and $R' = \sqrt{(R^2-2)/8}$. In this proof we take $R = R_0$. Let L be the differential operator defined by (2.27), and let $u \in \mathcal{S}_\infty(\boldsymbol{R}^{n''})$. Then we have

$$\begin{aligned} &\int\Big(\int e^{iU(y,\zeta)+iT(y,\eta)}p(\xi,w,\zeta,y,\eta)\hat{u}(\eta)\,d\eta\Big)\,dy \\ &= \int e^{iU(y,\zeta)+iT(y,\eta)}L^M(p(\xi,w,\zeta,y,\eta)\hat{u}(\eta))\,d\eta dy \\ &\equiv f_1(\xi,w,\zeta) + f_2(\xi,w,\zeta), \end{aligned}$$

where $M \ge n+1$ and

$$f_1(\xi,w,\zeta) = \int e^{iU(y,\zeta)+iT(y,\eta)}L^M(g^R(\zeta,\eta)p(\xi,w,\zeta,y,\eta)\hat{u}(\eta))\,d\eta dy.$$

We note that one can take $M = 0$ if p has a compact support with respect to y. It is obvious that

$$\begin{aligned} &\left|\partial_\xi^{\tilde\alpha}\partial_\zeta^{\tilde\gamma}\{e^{-\nu|\zeta|^2}L^M(g^R(\zeta,\eta)p(\xi,w,\zeta,y,\eta)\hat{u}(\eta))\}\right| \\ &\le C_{|\tilde\alpha|+|\tilde\gamma|,M}(T,p)\langle y\rangle^{-M}\langle\xi\rangle^{m_1}\langle\zeta\rangle^{-n-1}\exp[\delta_1\langle\xi\rangle - \langle\eta\rangle/2] \\ &\quad\times|\exp[(3c_2(T)(\delta_2)_+/c_3(U)+\delta_3+1)\langle\eta\rangle]\hat{u}(\eta)|_{S,M} \end{aligned}$$

for $0 < \nu \le 1$. Define a differential operator L_1 by

$$
\begin{aligned}
{}^tL_1 &= (1+|\nabla_\zeta U(w,\zeta)-\nabla_\zeta U(y,\zeta)|^2)^{-1} \\
&\quad \times\Big(\sum_{j=1}^n(\overline{\partial_{\zeta_j}U(y,\zeta)}-\overline{\partial_{\zeta_j}U(w,\zeta)})D_{\zeta_j}+1\Big).
\end{aligned}
$$

Noting that

$$
\begin{aligned}
&\left|\partial_\zeta^\alpha\{\partial_{\zeta_j}U(y,\zeta)-\partial_{\zeta_j}U(w,\zeta)\}\right| \\
&= \left|\partial_\zeta^\alpha\int_0^1(y-w)\cdot\{\partial_{\zeta_j}(\nabla_yU)(w+\theta(y-w),\zeta)\}\,d\theta\right| \\
&\le 2\sqrt{n}C(U)A_0^{|\alpha|+2}(|\alpha|+1)!\langle\zeta\rangle^{-|\alpha|}|y-w|,
\end{aligned}
$$

we have

$$
\begin{aligned}
&\left|\partial_\xi^{\tilde\alpha}L_1^N\{e^{-\nu|\zeta|^2}L^M(g^R(\zeta,\eta)p(\xi,w,\zeta,y,\eta)\hat u(\eta))\}\right| \\
&\le C_{|\tilde\alpha|,M,N}(T,U,p)\langle y\rangle^{-M}\langle y-w\rangle^{-N}\langle\xi\rangle^{m_1}\langle\zeta\rangle^{-n-1} \\
&\quad\times\exp[\delta_1\langle\xi\rangle-\langle\eta\rangle/2] \\
&\quad\times|\exp[(3c_2(T)(\delta_2)_+/c_3(U)+\delta_3+1)\langle\eta\rangle]\hat u(\eta)|_{S,M} \qquad (2.39)
\end{aligned}
$$

for $0 < \nu \le 1$. So Lebesgue's convergence theorem yields

$$
\begin{aligned}
F_1(\xi) &\equiv \lim_{\nu\downarrow 0}\int\Big(\int e^{-\nu|\zeta|^2}e^{iS(w,\xi)-iU(w,\zeta)}f_1(\xi,w,\zeta)\,d\zeta\Big)\,dw \\
&= \int e^{iS(w,\xi)-iU(w,\zeta)+iU(y,\zeta)+iT(y,\eta)}L_1^NL^M\{g^R(\zeta,\eta) \\
&\quad\times p(\xi,w,\zeta,y,\eta)\hat u(\eta)\}\,d\eta dyd\zeta dw,
\end{aligned}
$$

where $N \ge n+1$. Moreover, taking $M = N = |\alpha|+n+1$ we have

$$
\begin{aligned}
\left|D_\xi^\alpha F_1(\xi)\right| &\le C_{|\alpha|}(S,T,U,p)\langle\xi\rangle^{m_1}\exp[(\delta_1+c_1(S))\langle\xi\rangle] \\
&\quad\times\Big|\exp[\{3c_2(T)(\delta_2+c_1^+(U)+c_1^-(U))_+/c_3(U) \\
&\qquad+\delta_3+c_1(T)+1\}\langle\eta\rangle]\hat u(\eta)\Big|_{S,|\alpha|+n+1}. \qquad (2.40)
\end{aligned}
$$

Choose $\{\chi_j^{\varepsilon_0}(y)\}\subset C_0^\infty(\boldsymbol{R}^n)$ so that

$$
\begin{aligned}
&\chi_j^{\varepsilon_0}(y) = \begin{cases} 1 & \text{if } |y|\le\varepsilon_0/2, \\ 0 & \text{if } |y|\ge\varepsilon_0, \end{cases} \\
&\left|D_y^{\beta+\tilde\beta}\chi_j^{\varepsilon_0}\right| \le C_{\varepsilon_0,|\tilde\beta|}(C_*j/\varepsilon_0)^{|\beta|} \quad \text{if } |\beta|\le 2j.
\end{aligned}
$$

Let $\{\psi_j^R(\zeta)\}$ be the family defined before Lemma 2.2. 4 . Then we can write

$$f_2(\xi, w, \zeta) = \sum_{j=1}^{\infty}\{f_{2j}^1(\xi, w, \zeta) + f_{2j}^2(\xi, w, \zeta)\},$$

where

$$\begin{aligned} f_{2j}^1(\xi, w, \zeta) &= \psi_j^R(\zeta) \int e^{iU(y,\zeta)+iT(y,\eta)} \chi_j^{\varepsilon_0}(w-y) \\ &\quad \times (1 - g^R(\zeta, \eta)) p(\xi, w, \zeta, y, \eta) \hat{u}(\eta)\, d\eta dy, \\ f_{2j}^2(\xi, w, \zeta) &= \psi_j^R(\zeta) \int e^{iU(y,\zeta)+iT(y,\eta)} (1 - \chi_j^{\varepsilon_0}(w-y)) \\ &\quad \times L^M\{(1 - g^R(\zeta, \eta)) p(\xi, w, \zeta, y, \eta) \hat{u}(\eta)\}\, d\eta dy, \end{aligned}$$

Let K_1 be a differential operator defined by

$${}^tK_1 = |\nabla_y U(y, \zeta) + \nabla_y T(y, \eta)|^{-2} \sum_{k=1}^{n} (\overline{\partial_{y_k} U(y, \zeta)} + \overline{\partial_{y_k} T(y, \eta)}) D_{y_k}.$$

Then it follows from Lemma 2.1. 5 that

$$\begin{aligned} &\left|\partial_\xi^\alpha \partial_\eta^\rho K_1^j \{\chi_j^{\varepsilon_0}(w-y)(1 - g^R(\zeta, \eta)) p(\xi, w, \zeta, y, \eta)\}\right| \\ &\le C_{|\alpha|+|\rho|,\varepsilon_0,R}(T, U, p)(\Gamma(T, U, A_1, \varepsilon_0)/R)^j \langle\xi\rangle^{m_1-|\alpha|} \\ &\quad \times \langle\zeta\rangle^{m_2+(m_3)_+} \exp[\delta_1\langle\xi\rangle + \delta_2\langle\zeta\rangle + \delta_3\langle\eta\rangle] \end{aligned} \tag{2.41}$$

if $\langle\zeta\rangle \ge 2R(j-1)$, where

$$\begin{aligned} \Gamma(T, U, A_1, \varepsilon_0) &= 2^8 n A_0 c_3(U)^{-2} C(T, U) \\ &\quad \times \max\{C_*/\varepsilon_0 + A_1, 2^7 \cdot 15 n A_0^3 c_3(U)^{-2} C(T, U)^2\}, \\ C(T, U) &= C(U) + c_3(U) C(T)/(2c_2(T)). \end{aligned}$$

This yields

$$\begin{aligned} F_2^1(\xi) &\equiv \lim_{\nu \downarrow 0} \int \Big(\int e^{-\nu|\zeta|^2} e^{iS(w,\xi)-iU(w,\zeta)} \sum_{j=1}^{\infty} f_{2j}^1(\xi, w, \zeta)\, d\zeta\Big)\, dw \\ &= \int e^{iS(w,\xi)-iU(w,\zeta)+iU(y,\zeta)+iT(y,\eta)} \sum_{j=1}^{\infty} \psi_j^R(\zeta) \\ &\quad \times L^M K_1^j \{\chi_j^{\varepsilon_0}(w-y)(1 - g^R(\zeta, \eta)) p(\xi, w, \zeta, y, \eta) \hat{u}(\eta)\}\, d\eta dy d\zeta dw, \\ \left|D_\xi^\alpha F_2^1(\xi)\right| &\le C_{|\alpha|,M,\varepsilon_0,R}(S, T, U, p) \langle\xi\rangle^{m_1} \exp[(\delta_1 + c_1(S))\langle\xi\rangle] \\ &\quad \times \left|\exp[(\delta_3 + c_1(T) + 1)\langle\eta\rangle] \hat{u}(\eta)\right|_{S,M} \end{aligned} \tag{2.42}$$

if $R \geq 2e\Gamma(T, U, A_1, \varepsilon_0)$ and $\delta_2 + c_1^+(U) + c_1^-(U) \leq 1/(3R)$, where $M \geq |\alpha| + n + 1$. Here we have used the fact that $e^{-j} \leq \exp[-\langle\zeta\rangle/(3R)]$ if $\langle\zeta\rangle \leq 3Rj$. Let $\widetilde{L}_1$ be a differential operator defined by

$$ {}^t\widetilde{L}_1 = |\nabla_\zeta U(w,\zeta) - \nabla_\zeta U(y,\zeta)|^{-2} \sum_{k=1}^{n} (\overline{\partial_{\zeta_k} U(y,\zeta)} - \overline{\partial_{\zeta_k} U(w,\zeta)}) D_{\zeta_k}. $$

Applying the same arguments as for (2.39) and (2.41), we have

$$ \begin{aligned} &\left| \partial_\xi^\alpha \partial_\eta^\rho \widetilde{L}_1^{j+k+n+1} \{ e^{-\nu|\zeta|^2} \psi_j^R(\zeta)(1 - g^R(\zeta,\eta)) p(\xi, w, \zeta, y, \eta) \} \right| \\ &\quad \leq C_{|\alpha|+|\rho|,k,\varepsilon_0,R}(U,p) \langle y - w \rangle^{-k-n-1} (\Gamma(U, A_1, \varepsilon_0)/R)^j \\ &\qquad \times \langle\xi\rangle^{m_1 - |\alpha|} \langle\zeta\rangle^{m_2} \langle\eta\rangle^{m_3 - |\rho|} \exp[\delta_1\langle\xi\rangle + \delta_2\langle\zeta\rangle + \delta_3\langle\eta\rangle] \qquad (2.43) \end{aligned} $$

if $0 < \nu \leq 1$ and $|w - y| \geq \varepsilon_0/2$, where

$$ \begin{aligned} &\Gamma(U, A_1, \varepsilon_0) = 2^7 n^{3/2} A_0^2 \varepsilon_0^{-1} c_0(U)^{-2} C(U) \\ &\quad \times \max\{4(\sqrt{2} + \sqrt{3}) + \widehat{C} + \widehat{C}(T,U) + A_1, 2^5 \cdot 15 n^2 A_0^5 c_0(U)^{-2} C(U)^2\}. \end{aligned} $$

Here we have used (2.2), (2.21), (2.36) and (2.38). This yields

$$ \begin{aligned} F_2^2(\xi) &\equiv \lim_{\nu \downarrow 0} \int \left(\int e^{-\nu|\zeta|^2} e^{iS(w,\xi) - iU(w,\zeta)} \sum_{j=1}^{\infty} f_{2j}^2(\xi, w, \zeta)\, d\zeta \right) dw \\ &= \int e^{iS(w,\xi) - iU(w,\zeta) + iU(y,\zeta) + iT(y,\eta)} \sum_{j=1}^{\infty} (1 - \chi_j^{\varepsilon_0}(w - y)) \\ &\qquad \times L^M \widetilde{L}_1^{j+M} \{ \psi_j^R(\zeta)(1 - g^R(\zeta,\eta)) p(\xi, w, \zeta, y, \eta) \hat{u}(\eta) \}\, d\eta dy d\zeta dw, \\ &\left| D_\xi^\alpha F_2^2(\xi) \right| \leq C_{|\alpha|,M,\varepsilon_0,R}(S,T,U,p) \langle\xi\rangle^{m_1} \exp[(\delta_1 + c_1(S))\langle\xi\rangle] \\ &\qquad \times \left| \exp[(\delta_3 + c_1(T) + 1)\langle\eta\rangle] \hat{u}(\eta) \right|_{S,M} \qquad (2.44) \end{aligned} $$

if $R \geq 2e\Gamma(U, A_1, \varepsilon_0)$ and $\delta_2 + c_1^+(U) + c_1^-(U) \leq 1/(3R)$, where $M \geq |\alpha| + n + 1$. (2.40), (2.42) and (2.44) prove the lemma. □

We put

$$ \widetilde{\nabla}_y U(w, y, \zeta) := \int_0^1 (\nabla_y U)(w + \theta(y - w), \zeta)\, d\theta. $$

Fix $y, w \in \boldsymbol{R}^n$ so that $|y - w| \leq \varepsilon_0$, and put

$$ \begin{aligned} &\Gamma_w = \{\zeta \in \boldsymbol{R}^n \setminus \{0\};\ (w, \zeta) \in \Omega_1''\}, \\ &\mathcal{C}_{w,\delta} = \{\zeta \in \boldsymbol{C}^n;\ |\zeta - \tilde{\zeta}| < \delta\langle\tilde{\zeta}\rangle \quad \text{for some } \tilde{\zeta} \in \Gamma_w \text{ with } |\tilde{\zeta}| \geq 1\}, \end{aligned} $$

where $\delta > 0$. Let $\tilde{\Psi}_w(\zeta;\delta)$ be a function defined in $\boldsymbol{C}^n$ such that $\tilde{\Psi}_w(\zeta;\delta) = 1$ for $\zeta \in \mathcal{C}_{w,\delta}$ and $\tilde{\Psi}_w(\zeta;\delta) = 0$ for $\zeta \in \boldsymbol{C}^n \setminus \mathcal{C}_{w,\delta}$, where $\delta > 0$. We put

$$\Psi_w(\zeta;\delta) = \int_{\boldsymbol{C}^n} \tilde{\Psi}_w(z;\delta)\rho_\varepsilon(\langle\zeta\rangle^{-1}(\zeta - z))\langle\zeta\rangle^{-2n}\,dz^1dz^2,$$

where $\zeta = \zeta^1 + i\zeta^2 \in \boldsymbol{C}^n$, $z = z^1 + iz^2 \in \boldsymbol{C}^n$, $\zeta^1,\zeta^2,z^1,z^2 \in \boldsymbol{R}^n$, $\langle\zeta\rangle = \sqrt{1+|\zeta|^2}$, ρ_ε is as in the proof of Lemma 2.1.2 and $\varepsilon = \min\{\delta/4, 1/2\}$. Then, $0 \le \Psi_w(\zeta;\delta) \le 1$ for $\zeta \in \boldsymbol{C}^n$, $\Psi_w(\zeta;\delta) = 1$ for $\zeta \in \mathcal{C}_{w,\delta/2}$, $\Psi_w(\zeta;\delta) = 0$ for $\zeta \notin \mathcal{C}_{w,5\delta/2}$ and

$$\left|\partial_{\zeta^1}^\alpha\partial_{\zeta^2}^\beta\Psi_w(\zeta^1 + i\zeta^2;\delta)\right| \le C_{|\alpha|+|\beta|}\langle\zeta\rangle^{-|\alpha|-|\beta|}(1+\delta^{-1})^{|\alpha|+|\beta|}, \tag{2.45}$$

where the C_j are independent of δ (> 0). In fact, (2.45) is obvious. If $\zeta \in \mathcal{C}_{w,\delta/2}$ and $|\zeta - z| \le \varepsilon\langle\zeta\rangle$, then $|z - \tilde{\zeta}| \le |\zeta - z| + |\zeta - \tilde{\zeta}| \le \varepsilon\langle\tilde{\zeta}\rangle + (1+\varepsilon)|\zeta - \tilde{\zeta}| \le \delta\langle\tilde{\zeta}\rangle$ for some $\tilde{\zeta} \in \Gamma_w$ with $|\tilde{\zeta}| \ge 1$. This implies that $\Psi_w(\zeta;\delta) = 1$ for $\zeta \in \mathcal{C}_{w,\delta/2}$. If $\zeta \notin \mathcal{C}_{w,5\delta/2}$ and $|\zeta - z| \le \varepsilon\langle\zeta\rangle$, then $|z-\tilde{\zeta}| \ge |\zeta-\tilde{\zeta}| - |\zeta - z| \ge (1-\varepsilon)|\zeta-\tilde{\zeta}| - \varepsilon\langle\tilde{\zeta}\rangle \ge (5(1-\varepsilon)\delta/2-\varepsilon)\langle\tilde{\zeta}\rangle \ge \delta\langle\tilde{\zeta}\rangle$ for any $\tilde{\zeta} \in \Gamma_w$ with $|\tilde{\zeta}| \ge 1$. This gives $\Psi_w(\zeta;\delta) = 0$ for $\zeta \notin \mathcal{C}_{w,5\delta/2}$. By assumption we can regard $\tilde{\nabla}_y U(w,y,\zeta)$ ($= {}^tU_0\zeta + \tilde{\nabla}_y\tilde{U}(w,y,\zeta)$) as an analytic function of ζ in the set $\{\zeta \in \boldsymbol{C}^n;\ |\zeta - \tilde{\zeta}| < (\sqrt{n}A_0)^{-1}\langle\tilde{\zeta}\rangle$ for some $\tilde{\zeta} \in \Gamma_w$ with $|\tilde{\zeta}| \ge 1\}$. Moreover, we have

$$\left|\tilde{\nabla}_y\tilde{U}(w,y,\zeta)\right| \le c(U)(\langle\tilde{\zeta}\rangle + |\zeta - \tilde{\zeta}|) + 4n^{3/2}C(\tilde{U})A_0^3|\zeta - \tilde{\zeta}|^2/\langle\tilde{\zeta}\rangle, \tag{2.46}$$

if $\zeta \in \boldsymbol{C}^n$, $\tilde{\zeta} \in \Gamma_w$, $|\tilde{\zeta}| \ge 1$ and $|\zeta - \tilde{\zeta}| \le (2\sqrt{n}A_0)^{-1}\langle\tilde{\zeta}\rangle$. In fact,

$$\begin{aligned}\tilde{\nabla}_y\tilde{U}(w,y,\zeta) = \int_0^1 \Big\{&(\nabla_y\tilde{U})(w+\theta(y-w),\tilde{\zeta}) + \sum_{j=1}^n(\zeta_j - \tilde{\zeta}_j)\\ &\times(\partial_{\zeta_j}\nabla_y\tilde{U})(w+\theta(y-w),\tilde{\zeta})\\ &+ \sum_{|\alpha|\ge 2}\frac{(\zeta-\tilde{\zeta})^\alpha}{\alpha!}(\partial_\zeta^\alpha\nabla_y\tilde{U})(w+\theta(y-w),\tilde{\zeta})\Big\}\,d\theta,\end{aligned} \tag{2.47}$$

$$\begin{aligned}&\left|\tilde{\nabla}_y\tilde{U}(w,y,\zeta)\right|\\ &\quad\le c(U)(\langle\tilde{\zeta}\rangle + |\zeta-\tilde{\zeta}|) + \sum_{k=2}^\infty 2\sqrt{n}C(\tilde{U})A_0\langle\tilde{\zeta}\rangle(\sqrt{n}A_0|\zeta-\tilde{\zeta}|/\langle\tilde{\zeta}\rangle)^k.\end{aligned}$$

If $0 < \delta \le (5\sqrt{n}A_0)^{-1}$, then the map

$$\mathcal{Z}_\delta : \boldsymbol{C}^n(\simeq \boldsymbol{R}^{2n}) \ni \zeta \longmapsto \zeta + \Psi_w(\zeta;\delta)\,{}^tU_0^{-1}\tilde{\nabla}_y\tilde{U}(w,y,\zeta) \in \boldsymbol{C}^n(\simeq \boldsymbol{R}^{2n})$$

is well-defined.

Lemma 2.4. 2 (i) *There are* $\delta(U) > 0$ *and* $\gamma_j > 0$ $(j = 1, 2)$ *such that the map* $\mathcal{Z}_\delta$ *is a diffeomorphism on* $\boldsymbol{C}^n$ *if*

$$0 < \delta \le \delta(U), \quad c(U) \le \gamma_1 \delta / |U_0^{-1}| \quad \text{and} \quad c(U) \le \gamma_2 / |U_0^{-1}|. \tag{2.48}$$

(ii) *Assume that* (2.48) *is satisfied, and let* $Z_\delta(z; w, y)$ *be the inverse function (map) of* $\mathcal{Z}_\delta$, *i.e.,* $Z_\delta(\mathcal{Z}_\delta(\zeta); w, y) = \zeta$ *for* $\zeta \in \boldsymbol{C}^n$ *and* $\mathcal{Z}_\delta(Z_\delta(z; w, y)) = z$ *for* $z \in \boldsymbol{C}^n$. *We put*

$$\begin{aligned} &\widetilde{\Gamma}_{w,y,\delta} = \mathcal{Z}_\delta(\{\zeta \in \Gamma_w;\ |\zeta| \ge 1\}), \\ &\widetilde{\mathcal{C}}_{w,y,\delta} = \{z \in \boldsymbol{C}^n;\ |z - \tilde{z}| < \delta\langle\tilde{z}\rangle/8 \text{ for some } \tilde{z} \in \widetilde{\Gamma}_{w,y,\delta}\}. \end{aligned}$$

Then, modifying $\delta(U)$ *and* γ_j $(j = 1, 2)$ *if necessary, we have*

$$Z_\delta(\widetilde{\mathcal{C}}_{w,y,\delta}) (\equiv \{Z_\delta(z; w, y) \in \boldsymbol{C}^n;\ z \in \widetilde{\mathcal{C}}_{w,y,\delta}\}) \subset \mathcal{C}_{w,\delta/2}. \tag{2.49}$$

Moreover, $Z_\delta(z; w, y)$ *is analytic in* $z \in \widetilde{\mathcal{C}}_{w,y,\delta}$ *and there are* $C_0(U), A_0(U) > 0$ *such that*

$$\begin{aligned} &\left|\partial_z^\alpha D_w^\beta D_y^\lambda Z_\delta(z; w, y)\right| \le C_0(U) A_0(U)^{|\alpha|+|\beta|+|\lambda|-1} \\ &\quad \times (|\alpha| + |\beta| + |\lambda| - 1)!(|\alpha| + |\beta| + |\lambda|)^{-2} \langle Z_\delta(z; w, y)\rangle^{1-|\alpha|} \end{aligned} \tag{2.50}$$

for $|\alpha| + |\beta| + |\lambda| \ge 1$, $y, w \in \boldsymbol{R}^n$ *with* $|y - w| \le \varepsilon_0$ *and* $z \in \widetilde{\mathcal{C}}_{w,y,\delta}$.

Remark Although (2.50) is a well-known result on estimates for the inverse map of an analytic map, we shall give the proof.

Proof (i) By (2.46) we have

$$\left|\Psi_w(\zeta; \delta)^t U_0^{-1} \widetilde{\nabla}_y \widetilde{U}(w, y, \zeta)\right| \le \langle\zeta\rangle/2 \tag{2.51}$$

if

$$\begin{cases} c(U) \le (12|U_0^{-1}|)^{-1} \\ 0 < \delta \le \min\{(5\sqrt{n}A_0)^{-1}, 1/5, (40nC(\widetilde{U})A_0^2|U_0^{-1}|)^{-1}\}. \end{cases} \tag{2.52}$$

In fact, $|\zeta - \tilde{\zeta}| \le 5\delta\langle\zeta\rangle$ and $\langle\tilde{\zeta}\rangle \le 2\langle\zeta\rangle$ if $0 < \delta \le 1/5$, $|\tilde{\zeta}| \ge 1$ and $|\zeta - \tilde{\zeta}| \le 5\delta\langle\tilde{\zeta}\rangle/2$. (2.51) implies that $\mathcal{Z}_\delta$ is a proper mapping, *i.e.*, the inverse image of every compact subset of $\boldsymbol{C}^n$ is compact. Write

$$\begin{aligned} F(\zeta^1, \zeta^2) &\equiv (F_1(\zeta^1, \zeta^2), \cdots, F_{2n}(\zeta^1, \zeta^2)) \\ &= \Psi_w(\zeta; \delta)(\mathrm{Re}\ {}^t\widetilde{\nabla}_y \widetilde{U}(w, y, \zeta) U_0^{-1}, \mathrm{Im}\ {}^t\widetilde{\nabla}_y \widetilde{U}(w, y, \zeta) U_0^{-1}), \\ G(\zeta^1, \zeta^2) &= \Psi_w(\zeta; \delta)^t \widetilde{\nabla}_y \widetilde{U}(w, y, \zeta) U_0^{-1}, \end{aligned}$$

where $\zeta = \zeta^1 + i\zeta^2 \in \boldsymbol{C}^n$ and $\zeta^1, \zeta^2 \in \boldsymbol{R}^n$. It follows from (2.45), (2.46) and the assumptions on U that

$$\begin{aligned}
&\left|\left(\partial F/\partial(\zeta^1,\zeta^2)\right)(\zeta^1,\zeta^2)\right| \\
&\le \left(\left|\left(\partial G/\partial\zeta^1\right)(\zeta^1,\zeta^2)\right|^2 + \left|\left(\partial G/\partial\zeta^2\right)(\zeta^1,\zeta^2)\right|^2\right)^{1/2} \\
&\le \sqrt{2}\Big(\sqrt{n}(3c(U) + 50n^{3/2}C(\tilde{U})A_0^3\delta^2)C_1(1+\delta^{-1}) \\
&\qquad + c(U) + 30n^{3/2}C(\tilde{U})A_0^3\delta\Big)|U_0^{-1}|
\end{aligned}$$

if $0 < \delta \le \min\{(5\sqrt{n}A_0)^{-1}, 1/5\}$, where $|T|$ denotes the matrix norm of a matrix T. In fact,

$$\partial_{\zeta_j}\tilde{\nabla}_y\tilde{U}(w,y,\zeta) = (\partial_{\zeta_j}\tilde{\nabla}_y\tilde{U})(w,y,\tilde{\zeta}) + \sum_{|\alpha|\ge 1}\frac{(\zeta-\tilde{\zeta})^\alpha}{\alpha!}(\partial_\zeta^\alpha\partial_{\zeta_j}\tilde{\nabla}_y\tilde{U})(w,y,\tilde{\zeta})$$

if $\tilde{\zeta} \in \Gamma_w$, $|\tilde{\zeta}| \ge 1$ and $|\zeta - \tilde{\zeta}| \le 5\delta\langle\tilde{\zeta}\rangle/2$. Therefore, we have

$$\left|\left(\partial F/\partial(\zeta^1,\zeta^2)\right)(\zeta^1,\zeta^2)\right| \le 1/2$$

if

$$\left\{\begin{aligned}
&0 < \delta \le \min\{(5\sqrt{n}A_0)^{-1}, 1/5, (60\sqrt{2}n^{3/2}C(\tilde{U})A_0^2 \\
&\qquad\qquad \times(C_1 + 3A_0 + 5\sqrt{n}C_1A_0)|U_0^{-1}|)^{-1}\}, \\
&c(U) \le (6\sqrt{2}|U_0^{-1}|)^{-1}\min\{(3\sqrt{n}C_1+1)^{-1}, (3\sqrt{n}C_1)^{-1}\delta\},
\end{aligned}\right. \tag{2.53}$$

where C_1 is the constant in (2.45). This implies that the Jacobian of $\mathcal{Z}_\delta$ does not vanish if (2.53) is valid. The global implicit function theorem proves the assertion (i) (see, *e.g.*, [CH]). (ii) We note that

$$\begin{aligned}
&\tilde{\Gamma}_{w,y,\delta} = \{\zeta + {}^tU_0^{-1}\tilde{\nabla}_y\tilde{U}(w,y,\zeta);\ \zeta \in \Gamma_w \text{ and } |\zeta| \ge 1\}. \\
&\Gamma_w \cap \{|\zeta| \ge 1\} = Z_\delta(\tilde{\Gamma}_{w,y,\delta}), \\
&\frac{\partial Z_\delta}{\partial z}(z;w,y) = \left(I + {}^tU_0^{-1}\int_0^1 \frac{\partial^2\tilde{U}}{\partial y\partial\zeta}(w+\theta(y-w),\zeta)\,d\theta\right)^{-1}\Big|_{\zeta=Z_\delta(z;w,y)} \\
&\qquad\qquad \text{if } Z_\delta(z;w,y) \in \mathcal{C}_{w,\delta/2},
\end{aligned}$$

where I denotes the identity matrix of order n. By the proof of the assertion (i) we have

$$\left|\frac{\partial(\mathrm{Re}\, Z_\delta, \mathrm{Im}\, Z_\delta)}{\partial(z^1,z^2)}(z;w,y)\right| \le 2 \tag{2.54}$$

if (2.53) is valid, where $z = z^1 + iz^2$ and $z^1, z^2 \in \boldsymbol{R}^n$. Therefore, we have

$$|Z_\delta(z;w,y) - Z_\delta(\tilde{z};w,y)| \le 2|z-\tilde{z}| < \delta\langle\tilde{\zeta}\rangle/2 \tag{2.55}$$

if $z \in \boldsymbol{C}^n$, $\tilde{z} \in \tilde{\Gamma}_{w,y,\delta}$ and $|z-\tilde{z}| < \delta\langle\tilde{z}\rangle/8$, where $\tilde{\zeta} = Z_\delta(\tilde{z};w,y)$. In fact,

$$\langle\tilde{z}\rangle \le \langle\tilde{\zeta}\rangle + |U_0^{-1}||\tilde{\nabla}_y\tilde{U}(w,y,\tilde{\zeta})| \le 2\langle\tilde{\zeta}\rangle$$

if $c(U) \le |U_0^{-1}|^{-1}$. (2.55) yields (2.49). Now let us prove (2.50) and

$$\begin{aligned}
&\left|\partial_z^\alpha D_w^\beta D_y^\lambda \left(\partial_\zeta^{\alpha'} D_w^{\beta'} D_y^{\lambda'} \tilde{\nabla}_y \tilde{U}\right)(w,y,Z_\delta(z;w,y))\right| \\
&\le C_1(U) A_0(U)^{|\alpha|+|\beta|+|\lambda|-1} A_0'(U)^{|\alpha'|+|\beta'|+|\lambda'|} \\
&\quad \times (|\alpha|+|\beta|+|\lambda|+|\alpha'|+|\beta'|+|\lambda'|-1)! \\
&\quad \times (|\alpha|+|\beta|+|\lambda|)^{-2} \langle Z_\delta(z;w,y)\rangle^{1-|\alpha|-|\alpha'|} \qquad (2.56)\\
&\quad \text{for } |\alpha|+|\beta|+|\lambda| \ge 1,\ y,w \in \boldsymbol{R}^n \text{ with } |y-w| \le \varepsilon_0 \text{ and } z \in \tilde{\mathcal{C}}_{w,y,\delta},
\end{aligned}$$

together, by induction on $|\alpha|+|\beta|+|\lambda|$. We assume that (2.52) and (2.53) are valid and that $y, w \in \boldsymbol{R}^n$, $|y-w| \le \varepsilon_0$ and $z \in \tilde{\mathcal{C}}_{w,y,\delta}$. For $e, e' \in \boldsymbol{Z}_+^n$ with $|e|+|e'| = 1$ we have

$$\begin{aligned}
&\left(I + {}^tU_0^{-1} \frac{\partial(\tilde{\nabla}_y\tilde{U})}{\partial\zeta}(w,y,Z_\delta(z;w,y))\right) D_w^e D_y^{e'} Z_\delta(z;w,y) \\
&= -{}^tU_0^{-1}\left(D_w^e D_y^{e'} \tilde{\nabla}_y \tilde{U}\right)(w,y,Z_\delta(z;w,y)).
\end{aligned} \tag{2.57}$$

On the other hand, we have

$$\begin{aligned}
&\left|\partial_\zeta^\alpha D_w^\beta D_y^\lambda \tilde{\nabla}_y \tilde{U}(w,y,\zeta)\right| \\
&\le \int_0^1 \sum_\gamma \left|\frac{(\zeta-\tilde{\zeta})^\gamma}{\gamma!}\left(\partial_\zeta^{\gamma+\alpha} D_y^{\beta+\lambda} \nabla_y \tilde{U}\right)(w+\theta(y-w),\tilde{\zeta})\right| \theta^{|\lambda|}(1-\theta)^{|\beta|}\, d\theta \\
&\le (5/3)^2 \sqrt{n} C(\tilde{U}) (11A_0/5)^{|\alpha|} A_0^{|\beta|+|\lambda|+1} |\alpha|!|\beta|!|\lambda|! \langle\zeta\rangle^{1-|\alpha|}
\end{aligned} \tag{2.58}$$

if $|\zeta-\tilde{\zeta}| < \delta\langle\tilde{\zeta}\rangle/2$, $\tilde{\zeta} \in \Gamma_w$ and $|\tilde{\zeta}| \ge 1$. In fact, $10\langle\zeta\rangle/11 \le \langle\tilde{\zeta}\rangle \le 10\langle\zeta\rangle/9$ and $2\sqrt{n}A_0|\zeta-\tilde{\zeta}|/\langle\tilde{\zeta}\rangle \le 1/5$ if $|\zeta-\tilde{\zeta}| \le \delta\langle\tilde{\zeta}\rangle/2$,

$$\int_0^1 \theta^{|\lambda|}(1-\theta)^{|\beta|}\, d\theta = |\beta|!|\lambda|!/(|\beta|+|\lambda|+1)!.$$

From (2.54), (2.57) and (2.58) it follows that (2.50) is valid when $|\alpha| + |\beta| + |\lambda| = 1$ if $C_0(U) \ge 2\max\{1, 25\sqrt{n}C(\tilde{U})A_0^2|U_0^{-1}|/9\}$. For $e \in \boldsymbol{Z}_+^n$

with $|e| = 1$ we have, by (2.54) and (2.58),

$$\begin{aligned}&\left|\partial_z^e\left(\partial_\zeta^{\alpha'} D_w^{\beta'} D_y^{\lambda'} \tilde{\nabla}_y \tilde{U}\right)(w, y, Z_\delta(z; w, y))\right| \\ &\leq (110/9)nC(\tilde{U})A_0^2(22A_0/5)^{|\alpha'|}A_0^{|\beta'|+|\lambda'|} \\ &\quad \times(|\alpha'| + |\beta'| + |\lambda'|)!\langle Z_\delta(z; w, y)\rangle^{-|\alpha'|}.\end{aligned}$$

Similarly, for $e, e' \in Z_+^n$ with $|e| + |e'| = 1$ we have

$$\begin{aligned}&\left|D_w^e D_y^{e'}\left(\partial_\zeta^{\alpha'} D_w^{\beta'} D_y^{\lambda'} \tilde{\nabla}_y \tilde{U}\right)(w, y, Z_\delta(z; w, y))\right| \\ &\leq (5/3)^2\sqrt{n}C(\tilde{U})A_0^2(1 + 110nC(\tilde{U})A_0^2|U_0^{-1}|/9)(22A_0/5)^{|\alpha'|} \\ &\quad \times(2A_0)^{|\beta'|+|\lambda'|}(|\alpha'| + |\beta'| + |\lambda'|)!\langle Z_\delta(z; w, y)\rangle^{1-|\alpha'|}.\end{aligned}$$

So (2.56) is valid when $|\alpha|+|\beta|+|\lambda| = 1$ if $C_1(U) \geq ((5/3)^2\sqrt{n}C(\tilde{U})A_0^2 + |U_0^{-1}|^{-1})(1 + 110nC(\tilde{U})A_0^2|U_0^{-1}|/9)$ and $A_0'(U) \geq 22A_0/5$. Next suppose that (2.50) and (2.56) are valid when $|\alpha| + |\beta| + |\lambda| \leq \ell$, where $\ell \in \boldsymbol{N}$. Let $\alpha, \beta, \lambda \in Z_+^n$ satisfy $|\alpha| + |\beta| + |\lambda| = \ell$. Noting that

$$\left(I + {}^tU_0^{-1}\frac{\partial(\tilde{\nabla}_y\tilde{U})}{\partial\zeta}(w, y, Z_\delta(z; w, y))\right)\frac{\partial Z_\delta}{\partial z}(z; w, y) = I,$$

we have

$$\begin{aligned}\partial_z^\alpha D_w^\beta D_y^\lambda \frac{\partial Z_\delta}{\partial z}(z; w, y) &= -\left(I + {}^tU_0^{-1}\frac{\partial(\tilde{\nabla}_y\tilde{U})}{\partial\zeta}(w, y, Z_\delta(z; w, y))\right)^{-1} \\ &\quad \times \sum_{\substack{\alpha^1 \leq \alpha, \beta^1 \leq \beta, \lambda^1 \leq \lambda \\ |\alpha^1|+|\beta^1|+|\lambda^1| \geq 1}} \binom{\alpha}{\alpha^1}\binom{\beta}{\beta^1}\binom{\lambda}{\lambda^1} \\ &\quad \times {}^tU_0^{-1}\partial_z^{\alpha^1} D_w^{\beta^1} D_y^{\lambda^1} \frac{\partial(\tilde{\nabla}_y\tilde{U})}{\partial\zeta}(w, y, Z_\delta(z; w, y)) \\ &\quad \times \partial_z^{\alpha-\alpha^1} D_w^{\beta-\beta^1} D_y^{\lambda-\lambda^1} \frac{\partial Z_\delta}{\partial z}(z; w, y).\end{aligned}$$

This yields, for $e \in Z_+^n$ with $|e| = 1$,

$$\begin{aligned}&\left|\partial_z^{\alpha+e} D_w^\beta D_y^\lambda Z_\delta(z; w, y)\right| \leq C_0(U)A_0(U)^{|\alpha|+|\beta|+|\lambda|} \\ &\quad \times(|\alpha| + |\beta| + |\lambda|)!(|\alpha| + |\beta| + |\lambda| + 1)^{-2}\langle Z_\delta(z; w, y)\rangle^{-|\alpha|}I_{\alpha,\beta,\lambda},\end{aligned}$$

where

$$\begin{aligned}I_{\alpha,\beta,\lambda} &= 2\sqrt{n}|U_0^{-1}|C_1(U)A_0'(U)/A_0(U) \\ &\quad \times \sum_{k=1}^{|\alpha|+|\beta|+|\lambda|} (|\alpha| + |\beta| + |\lambda| + 1)^2 k^{-2}(|\alpha| + |\beta| + |\lambda| - k + 1)^{-2}.\end{aligned}$$

Since $\sum_{\mu=1}^{\infty}\mu^{-2} = \pi^2/6$, $I_{\alpha,\beta,\lambda} \le 1$ if $A_0(U) \ge 8\pi^2\sqrt{n}|U_0^{-1}|C_1(U)A_0'(U)/3$. Using (2.57), we have also, for $e, e' \in \mathbf{Z}_+^n$ with $|e|+|e'|=1$,

$$\begin{aligned}\left|D_w^{\beta+e}D_y^{\lambda+e'}Z_\delta(z;w,y)\right| &\le C_0(U)A_0(U)^{|\beta|+|\lambda|}\\ &\quad\times(|\beta|+|\lambda|)!(|\beta|+|\lambda|+1)^{-2}\langle Z_\delta(z;w,y)\rangle\end{aligned}$$

if $|\beta|+|\lambda|=\ell$ and

$$A_0(U) \ge 8|U_0^{-1}|C_1(U)A_0'(U)(\pi^2\sqrt{n}/3+1/C_0(U)). \tag{2.59}$$

Therefore, (2.50) is valid when $|\alpha|+|\beta|+|\lambda|=\ell+1$ if (2.59) is satisfied. It is easy to see that

$$\begin{aligned}&\left|\partial_z^{\alpha+e}D_w^{\beta}D_y^{\lambda}\left(\partial_\zeta^{\alpha'}D_w^{\beta'}D_y^{\lambda'}\tilde\nabla_y\tilde U\right)(w,y,Z_\delta(z;w,y))\right|\\ &\le \sum_{j=1}^{n}\left|\partial_z^{\alpha}D_w^{\beta}D_y^{\lambda}\left\{\left(\partial_\zeta^{\alpha'+e_j}D_w^{\beta'}D_y^{\lambda'}\tilde\nabla_y\tilde U\right)(w,y,Z_\delta(z;w,y))\partial_z^{e}Z_{\delta j}(z;w,y)\right\}\right|\\ &\le C_1(U)A_0(U)^{|\alpha|+|\beta|+|\lambda|}A_0'(U)^{|\alpha'|+|\beta'|+|\lambda'|}\\ &\quad\times(|\alpha|+|\beta|+|\lambda|+|\alpha'|+|\beta'|+|\lambda'|)!\\ &\quad\times(|\alpha|+|\beta|+|\lambda|+1)^{-2}\langle Z_\delta(z;w,y)\rangle^{-|\alpha|-|\alpha'|}\end{aligned}$$

if $|\alpha|+|\beta|+|\lambda|=\ell$, $|e|=1$ and

$$\begin{cases} A_0'(U)\ge 22A_0/5, \quad C_1(U)\ge 110nA_0^2C(\tilde U)C_0(U)/9\\ A_0(U)\ge 8\pi^2\sqrt{n}C_0(U)A_0'(U)/3,\end{cases}$$

where $Z_\delta(\cdot)=(Z_{\delta 1}(\cdot),\cdots,Z_{\delta n}(\cdot))$. Moreover, we can show that (2.56) is valid when $\alpha=0$ and $|\beta|+|\lambda|=\ell+1$ if

$$\begin{cases} A_0'(U)\ge 22A_0/5, \quad C_1(U)\ge 110nA_0^2C(\tilde U)C_0(U)/9\\ A_0(U)\ge 8A_0'(U)(1+\pi^2\sqrt{n}C_0(U)/3).\end{cases}$$

This proves (2.50). □

We take $\delta=\delta(U)$ and assume that

$$c(U)\le \min\{\gamma_1\delta,\gamma_2\}/|U_0^{-1}|.$$

Note that $Z_\delta(z;w,y)$ is independent of δ and

$$Z_\delta(z;w,y)+{}^tU_0^{-1}\tilde\nabla_y\tilde U(w,y,Z_\delta(z;w,y))=z$$

if $z \in \widetilde{C}_{w,y,\delta}$. We write $Z(z;w,y) = Z_\delta(z;w,y)$. Let $\kappa > 0$, and put

$$p_\kappa(\xi,w,\zeta,y,\eta;u) = \sum_{j=1}^{\infty} \psi_j^{\kappa R_0}(\eta) \sum_{|\alpha|\le j-1} u^\alpha \partial_\zeta^\alpha p(\xi,w,\zeta,y,\eta)/\alpha!$$

for $(\xi,w,\zeta,y,\eta,u) \in \boldsymbol{R}^{n'} \times \boldsymbol{R}^n \times \boldsymbol{R}^n \times \boldsymbol{R}^n \times \boldsymbol{R}^{n''} \times \boldsymbol{C}^n$. Then we have the following

Lemma 2.4. 3 *$p_\kappa(\xi,w,\zeta,y,\eta;u)$ is analytic in u,* supp $p_\kappa(\cdot;u) \subset$ supp *$p(\cdot)$ for $u \in \boldsymbol{C}^n$ and*

$$\begin{aligned}
&p_\kappa(\xi,w,\zeta,y,\eta;0) = p(\xi,w,\zeta,y,\eta),\\
&\left|\partial_\xi^{\alpha+\tilde\alpha} D_w^{\beta^1+\beta^2+\tilde\beta} \partial_\zeta^{\gamma+\tilde\gamma} D_y^{\lambda^1+\lambda^2+\tilde\lambda} \partial_\eta^{\rho+\tilde\rho} \partial_u^{\delta+\tilde\delta} p_\kappa(\xi,w,\zeta,y,\eta;u)\right|\\
&\le C_{|\tilde\alpha|+|\tilde\beta|+|\tilde\gamma|+|\tilde\delta|+|\tilde\lambda|,|\tilde\rho|,\kappa}((\widehat{C}/\kappa + A_1)/R_0)^{|\rho|}\\
&\quad \times (A_1/R_0)^{|\alpha|+|\beta^1|+|\beta^2|+|\gamma|+|\delta|+|\lambda^1|+|\lambda^2|}\\
&\quad \times \langle\xi\rangle^{m_1-|\tilde\alpha|+|\beta^1|}\langle\zeta\rangle^{m_2-|\tilde\gamma|-|\tilde\delta|+|\beta^2|+|\lambda^1|}\langle\eta\rangle^{m_3-|\tilde\rho|+|\lambda^2|+1}\\
&\quad \times \exp[\delta_1\langle\xi\rangle + \delta_2\langle\zeta\rangle + \delta_3\langle\eta\rangle + \sqrt{n}A_1\langle u\rangle/R_0], \qquad (2.60)\\
&\left|(\partial_{\zeta_j} - \partial_{u_j})\partial_\xi^{\alpha+\tilde\alpha} D_w^{\beta^1+\beta^2+\tilde\beta} \partial_\zeta^{\gamma+\tilde\gamma} D_y^{\lambda^1+\lambda^2+\tilde\lambda} \partial_\eta^{\rho+\tilde\rho} \partial_u^{\delta+\tilde\delta} p_\kappa(\xi,w,\zeta,y,\eta;u)\right|\\
&\le C_{|\tilde\alpha|+|\tilde\beta|+|\tilde\gamma|+|\tilde\delta|+|\tilde\lambda|,|\tilde\rho|}((\widehat{C}/\kappa + A_1)/R_0)^{|\rho|}\\
&\quad \times (A_1/R_0)^{|\alpha|+|\beta^1|+|\beta^2|+|\gamma|+|\lambda^1|+|\lambda^2|}\\
&\quad \times (eA_1/R_0)^{|\delta|}\langle\xi\rangle^{m_1-|\tilde\alpha|+|\beta^1|}\langle\zeta\rangle^{m_2-1-|\tilde\gamma|-|\tilde\delta|+|\beta^2|+|\lambda^1|}\langle\eta\rangle^{m_3-|\tilde\rho|+|\lambda^2|}\\
&\quad \times \exp[\delta_1\langle\xi\rangle + \delta_2\langle\zeta\rangle + (\delta_3 - 1/(3\kappa R_0))\langle\eta\rangle + e\sqrt{n}A_1\langle u\rangle/R_0], \quad (2.61)
\end{aligned}$$

if

$$\left\{\begin{array}{l}\langle\eta\rangle \le \kappa\langle\zeta\rangle, \quad |w-y| \le \varepsilon_0, \quad \langle\xi\rangle \ge R_0(|\alpha|+|\beta^1|),\\ \langle\zeta\rangle \ge 2R_0(|\gamma|+|\beta^2|+|\lambda^1|),\\ \langle\eta\rangle \ge \max\{1,3\kappa\}R_0(|\rho|+|\lambda^2|),\end{array}\right. \qquad (2.62)$$

where $\widehat{C}$ is the constant in (2.21).

Proof We have

$$\begin{aligned}
&\left|\partial_\xi^{\alpha+\tilde\alpha} D_w^{\beta^1+\beta^2+\tilde\beta} \partial_\zeta^{\gamma+\tilde\gamma} D_y^{\lambda^1+\lambda^2+\tilde\lambda} \partial_\eta^{\rho+\tilde\rho} \partial_u^{\delta+\tilde\delta} p_\kappa(\xi,w,\zeta,y,\eta;u)\right|\\
&\le \sum_{j\ge|\delta|+|\tilde\delta|+1} \chi_{F_j}(\eta) \sum_{\tilde\rho'\le\tilde\rho} \binom{\tilde\rho}{\tilde\rho'} 2\widehat{C}_{|\tilde\rho|-|\tilde\rho'|} C_{|\tilde\alpha|+|\tilde\beta|+|\tilde\gamma|+\tilde\delta|+|\tilde\lambda|+|\tilde\rho'|}\\
&\quad \times ((\widehat{C}/\kappa + A_1)/R_0)^{|\rho|}(A_1/R_0)^{|\alpha|+|\beta^1|+|\beta^2|+|\gamma|+|\lambda^1|+|\lambda^2|}\langle\xi\rangle^{m_1-|\tilde\alpha|+|\beta^1|}
\end{aligned}$$

$$\times\langle\zeta\rangle^{m_2-|\tilde\gamma|-|\tilde\delta|+|\beta^2|+|\lambda^1|}\langle\eta\rangle^{m_3-|\tilde\rho|+|\lambda^2|}\exp[\delta_1\langle\xi\rangle+\delta_2\langle\zeta\rangle+\delta_3\langle\eta\rangle]$$
$$\times\sum_{\mu\geq\delta+\tilde\delta,\,|\mu|\leq j-1}\left|u^{\mu-\delta-\tilde\delta}/(\mu-\delta-\tilde\delta)!\right|(A_1/R_0)^{|\mu|-|\tilde\delta|}$$

if (2.62) is valid, where $F_j=\{\eta\in \boldsymbol{R}^{n''};\ 2\kappa R_0(j-1)\leq\langle\eta\rangle\leq 3\kappa R_0 j\}$ and χ_F denotes the characteristic function of F. Noting that

$$\sum_{j\geq|\delta|+|\tilde\delta|+1}\chi_{F_j}(\eta)\sum_{\mu\geq\delta+\tilde\delta,\,|\mu|\leq j-1}\left|u^{\mu-\delta-\tilde\delta}/(\mu-\delta-\tilde\delta)!\right|(A_1/R_0)^{|\mu|-|\tilde\delta|}$$
$$\leq(5/2+\langle\eta\rangle/(4\kappa R_0))(A_1/R_0)^{|\delta|}\exp[\sqrt{n}A_1|u|/R_0],$$

we have (2.60). Similarly, we can prove (2.61). □

Choose $\tilde g^R(\xi,\eta)\in C^\infty(\boldsymbol{R}^{n'}\times\boldsymbol{R}^{n''})$ ($R\geq 2$) so that $0\leq\tilde g^R(\xi,\eta)\leq 1$ and

$$\tilde g^R(\xi,\eta)=\begin{cases}1 & \text{if } |\xi|\leq 2c_2(T)|\eta|/c_3(S) \text{ and } |\xi|^2+|\eta|^2\geq 8,\\ 0 & \text{if } |\xi|\geq 3c_2(T)|\eta|/c_3(S) \text{ or } |\xi|^2+|\eta|^2\leq 4,\end{cases}$$
$$\left|\partial_\xi^{\alpha+\tilde\alpha}\partial_\eta^{\rho+\tilde\rho}\tilde g^R(\xi,\eta)\right|\leq C_{|\tilde\alpha|+|\tilde\rho|}(\widehat C(S,T)/R)^{|\alpha|+|\rho|}\langle\xi\rangle^{-|\tilde\alpha|-|\tilde\rho|}$$
$$\text{if } \langle\xi\rangle\geq R|\alpha| \text{ and } \langle\eta\rangle\geq R|\rho|,$$

where $\widehat C(S,T)$ does not depend on R. Put

$$v(\zeta,\eta)={}^tU_0^{-1}\zeta-{}^tU_0^{-1}\,{}^tT_0\eta, \tag{2.63}$$
$$Z(\zeta,\eta,w,y)=Z(v(\zeta,\eta)-{}^tU_0^{-1}\widetilde\nabla_y\widetilde T(w,w+y,\eta);w,w+y),$$
$$p_\kappa(\xi,w,\eta)=\sum_{j=1}^\infty\tilde g^R(\xi,\eta)\psi_j^R(\eta)\sum_{|\gamma|<j}\frac{1}{\gamma!}\Big[(-\partial_\zeta)^\gamma D_y^\gamma$$
$$\times\Big\{p_\kappa(\xi,w,v(\zeta,\eta),w+y,\eta;Z(\zeta,\eta,w,y)-v(\zeta,\eta))|\det U_0|^{-1}$$
$$\times\det\frac{\partial Z}{\partial z}(v(\zeta,\eta)-{}^tU_0^{-1}\widetilde\nabla_y\widetilde T(w,w+y,\eta);w,w+y)\Big\}\Big]_{y=0,\zeta=0}.$$

We note that $p_\kappa(\xi,w,\eta)$ also depends on the choice of R.

Theorem 2.4.4 *There are positive constants $R(S,T,U,A_1,\varepsilon_0)$, $\kappa_0(S,T,U,A_1,\varepsilon_0)$, $\kappa(S,T,U,A_1,\varepsilon_0)$, $\delta(S,T,U,A_1,\varepsilon_0)$ and $\delta'(S,T,U,A_1,\varepsilon_0)$ such that*

$$p_{S,-U,U,T}(D_x,w,D_w,y,D_z)=p_{\kappa\,S,T}(D_x,y,D_z)+Q\quad\text{on } \mathcal{S}_\infty(\boldsymbol{R}^{n''}),$$

and $Q : \mathcal{S}_{\varepsilon_2}(\boldsymbol{R}^{n''}) \to \mathcal{S}_{\varepsilon_1}(\boldsymbol{R}^{n'})$ *and* $Q : \mathcal{S}_{-\varepsilon_2}{}'(\boldsymbol{R}^{n''}) \to \mathcal{S}_{-\varepsilon_1}{}'(\boldsymbol{R}^{n'})$ *are continuous if*

$$\begin{cases} R_0 \geq R(S,T,U,A_1,\varepsilon_0), \quad \kappa = \kappa_0(S,T,U,A_1,\varepsilon_0), \\ R = \kappa(S,T,U,A_1,\varepsilon_0)R_0, \\ \max\{\delta_1 + c_1(S) + \varepsilon_1, \delta_2, \delta_3 + c_1(T) - \varepsilon_2, c_1^+(U) + c_1^-(U), \\ \qquad c_1(T) + c_1^-(T)\} \leq \delta(S,T,U,A_1,\varepsilon_0)/R_0, \\ \max\{c(T), c(U)\} \leq \delta'(S,T,U,A_1,\varepsilon_0), \end{cases} \tag{2.64}$$

where $c_1^-(T)$ *is a constant satisfying* $\mathrm{Im}\, T(y,\eta) \leq c_1^-(T)|\eta| + C_1^-(T)$ *for* $(y,\eta) \in \Omega' \cap \{|\eta| \geq 1\}$ *with some constant* $C_1^-(T)$. *Moreover, there are positive constants* $C_j(S,T,U,p)$ ($j \in \boldsymbol{Z}_+$), $C(S,T,U,A_1,\varepsilon_0)$, $A(S,T,U,A_1,\varepsilon_0)$, $B(S,T,U,A_1,\varepsilon_0)$ *and* $\delta''(S,T,U,A_1,\varepsilon_0,\varepsilon)$ ($\varepsilon > 0$) *such that*

$$\begin{aligned} &\left|\partial_\xi^{\alpha+\tilde\alpha} D_y^{\beta+\tilde\beta} \partial_\eta^{\rho+\tilde\rho} p_\kappa(\xi,y,\eta)\right| \leq C_{|\tilde\alpha|+|\tilde\beta|+|\tilde\rho|}(S,T,U,p) \\ &\quad \times (A(S,T,U,A_1,\varepsilon_0)/R_0)^{|\alpha|+|\beta|+|\rho|} \langle\xi\rangle^{m_1-|\tilde\alpha|} \langle\eta\rangle^{m_2+m_3-|\tilde\rho|+|\beta|+2} \\ &\quad \times \exp[\delta_1\langle\xi\rangle + \delta_0(S,T,U,A_1,\delta_2,\delta_3,\varepsilon_0,\varepsilon,R_0)\langle\eta\rangle] \end{aligned} \tag{2.65}$$

if (2.64) *is satisfied,*

$$\max\{c(T), c(U)\} \leq \delta''(S,T,U,A_1,\varepsilon_0,\varepsilon), \tag{2.66}$$

$\varepsilon > 0$, $\langle\xi\rangle \geq R|\alpha|$ *and* $\langle\eta\rangle \geq C(S,T,U,A_1,\varepsilon_0)R_0(|\beta| + |\rho|)$, *where*

$$\begin{aligned} &\delta_0(S,T,U,A_1,\delta_2,\delta_3,\varepsilon_0,\varepsilon,R_0) = 4c_2(T)(\delta_2)_+/c_3(U) \\ &\quad -c_3(T)(\delta_2)_-/(4c_2(U)) + \delta_3 + \varepsilon/R_0 + B(S,T,U,A_1,\varepsilon_0)/R_0^2. \end{aligned}$$

Remark (i) It follows from the definition of p_κ that supp $p_\kappa \subset \{(\xi,y,\eta) \in \boldsymbol{R}^{n'} \times \boldsymbol{R}^n \times \boldsymbol{R}^{n''};\ |\xi| \leq 3c_2(T)|\eta|/c_3(S)$ and $|\xi|^2 + |\eta|^2 \geq 4\}$. So we can write $p_\kappa(\xi,y,\eta) = p_1(\xi,y,\eta) + p_2(\xi,y,\eta)$ so that $p_{2\,S,T}$ is a regularizer like Q, supp $p_1 \subset \{(\xi,y,\eta);\ c_3(T)|\eta|/(8c_2(S)) \leq |\xi| \leq 3c_2(T)|\eta|/c_3(S)$ and $|\xi|, |\eta| \geq 1\}$ and

$$\begin{aligned} &\left|\partial_\xi^{\alpha+\tilde\alpha} D_y^{\beta^1+\beta^2+\tilde\beta} \partial_\eta^{\rho+\tilde\rho} p_1(\xi,y,\eta)\right| \\ &\leq C_{|\tilde\alpha|+|\tilde\beta|+|\tilde\rho|}(A'(S,T,U,A_1,\varepsilon_0)/R_0)^{|\alpha|+|\beta^1|+|\beta^2|+|\rho|} \langle\xi\rangle^{m_1-|\tilde\alpha|+|\beta^1|} \\ &\qquad \times \langle\eta\rangle^{m_2+m_3-|\tilde\rho|+|\beta^2|+2} \exp[\delta_1\langle\xi\rangle + \delta_0(S,T,U,A_1,\delta_2,\delta_3,\varepsilon_0,\varepsilon,R_0)\langle\eta\rangle] \end{aligned}$$

if (2.64) and (2.66) are satisfied, $\varepsilon > 0$, $\langle\xi\rangle \geq C'(S,T,U,A_1,\varepsilon_0)R_0(|\alpha| + |\beta^1|)$ and $\langle\eta\rangle \geq C'(S,T,U,A_1,\varepsilon_0)R_0(|\beta^2| + |\rho|)$. (ii) Write $q(\eta,y,\zeta,w,\xi) = p(\xi,w,\zeta,y,\eta)$, $r(\xi,w,\zeta,y,\eta) = p(-\xi,w,\zeta,y,-\eta)$, $S'(w,\xi) = S(w,-\xi)$

and $T'(y,\eta) = T(y,-\eta)$. Since ${}^t q_{T,U,-U,S}(D_z, y, D_y, w, D_x) = r_{S',-U,U,T'}(D_x, w, D_w, y, D_z)$, applying Theorem 2.4. 4 to $r(\xi, w, \zeta, y, \eta)$ we can show that

$$q_{T,U,-U,S}(D_z, y, D_y, w, D_x) = q_{\kappa\, T,S}(D_z, y, D_x) + Q'$$

on $\mathcal{S}_\infty(\boldsymbol{R}^{n'})$ and $Q' : \mathcal{S}_{-\varepsilon_1}(\boldsymbol{R}^{n'}) \to \mathcal{S}_{-\varepsilon_2}(\boldsymbol{R}^{n''})$ and $Q' : \mathcal{S}'_{\varepsilon_1}(\boldsymbol{R}^{n'}) \to \mathcal{S}'_{\varepsilon_2}(\boldsymbol{R}^{n''})$ are continuous under the same conditions as in the theorem, where $q_\kappa(\eta, y, \xi) = p_\kappa(\xi, y, \eta)$.

The idea of the proof of Theorem 2.4. 4 is simple. However, the proof is a little long. So we shall give the proof in Appendix A.1.

Corollary 2.4. 5 *Modifying* $R(S,T,U,A_1,\varepsilon_0)$, $\kappa_0(S,T,U,A_1,\varepsilon_0)$, $\kappa(S,T,U,A_1,\varepsilon_0)$, $\delta'(S,T,U,A_1,\varepsilon_0)$, $C_j(S,T,U,p)$ ($j \in \boldsymbol{Z}_+$), $C(S,T,U,A_1,\varepsilon_0)$ *and* $A(S,T,U,A_1,\varepsilon_0)$ *in* Theorem 2.4. 4 *if necessary, we have the following:* (i) *In* Theorem 2.4. 4 $p_\kappa(\xi, w, \eta)$ *can be replaced by*

$$\begin{aligned} p'_\kappa(\xi, w, \eta) \equiv {} & \sum_{j=1}^{\infty} \tilde{g}^R(\xi,\eta)\psi_j^R(\eta) \sum_{|\gamma|<j} \frac{1}{\gamma!}\Big[(-\partial_\zeta)^\gamma D_y^\gamma \\ & \times \Big\{ p_\kappa(\xi, w, \mathrm{Re}\ Z(\zeta,\eta,w,y), w+y, \eta; i\,\mathrm{Im}\ Z(\zeta,\eta,w,y)) |\det U_0|^{-1} \\ & \times \det \frac{\partial Z}{\partial z}({}^tU_0^{-1}\zeta - {}^tU_0^{-1}\tilde{\nabla}_y T(w, w+y, \eta); w, w+y) \Big\}\Big]_{y=0,\zeta=0}. \end{aligned}$$

In particular, if $U(y,\zeta)$ *and* $T(y,\eta)$ *are real-valued, then* $p_\kappa(\xi, w, \eta)$ *can be replaced by*

$$\begin{aligned} p^0(\xi, w, \eta) \equiv {} & \sum_{j=1}^{\infty} \tilde{g}^R(\xi,\eta)\psi_j^R(\eta) \sum_{|\gamma|<j} \frac{1}{\gamma!}\Big[(-\partial_\zeta)^\gamma D_y^\gamma \\ & \times \Big\{ p(\xi, w, Z(\zeta,\eta,w,y), w+y, \eta) |\det U_0|^{-1} \\ & \times \det \frac{\partial Z}{\partial z}({}^tU_0^{-1}\zeta - {}^tU_0^{-1}\tilde{\nabla}_y T(w, w+y, \eta); w, w+y) \Big\}\Big]_{y=0,\zeta=0}. \end{aligned}$$

Moreover, we can take $\delta'(S,T,U,A_1,\varepsilon_0)$ *to be independent of* A_1, *and* (2.65) *is valid for* $\varepsilon = 0$ *without assuming* (2.66). (ii) *If* $p(\xi, w, \zeta, y, \eta)$ *satisfies, with some positive constant* ε'_0,

$$\begin{aligned} & \left|\partial_\xi^{\alpha+\tilde{\alpha}} D_w^{\beta^1+\beta^2+\tilde{\beta}} \partial_\zeta^\gamma D_y^{\lambda^1+\lambda^2+\tilde{\lambda}} \partial_\eta^{\rho+\tilde{\rho}} p(\xi, w, \zeta, y, \eta)\right| \\ & \le C_{|\tilde{\alpha}|+|\tilde{\beta}|+|\tilde{\lambda}|+|\tilde{\rho}|} (A_1/R_0)^{|\alpha|+|\beta^1|+|\beta^2|+|\lambda^1|+|\lambda^2|+|\rho|} A_1^{|\gamma|} |\gamma|! \\ & \quad \times \langle\xi\rangle^{m_1-|\tilde{\alpha}|+|\beta^1|} \langle\zeta\rangle^{m_2-|\gamma|+|\beta^2|+|\lambda^1|} \langle\eta\rangle^{m_3-|\tilde{\rho}|+|\lambda^2|} \\ & \quad \times \exp[\delta_1\langle\xi\rangle + \delta_2\langle\zeta\rangle + \delta_3\langle\eta\rangle] \end{aligned} \tag{2.67}$$

when $|w-y| \le \varepsilon_0'$, $|\zeta| \ge R_0$, $|\zeta + {}^tU_0^{-1}\,{}^tT_0\eta| \le \varepsilon_0'|\eta|$, $\langle\xi\rangle \ge R_0(|\alpha|+|\beta^1|)$, $\langle\zeta\rangle \ge R_0(|\beta^2|+|\lambda^1|)$ *and* $\langle\eta\rangle \ge R_0(|\rho|+|\lambda^2|)$, *then* $p(\xi,w,\zeta,y,\eta)$ *is analytic in* ζ *at* $\zeta = -{}^tU_0^{-1}\,{}^tT_0\eta$ *when* $|w-y| \le \varepsilon_0'$ *and* $|{}^tU_0^{-1}\,{}^tT_0\eta| \ge R_0$, *and* Theorem 2.4. 4 *is still valid with* $p_\kappa(\xi,w,\eta)$ *replaced by*

$$\begin{aligned}\tilde{p}^0(\xi,w,\eta) &\equiv \sum_{j=2}^{\infty} \tilde{g}^R(\xi,\eta)\psi_j^R(\eta) \sum_{|\gamma|<j} \frac{1}{\gamma!}\Big[(-\partial_\zeta)^\gamma D_y^\gamma \\ &\times\Big\{p(\xi,w,Z(\zeta,\eta,w,y),w+y,\eta)|\det U_0|^{-1} \\ &\times \det\frac{\partial Z}{\partial z}({}^tU_0^{-1}\zeta - {}^tU_0^{-1}\widetilde{\nabla}_y T(w,w+y,\eta);w,w+y)\Big\}\Big]_{y=0,\zeta=0}.\end{aligned}$$

Moreover, $\tilde{p}^0(\xi,w,\eta)$ *satisfies*

$$\begin{aligned}\Big|\partial_\xi^{\alpha+\tilde{\alpha}} D_y^{\beta+\tilde{\beta}} \partial_\eta^{\rho+\tilde{\rho}} \tilde{p}^0(\xi,y,\eta)\Big| &\le C_{|\tilde{\alpha}|+|\tilde{\beta}|+|\tilde{\rho}|}(S,T,U,p) \\ &\times (A'(S,T,U,A_1,\varepsilon_0)/R_0)^{|\alpha|+|\beta|+|\rho|}\langle\xi\rangle^{m_1-|\tilde{\alpha}|}\langle\eta\rangle^{m_2+m_3-|\tilde{\rho}|+|\beta|+1} \\ &\times \exp[\delta_1\langle\xi\rangle + \delta_2\langle{}^tU_0^{-1}\,{}^tT_0\eta\rangle + \delta_3\langle\eta\rangle] \qquad (2.68)\end{aligned}$$

if $\langle\xi\rangle \ge R|\alpha|$ *and* $\langle\eta\rangle \ge C(S,T,U,A_1,\varepsilon_0)R_0(|\beta|+|\rho|)$. (iii) *If* $p(\xi,w,\zeta,y,\eta)$ *satisfies*

$$\begin{aligned}&\Big|\partial_\xi^{\alpha+\tilde{\alpha}} D_w^{\beta^1+\beta^2+\tilde{\beta}} \partial_\zeta^{\gamma+\tilde{\gamma}} D_y^{\lambda} \partial_\eta^{\rho+\tilde{\rho}} p(\xi,w,\zeta,y,\eta)\Big| \\ &\le C_{|\tilde{\alpha}|+|\tilde{\beta}|+|\tilde{\gamma}|+|\tilde{\rho}|}(A_1/R_0)^{|\alpha|+|\beta^1|+|\beta^2|+|\gamma|+|\rho|} A_1^{|\lambda|}|\lambda|! \\ &\quad\times \langle\xi\rangle^{m_1-|\tilde{\alpha}|+|\beta^1|}\langle\zeta\rangle^{m_2-|\tilde{\gamma}|+|\beta^2|}\langle\eta\rangle^{m_3-|\tilde{\rho}|} \\ &\quad\times \exp[\delta_1\langle\xi\rangle + \delta_2\langle\zeta\rangle + \delta_3\langle\eta\rangle] \qquad (2.69)\end{aligned}$$

when $y = w$, $\zeta = -{}^tU_0^{-1}\,{}^tT_0\eta$, $\langle\xi\rangle \ge R_0(|\alpha|+|\beta^1|)$, $\langle\zeta\rangle \ge R_0(|\beta^2|+|\gamma|)$ *and* $\langle\eta\rangle \ge R_0|\rho|$, *then* Theorem 2.4. 4 *is still valid with* $p_\kappa(\xi,w,\eta)$ *replaced by*

$$\begin{aligned}p_\kappa^0(\xi,w,\eta) &\equiv \sum_{j=1}^{\infty} \psi_j^R(\eta) \sum_{|\gamma|<j} \frac{1}{\gamma!}\Big[(-\partial_\zeta)^\gamma D_y^\gamma \\ &\times\Big\{p_\kappa(\xi,w,v(\zeta,\eta),w+y,\eta;Z(\zeta,\eta,w,y)-v(\zeta,\eta))|\det U_0|^{-1} \\ &\times \det\frac{\partial Z}{\partial z}(v(\zeta,\eta) - {}^tU_0^{-1}\widetilde{\nabla}_y\widetilde{T}(w,w+y,\eta);w,w+y)\Big\}\Big]_{y=0,\zeta=0}.\end{aligned}$$

Moreover, $p_\kappa^0(\xi,w,\eta)$ *satisfies*

$$\Big|\partial_\xi^{\alpha+\tilde{\alpha}} D_y^{\beta^1+\beta^2+\tilde{\beta}} \partial_\eta^{\rho+\tilde{\rho}} p_\kappa^0(\xi,y,\eta)\Big|$$

$$\leq C_{|\tilde{\alpha}|+|\tilde{\beta}|+|\tilde{\rho}|}(A'(S,T,U,A_1,\varepsilon_0)/R_0)^{|\alpha|+|\beta^1|+|\beta^2|+|\rho|}$$
$$\times\langle\xi\rangle^{m_1-|\tilde{\alpha}|+|\beta^1|}\langle\eta\rangle^{m_2+m_3-|\tilde{\rho}|+|\beta^2|+2}$$
$$\times\exp[\delta_1\langle\xi\rangle+\delta_0(S,T,U,A_1,\delta_2,\delta_3,\varepsilon_0,\varepsilon,R_0)\langle\eta\rangle] \qquad (2.70)$$

if $\varepsilon>0$, $\langle\xi\rangle\geq C'(S,T,U,A_1,\varepsilon_0)R_0(|\alpha|+|\beta^1|)$, $\langle\eta\rangle\geq C'(S,T,U,A_1,\varepsilon_0)R_0$ $\times(|\beta^2|+|\rho|)$ *and* (2.66) *is valid.* (iv) *If* $p(\xi,w,\zeta,y,\eta)$ *satisfies*

$$\left|\partial_\xi^{\alpha+\tilde{\alpha}}D_w^{\beta^1+\beta^2+\tilde{\beta}}\partial_\zeta^\gamma D_y^\lambda\partial_\eta^{\rho+\tilde{\rho}}p(\xi,w,\zeta,y,\eta)\right|$$
$$\leq C_{|\tilde{\alpha}|+|\tilde{\beta}|+|\tilde{\rho}|}(A_1/R_0)^{|\alpha|+|\beta^1|+|\beta^2|+|\rho|}A_1^{|\gamma|+|\lambda|}|\gamma|!|\lambda|!$$
$$\times\langle\xi\rangle^{m_1-|\tilde{\alpha}|+|\beta^1|}\langle\zeta\rangle^{m_2-|\gamma|+|\beta^2|}\langle\eta\rangle^{m_3-|\tilde{\rho}|}\exp[\delta_1\langle\xi\rangle+\delta_2\langle\zeta\rangle+\delta_3\langle\eta\rangle]$$

when $|w-y|\leq\varepsilon_0'$, $|\zeta|\geq R_0$, $|\zeta+{}^tU_0^{-1}\,{}^tT_0\eta|\leq\varepsilon_0'|\eta|$, $\langle\xi\rangle\geq R_0(|\alpha|+|\beta^1|)$, $\langle\zeta\rangle\geq R_0|\beta^2|$ *and* $\langle\eta\rangle\geq R_0|\rho|$, *then* Theorem 2.4.4 *is still valid with* $p_\kappa(\xi,w,\eta)$ *replaced by*

$$p(\xi,w,\eta)\equiv\sum_{j=2}^{\infty}\psi_j^R(\eta)\sum_{|\gamma|<j}\frac{1}{\gamma!}\Big[(-\partial_\zeta)^\gamma D_y^\gamma$$
$$\times\Big\{p(\xi,w,Z(\zeta,\eta,w,y),w+y,\eta)|\det U_0|^{-1}$$
$$\times\det\frac{\partial Z}{\partial z}({}^tU_0^{-1}\zeta-{}^tU_0^{-1}\tilde{\nabla}_yT(w,w+y,\eta);w,w+y)\Big\}\Big]_{y=0,\zeta=0}.$$

Moreover, $p(\xi,w,\eta)$ *satisfies*

$$\left|\partial_\xi^{\alpha+\tilde{\alpha}}D_y^{\beta^1+\beta^2+\tilde{\beta}}\partial_\eta^{\rho+\tilde{\rho}}p(\xi,y,\eta)\right|\leq C_{|\tilde{\alpha}|+|\tilde{\beta}|+|\tilde{\rho}|}(S,T,U,p)$$
$$\times(A'(S,T,U,A_1,\varepsilon_0)/R_0)^{|\alpha|+|\beta^1|+|\beta^2|+|\rho|}\langle\xi\rangle^{m_1-|\tilde{\alpha}|+|\beta^1|}$$
$$\times\langle\eta\rangle^{m_2+m_3-|\tilde{\rho}|+|\beta^2|+1}\exp[\delta_1\langle\xi\rangle+\delta_2\langle{}^tU_0^{-1}\,{}^tT_0\eta\rangle+\delta_3\langle\eta\rangle]$$

if $\langle\xi\rangle\geq C'(S,T,U,A_1,\varepsilon_0)R_0(|\alpha|+|\beta^1|)$ *and* $\langle\eta\rangle\geq C'(S,T,U,A_1,\varepsilon_0)R_0\times$ $(|\beta^2|+|\rho|)$.

Corollary 2.4.5 will become evident if one reads the proof of Theorem 2.4.4 . So we shall give an outline of the proof in Appendix A.2. We can make the results in Theorem 2.4.4 a little more precise. In the case of pseudodifferential operators we shall give improved versions. Assume that $n'=n''=n$ and $p(\xi,w,\zeta,y,\eta)\in C^\infty(\boldsymbol{R}^n\times\boldsymbol{R}^n\times\boldsymbol{R}^n\times\boldsymbol{R}^n\times\boldsymbol{R}^n)$ satisfies (P–2) and (P–3). For $u\in\mathcal{S}_\infty(\boldsymbol{R}^n)$ we can define

$$p(D_x,w,D_w,y,D_y)u(x)$$
$$=\mathcal{F}_\xi^{-1}\Big[\lim_{\nu\downarrow 0}(2\pi)^{-2n}\int\Big(\int\Big(\int\Big(\int e^{-\nu|\zeta|^2}e^{iw\cdot(\zeta-\xi)+iy\cdot(\eta-\zeta)}$$
$$\times p(\xi,w,\zeta,y,\eta)\hat{u}(\eta)\,d\eta\Big)\,dy\Big)\,d\zeta\Big)\,dw\Big](x)$$

if $R_0 \gg 1$ and $\delta_2 \le 1/(3R_0)$ (see Lemma 2.4. 1).

Theorem 2.4. 6 (i) *Let $0 < \varepsilon \le 1$. There are symbols $p_{0,\varepsilon}(\xi, y, \eta)$ and $q_{0,\varepsilon}(\xi, y, \eta)$, positive constants $A_1' \equiv A_1'(A_1, \varepsilon)$, $A_1''(A_1, \varepsilon_0)$ and $R(A_1, \varepsilon_0, \varepsilon, \kappa)$ ($\kappa > 0$), which are polynomials of A_1 of degree 1, and $C > 0$ such that*

$$p(D_x, w, D_w, y, D_y) = p_{0,\varepsilon}(D_x, y, D_y) + q_{0,\varepsilon}(D_x, y, D_y) \quad \text{on } \mathcal{S}_\infty$$

if $R_0 \ge R(A_1, \varepsilon_0, \varepsilon, 1)$ and $\delta_2 \le 1/R_0$,

$$\begin{aligned} &\left|\partial_\xi^{\alpha+\tilde\alpha} D_y^{\beta^1+\beta^2+\tilde\beta} \partial_\eta^{\gamma+\tilde\gamma} p_{0,\varepsilon}(\xi, y, \eta)\right| \\ &\le C_{|\tilde\alpha|+|\tilde\beta|+|\tilde\gamma|,\varepsilon} (A_1/R_0)^{|\alpha|} ((3A_1/2 + \varepsilon/\delta)/R_0)^{|\beta^1|+|\beta^2|} \\ &\quad \times (A_1'/R_0)^{|\gamma|} \langle\xi\rangle^{m_1-|\tilde\alpha|+|\beta^1|} \langle\eta\rangle^{m_2+m_3-|\tilde\gamma|+|\beta^2|} \\ &\quad \times \exp[\delta_1\langle\xi\rangle + (\delta_2 + \varepsilon|\delta_2|/2 + \delta_3 + \delta)\langle\eta\rangle] \end{aligned} \tag{2.71}$$

if $\langle\xi\rangle \ge R_0(|\alpha| + |\beta^1|)$, $\langle\eta\rangle \ge 2R_0(|\beta^2| + |\gamma|)$, $\langle\eta\rangle \ge 4R_0|\gamma|$ and $\delta > 0$,

$$\begin{aligned} &\left|\partial_\xi^\alpha D_y^{\beta+\tilde\beta} \partial_\eta^\gamma q_{0,\varepsilon}(\xi, y, \eta)\right| \le C_{|\alpha|+|\tilde\beta|+|\gamma|,\varepsilon_0,\varepsilon,R_0} (\widetilde{A}_1''/R_0)^{|\beta|} \\ &\quad \times \langle\xi\rangle^{m_1-|\alpha|+|\beta|} \langle\eta\rangle^{m_3} \exp[\delta_1\langle\xi\rangle - \kappa\langle\eta\rangle/R_0] \end{aligned}$$

if $\kappa > 0$, $R_0 \ge R(A_1, \varepsilon_0, \varepsilon, \kappa)$, $\langle\xi\rangle \ge R_0|\beta|$ and $\max\{4(\delta_2)_+ + \delta_3, 4\delta_2 + 2|\delta_2| + 2\delta_3\} \le \kappa/R_0$, *where $\widetilde{A}_1'' = CR_0/\kappa + A_1''(A_1, \varepsilon_0)$. Put*

$$\begin{aligned} p(\xi, y, \eta) &= \sum_{j=0}^\infty \phi_j^{4R_0}(\eta) \sum_{|\gamma|=j} \frac{1}{\gamma!} \partial_\zeta^\gamma D_w^\gamma p(\xi, y, \eta+\zeta, y+w, \eta)\Big|_{w=0,\,\zeta=0}, \\ q_\varepsilon(\xi, y, \eta) &= p_{0,\varepsilon}(\xi, y, \eta) - p(\xi, y, \eta). \end{aligned}$$

Then we have

$$\begin{aligned} &\left|\partial_\xi^{\alpha+\tilde\alpha} D_y^{\beta+\tilde\beta} \partial_\eta^{\gamma+\tilde\gamma} p(\xi, y, \eta)\right| \le C_{|\tilde\alpha|+|\tilde\beta|+|\tilde\gamma|} (A_1/R_0)^{|\alpha|} \\ &\quad \times (2A_1/R_0)^{|\beta|} ((\widehat{C}/4 + 2A_1)/R_0)^{|\gamma|} \langle\xi\rangle^{m_1-|\tilde\alpha|} \langle\eta\rangle^{m_2+m_3-|\tilde\gamma|+|\beta|} \\ &\quad \times \exp[\delta_1\langle\xi\rangle + (\delta_2 + \delta_3 + nA_1^2/R_0^2)\langle\eta\rangle] \end{aligned} \tag{2.72}$$

if $\langle\xi\rangle \ge R_0|\alpha|$, $\langle\eta\rangle \ge 2R_0|\beta|$ and $\langle\eta\rangle \ge 6R_0|\gamma|$, and

$$\begin{aligned} &\left|\partial_\xi^\alpha D_y^{\beta+\tilde\beta} \partial_\eta^\gamma q_\varepsilon(\xi, y, \eta)\right| \le C_{|\alpha|+|\tilde\beta|+|\gamma|,\kappa} (5A_1/(2R_0))^{|\beta|} \langle\xi\rangle^{m_1-|\alpha|} \\ &\quad \times \langle\eta\rangle^{m_2+m_3-|\gamma|+|\beta|} \exp[\delta_1\langle\xi\rangle - \kappa\langle\eta\rangle/R_0] \end{aligned} \tag{2.73}$$

if $\kappa > 0$, $R_0 \ge R'(A_1, \varepsilon_0, \varepsilon, \kappa)$, $\langle \eta \rangle \ge 4R_0|\beta|$ *and* $\delta_2 + \varepsilon|\delta_2|/2 + \delta_3 \le \kappa/R_0$, *where* $R'(A_1, \varepsilon_0, \varepsilon, \kappa)$ *is a positive constant and a polynomial of* A_1 *of degree* 2. (ii) *Assume that there is an open conic set* Γ *in* $\boldsymbol{R}^n \times (\boldsymbol{R}^n \setminus \{0\})$ *such that*

$$\begin{aligned} &\left|\partial_\xi^{\alpha+\tilde{\alpha}} D_w^{\beta^1+\beta^2+\tilde{\beta}} \partial_\zeta^{\gamma+\tilde{\gamma}} D_y^\lambda \partial_\eta^{\rho+\tilde{\rho}} p(\xi, w, \zeta, y, \eta)\right| \\ &\le C_{|\tilde{\alpha}|+|\tilde{\beta}|+|\tilde{\gamma}|+|\tilde{\rho}|}(A_1/R_0)^{|\alpha|+|\beta^1|+|\beta^2|+|\gamma|+|\rho|} \\ &\quad \times B^{|\lambda|}|\lambda|!\langle \xi \rangle^{m_1-|\tilde{\alpha}|+|\beta^1|}\langle \zeta \rangle^{m_2-|\tilde{\gamma}|+|\beta^2|}\langle \eta \rangle^{m_3-|\tilde{\rho}|} \\ &\quad \times \exp[\delta_1\langle \xi \rangle + \delta_2\langle \zeta \rangle + \delta_3\langle \eta \rangle] \end{aligned} \tag{2.74}$$

if $(w, \eta) \in \Gamma$, $|w - y| \le \varepsilon_0$, $|\zeta - \eta| \le \varepsilon_0'|\eta|$, $\langle \xi \rangle \ge R_0(|\alpha| + |\beta^1|)$, $\langle \zeta \rangle \ge R_0(|\gamma| + |\beta^2|)$ *and* $\langle \eta \rangle \ge R_0|\rho|$, *where* $\varepsilon_0' > 0$. *Let* $0 < \varepsilon \le \min\{1, \varepsilon_0'\}$. *Then we have*

$$\begin{aligned} &\left|\partial_\xi^{\alpha+\tilde{\alpha}} D_y^{\beta^1+\beta^2+\tilde{\beta}} \partial_\eta^{\gamma+\tilde{\gamma}} p(\xi, y, \eta)\right| \le C_{|\tilde{\alpha}|+|\tilde{\beta}|+|\tilde{\gamma}|}(A_1/R_0)^{|\alpha|} \\ &\quad \times ((A_1 + 3B)/R_0)^{|\beta^1|+|\beta^2|}((\widehat{C}/4 + 2A_1)/R_0)^{|\gamma|}\langle \xi \rangle^{m_1-|\tilde{\alpha}|+|\beta^1|} \\ &\quad \times \langle \eta \rangle^{m_2+m_3-|\tilde{\gamma}|+|\beta^2|} \exp[\delta_1\langle \xi \rangle + (\delta_2 + \delta_3)\langle \eta \rangle] \end{aligned} \tag{2.75}$$

if $(y, \eta) \in \Gamma$, $R_0 \ge 8nA_1B$, $\langle \xi \rangle \ge R_0(|\alpha| + |\beta^1|)$, $\langle \eta \rangle \ge 2R_0|\beta^2|$ *and* $\langle \eta \rangle \ge 6R_0|\gamma|$. *Modifying* $\widetilde{A}_1''$ *there are* $R_j(B, \varepsilon_0, \varepsilon, \kappa) > 0$ ($j = 1, 2$, $\kappa > 0$) *such that the* $R_j(B, \varepsilon_0, \varepsilon, \kappa)$ *are polynomials of* B *of degree* 1 *and*

$$\begin{aligned} &\left|\partial_\xi^\alpha D_y^{\beta^1+\beta^2+\tilde{\beta}} \partial_\eta^\gamma q_\varepsilon(\xi, y, \eta)\right| \le C_{|\alpha|+|\tilde{\beta}|+|\gamma|, \varepsilon_0, \varepsilon, R_0}(\widetilde{A}_1''/R_0)^{|\beta^1|+|\beta^2|} \\ &\quad \times \langle \xi \rangle^{m_1-|\alpha|+|\beta^1|}\langle \eta \rangle^{m_2+m_3-|\gamma|+|\beta^2|} \exp[\delta_1\langle \xi \rangle - \kappa\langle \eta \rangle/R_0] \end{aligned} \tag{2.76}$$

if $(y, \eta) \in \Gamma$, $\kappa > 0$, $R_0 \ge R_1(B, \varepsilon_0, \varepsilon, \kappa)A_1$, $R_0 \ge R_2(B, \varepsilon_0, \varepsilon, \kappa)$, $\langle \xi \rangle \ge R_0|\beta^1|$, $\langle \eta \rangle \ge 4R_0|\beta^2|$ *and* $\delta_2 + \varepsilon|\delta_2|/2 + \delta_3 \le \kappa/R_0$.

Remark (i) Applying Theorem 2.3. 3 one can prove that $q_{0,\varepsilon}(D_x, y, D_y)$ is a regularizer like Q in Theorem 2.4. 4 if one takes κ (and, therefore, R_0) sufficiently large. Similarly, $q_\varepsilon(D_x, y, D_y)$ is a regularizer like Q in Theorem 2.4. 4 if (2.74) with $\Gamma = \boldsymbol{R}^n \times (\boldsymbol{R}^n \setminus \{0\})$ is satisfied and $\kappa \gg 1$. (ii) It follows from the proof of Theorem 2.4. 6 that the condition (P–3) on p can be replaced by the following condition (P–4):

(P–4) The estimates

$$\begin{aligned} &\left|\partial_\xi^\alpha D_w^{\beta+\tilde{\beta}} \partial_\zeta^\gamma D_y^{\lambda+\tilde{\lambda}} \partial_\eta^\rho p(\xi, w, \zeta, y, \eta)\right| \\ &\le C_{|\alpha|+|\tilde{\beta}|+|\gamma|+|\tilde{\lambda}|+|\rho|}(A_1/R_0)^{|\beta|+|\lambda|}\langle \xi \rangle^{m_1-|\alpha|+|\beta|}\langle \zeta \rangle^{m_2-|\gamma|} \\ &\quad \times \langle (\zeta, \eta) \rangle^{|\lambda|}\langle \eta \rangle^{m_3-|\rho|} \exp[\delta_1\langle \xi \rangle + \delta_2\langle \zeta \rangle + \delta_3\langle \eta \rangle] \end{aligned}$$

hold if $|w-y| \le \varepsilon_0$, $\langle\xi\rangle \ge R_0|\beta|$ and $\langle(\zeta,\eta)\rangle \ge R_0|\lambda|$.

We shall give the proof of Theorem 2.4.6 in Appendix A.3. Let $p(x,\xi,y,\eta) \in C^\infty(\boldsymbol{R}^n \times \boldsymbol{R}^n \times \boldsymbol{R}^n \times \boldsymbol{R}^n)$ satisfy the following estimates:

(P–2)′ The estimates

$$\begin{aligned}
&\left|D_x^{\beta+\tilde\beta}\partial_\xi^{\alpha+\tilde\alpha}D_y^{\lambda+\tilde\lambda}\partial_\eta^{\rho+\tilde\rho}p(x,\xi,y,\eta)\right| \\
&\le C_{|\tilde\alpha|+|\tilde\beta|+|\tilde\lambda|+|\tilde\rho|}(A_1/R_0)^{|\alpha|+|\beta|+|\lambda|+|\rho|}\langle\xi\rangle^{m_1-|\tilde\alpha|+|\beta|} \\
&\quad\times\langle\eta\rangle^{m_2-|\tilde\rho|+|\lambda|}\exp[\delta_1\langle\xi\rangle+\delta_2\langle\eta\rangle]
\end{aligned}$$

hold if $\langle\xi\rangle \ge R_0(|\alpha|+|\beta|)$ and $\langle\eta\rangle \ge R_0(|\lambda|+|\rho|)$.

(P–3)′ $\delta_1 = 0$ or there is $\varepsilon_0 > 0$ such that

$$\begin{aligned}
&\left|D_x^{\beta+\tilde\beta}\partial_\xi^{\alpha+\tilde\alpha}D_y^{\lambda^1+\lambda^2+\tilde\lambda}\partial_\eta^{\rho+\tilde\rho}p(x,\xi,y,\eta)\right| \\
&\le C_{|\tilde\alpha|+|\tilde\beta|+|\tilde\lambda|+|\tilde\rho|}(A_1/R_0)^{|\alpha|+|\beta|+|\lambda^1|+|\lambda^2|+|\rho|}\langle\xi\rangle^{m_1-|\tilde\alpha|+|\beta|+|\lambda^1|} \\
&\quad\times\langle\eta\rangle^{m_2-|\tilde\rho|+|\lambda^2|}\exp[\delta_1\langle\xi\rangle+\delta_2\langle\eta\rangle]
\end{aligned}$$

if $|x-y| < \varepsilon_0$, $\langle\xi\rangle \ge R_0(|\alpha|+|\beta|+|\lambda^1|)$ and $\langle\eta\rangle \ge R_0(|\lambda^2|+|\rho|)$.

For $u \in \mathcal{S}_\infty$ we can define

$$\begin{aligned}
p(x,D_x,y,D_y)u(x) &= \lim_{\nu\downarrow 0}(2\pi)^{-2n}\int e^{-\nu|\xi|^2}e^{ix\cdot\xi} \\
&\quad\times\Big(\int\Big(\int e^{iy\cdot(\eta-\xi)}p(x,\xi,y,\eta)\hat u(\eta)\,d\eta\Big)\,dy\Big)\,d\xi
\end{aligned}$$

if $R_0 \gg 1$ and $\delta_1 \le 1/(3R_0)$ (see Lemma 2.4.1). Similarly we have the following

Corollary 2.4.7 (i) *Let* $0 < \varepsilon \le 1$. *There are symbols* $p_{0,\varepsilon}(x,\xi)$ *and* $q_{0,\varepsilon}(x,\xi)$ *and* $R(A_1,\varepsilon_0,\varepsilon) > 0$ *such that* $R(A_1,\varepsilon_0,\varepsilon)$ *is a polynomial of* A_1 *of degree* 1,

$$p(x,D_x,y,D_y) = p_{0,\varepsilon}(x,D_x) + q_{0,\varepsilon}(x,D_x) \quad \text{on } \mathcal{S}_\infty$$

if $R_0 \ge R(A_1,\varepsilon_0,\varepsilon)$ *and* $\delta_1 \le 1/R_0$, *and*

$$\begin{aligned}
&\left|p_{0,\varepsilon(\beta+\tilde\beta)}^{(\alpha+\tilde\alpha)}(x,\xi)\right| \le C_{|\tilde\alpha|+|\tilde\beta|,\varepsilon}((\widehat{C}_*/(4\varepsilon)+5A_1/2)/R_0)^{|\alpha|+|\beta|} \\
&\quad\times\langle\xi\rangle^{m_1+m_2-|\tilde\alpha|+|\beta|}\exp[(\delta_1+\varepsilon|\delta_1|/2+\delta_2)\langle\xi\rangle]
\end{aligned}$$

if $\langle\xi\rangle \ge 4R_0(|\alpha|+|\beta|)$,

$$\left|q_{0,\varepsilon(\beta)}^{(\alpha)}(x,\xi)\right| \le C_{|\alpha|+|\beta|,\varepsilon}\exp[-\langle\xi\rangle/R_0]$$

if $R_0 \ge R(A_1,\varepsilon_0,\varepsilon)$ *and* $\max\{4(\delta_1)_+ + \delta_2, 3\delta_1/2\} \le 1/R_0$, *where* $R(A_1,\varepsilon_0,\varepsilon)$ *does not depend on* ε_0 *if* $\delta_1 = 0$. *Put*

$$p(x,\xi) = \sum_{j=0}^{\infty}\phi_j^{4R_0}(\xi)\sum_{|\gamma|=j}\frac{1}{\gamma!}\partial_\eta^\gamma D_y^\gamma p(x,\xi+\eta,x+y,\xi)\Big|_{y=0,\,\eta=0},$$

$$q_\varepsilon(x,\xi) = p_{0,\varepsilon}(x,\xi) - p(x,\xi).$$

Then we have

$$\left|p_{(\beta+\tilde\beta)}^{(\alpha+\tilde\alpha)}(x,\xi)\right| \le C_{|\tilde\alpha|+|\tilde\beta|}((\widehat{C}/4+2A_1)/R_0)^{|\alpha|}(2A_1/R_0)^{|\beta|} \times\langle\xi\rangle^{m_1+m_2-|\tilde\alpha|+|\beta|}\exp[(\delta_1+\delta_2+nA_1^2/R_0^2)\langle\xi\rangle]$$

if $\langle\xi\rangle \ge 6R_0|\alpha|$ *and* $\langle\xi\rangle \ge 2R_0|\beta|$, *and*

$$\left|q_{\varepsilon(\beta)}^{(\alpha)}(x,\xi)\right| \le C_{|\alpha|+|\beta|,\varepsilon}\langle\xi\rangle^{m_1+m_2-|\alpha|}\exp[-\langle\xi\rangle/R_0]$$

if $R_0 \ge R'(A_1,\varepsilon_0,\varepsilon)$ *and* $\delta_1+\varepsilon|\delta_1|/2+\delta_2 \le 1/R_0$, *where* $R'(A_1,\varepsilon_0,\varepsilon)$ *is a positive constant and a polynomial of* A_1 *of degree* 2 *and does not depend on* ε_0 *if* $\delta_1 = 0$. (ii) *Assume that there is an open conic set* Γ *in* $\boldsymbol{R}^n \times (\boldsymbol{R}^n\setminus\{0\})$ *such that*

$$\left|D_x^\beta\partial_\xi^{\alpha+\tilde\alpha}D_y^{\lambda+\tilde\lambda}\partial_\eta^{\rho+\tilde\rho}p(x,\xi,y,\eta)\right| \le C_{|\tilde\alpha|+|\tilde\lambda|+|\tilde\rho|}(A_1/R_0)^{|\alpha|+|\lambda|+|\rho|} \times B^{|\beta|}|\beta|!\langle\xi\rangle^{m_1-|\tilde\alpha|}\langle\eta\rangle^{m_2-|\tilde\rho|+|\lambda|}\exp[\delta_1\langle\xi\rangle+\delta_2\langle\eta\rangle] \tag{2.77}$$

if $(x,\eta)\in\Gamma$, $\langle\xi\rangle \ge R_0|\alpha|$ *and* $\langle\eta\rangle \ge R_0(|\lambda|+|\rho|)$,

$$\left|D_x^\beta\partial_\xi^{\alpha+\tilde\alpha}D_y^{\lambda^1+\lambda^2+\tilde\lambda}\partial_\eta^{\rho+\tilde\rho}p(x,\xi,y,\eta)\right| \le C_{|\tilde\alpha|+|\tilde\lambda|+|\tilde\rho|}(A_1/R_0)^{|\alpha|+|\lambda^1|+|\lambda^2|+|\rho|}B^{|\beta|}|\beta|!\langle\xi\rangle^{m_1-|\tilde\alpha|+|\lambda^1|} \times\langle\eta\rangle^{m_2-|\tilde\rho|+|\lambda^2|}\exp[\delta_1\langle\xi\rangle+\delta_2\langle\eta\rangle] \tag{2.78}$$

if $(x,\eta)\in\Gamma$, $|x-y|<\varepsilon_0$, $\langle\xi\rangle \ge R_0(|\alpha|+|\lambda^1|)$ *and* $\langle\eta\rangle \ge R_0(|\lambda^2|+|\rho|)$. *Then,*

$$\left|p_{0,\varepsilon(\beta)}^{(\alpha+\tilde\alpha)}(x,\xi)\right| \le C_{|\tilde\alpha|,\varepsilon}((\widehat{C}_*/(4\varepsilon)+2A_1)/R_0)^{|\alpha|}(B+\varepsilon/(2\delta))^{|\beta|} \times|\beta|!\langle\xi\rangle^{m_1+m_2-|\tilde\alpha|}\exp[(\delta_1+\varepsilon|\delta_1|/2+\delta_2+\delta)\langle\xi\rangle] \tag{2.79}$$

if $(x,\xi) \in \Gamma$, $\langle\xi\rangle \geq 4R_0|\alpha|$ *and* $\delta > 0$, *and*

$$\begin{aligned}\left|q_{0,\varepsilon(\beta)}^{(\alpha)}(x,\xi)\right| &\leq C_{|\alpha|,\varepsilon_0,\varepsilon,R_0}(B + 4R_0 + 12(1+\sqrt{2})/\varepsilon_0)^{|\beta|} \\ &\quad \times|\beta|!\langle\xi\rangle^{(m_1)_+ + m_2}\exp[-\langle\xi\rangle/R_0] \qquad (2.80)\end{aligned}$$

if $(x,\xi) \in \Gamma$, $R_0 \geq R(A_1, B, \varepsilon_0, \varepsilon)$ *and* $\max\{4(\delta_1)_+ + \delta_2, 2\delta_1, 4\delta_1 + 2|\delta_1| + 2\delta_2\} \leq 1/R_0$, *where* $R(A_1, B, \varepsilon_0, \varepsilon)$ *is a positive constant and a polynomial of* A_1 *and* B *of degree* 1. *Moreover, assume that*

$$\begin{aligned}\left|D_x^\beta \partial_\xi^{\alpha+\tilde{\alpha}} D_y^\lambda \partial_\eta^{\rho+\tilde{\rho}} p(x,\xi,y,\eta)\right| &\leq C_{|\tilde{\alpha}|+|\tilde{\rho}|}(A_1/R_0)^{|\alpha|+|\rho|}B^{|\beta|+|\lambda|} \\ &\quad \times|\beta|!|\lambda|!\langle\xi\rangle^{m_1-|\tilde{\alpha}|}\langle\eta\rangle^{m_2-|\tilde{\rho}|}\exp[\delta_1\langle\xi\rangle + \delta_2\langle\eta\rangle] \qquad (2.81)\end{aligned}$$

if $(x,\eta) \in \Gamma$, $|x-y| < \varepsilon_0$, $|\xi - \eta| \leq \varepsilon_0'|\eta|$, $\langle\xi\rangle \geq R_0|\alpha|$ *and* $\langle\eta\rangle \geq R_0|\rho|$, *where* $\varepsilon_0' > 0$. *Let* $0 < \varepsilon \leq \min\{1, \varepsilon_0'\}$. *Then we have*

$$\begin{aligned}\left|p_{(\beta)}^{(\alpha+\tilde{\alpha})}(x,\xi)\right| &\leq C_{|\tilde{\alpha}|}((\widehat{C}/4 + 2A_1)/R_0)^{|\alpha|}(2B)^{|\beta|}|\beta|! \\ &\quad \times\langle\xi\rangle^{m_1+m_2-|\tilde{\alpha}|}\exp[(\delta_1+\delta_2)\langle\xi\rangle] \qquad (2.82)\end{aligned}$$

if $(x,\xi) \in \Gamma$, $R_0 \geq 4nA_1B$ *and* $\langle\xi\rangle \geq 6R_0|\alpha|$, *and*

$$\begin{aligned}\left|q_{\varepsilon(\beta)}^{(\alpha)}(x,\xi)\right| &\leq C_{|\alpha|,\varepsilon,R_0}(2B + \varepsilon R_0 + 6(1+\sqrt{2})/\varepsilon_0)^{|\beta|} \\ &\quad \times|\beta|!\langle\xi\rangle^{m_1+m_2-|\alpha|}\exp[-\langle\xi\rangle/R_0] \qquad (2.83)\end{aligned}$$

if $(x,\xi) \in \Gamma$, $R_0 \geq e^{24}n(12/\varepsilon_0 + B)(\widehat{C}_*/\varepsilon + 4A_1)$ *and* $2\delta_1 + \varepsilon|\delta_1| + 2\delta_2 \leq 1/R_0$.

We shall give an outline of the proof of Corollary 2.4. 7 in Appendix A.4.

2.5 Product formula of Fourier integral operators II

To change the notations and the assumptions in Section 2.4 we begin a new section. Let Ω, Ω' and Ω'' be open conic subsets of $\boldsymbol{R}^{n''} \times (\boldsymbol{R}^{n'}\setminus\{0\})$, $\boldsymbol{R}^{n''} \times (\boldsymbol{R}^{n}\setminus\{0\})$ and $\boldsymbol{R}^{n} \times (\boldsymbol{R}^{n}\setminus\{0\})$, respectively, and let $S(w,\xi) \in \mathcal{P}(\Omega)$ and $T(w,\eta) \in \mathcal{P}(\Omega')$. We denote by $C(S)$, $A(S)$, $\cdots$ the constants C, A, $\cdots$ in ($\mathcal{P}$–0)–($\mathcal{P}$–4) for S. We also use the same notations $C(T)$,A(T), $\cdots$. We assume that $c_j(S), c_j(T) > 0$ ($j = 0, 3$). For simplicity we assume that $A_0 \equiv A(S) = B(S) = A(T) = B(T)$. Moreover, we assume that there are an $n \times n''$ real constant matrix T_0 and $c(T) \geq 0$ such that $\tilde{T}(w,\eta) \equiv T(w,\eta) - T_0 w \cdot \eta$ satisfies the following:

$(\mathcal{P}$–$5)'$ $\left|\tilde{T}^{(\rho)}_{(\beta)}(w,\eta)\right| \leq C(\tilde{T}) A_0^{|\beta|+|\rho|} |\beta|!|\rho|!\langle\eta\rangle^{1-|\rho|}$

in $\Omega' \cap \{|\eta| \geq 1\}$.

$(\mathcal{P}$–$6)'$ $|\mathrm{Im}\ \nabla_\eta T(w,\eta)| \leq c(T)$ in $\Omega' \cap \{|\eta| \geq 1\}$.

Then there are $c_1^{\pm}(T)$ and $C_1^{\pm}(T)$ such that

$$\pm \mathrm{Im}\ T(w,\eta) \geq -c_1^{\pm}(T)|\eta| - C_1^{\pm}(T) \quad \text{in } \Omega' \cap \{|\eta| \geq 1\}.$$

Let $U(y,\eta) \in C^\infty(\Omega'')$ be a symbol satisfying $(\mathcal{P}$–$0)$ and $(\mathcal{P}$–$1)$ with $A = B = A_0$ and the following conditions;

$(\mathcal{P}$–$1)'$ there is $c_0'(U) > 0$ such that

$$|\nabla_y U(y,\zeta) - \nabla_y U(y,\eta)| \geq c_0'(U)|\zeta - \eta|$$

for $(y,\zeta), (y,\eta) \in \Omega'' \cap \{|\eta| \geq 1\}$,

$(\mathcal{P}$–$2)'$ there are $c_1^{\pm}(U)$ and $C_1^{\pm}(U)$ such that

$$\pm \mathrm{Im}\ U(y,\eta) \geq -c_1^{\pm}(U)|\eta| - C_1^{\pm}(U) \quad \text{in } \Omega'' \cap \{|\eta| \geq 1\},$$

$(\mathcal{P}$–$5)''$ there are an $n \times n$ real constant nonsingular matrix U_0, a real-valued function $U_1(\eta) \in C^\infty(\tilde{\Omega}'')$ and $c(U) \geq 0$ such that

$$\left|\partial_\eta^\rho U_1(\eta)\right| \leq C(U_1) A_0^{|\rho|} |\rho|!\langle\eta\rangle^{1-|\rho|},$$
$$\left|\tilde{U}^{(\rho)}_{(\beta)}(y,\eta)\right| \leq C(\tilde{U}) A_0^{|\beta|+|\rho|} |\beta|!|\rho|!\langle\eta\rangle^{1-|\rho|},$$
$$|\nabla_\eta \tilde{U}(y,\eta)| + |(\partial^2 \tilde{U}/\partial\eta\partial y)(y,\eta)| \leq c(U)$$

if $(y,\eta) \in \Omega''$ and $|\eta| \geq 1$, where $\tilde{U}(y,\eta) = U(y,\eta) - U_0 y \cdot \eta - U_1(\eta)$, $\tilde{\Omega}'' = \{\eta \in \boldsymbol{R}^n \setminus \{0\};\ (y,\eta) \in \Omega'' \text{ for some } y \in \boldsymbol{R}^n\}$ and $C(U_1)$ and $C(\tilde{U})$ are positive constants.

Let $p(\xi, w, \zeta, y, \eta) \in C^\infty(\boldsymbol{R}^{n'} \times \boldsymbol{R}^{n''} \times \boldsymbol{R}^n \times \boldsymbol{R}^n \times \boldsymbol{R}^n)$ be a symbol satisfying the following:

(P–1) There are open conic subsets Ω_1' and Ω_1'' of Ω' and Ω'', respectively, and $\varepsilon_0 > 0$ such that

$$\{(w,\eta) \in \boldsymbol{R}^{n''} \times \boldsymbol{R}^n;\ |\zeta - \eta| \leq \varepsilon_0|\eta| \text{ for some } (w,\zeta) \in \Omega_1'\} \subset \Omega',$$
$$\{(y,\zeta) \in \boldsymbol{R}^n \times \boldsymbol{R}^n;\ |\zeta - \eta| \leq \varepsilon_0|\eta| \text{ for some } (y,\eta) \in \Omega_1''\} \subset \Omega''$$

and

$$\begin{aligned}\operatorname{supp} p \ \subset \ &\{(\xi, w, \zeta, y, \eta) \in \boldsymbol{R}^{n'} \times \Omega_1' \times \Omega_1''; \ (w, \xi) \in \Omega, \\ &(y, \zeta) \in \Omega'', \ |\xi| \geq 1, \ |\zeta| \geq 1 \text{ and } |\eta| \geq 1\}.\end{aligned}$$

(P–2) The estimates

$$\begin{aligned}&\left|\partial_\xi^{\alpha+\tilde\alpha} D_w^{\beta^1+\beta^2+\tilde\beta} \partial_\zeta^{\gamma+\tilde\gamma} D_y^{\lambda^1+\lambda^2+\tilde\lambda} \partial_\eta^{\rho+\tilde\rho} p(\xi, w, \zeta, y, \eta)\right| \\ &\leq C_{|\tilde\alpha|+|\tilde\beta|+|\tilde\gamma|+|\tilde\lambda|+|\tilde\rho|} (A_1/R_0)^{|\alpha|+|\beta^1|+|\beta^2|+|\gamma|+|\lambda^1|+|\lambda^2|+|\rho|} \langle\xi\rangle^{m_1-|\tilde\alpha|+|\beta^1|} \\ &\quad \times \langle\zeta\rangle^{m_2-|\tilde\gamma|+|\beta^2|+|\lambda^1|} \langle\eta\rangle^{m_3-|\tilde\rho|+|\lambda^2|} \exp[\delta_1\langle\xi\rangle + \delta_2\langle\zeta\rangle + \delta_3\langle\eta\rangle]\end{aligned}$$

hold if $\langle\xi\rangle \geq R_0(|\alpha| + |\beta^1|)$, $\langle\zeta\rangle \geq R_0(|\gamma| + |\beta^2| + |\lambda^1|)$ and $\langle\eta\rangle \geq R_0(|\rho| + |\lambda^2|)$, where $m_j, \delta_j \in \boldsymbol{R}$ ($j = 1, 2, 3$) and $R_0 \geq 1$.

For $u \in \mathcal{S}_\infty(\boldsymbol{R}^n)$ we can define

$$\begin{aligned}&p_{S,T,U,-U}(D_x, w, D_z, y, D_y)u(x) \\ &= \mathcal{F}_\xi^{-1}\Big[\lim_{\nu\downarrow 0}(2\pi)^{-2n} \int\Big(\int\Big(\int\Big(\int e^{-\nu|\zeta|^2} e^{iS(w,\xi)+iT(w,\zeta)+iU(y,\zeta)-iU(y,\eta)} \\ &\qquad \times p(\xi, w, \zeta, y, \eta)\hat u(\eta)\, d\eta\Big)\, dy\Big)\, d\zeta\Big)\, dw\Big](x).\end{aligned}$$

Lemma 2.5. 1 *Under the above hypotheses* $p_{S,T,U,-U}(D_x, w, D_z, y, D_y)$ *can be extended to a continuous linear operator from* $\mathcal{S}_\varepsilon(\boldsymbol{R}^n)$ *to* $\mathcal{S}_{-\delta}(\boldsymbol{R}^{n'})$ *if* $\varepsilon \geq \varepsilon(T, U, \delta_2, \delta_3)$, $\delta > \delta_1 + c_1(S)$, $R_0 \geq R(U, A_1)$ *and* $\delta_2 + c_1^+(T) + c_1^+(U) \leq 1/(3R_0)$.

Since one can prove Lemma 2.5. 1 by the same idea as in the proof of Lemma 2.4. 1 , we omit the proof of Lemma 2.5. 1 . We put

$$\tilde\nabla_\eta U(y, \eta, \zeta) := \int_0^1 (\nabla_\eta U)(y, \eta + \theta(\zeta - \eta))\, d\theta.$$

Fix $\eta, \zeta \in \boldsymbol{R}^n$ so that $|\eta - \zeta| \leq \varepsilon_0|\eta|$, $|\eta| \geq 1$ and $|\zeta| \geq 1$, and put

$$\begin{aligned}&\mathcal{U}_\eta = \{y \in \boldsymbol{R}^n; \ (y, \eta) \in \Omega_1''\}, \\ &\mathcal{U}_{\eta,\delta} = \{y \in \boldsymbol{C}^n; \ |y - \tilde y| < \delta \quad \text{for some } \tilde y \in \mathcal{U}_\eta\},\end{aligned}$$

where $\delta > 0$. Let $\Psi_\eta(y; \delta)$ be a function in $C^\infty(\boldsymbol{C}^n)$ such that $0 \leq \Psi_\eta(y; \delta) \leq 1$ and

$$\Psi_\eta(y; \delta) = \begin{cases} 1 & \text{if } y \in \mathcal{U}_{\eta,\delta}, \\ 0 & \text{if } y \notin \mathcal{U}_{\eta,2\delta}, \end{cases}$$

$$\left|\partial_{y^1}^\alpha \partial_{y^2}^\beta \Psi_\eta(y^1 + iy^2; \delta)\right| \leq C_{|\alpha|+|\beta|}\delta^{-|\alpha|-|\beta|} \quad \text{for } y^1, y^2 \in \boldsymbol{R}^n,$$

where the C_j are independent of δ. By assumption we can regard $U(y,\eta)$ and $\tilde{\nabla}_\eta U(y,\eta,\zeta)$ as analytic functions of y in the set $\{y \in \boldsymbol{C}^n;\ |y-\tilde{y}| < 1/(\sqrt{n}A_0)$ for some $\tilde{y} \in \mathcal{U}_\eta\}$. Moreover, we have

$$\left|\tilde{\nabla}_\eta \tilde{U}(y,\eta,\zeta)\right| \le c(U)(1+|y-\tilde{y}|) + 2n^{3/2}C(\tilde{U})A_0^3|y-\tilde{y}|^2 \qquad (2.84)$$

if $y \in \boldsymbol{C}^n$, $\tilde{y} \in \mathcal{U}_\eta$ and $|y-\tilde{y}| \le 1/(2\sqrt{n}A_0)$. If $0 < \delta \le 1/(4\sqrt{n}A_0)$, then the map

$$\mathcal{Z}_\delta : \boldsymbol{C}^n(\simeq \boldsymbol{R}^{2n}) \ni y \longmapsto y + \Psi_\eta(y;\delta)U_0^{-1}\tilde{\nabla}_\eta \tilde{U}(y,\eta,\zeta) \in \boldsymbol{C}^n(\simeq \boldsymbol{R}^{2n})$$

is well-defined.

Lemma 2.5. 2 (i) *There are $\delta(U) > 0$ and $\gamma_j > 0$ ($j = 1,2$) such that the map $\mathcal{Z}_\delta$ is a diffeomorphism on $\boldsymbol{C}^n$ if*

$$0 < \delta \le \delta(U), \quad c(U) \le \min\{\gamma_1\delta, \gamma_2\}/|U_0^{-1}|. \qquad (2.85)$$

(ii) *Assume that* (2.85) *is satisfied, and let $Z_\delta(z;\eta,\zeta)$ be the inverse function (map) of $\mathcal{Z}_\delta$. We put*

$$\mathcal{U}_{\eta,\zeta,\delta} = \mathcal{Z}_\delta(\mathcal{U}_\eta),$$
$$\tilde{\mathcal{U}}_{\eta,\zeta,\delta} = \{z \in \boldsymbol{C}^n;\ |z-\tilde{z}| < \delta/2 \text{ for some } \tilde{z} \in \mathcal{U}_{\eta,\zeta,\delta}\}.$$

Then, modifying $\delta(U)$ and γ_j ($j = 1,2$) if necessary, we have

$$Z_\delta(\tilde{\mathcal{U}}_{\eta,\zeta,\delta})(\equiv \{Z_\delta(z;\eta,\zeta) \in \boldsymbol{C}^n;\ z \in \tilde{\mathcal{U}}_{\eta,\zeta,\delta}\}) \subset \mathcal{U}_{\eta,\delta}.$$

Moreover, $Z_\delta(z;\eta,\zeta)$ is analytic in $z \in \tilde{\mathcal{U}}_{\eta,\zeta,\delta}$ and there are $C_0(U), A_0(U) > 0$ such that

$$\left|D_z^\beta \partial_\eta^\rho D_\zeta^\gamma Z_\delta(z;\eta,\zeta)\right| \le C_0(U)A_0(U)^{|\beta|+|\gamma|+|\rho|-1}$$
$$\times(|\beta|+|\gamma|+|\rho|-1)!(|\beta|+|\gamma|+|\rho|)^{-2}\langle\eta\rangle^{-|\gamma|-|\rho|}$$

for $|\beta|+|\gamma|+|\rho| \ge 1$, $\eta,\zeta \in \boldsymbol{R}^n$ with $|\eta-\zeta| \le \varepsilon_0|\eta|$, $|\eta| \ge 1$ and $|\zeta| \ge 1$ and $z \in \tilde{\mathcal{U}}_{\eta,\zeta,\delta}$.

The proof of Lemma 2.5. 2 is similar to that of Lemma 2.4. 2 . So we omit the proof. We take $\delta = \delta(U)$ and assume that

$$c(U) \le \min\{\gamma_1\delta, \gamma_2\}/|U_0^{-1}|.$$

Put

$$p(\xi, w, \zeta, y, \eta; u) = \sum_{j=1}^{\infty} \psi_j^{R_0}(\eta) \sum_{|\beta| \le j-1} u^\beta \partial_y^\beta p(\xi, w, \zeta, y, \eta)/\beta!$$

for $(\xi, w, \zeta, y, \eta, u) \in \boldsymbol{R}^{n'} \times \boldsymbol{R}^{n''} \times \boldsymbol{R}^n \times \boldsymbol{R}^n \times \boldsymbol{R}^n \times \boldsymbol{C}^n$. Choose $\tilde{g}^R(\xi, \eta) \in C^\infty(\boldsymbol{R}^{n'} \times \boldsymbol{R}^n)$ ($R \ge 2$) so that $0 \le \tilde{g}^R(\xi, \eta) \le 1$ and

$$\tilde{g}^R(\xi, \eta) = \begin{cases} 1 & \text{if } |\xi| \le 4c_2(T)|\eta|/c_3(S) \text{ and } |\xi|^2 + 2|\eta|^2 \ge 8, \\ 0 & \text{if } |\xi| \ge 5c_2(T)|\eta|/c_3(S) \text{ or } |\xi|^2 + 2|\eta|^2 \le 4, \end{cases}$$

$$\left|\partial_\xi^{\alpha+\tilde{\alpha}} \partial_\eta^{\rho+\tilde{\rho}} \tilde{g}^R(\xi, \eta)\right| \le C_{|\tilde{\alpha}|+|\tilde{\rho}|}(\widehat{C}(S,T)/R)^{|\alpha|+|\rho|} \langle \xi \rangle^{-|\tilde{\alpha}|-|\tilde{\rho}|}$$
$$\text{if } \langle \xi \rangle \ge R|\alpha| \text{ and } \langle \eta \rangle \ge R|\rho|,$$

where $\widehat{C}(S,T)$ does not depend on R. Taking $R = 8R_0$, we put

$$Z(z, \eta, \zeta, w) = Z_\delta(U_0^{-1}(z - \tilde{\nabla}_\eta U_1(\eta, \zeta) - \tilde{\nabla}_\eta T(w, \eta, \zeta)); \eta, \zeta),$$

$$p(\xi, w, \eta) = \sum_{j=1}^{\infty} \tilde{g}^R(\xi, \eta) \psi_j^R(\eta) \sum_{|\gamma| < j} \frac{1}{\gamma!} \Big[(-\partial_\zeta)^\gamma D_y^\gamma$$
$$\times \Big\{ p(\xi, w, \zeta, \mathrm{Re}\, Z(y, \eta, \zeta, w), \eta; i\, \mathrm{Im}\, Z(y, \eta, \zeta, w))$$
$$\times \mathrm{sgn}(\det U_0) \det \frac{\partial Z}{\partial z}(y, \eta, \zeta, w) \Big\} \Big]_{y=0, \zeta=\eta}.$$

Then we have the following

Theorem 2.5. 3 *There are positive constants $R(S, T, U, A_1, \varepsilon_0)$, $\delta(S, T, U)$ and $\delta'(T, U, A_1)$ such that*

$$p_{S,T,U,-U}(D_x, w, D_z, y, D_y) = p_{S,T}(D_x, w, D_y) + Q \quad \textit{on } \mathcal{S}_\infty(\boldsymbol{R}^n),$$

and $Q : \mathcal{S}_{\varepsilon_2}(\boldsymbol{R}^n) \to \mathcal{S}_{\varepsilon_1}(\boldsymbol{R}^{n'})$ and $Q : \mathcal{S}_{-\varepsilon_2}{}'(\boldsymbol{R}^n) \to \mathcal{S}_{-\varepsilon_1}{}'(\boldsymbol{R}^{n'})$ are continuous if

$$\begin{cases} R_0 \ge R(S, T, U, A_1, \varepsilon_0), \\ \max\{\delta_1 + c_1(S) + \varepsilon_1, \delta_2, \delta_3 + c_1^+(T) - \varepsilon_2, c_1^+(T) + c_1^+(U), \\ \quad c_1^-(T) + c_1^-(U)\} \le \delta(S, T, U)/R_0, \\ \max\{c(T), c(U)\} \le \delta'(T, U, A_1). \end{cases} \tag{2.86}$$

Moreover, there are positive constants $C_j(S, T, U, p)$ ($j \in \boldsymbol{Z}_+$), $A(S, T, U, A_1)$, $B(S, T, U, A_1)$ and $\delta''(U, A_1)$ such that

$$|\partial_\xi^{\alpha+\tilde{\alpha}} D_w^{\beta+\tilde{\beta}} \partial_\eta^{\rho+\tilde{\rho}} p(\xi, w, \eta)| \le C_{|\tilde{\alpha}|+|\tilde{\beta}|+|\tilde{\rho}|}(S, T, U, p)$$
$$\times (A(S, T, U, A_1)/R_0)^{|\alpha|+|\beta|+|\rho|} \langle \xi \rangle^{m_1 - |\tilde{\alpha}|} \langle \eta \rangle^{m_2 + m_3 - |\tilde{\rho}| + |\beta| + 2}$$
$$\times \exp[\delta_1 \langle \xi \rangle + (\delta_2 + \delta_3 + \varepsilon/R_0 + B(S, T, U, A_1)/R_0^2) \langle \eta \rangle] \tag{2.87}$$

if (2.86) *is satisfied,* $\varepsilon > 0$, $c(T) + c(U) \le \varepsilon\delta''(U, A_1)$, $\langle\xi\rangle \ge 8R_0|\alpha|$, $\langle\eta\rangle \ge 24R_0|\rho|$ *and* $\langle\eta\rangle \ge 8R_0(|\beta| + |\rho|)$. *If* $U(y,\eta)$ *and* $T(w,\eta)$ *are real-valued, then we can take* $\delta'(T, U, A_1)$ *to be independent of* A_1 *and* (2.87) *is valid for* $\varepsilon = 0$ *without the assumption* $c(T) + c(U) \le \varepsilon\delta''(U, A_1)$.

Remark Write $q(\eta, y, \zeta, w, \xi) = p(\xi, w, \zeta, y, \eta)$, and define $\rho : \mathcal{S}_\varepsilon' \mapsto \mathcal{S}_\varepsilon'$ by $\langle \rho u, \varphi\rangle = \langle u, \check{\varphi}\rangle$ for $\varphi \in \mathcal{S}_\varepsilon$, where $\check{\varphi} = \mathcal{F}_\xi^{-1}[\hat{\varphi}(-\xi)]$ and $\varepsilon \in \boldsymbol{R}$. Since ${}^t q_{-U,U,T,S}(D_y, z, D_z, w, D_x) = \rho \circ p_{S,T,U,-U}(D_x, w, D_z, y, D_y) \circ \rho$, we can show that

$$q_{-U,U,T,S}(D_y, z, D_z, w, D_x) = q_{T,S}(D_y, w, D_x) + Q'$$

on $\mathcal{S}_\infty(\boldsymbol{R}^{n'})$ and $Q' : \mathcal{S}_{-\varepsilon_1}(\boldsymbol{R}^{n'}) \to \mathcal{S}_{-\varepsilon_2}(\boldsymbol{R}^n)$ and $Q' : \mathcal{S}_{\varepsilon_1}'(\boldsymbol{R}^{n'}) \to \mathcal{S}_{\varepsilon_2}'(\boldsymbol{R}^n)$ are continuous under the same conditions as in the theorem, where $q(\eta, w, \xi) = p(\xi, w, \eta)$.

We can prove Theorem 2.5. 3 by the same argument as in the proof of Theorem 2.4. 4 . So we shall give an outline of the proof in Appendix A.5. We can also obtain the results corresponding to Corollary 2.4. 5 .

Corollary 2.5. 4 *If* $p(\xi, w, \zeta, y, \eta)$ *satisfies, with some positive constant* ε_0',

$$\begin{aligned}
&\left|\partial_\xi^{\alpha+\tilde\alpha} D_w^{\beta^1+\beta^2+\tilde\beta} \partial_\zeta^{\gamma+\tilde\gamma} D_y^\lambda \partial_\eta^{\rho+\tilde\rho} p(\xi, w, \zeta, y, \eta)\right| \\
&\le C_{|\tilde\alpha|+|\tilde\beta|+|\tilde\gamma|+|\tilde\rho|} (A_1/R_0)^{|\alpha|+|\beta^1|+|\beta^2|+|\gamma|+|\rho|} A_1^{|\lambda|} |\lambda|! \\
&\quad \times \langle\xi\rangle^{m_1-|\tilde\alpha|+|\beta^1|} \langle\zeta\rangle^{m_2-|\tilde\gamma|+|\beta^2|} \langle\eta\rangle^{m_3-|\tilde\rho|} \\
&\quad \times \exp[\delta_1\langle\xi\rangle + \delta_2\langle\zeta\rangle + \delta_3\langle\eta\rangle]
\end{aligned}$$

when $|y + U_0^{-1}\mathrm{Re}\,\nabla_\eta T(w,\eta) + U_0^{-1}\nabla_\eta U_1(\eta)| \le \varepsilon_0'$, $|\eta| \ge R_0$, $|\zeta - \eta| \le \varepsilon_0'|\eta|$, $\langle\xi\rangle \ge R_0(|\alpha| + |\beta^1|)$, $\langle\zeta\rangle \ge R_0(|\beta^2| + |\gamma|)$ *and* $\langle\eta\rangle \ge R_0|\rho|$, *then* $p(\xi, w, \zeta, y, \eta)$ *is analytic in* y *at* $y = -U_0^{-1}\mathrm{Re}\,\nabla_\eta T(w,\eta) - U_0^{-1}\nabla_\eta U_1(\eta)$ *when* $|\eta| \ge R_0$ *and* $|\zeta - \eta| \le \varepsilon_0'|\eta|$, *and* Theorem 2.5. 3 *is still valid with* $p(\xi, w, \eta)$ *replaced by*

$$\begin{aligned}
p^0(\xi, w, \eta) \equiv \sum_{j=2}^\infty \psi_j^{8R_0}(\eta) \sum_{|\gamma|<j} \frac{1}{\gamma!}\Big[(-\partial_\zeta)^\gamma D_y^\gamma \\
\times \Big\{p(\xi, w, \zeta, Z(y,\eta,\zeta,w), \eta)\, \mathrm{sgn}(\det U_0) \det \frac{\partial Z}{\partial z}(y, \eta, \zeta, w)\Big\}\Big]_{y=0,\zeta=\eta},
\end{aligned}$$

modifying $R(S, T, U, A_1, \varepsilon_0)$, $\delta(S, T, U)$ *and* $\delta'(T, U, A_1)$ *if necessary. Moreover,* $p^0(\xi, w, \eta)$ *satisfies*

$$\left|\partial_\xi^{\alpha+\tilde\alpha} D_y^{\beta^1+\beta^2+\tilde\beta} \partial_\eta^{\rho+\tilde\rho} p^0(\xi, y, \eta)\right| \le C_{|\tilde\alpha|+|\tilde\beta|+|\tilde\rho|}(S, T, U, p)$$

$$\times (A'(S,T,U,A_1)/R_0)^{|\alpha|+|\beta^1|+|\beta^2|+|\rho|} \langle\xi\rangle^{m_1-|\tilde{\alpha}|+|\beta^1|}$$
$$\times \langle\eta\rangle^{m_2+m_3-|\tilde{\rho}|+|\beta^2|+1} \exp[\delta_1\langle\xi\rangle + (\delta_2+\delta_3)\langle\eta\rangle]$$

if (2.86) *is satisfied,* $\langle\xi\rangle \geq R_0(|\alpha|+|\beta^1|)$ *and* $\langle\eta\rangle \geq 2R_0(|\beta^2|+|\rho|)$.

2.6 Pseudolocal properties

In this section we shall show that pseudodifferential operators have pseudolocal properties. Let $n, n' \in \boldsymbol{N}$, and let Ω be an open conic subset of $\boldsymbol{R}^n \times (\boldsymbol{R}^{n'} \setminus \{0\})$. Let $S(x,\eta) \in \mathcal{P}(\Omega; A_0, B_0, c_0(S), c_1(S), c_2(S), c_3(S))$. We assume that $c_j(S) > 0$ ($j = 0,3$) and that a symbol $p(x,\eta) \in C^\infty(\boldsymbol{R}^n \times \boldsymbol{R}^{n'})$ satisfies

$$\left|p^{(\alpha)}_{(\beta+\tilde{\beta})}(x,\eta)\right| \leq C_{|\alpha|+|\tilde{\beta}|}(B/R_0)^{|\beta|}\langle\eta\rangle^{m+|\beta|}e^{\delta\langle\eta\rangle}$$

if $\langle\eta\rangle \geq R_0|\beta|$, and $\operatorname{supp} p \subset \Omega \cap \{|\eta| \geq 1\}$, where $R_0 \geq 1$, $B \geq 0$ and $m, \delta \in \boldsymbol{R}$. It follows from Theorem 2.3. 3 that $p_S(x,D_y)$ maps $\mathcal{S}_{\varepsilon_1}{}'(\boldsymbol{R}^{n'})$ to $\mathcal{S}_{\varepsilon_2}{}'(\boldsymbol{R}^n)$ if $\nu > 1$, $\varepsilon_2 = \nu(\delta + c_1(S) + \varepsilon_1)_+/c_3(S)$, $\delta + c_1(S) + \varepsilon_1 \leq 1/R_0$, $R_0 \geq R_1(S,\nu)B$ and $R_0 \geq R_2(S,\nu)$, where $R_j(S,\nu)$ ($j = 1,2$) are positive constants depending on S and ν. We note that $p_S(x,D_y)$ maps $\mathcal{S}_{-\delta-c_1(S)}{}'(\boldsymbol{R}^{n'})$ to $\mathcal{S}'(\boldsymbol{R}^n)$ without any assumptions on R_0.

Theorem 2.6. 1 *Assume that X is an open subset of $\boldsymbol{R}^n$, $\varepsilon > 0$ and*

$$\overline{X}_\varepsilon \cap \operatorname{cl}(\{x \in \boldsymbol{R}^n;\ (x,\eta) \in \operatorname{supp} p \text{ for some } \eta\}) = \emptyset,$$

where X_ε denotes the ε-neighborhood of X and $\operatorname{cl}(S) = \overline{S}$. *Then there are $c_*(S) > 0$ and $R_j(S) > 0$ ($j = 1,2$) such that*

$$\operatorname{supp} p_S(x,D_y)u \cap X = \emptyset \quad \text{for } u \in \mathcal{S}_{\varepsilon_1}{}'$$

if

$$\begin{cases} R_0 \geq R_1(S)B, \quad R_0 \geq R_2(S), \\ \delta + c_1(S) + \varepsilon_1 < \min\{c_*(S)\varepsilon, 1/R_0\}. \end{cases} \tag{2.88}$$

Remark (i) If $\delta + c_1(S) + \varepsilon_1 = 0$, then $\operatorname{supp} p_S(x,D_y)u \cap X = \emptyset$ for $u \in \mathcal{S}_{\varepsilon_1}{}'$ without any conditions on R_0. (ii) If $n' = n$ and $S(x,\eta) = x\cdot\eta$, then one can take $R_1(S) = 2e\max\{n, 4\sqrt{n}\}$ and $R_2(S) = 0$. (iii) The condition (2.88) can be replaced by the condition

$$\begin{cases} R_0 \geq R_1(S)B, \\ \delta + c_1(S) + \varepsilon_1 < \min\{c_*(S)\varepsilon, 1/R_0, 1/R_2(S)\} \end{cases}$$

(see the remark of Theorem 2.3. 3).

Proof Assume that $\delta + c_1(S) + \varepsilon_1 > 0$ and $u \in \mathcal{S}_{\varepsilon_1}{}'$. Let us first prove that there are $\tilde{R}_j(S,\rho) > 0$ ($j = 1, 2$) such that

$$\begin{aligned} &\langle D_x\rangle^k \exp[-\rho\langle D_x\rangle] p_S(x, D_y)u(x) \\ &= (2\pi)^{-n}\langle \exp[-\varepsilon_1\langle\eta\rangle]\hat{u}(\eta), f_{k,\rho}(x,\eta)\rangle_\eta \end{aligned} \tag{2.89}$$

if $\rho > (\delta + c_1(S) + \varepsilon_1)/c_3(S)$, $R_0 \geq \tilde{R}_1(S,\rho)B$, $R_0 \geq \tilde{R}_2(S,\rho)$ and $\delta + c_1(S) + \varepsilon_1 < 1/R_0$, where $k = 0, 1$,

$$\begin{aligned} f_{k,\rho}(x,\eta) &= (2\pi)^{-n} \int \exp[i(x-w)\cdot\xi + iS(w,\eta) + \varepsilon_1\langle\eta\rangle] p(w,\eta) \\ &\qquad \times \langle x - w\rangle_B^{-2M} \langle D_\xi\rangle_B^{2M} (\langle\xi\rangle^k e^{-\rho\langle\xi\rangle})\, d\xi dw, \end{aligned}$$

$\langle x\rangle_B = \langle Bx/(4+4\sqrt{2})\rangle$ and $M \geq [n/2]+1$. If $u \in \mathcal{S}_\infty$, then we can easily prove that (2.89) is valid. Write

$$\begin{aligned} \langle\eta\rangle^\ell D_\eta^\gamma f_{k,\rho}(x,\eta) &= (2\pi)^{-n} \int \exp[i(x-w)\cdot\xi + iS(w,\eta) + \varepsilon_1\langle\eta\rangle] \\ &\qquad \times g_{k,\ell,\rho,\varepsilon_1,\gamma,M}(x,\eta,w,\xi)\, d\xi dw, \end{aligned}$$

where $\ell \in \mathbf{Z}_+$, $\gamma \in \mathbf{Z}_+^{n'}$, $M \geq [(n+|\gamma|)/2] + 1$ and

$$\begin{aligned} &g_{k,\ell,\rho,\varepsilon_1,\gamma,M}(x,\eta,w,\xi) = \sum_{\gamma'\leq\gamma} \binom{\gamma}{\gamma'} (-i)^{|\gamma|-|\gamma'|} t_{\varepsilon_1,\gamma'}(w,\eta) \\ &\qquad \times p^{(\gamma-\gamma')}(w,\eta)\langle\eta\rangle^\ell \langle x-w\rangle_B^{-2M}\langle D_\xi\rangle_B^{2M}(\langle\xi\rangle^k e^{-\rho\langle\xi\rangle}), \\ &t_{\varepsilon_1,\alpha}(w,\eta) = \exp[-iS(w,\eta) - \varepsilon_1\langle\eta\rangle] D_\eta^\alpha \exp[iS(w,\eta) + \varepsilon_1\langle\eta\rangle]. \end{aligned}$$

We can show by induction on $|\alpha|$ that

$$\left|\partial_\eta^\gamma D_w^\beta t_{\varepsilon_1,\alpha}(w,\eta)\right| \leq C_{|\alpha|,|\gamma|,\varepsilon_1}(S)(2B_0)^{|\beta|}|\beta|!\langle\eta\rangle^{-|\gamma|}\langle w\rangle^{|\alpha|} \tag{2.90}$$

in $\Omega \cap \{|\eta| \geq 1\}$ (see, also, Lemma 2.1. 7). Fix $\nu > 1$. If $|\xi| \geq c_3(S)|\eta|/\nu$, then

$$\begin{aligned} &\left|\exp[(c_1(S)+\varepsilon_1)\langle\eta\rangle] g_{k,\ell,\rho,\varepsilon_1,\gamma,M}(x,\eta,w,\xi)\right| \\ &\leq C_{\rho,\varepsilon_1,|\gamma|,M,B,\nu}(p,S)\langle\xi\rangle^{(m+\ell)_+ + k}\langle x\rangle^{|\gamma|} \\ &\qquad \times \langle x-w\rangle^{-(2M-|\gamma|)} \exp[(\nu(\delta + c_1(S) + \varepsilon_1)/c_3(S) - \rho)\langle\xi\rangle]. \end{aligned}$$

Therefore, putting

$$\begin{aligned} G^1_{k,\ell,\rho,\varepsilon_1,\gamma,M}(x,\eta) &\equiv \int_{|\xi|\geq c_3(S)|\eta|/\nu} \exp[i(x-w)\cdot\xi + iS(w,\eta) + \varepsilon_1\langle\eta\rangle] \\ &\qquad \times g_{k,\ell,\rho,\varepsilon_1,\gamma,M}(x,\eta,w,\xi)\, d\xi dw, \end{aligned}$$

we see that $G^1_{k,\ell,\rho,\varepsilon_1,\gamma,M}(x,\eta)$ is continuous and satisfies

$$\left|G^1_{k,\ell,\rho,\varepsilon_1,\gamma,M}(x,\eta)\right| \le C_{k,\ell,\rho,\varepsilon_1,|\gamma|,M,B,\nu}(p,S)\langle x\rangle^{|\gamma|} \tag{2.91}$$

if $\rho > \nu(\delta + c_1(S) + \varepsilon_1)/c_3(S)$. It is easy to see that

$$\left|D_w^\beta g_{k,\ell,\rho,\varepsilon_1,\gamma,M}(x,\eta,w,\xi)\right| \le C_{\rho,\varepsilon_1,|\gamma|,M,B}(p,S)$$
$$\times((B+2B_0)/R_0)^{|\beta|}\langle w\rangle^{|\gamma|}\langle x-w\rangle^{-2M}\langle\eta\rangle^{m+\ell+|\beta|}\langle\xi\rangle^k e^{\delta\langle\eta\rangle-\rho\langle\xi\rangle}$$

if $\langle\eta\rangle \ge R_0|\beta|$. Applying Lemma 2.1. 5 , we have

$$\left|K^j g_{k,\ell,\rho,\varepsilon_1,\gamma,M}(x,\eta,w,\xi)\right| \le C'_{\rho,\varepsilon_1,|\gamma|,M,B}(p,S)$$
$$\times(\Gamma(S,B,\nu)/R_0)^j\langle w\rangle^{|\gamma|}\langle x-w\rangle^{-2M}\langle\eta\rangle^{m+\ell}\langle\xi\rangle^k e^{\delta\langle\eta\rangle-\rho\langle\xi\rangle} \tag{2.92}$$

if $|\xi| \le c_3(S)|\eta|/\nu$, $j \in \boldsymbol{Z}_+$ and $R_0 j \le \langle\eta\rangle$ ($\le R_0(j+1)$), where

$$^tK = |\nabla_w S(w,\eta) - \xi|^{-2}\sum_{j=1}^n \left(\overline{\partial_{w_j}S(w,\eta)} - \xi_j\right)D_{w_j},$$
$$C(S,\nu) = 2C(S) + c_3(S)/(\nu B_0), \qquad c(S,\nu) = (1-1/\nu)^2 c_3(S)^2,$$
$$\Gamma(S,B,\nu)$$
$$= 2^5 n B_0 C(S,\nu)\max\{B+2B_0, 120nB_0^3C(S,\nu)^2/c(S,\nu)\}/c(S,\nu).$$

Therefore, putting

$$G^2_{k,\ell,\rho,\varepsilon_1,\gamma,M}(x,\eta) \equiv \int_{|\xi|\le c_3(S)|\eta|/\nu} \exp[i(x-w)\cdot\xi + iS(w,\eta) + \varepsilon_1\langle\eta\rangle]$$
$$\times g_{k,\ell,\rho,\varepsilon_1,\gamma,M}(x,\eta,w,\xi)\,d\xi dw,$$

we see that $G^2_{k,\ell,\rho,\varepsilon_1,\gamma,M}(x,\eta)$ is continuous and satisfies

$$\left|G^2_{k,\ell,\rho,\varepsilon_1,\gamma,M}(x,\eta)\right| \le C_{k,\ell,\rho,\varepsilon_1,|\gamma|,M,B,\nu,R_0}(p,S)\langle x\rangle^{|\gamma|} \tag{2.93}$$

if $R_0 \ge e\Gamma(S,B,\nu)$ and $\delta+c_1(S)+\varepsilon_1 < 1/R_0$. (2.93), together with (2.91), yields $f_{k,\rho}(x,\eta) \in C(\boldsymbol{R}^n_x; S(\boldsymbol{R}^n_\eta))$ if $\rho > (\delta + c_1(S) + \varepsilon_1)/c_3(S)$, $R_0 \ge e\Gamma(S,B,\nu_\rho)$ and $\delta+c_1(S)+\varepsilon_1 < 1/R_0$, where $\nu_\rho = c_3(S)\rho/(2(\delta+c_1(S)+\varepsilon_1))+1/2$. Therefore, (2.89) is valid with $\widetilde{R}_1(S,\rho) = 2^6 enB_0C(S,\nu_\rho)/c(S,\nu_\rho)$ and $\widetilde{R}_2(S,\rho) = 2^7 enB_0C(S,\nu_\rho)\max\{B_0, 30nB_0^3C(S,\nu_\rho)^2/c(S,\nu_\rho)\}/c(S,\nu_\rho)$. Moreover, the right-hand side of (2.89) is continuous in x. Choose $\{\chi_j(t)\} \subset C^\infty(\boldsymbol{R})$ so that $0 \le \chi_j(t) \le 1$, $\chi_j(t) = 0$ ($t \le 1/2$), $\chi_j(t) = 1$ ($t \ge 3/4$) and

$$\left|\partial_t^\mu \chi_j(t)\right| \le C_*(\widehat{C}_* j)^\mu \quad \text{if } \mu \le j.$$

Then, by induction we have

$$\left|\partial_\xi^\alpha \chi_j^{(\mu)}(\langle\xi\rangle/(c_3(S)\langle\eta\rangle))\right| \le C_*(\widehat{C}_* j)^\mu (\widehat{C}'_* j)^{|\alpha|}\langle\xi\rangle^{-|\alpha|}$$

if $|\alpha|+\mu \le j$, where $\widehat{C}'_* = (1+3\widehat{C}_*/4)(1+\sqrt{2})$. Write

$$\langle\eta\rangle^\ell D_\eta^\gamma f_{k,\rho}(x,\rho) = \sum_{j=1}^{\infty}\sum_{h=1}^{2} f_{k,\ell,\rho,\varepsilon_1,\gamma,M,j}^{R,h}(x,\eta),$$

where $R>0$, $\{\psi_j^R\}$ is the family defined before Lemma 2.2. 4 and

$$\begin{aligned} f_{k,\ell,\rho,\varepsilon_1,\gamma,M,j}^{R,1}(x,\eta) &= (2\pi)^{-n}\int \exp[i(x-w)\cdot\xi + iS(w,\eta)+\varepsilon_1\langle\eta\rangle] \\ &\quad \times g_{k,\ell,\rho,\varepsilon_1,\gamma,M}(x,\eta,w,\xi)\psi_j^R(\xi)\chi_j(\langle\xi\rangle/(c_3(S)\langle\eta\rangle))\,d\xi dw, \\ f_{k,\ell,\rho,\varepsilon_1,\gamma,M,j}^{R,2}(x,\eta) &= (2\pi)^{-n}\int \exp[i(x-w)\cdot\xi + iS(w,\eta)+\varepsilon_1\langle\eta\rangle] \\ &\quad \times g_{k,\ell,\rho,\varepsilon_1,\gamma,M}(x,\eta,w,\xi)\psi_j^R(\xi)(1-\chi_j(\langle\xi\rangle/(c_3(S)\langle\eta\rangle)))\,d\xi dw. \end{aligned}$$

We note that $f_{k,\ell,\rho,\varepsilon_1,\gamma,M,j}^{R,h}(x,\eta)$ ($h=1,2$, $j\in \boldsymbol{N}$) are analytic in $x\in\{x\in \boldsymbol{C}^n;\ \mathrm{Re}\ x\in X,\ |\mathrm{Im}\ x|^2<16(1+\sqrt{2})^2/B^2+\varepsilon^2\}$. In fact, Re $(1+B^2(x-w)^2/(4+4\sqrt{2})^2)>0$ if $x\in\boldsymbol{C}^n$, Re $x\in X$, $|\mathrm{Im}\ x|^2<16(1+\sqrt{2})^2/B^2+\varepsilon^2$, $w\in\boldsymbol{R}^n$ and $|\mathrm{Re}\ x-w|\ge\varepsilon$. Let $x\in\boldsymbol{C}^n$ satisfy Re $x\in X$, and let L be a differential operator defined by

$$L=(|\mathrm{Re}\ x-w|^2+|\mathrm{Im}\ x|^2)^{-1}\sum_{\mu=1}^{n}(w_\mu-\overline{x}_\mu)D_{\xi_\mu}.$$

($w\ne$ Re x). It is easy to see that

$$\begin{aligned} &\left|L^j\left\{\psi_j^R(\xi)\chi_j(\langle\xi\rangle/(c_3(S)\langle\eta\rangle))\langle D_\xi\rangle_B^{2M}(\langle\xi\rangle^k e^{-\rho\langle\xi\rangle})\right\}\right| \\ &\le C_{M,B,R}(A(\varepsilon)/R)^j j^k \exp[-\langle\xi\rangle/(3R)] \end{aligned}$$

if $A(\varepsilon)=e\sqrt{n}(\widehat{C}+\widehat{C}'_*+7(1+\sqrt{2}))/\varepsilon$, Re $x\in X$ and $(w,\zeta)\in\mathrm{supp}\ p$ for some $\zeta\in\boldsymbol{R}^{n'}$, since

$$\left|\partial_\xi^\alpha e^{-\rho\langle\xi\rangle}\right| \le (3(1+\sqrt{2}))^{|\alpha|}|\alpha|!\langle\xi\rangle^{-|\alpha|}e^{-\rho\langle\xi\rangle}\sum_{\mu=0}^{|\alpha|}(\rho\langle\xi\rangle)^\mu/\mu!.$$

Thus we have

$$\left|f_{k,\ell,\rho,\varepsilon_1,\gamma,M,j}^{R,1}(x,\eta)\right| \le C_{\ell,\varepsilon_1,|\gamma|,M,B,R,\varepsilon}(p,S)\langle\mathrm{Re}\ x\rangle^{|\gamma|}(A(\varepsilon)/R)^j \qquad (2.94)$$

if $\mathrm{Re}\, x \in X$, $|\mathrm{Im}\, x| \le \rho' \le 4(1+\sqrt{2})/B$ and $2(\delta+c_1(S)+\varepsilon_1)/c_3(S)+\rho' \le 1/(4R)$. It follows from Lemma 2.1.5 that

$$\begin{aligned}&\left|K^\mu g_{k,\ell,\rho,\varepsilon_1,\gamma,M}(x,\eta,w,\xi)\right| \le C_{\rho,\varepsilon_1,|\gamma|,M,B,\varepsilon}(p,S)\\ &\quad \times(\Gamma(S,B)/R_0)^\mu \langle w\rangle^{|\gamma|}\langle \mathrm{Re}\, x - w\rangle^{-2M}\langle\eta\rangle^{m+\ell}\langle\xi\rangle^k e^{\delta\langle\eta\rangle-\rho\langle\xi\rangle}\end{aligned} \tag{2.95}$$

if $R_0 \ge 2\sqrt{2}$, $\mu \in \boldsymbol{Z}_+$, $\langle\eta\rangle \ge R_0\mu$, $\mathrm{Re}\, x \in X$, $|\mathrm{Im}\, x| \le 4(1+\sqrt{2})/B$ and $\chi_j(\langle\xi\rangle/(c_3(S)\langle\eta\rangle)) \ne 1$, where

$$\begin{aligned}&C'(S) = 2C(S) + 3c_3(S)/(4B_0),\\ &\Gamma(S,B)\\ &= 2^{11}nB_0C'(S)\max\{B+2B_0, 2^9\cdot 15nB_0^3C'(S)^2/c_3(S)^2\}/c_3(S)^2\end{aligned}$$

(see, also, (2.92)). Here we have used the fact that $|\nabla_w S(w,\eta) - \xi| \ge c_3(S)\langle\eta\rangle/8$ if $(w,\eta) \in \Omega$, $\langle\eta\rangle \ge R_0 \ge 2\sqrt{2}$ and $\chi_j(\langle\xi\rangle/(c_3(S)\langle\eta\rangle)) \ne 1$. (2.95) yields

$$\begin{aligned}&\left|f^{R,2}_{k,\ell,\rho,\varepsilon_1,\gamma,M,j}(x,\eta)\right| \le C_{\rho,\varepsilon_1,|\gamma|,M,B,R,\varepsilon}(p,S)\\ &\quad\times\langle \mathrm{Re}\, x\rangle^{|\gamma|}\langle\eta\rangle^{m+\ell+k+n+2}j^{-2}(e\Gamma(S,B)/R_0)^\mu\\ &\quad\times\exp[(\delta+c_1(S)+\varepsilon_1+3c_3(S)(\rho'-\rho)_+/4-1/R_0)\langle\eta\rangle]\end{aligned} \tag{2.96}$$

if $R_0 \ge 2\sqrt{2}$, $\mu \in \boldsymbol{Z}_+$, $R_0\mu \le \langle\eta\rangle \le R_0(\mu+1)$, $\mathrm{Re}\, x \in X$ and $|\mathrm{Im}\, x| \le \rho' \le 4(1+\sqrt{2})/B$. Therefore, it follows from (2.94) and (2.96) that $\sum_{j=1}^\infty\sum_{h=1}^2 f^{R,h}_{k,\ell,\rho,\varepsilon_1,\gamma,M,j}(x,\eta)$ converges uniformly in $K\times\boldsymbol{R}^{n'}$ for every compact subset K of $\{x\in\boldsymbol{C}^n;\ \mathrm{Re}\, x\in X \text{ and } |\mathrm{Im}\, x|\le\rho'\}$ if

$$\begin{cases} R_0 \ge \max\{2\sqrt{2}, 2e\Gamma(S,B), e\Gamma(S,B,3/2), R_1(S,2)B, R_2(S,2)\},\\ 0<\delta+c_1(S)+\varepsilon_1\\ <\min\{1/R_0, c_3(S)/(2^5A(\varepsilon)), 2(1+\sqrt{2})c_3(S)/B\}\end{cases} \tag{2.97}$$

and R, ρ and ρ' satisfy

$$\begin{cases} 2(\delta+c_1(S)+\varepsilon_1)/c_3(S) < \rho < \rho' \le 4(1+\sqrt{2})/B,\\ \delta+c_1(S)+\varepsilon_1+3c_3(S)(\rho'-\rho)/4 \le 1/R_0,\\ R\ge 2A(\varepsilon),\\ 2(\delta+c_1(S)+\varepsilon_1)/c_3(S)+\rho' \le 1/(4R).\end{cases} \tag{2.98}$$

Here we remark that there are R, ρ and ρ' satisfying (2.98) if (2.97) is satisfied. In particular, $\langle\eta\rangle^\ell D^\gamma_\eta f_{k,\rho}(x,\eta)$ can be regarded as an analytic function of x in $\{x\in\boldsymbol{C}^n;\ \mathrm{Re}\, x\in X \text{ and } |\mathrm{Im}\, x|\le\rho'\}$ for $\eta\in\boldsymbol{R}^{n'}$ and satisfies the estimate

$$\left|\langle\eta\rangle^\ell D^\gamma_\eta f_{k,\rho}(x,\eta)\right| \le C_{\rho,\varepsilon_1,|\gamma|,B,R,\varepsilon,\ell}(p,S)\langle\mathrm{Re}\, x\rangle^{|\gamma|} \tag{2.99}$$

there, if (2.97) and (2.98) are valid. Now we assume that (2.97) and (2.98) are satisfied. Let $\{u_j\}$ be a sequence in $\mathcal{S}_\infty$ such that $u_j \to u$ in $\mathcal{S}_{\varepsilon_1}{}'$. (2.89) and (2.99) show that $\{\langle D_x\rangle^k \exp[-\rho\langle D_x\rangle] p_S(x, D_y)u_j(x)\}$ converges uniformly to $\langle D_x\rangle^k \exp[-\rho\langle D_x\rangle] p_S(x, D_y)u(x)$ in $\{x \in \boldsymbol{C}^n;\ \mathrm{Re}\, x \in X \cap K$ and $|\mathrm{Im}\, x| \le \rho'\}$ for any compact subset K of $\boldsymbol{R}^n$. Put

$$V_j(x, x_{n+1}) = \mathcal{H}(p_S(x, D_y)u_j(x))(x, x_{n+1}),$$
$$V(x, x_{n+1}) = \mathcal{H}(p_S(x, D_y)u(x))(x, x_{n+1}),$$

where $\mathcal{H}$ denotes the transformation defined in Section 1.1. Since $p_S(x, D_y)u \in \mathcal{S}_{2(\delta+c_1(S)+\varepsilon_1)/c_3(S)}{}'$, $V(x, x_{n+1})$ is real analytic for $|x_{n+1}| > 2(\delta + c_1(S) + \varepsilon_1)/c_3(S)$ and satisfies $(1 - \Delta_{x,x_{n+1}})V(x, x_{n+1}) = 0$ there. On the other hand, we have proved that $\langle D_x\rangle^k V(x, \rho)$ ($k = 0, 1$) can be continued analytically to $\{x \in \boldsymbol{C}^n;\ |\mathrm{Im}\, x| < \rho'\}$. Applying Lemma 1.2. 4 to the Cauchy problem

$$\begin{cases} (1 - \Delta_{x,x_{n+1}})v(x, x_{n+1}) = 0, \\ v(x, \rho) = V(x, \rho), \quad (\partial v/\partial x_{n+1})(x, \rho) = -\langle D_x\rangle V(x, \rho), \end{cases}$$

we can show that $V(x, x_{n+1})$ can be continued analytically from $\boldsymbol{R}^n \times (0, \infty)$ to a neighborhood of $X \times [0, \infty)$. Similarly, the $V_j(x, x_{n+1})$ can be regarded as analytic functions in a neighborhood of $X \times [0, \infty)$. Moreover, $V_j(x, +0) = p_S(x, D_y)u_j(x)$ in $\mathcal{S}'$ ($j \in \boldsymbol{N}$). Since $\{\langle D_x\rangle^k V_j(x, \rho)\}$ ($k = 0, 1$) converge to $\langle D_x\rangle^k V(x, \rho)$ locally uniformly in $\{x \in \boldsymbol{C}^n;\ \mathrm{Re}\, x \in X$ and $|\mathrm{Im}\, x| \le \rho'\}$, respectively, applying the same arguments as in the proofs of Lemmas 1.2. 3 and 1.2. 4 we can show that $\{V_j(x, +0)\}$ converges to $V(x, +0)$ locally uniformly in X. Since $V_j(x, +0) \in \mathcal{S}'$ and $\mathrm{supp}\, V_j(x, +0) \subset \mathrm{cl}(\{x \in \boldsymbol{R}^n;\ (x, \eta) \in \mathrm{supp}\, p$ for some $\eta\})$, we have $V(x, +0) = 0$ for $x \in X$ and, therefore, $\mathrm{supp}\, p_S(x, D_y)u \cap X = \emptyset$. □

Corollary 2.6. 2 *Assume that* $\mathrm{supp}\, p \subset \Omega \cap \{|\eta| \ge 1\}$ *and for any* $\varepsilon > 0$,

$$\left|p^{(\alpha)}_{(\beta+\tilde{\beta})}(x, \eta)\right| \le C_{|\alpha|+|\tilde{\beta}|,\varepsilon}(p)(B/R_0)^{|\beta|}\langle\eta\rangle^{|\beta|}e^{\varepsilon\langle\eta\rangle}$$

if $\langle\eta\rangle \ge R_0|\beta|$, *and that* $c_1(S) = 0$. *Then there is* $R(S) > 0$ *such that* $p_S(x, D_y)$ *maps* $\mathcal{F}_0$ *to* $\mathcal{F}_0$ *continuously and*

$$\mathrm{supp}\, p_S(x, D_y)u \subset \mathrm{cl}(\{x \in \boldsymbol{R}^n;\ (x, \eta) \in \mathrm{supp}\, p \textit{ for some } \eta\})$$

for $u \in \mathcal{F}_0$ *if* $R_0 \ge R(S)B$.

Corollary 2.6. 3 *Let $p(\xi, y, \eta)$ be a symbol in $S^{m_1,m_2,\delta_1,\delta_2}(R_0, A)$. We assume that X is an open subset of $\mathbf{R}^n$, $\varepsilon > 0$ and that $p(\xi, y, \eta) = 0$ if $y \in X_\varepsilon$, $|\xi - \eta| \le \varepsilon_0|\eta|$ and $\langle\xi\rangle \ge R_0$, where ε_0 is a positive constant. Then there are $R_0(A, \varepsilon_0, \varepsilon) > 0$ and $\delta(\varepsilon) > 0$ such that $R_0(A, \varepsilon_0, \varepsilon)$ is a polynomial of A of degree 1 and $p(D_x, y, D_y)u$ is analytic in X for $u \in \mathcal{F}_0$ if $R_0 \ge R_0(A, \varepsilon_0, \varepsilon)$, $\max\{4(\delta_1)_+ + \delta_2, 2\delta_1, 4\delta_1 + 2|\delta_1| + 2\delta_2, 4\delta_1 + 2|\delta_1| + 4\delta_2\} < 1/R_0$ and $\delta_1 + |\delta_1|/2 + \delta_2 < \delta(\varepsilon)$.*

Proof From Corollary 2.4. 7 there are symbols $p(x, \xi)$ and $q(x, \xi)$ and $R(A, \varepsilon_0) > 0$ such that $R(A, \varepsilon_0)$ is a polynomial of A of degree 1 and

$$\begin{aligned}
&p(D_x, y, D_y) = p(x, D) + q(x, D) \quad \text{on } \mathcal{S}_\infty,\\
&p(x, \xi) \in S^{m_1+m_2,\delta}(4R_0, \widehat{C}_*/\varepsilon_0' + 10A),\\
&\left|q^{(\alpha)}_{(\beta)}(x, \xi)\right|\\
&\le C_{|\alpha|,\varepsilon_0,R_0}(4R_0 + 1)^{|\beta|}|\beta|!\langle\xi\rangle^{(m_1)_+ + m_2}\exp[-\langle\xi\rangle/R_0]
\end{aligned}$$

if

$$\left\{\begin{array}{l} R_0 \ge R(A, \varepsilon_0),\\ \max\{4(\delta_1)_+ + \delta_2, 2\delta_1, 4\delta_1 + 2|\delta_1| + 2\delta_2\} \le 1/R_0, \end{array}\right. \tag{2.100}$$

where $\delta = \delta_1 + |\delta_1|/2 + \delta_2$ and $\varepsilon_0' = \min\{1, \varepsilon_0\}$. Moreover, there is $R(A, \varepsilon_0, \varepsilon) > 0$ such that

$$\begin{aligned}
\left|p^{(\alpha)}_{(\beta)}(x, \xi)\right| \le\ & C_{|\alpha|,\varepsilon_0,R_0}(R_0 + 1)^{|\beta|}|\beta|!\\
&\times\langle\xi\rangle^{m_1+m_2-|\alpha|}\exp[-\langle\xi\rangle/R_0]
\end{aligned}$$

if $x \in X_{\varepsilon/2}$ and

$$R_0 \ge R(A, \varepsilon_0, \varepsilon), \qquad 2\delta \le 1/R_0. \tag{2.101}$$

Assume that (2.100) and (2.101) are satisfied. Theorem 2.3. 3 implies that $q(x, D)u$ is analytic if $u \in \mathcal{F}_0$. Choose $\psi^R(x, \xi) \in S^0(R, C_*, C(\varepsilon))$ ($R \ge 4$) so that $\psi^R(x, \xi) = 1$ for $x \in X_{\varepsilon/4}$ and supp $\psi^R \subset X_{\varepsilon/2} \times \mathbf{R}^n$. We put

$$\begin{aligned}
p_1^R(x, \xi) &= \psi^R(x, \xi)p(x, \xi),\\
p_2^R(x, \xi) &= (1 - \psi^R(x, \xi))p(x, \xi).
\end{aligned}$$

Then we have

$$\begin{aligned}
&\left|p^{R(\alpha)}_{1(\beta+\tilde\beta)}(x, \xi)\right| \le C_{|\alpha|+|\tilde\beta|,\varepsilon_0,R_0,R}\\
&\quad\times((2R_0 + 2 + C(\varepsilon))/R)^{|\beta|}\langle\xi\rangle^{m_1+m_2-|\alpha|+|\beta|}\exp[-\langle\xi\rangle/R_0]
\end{aligned}$$

if $\langle\xi\rangle \ge R|\beta|$, and

$$\left|p_{2(\beta+\tilde{\beta})}^{R(\alpha)}(x,\xi)\right| \le C_{|\alpha|+|\tilde{\beta}|} \times((\widehat{C}_*/\varepsilon_0' + 10A)/(4R_0) + C(\varepsilon)/R)^{|\beta|}\langle\xi\rangle^{m_1+m_2+|\beta|}\exp[\delta\langle\xi\rangle]$$

if $R \ge 4R_0$ and $\langle\xi\rangle \ge R|\beta|$. Since $p_1^R(x,D)u = (2\pi)^{-n}\langle e^{-\rho\langle\xi\rangle}\hat{u}(\xi), e^{ix\cdot\xi+\rho\langle\xi\rangle}p_1^R(x,\xi)\rangle$ for $u \in \mathcal{F}_0$ and $0 < \rho < 1/R_0$, $p_1^R(x,D)u$ is analytic in X if $u \in \mathcal{F}_0$. Applying Theorem 2.6.1 we have supp $p_2^R(x,D)u \cap X = \emptyset$ if $u \in \mathcal{F}_0$, $R \ge 4R_0$, $R \ge 2e\max\{n, 4\sqrt{n}\}(R(\widehat{C}_*/\varepsilon_0' + 10A)/(4R_0) + C(\varepsilon))$ and $\delta < \min\{c_*\varepsilon/4, 1/R\}$, which proves the corollary. □

Let X, X_j ($j = 1,2$) be open bounded subsets of $\mathbf{R}^n$ such that $X_1 \Subset X_2 \Subset X$, and let $p(x,\xi)$ be a symbol in $S^{m,\delta}(R_0,A,B)$ such that supp $p(x,\xi) \subset X_2 \times \mathbf{R}^n$,

$$\left|p_{(\beta)}^{(\alpha)}(x,\xi)\right| \le C_{|\alpha|}(B_1/R_1)^{|\beta|}\langle\xi\rangle^{m+|\beta|}\exp[\delta_1\langle\xi\rangle]$$

if $\langle\xi\rangle \ge R_1|\beta|$, and

$$\left|p_{(\beta)}^{(\alpha+\tilde{\alpha})}(x,\xi)\right| \le C_{|\tilde{\alpha}|}(A/R_0)^{|\alpha|}B'^{|\beta|}|\beta|!\langle\xi\rangle^{m-|\tilde{\alpha}|}\exp[\delta\langle\xi\rangle]$$

if $x \in X_1$, $\langle\xi\rangle \ge R_0|\alpha|$ and $|\xi| \ge 1$, where $\delta \ge \delta_1 \ge 0$ and $R_0, R_1 \ge 1$. Assume that $u \in \mathcal{F}_{\delta_0}$ is analytic in a neighborhood of $\overline{X}$, where $\delta_0 \ge 0$. Put $u_\rho = e^{-\rho\langle D\rangle}u$ for $0 < \rho \le 1+\delta_0$. By definition $u_\rho(x)$ can be regarded as a function in $C^\infty(\overline{X})$ by analytic continuation, and there are positive constants $C(u)$ and $A(u)$ such that

$$\left|D^\beta u_\rho(x)\right| \le C(u)A(u)^{|\beta|}|\beta|! \tag{2.102}$$

for $x \in X$ and $0 < \rho \le 1+\delta_0$. Then we have the following

Lemma 2.6.4 *There are $c_j > 0$ ($1 \le j \le 3$) such that $p(x,D)u$ is analytic in X_1 if $u \in \mathcal{F}_{\delta_0}$ satisfies* (2.102), *$R_1 > 2enB_1$, $R_0 \ge 12e\sqrt{n}A/\varepsilon$, $\delta_0 + \delta_1 < 1/R_1$ and $2\delta_0 + \delta < \min\{c_1, c_2\varepsilon, c_3/A(u)\}$, where the c_j depend only on n and $\varepsilon =$ dis$(X_2, \mathbf{R}^n \setminus X)$.*

Proof Let X_3 be an open subset of $\mathbf{R}^n$ such that $X_2 \Subset X_3 \Subset X$, dis$(X_2, \mathbf{R}^n \setminus X_3) \ge \varepsilon/3$ and dis$(X_3, \mathbf{R}^n \setminus X) \ge \varepsilon/3$. We choose a family $\{\chi_j\}_{j\in N}$ of functions in $C_0^\infty(X)$ so that $\chi_j(x) = 1$ in X_3 and

$$\left|D^\beta\chi_j(x)\right| \le C(C_*j/\varepsilon)^{|\beta|} \quad \text{for } |\beta| \le j, \tag{2.103}$$

where C_* (>0) depends only on n. Let $u \in \mathcal{F}_{\delta_0}$ satisfy (2.102). By Lemma 1.1.3 there is $\ell \in \boldsymbol{Z}_+$ such that

$$|u_\rho(x)| \le C_\rho(1+|x|)^\ell \quad \text{for } \rho > \delta_0, \tag{2.104}$$

where C_ρ is a positive constant depending on u and ρ. Fix $\rho > \delta_0$ so that $\rho \le \delta_0 + 1$. Since $\sum_{j=1}^\infty \psi_j^R(D)u = u$ in $\mathcal{F}_{\delta_0}$, it follows from Theorem 2.3. 3 (and its remark) that

$$p(x,D)u = \sum_{j=1}^\infty p(x,D)\psi_j^R(D)e^{\rho\langle D\rangle}\{(1-\chi_j)u_\rho + \chi_j u_\rho\} \tag{2.105}$$

in $\mathcal{F}_{2(\delta_0+\delta_1)}$ if $R_1 > 2enB_1$ and $\delta_0 + \delta_1 < 1/R_1$, where $R > 0$. By (2.102) we have

$$\left|D^\beta(\chi_j(x)u_\rho(x))\right| \le CC(u)((C_*/\varepsilon + A(u))j)^{|\beta|} \quad \text{if } |\beta| \le j.$$

This yields

$$|\mathcal{F}[\chi_j u_\rho](\xi)| \le C'(u)(1+\sqrt{n}(C_*/\varepsilon + A(u))j)^j \langle\xi\rangle^{-j}. \tag{2.106}$$

Therefore, we have

$$\begin{aligned} &\left|p(x,D)\psi_j^R(D)e^{\rho\langle D\rangle}(\chi_j u_\rho)(x)\right| \\ &\le \begin{cases} C_R(u)j^{m+n}((1+\sqrt{n}(C_*/\varepsilon + A(u)))\exp[3(\rho+\rho_0+\delta)R]/R)^j \\ \quad \text{if } \rho_0 \le 1/(2\sqrt{n}B'),\ \mathrm{Re}\, x \in X_1 \text{ and } |\mathrm{Im}\, x| \le \rho_0, \\ C_R(u)j^{m+n}((1+\sqrt{n}(C_*/\varepsilon + A(u)))\exp[3(\rho+\delta)R]/R)^j \\ \quad \text{if } x \in \boldsymbol{R}^n. \end{cases} \end{aligned} \tag{2.107}$$

Write

$$\begin{aligned} &p(x,D)\psi_j^R(D)e^{\rho\langle D\rangle}((1-\chi_j)u_\rho)(x) \\ &= (2\pi)^{-n}\int e^{i(x-y)\cdot\xi}\langle x-y\rangle^{-2M}L^j\langle D_\xi\rangle^{2M} \\ &\qquad \times (p(x,\xi)\psi_j^R(\xi)e^{\rho\langle\xi\rangle})(1-\chi_j(y))u_\rho(y)\,dyd\xi, \end{aligned}$$

where $M \ge [(n+\ell)/2]+1$ and $L = |x-y|^{-2}\sum_{k=1}^n (y_k - x_k)D_{\xi_k}$. Since

$$\begin{aligned} &\left|L^j\langle D_\xi\rangle^{2M}(p(x,\xi)\psi_j^R(\xi)e^{\rho\langle\xi\rangle})\right| \\ &\le C_{M,R}(3\sqrt{n}(A/R_0 + \widehat{C}/R + 6(1+\sqrt{2})/R)/\varepsilon)^j e^{(2\rho+\delta)\langle\xi\rangle}\langle\xi\rangle^m \end{aligned}$$

if $y \notin X_3$ and $x \in \boldsymbol{R}^n$ or if $y \notin X_3$, Re $x \in X_1$ and $|\text{Im}\, x| \le 1/(2\sqrt{n}B')$, we have

$$\left|p(x,D)\psi_j^R(D)e^{\rho\langle D\rangle}((1-\chi_j)u_\rho)(x)\right| \le \begin{cases} C_{\rho,R}(u)j^{m+n}\{3\sqrt{n}(AR/R_0+\widehat{C}+6(1+\sqrt{2})) \\ \qquad \times \exp[3(2\rho+\rho_0+\delta)R]/(\varepsilon R)\}^j \\ \quad \text{if } \rho_0 \le 1/(2\sqrt{n}B'),\ \text{Re}\, x \in X_1 \text{ and } |\text{Im}\, x| \le \rho_0, \\ C_{\rho,R}(u)j^{m+n}\{3\sqrt{n}(AR/R_0+\widehat{C}+6(1+\sqrt{2})) \\ \qquad \times \exp[3(2\rho+\delta)R]/(\varepsilon R)\}^j \\ \quad \text{if } x \in \boldsymbol{R}^n. \end{cases} \tag{2.108}$$

It follows from (2.107) and (2.108) that the right-hand side of (2.105) converges uniformly in $\boldsymbol{R}^n$ if $2\rho+\delta \le 1/(3R)$,

$$\begin{cases} R \ge \max\{2e(1+\sqrt{n}(C_*/\varepsilon + A(u))), \\ \qquad 12e\sqrt{n}(\widehat{C}+6(1+\sqrt{2}))/\varepsilon\}, \\ R_0 \ge 12e\sqrt{n}A/\varepsilon. \end{cases} \tag{2.109}$$

In particular, $p(x,D)u(x)$ is continuous in $\boldsymbol{R}^n$. We note that one can easily prove that $p(x,D)u \in C^\infty(\boldsymbol{R}^n)$. Similarly, the right-hand side of (2.105) converges uniformly in $\{x \in \boldsymbol{C}^n;\ \text{Re}\, x \in X_1 \text{ and } |\text{Im}\, x| \le \rho_0\}$ if $\rho_0 \le 1/(2\sqrt{n}B')$, $2\rho+\rho_0+\delta \le 1/(3R)$ and (2.109) is valid. Therefore, $p(x,D)u(x)$ is analytic in $\{x \in \boldsymbol{C}^n;\ \text{Re}\, x \in X_1 \text{ and } |\text{Im}\, x| < \rho_0\}$ if $\rho_0 \le 1/(2\sqrt{n}B')$, $2\rho+\rho_0+\delta \le 1/(3R)$ and (2.109) is valid. This completes the proof. □

Theorem 2.6. 5 (i) *Let X and X_j $(j=1,2)$ be bounded open subsets of $\boldsymbol{R}^n$ such that $X_1 \Subset X_2 \Subset X$. Assume that $p(x,\xi)$ is a symbol satisfying*

$$\left|p^{(\alpha)}_{(\beta+\tilde{\beta})}(x,\xi)\right| \le C_{|\alpha|+|\tilde{\beta}|}(A/R_0)^{|\beta|}\langle\xi\rangle^{m+|\beta|}e^{\delta\langle\xi\rangle}$$

if $\langle\xi\rangle \ge R_0|\beta|$ and that

$$\left|p^{(\alpha+\tilde{\alpha})}_{(\beta)}(x,\xi)\right| \le C_{|\tilde{\alpha}|}(A_1/R_1)^{|\alpha|}B^{|\beta|}|\beta|!\langle\xi\rangle^{m-|\tilde{\alpha}|}\exp[\delta_1\langle\xi\rangle]$$

if $x \in X$ and $\langle\xi\rangle \ge R_1|\alpha|$, where $m \in \boldsymbol{R}$, $\delta \ge 0$ and $\delta_1 \ge 0$. Moreover, we assume that $u \in \mathcal{F}_0$ is analytic in a neighborhood of $\overline{X}_2$. Put $\varepsilon =$ $\text{dis}(X_1, \boldsymbol{R}^n \setminus X_2)$. Then there are $c_0 > 0$ and $\delta_1(\varepsilon,u) > 0$ such that $p(x,D)u$ is analytic in X_1 if $R_0 \ge 4e\max\{n, 4\sqrt{n}\}A$, $R_1 \ge 72e\sqrt{n}A_1/\varepsilon$, $\delta < 1/R_0$, $\delta < c_0\varepsilon$ and $\delta_1 < \delta_1(\varepsilon,u)$. (ii) *Let $x^0 \in \boldsymbol{R}^n$, and let X be a neighborhood of x^0. Assume that $p(x,\xi) \in S^+(R_0,A)$, and that*

$$\left|p^{(\alpha+\tilde{\alpha})}_{(\beta)}(x,\xi)\right| \le C_{|\tilde{\alpha}|,\delta}B_\delta^{|\alpha|+|\beta|}|\alpha|!|\beta|!\langle\xi\rangle^{-|\alpha|-|\tilde{\alpha}|}e^{\delta\langle\xi\rangle}$$

if $\delta > 0$, $x \in X$ and $\langle\xi\rangle \geq R_\delta|\alpha|$, where B_δ and R_δ are positive constants depending on δ (> 0). Moreover, we assume that $u \in \mathcal{F}_0$ is analytic at x^0. Then, $p(x,D)u$ is analytic at x^0 if $R_0 \geq 4e\max\{n, 4\sqrt{n}\}A$.

Proof The assertion (ii) easily follows from the assertion (i). So we shall prove the assertion (i). It follows from Theorem 2.3. 3 that $p(x,D)u \in \mathcal{F}_{2\delta}$ if $R_0 > 2enA$ and $\delta < 1/R_0$. Choose open subsets Y_j ($j = 1,2$) of X so that $X_1 \subset\subset Y_1 \subset\subset Y_2 \subset\subset X_2$ and $\mathrm{dis}(X_1, \boldsymbol{R}^n \setminus Y_1) = \mathrm{dis}(Y_1, \boldsymbol{R}^n \setminus Y_2) = \varepsilon/3$. By assumption there are positive constants $C(u)$ and $A(u)$ such that

$$\left|D^\beta u_\rho(x)\right| \leq C(u)A(u)^{|\beta|}|\beta|!$$

for $x \in X_2$ and $0 < \rho \leq 1$. Now we assume that $R_0 > 2enA$ and $\delta < 1/R_0$. From Proposition 2.2. 3 and its proof we can choose a symbol $\Phi^R(x,\xi) \in S^0(R, C_*, \hat{c}/\varepsilon)$ ($R \geq 4$) such that $0 \leq \Phi^R \leq 1$, $\mathrm{supp}\, \Phi^R(x,\xi) \subset Y_2 \times \boldsymbol{R}^n$, $\Phi^R(x,\xi) \equiv \tilde{\Phi}^R(\xi)$ for $x \in Y_1$ and $\tilde{\Phi}^R(\xi) = 1$ if $\langle\xi\rangle \geq R$, where $\hat{c} > 0$. Put

$$\begin{aligned}
p_1(x,\xi) &= \Phi^R(x,\xi)p(x,\xi), \\
p_2(x,\xi) &= (1 - \Phi^R(x,\xi))\phi_1^R(\xi)p(x,\xi), \\
p_3(x,\xi) &= (1 - \Phi^R(x,\xi))(1 - \phi_1^R(\xi))p(x,\xi),
\end{aligned}$$

where $\phi_1^R(\xi)$ is the symbol defined in Section 2.2. Since $\mathrm{supp}\, p_3 \subset \{(x,\xi) \in \boldsymbol{R}^n \times \boldsymbol{R}^n;\ \langle\xi\rangle \leq 3R\}$ and $p_3(x,\xi) = (1 - \tilde{\Phi}^R(\xi))(1 - \phi_1^R(\xi))p(x,\xi)$ for $x \in Y_1$, $p_3(x,D)u$ is analytic in Y_1. It is easy to see that $p_2(x,\xi) = 0$ for $x \in Y_1$ and

$$\left|p_{2(\beta+\tilde{\beta})}^{(\alpha)}(x,\xi)\right| \leq C_{|\alpha|+|\tilde{\beta}|}(A/R_0 + \hat{c}/(\varepsilon R))^{|\beta|}\langle\xi\rangle^{m+|\beta|}e^{\delta\langle\xi\rangle}$$

if $R \geq R_0$ and $\langle\xi\rangle \geq R|\beta|$. From Theorem 2.6. 1 and its remark we can see that $\mathrm{supp}\, p_2(x,D)u \cap X_1 = \emptyset$ if $R_0 \geq 4e\max\{n, 4\sqrt{n}\}A$, $R \geq R_0$, $R \geq 4e\hat{c}\max\{n, 4\sqrt{n}\}/\varepsilon$ and $\delta < \min\{c_*\varepsilon/3, 1/R\}$, where c_* is equal to the constant $c_*(S)$ in Theorem 2.6. 1 with $n = n'$ and $S(x,\eta) = x\cdot\eta$. On the other hand $p_1(x,\xi)$ belongs to $S^{m,\delta_1}(\tilde{R}, \tilde{R}A_1/R_1 + \tilde{R}C_*/R, B + \tilde{R}\hat{c}/(\varepsilon R))$ for $\tilde{R} \geq \max\{R, R_1\}$ and satisfies

$$\left|p_{1(\beta)}^{(\alpha)}(x,\xi)\right| \leq C_{|\alpha|}(A/R_0 + \hat{c}/(\varepsilon R))^{|\beta|}\langle\xi\rangle^{m+|\beta|}e^{\delta\langle\xi\rangle}$$

if $R \geq R_0$ and $\langle\xi\rangle \geq R|\beta|$, and

$$\left|p_{1(\beta)}^{(\alpha+\tilde{\alpha})}(x,\xi)\right| \leq C_{|\tilde{\alpha}|}(A_1/R_1 + C_*/R)^{|\alpha|}B^{|\beta|}|\beta|!\langle\xi\rangle^{m-|\tilde{\alpha}|}\exp[\delta_1\langle\xi\rangle]$$

if $x \in Y_1$ and $\langle\xi\rangle \geq \max\{R, R_1\}|\alpha|$. Therefore, it follows from Lemma 2.6. 4 that $p_1(x,D)u$ is analytic in Y_1 if $R_0 \geq 4enA$, $R > \max\{R_0, 4en\hat{c}/\varepsilon,$

$72e\sqrt{n}C_*/\varepsilon\}$. $R_1 \geq 72e\sqrt{n}A_1/\varepsilon$, $\delta < 1/R$ and $\delta_1 < \min\{c_1, c_2\varepsilon/3, c_3/A(u)\}$, where the c_j are the constants in Lemma 2.6.4. Taking $c_0 = \min\{c_*/3, 1/(4en\hat{c}), 1/(72e\sqrt{n}C_*)\}$ and $\delta_1(\varepsilon,u) = \min\{c_1, c_2\varepsilon/3, c_3/A(u)\}$ we have the assertion (i). □

Corollary 2.6.6 *Let X and X_1 be open bounded subsets of $\boldsymbol{R}^n$ such that $X_1 \subset\subset X$, and let $p(\xi, y, \eta) \in S^{0,0,\delta_1,\delta_2}(R_0, A)$. Assume that $u \in \mathcal{F}_0$ is analytic in X. We put $\varepsilon = \mathrm{dis}(X_1, \boldsymbol{R}^n \setminus X)$. Then there are positive constants $R(A,\varepsilon)$ and $\delta(\varepsilon,u)$ such that $R(A,\varepsilon)$ is a polynomial of A of degree 1, and $p(D_x, y, D_y)u$ is analytic in X_1 if $R_0 \geq R(A,\varepsilon)$, $\max\{4\delta_1 + 2|\delta_1| + 4\delta_2, 4(\delta_1)_+ + \delta_2, 2\delta_1, 4\delta_1 + 2|\delta_1| + 2\delta_2\} < 1/R_0$ and $\delta_1 + |\delta_1|/2 + \delta_2 < \delta(\varepsilon,u)$.*

Proof From Corollary 2.4.7 there are symbols $p_0(x,\xi)$ and $q_0(x,\xi)$ and $C_j > 0$ ($j = 1,2$) such that $p(D_x, y, D_y) = p_0(x,D) + q_0(x,D)$ on $\mathcal{S}_\infty$, $p_0(x,\xi) \in S^{0,\delta}(4R_0, 10A + \widehat{C}_*)$,

$$\begin{aligned} \left|p_{0(\beta)}^{(\alpha+\gamma)}(x,\xi)\right| &\leq C_{|\gamma|}((2A + \widehat{C}_*/4)/R_0)^{|\alpha|}(2\delta')^{-|\beta|} \\ &\quad \times |\beta|!\langle\xi\rangle^{-|\gamma|}e^{(\delta+\delta')\langle\xi\rangle} \end{aligned}$$

if $\langle\xi\rangle \geq 4R_0$ and $\delta' > 0$, and

$$\left|q_{0(\beta)}^{(\alpha)}(x,\xi)\right| \leq C_{|\alpha|,R_0}(4R_0+1)^{|\beta|}|\beta|!\exp[-\langle\xi\rangle/R_0]$$

if

$$\left\{\begin{array}{l} R_0 \geq C_1 A + C_2, \\ \max\{4(\delta_1)_+ + \delta_2, 2\delta_1, 4\delta_1 + 2|\delta_1| + 2\delta_2\} \leq 1/R_0, \end{array}\right. \tag{2.110}$$

where $\delta = \delta_1 + |\delta_1|/2 + \delta_2$. Theorem 2.3.3 yields $q_0(x,D)u \in \mathcal{S}'_{-\delta'}$ if $\delta' \leq (2en(4R_0+1))^{-1}$ and (2.110) is valid, since

$$\begin{aligned} \left|\partial_\xi^\alpha D_y^\beta \partial_\eta^\gamma q_0(y,\eta)\right| &\leq C_{|\alpha|+|\gamma|,R_0}((4R_0+1)/R)^{|\beta|} \\ &\quad \times \langle\xi\rangle^{|\beta|}\exp[-\langle\eta\rangle/R_0] \end{aligned}$$

if $R > 0$ and $\langle\xi\rangle \geq R|\beta|$. On the other hand Theorem 2.6.5 implies that $p_0(x,D)u$ is analytic in X_1 if $R_0 \geq e\max\{n, 4\sqrt{n}\}(10A + \widehat{C}_*)$, $R_0 \geq 18e\sqrt{n}(8A + \widehat{C}_*)/\varepsilon$ and $\delta < \min\{1/(4R_0), c_0\varepsilon, \delta_1(\varepsilon,u)\}$, where c_0 and $\delta_1(\varepsilon,u)$ are the constants in Theorem 2.6.5. □

Theorem 2.6.7 *Let X and X_1 be open bounded subsets of $\boldsymbol{R}^n$ such that $X_1 \subset\subset X$, and put $\varepsilon = \mathrm{dis}(X_1, \boldsymbol{R}^n \setminus X)$. Let $\delta_0 \geq 0$, and assume that*

$u \in \mathcal{F}_{\delta_0}$ and u is analytic in a neighborhood of $\overline{X}$. Moreover, we assume that a symbol $p(\xi, y, \eta)$ satisfies $\operatorname{supp} p \subset \boldsymbol{R}^n \times X_1 \times \boldsymbol{R}^n$ *and*

$$\left|\partial_\xi^\alpha D_y^{\beta+\tilde{\beta}} \partial_\eta^{\gamma+\tilde{\gamma}} p(\xi, y, \eta)\right| \leq C_{|\alpha|+|\tilde{\beta}|+|\tilde{\gamma}|}(A/R_0)^{|\beta|+|\gamma|} \times \langle\xi\rangle^{m_1-|\alpha|+|\beta|}\langle\eta\rangle^{m_2-|\tilde{\gamma}|} \exp[\delta_1\langle\xi\rangle + \delta_2\langle\eta\rangle]$$

if $\langle\xi\rangle \geq R_0|\beta|$ and $\langle\eta\rangle \geq R_0|\gamma|$. Then there are positive constants $\delta(\varepsilon, u)$ and $\delta_j(\varepsilon, u)$ ($j = 1, 2$) such that $p(D_x, y, D_y)u$ is well-defined and $p(D_x, y, D_y)u \in \mathcal{S}'_{-\delta}$, i.e., $\sum_{j=1}^\infty p(D_x, y, D_y)\psi_j^R(D)u$ is convergent in $\mathcal{S}'_{-\delta}$ for $R \gg 1$, if $R_0 \geq 4e\sqrt{n}\max\{1, 2/\varepsilon\}A$, $2\delta_1 + (\delta_0 + \delta_2)_+ < 1/R_0$, $\delta \leq \min\{1/(2R_0), \delta(\varepsilon, u)\}$, $\delta_1 \leq \delta_1(\varepsilon, u)$ and $\delta_0 + \delta_2 \leq \delta_2(\varepsilon, u)$.

Remark From the proof of Theorem 2.6. 7 we have

$$|\mathcal{F}[p(D_x, y, D_y)u](\xi)| \leq C_{R_0,\delta}(u, p)\langle\xi\rangle^{m_1} e^{-\delta\langle\xi\rangle}$$

if R_0, δ and δ_j ($j = 0, 1, 2$) satisfy the above conditions.

Proof We may assume that $u_\rho \equiv \exp[-\rho\langle D\rangle]u$ ($\rho > \delta_0$) satisfies (2.102) and (2.104). Let X_2 be an open subset of X such that $X_1 \subset\subset X_2 \subset\subset X$ and $\operatorname{dis}(X_1, \boldsymbol{R}^n \setminus X_2) = \varepsilon/2$. We choose a family $\{\chi_j\}_{j\in N}$ of $C_0^\infty(X)$ so that $\chi_j(x) = 1$ in X_2 and (2.103) is satisfied. Then we have (2.106) and

$$\begin{aligned}
&\left|\mathcal{F}\left[p(D_x, y, D_y)\psi_j^R(D)e^{\rho\langle D\rangle}(\chi_j u_\rho)\right](\xi)\right| \\
&= (2\pi)^{-n}\left|\int e^{-iy\cdot(\xi-\eta)} p(\xi, y, \eta)\psi_j^R(\eta)e^{\rho\langle\eta\rangle}\mathcal{F}[\chi_j u_\rho](\eta)\, d\eta dy\right| \\
&\leq \int_{\Omega_j^R\cap\{|\eta|\geq|\xi|/2\}} C''(u)\langle\xi\rangle^{m_1}\langle\eta\rangle^{m_2}\exp[\delta_1\langle\xi\rangle + (\rho + \delta_2)\langle\eta\rangle] \\
&\qquad \times(1 + \sqrt{n}(C_*/\varepsilon + A(u))j)^j\langle\eta\rangle^{-j}\, d\eta dy \\
&\quad + \int_{\Omega_j^R\cap\{|\eta|\leq|\xi|/2\}} C'(u)|K^k p(\xi, y, \eta)|e^{\rho\langle\eta\rangle}(1 + \sqrt{n}(C_*/\varepsilon + A(u))j)^j \\
&\qquad \times\langle\eta\rangle^{-j}\, d\eta dy
\end{aligned}$$

if $R_0 k \leq \langle\xi\rangle \leq R_0(k+1)$, where $R > 0$, $\Omega_j^R = \{(y, \eta) \in X_1 \times \boldsymbol{R}^n;\ 2R(j-1) \leq \langle\eta\rangle \leq 3Rj\}$ and

$$K = |\xi - \eta|^{-2}\sum_{\mu=1}^n(\xi_\mu - \eta_\mu)D_{y_\mu}.$$

Therefore, we have

$$\begin{aligned}
&\left|\mathcal{F}\left[p(D_x, y, D_y)\psi_j^R(D)e^{\rho\langle D\rangle}(\chi_j u_\rho)\right](\xi)\right| \\
&\leq C_{R,R_0}(u)j^{n+m_2}2^{-j}\langle\xi\rangle^{m_1}e^{-\delta\langle\xi\rangle}
\end{aligned}$$

if $R \geq 2e(1+\sqrt{n}(C_*/\varepsilon+A(u)))$, $R_0 \geq 2e\sqrt{n}A$, $\rho+\delta_2+2(\delta_1+\delta)_+ \leq 1/(3R)$, $\delta_1 \leq 1/(2R_0)$ and $\delta \leq 1/(2R_0)$. It is easy to see that

$$\begin{aligned}&\mathcal{F}\Big[p(D_x,y,D_y)\psi_j^R(D)e^{\rho\langle D\rangle}((1-\chi_j)u_\rho)\Big](\xi)\\&= \langle (1-\chi_j(w))u_\rho(w), f_{j,\rho}^R(w,\xi)\rangle_w,\end{aligned}$$

where $M = [(\ell+n)/2]+1$ and

$$\begin{aligned}f_{j,\rho}^R(w,\xi) = (2\pi)^{-n}\int e^{-iy\cdot(\xi-\eta)-iw\cdot\eta}\langle y-w\rangle_A^{-2M}\\ \times\langle D_\eta\rangle_A^{2M}\Big\{p(\xi,y,\eta)\psi_j^R(\eta)e^{\rho\langle\eta\rangle}\Big\}\,d\eta dy.\end{aligned}$$

Since $\langle w\rangle_A \leq \sqrt{2}\langle y-w\rangle_A\langle y\rangle_A$ and $\langle w\rangle_A^{-2M}u_\rho(w) \in L^1$, we have

$$\begin{aligned}&\Big|\mathcal{F}\Big[p(D_x,y,D_y)\psi_j^R(D)e^{\rho\langle D\rangle}((1-\chi_j)u_\rho)\Big](\xi)\Big|\\&\leq C'_{\rho,A}\sup_{w\notin X_2}\Big|\langle w\rangle_A^{2M}f_{j,\rho}^R(w,\xi)\Big|.\end{aligned}$$

Choose $g^R(\xi,\eta) \in C^\infty(\boldsymbol{R}^n\times\boldsymbol{R}^n)$ ($R \geq 2$) so that $g^R(\xi,\eta) = 1$ if $|\eta| \leq |\xi|/4$ and $|\xi| \geq R/2$, $g^R(\xi,\eta) = 0$ if $|\eta| \geq |\xi|/2$ or $|\xi| \leq R/4$, and $|\partial_\eta^\gamma g^R(\xi,\eta)| \leq C(C_*/R)^{|\gamma|}$ if $\langle\eta\rangle \geq R|\gamma|$. We can write

$$\begin{aligned}\langle w\rangle_A^{2M}f_{j,\rho}^R(w,\xi) &= (2\pi)^{-n}\int e^{-iy\cdot(\xi-\eta)-iw\cdot\eta}\langle w\rangle_A^{2M}\\ &\quad\times K^k\Big[\langle y-w\rangle_A^{-2M}g^R(\xi,\eta)\langle D_\eta\rangle_A^{2M}\Big\{p(\xi,y,\eta)\psi_j^R(\eta)e^{\rho\langle\eta\rangle}\Big\}\Big]\,d\eta dy\\ &\quad+(2\pi)^{-n}\int e^{-iy\cdot(\xi-\eta)-iw\cdot\eta}\langle y-w\rangle_A^{-2M}L^j\Big[(1-g^R(\xi,\eta))\\ &\quad\times\langle D_\eta\rangle_A^{2M}\Big\{p(\xi,y,\eta)\psi_j^R(\eta)e^{\rho\langle\eta\rangle}\Big\}\Big]\,d\eta dy\\ &\equiv I_{j,\rho,k}^{R,1}(w,\xi)+I_{j,\rho}^{R,2}(w,\xi)\end{aligned}$$

if $w \notin X_2$, $k \in \boldsymbol{Z}_+$ and $R_0 k \leq \langle\xi\rangle \leq R_0(k+1)$, where

$$L = |y-w|^{-2}\sum_{\mu=1}^n (w_\mu - y_\mu)D_{\eta_\mu}.$$

Then,

$$\Big|I_{j,\rho,k}^{R,1}(w,\xi)\Big| \leq C_{R_0}\langle\xi\rangle^{m_1}\exp[-\langle\xi\rangle/(2R_0)]j^{-2}$$

if $w \notin X_2$, $R_0 k \leq \langle\xi\rangle \leq R_0(k+1)$, $R_0 \geq 4e\sqrt{n}A$ and $2\delta_1+(\rho+\delta_2)_+ \leq 1/R_0$, and

$$\Big|I_{j,\rho}^{R,2}(w,\xi)\Big| \leq C_R\langle\xi\rangle^{m_1}\exp[(\delta_1+(\rho+\delta_2)/4-1/(12R))\langle\xi\rangle]j^{m_2+n}2^{-j}$$

if $w \notin X_2$, $R_0 \geq 8e\sqrt{n}A/\varepsilon$, $R \geq 8e\sqrt{n}(C_* + \widehat{C} + 6(1+\sqrt{2}))/\varepsilon$ and $\rho + \delta_2 \leq 1/(3R)$. Therefore, we have

$$\begin{aligned}&\left|\mathcal{F}\left[p(D_x, y, D_y)\psi_j^R(D)e^{\rho\langle D\rangle}((1-\chi_j)u_\rho)\right](\xi)\right| \\ &\leq C_{\rho,A,R_0,R}\langle\xi\rangle^{m_1}\exp[-\delta\langle\xi\rangle]j^{-2}\end{aligned}$$

if $R_0 \geq 4e\sqrt{n}\max\{1, 2/\varepsilon\}A$, $R \geq 8e\sqrt{n}(C_* + \widehat{C} + 6(1+\sqrt{2}))/\varepsilon$, $\delta \leq 1/(2R_0)$, $2\delta_1 + (\rho+\delta_2)_+ \leq 1/R_0$, $\rho + \delta_2 \leq 1/(3R)$ and $\delta \leq \delta_1 + (\rho + \delta_2)/4 - 1/(12R)$. This proves the theorem. □

2.7 Pseudodifferential operators in $\mathcal{B}$

By Theorem 2.6. 1 and 2.6. 5 we can define pseudodifferential operators acting on $\mathcal{B}$. Let X be an open subset of $\boldsymbol{R}^n$. We say that $p(x,\xi) \in PS^+_{loc}(X)$ if $p(x,\xi)$ satisfies the following; for any compact subset K of X there are $A_K > 0$, $R_K \geq 0$ and $C_{K,j,\delta} > 0$ ($\delta > 0$) such that

$$\left|p^{(\alpha+\tilde{\alpha})}_{(\beta)}(x,\xi)\right| \leq C_{K,|\tilde{\alpha}|,\delta}A_K^{|\alpha|+|\beta|}|\alpha|!|\beta|!\langle\xi\rangle^{-|\alpha|}e^{\delta\langle\xi\rangle}$$

if $x \in K$, $\langle\xi\rangle \geq R_K|\alpha|$ and $\delta > 0$. Now assume that $p(x,\xi) \in PS^+_{loc}(X)$. Let U be an open subset of X such that $U \Subset X$, and choose a compact subset K of X so that $U \Subset K \Subset X$. It follows from Proposition 2.2. 3 that there are symbols $\Phi^R(x,\xi) \in S^{0,0}(R, C_*, C(U,K))$ ($R \geq 4$) satisfying $0 \leq \Phi^R(x,\xi) \leq 1$, supp $\Phi^R(x,\xi) \subset K \times \boldsymbol{R}^n$ and $\Phi^R(x,\xi) = 1$ for $(x,\xi) \in U \times \boldsymbol{R}^n$. Put $\tilde{p}^R(x,\xi) = \Phi^R(x,\xi)p(x,\xi)$. Then we have $\tilde{p}^R(x,\xi) \in S^+(R, A_K + C_*, A_K + C(U,K))$ for $R \geq \max\{4, R_K\}$. Let $u \in \mathcal{B}(U)$, and choose $v \in \mathcal{A}'(\overline{U})$ so that $v|_U = u$. Then $\tilde{p}^R(x,\xi)v$ ($\in \mathcal{F}_0$) determines an element $(\tilde{p}^R(x,D)v)|_U$ of $\mathcal{B}(U)$ and, therefore, an element of $\mathcal{B}(U)/\mathcal{A}(U)$ if $R \gg 1$. Theorem 2.6. 1 (or Corollary 2.6. 2) implies that $(\tilde{p}^R(x,\xi)v)|_U$ does not depend on the choice of $\Phi^R(x,\xi)$ for $R \gg 1$. Moreover, it follows from Theorem 2.6. 5 that $\tilde{p}^R(x,D)w$ is analytic in U if $R \gg 1$, $w \in \mathcal{F}_0$ and supp $w \cap U = \emptyset$. This implies that $(\tilde{p}^R(x,D)v)|_U$ as an elemant of $\mathcal{B}(U)/\mathcal{A}(U)$ is uniquely determined by u and does not depend on the choice of v. Therefore, we can define $p(x,D)$ as $p(x,D):\mathcal{B}(U) \to \mathcal{B}(U)/\mathcal{A}(U)$. Moreover, it follows from Theorem 2.6. 5 and Corollary 2.6. 2 that

$$p(x,D)(u|_U) - (p(x,D)u)|_U \in \mathcal{A}(U) \quad \text{for } u \in \mathcal{B}(V),$$

where V is an open subset of X satisfying $U \subset V \Subset X$. So we can define $p(x,D)$ as $p(x,D): \mathcal{B}_X \to \mathcal{B}_X/\mathcal{A}_X$, which is a sheaf homomorphism. By

Theorem 2.6. 5 we can also define $p(x,D) : \mathcal{B}_X/\mathcal{A}_X \to \mathcal{B}_X/\mathcal{A}_X$, which is a sheaf homomorphism.

Theorem 2.7. 1 *Let $p(x,D) = \sum_{|\alpha|\le m} a_\alpha(x)D^\alpha$ be a differential operator in Ω, where Ω is an open subset of $\boldsymbol{R}^n$ and $a_\alpha(x) \in \mathcal{A}(\Omega)$. Let K and U_j ($j = 0,1$) be a compact subset and open subsets of Ω, respectively, such that $K \subset\subset U_0 \subset\subset U_1 \subset\subset \Omega$, and choose $\Phi^R(x,\xi) \in S^0(R, C_*, C(U_0,U_1))$ so that* supp $\Phi^R \subset U_1 \times \boldsymbol{R}^n$ *and $\Phi^R(x,\xi) = 1$ for $(x,\xi) \in U_0 \times \boldsymbol{R}^n$. We put $\tilde{p}^R(x,\xi) = \Phi^R(x,\xi)p(x,\xi)$. Then for any open set U with $U \subset\subset U_0$ there is $R(K,U,U_0,U_1,\Omega,p) > 0$ such that*

$$(\tilde{p}^R(x,D)u)|_U = (p(x,D)u)|_U \quad \text{in } \mathcal{B}(U) \tag{2.111}$$

if $u \in \mathcal{A}'(K)$ and $R \ge R(K,U,U_0,U_1,\Omega,p)$. Here $p(x,D)u$ ($\in \mathcal{A}'(K)$) on the right-hand side of (2.111) *is defined by $(p(x,D)u)(\varphi) = u({}^tp(x,D)\varphi)$ for $\varphi \in \mathcal{A}$, where ${}^tp(x,D)\varphi(x) = \sum_{|\alpha|\le m}(-D)^\alpha(a_\alpha(x)\varphi(x))$. Namely, two definitions of $p(x,D)u$ ($\in \mathcal{B}(U)$) for $u \in \mathcal{B}(U)$ are consistent.*

Proof It suffices to prove the theorem in the case where $p(x,\xi) = a(x) \in \mathcal{A}(\Omega)$. Let U_j ($j = 2,3$) be open subsets of Ω such that $U_1 \subset\subset U_2 \subset\subset U_3 \subset \subset \Omega$, and choose $\Psi(x) \in C_0^\infty(U_2)$ and $\widetilde{\Psi}^R(x,\xi) \in S^0(R, C_*, C(U_2,U_3))$ so that $\Psi(x) = 1$ for $x \in U_1$, supp $\widetilde{\Psi}^R \subset U_3 \times \boldsymbol{R}^n$ and $\widetilde{\Psi}^R(x,\xi) = 1$ for $x \in U_2 \times \boldsymbol{R}^n$. Put

$$a^R(x,\xi) = \Phi^R(x,\xi)a(x), \qquad b^R(x,\xi) = \widetilde{\Psi}^R(x,\xi)(1 - \Phi^R(x,\xi))a(x).$$

Then it follows from Theorem 2.3. 3 and Corollary 2.6. 2 that $a^R(x,D)u$, $b^R(x,D)u \in \mathcal{F}_0$, supp $a^R(x,D)u \subset \overline{U}_1$ and supp $b^R(x,D)u \subset U_3 \setminus U_0$ if $u \in \mathcal{A}'(K)$ (or $\in \mathcal{F}_0$) and $R \ge R_0(C(U_0,U_1) + C(U_2,U_3) + B(a))$, where R_0 is a positive constant and $|D^\beta a(x)| \le CB(a)^{|\beta|}|\beta|!$ for $x \in U_3$. Moreover, Lemma 2.6. 4 and its proof imply that $b^R(x,D)u \in C^\infty(\boldsymbol{R}^n)$ and $b^R(x,D)u$ is analytic in $U_2\setminus U_1$ if $u \in \mathcal{A}'(K)$ and $R \ge 2en(C(U_0,U_1)+ C(U_2,U_3) + B(a))$. Now assume that $R \ge \max\{R_0, 2en\}(C(U_0,U_1) + C(U_2,U_3) + B(a))$. Then we can define the linear operator $A : \mathcal{A}'(K) \to \mathcal{F}_0$ by

$$Au = a^R(x,D)u + \Psi(x)b^R(x,D)u.$$

Note that $(Au)|_{U_0} = (a^R(x,D)u)|_{U_0}$ in $\mathcal{B}(U)$ for $u \in \mathcal{A}'(K)$. Let $u \in \mathcal{A}'(K)$ and $\varphi \in \mathcal{S}_\infty$. It is obvious that

$$\langle a^R(x,D)u, \varphi\rangle = \langle u, {}^ta^R(x,D)\varphi\rangle = u({}^ta^R(x,D)\varphi).$$

From the proof of Lemma 2.6. 4 we see that $b^R(x,D)u = \sum_{j=1}^\infty b^R(x,D) \times \psi_j^R(D)u$ converges uniformly on $\boldsymbol{R}^n$. Therefore, we have

$$\langle \Psi(x) b^R(x,D)u, \varphi\rangle = \sum_{j=1}^\infty \langle b^R(x,D)\psi_j^R(D)u, \Psi(x)\varphi(x)\rangle.$$

Since $\tilde{\psi}_j^R(D)u \in \mathcal{S}\ (\subset \mathcal{S}')$, $\Psi(x)\varphi(x) \in \mathcal{S}$ and ${}^t\psi_j^R(D)\,{}^tb^R(x,D)(\Psi(x)\varphi(x)) \in \mathcal{S}_\infty \subset \mathcal{A}$, we have also

$$\begin{aligned}&\langle b^R(x,D)\psi_j^R(D)u, \Psi(x)\varphi(x)\rangle = \langle \tilde{\psi}_j^R(D)u, {}^t\psi_j^R(D)\,{}^tb^R(x,D)(\Psi(x)\varphi(x))\rangle\\ &= \langle u, {}^t\psi_j^R(D)\,{}^tb^R(x,D)(\Psi(x)\varphi(x))\rangle = u({}^t\psi_j^R(D)\,{}^tb^R(x,D)(\Psi(x)\varphi(x))),\end{aligned}$$

where $\tilde{\psi}_j^R(\xi) \in C_0^\infty(\boldsymbol{R}^n)$ satisfies $\tilde{\psi}_j^R(\xi) = 1$ in supp $\psi_j^R(\xi)$. Let U be an open subset of U_0 such that $K \subset\subset U \subset\subset U_0$. Noting that ${}^t\psi_j^R(D)\,{}^tb^R(x,D)(\Psi(x)\varphi(x))$ can be regarded as an entire analytic function, we have

$$\begin{aligned}&{}^t\psi_j^R(D)\,{}^tb^R(x,D)(\Psi(x)\varphi(x))\\ &= (2\pi)^{-n}\int e^{i(x-y)\cdot\xi} L^j\{\psi_j^R(-\xi)(1-\Phi^R(y,-\xi))\}\Psi(y)\varphi(y)\,dyd\xi\end{aligned}$$

for $x \in \boldsymbol{C}^n$ with Re $x \in \overline{U}$, where $L = |x-y|^{-2}\sum_{k=1}^n (y_k - \overline{x}_k)D_{\xi_k}$. This yields

$$\begin{aligned}\left|{}^t\psi_j^R(D)\,{}^tb^R(x,D)(\Psi(x)\varphi(x))\right| \le\ & C_R(e\sqrt{n}(\widehat{C}+C_*)/(\varepsilon R))^j j^n\\ &\times \exp[(|\mathrm{Im}\ x| - 1/(3R))\langle\xi\rangle]\end{aligned}$$

if Re $x \in \overline{U}$, where $\varepsilon = \mathrm{dis}(U, \boldsymbol{R}^n\setminus U_0)$. Therefore, $\sum_{j=1}^\infty {}^t\psi_j^R(D)\,{}^tb^R(x,D) \times(\Psi(x)\varphi(x))$ converges uniformly on $\{x \in \boldsymbol{C}^n;$ Re $x \in \overline{U}$ and $|\mathrm{Im}\ x| \le 1/(3R)\}$ if $R \ge 2e\sqrt{n}(\widehat{C}+C_*)/\varepsilon$. On the other hand, we have

$$\sum_{j=1}^\infty {}^t\psi_j^R(D)\,{}^tb^R(x,D)(\Psi(x)\varphi(x)) = {}^tb^R(x,D)(\Psi(x)\varphi(x))$$

in $\mathcal{S}$. Thus we have

$$\begin{aligned}&\langle \Psi(x)b^R(x,D)u,\varphi\rangle = u({}^tb^R(x,D)(\Psi(x)\varphi(x))),\\ &\langle Au,\varphi\rangle = u({}^ta^R(x,D)\varphi + {}^tb^R(x,D)(\Psi(x)\varphi(x))),\end{aligned}$$

if $R \ge 2e\sqrt{n}(\widehat{C}+C_*)/\varepsilon$. This gives $Au = a(x)u$ in $\mathcal{A}'(K)\ (\subset \mathcal{F}_0)$ if $R \ge 2e\sqrt{n}(\widehat{C}+C_*)/\varepsilon$, since $a^R(x,\xi) + \Psi(x)b^R(x,\xi) = \Psi(x)a(x)$. So we have

$$(a^R(x,D)u)|_U = (Au)|_U = (a(x)u)|_U \quad \text{in } \mathcal{B}(U)$$

if $R \ge R(K,U,U_0,U_1,\Omega,a)$, where $R(K,U,U_0,U_1,\Omega,a)$ is a positive constant. □

Theorem 2.7.2 *Let X be an open subset of $\mathbf{R}^n$, and let $p_j(x,\xi) \in PS^+(X \times \mathbf{R}^n; R_0, A, B)$ ($j = 1,2$). We fix $R_1 \geq R_0 + enAB$ and put*

$$p(x,\xi) = \sum_{j=0}^{\infty} \phi_j^{R_1}(\xi) \sum_{|\gamma|=j} \frac{1}{\gamma!} p_1^{(\gamma)}(x,\xi) p_{2(\gamma)}(x,\xi).$$

Then we have

$$p(x,\xi) \in PS^+(X \times \mathbf{R}^n; 2R_1, A', B) \tag{2.112}$$

and $p_1(x,D)p_2(x,D)u - p(x,D)u \in \mathcal{A}(U)$ for $u \in \mathcal{B}(U)$, where $A' = A + 3\widehat{C}$ and U is an open subset of X.

Remark When $p_j(x,\xi) \in PS^+_{loc}(X)$ ($j = 1,2$), the above theorem is still valid if $p(x,D)$ is replaced by a family of operators. We omit the details as we need the sheaf of pseudodifferential operators.

Proof Since

$$\begin{aligned}
&\Big|\partial_\xi^{\alpha+\tilde{\alpha}} D_x^\beta \Big\{ \sum_{|\gamma|=j} \frac{1}{\gamma!} p_1^{(\gamma)}(x,\xi) p_{2(\gamma)}(x,\xi) \Big\}\Big| \\
&\leq C_{|\tilde{\alpha}|,\delta} (4nAB)^j (2A)^{|\alpha|} (2B)^{|\beta|} j! |\alpha|! |\beta|! \langle\xi\rangle^{-|\alpha|-|\tilde{\alpha}|-j} e^{\delta\langle\xi\rangle}
\end{aligned}$$

for $j \in \mathbf{Z}_+$ and $(x,\xi) \in X \times \mathbf{R}^n$ with $\langle\xi\rangle \geq R_0(|\alpha|+j)$, from Lemma 2.2.4 we have (2.112). Let U be an open subset of X such that $U \Subset X$, and choose compact subsets K_j ($j = 0,1,2$) of X and $\varepsilon_0 > 0$ so that $U \Subset K_0 \Subset K_1 \Subset (K_1)_{\varepsilon_0} \Subset K_2 \Subset X$, where $(K_1)_{\varepsilon_0}$ denotes the ε_0-neighborhood of K_1. Then, by Proposition 2.2.3 there are symbols $\Phi_j^R(x,\xi) \in S^{0,0}(R, C_*, B_j)$ ($j = 1,2$, $R \geq 4$) satisfying $0 \leq \Phi_j^R(x,\xi) \leq 1$ ($j = 1,2$), supp $\Phi_1^R \subset K_1 \times \mathbf{R}^n$, $\Phi_1^R(x,\xi) = 1$ for $(x,\xi) \in K_0 \times \mathbf{R}^n$, supp $\Phi_2^R \subset K_2 \times \mathbf{R}^n$ and $\Phi_2^R(x,\xi) = 1$ for $(x,\xi) \in (K_1)_{\varepsilon_0} \times \mathbf{R}^n$, where B_1 and B_2 depend on $\mathrm{dis}(K_0, \mathbf{R}^n \setminus K_1)$ and $\mathrm{dis}((K_1)_{\varepsilon_0}, \mathbf{R}^n \setminus K_2)$, respectively. Put $\tilde{p}_j^R(x,\xi) = p_j(x,\xi)\Phi_j^R(x,\xi)$ ($j = 1,2$). For $u \in \mathcal{B}(U)$ we defined $p_1(x,D)p_2(x,D)u$ ($\in \mathcal{B}(U)/\mathcal{A}(U)$) by the residue class of $(\tilde{p}_1^R(x,D)\tilde{p}_2^R(x,D)v)|_U$ in $\mathcal{B}(U)/\mathcal{A}(U)$, where $R \gg 1$ and $v \in \mathcal{A}(\overline{U})$ satisfies $v|_U = u$. $p^R(x,\xi,y,\eta) \equiv \tilde{p}_1^R(x,\xi)\tilde{p}_2^R(y,\eta)$ satisfies

$$\begin{aligned}
&\Big|D_x^{\beta+\tilde{\beta}} \partial_\xi^{\alpha+\tilde{\alpha}} D_y^{\lambda+\tilde{\lambda}} \partial_\eta^{\rho+\tilde{\rho}} p^R(x,\xi,y,\eta)\Big| \leq C_{|\tilde{\alpha}|+|\tilde{\beta}|+|\tilde{\lambda}|+|\tilde{\rho}|,\delta} \\
&\quad \times ((A+C_*)/R)^{|\alpha|+|\rho|} ((B+B_1)/R)^{|\beta|} ((B+B_2)/R)^{|\lambda|} \\
&\quad \times \langle\xi\rangle^{-|\tilde{\alpha}|+|\beta|} \langle\eta\rangle^{-|\tilde{\rho}|+|\lambda|} e^{\delta\langle\xi\rangle+\delta\langle\eta\rangle}
\end{aligned}$$

if $\delta > 0$, $\langle\xi\rangle \geq R(|\alpha| + |\beta|)$ and $\langle\eta\rangle \geq R(|\lambda| + |\rho|)$,

$$\begin{aligned}\left|D_x^{\beta+\tilde\beta}\partial_\xi^{\alpha+\tilde\alpha}D_y^\lambda\partial_\eta^{\rho+\tilde\rho}p^R(x,\xi,y,\eta)\right| &\leq C_{|\tilde\alpha|+|\tilde\beta|+|\tilde\rho|,\delta}\\ &\times((A+C_*)/R)^{|\alpha|}((B+B_1)/R)^{|\beta|}B^{|\lambda|}A^{|\rho|}|\lambda|!|\rho|!\\ &\times\langle\xi\rangle^{-|\tilde\alpha|+|\beta|}\langle\eta\rangle^{-|\rho|-|\tilde\rho|}e^{\delta\langle\xi\rangle+\delta\langle\eta\rangle}\end{aligned}$$

if $\delta > 0$, $|x-y| < \varepsilon_0$, $\langle\xi\rangle \geq R(|\alpha| + |\beta|)$ and $\langle\eta\rangle \geq R|\rho|$,

$$\begin{aligned}\left|D_x^{\beta}\partial_\xi^{\alpha+\tilde\alpha}D_y^{\lambda+\tilde\lambda}\partial_\eta^{\rho+\tilde\rho}p^R(x,\xi,y,\eta)\right| &\leq C_{|\tilde\alpha|+|\tilde\lambda|+|\tilde\rho|,\delta}\\ &\times A^{|\alpha|}B^{|\beta|}((B+B_2)/R)^{|\lambda|}((A+C_*)/R)^{|\rho|}|\alpha|!|\beta|!\\ &\times\langle\xi\rangle^{-|\alpha|-|\tilde\alpha|}\langle\eta\rangle^{-|\tilde\rho|+|\lambda|}e^{\delta\langle\xi\rangle+\delta\langle\eta\rangle}\end{aligned}$$

if $\delta > 0$, $x \in K_0$, $\langle\xi\rangle \geq R|\alpha|$ and $\langle\eta\rangle \geq R(|\lambda| + |\rho|)$, and

$$\begin{aligned}\left|D_x^{\beta}\partial_\xi^{\alpha+\tilde\alpha}D_y^{\lambda}\partial_\eta^{\rho+\tilde\rho}p^R(x,\xi,y,\eta)\right| &\leq C_{|\tilde\alpha|+|\tilde\rho|,\delta}A^{|\alpha|+|\rho|}B^{|\beta|+|\lambda|}\\ &\times|\alpha|!|\beta|!|\lambda|!|\rho|!\langle\xi\rangle^{-|\alpha|-|\tilde\alpha|}\langle\eta\rangle^{-|\rho|-|\tilde\rho|}e^{\delta\langle\xi\rangle+\delta\langle\eta\rangle}\end{aligned}$$

if $\delta > 0$, $x \in K_0$, $|x-y| < \varepsilon_0$, $\langle\xi\rangle \geq R|\alpha|$ and $\langle\eta\rangle \geq R|\rho|$, where $R \geq \max\{4, R_0\}$. It follows from Corollary 2.4. 7 that there are a symbol $q^R(x,\xi)$ and $R(A,B,B_1,B_2,\varepsilon_0) > 0$ such that

$$\begin{aligned}&p^R(x,D_x,y,D_y) = \tilde p_0^R(x,D) + q^R(x,D) \quad \text{on } \mathcal{S}_\infty,\\ &\left|q_{(\beta)}^{R(\alpha)}(x,\xi)\right| \leq C_{|\alpha|+|\beta|}\exp[-\langle\xi\rangle/R]\end{aligned}$$

if $R \geq R(A,B,B_1,B_2,\varepsilon_0)$, and

$$\left|q_{(\beta)}^{R(\alpha)}(x,\xi)\right| \leq C_{|\alpha|,R}(2B+4R+12(1+\sqrt{2})/\varepsilon_0)^{|\beta|}|\beta|!\exp[-\langle\xi\rangle/R]$$

if $x \in K_0$ and $R \geq R(A,B,B_1,B_2,\varepsilon_0)$, where

$$\tilde p_0^R(x,\xi) = \sum_{j=0}^\infty \phi_j^{4R}(\xi)\sum_{|\gamma|=j}\frac{1}{\gamma!}\tilde p_1^{R(\gamma)}(x,\xi)\tilde p_{2(\gamma)}^R(x,\xi).$$

It is obvious that $q^R(x,D)v$ is analytic in U for $v \in \mathcal{F}_0$ if $R \geq R(A,B,B_1,B_2,\varepsilon_0)$. Choose a compact subset K of X and a symbol $\Phi^R(x,\xi) \in S^{0,0}(R,C_*,C(K,K_0))$ ($R \geq 4$) so that $U \subset\subset K \subset\subset K_0$, $0 \leq \Phi^R(x,\xi) \leq 1$, supp $\Phi^R \subset K_0 \times \mathbf{R}^n$ and $\Phi^R(x,\xi) = 1$ for $(x,\xi) \in K \times \mathbf{R}^n$. Put

$$\begin{aligned}&q_1^R(x,\xi) = (\tilde p_0^R(x,\xi) - p(x,\xi))\Phi^R(x,\xi),\\ &q_2^R(x,\xi) = \tilde p_0^R(x,\xi)(1-\Phi^R(x,\xi)).\end{aligned}$$

Then we have

$$q_1^R(x,\xi) = \sum_{j=0}^{\infty}(\phi_j^{4R}(\xi) - \phi_j^{R_1}(\xi))\Phi^R(x,\xi) \sum_{|\gamma|=j} \frac{1}{\gamma!} p_1^{(\gamma)}(x,\xi)p_{2(\gamma)}(x,\xi),$$

$$\left|q_{1(\beta)}^{R(\alpha)}(x,\xi)\right| \le C_{|\alpha|+|\beta|} \exp[-\langle\xi\rangle/(24R)]$$

if $4R \ge R_1$ and $R_1 \ge enAB/2$, and

$$\left|q_{1(\beta)}^{R(\alpha)}(x,\xi)\right| \le C_{|\alpha|}(2B)^{|\beta|}|\beta|! \exp[-\langle\xi\rangle/(24R)]$$

if $x \in K$, $4R \ge R_1$ and $R_1 \ge enAB$. Therefore, $q_1^R(x,D)v$ is analytic in U for $v \in \mathcal{F}_0$ if $R_1 \ge enAB$ and $R \gg 1$. On the other hand, we have $\operatorname{supp} q_2^R \subset \overline{(\boldsymbol{R}^n \setminus K)} \times \boldsymbol{R}^n$ and

$$\begin{aligned} \left|q_{2(\beta+\tilde\beta)}^{R(\alpha)}(x,\xi)\right| &\le C_{|\alpha|+|\tilde\beta|,\delta}((3B/8 + B_1 + C(K,K_0))/R)^{|\beta|} \\ &\quad \times\langle\xi\rangle^{-|\tilde\alpha|+|\beta|}e^{\delta\langle\xi\rangle} \end{aligned}$$

if $\delta > 0$, $R \ge \max\{4, R_0, n(A/2 + 4C_*)B\}$ and $\langle\xi\rangle \ge R|\beta|$. So Corollary 2.6.2 yields $\operatorname{supp} q_2^R(x,D)v \cap K_0 = \emptyset$ for $v \in \mathcal{F}_0$ if $R \gg 1$. So $\tilde p_1^R(x,D)\tilde p_2^R(x,D)v - \tilde p^R(x,D)v$ is analytic in U for $v \in \mathcal{F}_0$ if $R_1 \ge enAB$ and $R \gg 1$, where $\tilde p^R(x,\xi) = p(x,\xi)\Phi^R(x,\xi)$. This proves the theorem. □

2.8 Parametrices of elliptic operators

Let X be an open subset of $\boldsymbol{R}^n$, and let $p(x,\xi) \in PS^m(X \times \boldsymbol{R}^n; R_0, A)$, where $m \in \boldsymbol{R}$, $R_0 \ge 0$ and $A > 0$. We say that $p(x,\xi)$ is elliptic at x^0 ($\in X$) if there are a neighborhood U of x^0 in X and positive constants C and c such that

$$|p(x,\xi)| \ge c\langle\xi\rangle^m \quad \text{if } x \in U \text{ and } |\xi| \ge C.$$

Moreover, we say that $p(x,\xi)$ is elliptic in U if $p(x,\xi)$ is elliptic at every x in U, where U is an open subset of X. Applying Theorem 2.7.2 we can construct parametrices for elliptic operators.

Theorem 2.8.1 *Assume that there are positive constants C and c such that*

$$|p(x,\xi)| \ge c\langle\xi\rangle^m \quad \text{if } x \in X \text{ and } |\xi| \ge C.$$

Then there are positive constants R' and A' and a symbol $q(x,\xi) \in PS^{-m}(X \times \boldsymbol{R}^n; R', A')$ such that $q(x,D)p(x,D)u - u \in \mathcal{A}(X)$ and $p(x,D) \times q(x,D)u - u \in \mathcal{A}(X)$ for $u \in \mathcal{B}(X)$, i.e., $p(x,D) : \mathcal{B}_X/\mathcal{A}_X \to \mathcal{B}_X/\mathcal{A}_X$ is a sheaf isomorphism and $p(x,D)^{-1} = q(x,D)$ in $\mathcal{B}_X/\mathcal{A}_X$.

Proof Put

$$q_0(x,\xi) = 1/p(x,\xi),$$
$$q_j(x,\xi) = -q_0(x,\xi)\sum_{k=0}^{j-1}\sum_{|\gamma|=j-k}\frac{1}{\gamma!}q_k^{(\gamma)}(x,\xi)p_{(\gamma)}(x,\xi)$$
$$(\, j = 1,2,\cdots)$$

for $x \in X$ and $\xi \in \boldsymbol{R}^n$ with $|\xi| \geq C$. Then, by induction we can prove

$$\left|q_{0(\beta)}^{(\alpha+\tilde{\alpha})}(x,\xi)\right| \leq C_{|\tilde{\alpha}|}A_1^{|\alpha|+|\beta|}|\alpha|!|\beta|!\langle\xi\rangle^{-m-|\alpha|-|\tilde{\alpha}|} \tag{2.113}$$

for $(x,\xi) \in X \times \boldsymbol{R}^n$ with $|\xi| \geq C$ and $\langle\xi\rangle \geq R_0|\alpha|$, where $A_1 = \max\{2, 24 \times C_0(p)/c\}A$ and $C_0(p) = \sup_{x\in X, |\xi|\geq C}|p(x,\xi)|\langle\xi\rangle^{-m}$ (see, also, Lemma 2.1. 4). It follows from (2.113) that

$$\left|q_{j(\beta)}^{(\alpha+\tilde{\alpha})}(x,\xi)\right| \leq C_{|\tilde{\alpha}|}(2A_1)^{|\alpha|+|\beta|}B^j(|\alpha|+j)!|\beta|!\langle\xi\rangle^{-m-|\alpha|-|\tilde{\alpha}|-j}$$

for $(x,\xi) \in X \times \boldsymbol{R}^n$ with $|\xi| \geq C$, $j \in \boldsymbol{N}$ and $\langle\xi\rangle \geq R_0(|\alpha|+j)$, where $B = 2^8 nAA_1$. Assume that $R \geq \max\{1, C, B/2, R_0\}$, and put

$$q(x,\xi) = \phi_1^R(\xi)q_0(x,\xi) + \sum_{j=1}^{\infty}\phi_j^R(\xi)q_j(x,\xi).$$

Then, by Lemma 2.2. 4 we have $q(x,\xi) \in PS^{-m}(X \times \boldsymbol{R}^n; 2R, A', 2A_1)$, where $A' = 2A_1 + 3\widehat{C}$. A simple calculation yields

$$\sum_{j=0}^{\infty}\phi_j^{R_1}(\xi)\sum_{|\gamma|=j}\frac{1}{\gamma!}q^{(\gamma)}(x,\xi)p_{(\gamma)}(x,\xi) - 1$$
$$= r_1^{R,R_1}(x,\xi) + r_2^{R,R_1}(x,\xi), \tag{2.114}$$

where $R_1 \geq 2R + 2enA_1A'$ and

$$r_1^{R,R_1}(x,\xi)$$
$$= \phi_1^R(\xi) - 1 + \sum_{\gamma>0}\frac{1}{\gamma!}(\phi_1^R(\xi)\phi_{|\gamma|}^{R_1}(\xi) - \phi_{|\gamma|}^R(\xi))q_0^{(\gamma)}(x,\xi)p_{(\gamma)}(x,\xi)$$
$$+\sum_{j=1}^{\infty}\sum_{\gamma>0}\frac{1}{\gamma!}(\phi_j^R(\xi)\phi_{|\gamma|}^{R_1}(\xi) - \phi_{j+|\gamma|}^R(\xi))q_j^{(\gamma)}(x,\xi)p_{(\gamma)}(x,\xi),$$
$$r_2^{R,R_1}(x,\xi) = \sum_{\gamma>0}\phi_{|\gamma|}^{R_1}(\xi)\sum_{\gamma'<\gamma}\frac{1}{\gamma'!(\gamma-\gamma')!}$$
$$\times\Big\{\phi_1^{R(\gamma-\gamma')}(\xi)q_0^{(\gamma')}(x,\xi) + \sum_{j=1}^{\infty}\phi_j^{R(\gamma-\gamma')}(\xi)q_j^{(\gamma')}(x,\xi)\Big\}p_{(\gamma)}(x,\xi).$$

Noting that

$$\begin{aligned}&\operatorname{supp}\,(\phi_1^R(\xi)\phi_k^{R_1}(\xi)-\phi_k^R(\xi))\subset\{\xi\in\boldsymbol{R}^n;\ 2Rk\le\langle\xi\rangle\le 3R_1k\},\\&\operatorname{supp}\,(\phi_j^R(\xi)\phi_k^{R_1}(\xi)-\phi_{j+k}^R(\xi))\\&\subset\{\xi\in\boldsymbol{R}^n;\ 2R\max\{j,k\}\le\langle\xi\rangle\le 3R_1(j+k)\}\end{aligned}$$

for $j,k\in\boldsymbol{N}$, we have

$$\left|r_{1(\beta)}^{R,R_1(\alpha)}(x,\xi)\right|\le C_{|\alpha|}(2A_1)^{|\beta|}|\beta|!\langle\xi\rangle^{-m-|\alpha|}\exp[-\langle\xi\rangle/(3R_1)] \tag{2.115}$$

if $x\in X$, $R\ge 2^8enAA_1$ (and $R_1\ge 2R+2enA_1A'$). Since

$$\operatorname{supp}\,\phi_k^{R_1}(\xi)\cap\operatorname{supp}\,d\phi_j^R(\xi)\subset\{\xi\in\boldsymbol{R}^n;\ 2R\max\{j,k\}\le\langle\xi\rangle\le 3Rj\}$$

for $j\in\boldsymbol{N}$ and $k\in\boldsymbol{Z}_+$, we have also

$$\left|r_{2(\beta)}^{R,R_1(\alpha)}(x,\xi)\right|\le C_{|\alpha|}(2A_1)^{|\beta|}|\beta|!\langle\xi\rangle^{-m-|\alpha|}\exp[-\langle\xi\rangle/(3R)] \tag{2.116}$$

if $x\in X$, $R\ge 2^8enAA_1$ and $R\ge 4nA(2A_1+\widehat{C})$. Therefore, it follows from Theorem 2.7. 2 and (2.114)–(2.116) that $q(x,D)p(x,D)u-u\in\mathcal{A}(X)$ for $u\in\mathcal{B}(X)$. Similarly we can construct a symbol $\tilde{q}(x,\xi)\in PS^{-m}(X\times\boldsymbol{R}^n;R',A')$ such that $p(x,D)\tilde{q}(x,D)u-u\in\mathcal{A}(X)$ for $u\in\mathcal{B}(X)$, which proves the theorem. □

Chapter 3

Analytic wave front sets and microfunctions

In Section 3.1 we shall give a new definition of the analytic wave front set, which coincides with usual definitions. Our definition is very similar to that of the wave front set (with respect to C^∞-singularities) of distributions. Action of Fourier integral operators on the analytic wave front sets will be described in Section 3.2. In particular, it will be proved that pseudodifferential operators have pseudolocal properties in a microlocal sense. Several operations (pull-back, tensor product, push-forward and so forth) on hyperfunctions will be defined in Section 3.4. A generalization of the Holmgren uniqueness theorem due to Kashiwara and Kawai will be proved in Section 3.5. The so-called watermelon-slicing theorem will be also proved, following [Hr5]. In Section 3.6 we shall define the sheaf $\mathcal{C}$ of microfunctions and show that $\mathcal{C}$ is flabby. Moreover, we shall define Fourier integral operators with analytic symbols acting on $\mathcal{C}$ and prove that elliptic pseudodifferential operators give sheaf isomorphisms on $\mathcal{C}$.

3.1 Analytic wave front sets

There are several definitions of analytic wave front set (or singular spectrum), which are equivalent to each other. Here we adopt the following definition.

Definition 3.1.1 (i) Let $u \in \mathcal{F}_0$. The analytic wave front set $WF_A(u) \subset T^*\boldsymbol{R}^n \setminus 0$ ($\simeq \boldsymbol{R}^n \times (\boldsymbol{R}^n \setminus \{0\})$) is defined as follows: $(x^0, \xi^0) \in T^*\boldsymbol{R}^n \setminus 0$ does not belong to $WF_A(u)$ if there are a conic neighborhood Γ of ξ^0,

$R_0 > 0$ and $\{g^R(\xi)\}_{R \geq R_0} \subset C^\infty(\boldsymbol{R}^n)$ such that $g^R(\xi) = 1$ in $\Gamma \cap \{\langle \xi \rangle \geq R\}$,

$$\left| \partial_\xi^{\alpha+\tilde{\alpha}} g^R(\xi) \right| \leq C_{|\tilde{\alpha}|} (C/R)^{|\alpha|} \langle \xi \rangle^{-|\tilde{\alpha}|} \tag{3.1}$$

if $\langle \xi \rangle \geq R|\alpha|$, and $g^R(D)u$ is analytic at x^0 for $R \geq R_0$, where C is a positive constant independent of R. (ii) Let X be an open subset of $\boldsymbol{R}^n$, and let $u \in \mathcal{B}(X)$ and $(x^0, \xi^0) \in T^*X \setminus 0$ ($\simeq X \times (\boldsymbol{R}^n \setminus \{0\})$). Then we say that $(x^0, \xi^0) \notin WF_A(u)$ ($\subset T^*X \setminus 0$) if there are a bounded open neighborhood U of x^0 and $v \in \mathcal{A}'(\overline{U})$ such that $v|_U = u|_U$ in $\mathcal{B}(U)$ and $(x^0, \xi^0) \notin WF_A(v)$.

Remark $WF_A(u)$ for $u \in \mathcal{B}(X)$ is well-defined. In fact, it follows from Theorem 2.6. 5 that for any $v \in \mathcal{A}'$ with $x^0 \notin \operatorname{supp} v$ there is $R_1 > 0$ such that $g^R(D)v$ is analytic at x^0 if $R \geq R_1$, where $\{g^R(\xi)\}_{R \geq R_0}$ is a family of symbols satisfying (3.1).

Proposition 3.1. 2 (i) *Let $u \in \mathcal{F}_0$ and $(x^0, \xi^0) \in T^*\boldsymbol{R}^n \setminus 0$, and let Γ be a conic neighborhood of ξ^0. Assume that a symbol $g(\xi)$ satisfies $|\partial_\xi^\alpha g(\xi)| \leq C_{|\alpha|}$ for any $\xi \in \boldsymbol{R}^n$ and $\alpha \in \boldsymbol{Z}_+^n$ and, with $R_1 > 0$, $g(\xi) = 1$ in $\Gamma \cap \{\langle \xi \rangle \geq R_1\}$, and that $g(D)u$ is analytic in a neighborhood U of x^0. Let $\tilde{U}$ ($\subset\subset U$) be a neighborhood of x^0, and put $\varepsilon = \operatorname{dis}(\tilde{U}, \boldsymbol{R}^n \setminus U)$. Then $g^R(D)u$ is analytic in $\tilde{U}$ for $R \geq 72e\sqrt{n}C/\varepsilon$ if $g^R(\xi)$ satisfies (3.1) with some $C > 0$ and* $\operatorname{supp} g^R \subset \Gamma$. *In particular, we have $(x^0, \xi^0) \notin WF_A(u)$.* (ii) *Let $u \in \mathcal{F}_0$, $x^0 \in \boldsymbol{R}^n$, and let Γ be an open conic subset of $\boldsymbol{R}^n \setminus \{0\}$. Assume that $\{x^0\} \times \overline{\Gamma} \cap WF_A(u) = \emptyset$. Then there are a symbol $g(\xi)$ and $R_1 > 0$ such that $|\partial_\xi^\alpha g(\xi)| \leq C_{|\alpha|}$, $g(\xi) = 1$ in $\Gamma \cap \{\langle \xi \rangle \geq R_1\}$ and $g(D)u$ is analytic at x^0. In particular, u is analytic at x^0 if and only if $(x^0, \xi) \notin WF_A(u)$ for any $\xi \in \boldsymbol{R}^n \setminus \{0\}$.*

Proof (i) Let $g^R(\xi)$ be a symbol satisfying (3.1) and $\operatorname{supp} g^R \subset \Gamma$. Note that $\operatorname{supp} (g^R(\xi)g(\xi) - g^R(\xi)) \subset \{\xi \in \boldsymbol{R}^n;\ \langle \xi \rangle \leq R_1\}$ and that $g^R(D)g(D)u - g^R(D)u$ is (entire) analytic. It follows from Theorem 2.6. 5 that $g^R(D)g(D) \times u$ is analytic in $\tilde{U}$ for $R \geq 72e\sqrt{n}C/\varepsilon$. This proves the assertion (i). (ii) It suffices to prove the first part of the assertion (ii). By definition, for each $\eta \in \overline{\Gamma} \cap S^{n-1}$ there are a conic neighborhood Γ_η of η, a neighborhood U_η of x^0, $R_\eta > 0$ and a symbol $g_\eta(\xi)$ such that $g_\eta(\xi) = 1$ in $\Gamma_\eta \cap \{\langle \xi \rangle \geq R_\eta\}$, $|\partial_\xi^\alpha g_\eta(\xi)| \leq C_{\eta,|\alpha|}$, and $g_\eta(D)u$ is analytic in U_η. Since $\overline{\Gamma} \cap S^{n-1}$ is compact, there are $\eta_1, \cdots, \eta_N \in \overline{\Gamma} \cap S^{n-1}$ such that $\Gamma \subset\subset \bigcup_{j=1}^N \Gamma_{\eta_j}$. Let U be an open neighborhood of x^0 such that $U \subset \subset \bigcap_{j=1}^N U_{\eta_j}$. By Proposition 2.2. 3 there are symbols $g_j^R(\xi)$ ($1 \leq j \leq N$,

$R \geq 4$) such that $\operatorname{supp} g_j^R \subset \Gamma_{\eta_j} \cap \{\langle\xi\rangle \geq R/2\}$, $\sum_{j=1}^N g_j^R(\xi) = 1$ for $\xi \in \Gamma$ with $\langle\xi\rangle \geq R$ and

$$\left|\partial_\xi^{\alpha+\tilde{\alpha}} g_j^R(\xi)\right| \leq C_{|\tilde{\alpha}|}(C/R)^{|\alpha|}\langle\xi\rangle^{-|\tilde{\alpha}|}$$

if $\langle\xi\rangle \geq R|\alpha|$. By the assertion (i) there is $R_0 > 0$ such that $g_j^R(D)u$ ($1 \leq j \leq N$) are analytic in U if $R \geq R_0$. If we put $g(\xi) \equiv \sum_{j=1}^N g_j^{R_0}(\xi)$ and $R_1 = R_0$, then $g(\xi)$ has the desired properties. □

Hörmander gave a definition of $WF_A(u)$ in [Hr5]. We shall give his definition of $WF_A(u)$ and prove that his definition coincides with ours. Define

$$I(\xi) := \int_{|\omega|=1} e^{-\omega\cdot\xi}\, dS_\omega \quad \text{for } \xi \in \boldsymbol{C}^n,$$

where dS_ω denotes the surface element on S^{n-1}. Then we have

$$I(\xi) = \begin{cases} 2\cosh\xi & (\,n = 1), \\ I_0(\sqrt{\xi^2}) & (\,n > 1), \end{cases}$$

where

$$I_0(\tau) = c_{n-1}\int_{-1}^{1} (1-t^2)^{(n-3)/2} e^{t\tau}\, dt$$

and c_{n-1} denotes the area of S^{n-2}, *i.e.*,

$$c_{n-1} = 2\pi^{(n-1)/2}/\Gamma((n-1)/2).$$

It is easy to see that

$$I_0(\tau) = c(n)\tau^{-(n-2)/2} J_{(n-2)/2}(i\tau),$$

where $J_\nu(z)$ denotes the Bessel function of order ν and

$$c(n) = (2\pi)^{n/2} i^{-(n-2)/2}, \quad \arg c(n) = -\pi(n-2)/4.$$

Lemma 3.1.3 (i) *$I(\xi)$ is an entire analytic function in $\boldsymbol{C}^n$. Moreover, we have $\xi^2\ (= \xi\cdot\xi) < 0$ if $I(\xi) = 0$, and*

$$I(\xi) \geq \begin{cases} 1 & \text{if } n = 1 \text{ and } \xi \in \boldsymbol{R}, \\ 2\pi^{n/2}/\Gamma(n/2) & \text{if } n > 1 \text{ and } \xi \in \boldsymbol{R}^n. \end{cases}$$

(ii) *For any $\varepsilon > 0$ we have*

$$I_0(\tau) = (2\pi)^{(n-1)/2} e^\tau \tau^{-(n-1)/2}(1 + O(1/\tau))$$

as $\tau \to \infty$ *in the sector* $|\arg \tau| \le \pi/2 - \varepsilon$. *Moreover, there is* $C > 0$ *such that*

$$|I_0(\tau)| \le C(1+|\tau|)^{-(n-1)/2} e^{|\mathrm{Re}\,\tau|} \quad \textit{for } \tau \in \boldsymbol{C}.$$

(iii) *There are positive constants* A_k $(1 \le k \le 3)$ *and* C_k $(0 \le k \le 3)$ *such that*

$$\left|\partial_\tau^j \left(e^{-\tau} I_0(\tau)\right)\right| \le C_0 2^j j! \tau^{-(n-1)/2-j} \tag{3.2}$$

for $\tau \ge 1$*, and*

$$\left|\partial_\xi^\alpha \left\{e^{\langle\xi\rangle} I(\xi)^{-1}\right\}\right| \le C_1 A_1^{|\alpha|} |\alpha|! \langle\xi\rangle^{(n-1)/2-|\alpha|}, \tag{3.3}$$

$$\left|\partial_\xi^\alpha I(\xi)^{-1}\right| \le \begin{cases} C_2 (A_2/\rho)^{|\alpha|} |\alpha|! \langle\xi\rangle^{(n-1)/2-|\alpha|} e^{(\rho-1)\langle\xi\rangle} \\ \qquad\qquad \textit{if } 0 < \rho \le 1, \\ C_3 A_3^{|\alpha|} |\alpha|! \langle\xi\rangle^{(n-1)/2} e^{-\langle\xi\rangle} \end{cases} \tag{3.4}$$

for $\xi \in \boldsymbol{R}^n$.

Proof (i) It is obvious that $I(\xi)$ is entire analytic. Since the zeros of $J_\nu(z)$ are all real when $\nu \ge -1$ and

$$I_0(\tau) = c_{n-1}\sqrt{\pi}\,\Gamma((n-1)/2) \sum_{k=0}^{\infty} (\tau^2/4)^k / (k!\Gamma(n/2+k)),$$

we obtain the assertion (i) (see, *e.g.*, [Ol]). (ii) From the asymptotic behaviors of the Bessel functions we have

$$\begin{aligned} I_0(\tau) &= 2(2\pi)^{(n-1)/2} \tau^{-(n-1)/2} i^{-(n-1)/2} \\ &\quad \times \Big[\cos(i\tau - (n-1)\pi/4)\,(1 + O(1/\tau^2)) \\ &\quad - \sin(i\tau - (n-1)\pi/4)\, O(1/\tau)\Big] \end{aligned} \tag{3.5}$$

as $\tau \to \infty$ in the sector $-3\pi/2 + \varepsilon < \arg \tau < \pi/2 - \varepsilon$, where $\varepsilon > 0$ (see, *e.g.*, [Ol]). We note that (3.5) is valid as $\tau \to \infty$ in $\boldsymbol{C}$, since $I_0(-\tau) = I_0(\tau)$ and $I_0(\tau)$ is entire analytic in $\boldsymbol{C}$. This proves the assertion (ii). (iii) We have

$$\begin{aligned} \partial_\tau^j \left(e^{-\tau} I_0(\tau)\right) &= c_{n-1} \tau^{-(n-1)/2-j} \int_0^{2\tau} (-1)^j s^{(n-3)/2+j} \\ &\quad \times (2 - s/\tau)^{(n-3)/2} e^{-s}\, ds. \end{aligned}$$

Therefore, a simple calculation yields

$$\left|\partial_\tau^j \left(e^{-\tau} I_0(\tau)\right)\right| \le c_{n-1} 2^{(n-3)/2} \Gamma((n-1)/2 + j) \tau^{-(n-1)/2-j}$$

if $n \ge 3$ and $\tau \ge 1$. For $n = 2$ we have

$$\begin{aligned}\left|\partial_\tau^j\left(e^{-\tau}I_0(\tau)\right)\right| &\le c_1\tau^{-1/2-j}\Big\{\int_0^\tau s^{-1/2+j}e^{-s}\,ds\\ &\quad+\int_\tau^{2\tau}(2\max\{1/2,j-1/2\}/e)^{j-1/2}(2-s/\tau)^{-1/2}e^{-s/2}\,ds\Big\}\\ &\le c_1\tau^{-1/2-j}\Big\{\Gamma(1/2+j)\\ &\qquad+\int_0^1(2\max\{1/2,j-1/2\}/e)^{j-1/2}t^{-1/2}\tau e^{-\tau/2}\,dt\Big\}\\ &\le c_1\tau^{-1/2-j}\Big\{\Gamma(1/2+j)+4e^{-1}(2\max\{1/2,j-1/2\}/e)^{j-1/2}\Big\}\end{aligned}$$

for $\tau \ge 1$. Here we have used the estimates

$$\sup_{s\ge1} s^a e^{-s/2} \le \begin{cases} e^{-1/2} & \text{if } a \le 0,\\ (2a/e)^a & \text{if } a > 0.\end{cases}$$

Since $\Gamma(k) \sim e^{-k}k^{k-1/2}$ and $j! \sim e^{-j}j^{j+1/2}$, we have

$$\Gamma((n-1)/2+j) \le C_\varepsilon(1+\varepsilon)^j j!$$

for $\varepsilon > 0$ and $j \in \boldsymbol{Z}_+$, and

$$(2\max\{1/2,j-1/2\}/e)^{j-1/2} \le C2^j j!.$$

This proves (3.2). If $|\xi| \le 1$, then it is obvious that (3.3) is valid. Assume that $|\xi| \ge 1$. If $n = 1$, then we have $e^{|\xi|}I(\xi)^{-1} = (1+e^{-2|\xi|})^{-1}$ and, by induction,

$$\left|\partial_\xi^j(e^{|\xi|}I(\xi)^{-1})\right| \le A_1'^{\,j} j!|\xi|^{-j}. \tag{3.6}$$

Now assume that $n \ge 2$, and put $f(\tau) = e^{-\tau}I_0(\tau)$. Then, by induction on $|\alpha|$ we can show that

$$\left|\partial_\xi^\alpha f^{(j)}(|\xi|)\right| \le C_0 2^j A_0^{|\alpha|}(j+|\alpha|)!|\xi|^{-(n-1)/2-j-|\alpha|}.$$

Therefore, by induction we can also prove that

$$\left|\partial_\xi^\alpha f(|\xi|)^{-1}\right| \le A_1'^{|\alpha|}|\alpha|!|f(|\xi|)|^{-1}|\xi|^{-|\alpha|}. \tag{3.7}$$

On the other hand, we have

$$\left|\partial_\xi^\alpha e^{\langle\xi\rangle-|\xi|}\right| \le (8(1+\sqrt{2}))^{|\alpha|}|\alpha|!\langle\xi\rangle^{-|\alpha|} \tag{3.8}$$

for $|\xi| \geq 1$. The assertions (i) and (ii) and (3.6)–(3.8) give (3.3). Since $I(\xi)^{-1} = e^{-|\xi|}(e^{|\xi|}I(\xi)^{-1})$ and

$$\left|\partial_\xi^\alpha e^{-|\xi|}\right| \leq (\max\{2, 4/\rho\}(1+\sqrt{2}))^{|\alpha|}|\alpha|!\langle\xi\rangle^{-|\alpha|} \times e^{-|\xi|}\sum_{\mu=0}^{|\alpha|}(\rho\langle\xi\rangle)^\mu/\mu! \quad \text{if } \rho > 0,$$

the assertions (i) and (ii) and (3.6) and (3.7) give (3.4). □

From Lemma 3.1. 3 we can define

$$K(z) := (2\pi)^{-n}\int_{R^n} e^{iz\cdot\xi}/I(\xi)\, d\xi$$

for $z \in \Omega \equiv \{z \in C^n;\ |\text{Im } z| < 1\}$.

Lemma 3.1. 4 (Lemma 8.4.10 in Hörmander [Hr5]) *$K(z)$ can be continued analytically to an analytic function in the connected open set $\tilde{\Omega} \equiv \{z \in C^n;\ z^2 \notin (-\infty, -1]\}$ $(\supset \Omega)$. For any closed cone $\Gamma \subset \tilde{\Omega}$ satisfying $z^2 \notin (-\infty, 0]$ for every $z \in \Gamma \setminus \{0\}$ there is $c > 0$ such that $K(z) = O(e^{-c|z|})$ as $z \to \infty$ in Γ. Moreover, we have*

$$|K(x+iy)| \leq K(iy) \quad \text{for } x, y \in R^n \text{ with } |y| < 1$$

and

$$K(iy) = (2\pi)^{-n}(n-1)!(1-|y|)^{-n}(1+O(1-|y|))$$

as $|y| \uparrow 1$ for $y \in R^n$.

We define

$$K(\cdot - i\omega) * u(x) = \mathcal{F}^{-1}[e^{\omega\cdot\xi}I(\xi)^{-1}\hat{u}(\xi)](x)$$

for $u \in \cup_{\varepsilon \in R} \mathcal{S}_\varepsilon{}'$ and $\omega \in R^n$ with $|\omega| \leq 1$. If $u \in \mathcal{A}'(R^n)$, then

$$K(\cdot - i\omega) * u(x) = u_y(K(x-y-i\omega)) \quad \text{for } \omega \in R^n \text{ with } |\omega| < 1$$

and

$$K(\cdot - i\omega) * u(x) = \lim_{r \to 1-0} u_y(K(x-y-ir\omega)) \quad \text{in } \mathcal{F}_0$$

for $\omega \in S^{n-1}$.

Lemma 3.1. 5 *For* $\omega \in \boldsymbol{R}^n$ *with* $|\omega| \le 1$ $K(\cdot - i\omega)*$ *maps continuously* $\mathcal{S}_\varepsilon$ *and* $\mathcal{S}_\varepsilon{}'$ *to* $\mathcal{S}_{\varepsilon+1-|\omega|}$ *and* $\mathcal{S}_{\varepsilon-1+|\omega|}{}'$, *respectively, where* $\varepsilon \in \boldsymbol{R}$. *If* $u \in \mathcal{S}_\varepsilon$
(*resp.* $u \in \mathcal{S}_\varepsilon{}'$), *then* $K(\cdot - i\omega) * u(x) \in C(\boldsymbol{B}; \mathcal{S}_\varepsilon)$ (*resp.* $\in C(\boldsymbol{B}; \mathcal{S}_\varepsilon{}')$), *where* $\boldsymbol{B} = \{\omega \in \boldsymbol{R}^n;\ |\omega| \le 1\}$. *Moreover, we have*

$$u = \int_{S^{n-1}} K(\cdot - i\omega) * u(x)\, dS_\omega \quad \text{in } \mathcal{S}_\varepsilon \ (\textit{ resp. } \mathcal{S}_\varepsilon{}') \tag{3.9}$$

$$u = \lim_{r \to 1-0} \int_{S^{n-1}} K(\cdot - ir\omega) * u(x)\, dS_\omega \quad \text{in } \mathcal{S}_\varepsilon \ (\textit{ resp. } \mathcal{S}_\varepsilon{}') \tag{3.10}$$

if $u \in \mathcal{S}_\varepsilon$ (*resp.* $\mathcal{S}_\varepsilon{}'$).

Proof It follows from Lemma 3.1. 3 that

$$\widehat{\mathcal{S}}_\varepsilon \ni v(\xi) \longmapsto e^{\omega\cdot\xi} v(\xi)/I(\xi) \in \widehat{\mathcal{S}}_{\varepsilon+1-|\omega|},$$
$$\widehat{\mathcal{S}}_\varepsilon{}' \ni v(\xi) \longmapsto e^{\omega\cdot\xi} v(\xi)/I(\xi) \in \widehat{\mathcal{S}}_{\varepsilon-1+|\omega|}{}'$$

are continuous. Since

$$\left|\partial_\xi^\alpha (e^{\omega\cdot\xi - \langle\xi\rangle} - e^{\omega'\cdot\xi - \langle\xi\rangle})\right| \le C_{|\alpha|}(1+|\xi|)|\omega - \omega'| \tag{3.11}$$

for $\omega, \omega' \in \boldsymbol{B}$, $e^{\omega\cdot\xi}\hat{u}(\xi)/I(\xi) \in C(\boldsymbol{B}; \widehat{\mathcal{S}}_\varepsilon)$ (*resp.* $\in C(\boldsymbol{B}; \widehat{\mathcal{S}}_\varepsilon{}')$) if $u \in \mathcal{S}_\varepsilon$ (*resp.* $\in \mathcal{S}_\varepsilon{}'$). This proves the first part of the lemma. In order to prove (3.9) it is sufficient to show that

$$\hat{u}(\xi) = \int_{S^{n-1}} e^{\omega\cdot\xi}\hat{u}(\xi)/I(\xi)\, dS_\omega \tag{3.12}$$

in $\widehat{\mathcal{S}}_\varepsilon$ (*resp.* $\widehat{\mathcal{S}}_\varepsilon{}'$) if $u \in \mathcal{S}_\varepsilon$ (*resp.* $\mathcal{S}_\varepsilon{}'$). Using (3.11), we can see that the Riemann sum of the integral

$$\langle\xi\rangle^{-1} \int_{S^{n-1}} e^{\omega\cdot\xi - \langle\xi\rangle}\, dS_\omega \ \left(= \langle\xi\rangle^{-1} e^{-\langle\xi\rangle} I(\xi)\right)$$

converges in $\mathcal{B}^\infty(\boldsymbol{R}^n_\xi)$. Here $\mathcal{B}^\infty(\boldsymbol{R}^n)$ denotes the set

$$\{f \in C^\infty(\boldsymbol{R}^n);\ \sup_{x \in R^n} |D^\alpha f(x)| < \infty \text{ for any } \alpha \in \boldsymbol{Z}_+^n\}$$

and we introduce the natural topology to $\mathcal{B}^\infty(\boldsymbol{R}^n)$. Noting that $\langle\xi\rangle e^{\langle\xi\rangle} \times I(\xi)^{-1}\hat{u}(\xi) \in \widehat{\mathcal{S}}_\varepsilon$ (*resp.* $\in \widehat{\mathcal{S}}_\varepsilon{}'$) if $u \in \mathcal{S}_\varepsilon$ (*resp.* $\in \mathcal{S}_\varepsilon{}'$), we have (3.12) and, therefore, (3.9). Similarly, we can show that

$$\langle\xi\rangle^{-1} \int_{S^{n-1}} e^{r\omega\cdot\xi - \langle\xi\rangle}\, dS_\omega \longrightarrow \langle\xi\rangle^{-1} \int_{S^{n-1}} e^{\omega\cdot\xi - \langle\xi\rangle}\, dS_\omega$$

in $\mathcal{B}^\infty(\boldsymbol{R}^n_\xi)$ as $r \to 1-0$, which proves (3.10). □

From Lemmas 1.1.3 and 3.1.5 it follows that

$$U(z) \equiv K * u(z) \ \left(= \mathcal{F}^{-1}[e^{-\mathrm{Im}\, z\cdot\xi} I(\xi)^{-1}\hat{u}(\xi)](\mathrm{Re}\ z)\right)$$

is analytic in Ω. Hörmander defined $WF_A(u)$ for $u \in \mathcal{A}'(\boldsymbol{R}^n)$ in [Hr5] as follows: $(x^0,\xi^0) \in T^*\boldsymbol{R}^n \setminus 0$ does not belong to $WF_A(u)$ if $U(z)$ can be continued analytically to a neighborhood of $x^0 - i\xi^0/|\xi^0|$. In this section we shall prove the following

Theorem 3.1.6 *Assume that $u \in \mathcal{F}_0$ and $(x^0,\xi^0) \in T^*\boldsymbol{R}^n \setminus 0$. Then, $(x^0,\xi^0) \notin WF_A(u)$ if and only if $U(z) \equiv K * u(z)$ can be continued analytically to a neighborhood of $x^0 - i\xi^0/|\xi^0|$, i.e., Our definition of $WF_A(u)$ coincides with Hörmander's.*

Put $a(\xi;\omega) = e^{\omega\cdot\xi}I(\xi)^{-1}$, $\tilde{a}(\xi) = e^{\langle\xi\rangle}I(\xi)^{-1}$ and $S(x,\xi;\omega) = x\cdot\xi + i(\langle\xi\rangle - \omega\cdot\xi)$ for $\xi \in \boldsymbol{R}^n$ and $\omega \in \boldsymbol{R}^n$ with $|\omega| \le 1$. Then we have

$$K(\cdot - i\omega) * u(x) = a(D;\omega)u(x) = \tilde{a}_S(D)u(x)$$

for $u \in \mathcal{F}_0$ and $\omega \in \boldsymbol{R}^n$ with $|\omega| \le 1$. To prove Theorem 3.1.6 we need the following

Lemma 3.1.7 *Let $x^0 \in \boldsymbol{R}^n$, and assume that $v \in \mathcal{F}_0$ is analytic at x^0. Then $K(\cdot - i\omega) * v(x)$ is analytic at x^0 if $\omega \in \boldsymbol{R}^n$ and $|\omega| \le 1$.*

Proof The proof is very similar to that of Theorem 2.6.5. Let X and X_j ($j = 1,2$) be bounded open neighborhoods of x^0 such that $X_1 \Subset X_2 \Subset X$ and

$$|D^\beta v_\rho(x)| \le C(v)A(v)^{|\beta|}|\beta|!$$

for $x \in X$ and $0 < \rho \le 1$, where $v_\rho(x) = e^{-\rho\langle D\rangle}v(x)$ and $C(v)$ and $A(v)$ are positive constants independent of ρ. Let $\Phi^R(x,\xi)$ ($R \ge 4$) be a symbol in the proof of Theorem 2.6.5. we put

$$\begin{aligned}
&a_1^R(x,\xi;\omega) = \Phi^R(x,\xi)a(\xi;\omega),\\
&\tilde{a}_1^R(x,\xi) = \Phi^R(x,\xi)\tilde{a}(\xi),\\
&a_2^R(x,\xi;\omega) = (1-\Phi^R(x,\xi))\phi_1^R(\xi)a(\xi;\omega),\\
&a_3^R(x,\xi;\omega) = (1-\Phi^R(x,\xi))(1-\phi_1^R(\xi))a(\xi;\omega).
\end{aligned}$$

It is obvious that $a_3^R(x,D;\omega)v$ is analytic in X_1. It follows from Corollary 2.6.2 and (3.4) that $x^0 \notin \mathrm{supp}\ a_2^R(x,D;\omega)v$ if $R \ge 2e\max\{n, 4\sqrt{n}\}C(X_1,X_2)$. Note that $S(x,\xi;\omega) \in \mathcal{P}(\boldsymbol{R}^n\times(\boldsymbol{R}^n\setminus\{0\}); 1+\sqrt{2}, B, 1, 0, 1, 1)$ for any $B > 0$, $\tilde{a}_1^R(x,\xi) \in S^{(n-1)/2}(R, C_* + 2A_1, C(X_1, X_2))$, supp $\tilde{a}_1^R(x,\xi) \subset X_2 \times \boldsymbol{R}^n$ and

$$\left|\tilde{a}_{1(\beta)}^{R(\alpha+\tilde{\alpha})}(x,\xi)\right| \le \begin{cases} 0 & (\beta > 0), \\ C_{|\tilde{\alpha}|}(2A_1)^{|\alpha|}|\alpha|!\langle\xi\rangle^{(n-1)/2-|\alpha|-|\tilde{\alpha}|} & (\beta = 0) \end{cases}$$

if $x \in X_1$ and $\langle\xi\rangle \ge R|\alpha|$, where $\omega \in \boldsymbol{R}^n$ and $|\omega| \le 1$. In order to prove that $a_1^R(x, D; \omega)v$ ($= \tilde{a}_{1S}^R(x, D)v$) is analytic at x^0 if $R \gg 1$, we have only to repeat the argument in the proof of Lemma 2.6. 4 , replacing the phase function $x \cdot \xi$ with $S(x,\xi)$. We omit the details as we shall prove general results on effects of Fourier integral operators on analytic wave front sets
(see Proposition 3.2. 3 and Theorems 3.2. 4 and 3.2. 5). □

Lemma 3.1. 8 *Assume that $u \in \mathcal{F}_0$ and $(x^0, \omega^0) \in \boldsymbol{R}^n \times S^{n-1}$, and put $U(z) \equiv K * u(z)$. Then $U(z)$ can be continued analytically from Ω to a neighborhood of $x^0 - i\omega^0$ if and only if $U(x - i\omega^0)$ ($\in \mathcal{F}_0$) is analytic at x^0.*

Proof First assume that $U(x - i\omega^0)$ is analytic at x^0. Put

$$V(x, x_{n+1}; \omega) = \mathcal{H}(U(\cdot - i\omega))(x, x_{n+1})$$
$$\Big(= (\text{sgn } x_{n+1}) \exp[-|x_{n+1}|\langle D\rangle]U(x - i\omega)/2\Big)$$

for $x_{n+1} \in \boldsymbol{R} \setminus \{0\}$ and $\omega \in \boldsymbol{R}^n$ with $|\omega| \le 1$. $V(\text{Re } z, x_{n+1}; -\text{Im } z)$ is an analytic function of (z, x_{n+1}) in $\Omega \times (\boldsymbol{R} \setminus \{0\})$. In fact, we have

$$\exp[iz \cdot \xi - x_{n+1}\langle\xi\rangle]I(\xi)^{-1} \in \widehat{\mathcal{S}}_\varepsilon,$$
$$V(\text{Re } z, x_{n+1}, -\text{Im } z) = (2\pi)^{-n}\langle\hat{u}(\xi), \exp[iz \cdot \xi - x_{n+1}\langle\xi\rangle]I(\xi)^{-1}\rangle_\xi/2$$

if $x_{n+1} > \varepsilon > 0$ and $z \in \Omega$. So we write

$$\widetilde{V}(z, x_{n+1}) = V(\text{Re } z, x_{n+1}; -\text{Im } z).$$

By definition $V(x, x_{n+1}; \omega^0)$ can be continued analytically from $\boldsymbol{R}^n \times (0, \infty)$ to a neighborhood of $(x^0, 0)$ in $\boldsymbol{R}^{n+1}$. Therefore, $\widetilde{V}(z, x_{n+1})$ can be continued analytically to a neighborhood of $(x^0 - i\omega^0, 0)$ in $\boldsymbol{C}^n \times \boldsymbol{R}$. Since $\widetilde{V}(z, +0) = U(z)/2$ for $z \in \Omega$, $U(z)$ can be continued analytically from Ω to a neighborhood of $x^0 - i\omega^0$. Conversely we assume that $U(z)$ can be continued analytically from Ω to a neighborhood of $x^0 - i\omega^0$. Choose $R > 1$ so that $|x^0| < R - 1$. It follows from Theorem 1.3. 3 that there are $u_1 \in \mathcal{A}'(\boldsymbol{R}^n)$ and $u_2 \in \mathcal{F}_0$ such that $u = u_1 + u_2$, supp $u_1 \subset \{x \in \boldsymbol{R}^n; |x| \le R\}$ and supp $u_2 \subset \{x \in \boldsymbol{R}^n; |x| \ge R\}$. Put

$$U_j(x - i\omega) = K(\cdot - i\omega) * u_j(x) \in \mathcal{F}_0 \quad \text{for } \omega \in \boldsymbol{R}^n \text{ with } |\omega| \le 1$$

($j = 1, 2$). By Lemma 3.1.7 $U_2(x - i\omega)$ is analytic in $\{x \in \boldsymbol{R}^n; |x| < R\}$ for $\omega \in S^{n-1}$. So $U_2(x - i\omega^0)$ can be continued analytically to a neighborhood of x^0 as we have just proved. Therefore, there is $\delta > 0$ such that $U_1(x - i\omega^0)$ can be continued analytically to a neighborhood of $\{x \in \boldsymbol{R}^n; |x - x^0| \le \delta\}$. Since $u_1 \in \mathcal{A}'(\boldsymbol{R}^n)$, we have

$$\begin{aligned} U_1(x - i\omega) &= u_{1y}(K(x - y - i\omega)) \quad \text{for } |\omega| < 1, \\ U_1(x - i\omega) &= \lim_{r \to 1-0} u_{1y}(K(x - y - ir\omega)) \quad \text{in } \mathcal{F}_0 \text{ for } |\omega| \le 1. \end{aligned}$$

We have also for $\varphi \in \mathcal{S}_\infty$ and $\omega \in \boldsymbol{R}^n$ with $|\omega| \le 1$

$$\langle U_1(x - i\omega), \varphi \rangle = \lim_{r \to 1-0} \int U_1(x - ir\omega)\varphi(x)\, dx.$$

By Lemma 3.1.4 $K(x - y - i\omega)$ is an analytic function of y in a complex neighborhood of $\{y \in \boldsymbol{R}^n; |y| < R + 1\}$ for $x \in \boldsymbol{R}^n$ with $|x| \ge R + 1$ and $\omega \in \boldsymbol{R}^n$ with $|\omega| \le 1$. Moreover, there are positive constants C and c such that

$$|K(x - y - i\omega)| \le Ce^{-c|x|}$$

if $x \in \boldsymbol{R}^n$, $|x| \ge R + 2$, $\omega \in \boldsymbol{R}^n$, $|\omega| \le 1$, $y \in \boldsymbol{C}^n$, $|\mathrm{Re}\, y| < R + 1$ and $|\mathrm{Im}\, y| < 1/2$. Hence $U_1(x - i\omega)$ is an analytic function of x in $\{x \in \boldsymbol{R}^n; |x| \ge R + 1\}$ for $\omega \in \boldsymbol{R}^n$ with $|\omega| \le 1$, and

$$|U_1(x - i\omega)| \le C(u)e^{-c|x|}$$

for $x \in \boldsymbol{R}^n$ with $|x| \ge R + 2$ and $\omega \in \boldsymbol{R}^n$ with $|\omega| \le 1$. Choose $\chi(x) \in C_0^\infty(\boldsymbol{R}^n)$ so that $\chi(x) = 1$ if $|x - x^0| \ge \delta$ and $|x| \le R + 2$, supp $\chi \subset \{x \in \boldsymbol{R}^n; |x - x^0| \ge \delta/2 \text{ and } |x| \le R + 3\}$ and $0 \le \chi(x) \le 1$. Since $\mathcal{S}_\infty \subset \mathcal{A} \cap \mathcal{S}$ and $U_1(z)$ is analytic in Ω, (Cauchy-) Stokes' formula yields

$$\begin{aligned} &\int U_1(x - ir\omega^0)\varphi(x)\, dx \\ &= \int_{\gamma(\omega^0, \varepsilon)} U_1(z - ir\omega^0)\varphi(z)\, dz_1 \wedge \cdots \wedge dz_n \end{aligned}$$

for $\varphi \in \mathcal{S}_\infty$, $0 \le r < 1$ and $0 < \varepsilon < 1$, where $\gamma(\omega^0, \varepsilon) = \{x + i\varepsilon\chi(x)\omega^0; x \in \boldsymbol{R}^n\}$. Therefore, we have

$$\langle U_1(x - i\omega^0), \varphi \rangle = \int_{\gamma(\omega^0, \varepsilon)} U_1(z - i\omega^0)\varphi(z)\, dz_1 \wedge \cdots \wedge dz_n$$

for $\varphi \in \mathcal{S}_\infty$. Write

$$U_1(x - i\omega^0) = U_{11}(x) + U_{12}(x) + U_{13}(x) \quad \text{in } \mathcal{F}_0,$$

where U_{1j} ($1 \le j \le 3$) are defined by

$$\langle U_{11}, \varphi\rangle = \int_{|x|\ge R+3} U_1(x - i\omega^0)\varphi(x)\, dx,$$
$$\langle U_{12}, \varphi\rangle = \int_{|x-x^0|\le\delta/2} U_1(x - i\omega^0)\varphi(x)\, dx,$$
$$\langle U_{13}, \varphi\rangle = \int_{\gamma'(\omega^0,\varepsilon)} U_1(z - i\omega^0)\varphi(z)\, dz_1 \wedge \cdots \wedge dz_n,$$

for $\varphi \in \mathcal{S}_\infty$. Here $\gamma'(\omega^0,\varepsilon)$ denotes the chain $\gamma(\omega^0,\varepsilon)\cap\{z \in \boldsymbol{C}^n;\ |\text{Re } z| \le R+3 \text{ and } |\text{Re } z - x^0| \ge \delta/2\}$. Then it is obvious that $U_{11}(x) = U_1(x - i\omega^0)\chi_{\{|x|\ge R+3\}}(x) \in L^1(\boldsymbol{R}^n)$, supp $U_{11} \subset \{x \in \boldsymbol{R}^n;\ |x| \ge R+3\}$, $U_{12}(x) = U_1(x - i\omega^0)\chi_{\{|x-x^0|\le\delta/2\}}(x)$ and U_{12} is analytic at x^0, where χ_F denotes the characteristic function of F. Moreover, $U_{13} \in \mathcal{A}'(\boldsymbol{R}^n)$ and supp $U_{13} \subset \{x \in \boldsymbol{R}^n;\ |x| \le R+3 \text{ and } |x - x^0| \ge \delta/2\}$. In fact, U_{13} is an analytic functional carried by $\gamma'(\omega^0,\varepsilon)$ for $0 < \varepsilon < 1$. Therefore, $U_1(x - i\omega^0)$ is analytic at x^0, which implies that $U(x - i\omega^0)$ is analytic at x^0. □

Proof of Theorem 3.1. 6 First assume that $(x^0,\xi^0) \notin WF_A(u)$. By definition there are a conic neighborhood Γ of ξ^0, $R_0 > 0$ and $\{g^R(\xi)\}_{R\ge R_0} \subset C^\infty(\boldsymbol{R}^n)$ such that $g^R(\xi) = 1$ in $\Gamma \cap \{\langle\xi\rangle \ge R\}$, $g^R(\xi)$ satisfies (3.1) and $g^R(D)u$ is analytic at x^0 for $R \ge R_0$. Write

$$U(x - i\xi^0/|\xi^0|) = U_1^R(x) + U_2^R(x),$$

where

$$\begin{aligned} U_1^R(x) &= \mathcal{F}^{-1}[\exp[\xi^0\cdot\xi/|\xi^0|]I(\xi)^{-1}g^R(\xi)\hat{u}(\xi)](x) \\ \Big(&= g^R(D)U(x - i\xi^0/|\xi^0|)\Big), \\ U_2^R(x) &= \mathcal{F}^{-1}[\exp[\xi^0\cdot\xi/|\xi^0|]I(\xi)^{-1}(1 - g^R(\xi))\hat{u}(\xi)](x) \\ \Big(&= (1 - g^R(D))U(x - i\xi^0/|\xi^0|)\Big). \end{aligned}$$

There are positive constants C and δ such that

$$\exp[\xi^0\cdot\xi/|\xi^0| - \langle\xi\rangle] \le Ce^{-\delta\langle\xi\rangle} \quad \text{for } \xi \notin \Gamma.$$

This gives $U_2^R \in \mathcal{S}_{-\delta/2}'$ and, therefore, U_2^R is analytic at x^0. Since

$$U_1^R(x) = a(D;\xi^0/|\xi^0|)(g^R(D)u),$$

by Lemma 3.1. 7 U_1^R is analytic at x^0 for $R \geq R_0$. Therefore, it follows from Lemma 3.1. 8 that $U(z)$ can be continued analytically to a neighborhood of $x^0 - i\xi^0/|\xi^0|$. Conversely we assume that $U(z)$ can be continued analytically to a neighborhood of $x^0 - i\xi^0/|\xi^0|$. Then, by Lemma 3.1. 8 there are $\delta > 0$ and a neighborhood U_0 of x^0 such that $U(x - i\omega)$ is analytic at every $x \in U_0$ if $\omega \in S^{n-1}$ and $|\omega - \xi^0/|\xi^0|| < \delta$. Let $\{g^R(\xi)\}_{R \geq 2} \subset C^\infty(\boldsymbol{R}^n)$ be a family of symbols such that $g^R(\xi) = 1$ if $|\xi/|\xi| - \xi^0/|\xi^0|| < \delta/4$, $g^R(\xi) = 0$ if $|\xi/|\xi| - \xi^0/|\xi^0|| > \delta/2$ and $|\xi| \geq 1$, and

$$\left|\partial_\xi^{\alpha+\tilde{\alpha}} g^R(\xi)\right| \leq C_{|\tilde{\alpha}|}(C_*/(\delta R))^{|\alpha|}\langle\xi\rangle^{-|\tilde{\alpha}|}$$

if $\langle\xi\rangle \geq R|\alpha|$. Choose a neighborhood U_1 of x^0 so that $U_1 \subset\subset U_0$. By Theorem 2.6. 5 there is $R_\delta \geq 1$ such that $g^R(D)U(x - i\omega)$ ($= K * (g^R(D)u)$ $(x - i\omega)$) is analytic at every $x \in U_1$ if $R \geq R_\delta$, $\omega \in S^{n-1}$ and $|\omega - \xi^0/|\xi^0|| < \delta$. On the other hand, there is $c > 0$ such that

$$|e^{\omega\cdot\xi} I(\xi)^{-1} g^R(\xi)| \leq C \exp[-c\delta^2\langle\xi\rangle]$$

if $\omega \in S^{n-1}$ and $|\omega - \xi^0/|\xi^0|| \geq \delta$. Thus $K * (g^R(D)u)(x - i\omega)$ is analytic in $\boldsymbol{R}^n$ if $\omega \in S^{n-1}$ and $|\omega - \xi^0/|\xi^0|| \geq \delta$. It follows from Lemma 3.1. 5 that

$$g^R(D)u = \int_{S^{n-1}} K * (g^R(D)u)(x - i\omega)\, dS_\omega \quad \text{in } \mathcal{F}_0.$$

We put

$$F^R(x, x_{n+1}; \omega) = \mathcal{H}(K * (g^R(D)u)(\cdot - i\omega))(x, x_{n+1}),$$
$$F^R(x, x_{n+1}) = \mathcal{H}(g^R(D)u)(x, x_{n+1})$$

for $x_{n+1} \in \boldsymbol{R} \setminus \{0\}$. Then, we have

$$F^R(x, x_{n+1}) = \int_{S^{n-1}} F^R(x, x_{n+1}; \omega)\, dS_\omega \quad \text{in } \mathcal{S}_{-\varepsilon}' \tag{3.13}$$

for $|x_{n+1}| > \varepsilon > 0$. Since $K * (g^R(D)u)(x - i\omega)$ is analytic in U_1 if $R \geq R_\delta$ and $\omega \in S^{n-1}$, for any $\omega \in S^{n-1}$ there is $\delta_\omega > 0$ such that $F^R(x, x_{n+1}; \omega)$ can be continued analytically from $\boldsymbol{R}^n \times (0, \infty)$ to the set $\{(x, x_{n+1}) \in \boldsymbol{C}^n \times \boldsymbol{R};\ |x - x^0| < \delta_\omega \text{ and } x_{n+1} > -\delta_\omega\}$ if $R \geq R_\delta$. Noting that

$$\begin{aligned}
&F^R(x, x_{n+1}; \omega) \\
&= (2\pi)^{-n}\langle g^R(\xi)\hat{u}(\xi), \exp[(x - i\omega)\cdot\xi - x_{n+1}\langle\xi\rangle] I(\xi)^{-1}\rangle_\xi/2, \\
&\exp[(x - i\omega)\cdot\xi - x_{n+1}\langle\xi\rangle] I(\xi)^{-1} \in \widehat{\mathcal{S}}_\varepsilon(\boldsymbol{R}^n_\xi)
\end{aligned}$$

if $x \in \boldsymbol{R}^n$, $x_{n+1} > \varepsilon > 0$ and $\omega \in S^{n-1}$, we have

$$F^R(x, x_{n+1}; \omega^1) = F^R(x - i(\omega^1 - \omega), x_{n+1}; \omega)$$

if $x \in \boldsymbol{R}^n$, $x_{n+1} > 0$, $|x - x^0| < \delta_\omega/2$, $\omega, \omega^1 \in S^{n-1}$ and $|\omega - \omega^1| < \delta_\omega/2$. This implies that $F^R(x, x_{n+1}; \omega^1)$ can be continued analytically from $\boldsymbol{R}^n \times (0, \infty)$ to the set $\{(x, x_{n+1}) \in \boldsymbol{C}^n \times \boldsymbol{R};\ |x - x^0| < \delta_\omega/2$ and $x_{n+1} > -\delta_\omega\}$ if $R \geq R_\delta$, $\omega, \omega^1 \in S^{n-1}$ and $|\omega - \omega^1| < \delta_\omega/2$. Since S^{n-1} is compact, there are $N \in \boldsymbol{N}$ and $\omega^j \in S^{n-1}$ ($1 \leq j \leq N$) such that

$$S^{n-1} = \bigcup_{j=1}^N \{\omega \in S^{n-1};\ |\omega - \omega^j| < \delta_{\omega^j}/2\}.$$

Put $\delta_0 = \min_{1 \leq j \leq N} \delta_{\omega^j}$. Then $F^R(x, x_{n+1}; \omega)$ can be continued analytically from $\boldsymbol{R}^n \times (0, \infty)$ to the set $\{(x, x_{n+1}) \in \boldsymbol{C}^n \times \boldsymbol{R};\ |x - x^0| < \delta_0/2$ and $x_{n+1} > -\delta_0\}$ if $R \geq R_\delta$ and $\omega \in S^{n-1}$, and

$$F^R(x, x_{n+1}; \omega) = F^R(x - i(\omega - \omega'), x_{n+1}; \omega')$$

if $R \geq R_\delta$, $x \in \boldsymbol{C}^n$, $|x - x^0| < \delta_0/4$, $x_{n+1} > -\delta_0$, $\omega, \omega' \in S^{n-1}$ and $|\omega - \omega'| < \delta_0/4$. This, together with (3.13), show that $F^R(x, x_{n+1})$ can be continued analytically from $\boldsymbol{R}^n \times (0, \infty)$ to the set $\{(x, x_{n+1}) \in \boldsymbol{R}^n \times \boldsymbol{R};\ |x - x^0| < \delta_0/4$ and $x_{n+1} > -\delta_0\}$ if $R \geq R_\delta$. This implies that $g^R(D)u$ is analytic at x^0 if $R \geq R_\delta$, *i.e.*, $(x^0, \xi^0) \notin WF_A(u)$. □

Let $c \in \boldsymbol{R}$, and put

$$a_c(\xi; \omega) = e^{c(\omega \cdot \xi - \langle \xi \rangle)}.$$

Theorem 3.1. 9 *Let $(x^0, \xi^0) \in T^*\boldsymbol{R}^n \setminus 0$ and $u \in \mathcal{F}_0$, and assume that $c > 0$. Then $(x^0, \xi^0) \notin WF_A(u)$ if and only if $a_c(D; \xi^0/|\xi^0|)u$ is analytic at x^0.*

Remark Theorem 3.1. 9 is very similar to Theorem 3.1. 6 , and we can also prove Theorem 3.1. 6 by the same argument as in the proof of Theorem 3.1. 9 below.

Proof Applying the same argument as in the proof of Theorem 3.1. 6 we can prove that $a_c(D; \xi^0/|\xi^0|)u$ is analytic at x^0 if $(x^0, \xi^0) \notin WF_A(u)$. Now we assume that $a_c(D; \xi^0/|\xi^0|)u$ is analytic at x^0. Let $\{g^{R,\varepsilon}\}_{R \geq 2, \varepsilon > 0} \subset C^\infty(\boldsymbol{R}^n)$ be a family of symbols such that $g^{R,\varepsilon}(\xi) = 1$ if $|\xi/|\xi| - \xi^0/|\xi^0|| < \varepsilon$, $g^{R,\varepsilon}(\xi) = 0$ if $|\xi/|\xi| - \xi^0/|\xi^0|| > 2\varepsilon$ and $|\xi| \geq 1$, and

$$\left|\partial_\xi^{\alpha + \tilde{\alpha}} g^{R,\varepsilon}(\xi)\right| \leq C_{|\tilde{\alpha}|}(C_*/(\varepsilon R))^{|\alpha|} \langle \xi \rangle^{-|\tilde{\alpha}|}$$

if $\langle\xi\rangle \ge R|\alpha|$, where C_* is independent of ε and R. We put

$$v(x) = a_c(D;\xi^0/|\xi^0|)u,$$
$$b^{R,\varepsilon}(\xi) = g^{R,\varepsilon}(\xi)a_{-c}(\xi;\xi^0/|\xi^0|).$$

Then we have $g^{R,\varepsilon}(D)u = b^{R,\varepsilon}(D)v(x)$. Since

$$\left|\partial_\xi^\alpha\{c(\langle\xi\rangle - \xi^0\cdot\xi/|\xi^0|)\}\right|$$
$$\le \begin{cases} 3c\varepsilon \quad \text{if } |\alpha| = 1,\ |\xi| \ge \varepsilon^{-1/2} \text{ and } |\xi/|\xi| - \xi^0/|\xi^0|| \le 2\varepsilon, \\ c(1+\sqrt{2})^{|\alpha|}|\alpha|!\langle\xi\rangle^{1-|\alpha|} \quad \text{if } |\alpha| \ge 1, \end{cases}$$

(2.17) with $X(x,\xi) = 2/\varepsilon$ and $\rho(x,\xi) = c\varepsilon^2$ yields

$$\left|\partial_\xi^\alpha a_{-c}(\xi;\xi^0/|\xi^0|)\right| \le A^{|\alpha|}|\alpha|!\langle\xi\rangle^{-|\alpha|}\exp[c(\langle\xi\rangle - \xi^0\cdot\xi/|\xi^0| + \varepsilon^2\langle\xi\rangle)]$$

if $0 < \varepsilon \le 1$, $|\xi/|\xi| - \xi^0/|\xi^0|| \le 2\varepsilon$, $|\xi| \ge \varepsilon^{-1/2}$, $\rho > 0$ and $A \ge 24(1+\sqrt{2})/\varepsilon$. Noting that

$$\langle\xi\rangle - \xi^0\cdot\xi/|\xi^0| \le 1 + 2\varepsilon^2|\xi| \quad \text{if } |\xi/|\xi| - \xi^0/|\xi^0|| \le 2\varepsilon,$$

we have

$$\left|\partial_\xi^{\alpha+\tilde\alpha} b^{R,\varepsilon}(\xi)\right| \le C_{|\tilde\alpha|}((48(1+\sqrt{2}) + C_*)/(\varepsilon R))^{|\alpha|}\langle\xi\rangle^{-|\tilde\alpha|}\exp[3c\varepsilon^2\langle\xi\rangle]$$

if $0 < \varepsilon \le 1$, $|\xi| \ge \varepsilon^{-1/2}$ and $\langle\xi\rangle \ge R|\alpha|$. Applying the same argument as in the proof of Theorem 2.6. 5 , we will prove that $g^{R,\varepsilon}(D)u$ is analytic at x^0 for some R and ε. Let X and X_j ($j = 1,2$) be bounded open neighborhoods of x^0 such that $X_1 \Subset X_2 \Subset X$ and

$$|D^\beta v_\rho(x)| \le C(v)A(v)^{|\beta|}|\beta|!$$

for $x \in X$ and $0 < \rho \le 1$, where $v_\rho(x) = e^{-\rho\langle D\rangle}v(x)$ and $C(v)$ and $A(v)$ are positive constants independent of ρ. We choose $\delta > 0$ so that $\mathrm{dis}(\{x^0\}, \boldsymbol{R}^n \setminus X_1) \ge \delta$, $\mathrm{dis}(X_1, \boldsymbol{R}^n \setminus X_2) \ge \delta$ and $\mathrm{dis}(X_2, \boldsymbol{R}^n \setminus X) \ge \delta$. Let $\Phi^R(x,\xi)$ ($R \ge 4$) be a symbol in the proof of Theorem 2.6. 5 . We put

$$b_1^{R,\varepsilon}(x,\xi) = \Phi^R(x,\xi)b^{R,\varepsilon}(\xi),$$
$$b_2^{R,\varepsilon}(x,\xi) = (1-\Phi^R(x,\xi))\phi_1^R(\xi)b^{R,\varepsilon}(\xi),$$
$$b_3^{R,\varepsilon}(x,\xi) = (1-\Phi^R(x,\xi))(1-\phi_1^R(\xi))b^{R,\varepsilon}(\xi).$$

It is obvious that $b_3^{R,\varepsilon}(x,D)v$ is analytic in X_1. Since

$$\left|b_{2(\beta+\tilde{\beta})}^{R,\varepsilon(\alpha)}(x,\xi)\right| \le C_{|\alpha|+|\tilde{\beta}|}(C(X_1,X_2)/R)^{|\beta|}\langle\xi\rangle^{|\beta|-|\alpha|}\exp[3c\varepsilon^2\langle\xi\rangle]$$

if $0<\varepsilon\le 1$, $R\ge\max\{4,\varepsilon^{-1/2}\}$ and $\langle\xi\rangle\ge R|\beta|$, Theorem 2.6. 1 yields $x^0\notin \operatorname{supp} b_2^{R,\varepsilon}(x,D)v$ if $0<\varepsilon\le 1$, $R\ge\max\{4,\varepsilon^{-1/2},2enC(X_1,X_2),$ $8e\sqrt{n}C(X_1,X_2)\}$ and $3c\varepsilon^2<\min\{c_*\delta/2,1/R\}$, where c_* is equal to the constant $c_*(S)$ with $S=x\cdot\xi$ in Theorem 2.6. 1 . We may assume that $C(X_1,X_2)\le C_*'/\delta$ and C_*' is independent of δ, X_1 and X_2. It is obvious that

$$\begin{aligned}\left|b_{1(\beta+\tilde{\beta})}^{R,\varepsilon(\alpha+\tilde{\alpha})}(x,\xi)\right| &\le C_{|\tilde{\alpha}|+|\tilde{\beta}|}(((48(1+\sqrt{2})+C_*)/\varepsilon+C_*)/R)^{|\alpha|}\\ &\quad\times(C(X_1,X_2)/R)^{|\beta|}\langle\xi\rangle^{|\beta|-|\tilde{\alpha}|}\exp[3c\varepsilon^2\langle\xi\rangle]\end{aligned}$$

if $0<\varepsilon\le 1$, $R\ge\max\{4,\varepsilon^{-1/2}\}$ and $\langle\xi\rangle\ge R(|\alpha|+|\beta|)$,

$$\begin{aligned}\left|b_{1(\beta)}^{R,\varepsilon(\alpha+\tilde{\alpha})}(x,\xi)\right| &\le C_{|\tilde{\alpha}|}(((48(1+\sqrt{2})+C_*)/\varepsilon+C_*)/R)^{|\alpha|}\\ &\quad\times B^{|\beta|}|\beta|!\langle\xi\rangle^{-|\tilde{\alpha}|}\exp[3c\varepsilon^2\langle\xi\rangle]\end{aligned}$$

if $0<\varepsilon\le 1$, $R\ge\max\{4,\varepsilon^{-1/2}\}$, $x\in X_1$, $\langle\xi\rangle\ge R|\alpha|$ and $B>0$, and $\operatorname{supp} b_1^{R,\varepsilon}\subset X_2\times\boldsymbol{R}^n$. Applying Lemma 2.6. 4 , we can see that $b_1^{R,\varepsilon}(x,D)v$ is analytic in X_1 if $0<\varepsilon\le 1$, $R\ge\max\{4,\varepsilon^{-1/2},2enC_*'/\delta,12e\sqrt{n}((48(1+$ $\sqrt{2})+C_*)/\varepsilon+C_*)/\delta\}$ and $3c\varepsilon^2<\min\{c_1,c_2\delta,c_3A(v),1/R\}$, where c_j $(1\le j\le 3)$ are the constants in Lemma 2.6. 4 . Therefore, $g^{R,\varepsilon}(D)u$ is analytic at x^0 if

$$\left\{\begin{array}{l}0<\varepsilon<\min\{1,\sqrt{c_1/(3c)},\sqrt{c_2\delta/(3c)},\sqrt{c_3A(v)/(3c)}\},\\ 1/(3c\varepsilon^2)>R\ge\max\{4,\varepsilon^{-1/2},2enC_*'/\delta,8e\sqrt{n}C_*'/\delta,\\ \qquad\qquad 12e\sqrt{n}((48(1+\sqrt{2})+C_*)/\varepsilon+C_*)/\delta\}.\end{array}\right. \tag{3.14}$$

Since there are R and ε satisfying (3.14), Proposition 3.1. 2 yields (x^0,ξ^0) $\notin WF_A(u)$. □

From Theorem 3.1. 6 and the results in [Hr5] one can give another equivalent definition of $WF_A(u)$ for $u\in\mathcal{A}'(\boldsymbol{R}^n)$ using the FBI (Fourier-Bros-Iagolnitzer) transformation. The FBI transform $T_\lambda u(z)$ of $u\in\mathcal{A}'(\boldsymbol{R}^n)$ is defined by

$$T_\lambda u(z)=u_y(\exp[-\lambda(z-y)^2/2])$$

for $z\in\boldsymbol{C}^n$ and $\lambda>0$ (see Sjöstrand [Sj]). Let $u\in\mathcal{A}'(\boldsymbol{R}^n)$ and $(x^0,\xi^0)\in T^*\boldsymbol{R}^n\backslash 0$. Then $(x^0,\xi^0)\notin WF_A(u)$ if and only if there are a neighborhood U of $x^0-i\xi^0/|\xi^0|$ and positive constants C and c such that

$$|T_\lambda u(z)|\le Ce^{\lambda(1/2-c)}\quad\text{for } z\in V \text{ and } \lambda>0.$$

We note that

$$T_\lambda u(x-i\xi)=\exp[i\lambda x\cdot\xi+\lambda\xi^2/2]\mathcal{F}_y\Big[\exp[-\lambda(x-y)^2/2]u(y)\Big](\lambda\xi).$$

3.2 Action of Fourier integral operators on wave front sets

Let Ω and Ω' be open conic subsets of $\boldsymbol{R}^n\times(\boldsymbol{R}^{n'}\setminus\{0\})$ and $\boldsymbol{R}^n\times(\boldsymbol{R}^{n''}\setminus\{0\})$, respectively, and let $S(y,\xi)\in\mathcal{P}(\Omega;A_0,B_0,c_0(S),0,c_2(S),c_3(S))$, $T(y,\eta)\in\mathcal{P}(\Omega';A_0,B_0,c_0(T),0,c_2(T),c_3(T))$ and $p(\xi,y,\eta)\in S^+(\boldsymbol{R}^{n'}\times\boldsymbol{R}^n\times\boldsymbol{R}^{n''};R_0,A)$. We assume that $S(y,\xi)$ and $T(y,\eta)$ is positively homogeneous of degree 1 in ξ and η, respectively, $c_j(S)>0$ ($j=0,2$) and $\operatorname{supp} p\subset\{(\xi,y,\eta)\in\boldsymbol{R}^{n'}\times\Omega';\ (y,\xi)\in\Omega,\ |\xi|\geq1 \text{ and } |\eta|\geq1\}$. We may assume that $S(y,\xi)$ and $T(y,\eta)$ are defined and analytic in $\overline{\Omega}\cap\{|\xi|>0\}$ and $\overline{\Omega'}\cap\{|\eta|>0\}$, respectively, and that $S(y,0)=0$ and $T(y,0)=0$. It follows from Theorem 2.3.3 that there is $R(S,T)>0$ such that $p_{S,T}(D_x,y,D_z)$ maps continuously $\mathcal{F}_0(\boldsymbol{R}^{n''})$ to $\mathcal{F}_0(\boldsymbol{R}^{n'})$ if $c_3(T)>0$ and $R_0\geq R(S,T)A$. It also follows from Lemma A.1.7 that $p_{S,T}(D_x,y,D_z)$ maps continuously $\mathcal{S}_{-\varepsilon}'(\boldsymbol{R}^{n''})$ to $\mathcal{F}_0(\boldsymbol{R}^{n'})$ if $\varepsilon>0$. We define

$$\begin{aligned}\mathcal{M}(x^0,\xi^0)=\{(\lambda\xi^0,y,\lambda\eta);\ &(y,\xi^0)\in\overline{\Omega},\ (y,\eta)\in\overline{\Omega'},\\ &x^0=-\nabla_\xi S(y,\xi^0),\ \nabla_y S(y,\xi^0)+\nabla_y T(y,\eta)=0 \text{ and } \lambda>0\}\end{aligned}$$

for $(x^0,\xi^0)\in T^*\boldsymbol{R}^{n'}\setminus0$. We note that $\mathcal{M}(x^0,\xi^0)$ is a closed conic subset of $(\boldsymbol{R}^{n'}\setminus\{0\})\times\boldsymbol{R}^n\times\boldsymbol{R}^{n''}$ and $\mathcal{M}(x^0,\xi^0)\cap S^{n'-1}\times\boldsymbol{R}^n\times\boldsymbol{R}^{n''}$ is compact.

Proposition 3.2.1 *Let $(x^0,\xi^0)\in T^*\boldsymbol{R}^{n'}\setminus0$, and assume that $c_3(T)>0$ and that there are a conic neighborhood $\mathcal{U}$ of $\mathcal{M}(x^0,\xi^0)$ and $c>0$ satisfying*

$$\Big|e^{iS(y,\xi)+iT(y,\eta)}\partial_\eta^\gamma p(\xi,y,\eta)\Big|\leq C_{|\gamma|}(p)e^{-c\langle\xi\rangle}$$

for $(\xi,y,\eta)\in\mathcal{U}$. Then there is $R(S,T)>0$ such that

$$(x^0,\xi^0)\notin WF_A(p_{S,T}(D_x,y,D_z)u)\quad \textit{for } u\in\mathcal{F}_0(\boldsymbol{R}^{n''})$$

if $R_0\geq R(S,T)A$.

Remark $R(S,T)$ depends on (x^0,ξ^0) and $\mathcal{U}$.

Proof Let $\mathcal{U}_0$ ($\subset\subset\mathcal{U}$) be an open conic neighborhood of $\mathcal{M}(x^0,\xi^0)$. Then there are $\varepsilon>0$ and a conic neighborhood Γ of ξ^0 such that

$$|x^0+\nabla_\xi S(y,\xi)|^2/4+|\nabla_y S(y,\xi)+\nabla_y T(y,\eta)|^2/|\xi|^2\geq\varepsilon^2\langle y\rangle^2 \qquad (3.15)$$

if $(\xi, y, \eta) \in \Gamma \times \Omega' \setminus \mathcal{U}_0$ and $(y, \xi) \in \Omega$. In fact, suppose that there are a sequence $\{(\xi^j, y^j, \eta^j)\} \subset S^{n'-1} \times \Omega' \setminus \mathcal{U}_0$ such that $(y^j, \xi^j) \in \Omega$, and $\xi^j \to \xi^0/|\xi^0|$, $|x^0 + \nabla_\xi S(y^j, \xi^j)|/\langle y^j \rangle \to 0$ and $|\nabla_y S(y^j, \xi^j) + \nabla_y T(y^j, \eta^j)|/\langle y^j \rangle \to 0$ as $j \to \infty$. Since $c_0(S) > 0$ and $|x^0 + \nabla_\xi S(y^j, \xi^j)|/\langle y^j \rangle \geq \sqrt{c_0(S)} - (1 + |x^0|)/\langle y^j \rangle$, $\{|y^j|\}$ is bounded and we may suppose that there is $y^0 \in \boldsymbol{R}^n$ satisfying $y^j \to y^0$ as $j \to \infty$. Then we have $x^0 = -\nabla_\xi S(y^0, \xi^0)$. Since $c_3(T)|\eta^j| \leq |\nabla_y T(y^j, \eta^j)|$ and $|\nabla_y S(y^j, \xi^j)| \leq c_2(S)$, $\{|\eta^j|\}$ is bounded and we may suppose that there is $\eta^0 \in \boldsymbol{R}^{n''}$ satisfying $\eta^j \to \eta^0$ as $j \to \infty$. Then we have $\nabla_y S(y^0, \xi^0/|\xi^0|) + \nabla_y T(y^0, \eta^0) = 0$ and, therefore, $(\xi^0/|\xi^0|, y^0, \eta^0) \in \mathcal{M}(x^0, \xi^0)$, which is a contradiction. This gives (3.15). Let Γ_0 ($\subset\subset \Gamma$) be a conic neighborhood of ξ^0. Choose $g^R(\xi) \in C^\infty(\boldsymbol{R}^{n'})$ and $\Phi^R(\xi, y, \eta) \in C^\infty(\boldsymbol{R}^{n'} \times \boldsymbol{R}^n \times \boldsymbol{R}^{n''})$ ($R \geq 4$) such that $g^R(\xi) = 1$ in $\Gamma_0 \cap \{\langle \xi \rangle \geq 2R\}$, $\operatorname{supp} g^R \subset \Gamma \cap \{|\xi| \geq R\}$,

$$\left|\partial_\xi^{\alpha+\tilde{\alpha}} g^R(\xi)\right| \leq C_{|\tilde{\alpha}|}(\Gamma, \Gamma_0)(C(\Gamma, \Gamma_0)/R)^{|\alpha|}$$

if $\langle \xi \rangle \geq R|\alpha|$, $\Phi^R(\xi, y, \eta) = 1$ in $\mathcal{U}_0 \cap \{\langle \xi \rangle \geq R\}$, $\operatorname{supp} \Phi^R \subset \mathcal{U}$ and $\Phi^R(\xi, y, \eta) \in S^{0,0}(\boldsymbol{R}^{n'} \times \boldsymbol{R}^n \times \boldsymbol{R}^{n''}; R, C(\mathcal{U}, \mathcal{U}_0))$ (see Lemma 2.1. 3 and Proposition 2.2. 3). We put

$$p_0^R(\xi, y, \eta) = \Phi^R(\xi, y, \eta) p(\xi, y, \eta),$$
$$p_1^R(\xi, y, \eta) = (1 - \Phi^R(\xi, y, \eta)) p(\xi, y, \eta).$$

We can assume without loss of generality that $c_3(T)|\eta|/2 \leq c_2(S)|\xi|$ if $(\xi, y, \eta) \in \mathcal{U}$, since $c_3(T)|\eta| \leq c_2(S)|\xi|$ if $(\xi, y, \eta) \in \mathcal{M}(x^0, \xi^0)$. Then it is obvious that

$$\begin{aligned} &\left|e^{iS(y,\xi)+iT(y,\eta)} \partial_\eta^\gamma p_0^R(\xi, y, \eta)\right| \\ &\leq C_{|\gamma|}(p, \mathcal{U}, \mathcal{U}_0) \exp[-c\langle \xi \rangle/2 - cc_3(T)\langle \eta \rangle/(4c_2(S))]. \end{aligned}$$

This, together with Lemma A.1. 7 , yields $p^R_{0\,S,T}(D_x, y, D_z)u \in \mathcal{S}'_{-c/2}$ for $u \in \mathcal{F}_0(\boldsymbol{R}^{n''})$, *i.e.*, $p^R_{0\,S,T}(D_x, y, D_z)u$ is analytic if $u \in \mathcal{F}_0(\boldsymbol{R}^{n''})$ (see the proof of Lemma A.1. 7 and Lemma 1.1. 3). Next consider $p^R_{1\,S,T}(D_x, y, D_z)$. The proof below is very similar to that of Theorem 2.6. 1 . It is obvious that $p_1^R(\xi, y, \eta) \in S^+(\boldsymbol{R}^{n'} \times \boldsymbol{R}^n \times \boldsymbol{R}^{n''}; R, RA/R_0 + C(\mathcal{U}, \mathcal{U}_0))$. By Theorem 2.3. 3 there are $\tilde{R}_j(S, T) > 0$ ($j = 1, 2$) such that $p^R_{1\,S,T}(D_x, y, D_z)$ maps continuously $\mathcal{F}_0(\boldsymbol{R}^{n''})$ to $\mathcal{F}_0(\boldsymbol{R}^{n'})$ if

$$\begin{cases} R \geq R_0 \geq \tilde{R}_1(S, T)A, \\ R \geq \max\{\tilde{R}_1(S, T)C(\mathcal{U}, \mathcal{U}_0), \tilde{R}_2(S, T)\}. \end{cases} \tag{3.16}$$

Therefore, we have

$$
\begin{aligned}
&\langle D_x\rangle^\nu \exp[-\rho\langle D_x\rangle] g^R(D_x) p^R_{1\,S,T}(D_x,y,D_z)u \\
&= \sum_{k=1}^{\infty} \langle D_x\rangle^\nu \exp[-\rho\langle D_x\rangle] g^R(D_x) p^R_{1\,S,T}(D_x,y,D_z)\psi_k^R(D_z)u \\
&= \sum_{k=1}^{\infty}\Big(\sum_{j=1}^{\infty} \langle D_x\rangle^\nu \exp[-\rho\langle D_x\rangle] \psi_j^R(D_x) g^R(D_x) p^R_{1\,S,T}(D_x,y,D_z)\psi_k^R(D_z)u\Big)
\end{aligned}
$$

in $\mathcal{F}_0$ for $u \in \mathcal{F}_0(\boldsymbol{R}^{n''})$, $\nu = 0,1$ and $0 < \rho \le 1$ if (3.16) is valid. Here $\{\psi_j^R(\xi)\}$ and $\{\psi_j^R(\eta)\}$ are the families of symbols in Section 2.2. We use the same notations although the $\psi_j^R(\xi)$ and the $\psi_j^R(\eta)$ are different if $n' \neq n''$. Thus a usual calculation in $\mathcal{S}'$ gives

$$
\begin{aligned}
&\langle D_x\rangle^\nu \exp[-\rho\langle D_x\rangle] g^R(D_x) p^R_{1\,S,T}(D_x,y,D_z)u \\
&= \sum_{k=1}^{\infty}\Big(\sum_{j=1}^{\infty} \langle e^{-\delta\langle\eta\rangle}\hat{u}(\eta), f^R_{\nu,\delta,j,k}(x,\eta;\rho)\rangle_\eta\Big)
\end{aligned}
\tag{3.17}
$$

for $u \in \mathcal{F}_0(\boldsymbol{R}^{n''})$, $\nu = 0,1$, $0 < \rho \le 1$ and $0 < \delta \le 1$ if (3.16) is valid, where $M \ge n+1$ and

$$
\begin{aligned}
{}^tL_1 &= \Big(1 + (x + \nabla_\xi S(y,\xi)) \cdot (x + \overline{\nabla_\xi S(y,\xi)})\Big)^{-1} \\
&\quad \times \Big(\sum_{\mu=1}^{n'} (x_\mu + \overline{\partial_{\xi_\mu} S(y,\xi)}) D_{\xi_\mu} + 1\Big), \\
f^R_{\nu,\delta,j,k}(x,\eta;\rho) &= (2\pi)^{-n'-n''} \int e^{ix\cdot\xi + iS(y,\xi) + iT(y,\eta) + \delta\langle\eta\rangle} \\
&\quad \times \psi_k^R(\eta) L_1^M \Big(\langle\xi\rangle^\nu e^{-\rho\langle\xi\rangle} \psi_j^R(\xi) g^R(\xi) p_1^R(\xi,y,\eta)\Big)\, d\xi\, dy.
\end{aligned}
$$

Let $\nu = 0,1$, $0 < \rho \le 1$ and $0 < \delta \le 1$, and assume that (3.16) is valid. We note that $f^R_{\nu,\delta,j,k}(x,\eta;\rho)$ is analytic in x when $|\mathrm{Im}\, x| < 1$, since

$$
\mathrm{Re}\, (x + \nabla_\xi S(y,\xi)) \cdot (x + \overline{\nabla_\xi S(y,\xi)})) = |\mathrm{Re}\, x + \nabla_\xi S(y,\xi)|^2 - |\mathrm{Im}\, x|^2.
$$

Let us first consider the case where $j,k \in \boldsymbol{N}$ and $c_3(T)(2R(k-1)-1) \ge 6c_2(S)Rj$. Then we have $c_3(T)|\eta| \ge 2c_2(S)|\xi|$ if $\psi_j^R(\xi)\psi_k^R(\eta) \neq 0$. Let K be a differential operator defined by

$$
{}^tK = |\nabla_y S(y,\xi) + \nabla_y T(y,\eta)|^{-2} \sum_{\mu=1}^{n} \Big(\overline{\partial_{y_\mu} S(y,\xi)} + \overline{\partial_{y_\mu} T(y,\eta)}\Big) D_{y_\mu}. \tag{3.18}
$$

It follows from Lemma 2.1. 5 that

$$\begin{aligned}&\left|\partial_\xi^\alpha \partial_\eta^\gamma K^k\left\{\psi_k^R(\eta)\langle\xi\rangle^\nu e^{-\rho\langle\xi\rangle}\psi_j^R(\xi)g^R(\xi)p_1^R(\xi,y,\eta)\right\}\right|\\ &\le C_{|\alpha|+|\gamma|,\delta'}(p,S,T,\mathcal{U},\mathcal{U}_0,\Gamma,\Gamma_0)(\Gamma(S,T,A,\mathcal{U},\mathcal{U}_0,R/R_0)/R)^k\\ &\quad\times\langle\xi\rangle^{\nu-|\alpha|}\langle\eta\rangle^{-|\gamma|}e^{\delta'\langle\xi\rangle+\delta'\langle\eta\rangle}\end{aligned}$$

if $\delta' > 0$, where

$$\begin{aligned}\Gamma(S,T,A,\mathcal{U},\mathcal{U}_0,R/R_0) &= 2^7 nB_0C(S,T)\max\{AR/R_0+C(\mathcal{U},\mathcal{U}_0),\\ &\qquad 480nC(S,T)^2B_0^3\}/c_3(T),\\ C(S,T) &= C(S)/c_2(S)+2C(T)/c_3(T).\end{aligned}$$

Here we have used the estimates that

$$\begin{aligned}\left|\partial_\xi^\alpha\partial_\eta^\gamma D_y^\beta(S(y,\xi)+T(y,\eta))\right| &\le 2(C(S)\langle\xi\rangle+C(T)\langle\eta\rangle)\\ &\quad\times A_0^{|\alpha|+|\gamma|}B_0^{|\beta|}|\alpha|!|\beta|!|\gamma|!\langle\xi\rangle^{-|\alpha|}\langle\eta\rangle^{-|\gamma|}\end{aligned}$$

if $|\beta| \ge 1$, $|\xi| \ge 1$ and $|\eta| \ge 1$, and

$$|\nabla_y S(y,\xi)+\nabla_y T(y,\eta)| \ge c_3(T)|\eta|/2$$

if $\psi_j^R(\xi)\psi_k^R(\eta) \ne 0$. We can write

$$\begin{aligned}\langle\eta\rangle^\ell D_\eta^\gamma f_{\nu,\delta,j,k}^R(x,\eta;\rho) &= (2\pi)^{-n'-n''}\int e^{ix\cdot\xi+iS(y,\xi)+iT(y,\eta)+\delta\langle\eta\rangle}\\ &\quad\times\langle\eta\rangle^\ell\sum_{\gamma'\le\gamma}\binom{\gamma}{\gamma'}t_{\delta,\gamma-\gamma'}(y,\eta)D_\eta^{\gamma'}L_1^M K^k\\ &\quad\times\left\{\psi_k^R(\eta)\langle\xi\rangle^\nu e^{-\rho\langle\xi\rangle}\psi_j^R(\xi)g^R(\xi)p_1^R(\xi,y,\eta)\right\}d\xi dy,\end{aligned}$$

where $M \ge |\gamma|+n+1$ and

$$t_{\delta,\alpha}(y,\eta) = e^{-iT(y,\eta)-\delta\langle\eta\rangle}D_\eta^\alpha e^{iT(y,\eta)+\delta\langle\eta\rangle}.$$

Therefore, we have

$$\begin{aligned}&\left|\langle\eta\rangle^\ell D_\eta^\gamma f_{\nu,\delta,j,k}^R(x,\eta;\rho)\right| \le C_{\delta,|\gamma|,\ell,\delta',R}(p,S,T,\mathcal{U},\mathcal{U}_0,\Gamma,\Gamma_0)\\ &\quad\times j^{-2}k^{-2}\langle\mathrm{Re}\, x\rangle^{|\gamma|}\exp[(\delta+\delta'+c_3(T)(\rho_1+\delta')/(2c_2(S))-1/(3R))\langle\eta\rangle]\end{aligned}$$

if $\ell \in \boldsymbol{Z}_+$, $\gamma \in \boldsymbol{Z}_+^{n''}$, $\delta' > 0$, $x \in \boldsymbol{C}^n$, $|\mathrm{Im}\, x| \le \rho_1 \le 1/2$ and $R \ge 2e\Gamma(S,T,A,\mathcal{U},\mathcal{U}_0,R/R_0)$. Moreover, $\langle e^{-\delta\langle\eta\rangle}\hat{u}(\eta), f_{\nu,\delta,j,k}^R(x,\eta;\rho)\rangle_\eta$ is analytic in x and

$$\begin{aligned}&\left|\langle e^{-\delta\langle\eta\rangle}\hat{u}(\eta), f_{\nu,\delta,j,k}^R(x,\eta;\rho)\rangle_\eta\right|\\ &\le C_{\delta,R,r}(p,S,T,\mathcal{U},\mathcal{U}_0,\Gamma,\Gamma_0,u)j^{-2}k^{-2} \qquad (3.19)\end{aligned}$$

if $u \in \mathcal{F}_0(\boldsymbol{R}^{n''})$, $x \in \boldsymbol{C}^n$, $|\mathrm{Re}\, x| \le r$, $|\mathrm{Im}\, x| \le \rho_1 \le 1/2$, $R \ge 2e\Gamma(S,T,A,\mathcal{U},\mathcal{U}_0,R/R_0)$ and $\delta + c_3(T)\rho_1/(2c_2(S)) < 1/(3R)$. Next consider the case where $j,k \in \boldsymbol{N}$ and $c_3(T)(2R(k-1)-1) < 6c_2(S)Rj$. Then we have $2c_3(T)\langle\eta\rangle \le 9c_2(S)\langle\xi\rangle(1+\kappa(S,T)R/\langle\xi\rangle)$ with some constant $\kappa(S,T)$ if $\psi_j^R(\xi)\psi_k^R(\eta) \ne 0$. Let L_2 be a differential operator defined by

$$
{}^tL_2 = \left(|\nabla_y S(y,\xi) + \nabla_y T(y,\eta)|^2 + \langle\xi\rangle^2|x + \nabla_\xi S(y,\xi)|^2\right)^{-1}
$$
$$
\times\Big\{\sum_{\mu=1}^{n}\left(\overline{\partial_{y_\mu}S(y,\xi)} + \overline{\partial_{y_\mu}T(y,\eta)}\right)D_{y_\mu} + \sum_{\mu=1}^{n'}\langle\xi\rangle^2\left(\overline{x}_\mu + \overline{\partial_{\xi_\mu}S(y,\xi)}\right)D_{\xi_\mu}\Big\}.
$$

We can write

$$
\begin{aligned}
f_{\nu,\delta,j,k}^R(x,\eta;\rho) &= (2\pi)^{-n'-n''}\int e^{ix\cdot\xi+iS(y,\xi)+iT(y,\eta)+\delta\langle\eta\rangle} \\
&\quad \times\psi_k^R(\eta)L_2^{j+M}\left(\langle\xi\rangle^\nu e^{-\rho\langle\xi\rangle}\psi_j^R(\xi)g^R(\xi)p_1^R(\xi,y,\eta)\right)d\xi dy.
\end{aligned}
$$

It follows from (3.15) that

$$
\begin{aligned}
&|\nabla_y S(y,\xi) + \nabla_y T(y,\eta)|^2 + \langle\xi\rangle^2|x + \nabla_\xi S(y,\xi)|^2 \\
&\ge \left(\varepsilon^2\langle y\rangle^2 + |x + \nabla_\xi S(y,\xi)|^2\right)|\xi|^2/2
\end{aligned}
$$

if $(\xi,y,\eta) \in \Gamma\times\Omega'\setminus\mathcal{U}_0$, $(y,\xi)\in\Omega$ and $|x - x^0| \le \varepsilon$. Since

$$
\begin{aligned}
&\left|\partial_\xi^\alpha\partial_\eta^\gamma D_y^\beta(x\cdot\xi + S(y,\xi) + T(y,\eta))\right| \\
&\le (2C(S)\langle y\rangle\langle\xi\rangle + 2C(T)\langle\eta\rangle + |x + \nabla_\xi S(y,\xi)|\langle\xi\rangle/A_0) \\
&\quad \times A_0^{|\alpha|+|\gamma|}B_0^{|\beta|}|\alpha|!|\beta|!|\gamma|!\langle\xi\rangle^{-|\alpha|}\langle\eta\rangle^{-|\gamma|}
\end{aligned}
$$

if $|\alpha|+|\beta| \ge 1$, $(y,\xi)\in\Omega$, $(y,\eta)\in\Omega'$, $|\xi|\ge 1$ and $|\eta| \ge 1$, and

$$
\begin{aligned}
&\left|\partial_\xi^{\alpha+\tilde\alpha}\partial_\eta^\gamma D_y^{\beta+\tilde\beta}\left\{\langle\xi\rangle^\nu e^{-\rho\langle\xi\rangle}\psi_j^R(\xi)g^R(\xi)p_1^R(\xi,y,\eta)\right\}\right| \\
&\le C_{|\tilde\alpha|+|\tilde\beta|+|\gamma|,\delta'}(p,\mathcal{U},\mathcal{U}_0,\Gamma,\Gamma_0)(A_1/R)^{|\alpha|}(A_2/R)^{|\beta|} \\
&\quad \times\langle\xi\rangle^{\nu+|\beta|-|\tilde\alpha|}\langle\eta\rangle^{-|\gamma|}e^{\delta'\langle\xi\rangle+\delta'\langle\eta\rangle}
\end{aligned}
$$

if $\delta' > 0$ and $\langle\xi\rangle \ge R(|\alpha|+|\beta|)$, where $A_1 \equiv A_1(A,\mathcal{U},\mathcal{U}_0,\Gamma,\Gamma_0,R/R_0) = RA/R_0 + \widehat{C} + C(\Gamma,\Gamma_0) + C(\mathcal{U},\mathcal{U}_0) + 6(1+\sqrt{2})$ and $A_2 \equiv A_2(A,\mathcal{U},\mathcal{U}_0,R/R_0) = RA/R_0 + C(\mathcal{U},\mathcal{U}_0)$, Lemma 2.1.5 yields

$$
\begin{aligned}
&\left|\partial_\eta^\gamma L_2^{j+M}\left\{\psi_k^R(\eta)\langle\xi\rangle^\nu e^{-\rho\langle\xi\rangle}\psi_j^R(\xi)g^R(\xi)p_1^R(\xi,y,\eta)\right\}\right| \\
&\le C_{|\gamma|,M,\varepsilon,\delta',R}(p,S,T,\mathcal{U},\mathcal{U}_0,\Gamma,\Gamma_0)(\Gamma(S,T,A,\mathcal{U},\mathcal{U}_0,\Gamma,\Gamma_0,\varepsilon,R/R_0)/R)^j \\
&\quad \times\langle y\rangle^{-M}\langle\xi\rangle^{\nu-M}\langle\eta\rangle^{-|\gamma|}e^{\delta'\langle\xi\rangle+\delta'\langle\eta\rangle}
\end{aligned}
$$

if $\delta' > 0$, $x \in \boldsymbol{C}^n$ and $|x - x^0| \le \varepsilon$, where

$$\begin{aligned}
&\Gamma(S,T,A,\mathcal{U},\mathcal{U}_0,\Gamma,\Gamma_0,\varepsilon,R/R_0) = 32(C_1(S,T)/\varepsilon^2 + 1/(\varepsilon A_0)) \\
&\quad \times \Big[2nB_0 \max\Big\{RA/R_0 + C(\mathcal{U},\mathcal{U}_0), \\
&\qquad\qquad 480C(S,T,\varepsilon)(nB_0^2 + n'(A_0+1/2)^2)B_0\Big\} \\
&\qquad +n'(\sqrt{2A_0} + 1/\sqrt{2A_0})^2 \max\Big\{A_1(A,\mathcal{U},\mathcal{U}_0,\Gamma,\Gamma_0,R/R_0), \\
&\qquad\qquad 480C(S,T,\varepsilon)(nB_0^2 + n'(A_0+1/2)^2)A_0\Big\}\Big], \\
&C_1(S,T) = 2C(S) + 9c_2(S)C(T)/c_3(T), \\
&C(S,T,\varepsilon) = C_1(S,T)^2/\varepsilon^2 + A_0^{-2}.
\end{aligned}$$

Therefore, we have

$$\left|\langle\eta\rangle^\ell D_\eta^\gamma f^R_{\nu,\delta,j,k}(x,\eta;\rho)\right| \le C_{\delta,|\gamma|,\ell,\varepsilon,R}(p,S,T,\mathcal{U},\mathcal{U}_0,\Gamma,\Gamma_0)j^{-2}k^{-2}$$

if $\ell \in \boldsymbol{Z}_+$, $\gamma \in \boldsymbol{Z}_+^{n''}$, $x \in \boldsymbol{C}^n$, $|x - x^0| \le \varepsilon$, $|\mathrm{Im}\, x| \le \rho_1$ and

$$\begin{cases} R \ge 2e\Gamma(S,T,A,\mathcal{U},\mathcal{U}_0,\Gamma,\Gamma_0,\varepsilon,R/R_0), \\ 9c_2(S)\delta/(2c_3(T)) + \rho_1 < 1/(3R). \end{cases} \tag{3.20}$$

Moreover, $\langle e^{-\delta\langle\eta\rangle}\hat{u}(\eta), f^R_{\nu,\delta,j,k}(x,\eta;\rho)\rangle_\eta$ is analytic in x and

$$\begin{aligned}
&\left|\langle e^{-\delta\langle\eta\rangle}\hat{u}(\eta), f^R_{\nu,\delta,j,k}(x,\eta;\rho)\rangle_\eta\right| \\
&\le C_{\delta,\varepsilon,R}(p,S,T,\mathcal{U},\mathcal{U}_0,\Gamma,\Gamma_0,u)j^{-2}k^{-2}
\end{aligned} \tag{3.21}$$

if $x \in \boldsymbol{C}^n$, $|x - x^0| \le \varepsilon$, $|\mathrm{Im}\, x| \le \rho_1 \le 1/2$ and (3.20) is valid. We put

$$V(x,x_{n+1}) = \mathcal{H}(g^R(D)p^R_{1\,S,T}(D_x,y,D_z)u)(x,x_{n+1}),$$

and assume that

$$\begin{aligned}
&R \ge 2e\max\{\Gamma(S,T,A,\mathcal{U},\mathcal{U}_0,R/R_0), \Gamma(S,T,A,\mathcal{U},\mathcal{U}_0,\Gamma,\Gamma_0,\varepsilon,R/R_0)\}, \\
&0 < \rho_1 < \min\{1/2, 2c_2(S)/(3c_3(T)R), 1/(3R)\}.
\end{aligned}$$

Then it follows from (3.17), (3.19) and (3.21) that $\langle D_x\rangle^\nu V(x,\rho)$ ($\nu = 0,1$) can be continued analytically to $\{x \in \boldsymbol{C}^n;\ |x - x^0| < \varepsilon$ and $|\mathrm{Im}\, x| < \rho_1\}$. Applying Lemma 1.2.4 to the Cauchy problem

$$\begin{cases} (1 - \Delta_{x,x_{n+1}})v(x,x_{n+1}) = 0, \\ v(x,\rho) = V(x,\rho), \quad (\partial v/\partial x_{n+1})(x,\rho) = -\langle D_x\rangle V(x,\rho), \end{cases}$$

we can show that $V(x, x_{n+1})$ can be continued analytically from $\boldsymbol{R}^n \times (0,\infty)$ to a neighborhood of $\{x^0\} \times [0,\infty)$. This implies that $(x^0,\xi^0) \notin WF_A(p^R_{1\,S,T}(D_x, y, D_z)u)$ and, therefore, $(x^0,\xi^0) \notin WF_A(p_{S,T}(D_x, y, D_z) \times u)$. □

Corollary 3.2. 2 *Let $q(\xi, y, \eta)$ be a symbol in $C^\infty(\boldsymbol{R}^{n'} \times \boldsymbol{R}^n \times \boldsymbol{R}^{n''})$ such that* $\operatorname{supp} q \subset \{(\xi, y, \eta) \in \boldsymbol{R}^{n'} \times \Omega';\ (y,\xi) \in \Omega,\ |\xi| \geq 1$ *and* $|\eta| \geq 1\}$, *and*

$$\left|\partial_\xi^{\alpha+\tilde{\alpha}} D_y^\beta \partial_\eta^\gamma q(\xi, y, \eta)\right| \leq C_{|\tilde{\alpha}|+|\gamma|}(q)(A/R_0)^{|\alpha|+|\beta|}\langle\eta\rangle^{|\beta|} e^{\delta\langle\xi\rangle+\delta\langle\eta\rangle}$$

if $\langle\xi\rangle \geq R_0|\alpha|$ and $\langle\eta\rangle \geq R_0|\beta|$, where $\delta \geq 0$. Let $x^0 \in \boldsymbol{R}^n$, and assume that $c_3(T) > 0$ and that there are bounded open subsets U_0 and U of $\boldsymbol{R}^n$, $c > 0$ and $\varepsilon > 0$ such that $U_0 \Subset U$,

$$\left|e^{iS(y,\xi)+iT(y,\eta)} \partial_\eta^\gamma q(\xi, y, \eta)\right| \leq C'_{|\gamma|}(q) e^{-c\langle\xi\rangle - c\langle\eta\rangle}$$

for $(\xi, y, \eta) \in \boldsymbol{R}^{n'} \times U \times \boldsymbol{R}^{n''}$, and

$$|x^0 + \nabla_\xi S(y,\xi)| \geq \varepsilon\langle y\rangle$$

for $(y,\xi) \in \Omega \setminus (U_0 \times (\boldsymbol{R}^{n'} \setminus \{0\}))$. Then there are positive constants $R(S,T,\varepsilon)$, $c(S,T)$ and $\delta(S,T,U,U_0,\varepsilon)$ such that $q_{S,T}(D_x, y, D_z)u \in \mathcal{F}_{(1+2c_2(S)/c_3(T))\delta}(\boldsymbol{R}^{n'})$ and $q_{S,T}(D_x, y, D_z)u$ is analytic at x^0 for $u \in \mathcal{F}_0(\boldsymbol{R}^{n''})$ if $R_0 \geq R(S,T,\varepsilon)A$ and $\delta < \min\{c(S,T)/R_0, \delta(S,T,U,U_0,\varepsilon)\}$.

Proof Choose $\Phi^R(\xi, y, \eta) \in S^{0,0}(\boldsymbol{R}^{n'} \times \boldsymbol{R}^n \times \boldsymbol{R}^{n''}; R, C_*, C(U, U_0), C_*)$ ($R \geq 4$) so that $\Phi^R(\xi, y, \eta) = 1$ for $y \in U_0$, and $\Phi^R(\xi, y, \eta) = 0$ for $y \notin U$. We put

$$\begin{aligned} q_0^R(\xi, y, \eta) &= \Phi^R(\xi, y, \eta) q(\xi, y, \eta), \\ q_1^R(\xi, y, \eta) &= (1 - \Phi^R(\xi, y, \eta)) q(\xi, y, \eta). \end{aligned}$$

Lemma A.1. 7 and its proof show that $q^R_{0\,S,T}(D_x, y, D_z)u \in \mathcal{S}'_{-c}$ for $u \in \mathcal{F}_0(\boldsymbol{R}^{n''})$. By Theorem 2.3. 3 there are $R_j(S,T) > 0$ ($j = 0,1,2$) that $q^R_{1\,S,T}(D_x, y, D_z)$ maps continuously from $\mathcal{F}_0(\boldsymbol{R}^{n''})$ to $\mathcal{F}_{(1+2c_2(S)/c_3(T))\delta}(\boldsymbol{R}^{n'})$

$$\begin{cases} R_0 \geq R_0(S,T)A, \qquad R \geq \max\{R_0, R_1(S,T)\}, \\ \delta < \min\{1/R, 1/R_2(S,T)\}. \end{cases} \tag{3.22}$$

We can apply the same argument as in the proof of Proposition 3.2. 1, replacing $g^R(\xi)p_1^R(\xi,y,\eta)$ and ${}^tL_2$ by $q_1^R(\xi,y,\eta)$ and

$$ {}^t\widetilde{L}_1 \equiv |x+\nabla_\xi S(y,\xi)|^{-2}\sum_{\mu=1}^{n'}\Big(\overline{x}_\mu+\overline{\partial_{\xi_\mu}S(y,\xi)}\Big)D_{\xi_\mu}, $$

respectively. Then we can show that $q^R_{1\,S,T}(D_x,y,D_z)u$ is analytic at x^0 for $u\in\mathcal{F}_0(\boldsymbol{R}^{n''})$ if $R_0\ge R(S,T,\varepsilon)A$, $R\ge\max\{R_0,R'(S,T,U,U_0,\varepsilon)\}$ and $\delta<\min\{c'(S,T)/R,\delta'(S,T)\}$, where $R(S,T,\varepsilon)$, $R'(S,T,\varepsilon)$, $c'(S,T)$ and $\delta'(S,T)$ are positive constants. In fact, we have

$$ \begin{aligned} &\langle D\rangle^\nu e^{-\rho\langle D\rangle}q^R_{1\,S,T}(D_x,y,D_z)u\\ &=\sum_{k=1}^\infty\sum_{j=1}^\infty\langle\exp[-\delta_1\langle\eta\rangle]\hat{u}(\eta),f^R_{\nu,\delta_1,j,k}(x,\eta;\rho)\rangle_\eta \end{aligned} $$

for $u\in\mathcal{F}_0(\boldsymbol{R}^{n''})$, $\nu=0,1$, $\delta<\rho\le\delta+1$ and $0<\delta_1\le 1$ if (3.22) is valid, where $M\ge n+1$ and

$$ \begin{aligned} f^R_{\nu,\delta_1,j,k}(x,\eta;\rho)&=(2\pi)^{-n'-n''}\int e^{ix\cdot\xi+iS(y,\xi)+iT(y,\eta)+\delta_1\langle\eta\rangle}\\ &\quad\times\psi_k^R(\eta)L_1^M\Big(\langle\xi\rangle^\nu e^{-\rho\langle\xi\rangle}\psi_j^R(\xi)q_1^R(\xi,y,\eta)\Big)\,d\xi dy. \end{aligned} $$

Replacing L_2 by $\widetilde{L}_1$ and taking $\rho_1>\rho>\delta$, we can repeat the argument in the proof of Proposition 3.2. 1 . □

We define the essential cone support $ECS(p)$ ($\subset X\equiv\{(\xi,y,\eta)\in\boldsymbol{R}^{n'}\times\boldsymbol{R}^n\times\boldsymbol{R}^{n''};\ |\xi|+|\eta|\neq 0\}$) of $p(\xi,y,\eta)$ as follows: $(\xi^0,x^0,\eta^0)\in X$ does not belong to $ECS(p)$ if there are a conic neighborhood $\mathcal{U}$ of (ξ^0,x^0,η^0) in X, $c>0$ and $C_j>0$ ($j\in\boldsymbol{Z}_+$) such that

$$ \left|\partial_\xi^\alpha\partial_\eta^\gamma p(\xi,y,\eta)\right|\le C_{|\alpha|+|\gamma|}e^{-c\langle\xi\rangle-c\langle\eta\rangle}\quad\text{for } (\xi,y,\eta)\in\mathcal{U}. $$

It is obvious that $ECS(p)$ is a closed conic subset of X.

Proposition 3.2. 3 *Let $(x^0,\xi^0)\in T^*\boldsymbol{R}^{n'}\setminus 0$, and assume that $c_3(T)>0$ and that*

$$ \eta\neq 0\quad \textit{if } (\xi,y,\eta)\in\mathcal{M}(x^0,\xi^0)\cap ECS(p) \textit{ and } \operatorname{Im} S(y,\xi^0)=0. \tag{3.23} $$

(i) *We put*

$$ \begin{aligned} \mathcal{N}(x^0,\xi^0)\equiv\{&(y,\lambda\eta)\in\boldsymbol{R}^n\times(\boldsymbol{R}^{n''}\setminus\{0\});\ (\xi^0,y,\eta)\in ECS(p),\\ &(y,\xi^0)\in\overline{\Omega},\ (y,\eta)\in\overline{\Omega'},\ \operatorname{Im} S(y,\xi^0)=0,\ \operatorname{Im} T(y,\eta)=0,\\ &x^0=-\nabla_\xi S(y,\xi^0),\ \nabla_y S(y,\xi^0)+\nabla_y T(y,\eta)=0 \textit{ and } \lambda>0\}. \end{aligned} $$

Then $\mathcal{N}(x^0,\xi^0)$ *is a closed conic subset of* $\mathbf{R}^n \times (\mathbf{R}^{n''}\setminus\{0\})$ *and* $\mathcal{N}(x^0,\xi^0)\cap(\mathbf{R}^n \times S^{n''-1})$ *is compact.* (ii) *Let* $u \in \mathcal{F}_0(\mathbf{R}^{n''})$ *satisfy*

$$WF_A(u)\cap\{(\nabla_\eta T(y,\eta),\eta);\ (y,\eta)\in\mathcal{N}(x^0,\xi^0)\}=\emptyset. \tag{3.24}$$

Then there is $R(S,T,u,x^0,\xi^0)>0$ *such that* $(x^0,\xi^0)\notin WF_A(p_{S,T}(D_x,y,D_z)u)$ *if* $R_0\geq R(S,T,u,x^0,\xi^0)A$.

Remark (i) If $\nabla_y S(y,\xi^0)\neq 0$ when $(\xi^0,y,\eta)\in ECS(p)$, $(y,\xi^0)\in\overline{\Omega}$, $(y,\eta)\in\overline{\Omega}'$, Im $S(y,\xi^0)=$ Im $T(y,\eta)=0$ and $x^0=-\nabla_\xi S(y,\xi^0)$, then (3.23) is valid. (ii) By assumption, Im $\nabla_\xi S(y,\xi)=0$ and Im $\nabla_y S(y,\xi)=0$ if $(y,\xi)\in\Omega$ and Im $S(x,\xi)=0$. (iii) It is obvious that $\mathcal{N}(x^0,\xi^0)=\{(y,\eta)\in\mathbf{R}^n\times\mathbf{R}^{n''};\ (\xi,y,\eta)\in\mathcal{M}(x^0,\xi^0)\cap ECS(p)$ and Im $S(y,\xi^0)=$ Im $T(y,\eta)=0\}\subset\mathbf{R}^n\times(\mathbf{R}^{n'}\setminus\{0\})$.

Proof (i) Let $\{(y^j,\eta^j)\}$ be a sequence in $\mathcal{N}(x^0,\xi^0)$ such that $y^j\to y$ and $\eta^j\to\eta$ as $j\to\infty$ for some $y\in\mathbf{R}^n$ and $\eta\in\mathbf{R}^{n''}\setminus\{0\}$. Then there are $\lambda_j>0$ ($j\in\mathbf{N}$) such that $\nabla_y S(y^j,\xi^0)+\nabla_y T(y^j,\lambda_j^{-1}\eta^j)=0$ ($j\in\mathbf{N}$). By assumption we have

$$|\eta^j|/\lambda_j\leq|\nabla_y S(y^j,\xi^0)|/c_3(T)\leq c_2(S)|\xi^0|/c_3(T)$$

and, therefore, $\{\lambda_j^{-1}\eta^j\}$ is a bounded sequence. So we may assume that $\lambda_j^{-1}\eta^j\to\tilde{\eta}$ as $j\to\infty$ for some $\tilde{\eta}\in\mathbf{R}^{n''}$. Then we have $(\xi^0,y,\tilde{\eta})\in\mathcal{M}(x^0,\xi^0)\cap ECS(p)$ and Im $S(y,\xi^0)=$ Im $T(y,\tilde{\eta})=0$. This implies that $\tilde{\eta}\neq 0$. Since $\lambda_j^{-1}=|\lambda_j^{-1}\eta^j|/|\eta^j|\to|\tilde{\eta}|/|\eta|$ (>0) as $j\to\infty$, we have $\lim_{j\to\infty}\lambda_j=|\eta|/|\tilde{\eta}|\equiv\lambda$ (>0), $\eta=\lambda\tilde{\eta}$ and $(y,\eta)\in\mathcal{N}(x^0,\xi^0)$. Therefore, $\mathcal{N}(x^0,\xi^0)$ is closed in $\mathbf{R}^n\times(\mathbf{R}^{n''}\setminus\{0\})$. If $(y,\eta)\in\mathcal{N}(x^0,\xi^0)$, then we have $1+|x^0|^2=1+|\nabla_\xi S(y,\xi^0)|^2\geq c_0(S)\langle y\rangle^2$. This proves the assertion (i). (ii) Let $(\hat{y},\hat{\eta})\in\mathcal{N}(x^0,\xi^0)\cap(\mathbf{R}^n\times S^{n''-1})$. (3.24) implies that there are an open bounded neighborhood $U_{(\hat{y},\hat{\eta})}$ of $\nabla_\eta T(\hat{y},\hat{\eta})$ in $\mathbf{R}^{n''}$, a conic neighborhood $\Gamma_{(\hat{y},\hat{\eta})}$ of $\hat{\eta}$, $R(\hat{y},\hat{\eta})\geq 1$ and a family $\{g^R_{(\hat{y},\hat{\eta})}(\eta)\}_{R\geq R(\hat{y},\hat{\eta})}$ of symbols such that $g^R_{(\hat{y},\hat{\eta})}(\eta)=1$ in $\Gamma_{(\hat{y},\hat{\eta})}\cap\{\langle\eta\rangle\geq R/2\}$,

$$\left|\partial_\eta^{\alpha+\tilde{\alpha}}g^R_{(\hat{y},\hat{\eta})}(\eta)\right|\leq C_{|\tilde{\alpha}|}(u,x^0,\xi^0)(C(u,x^0,\xi^0)/R)^{|\alpha|}\langle\eta\rangle^{-|\tilde{\alpha}|}$$

if $\langle\eta\rangle\geq R|\alpha|$, and $g^R_{(\hat{y},\hat{\eta})}(D)u$ is analytic in a neighborhood of $\overline{U}_{(\hat{y},\hat{\eta})}$ for $R\geq R(\hat{y},\hat{\eta})$. Let $U^j_{(\hat{y},\hat{\eta})}$ be neighborhoods of $\nabla_\eta T(\hat{y},\hat{\eta})$ such that $U^1_{(\hat{y},\hat{\eta})}\Subset U^2_{(\hat{y},\hat{\eta})}\Subset U_{(\hat{y},\hat{\eta})}$. Then there are a bounded neighborhood $V_{(\hat{y},\hat{\eta})}$ of $\hat{y}$ and a conic neighborhood $\Gamma^1_{(\hat{y},\hat{\eta})}$ of $\hat{\eta}$ such that $\Gamma^1_{(\hat{y},\hat{\eta})}\Subset\Gamma_{(\hat{y},\hat{\eta})}$ and

$$\nabla_\eta T(y,\eta)\in U^1_{(\hat{y},\hat{\eta})}\quad\text{for } y\in V_{(\hat{y},\hat{\eta})}\text{ and }\eta\in\Gamma^1_{(\hat{y},\hat{\eta})}.$$

Since $\nabla_y S(\hat{y},\xi^0) \neq 0$, there are a neighborhood $V_{\hat{y}}$ of $\hat{y}$, a conic neighborhood $\widetilde{\Gamma}_{\hat{y}}$ of ξ^0 and $c(\hat{y}) > 0$ such that

$$|\nabla_y S(y,\xi)| \geq c(\hat{y})|\xi| \quad \text{for } y \in V_{\hat{y}} \text{ and } \xi \in \widetilde{\Gamma}_{\hat{y}}.$$

Let $V^1_{(\hat{y},\hat{\eta})}$ ($\Subset V_{(\hat{y},\hat{\eta})} \cap V_{\hat{y}}$) be a neighborhood of $\hat{y}$, and let $\Gamma^2_{(\hat{y},\hat{\eta})}$ ($\subset \subset \Gamma^1_{(\hat{y},\hat{\eta})}$) be a conic neighborhood of $\hat{\eta}$. Since $\mathcal{N}(x^0,\xi^0) \cap (\boldsymbol{R}^n \times S^{n''-1})$ is compact, there are $N \in \boldsymbol{N}$ and $(y^j,\eta^j) \in \mathcal{N}(x^0,\xi^0) \cap \boldsymbol{R}^n \times S^{n''-1}$ ($1 \leq j \leq N$) such that $\mathcal{N}(x^0,\xi^0) \Subset \cup_{j=1}^N V^1_{(y^j,\eta^j)} \times \Gamma^2_{(y^j,\eta^j)} \equiv \mathcal{C}$. Choose $\Phi_j^R(\xi,y,\eta) \in S^{0,0}(\boldsymbol{R}^{n'} \times \boldsymbol{R}^n \times \boldsymbol{R}^{n''}; R, C_*, \widehat{C}, \widehat{C})$ ($1 \leq j \leq N$, $R \geq 4$) so that $0 \leq \Phi_j^R(\xi,y,\eta) \leq 1$ and

$$\text{supp } \Phi_j^R \subset \boldsymbol{R}^{n'} \times (V_{(y^j,\eta^j)} \cap V_{y^j}) \times \Gamma^1_{(y^j,\eta^j)} \cap \{\langle\eta\rangle \geq R/2\},$$

$$\sum_{j=1}^N \Phi_j^R(\xi,y,\eta) = 1 \quad \text{if } (y,\eta) \in \mathcal{C} \text{ and } \langle\eta\rangle \geq R,$$

where $\widehat{C}$ also depends on $\{V_{(y^j,\eta^j)} \cap V_{y^j}\}$, $\{\Gamma^1_{(y^j,\eta^j)}\}$, $\{V^1_{(y^j,\eta^j)}\}$ and $\{\Gamma^2_{(y^j,\eta^j)}\}$ (see Proposition 2.2. 3). We put $\widetilde{\Gamma} = \bigcap_{j=1}^N \widetilde{\Gamma}_{y^j}$, $\tilde{c} = \min_{1\leq j\leq N} c(y^j)$ and $R_1 = \max\{4, R(y^1,\eta^1), \cdots, R(y^N,\eta^N)\}$. Let $\widetilde{\Gamma}^1$ ($\Subset \widetilde{\Gamma}$) be a conic neighborhood of ξ^0, and choose $\{\tilde{g}^R(\xi)\}_{R\geq R_1} \subset C^\infty(\boldsymbol{R}^{n'})$ so that supp $\tilde{g}^R \subset \widetilde{\Gamma} \cap \{|\xi| \geq 1\}$, $\tilde{g}^R(\xi) = 1$ in $\widetilde{\Gamma}^1 \cap \{\langle\xi\rangle \geq R\}$, and $|\partial_\xi^{\alpha+\tilde{\alpha}} \tilde{g}^R(\xi)| \leq C_{|\tilde{\alpha}|}(\widehat{C}'/R)^{|\alpha|}\langle\xi\rangle^{-|\tilde{\alpha}|}$ if $\langle\xi\rangle \geq R|\alpha|$, where $\widehat{C}'$ also depends on $\widetilde{\Gamma}$ and $\widetilde{\Gamma}^1$. Put

$$p_0^R(\xi,y,\eta) = (1 - \sum_{j=1}^N \tilde{g}^R(\xi)\Phi_j^R(\xi,y,\eta))p(\xi,y,\eta),$$

$$p_\ell^R(\xi,y,\eta) = \tilde{g}^R(\xi)\Phi_\ell^R(\xi,y,\eta))p(\xi,y,\eta) \quad (1 \leq \ell \leq N).$$

If $\xi \in \widetilde{\Gamma}^1$, $(y,\eta) \in \mathcal{C}$, $\langle\xi\rangle \geq R$ and $\langle\eta\rangle \geq R$, then $p_0^R(\xi,y,\eta) = 0$. Since $\widetilde{\Gamma}^1 \times \mathcal{C}$ ($\supset \widetilde{\Gamma}^1 \times \mathcal{N}(x^0,\xi^0)$) is a conic neighborhood of $\mathcal{M}(x^0,\xi^0) \cap \{(\xi,y,\eta) \in ECS(p);\ \text{Im } S(y,\xi) = 0 \text{ and Im } T(y,\eta) = 0\}$, for any $(\xi^0,\hat{y},\hat{\eta}) \in \mathcal{M}(x^0,\xi^0) \setminus \widetilde{\Gamma}^1 \times \mathcal{C}$ there are a conic neighborhood $\mathcal{U}_{(\hat{y},\hat{\eta})}$ of $(\xi^0,\hat{y},\hat{\eta})$ in $(\boldsymbol{R}^{n'} \setminus \{0\}) \times \boldsymbol{R}^n \times \boldsymbol{R}^{n''}$ and $c(\hat{y},\hat{\eta}) > 0$ such that

$$\left|e^{iS(y,\xi)+iT(y,\eta)}\partial_\eta^\gamma p(\xi,y,\eta)\right| \leq C_{|\gamma|}(p)\exp[-c(\hat{y},\hat{\eta})\langle\xi\rangle]$$

for $(\xi,y,\eta) \in \mathcal{U}_{(\hat{y},\hat{\eta})}$. Here we have used the assumption (3.23). Noting that $\mathcal{M}(x^0,\xi^0) \cap S^{n'-1} \times \boldsymbol{R}^n \times \boldsymbol{R}^{n''}$ is compact, we can show that there are a conic neighborhood $\mathcal{U}$ of $\mathcal{M}(x^0,\xi^0)$ and $c > 0$ satisfying

$$\left|e^{iS(y,\xi)+iT(y,\eta)}\partial_\eta^\gamma p_0^R(\xi,y,\eta)\right| \leq C_{|\gamma|}(p)e^{-c\langle\xi\rangle}$$

for $(\xi, y, \eta) \in \mathcal{U}$. Therefore, it follows from Proposition 3.2.1 that there are $R_j(S,T,u,x^0,\xi^0) > 0$ ($j = 1,2$) such that $(x^0,\xi^0) \notin WF_A(p^R_{0\,S,T}(D_x, y, D_z)u)$ if $R_0 \geq R_1(S,T,u,x^0,\xi^0)A$, $R \geq R_1(S,T,u,x^0,\xi^0)(C_* + \widehat{C} + \widehat{C}')$ and $R \geq R_2(S,T,u,x^0,\xi^0)$. We note that

$$p^R_\ell(\xi,y,\eta) g^R_{(y^\ell,\eta^\ell)}(\eta) = p^R_\ell(\xi,y,\eta) \quad (1 \leq \ell \leq N).$$

Put $u_{\ell,\rho}(z) = e^{-\rho\langle D\rangle} g^R_{(y^\ell,\eta^\ell)}(D)u(z)$ ($1 \leq \ell \leq N$, $0 < \rho \leq 1$). Then there are positive constants $C(u,x^0,\xi^0)$ and $A(u,x^0,\xi^0)$ such that

$$\left|D^\lambda u_{\ell,\rho}(z)\right| \leq C(u,x^0,\xi^0) A(u,x^0,\xi^0)^{|\lambda|} |\lambda|!$$

for $z \in U_{(y^\ell,\eta^\ell)}$, $0 < \rho \leq 1$ and $1 \leq \ell \leq N$. Choose $\{\chi_{\ell,k}(z)\}_{k \in N} \subset C_0^\infty(U_{(y^\ell,\eta^\ell)})$ ($1 \leq \ell \leq N$) so that $\chi_{\ell,k}(z) = 1$ in $U^2_{(y^\ell,\eta^\ell)}$ and

$$\left|D^\lambda \chi_{\ell,k}(z)\right| \leq \widehat{C}_0 (\widehat{C}_1 k)^{|\lambda|} \quad \text{for } |\lambda| \leq k.$$

Then we have, for $0 < \rho \leq 1$,

$$\begin{aligned} &\left|\mathcal{F}[\chi_{\ell,k}(z) u_{\ell,\rho}(z)](\eta)\right| \\ &\leq C'(u,x^0,\xi^0)(1 + \sqrt{n''}(\widehat{C}_1 + A(u,x^0,\xi^0))k)^k \langle\eta\rangle^{-k} \end{aligned} \tag{3.25}$$

Fix $\ell \in N$ so that $1 \leq \ell \leq N$. Since $\sum_{k=1}^\infty \psi^R_k(D)u = u$ in $\mathcal{F}_0(\boldsymbol{R}^{n''})$, it follows from Theorem 2.3.3 that

$$\begin{aligned} p^R_{\ell\,S,T}(D_x,y,D_z)u = &\sum_{k=1}^\infty p^R_{\ell\,S,T}(D_x,y,D_z)\psi^R_k(D_z) \\ &\times e^{\rho\langle D_z\rangle}\{\chi_{\ell,k}(z)u_{\ell,\rho}(z) + (1 - \chi_{\ell,k}(z))u_{\ell,\rho}(z)\} \end{aligned}$$

in $\mathcal{F}_0(\boldsymbol{R}^{n'})$ if $R_0 \geq R_1(S,T)A$, $R \geq R_1(S,T)\widehat{C}$ and $R \geq R_2(S,T)$, where $R_j(S,T)$ ($j = 1,2$) are some positive constants. (3.25) yields

$$\left|\mathcal{F}\left[\psi^R_k(D_z) e^{\rho\langle D_z\rangle}(\chi_{\ell,k}(z)u_{\ell,\rho}(z))\right](\eta)\right| \leq C_R(u) 2^{-k} e^{-\delta\langle\eta\rangle}$$

if $0 < \rho \leq 1$, $R \geq 2e(1+\sqrt{n''}(\widehat{C}_1 + A(u,x^0,\xi^0)))$, $\delta \geq 0$ and $\rho + \delta \leq 1/(3R)$. Hence we have

$$\sum_{k=1}^\infty \psi^R_k(D_z) e^{\rho\langle D_z\rangle}(\chi_{\ell,k}u_{\ell,\rho}) \in \mathcal{S}'_{-\delta}$$

if $0 < \rho \leq 1$, $R \geq 2e(1 + \sqrt{n''}(\widehat{C}_1 + A(u,x^0,\xi^0)))$, $\delta \geq 0$ and $\rho + \delta \leq 1/(3R)$. Recall that $|\nabla_y S(y,\xi)| \geq \tilde{c}|\xi|$ in supp p^R_ℓ, and that $c_0(S) > 0$ and

$c_2(T) > 0$. By Theorem 2.3. 3 (or its proof) there are positive constants $R_1'(S,T)$ and $R(S,T,u,x^0,\xi^0)$ such that

$$\sum_{k=1}^{\infty} p^R_{\ell S,T}(D_x,y,D_z)\psi_k^R(D_z)e^{\rho\langle D_z\rangle}(\chi_{\ell,k}(z)u_{\ell,\rho}(z)) \in \mathcal{S}'_{-\bar{c}\delta/(2c_2(T))}$$

if $0<\rho\le 1$, $\delta\ge 0$, $\rho+\delta\le 1/(3R)$, $\bar{c}\delta/(2c_2(T))<1/R$, $R_0\ge R_1'(S,T)A$ and $R\ge R(S,T,u,x^0,\xi^0)$. This implies that $\sum_{k=1}^{\infty} p^R_{\ell S,T}(D_x,y,D_z)\psi_k^R(D_z)e^{\rho\langle D_z\rangle}(\chi_{\ell,k}u_{\ell,\rho})$ is analytic if $\rho<1/(3R)$, $R_0\ge R_1'(S,T)A$ and $R\ge R(S,T,u,x^0,\,\xi^0)$. Let L be a differential operator defined by

$$\begin{aligned} {}^tL &= |\nabla_\eta T(y,\eta)-z-i\rho\eta/\langle\eta\rangle|^{-2} \\ &\quad\times\sum_{\mu=1}^{n''}\Big(\overline{\partial_{\eta_\mu}T(y,\eta)}-z_\mu+i\rho\eta_\mu/\langle\eta\rangle\Big)D_{\eta_\mu} \end{aligned}$$

for $y\in V_{(y^\ell,\eta^\ell)}$, $\eta\in\Gamma^1_{(y^\ell,\eta^\ell)}$ and $z\notin U^2_{(y^\ell,\eta^\ell)}$. By Lemma 1.1. 3 $u_{\ell,\rho}\in \mathcal{S}'\cap C^\infty$ for $\rho>0$ and there is $\ell_0\in \boldsymbol{Z}_+$ such that

$$|u_{\ell,\rho}(z)|\le C_\rho(1+|z|)^{\ell_0} \quad \text{for } \rho>0,$$

where C_ρ is a positive constant depending on ρ, u and (x^0,ξ^0). Choose $\{\phi_k(t)\}\subset C^\infty(\boldsymbol{R})$ so that $0\le\phi_k(t)\le 1$, $\phi_k(t)=1$ ($|t|\le 2$), $\phi_k(t)=0$ ($|t|\ge 3$) and

$$\left|\partial_t^{\nu+\tilde{\nu}}\phi_k(t)\right|\le C_{\tilde{\nu}}(\widehat{C}_*k)^\nu \quad \text{if } \nu\le k.$$

Then, by induction we have

$$\left|\partial_\eta^{\gamma+\tilde{\gamma}}\phi_k^{(\nu+\tilde{\nu})}(\bar{c}|\xi|/(c_2(T)\langle\eta\rangle))\right|\le C_{|\tilde{\gamma}|,\tilde{\nu}}(\widehat{C}_*k)^\nu(\widehat{C}_*'k)^{|\gamma|}\langle\eta\rangle^{-|\gamma|-|\tilde{\gamma}|}$$

if $|\gamma|+\nu\le k$, where $\widehat{C}_*'=8(1+\sqrt{2})\max\{1,12\widehat{C}_*\}$. Noting that there is $c>0$ such that $|\mathrm{Re}\,\nabla_\eta T(y,\eta)-z|\ge c\langle z\rangle$ if $y\in V_{(y^\ell,\eta^\ell)}$, $\eta\in\Gamma^1_{(y^\ell,\eta^\ell)}$ and $z\notin U^2_{(y^\ell,\eta^\ell)}$, we can write

$$\begin{aligned} &\mathcal{F}\Big[p^R_{\ell S,T}(D_x,y,D_z)\psi_k^R(D_z)e^{\rho\langle D_z\rangle}((1-\chi_{\ell,\rho}(z))u_{\ell,\rho}(z))\Big](\xi) \\ &= F^{1,R}_{\ell,k,\rho}(\xi)+F^{2,R}_{\ell,k,\rho}(\xi), \end{aligned} \tag{3.26}$$

where

$$F^{1,R}_{\ell,k,\rho}(\xi)=(2\pi)^{-n''}\int e^{iS(y,\xi)+iT(y,\eta)-iz\cdot\eta+\rho\langle\eta\rangle}$$

$$\times L^{\ell_0+n''+1+k}\Big\{p_\ell^R(\xi,y,\eta)\psi_k^R(\eta)\phi_k(\tilde{c}|\xi|/(c_2(T)\langle\eta\rangle))\Big\}$$
$$\times(1-\chi_{\ell,k}(z))u_{\ell,\rho}(z)\,dzd\eta dy,$$
$$F_{\ell,k,\rho}^{2,R}(\xi)=(2\pi)^{-n''}\int e^{iS(y,\xi)+iT(y,\eta)-iz\cdot\eta+\rho\langle\eta\rangle}$$
$$\times L^{\ell_0+n''+1}\Big\{p_\ell^R(\xi,y,\eta)\psi_k^R(\eta)(1-\phi_k(\tilde{c}|\xi|/(c_2(T)\langle\eta\rangle)))\Big\}$$
$$\times(1-\chi_{\ell,k}(z))u_{\ell,\rho}(z)\,dzd\eta dy.$$

Since

$$\Big|D_y^{\beta+\tilde{\beta}}\partial_\eta^{\gamma+\tilde{\gamma}}\Big\{p_\ell^R(\xi,y,\eta)\psi_k^R(\eta)\phi_k(\tilde{c}|\xi|/(c_2(T)\langle\eta\rangle))\Big\}\Big|$$
$$\le C_{|\tilde{\beta}|+|\tilde{\gamma}|,\delta',R}(p)(A/R_0+\widehat{C}/R)^{|\beta|}(A/R_0+(\widehat{C}+\widehat{C}'_*)/R)^{|\gamma|}$$
$$\times\langle\xi\rangle^{|\beta|}\langle\eta\rangle^{-|\tilde{\gamma}|}e^{\delta'\langle\xi\rangle+\delta'\langle\eta\rangle}$$

if $\delta'>0$, $R\ge R_0$, $\langle\xi\rangle\ge R|\beta|$ and $|\gamma|\le k$, and

$$\Big|\partial_\eta^\alpha(T(y,\eta)-z\cdot\eta-i\rho\langle\eta\rangle)\Big|$$
$$\le(2C(T)\langle y\rangle+|z|/\widetilde{A}_0+1)\widetilde{A}_0^{|\alpha|}|\alpha|!\langle\eta\rangle^{1-|\alpha|}$$

if $(y,\eta)\in\Omega'$, $0\le\rho\le1$ and $\widetilde{A}_0=\max\{A_0,1+\sqrt{2}\}$, Lemma 2.1. 5 gives

$$\Big|L^{\ell_0+n''+1+k}\Big\{p_\ell^R(\xi,y,\eta)\psi_k^R(\eta)\phi_k(\tilde{c}|\xi|/(c_2(T)\langle\eta\rangle))\Big\}\Big|$$
$$\le C_{\ell_0+n''+1,\delta',R}(p,T)(\Gamma_\ell(T,A,R/R_0)/R)^k\langle z\rangle^{-\ell_0-n''-1-k}$$
$$\times\langle\eta\rangle^{-\ell_0-n''-1}e^{\delta'\langle\xi\rangle+\delta'\langle\eta\rangle}$$

if $\delta'>0$ and $z\notin U^2_{(y^\ell,\eta^\ell)}$, where $C(T,V_{(y^\ell,\eta^\ell)})$ is a positive constant and

$$\Gamma_\ell(T,A,R/R_0)=32n''(1+C(T,V_{(y^\ell,\eta^\ell)})\widetilde{A}_0)$$
$$\times\max\Big\{AR/R_0+\widehat{C}+\widehat{C}'_*,120n''(1+C(T,V_{(y^\ell,\eta^\ell)})\widetilde{A}_0)^2\widetilde{A}_0/c^2\Big\}/c^2.$$

Therefore, we have

$$\Big|F_{\ell,k,\rho}^{1,R}(\xi)\Big|\le C_{\rho,\delta',R}\,e^{-\delta\langle\xi\rangle}2^{-k}\tag{3.27}$$

if $\delta'>0$, $\delta\ge0$, $R\ge2e\Gamma_\ell(T,A,R/R_0)$ and $3c_2(T)(\delta+\delta')/\tilde{c}+\rho+\delta'\le 1/(3R)$. Write

$$F_{\ell,k,\rho}^{2,R}(\xi)=(2\pi)^{-n''}\int e^{iS(y,\xi)+iT(y,\eta)-iz\cdot\eta+\rho\langle\eta\rangle}$$
$$\times L^{\ell_0+n''+1}K^\mu\Big\{p_\ell^R(\xi,y,\eta)\psi_k^R(\eta)(1-\phi_k(\tilde{c}|\xi|/(c_2(T)\langle\eta\rangle)))\Big\}$$
$$\times(1-\chi_{\ell,k}(z))u_{\ell,\rho}(z)\,dzd\eta dy$$

for $\langle\xi\rangle \geq R\mu$ and $\mu \in \boldsymbol{Z}_+$, where $R \geq R_0$ and K is the differential operator defined by (3.18). Applying Lemma 2.1.5 , we have

$$\begin{aligned}&\left|\partial_\eta^\gamma K^\mu\left\{p_\ell^R(\xi,y,\eta)\psi_k^R(\eta)(1-\phi_k(\tilde{c}|\xi|/(c_2(T)\langle\eta\rangle)))\right\}\right|\\&\leq C_{|\gamma|,\delta'}(p,S,T)(\widetilde{\Gamma}_\ell(S,T,A,R/R_0)/R)^\mu\langle\eta\rangle^{-|\gamma|}e^{\delta'\langle\xi\rangle+\delta'\langle\eta\rangle}\end{aligned}$$

if $\delta' > 0$, $R \geq R_0$, $\mu \in \boldsymbol{Z}_+$ and $\langle\xi\rangle \geq R\mu$, where

$$\begin{aligned}&\widetilde{\Gamma}_\ell(S,T,A,R/R_0) = 2^8 nC(S,T)B_0\\&\quad\times\max\{AR/R_0+\widehat{C}, 2^7\cdot 15nC(S,T)^2B_0^3\}/\tilde{c},\\&C(S,T) = C(S)/\tilde{c}+C(T)/(2c_2(T)).\end{aligned}$$

Here we have used the estimates that

$$\left|D_y^\beta\partial_\eta^\gamma(S(y,\xi)+T(y,\eta))\right| \leq 2(C(S)|\xi|+C(T)\langle\eta\rangle)A_0^{|\gamma|}B_0^{|\beta|}\langle\eta\rangle^{-|\gamma|}$$

if $|\beta| \geq 1$, and

$$|\nabla_y S(y,\xi)+\nabla_y T(y,\eta)| \geq \tilde{c}|\xi|/2$$

if $\phi_k(\tilde{c}|\xi|/(c_2(T)\langle\eta\rangle)) \neq 1$. Therefore, we have

$$\left|F_{\ell,k,\rho}^{2,R}(\xi)\right| \leq C_{\rho,\delta',R}\,e^{-\delta\langle\xi\rangle}k^{-2} \tag{3.28}$$

if $\delta' > 0$, $\delta \geq 0$, $R \geq R_0$, $R \geq 2e\widetilde{\Gamma}_\ell(S,T,A,R/R_0)$ and $\delta+\delta'+\tilde{c}(\rho+\delta')/(2c_2(T)) \leq 1/R$. (3.26) – (3.28) gives

$$\sum_{k=1}^\infty p_{\ell\, S,T}^R(D_x,y,D_z)\psi_k^R(D_z)e^{\rho\langle D_z\rangle}((1-\chi_{\ell,k})u_{\ell,\rho}) \in \mathcal{S}'_{-\delta}$$

if $\delta \geq 0$, $R \geq R_0$, $R \geq 2e\max\{\Gamma_\ell(T,A,R/R_0), \widetilde{\Gamma}_\ell(S,T,A,R/R_0)\}$, $3c_2(T)\times\delta\tilde{c}^{-1}+\rho < 1/(3R)$ and $\delta+\tilde{c}\rho/(2c_2(T)) < 1/R$. Therefore, $p_{\ell\, S,T}^R(D_x,y,D_z)u$ is analytic if $R_0 \geq R_1'(S,T)A$, $R \geq R_0$, $R \geq R(S,T,u,x^0,\xi^0)$ and $R \geq 2e\max\{\Gamma_\ell(T,A,R/R_0), \widetilde{\Gamma}_\ell(S,T,A,R/R_0)\}$. □

It is not necessary to assume that $c_3(T) > 0$ in Proposition 3.2. 3 . Instead, we assume that

(A–1) there are positive constant C_T and c such that

$$|\nabla_y T(y,\eta)| \geq c|\eta| \quad \text{if } (y,\eta) \in \overline{\Omega}' \text{ and } |y| \geq C_T,$$

and that

(A–2) $|\nabla_y S(y,\xi)| \neq 0$ if $(y,\xi) \in \overline{\Omega}$, $\xi \neq 0$, $\operatorname{Im} S(y,\xi) = 0$, and $(y,\eta) \in \mathcal{T}$ for some $\eta \in \boldsymbol{R}^{n''}\setminus\{0\}$ with $(y,\eta) \in \overline{\Omega}'$, where $\mathcal{T} = \{(y,\eta) \in \overline{\Omega}';\ \eta \neq 0$ and $\nabla_y T(y,\eta) = 0\}$.

Theorem 3.2.4 *We assume that (A–1) and (A–2) are valid, and that* $u \in \mathcal{F}_0(\boldsymbol{R}^{n''})$ *satisfies*

$$WF_A(u) \cap \{(\nabla_\eta T(y,\eta),\eta);\ (y,\eta) \in \mathcal{T} \text{ and } \mathrm{Im}\, T(y,\eta)=0\} = \emptyset.$$

(i) *There is* $R(S,T,u)>0$ *such that* $p_{S,T}(D_x,y,D_z)u$ *is well-defined, that is, we can define* $p_{S,T}(D_x,y,D_z)u$ *by*

$$p_{S,T}(D_x,y,D_z)u = \lim_{\varepsilon\downarrow 0} p_{S,T}(D_x,y,D_z)(\exp[-\varepsilon\langle D_z\rangle]u) \quad \text{in } \mathcal{F}_0(\boldsymbol{R}^{n'}),$$

if $R_0 \ge R(S,T,u)A$. (ii) *Let* $(x^0,\xi^0) \in T^*\boldsymbol{R}^{n'}\setminus 0$, *and assume that* (3.23) *is valid. Moreover, we assume that*

$$WF_A(u) \cap \{(\nabla_\eta T(y,\eta),\eta) \in T^*\boldsymbol{R}^{n''}\setminus 0;\ (y,\eta)\in \mathcal{N}(x^0,\xi^0)\} = \emptyset,$$

where $\mathcal{N}(x^0,\xi^0)$ *is defined in* Proposition 3.2.3 . *Then there is* $R(S,T,u,x^0,\xi^0)>0$ *such that* $R(S,T,u,x^0,\xi^0)\ge R(S,T,u)$ *and*

$$(x^0,\xi^0)\notin WF_A(p_{S,T}(D_x,y,D_z)u) \quad \text{if } R_0 \ge R(S,T,u,x^0,\xi^0)A.$$

Proof There are a conic neighborhood $\mathcal{U}_1$ of $\mathcal{T}$ and a conic neighborhood $\mathcal{U}_2$ of $\{(y,\eta)\in\overline{\Omega}';\ \eta\neq 0 \text{ and } \mathrm{Im}\,T(y,\eta)=0\}$ and $c_1>0$ such that $\mathcal{U}_1 \subset \{|y|\le C_T+1\}$,

$$WF_A(u)\cap\{(\nabla_\eta T(y,\eta),\eta);\ (y,\eta)\in\overline{\Omega}'\cap\overline{\mathcal{U}}_1\cap\overline{\mathcal{U}}_2 \text{ and } \eta\neq 0\}=\emptyset,$$
$$|\nabla_y S(y,\xi)|\ge 2c_1|\xi| \quad \text{if } (y,\eta)\in\mathcal{U}_1,\ (y,\xi)\in\overline{\Omega} \text{ and } \mathrm{Im}\,S(y,\xi)=0.$$

So we can choose conic neighborhoods $\mathcal{V}_j$ ($j=1,2$) of $\{(y,\xi)\in\overline{\Omega};\ y\in Y_1,\ \xi\neq 0 \text{ and } \mathrm{Im}\,S(y,\xi)=0\}$ so that $\mathcal{V}_1 \subset\subset \mathcal{V}_2$, $\mathcal{V}_2\cap\boldsymbol{R}^n\times S^{n'-1}$ is bounded and

$$|\nabla_y S(y,\xi)|\ge c_1|\xi| \quad \text{for } (y,\xi)\in\mathcal{V}_2\cap\overline{\Omega},$$

where $Y_1=\{y\in\boldsymbol{R}^n;\ (y,\eta)\in\overline{\mathcal{U}}_1 \text{ for some } \eta\}$. Then there is $c_2>0$ such that

$$\mathrm{Im}\,S(y,\xi)\le -c_2|\xi| \quad \text{if } (y,\xi)\in\overline{\Omega}\setminus\mathcal{V}_1,\ y\in Y_1 \text{ and } \xi\neq 0.$$

Let $\mathcal{U}_{1,0}$ ($\subset\subset \mathcal{U}_1$) be a conic neighborhood of $\mathcal{T}$, and choose $\Psi_1^R(\xi,y,\eta)\in S^{0,0}(\boldsymbol{R}^{n'}\times\boldsymbol{R}^n\times\boldsymbol{R}^{n''};R,C_*,C(\mathcal{U}_1,\mathcal{U}_{1,0}),C(\mathcal{U}_1,\mathcal{U}_{1,0}))$ ($R\ge 4$) so that $0\le\Psi_1^R(\xi,y,\eta)\le 1$, supp $\Psi_1^R\subset\boldsymbol{R}^{n'}\times\mathcal{U}_1$ and $\Psi_1^R(\xi,y,\eta)=1$ if $(y,\eta)\in\mathcal{U}_{1,0}$ and $|\eta|\ge 1$. Put

$$\begin{aligned} p_0^R(\xi,y,\eta) &= (1-\Psi_1^R(\xi,y,\eta))p(\xi,y,\eta),\\ p_1^R(\xi,y,\eta) &= \Psi_1^R(\xi,y,\eta)p(\xi,y,\eta). \end{aligned}$$

Since $p_0^R(\xi, y, \eta) = 0$ for $(y, \eta) \in \overline{\mathcal{U}}_{1,0}$, we can apply Proposition 3.2.3 to $p_0^R(\xi, y, \eta)$ replacing Ω' with $\Omega' \setminus \overline{\mathcal{U}}_{1,0}$. In particular, it is obvious that

$$\lim_{\varepsilon \downarrow 0} p^R_{0\,S,T}(D_x, y, D_z)(\exp[-\varepsilon \langle D_z \rangle]u) = p^R_{0\,S,T}(D_x, y, D_z)u \quad \text{in } \mathcal{F}_0(\boldsymbol{R}^{n'}).$$

Therefore, it suffices to prove that there is $R(S, T, u) > 0$ such that $p^R_{1\,S,T}(D_x, y, D_z)u$ is well-defined and analytic if $R_0 \geq R(S, T, u)A$ and $R \gg 1$. Let $\Psi_2^R(\xi, y, \eta)$ ($R \geq 4$) be symbols in $S^{0,0}(\boldsymbol{R}^{n'} \times \boldsymbol{R}^n \times \boldsymbol{R}^{n''}; R, C(\mathcal{V}_2, \mathcal{V}_1), C(\mathcal{V}_2, \mathcal{V}_1), C_*)$ such that $0 \leq \Psi_2^R(\xi, y, \eta) \leq 1$, $\operatorname{supp} \Psi_2^R \subset \{(\xi, y, \eta) \in \boldsymbol{R}^{n'} \times \boldsymbol{R}^n \times \boldsymbol{R}^{n''};\ (y, \xi) \in \mathcal{V}_2\}$ and $\Psi_2^R(\xi, y, \eta) = 1$ if $(y, \xi) \in \mathcal{V}_1$ and $|\xi| \geq 1$. We put

$$\begin{aligned} p^R_{1,1}(\xi, y, \eta) &= (1 - \Psi_2^R(\xi, y, \eta))p_1^R(\xi, y, \eta) \\ p^R_{1,2}(\xi, y, \eta) &= \Psi_2^R(\xi, y, \eta)p_1^R(\xi, y, \eta) \end{aligned}$$

Let $g^R(\xi, \eta)$ ($R \geq 4$) be symbols such that $0 \leq g^R(\xi, \eta) \leq 1$, $g^R(\xi, \eta) = 0$ if $|\xi| \leq 2c_2(T)|\eta|/c_1$, $g^R(\xi, \eta) = 1$ if $|\xi| \geq 3c_2(T)|\eta|/c_1$ and $\langle \xi \rangle \geq R$, and

$$\left| \partial_\xi^\alpha \partial_\eta^{\gamma + \tilde{\gamma}} g^R(\xi, \eta) \right| \leq C_{|\alpha| + |\tilde{\gamma}|}(\widehat{C}_1/R)^{|\gamma|} \langle \eta \rangle^{-|\tilde{\gamma}|}$$

if $\langle \eta \rangle \geq R|\gamma|$, and put

$$\begin{aligned} p^R_{1,1,0}(\xi, y, \eta) &= g^R(\xi, \eta)p^R_{1,1}(\xi, y, \eta), \\ p^R_{1,2,0}(\xi, y, \eta) &= g^R(\xi, \eta)p^R_{1,2}(\xi, y, \eta). \end{aligned}$$

Since $\operatorname{Im} S(y, \xi) \leq -c_2|\xi|$ if $(\xi, y, \eta) \in \operatorname{supp} p^R_{1,1,0}$, and

$$\begin{aligned} &\left| \partial_\xi^\alpha \partial_\eta^\gamma p^R_{1,1,0}(\xi, y, \eta) \right| \\ &\leq C_{|\alpha| + |\gamma|, \delta} \exp[(\delta + c_2/2)\langle \xi \rangle + (\delta - c_2 c_2(T)/c_1)\langle \eta \rangle] \end{aligned}$$

if $\delta > 0$, Lemma A.1.7 yields

$$\lim_{\varepsilon \downarrow 0} p^R_{1,1,0\,S,T}(D_x, y, D_z)(\exp[-\varepsilon \langle D_z \rangle]u) = p^R_{1,1,0\,S,T}(D_x, y, D_z)u$$

in $\mathcal{S}'_{-c_2/4}(\boldsymbol{R}^{n'})$ ($\subset \mathcal{F}_0(\boldsymbol{R}^{n'})$). In particular, $p^R_{1,1,0\,S,T}(D_x, y, D_z)u$ is analytic. Note that $|\nabla_y S(y, \xi) + \nabla_y T(y, \eta)| \geq c_1|\xi|/2 \geq c_2(T)|\eta|$ for $(\xi, y, \eta) \in \operatorname{supp} p^R_{1,2,0}$. We can write

$$\begin{aligned} &\mathcal{F}_x^{-1}\left[p^R_{1,2,0\,S,T}(D_x, y, D_z)(\exp[-\varepsilon \langle D_z \rangle]u)(x) \right](\xi) \\ &= (2\pi)^{-n''} \langle e^{-\varepsilon \langle \eta \rangle} \hat{u}(\eta), g^R(\xi, \eta) \int e^{iS(y,\xi) + iT(y,\eta)} K^\mu p^R_{1,2}(\xi, y, \eta)\, dy \rangle_\eta \end{aligned}$$

if $\mu \in \boldsymbol{Z}_+$, $\langle\xi\rangle \ge \max\{R_0, R\}\mu$ and $\varepsilon \ge 0$. Applying Lemma 2.1. 5 , we have

$$\left|\partial_\eta^\gamma K^\mu p_{1,2}^R(\xi, y, \eta)\right| \le C_{|\gamma|,\delta}(p, S, T)$$
$$\times(\Gamma(S, T, A, \mathcal{U}_1, \mathcal{U}_{1,0}, \mathcal{V}_2, \mathcal{V}_1, R/R_0)/R)^\mu \langle\eta\rangle^{-|\gamma|} e^{\delta\langle\xi\rangle+\delta\langle\eta\rangle}$$

if $\delta > 0$, $R \ge R_0$, $\mu \in \boldsymbol{Z}_+$, $\langle\xi\rangle \ge \mu R$ and $g^R(\xi, \eta) \ne 0$, where

$$\Gamma(S, T, A, \mathcal{U}_1, \mathcal{U}_{1,0}, \mathcal{V}_2, \mathcal{V}_1, R/R_0) = 2^8 n C(S, T) B_0$$
$$\times \max\{AR/R_0 + C(\mathcal{U}_1, \mathcal{U}_{1,0}) + C(\mathcal{V}_2, \mathcal{V}_1), 2^7 \cdot 15 n C(S, T)^2 B_0^3\}/c_1,$$
$$C(S, T) = C(S)/c_1 + C(T)/(2c_2(T)).$$

This gives

$$\lim_{\varepsilon\downarrow 0} p_{1,2,0\,S,T}^R(D_x, y, D_z)(\exp[-\varepsilon\langle D_z\rangle]u) = p_{1,2,0\,S,T}^R(D_x, y, D_z)u$$

in $\mathcal{S}'_{-1/(2R)}$ if $R \ge R_0$ and $R \ge e\Gamma(S, T, A, \mathcal{U}_1, \mathcal{U}_{1,0}, \mathcal{V}_2, \mathcal{V}_1, R/R_0)$. Let $\mathcal{U}_{2,0}$ ($\subset\subset \mathcal{U}_2 \cap \{|y| \le C_T + 2\}$) be a conic neighborhood of $\{(y, \eta) \in \overline{\Omega}'$; $\eta \ne 0$, $\mathrm{Im}\, T(y, \eta) = 0$ and $|y| \le C_T + 1\}$, and choose $\Psi_3^R(\xi, y, \eta) \in S^{0,0}(\boldsymbol{R}^{n'} \times \boldsymbol{R}^n \times \boldsymbol{R}^{n''}; R, C_*, C(\mathcal{U}_2, \mathcal{U}_{2,0}, T), C(\mathcal{U}_2, \mathcal{U}_{2,0}, T))$ ($R \ge 4$) so that $0 \le \Psi_3^R(\xi, y, \eta) \le 1$, $\mathrm{supp}\, \Psi_3^R \subset \boldsymbol{R}^{n'} \times \mathcal{U}_2 \cap \{|y| \le C_T + 2\}$ and $\Psi_3^R(\xi, y, \eta) = 1$ if $(y, \eta) \in \mathcal{U}_{2,0}$ and $|\eta| \ge 1$. Put

$$\begin{aligned} p_2^R(\xi, y, \eta) &= (1 - g^R(\xi, \eta))(1 - \Psi_3^R(\xi, y, \eta))p_1^R(\xi, y, \eta), \\ p_3^R(\xi, y, \eta) &= (1 - g^R(\xi, \eta))\Psi_3^R(\xi, y, \eta)p_1^R(\xi, y, \eta). \end{aligned}$$

Note that $p_1^R(\xi, y, \eta) = p_{1,1,0}^R(\xi, y, \eta) + p_{1,2,0}^R(\xi, y, \eta) + p_2^R(\xi, y, \eta) + p_3^R(\xi, y, \eta)$. It is obvious that there is $c_3 > 0$ such that

$$\mathrm{Im}\, T(y, \eta) \le -c_3|\eta| \quad \text{if } (y, \eta) \in \overline{\Omega}' \setminus \mathcal{U}_{2,0},\ |y| \le C_T + 1 \text{ and } \eta \ne 0.$$

Therefore, applying Lemma A.1. 7 we have

$$\lim_{\varepsilon\downarrow 0} p_{2\,S,T}^R(D_x, y, D_z)(\exp[-\varepsilon\langle D_z\rangle]u) = p_{2\,S,T}^R(D_x, y, D_z)u$$

in $\mathcal{S}'_{-c_1c_3/(12c_2(T))}$. Let us consider $p_3^R(\xi, y, \eta)$. Repeating the same argument as in the proof of Proposition 3.2. 3 (ii), we can show that there are $N \in \boldsymbol{N}$, $R_1 \ge 1$, $(y^j, \eta^j) \in \overline{\Omega}' \cap \overline{\mathcal{U}}_1 \cap \overline{\mathcal{U}}_2 \cap (\boldsymbol{R}^n \times S^{n''-1})$, bounded open neighborhoods U_j^1 and U_j of $\nabla_\eta T(y^j, \eta^j)$ in $\boldsymbol{R}^{n''}$, conic neighborhoods Γ_j^1, Γ_j^2 and Γ_j of η^j, bounded neighborhoods V_j^1 and V_j of y^j and a family $\{g_j^R(\eta)\}_{R\ge R_1}$ of symbols ($1 \le j \le N$) such that $U_j^1 \subset\subset U_j$, $\Gamma_j^2 \subset\subset \Gamma_j^1 \subset\subset \Gamma_j$, $V_j^1 \subset\subset V_j$, $g_j^R(\eta) = 1$ if $\eta \in \Gamma_j$ and $\langle\eta\rangle \ge R/2$,

$$\left|\partial_\eta^{\alpha+\tilde{\alpha}} g_j^R(\eta)\right| \le C_{|\tilde{\alpha}|}(u)(\widehat{C}(u)/R)^{|\alpha|}\langle\eta\rangle^{-|\tilde{\alpha}|} \quad \text{if } \langle\eta\rangle \ge R|\alpha|,$$

$g_j^R(D)u$ is analytic in a neighborhood of $\overline{U}_j$,

$$\nabla_\eta T(y,\eta) \in U_j^1 \quad \text{for } y \in V_j \text{ and } \eta \in \Gamma_j^1$$

and

$$\overline{\Omega}' \cap \overline{\mathcal{U}}_1 \cap \overline{\mathcal{U}}_2 \setminus 0 \subset\subset \bigcup_{k=1}^N V_k^1 \times \Gamma_k^2 \equiv \mathcal{C},$$

where $1 \le j \le N$ and $R \ge R_1$. Choose $\Phi_j^R(\xi, y, \eta) \in S^{0,0}(\boldsymbol{R}^{n'} \times \boldsymbol{R}^n \times \boldsymbol{R}^{n''}; R, C_*, \widehat{C}_*, \widehat{C}_*)$ ($1 \le j \le N$, $R \ge 4$) so that $0 \le \Phi_j^R(\xi, y, \eta) \le 1$ and

$$\operatorname{supp} \Phi_j^R \subset \boldsymbol{R}^{n'} \times V_j \times \Gamma_j^1 \cap \{\langle\eta\rangle \ge R/2\},$$
$$\sum_{j=1}^N \Phi_j^R(\xi, y, \eta) = 1 \quad \text{if } (y,\eta) \in \mathcal{C} \text{ and } \langle\eta\rangle \ge R,$$

where $\widehat{C}_*$ also depends on $\{V_j \times \Gamma_j^1\}$ and $\{V_j^1 \times \Gamma_j^2\}$. We put

$$\begin{aligned} p_{3,0}^R(\xi,y,\eta) &= \Big(1 - \sum_{j=1}^N \Phi_j^R(\xi,y,\eta)\Big) p_3^R(\xi,y,\eta), \\ p_{3,\ell}^R(\xi,y,\eta) &= \Phi_\ell^R(\xi,y,\eta) p_3^R(\xi,y,\eta) \quad (1 \le \ell \le N). \end{aligned}$$

Since $\operatorname{supp} p_{3,0}^R(\xi,y,\eta) \subset \{(\xi,y,\eta);\ \langle\eta\rangle \le R$ and "$\langle\xi\rangle \le R$ or $|\xi| \le 3c_2(T) \times|\eta|/c_1$"$\} \subset \{(\xi,y,\eta);\ \langle\eta\rangle \le R$ and $|\xi| \le \max\{1, 3c_2(T)/c_1\}R\}$, we have

$$\lim_{\varepsilon\downarrow 0} p_{3,0\,S,T}^R(D_x, y, D_z)(\exp[-\varepsilon\langle D_z\rangle]u) = p_{3,0\,S,T}^R(D_x,y,D_z)u$$

in $\mathcal{S}_{-a}'$ for any $a \in \boldsymbol{R}$. We note that $p_{3,\ell}^R(\xi,y,\eta) = p_{3,\ell}^R(\xi,y,\eta) g_\ell^R(\eta)$. Fix $\ell \in \boldsymbol{N}$ so that $1 \le \ell \le N$. Put $u_{\ell,\rho}(z) = e^{-\rho\langle D\rangle} g_\ell^R(D)u(z)$ ($0 < \rho \le 1$). Then there are positive constants $C(u)$ and $A(u)$ such that

$$\left|D_z^\lambda u_{\ell,\rho}(z)\right| \le C(u)A(u)^{|\lambda|}|\lambda|!$$

for $z \in U_\ell$ and $0 < \rho \le 1$. Let U_ℓ^2 be an open neighborhood of $\nabla_\eta T(y^\ell, \eta^\ell)$ such that $U_\ell^1 \subset\subset U_\ell^2 \subset\subset U_\ell$. We choose $\{\chi_{\ell,k}(z)\}_{k\in N} \subset C_0^\infty(U_\ell)$ so that $\chi_{\ell,k}(z) = 1$ in U_ℓ^2 and

$$\left|D^\lambda \chi_{\ell,k}(z)\right| \le \widehat{C}_0(\widehat{C}_1 k)^{|\lambda|} \quad \text{for } |\lambda| \le k.$$

Then we have, for $0 < \rho \le 1$,

$$\left|\mathcal{F}[\chi_{\ell,k}(z)u_{\ell,\rho}(z)](\eta)\right| \le C'(u)(1+\sqrt{n''}(\widehat{C}_1 + A(u))k)^k \langle\eta\rangle^{-k}$$

(see, also, (3.25)). Since $\sum_{k=1}^{\infty} e^{-\varepsilon\langle D\rangle}\psi_k^R(D)g_\ell^R(D)u = e^{-\varepsilon\langle D\rangle}g_\ell^R(D)u$ in $\mathcal{F}_{-\varepsilon}(\boldsymbol{R}^{n''})$ ($=\bigcap_{\delta>-\varepsilon} \mathcal{S}_\delta'$), it follows from Lemma A.1. 7 that

$$p_{3,\ell\,S,T}^R(D_x,y,D_z)(\exp[-\varepsilon\langle D_z\rangle]u) = \sum_{k=1}^{\infty} p_{3,\ell\,S,T}^R(D_x,y,D_z)$$
$$\times\psi_k^R(D_z)\exp[(\rho-\varepsilon)\langle D_z\rangle](\chi_{\ell,k}(z)u_{\ell,\rho}(z) + (1-\chi_{\ell,k}(z))u_{\ell,\rho}(z))$$

in $\mathcal{F}_0(\boldsymbol{R}^{n'})$ if $0<\varepsilon\le 1$ and $0<\rho\le 1$. Since $|\xi| \le \max\{R, 3c_2(T)|\eta|/c_1\}$ if $p_{3,\ell}^R(\xi,y,\eta)\neq 0$, we have

$$\begin{aligned}&\left|\partial_\xi^\alpha\partial_\eta^\gamma p_{3,\ell}^R(\xi,y,\eta)\right| \\ &\le C_{|\alpha|+|\gamma|,R,\delta}\exp[(\delta-a)\langle\xi\rangle + (\delta+3c_2(T)a/c_1)\langle\eta\rangle] \qquad (3.29)\end{aligned}$$

if $\delta>0$ and $a\ge 0$. On the other hand, we have

$$\left|\mathcal{F}\left[\psi_k^R(D_z)(\chi_{\ell,k}(z)u_{\ell,\rho}(z))\right](\eta)\right| \le C_R(u)2^{-k}e^{-\langle\eta\rangle/(3R)}$$

if $0<\rho\le 1$ and $R\ge 2e(1+\sqrt{n''}(\widehat{C}_1 + A(u)))$. This implies that

$$\sum_{k=1}^{\infty}\psi_k^R(D_z)\exp[(\rho-\varepsilon)\langle D_z\rangle](\chi_{\ell,k}(z)u_{\ell,\rho}(z))$$
$$\longrightarrow \sum_{k=1}^{\infty}\psi_k^R(D_z)\exp[\rho\langle D_z\rangle](\chi_{\ell,k}(z)u_{\ell,\rho}(z))$$

in $\mathcal{S}'_{-1/(6R)}$ as $\varepsilon\downarrow 0$ if $R\ge 2e(1+\sqrt{n''}(\widehat{C}_1+A(u)))$ and $0<\rho\le 1/(6R)$. Lemma A.1. 7 , together with (3.29), yields

$$\sum_{k=1}^{\infty} p_{3,\ell\,S,T}^R(D_x,y,D_z)\psi_k^R(D_z)\exp[(\rho-\varepsilon)\langle D_z\rangle](\chi_{\ell,k}(z)u_{\ell,\rho}(z))$$
$$\longrightarrow \sum_{k=1}^{\infty} p_{3,\ell\,S,T}^R(D_x,y,D_z)\psi_k^R(D_z)\exp[\rho\langle D_z\rangle](\chi_{\ell,k}(z)u_{\ell,\rho}(z))$$

in $\mathcal{S}'_{-c_1/(36c_2(T)R)}$ as $\varepsilon\downarrow 0$ if $R\ge 2e(1+\sqrt{n''}(\widehat{C}_1+A(u)))$ and $0<\rho\le 1/(6R)$. If one applies the same argument as for $F_{\ell,k,\rho}^{1,R}(\xi)$ in (3.26) with $p_\ell^R(\xi,y,\eta)\phi_k(\tilde{c}|\xi|/(c_2(T)\langle\eta\rangle))$ and ρ replaced by $p_{3,\ell}^R(\xi,y,\eta)$ and $\rho-\varepsilon$, respectively, then one can easily prove that there are positive constants $R_0(T)$ and $R(T,u)$ such that

$$\sum_{k=1}^{\infty} p^R_{3,\ell\, S,T}(D_x,y,D_z)\psi^R_k(D_z)\exp[(\rho-\varepsilon)\langle D_z\rangle]((1-\chi_{\ell,k}(z))u_{\ell,\rho}(z))$$
$$\longrightarrow \sum_{k=1}^{\infty} p^R_{3,\ell\, S,T}(D_x,y,D_z)\psi^R_k(D_z)\exp[\rho\langle D_z\rangle]((1-\chi_{\ell,k}(z))u_{\ell,\rho}(z))$$

in $\mathcal{S}'_{-c_1/(18c_2(T)R)}$ as $\varepsilon \downarrow 0$ if $R \geq R_0$, $R_0 \geq R_0(T)A$ and $R \geq R(T,u)$. Therefore, $p^R_{3,\ell}(D_x,y,D_z)u$ is well-defined and analytic if $R_0 \geq R_0(T)A$ and $R \gg 1$, which completes the proof. □

Theorem 3.2.5 *Assume that (A–1) is valid and that $a(x,\eta) \in S^+(\boldsymbol{R}^n \times \boldsymbol{R}^{n''}; R_0, A)$ and* supp $a \subset \Omega' \cap \{|\eta| \geq 1\}$. *Let X be an open subset of $\boldsymbol{R}^n$ such that $Y \equiv \{y \in \boldsymbol{R}^n;\ (y,\eta) \in \mathcal{T}$ for some $\eta\} \subset X$. We assume that*

$$\left|a^{(\alpha+\tilde\alpha)}_{(\beta)}(x,\eta)\right| \leq C_{|\tilde\alpha|,\varepsilon}(A/R_0)^{|\alpha|}B^{|\beta|}|\beta|!e^{\varepsilon\langle\eta\rangle}$$

if $\varepsilon > 0$, $x \in X$ and $\langle\eta\rangle \geq R_0|\alpha|$, and that $u \in \mathcal{F}_0(\boldsymbol{R}^{n''})$ satisfies

$$WF_A(u) \cap \{(\nabla_\eta T(y,\eta),\eta);\ (y,\eta) \in \mathcal{T} \text{ and } \mathrm{Im}\, T(y,\eta) = 0\} = \emptyset.$$

(i) *There is $R(T,u) > 0$ such that $a_T(x,D_z)u$ is well-defined, that is, $a_T(x,D_z)u$ can be defined by*

$$a_T(x,D_z)u = \lim_{\varepsilon\downarrow 0} a_T(x,D_z)(\exp[-\varepsilon\langle D_z\rangle]u) \quad \text{in } \mathcal{F}_0(\boldsymbol{R}^n)$$

if $R_0 \geq R(T,u)A$. (ii) *Let $(x^0,\xi^0) \in T^*X \setminus 0$, and assume that*

$$WF_A(u) \cap \{(\nabla_\eta T(x^0,\eta),\eta);\ \eta \in \widetilde{\mathcal{N}}(x^0,\xi^0)\} = \emptyset, \tag{3.30}$$

where

$$\begin{aligned}\widetilde{\mathcal{N}}(x^0,\xi^0) \;=\; &\{\lambda\eta \in \boldsymbol{R}^{n''}\setminus\{0\};\ (x^0,\eta) \in ECS(a),\\ &\mathrm{Im}\, T(x^0,\eta) = 0,\ \xi^0 = \nabla_x T(x^0,\eta) \text{ and } \lambda > 0\}\end{aligned}$$

and the essential cone support $ECS(a)$ ($\subset \boldsymbol{R}^n \times (\boldsymbol{R}^{n''}\setminus\{0\})$) is defined in the same manner as for $p(\xi,y,\eta)$. Then there is $R(T,u,x^0,\xi^0) > 0$ such that $R(T,u,x^0,\xi^0) \geq R(T,u)$ and $(x^0,\xi^0) \notin WF_A(a_T(x,D_z)u)$ if $R_0 \geq R(T,u,x^0,\xi^0)A$.

Proof Let U be a neighborhood of Y such that $U \subset\subset X$. Choose $\Phi^R(\xi, y, \eta) \in S^{0,0}(\mathbf{R}^n \times \mathbf{R}^n \times \mathbf{R}^{n''}; R, C_*, C(X, U), C_*)$ ($R \geq 4$) so that $\Phi^R(\xi, y, \eta) = 1$ for $y \in U$, and $\Phi^R(\xi, y, \eta) = 0$ for $y \notin X$. We put

$$\begin{aligned} a^R(\xi, y, \eta) &= \Phi^R(\xi, y, \eta) a(y, \eta), \\ q^R(\xi, y, \eta) &= (1 - \Phi^R(\xi, y, \eta)) a(y, \eta). \end{aligned}$$

Then we have $a^R(\xi, y, \eta) \in S^+(\mathbf{R}^n \times \mathbf{R}^n \times \mathbf{R}^{n''}; R, C_*, B + C(X, U), AR/R_0 + C_*)$ if $R \geq R_0$. It follows from Theorem 3.2. 4 that there is $R(T, u) > 0$ such that $a_T^R(D_x, y, D_z)u$ is well-defined if $R_0 \geq R(T, u)A$ and $R \gg 1$. Note that

$$\begin{aligned} &\left|\partial_\xi^{\alpha+\tilde{\alpha}} D_y^\beta \partial_\eta^\gamma q^R(\xi, y, \eta)\right| \\ &\leq C_{|\tilde{\alpha}|+|\gamma|, \varepsilon} (C_*/R)^{|\alpha|} (A/R_0 + C(X, U)/R)^{|\beta|} \langle \eta \rangle^{|\beta|} e^{\varepsilon \langle \xi \rangle + \varepsilon \langle \eta \rangle} \end{aligned}$$

if $\varepsilon > 0$, $R \geq R_0$, $\langle \xi \rangle \geq R|\alpha|$ and $\langle \eta \rangle \geq R|\beta|$, and that $q^R(\xi, y, \eta) = 0$ for $y \in U$. Moreover, there is $c > 0$ such that

$$|\nabla_y T(y, \eta)| \geq c|\eta| \quad \text{if } q^R(\xi, y, \eta) \neq 0.$$

It follows from Theorem 2.3. 3 that there is $R(T) > 0$ such that

$$q_T^R(D_x, y, D_z)u = \lim_{\varepsilon \downarrow 0} q_T^R(D_x, y, D_z)(\exp[-\varepsilon \langle D_z \rangle]u) \quad \text{in } \mathcal{F}_0(\mathbf{R}^n)$$

if $R_0 \geq R(T)A$ and $R \gg 1$. This proves the assertion (i). Next let $(x^0, \xi^0) \in T^*X \setminus 0$, and assume that (3.30) is valid. We may assume that $x^0 \in U$. By Theorem 3.2. 4 there is $R(T, u, x^0, \xi^0) > 0$ such that $R(T, u, x^0, \xi^0) \geq R(T, u)$ and $(x^0, \xi^0) \notin WF_A(a_T^R(D_x, y, D_z)u)$ if $R_0 \geq R(T, u, x^0, \xi^0)A$ and $R \gg 1$. Moreover, from Corollary 3.2. 2 we can see that $q_T^R(D_x, y, D_z)u$ is analytic at x^0 if $R_0 \geq R(T, x^0, U)A$ and $R \gg 1$, where $R(T, x^0, U)$ is a positive constant. This proves the assertion (ii). □

Lemma 3.2. 6 *Let $(x^0, \xi^0) \in T^*\mathbf{R}^n \backslash 0$, and let $U \times \Omega$ be a conic neighborhood of (x^0, ξ^0). We assume that $p(\xi, y, \eta) \in S^{m_1, m_2}(R, A)$ and $p(\xi, y, \eta) = 1$ for $(\xi, y, \eta) \in \Omega \times U \times \Omega$ with $\langle \xi \rangle \geq R_0$ and $\langle \eta \rangle \geq R_0$. Then there is $R_0(U, \Omega, x^0, \xi^0) > 0$ such that $(x^0, \xi^0) \notin WF_A(u)$ if $u \in \mathcal{F}_0$, $R_0 \geq R_0(U, \Omega, x^0, \xi^0)A$ and $(x^0, \xi^0) \notin WF_A(p(D_x, y, D_y)u)$.*

Proof Assume that $R_0 \geq 2enA$, $u \in \mathcal{F}_0$ and $(x^0, \xi^0) \notin WF_A(p(D_x, y, D_y) \times u)$. It follows from Proposition 3.1. 2 that there are conic neighborhoods $U_j \times \Omega_j$ ($j = 1, 2$) of (x^0, ξ^0), $R_1 \equiv R_1(u, x^0, \xi^0) > 0$ and $\{g^R(\xi)\}_{R \geq R_1} \subset$

$C^\infty(\boldsymbol{R}^n)$ such that $U_1 \times \Omega_1 \subset\subset U_2 \times \Omega_2 \subset\subset U \times \Omega$, $\mathrm{dis}(U_1, \boldsymbol{R}^n \setminus U_2) \ge \varepsilon/3$, $\mathrm{dis}(U_2, \boldsymbol{R}^n \setminus U) \ge \varepsilon/3$, $\mathrm{dis}(\Omega_1 \cap S^{n-1}, \boldsymbol{R}^n \setminus \Omega_2) \ge \varepsilon'/3$, $\mathrm{dis}(\Omega_2 \cap S^{n-1}, \boldsymbol{R}^n \setminus \Omega) \ge \varepsilon'/3$, $g^R(\xi) = 1$ for $\xi \in \Omega_1$ with $\langle\xi\rangle \ge 2R$, $\mathrm{supp}\, g^R \subset \Omega_2 \cap \{\langle\xi\rangle \ge R\}$,

$$\left|\partial_\xi^{\alpha+\gamma} g^R(\xi)\right| \le C_{|\gamma|}(C(\varepsilon')/R)^{|\alpha|}\langle\xi\rangle^{-|\gamma|} \quad \text{if } \langle\xi\rangle \ge R|\alpha|$$

and $g^R(D)p(D_x, y, D_y)u$ is analytic in U_1 for $R \ge R_1$, where $\varepsilon = \mathrm{dis}(\{x^0\}, \boldsymbol{R}^n \setminus U)$, $\varepsilon' = \min\{1, \mathrm{dis}(\{\xi^0/|\xi^0|\}, \boldsymbol{R}^n \setminus \Omega)\}$ and $C(\varepsilon')$ is a positive constant. Applying Corollary 2.4. 7 we can write

$$\begin{aligned}
&g^R(D)p(D_x, y, D_y) = a(x, D) + r(x, D) \quad \text{on } \mathcal{S}_\infty,\\
&a(x,\xi) \in S^{m_1+m_2}(4R, 10RA/R_0 + C'(\varepsilon')),\\
&\left|r_{(\beta)}^{(\alpha)}(x,\xi)\right| \le C_{|\alpha|,\varepsilon',R}(5R+1)^{|\beta|}|\beta|!\langle\xi\rangle^{(m_1)_+ + m_2} e^{-\langle\xi\rangle/R}
\end{aligned}$$

if $R \ge R_0$, $R_0 \ge R_0(\varepsilon')A$ and $R \ge R(\varepsilon')$, where $C'(\varepsilon')$, $R_0(\varepsilon')$ and $R(\varepsilon')$ are positive constants. Moreover, we have

$$\begin{aligned}
\left|\partial_\xi^\alpha D_x^\beta (a(x,\xi) - g^R(\xi))\right| \;&\le\; C_{|\alpha|,\varepsilon',R}(4R + C''(\varepsilon))^{|\beta|}|\beta|!\\
&\quad \times \langle\xi\rangle^{m_1+m_2-|\alpha|} e^{-\langle\xi\rangle/R}
\end{aligned}$$

if $x \in U_2$, $R \ge R_0$, $R_0 \ge R_0'(\varepsilon)A$ and $R \ge R(\varepsilon, \varepsilon')$, where $C''(\varepsilon)$, $R_0'(\varepsilon)$ and $R(\varepsilon, \varepsilon')$ are positive constants. Therefore, it follows from Theorems 2.3. 3 and 2.6. 1 that $g^R(D)u$ is analytic at x^0 if $R \ge R_0$, $R_0 \ge R_0(\varepsilon, \varepsilon')A$ and $R \ge R(\varepsilon, \varepsilon', u, x^0, \xi^0)$, where $R_0(\varepsilon, \varepsilon')$ and $R(\varepsilon, \varepsilon', u, x^0, \xi^0)$ are positive constants. □

Theorem 3.2. 7 *Let $p(\xi, y, \eta)$ be a symbol in $S^{m_1,m_2}(R_0, A)$. Let Γ_0 and Γ be open conic subsets of $\boldsymbol{R}^n \times (\boldsymbol{R}^n \setminus \{0\})$ such that $\Gamma_0 \subset\subset \Gamma$. We assume that there are a symbol $\tilde{p}(x,\xi) \in PS^{m_2}(\Gamma; R_0, A')$ and positive constants C and c such that $p(\xi, y, \eta) = \tilde{p}(y, \eta)$ for $(y, \eta) \in \Gamma$ with $|\eta| \ge C$ and*

$$|\tilde{p}(x,\xi)| \ge c\langle\xi\rangle^{m_2} \quad \text{if } (x,\xi) \in \Gamma \text{ and } |\xi| \ge C.$$

Then there are positive constants $R_0(A, A', C_0/c, \Gamma_0, \Gamma)$, $A(A', C_0/c, \Gamma_0, \Gamma)$ and C_1 and a symbol $q(\xi, y, \eta) \in S^{0,-m_2}(2R_0, A(A', C_0/c, \Gamma_0, \Gamma))$ such that $(q(D_x, y, D_y)p(D_x, y, D_y) - 1)\varphi(D_x, y, D_y)$ and $\varphi(D_x, y, D_y)(q(D_x, y, D_y)p(D_x, y, D_y) - 1)$ map continuously $\mathcal{S}_{-\delta}$ (resp. $\mathcal{S}_\delta'$) to $\mathcal{S}_\delta$ (resp. $\mathcal{S}_{-\delta}'$) if $R_0 \ge R_0(A, A', C_0/c, \Gamma_0, \Gamma)$, $\varphi(\xi, y, \eta) \in S^{m_1', m_2'}(R, B)$, $\varphi(\xi, y, \eta) = 0$ for $(y, \eta) \notin \Gamma_0$, $R \ge 6R_0$, $R \ge R(B, \Gamma_0, \Gamma)$ and $\delta \le \min\{1/(96R_0), 1/R, enC_1/R\}$, where $C_0 = \sup_{(x,\xi)\in\Gamma, |\xi|\ge C} |\tilde{p}(x,\xi)|\langle\xi\rangle^{-m_2}$ and $R(B, \Gamma_0, \Gamma)$ is a polynomial of B of degree 1.

Remark The theorem implies that $p(D_x, y, D_y)$ has a left microlocal parametrix in Γ_0. One can similarly construct a right microlocal parametrix of p in Γ_0.

Proof Put

$$q_0(x,\xi) = 1/\tilde{p}(x,\xi),$$
$$q_j(x,\xi) = -q_0(x,\xi) \sum_{k=0}^{j-1} \sum_{|\gamma|=j-k} \frac{1}{\gamma!} q_k^{(\gamma)}(x,\xi)\tilde{p}_{(\gamma)}(x,\xi) \qquad (j = 1, 2, \cdots)$$

for $(x,\xi) \in \Gamma$ with $|\xi| \geq C$. We assume that $R_0 \geq \max\{2, C, 2^7 nA'A_1\}$, and put

$$\tilde{q}(x,\xi) = \phi_1^{R_0}(\xi) q_0(x,\xi) + \sum_{j=1}^{\infty} \phi_j^{R_0}(\xi) q_j(x,\xi),$$

where $A_1 = \max\{2, 24C_0/c\}A'$. By the same argument as in the proof of Theorem 2.8. 1 we have

$$\tilde{q}(x,\xi) \in PS^{-m_2}(\Gamma; 2R_0, 2A_1 + 3\widehat{C}, 2A_1).$$

Let Γ_j ($1 \leq j \leq 3$) be open conic subsets of Γ such that $\Gamma_0 \subset\subset \Gamma_1 \subset\subset \Gamma_2 \subset \subset \Gamma_3 \subset\subset \Gamma$, and choose $\psi(\xi, y, \eta) \in S^{0,0}(2R_0, C_*, C(\Gamma_2, \Gamma_3), C(\Gamma_2, \Gamma_3))$ so that $\psi(\xi, y, \eta) = 1$ for $(y, \eta) \in \Gamma_2$ with $\langle\eta\rangle \geq 2R_0$ and $\psi(\xi, y, \eta) = 0$ if $(y, \eta) \notin \Gamma_3$ or $\langle\eta\rangle \leq R_0$. We put $q(\xi, y, \eta) = \psi(\xi, y, \eta)\tilde{q}(y, \eta)$. Then, $q(\xi, y, \eta) \in S^{0,-m_2}(2R_0, C_*, 2A_1 + 3\widehat{C} + C(\Gamma_2, \Gamma_3), 2A_1 + C(\Gamma_2, \Gamma_3))$. Put

$$e(\xi, y, \eta) = \sum_{j=0}^{\infty} \phi_j^{8R_0}(\eta) \sum_{|\gamma|=j} \frac{1}{\gamma!} \partial_\eta^\gamma q(\xi, y, \eta) \cdot \tilde{p}_{(\gamma)}(y, \eta).$$

By Theorem 2.4. 6 there are a symbol $r(\xi, y, \eta)$ and positive constants $R(A, A', C_0/c, \Gamma_2, \Gamma_3, \Gamma, \kappa)$ ($\kappa > 0$), $A_1' \equiv A_1'(A, A', C_0/c, \Gamma_2, \Gamma_3)$ and $A_2 \equiv A_2(A, A', C_0/c, \Gamma_2, \Gamma_3, \Gamma)$, which are polynomials of A and A' of degree 1, and $C_1 > 0$ such that $e(\xi, y, \eta) \in S^{0,m_1}(6R_0, A_1')$ and

$$q(D_x, y, D_y)p(D_x, y, D_y) = e(D_x, y, D_y) + r(D_x, y, D_y) \quad \text{on } \mathcal{S}_\infty,$$
$$\left|\partial_\xi^\alpha D_y^{\beta+\tilde{\beta}} \partial_\eta^\gamma r(\xi, y, \eta)\right| \leq C_{|\alpha|+|\tilde{\beta}|+|\gamma|, R_0}(\Gamma_3, \Gamma)$$
$$\times ((C_1 R_0/\kappa + A_2)/R_0)^{|\beta|} \langle\xi\rangle^{-|\alpha|+|\beta|} \langle\eta\rangle^{(m_1-m_2)_+ + m_2} \exp[-\kappa\langle\eta\rangle/R_0]$$

if $\kappa > 0$, $R_0 \geq R(A, A', C_0/c, \Gamma_2, \Gamma_3, \Gamma, \kappa)$ and $\langle\xi\rangle \geq 2R_0|\beta|$. We can see that $e(\xi, y, \eta)$ does not depend on ξ if $(y, \eta) \in \Gamma_2$ and $\langle\eta\rangle \geq 2R_0$, and that

$$\left|D_y^\beta \partial_\eta^\gamma \{e(\xi, y, \eta) - 1\}\right| \leq C_{|\gamma|}(2A_1)^{|\beta|}|\beta|!\langle\eta\rangle^{-m_2-|\gamma|} \exp[-\langle\eta\rangle/(24R_0)]$$

if $(y,\eta)\in\Gamma_2$, $\langle\eta\rangle\geq 2R_0$ and

$$R_0 \geq en\max\{2A_1+\widehat{C}, 2^8 A'\}A_1 \tag{3.31}$$

(see the proof of Theorem 2.8. 1). Assume that $\kappa>0$, $R_0\geq R(A,A',C_0/c,\Gamma_2,\Gamma_3,\Gamma,\kappa)$ and that (3.31) is satisfied, and choose $\psi_1(\xi,y,\eta)\in S^{0,0}(2R_0,C_*,C(\Gamma_1,\Gamma_2),C(\Gamma_1,\Gamma_2))$ so that $\psi_1(\xi,y,\eta)=1$ for $(y,\eta)\in\Gamma_1$ with $\langle\eta\rangle\geq 4R_0$ and $\psi_1(\xi,y,\eta)=0$ if $(y,\eta)\notin\Gamma_2$ or $\langle\eta\rangle\leq 2R_0$. We put

$$\begin{aligned}
f_1(\xi,y,\eta)&=(e(\xi,y,\eta)-1)\psi_1(\xi,y,\eta)+r(\xi,y,\eta),\\
f_2(\xi,y,\eta)&=(e(\xi,y,\eta)-1)(1-\psi_1(\xi,y,\eta))\phi_2^{2R_0}(\eta),\\
f_3(\xi,y,\eta)&=(e(\xi,y,\eta)-1)(1-\psi_1(\xi,y,\eta))(1-\phi_2^{2R_0}(\eta)).
\end{aligned}$$

Let $\varphi(\xi,y,\eta)$ be a symbol in $S^{m_1',m_2'}(R,B)$ satisfying $\varphi(\xi,y,\eta)=0$ for $(y,\eta)\notin\Gamma_0$. From Theorem 2.3. 3 it is easy to see that $f_1(D_x,y,D_y)\varphi(D_x,y,D_y)$ and $\varphi(D_x,y,D_y)f_1(D_x,y,D_y)$ map continuously $\mathcal{S}_{-\delta}$ (resp. $\mathcal{S}_\delta{}'$) to $\mathcal{S}_\delta$ (resp. $\mathcal{S}'_{-\delta}$) if $\kappa=2enC_1$, $R_0\geq en\max\{2A_1+C(\Gamma_1,\Gamma_2),4A_2\}$, $R\geq 2enB$ and $\delta\leq\min\{1/R,1/(96R_0),enC_1/R_0\}$. By Theorem 2.4. 6 there are a symbol $s(\xi,y,\eta)$ and positive constants $R_0(A,A',C_0/c,\Gamma_0,\Gamma_1,\Gamma_2,\Gamma_3,\kappa')$, $R(B,\Gamma_0,\Gamma_1,\kappa')$ ($\kappa'>0$) and $A_2'\equiv A_2'(A,A',C_0/c,B,\Gamma_0,\Gamma_1,\Gamma_2,\Gamma_3)$, which are polynomials of A, A' and B of degree 1, and $C_1>0$ such that

$$\begin{aligned}
&f_2(D_x,y,D_y)\varphi(D_x,y,D_y)=s(D_x,y,D_y)\quad\text{on } \mathcal{S}_\infty,\\
&\left|\partial_\xi^\alpha D_y^{\beta+\tilde\beta}\partial_\eta^\gamma s(\xi,y,\eta)\right|\leq C_{|\alpha|+|\tilde\beta|+|\gamma|,R}(\Gamma_1,\Gamma_2)\\
&\quad\times((C_1R/\kappa'+A_2')/R)^{|\beta|}\langle\xi\rangle^{-|\alpha|+|\beta|}\langle\eta\rangle^{(m_1+m_1')_++m_2'}e^{-\kappa\langle\eta\rangle/R}
\end{aligned}$$

if $\kappa'>0$, $\langle\xi\rangle\geq R|\beta|$ and

$$\begin{cases} R_0\geq R_0(A,A',C_0/c,\Gamma_0,\Gamma_1,\Gamma_2,\Gamma_3,\kappa'),\\ R\geq 6R_0,\qquad R\geq R(B,\Gamma_0,\Gamma_1,\kappa'),\end{cases} \tag{3.32}$$

since supp $f_2\subset\{(\xi,y,\eta);\ (y,\eta)\notin\Gamma_1\}$. Theorem 2.3. 3 proves that $f_2(D_x,y,D_y)\varphi(D_x,y,D_y)$ maps continuously $\mathcal{S}_{-\delta}$ (resp. $\mathcal{S}_\delta{}'$) to $\mathcal{S}_\delta$ (resp. $\mathcal{S}'_{-\delta}$) if (3.32) is satisfied, $\kappa'=4enC_1$, $R\geq 4enA_2'$ and $\delta\leq R^{-1}\min\{1,en\times C_1\}$. Similarly we can show that $\varphi(D_x,y,D_y)f_2(D_x,y,D_y)$ maps continuously $\mathcal{S}_{-\delta}$ (resp. $\mathcal{S}_\delta{}'$) to $\mathcal{S}_\delta$ (resp. $\mathcal{S}'_{-\delta}$) if $R_0\geq R_0'(A,A',C_0/c,\Gamma_0,\Gamma_1,\Gamma_2,\Gamma_3)$, $R\geq 6R_0$, $R\geq R'(B,\Gamma_0,\Gamma_1)$ and $\delta\leq\min\{1,enC_1\}/R$. Since $f_3(\xi,y,\eta)\in S^{0,0,0,-\varepsilon}(6R_0,A_1'+3C_*,A_1'+3C(\Gamma_1,\Gamma_2),A_1'+3C(\Gamma_1,\Gamma_2)+3\widehat{C})$ for any $\varepsilon\in\boldsymbol{R}$, $f_3(D_x,y,D_y)\varphi(D_x,y,D_y)$ and $\varphi(D_x,y,D_y)f_3(D_x,y,D_y)$ have the same properties as $f_1(D_x,y,D_y)\varphi(D_x,y,D_y)$. This proves the theorem. □

Theorem 3.2.8 *Assume that $p(\xi,y,\eta)$ satisfies the assumptions in* Theorem 3.2.7 *with $\Gamma = U\times\Omega$, where U is an open subset of $\boldsymbol{R}^n$ and Ω is an open cone in $\boldsymbol{R}^n\setminus\{0\}$. Let $\Gamma' \equiv U'\times\Omega'$ be an open conic subset of Γ such that $\Gamma' \subset\subset \Gamma$. Then there is $R_0(\Gamma',\Gamma) > 0$ such that $(x^0,\xi^0) \notin WF_A(u)$ if $u \in \mathcal{F}_0$, $R_0 \ge R_0(\Gamma',\Gamma)A$ and $(x^0,\xi^0) \in \Gamma'\setminus WF_A(p(D_x,y,D_y)u)$.*

Proof Let $\Gamma_j \equiv U_j\times\Omega_j$ ($j=0,1$) be open conic subsets of Γ satisfying $\Gamma' \subset\subset \Gamma_0 \subset\subset \Gamma_1 \subset\subset \Gamma$, and choose $\Phi^R(\xi,y,\eta) \in S^{0,0}(R,C(\Gamma_1,\Gamma))$ and a symbol $g^R(\xi)$ ($R\ge 4$) so that $\Phi^R(\xi,y,\eta)=1$ for $(y,\eta)\in\Gamma_1$ with $\langle\eta\rangle \ge R$, $\operatorname{supp}\Phi^R \subset \boldsymbol{R}^n\times\Gamma\cap\{|\eta|\ge R/2\}$, $g^R(\xi)=1$ for $\xi\in\Omega'$ with $\langle\xi\rangle\ge 4R$, $\operatorname{supp} g^R \subset \Omega_0\cap\{\langle\xi\rangle \ge 2R\}$ and $|\partial_\xi^{\alpha+\gamma} g^R(\xi)| \le C_{|\gamma|}(C(\Omega',\Omega_0)/R)^{|\alpha|}$ if $\langle\xi\rangle \ge R|\alpha|$. We put

$$\begin{aligned} p_1^R(\xi,y,\eta) &= \Phi^R(\xi,y,\eta)p(\xi,y,\eta),\\ p_2^R(\xi,y,\eta) &= g^R(\xi)(1-\Phi^R(\xi,y,\eta))p(\xi,y,\eta),\\ p_3^R(\xi,y,\eta) &= (1-g^R(\xi))(1-\Phi^R(\xi,y,\eta))p(\xi,y,\eta). \end{aligned}$$

It is obvious that $p_1^R(\xi,y,\eta) \in S^{0,m_2}(R,A'+C(\Gamma_1,\Gamma))$ and $p_2^R(\xi,y,\eta) \in S^{m_1,m_2}(R,RA/R_0+C(\Gamma_1,\Gamma)+C(\Omega',\Omega_0))$ if $R\ge R_0$. Let $\varphi(\xi,y,\eta) \in S^{0,0}(R_1,C(\Gamma',\Gamma_0))$ satisfy $\varphi(\xi,y,\eta)=1$ for $(y,\eta)\in\Gamma'$ with $\langle\eta\rangle\ge R_1$ and $\operatorname{supp}\varphi \subset \boldsymbol{R}^n\times\Gamma_0$. From Theorem 3.2.7 there are positive constants $R(A',C_0/c,\Gamma_0,\Gamma_1,\Gamma)$, $R_1(\Gamma',\Gamma_0,\Gamma_1)$ and $A(A',C_0/c,\Gamma_0,\Gamma_1)$ and a symbol $q(\xi,y,\eta)\in S^{0,-m_2}(2R,\ A(A',C_0/c,\Gamma_0,\Gamma_1))$ such that $\varphi(D_x,y,D_y)(q(D_x,y,D_y)p_1^R(D_x,y,D_y)-1)u$ is analytic if $u\in\mathcal{F}_0$, $R\ge R(A',C_0/c,\Gamma_0,\Gamma_1,\Gamma)$, $R_1\ge 6R$ and $R_1\ge R_1(\Gamma',\Gamma_0,\Gamma_1)$. It follows from Theorem 3.2.4 that for any $u\in\mathcal{F}_0$ with $(x^0,\xi^0)\notin WF_A(p_1^R(D_x,y,D_y)u)$ there are positive constants $R(u,x^0,\xi^0)$ and $R_1(u,x^0,\xi^0)$ such that

$$(x^0,\xi^0) \notin WF_A(\varphi(D_x,y,D_y)q(D_x,y,D_y)p_1^R(D_x,y,D_y)u)$$

if $R\ge R(u,x^0,\xi^0)$ and $R_1\ge R_1(u,x^0,\xi^0)$. Since there is $\varepsilon_0(\Omega_0,\Omega_1)>0$ such that $p_2^R(\xi,y,\eta)=0$ if $y\in U_1$ and $|\xi-\eta|\le\varepsilon_0(\Omega_0,\Omega_1)|\eta|$, by Corollary 2.4.7 $p_2^R(D_x,y,D_y)u$ is analytic in U' for $u\in\mathcal{F}_0$ if $R_0\ge R_0(\Omega_0,\Omega_1,U',U_1)A$ and $R\ge R(\Gamma',\Gamma_0,\Gamma_1,\Gamma)$, where $R_0(\Omega_0,\Omega_1,U',U_1)>0$ and $R(\Gamma',\Gamma_0,\Gamma_1,\Gamma)>0$. It is obvious that $(x^0,\xi^0)\notin WF_A(p_3^R(D_x,y,D_y)u)$ for $u\in\mathcal{F}_0$ and $\xi^0\in\Omega'$. Therefore, we have $(x^0,\xi^0)\notin WF_A(\varphi(D_x,y,D_y)u)$ if $u\in\mathcal{F}_0$, $(x^0,\xi^0)\in\Gamma'\setminus WF_A(p(D_x,y,D_y)u)$, $R_0\ge R_0(\Omega_0,\Omega_1,U',U_1)A$, $R\gg 1$ and $R_1\gg 1$. Thus Lemma 3.2.6 (or its proof) proves the theorem. □

3.3 The boundary values of analytic functions

One can regard hyperfunctions as the boundary values of analytic functions. This interpretation makes various arguments on hyperfunctions clearer and simpler. Here we shall give several fundamental results on "the boundary values of analytic functions," following Hörmander [Hr5].

Theorem 3.3. 1 (Theorem 9.3.3 in [Hr5]) *Let X be an open subset of $\mathbf{R}^n$, and let Γ be an open connected cone in $\mathbf{R}^n \setminus \{0\}$. Moreover, let Z be an open subset of $\mathbf{C}^n$ such that for every open set $X_1 \subset\subset X$ and closed convex cone $\Gamma_1 \subset \Gamma \cup \{0\}$ there is $\gamma > 0$ satisfying*

$$Z \supset \{z \in \mathbf{C}^n;\ \mathrm{Re}\, z \in X_1,\ \mathrm{Im}\, z \in \Gamma_1 \text{ and } 0 < |\mathrm{Im}\, z| < \gamma\}. \tag{3.33}$$

Let f be an analytic function in Z. (i) *For every open subset X_1 of X there is a unique $f_{X_1} \in \mathcal{B}(X_1)$ satisfying the following: If Γ_1 is a closed convex cone in $\Gamma \cup \{0\}$ and* (3.33) *is valid with some $\gamma > 0$, then for any neighborhood U of ∂X_1 in $\mathbf{R}^n$ there is $\gamma_0 > 0$ such that $\gamma_0 < \gamma$ and the analytic functional*

$$v:\ \mathcal{A} \ni \phi \longmapsto u(\phi) - \int_{X_1} f(x+iy)\phi(x+iy)\, dx \in \mathbf{C}$$

satisfies supp $v \subset U$ *for $y \in \Gamma_1$ with $0 < |y| < \gamma_0$, where $u \in \mathcal{A}'(\overline{X}_1)$ and the residue class $u|_{X_1}$ in $\mathcal{B}(X_1)$ of u is f_{X_1}.* (ii) *We have $f_{X_1}|_{X_2} = f_{X_2}$ for open subsets X_1 and X_2 of X with $X_2 \subset X_1 \subset\subset X$, and, therefore, there is a unique $f_X \in \mathcal{B}(X)$ such that $f_X|_{X_1} = f_{X_1}$ for every open subset X_1 of X.* (iii) *If X_1 is a bounded open subset of X, Γ_1 is a closed convex cone in $\Gamma \cup \{0\}$,* (3.33) *is valid with some $\gamma > 0$ and there are $N \in \mathbf{R}$ and $C > 0$ satisfying*

$$|f(x+iy)| \le C|y|^{-N} \quad \textit{for } x \in X_1 \textit{ and } y \in \Gamma_1 \textit{ with } 0 < |y| < \gamma,$$

then f_{X_1} can be regarded as an element of $\mathcal{D}'(X_1)$ and

$$f(x+iy) \longrightarrow f_{X_1}(x) \quad \textit{in } \mathcal{D}'(X_1) \textit{ as } y \to 0 \textit{ in } \Gamma_1.$$

(iv) *If X_1 is a bounded open subset of X and f can be continued analytically to a neighborhood of ∂X_1, then there is $u \in \mathcal{A}'(\overline{X}_1)$ such that*

$$\int_{X_1} f(x+iy)\phi(x)\, dx \longrightarrow u(\phi) \quad \textit{as } y \to 0 \textit{ in } \Gamma_1$$

for any $\phi \in \mathcal{A}$ and $u|_{X_1} = f_{X_1}$, where Γ_1 is a closed convex cone in $\Gamma \cup \{0\}$. (v) *If $f_{X_1} = 0$ in $\mathcal{B}(X_1)$ for some non-empty open subset X_1 of*

X, and Z is connected, then $f \equiv 0$ in Z and $f_X = 0$ in $\mathcal{B}(X)$. (vi) *We have*

$$WF_A(f_X) \subset X \times (\Gamma^* \setminus \{0\}),$$

where $\Gamma^* = \{\xi \in \boldsymbol{R}^n;\ y \cdot \xi \geq 0 \text{ for any } y \in \Gamma\}$.

Remark If Z ($\subset \boldsymbol{C}^n$) has the properties in the theorem and f is analytic in Z, then f is called Γ analytic at X. $f_X \in \mathcal{B}(X)$ defined in the assertion (ii) is called the boundary value of f from Γ, and we also denote f_X by $b_\Gamma f$.

Proof By Proposition 1.2.6 the assertion (i) easily follows from the corresponding assertion of Theorem 9.3.3 in [Hr5]. Although the proof was given in [Hr5], we shall give an alternative proof of the assertions (v) and (vi). Let $X_1 \subset\subset X$ be a non-empty open set, and let Γ_1 be a closed convex cone in $\Gamma \cup \{0\}$. We choose $\gamma > 0$ so that (3.33) is satisfied. Define $u_y \in \mathcal{A}'(\overline{X}_1^{|y|})$ for $y \in \Gamma_1 \setminus \{0\}$ with $|y| < \gamma$ by

$$u_y(\phi) = \int_{X_1} f(x+iy)\phi(x+iy)\,dx \quad \text{for } \phi \in \mathcal{A},$$

where $\overline{X}_1^{|y|} = \{z \in \boldsymbol{C}^n;\ \mathrm{Re}\, z \in \overline{X}_1 \text{ and } |\mathrm{Im}\, z| \leq |y|\}$. Assume that $y \in \Gamma_1 \setminus \{0\}$ and $|y| < \gamma$. We put

$$U_y(x, x_{n+1}) = \mathcal{H}(u_y)(x, x_{n+1}) \quad \text{for } |x_{n+1}| > |y|.$$

Then

$$U_y(x, x_{n+1}) = \int_{X_1} f(t+iy) P_0(x - t - iy, x_{n+1})\,dt$$

(see Section 1.1). Choose $u \in \mathcal{A}'(\overline{X}_1)$ so that $u|_{X_1} = f_{X_1}$, and let $X_2 \subset\subset X_1$ be an open set. By the assertion (i) there is $\gamma_0 > 0$ such that

$$\mathrm{supp}\,(u - u_y) \subset \boldsymbol{R}^n \setminus X_2 \quad \text{if } |y| < \gamma_0. \tag{3.34}$$

We put

$$\begin{aligned} v_y(x) &= f(x+iy)\chi_{X_1}(x) \in \mathcal{S}', \\ V_y(x, x_{n+1}) &= \mathcal{H}(v_y)(x, x_{n+1}) \quad \text{for } x_{n+1} \neq 0, \end{aligned}$$

where $\chi_A(x)$ denotes the characteristic function of A. First assume that $f_{X_1} = 0$ in $\mathcal{B}(X_1)$. Then we have $\mathrm{supp}\, u_y \subset \boldsymbol{R}^n \setminus X_2$ if $|y| < \gamma_0$, and, therefore, $U_y(x, x_{n+1})$ can be continued analytically to $X_2 \times \boldsymbol{R}$ if $|y| < \gamma_0$. We denote its analytic continuation by $\widetilde{U}_y(x, x_{n+1})$. We have $\widetilde{U}_y(x, 0) = 0$

in X_2 and $(1-\Delta_{x,x_{n+1}})\widetilde{U}_y(x,x_{n+1}) = 0$ in $X_2 \times \boldsymbol{R}$ if $|y| < \gamma_0$. Let $X_3 \subset\subset X_2$ be an open set. It follows from Lemma 1.2.4 that there is $\varepsilon > 0$ such that $\widetilde{U}_y(x,x_{n+1})$ can be regarded as an analytic function in $\{z \in \boldsymbol{C}^n;\ \mathrm{Re}\, z \in X_3 \text{ and } |\mathrm{Im}\, z| < \varepsilon\} \times \boldsymbol{R}$ if $|y| < \gamma_0$. Here ε does not depend on the choice of y. So we have

$$\widetilde{U}_y(x+iy,x_{n+1}) = \int_{X_1} f(t+iy)P_0(x-t,x_{n+1})\,dt$$

if $x \in X_3$, $|y| < \min\{\varepsilon,\gamma_0\}$ and $|x_{n+1}| \gg |y|$. This yields

$$V_y(x,x_{n+1}) = \widetilde{U}_y(x+iy,x_{n+1})$$

for $x \in X_3$ and $x_{n+1} \neq 0$ if $|y| < \min\{\varepsilon,\gamma_0\}$. Lemma 1.2.7 implies that $V_y(x,\pm 0) = \pm f(x+iy)/2$ for $x \in X_1$. Since $\widetilde{U}_y(x,0) = 0$ in X_2, we have $\widetilde{U}_y(x+iy,0) = 0$ for $x \in X_3$ if $|y| < \min\{\varepsilon,\gamma_0\}$. So we have $f(x+iy) = 0$ for $x \in X_3$ if $|y| < \min\{\varepsilon,\gamma_0\}$, which proves the assertion (v). Next let us prove the assertion (vi). Let $\xi^0 \in S^{n-1}\setminus\Gamma^*$. Then there are $y^0 \in \Gamma\cap S^{n-1}$, a conic neighborhood $\mathcal{C}_0$ of ξ^0 and $c > 0$ such that $y^0\cdot\xi \le -c|\xi|$ for $\xi \in \mathcal{C}_0$. We take $\Gamma_1 = \{\lambda y^0;\ \lambda > 0\}$, *i.e.*, we assume that $y = sy^0$, $0 < s < \gamma$ and that $X_1 + i\{sy^0\} \subset Z$ for $0 < s < \gamma$. It is obvious that

$$\mathcal{F}[u_y](\xi) = \int_{X_1} f(x+iy)e^{-i(x+iy)\cdot\xi}\,dx = e^{y\cdot\xi}\mathcal{F}[v_y](\xi).$$

Let $\mathcal{C}_1 \subset\subset \mathcal{C}_0$ be a conic neighborhood of ξ^0, and let $\{g^R(\xi)\}_{R\ge 2} \subset C^\infty(\boldsymbol{R}^n)$ be a family of symbols such that $g^R(\xi) = 1$ in $\mathcal{C}_1$, $\mathrm{supp}\, g^R \subset \mathcal{C}_0 \cup \{\langle\xi\rangle \ge 1\}$ and

$$\left|\partial_\xi^{\alpha+\tilde{\alpha}} g^R(\xi)\right| \le C_{|\tilde{\alpha}|}(C/R)^{|\alpha|}\langle\xi\rangle^{-|\tilde{\alpha}|} \quad \text{if } \langle\xi\rangle \ge R|\alpha|.$$

Since

$$g^R(D_x)u_y(x) = \mathcal{F}_\xi^{-1}\left[e^{-c|y|\langle\xi\rangle}g^R(\xi)e^{y\cdot\xi+c|y|\langle\xi\rangle}\mathcal{F}[v_y](\xi)\right](x)$$

and $g^R(\xi)\exp[y\cdot\xi + c|y|\langle\xi\rangle]\mathcal{F}[v_y](\xi) \in L^\infty(\boldsymbol{R}^n) \subset \mathcal{S}'$, $g^R(D_x)u_y(x)$ is an analytic function of x in $\boldsymbol{R}^n$. Let $X_3 \subset\subset X_2$ be an open set. Put $\varepsilon = \mathrm{dis}(X_3, \boldsymbol{R}^n \setminus X_2)/2 > 0$, $w_y(x) = u(x) - u_y(x) \in \mathcal{A}'(\overline{X}_1^{|y|})$ and $W_y(x,x_{n+1}) = \mathcal{H}(w_y)(x,x_{n+1})$ for $|x_{n+1}| > |y|$. Then, by Lemma 1.2.4 and (3.34) we can see that $W_y(x,x_{n+1})$ can be continued analytically to $\{z \in \boldsymbol{C}^n;\ |\mathrm{Re}\, z - x| + |\mathrm{Im}\, z| < 2\varepsilon \text{ for some } x \in X_3\} \times \boldsymbol{R}$ if $|y| < \gamma_0$. We also denote its analytic continuation by $W_y(x,x_{n+1})$. Therefore, there is $C_{|y|} > 0$ such that

$$\left|D_x^\beta W_y(x,\rho)\right| \le C_{|y|}(2\sqrt{n}/\varepsilon)^{|\beta|}|\beta|!$$

if $x \in \boldsymbol{R}^n$, $|x-\tilde{x}| < \varepsilon$ for some $\tilde{x} \in X_3$, $0 < \rho < 2$ and $|y| < \gamma_0$. It follows from Lemma 2.6. 4 that there are $R_0 > 0$ and $c_0 > 0$ such that $g^R(D_x)w_y(x)$ is analytic in X_3 if $R \geq CR_0/\varepsilon$, $|y| < \gamma_0$ and $|y| < c_0 \min\{1, \varepsilon\}$. So $g^R(D)u(x)$ is analytic in X_3, *i.e.*, $WF_A(u) \cap X_3 \times \{\xi^0\} = \emptyset$. This proves the assertion (vi). □

We define for $u \in \mathcal{F}_0$ and $S \subset S^{n-1}$

$$u_S(x) := \int_S U(x - i\omega)\, dS_\omega \quad \text{in } \mathcal{F}_0,$$

where $U(x-i\omega) = K(\cdot - i\omega) * u(x) \in C(S^{n-1}_\omega; \mathcal{F}_0(\boldsymbol{R}^n_x))$ and $K(z)$ is as defined in Section 3.1 (see Lemma 3.1. 5).

Lemma 3.3. 2 *Let $u \in \mathcal{F}_0$ and $S \subset S^{n-1}$. Then we have*

$$WF_A(u_S) \subset \{(x,\xi) \in WF_A(u);\ \xi/|\xi| \in \overline{S}\}.$$

Proof It is easy to see that

$$\mathcal{F}[u_S](\xi) = \int_S e^{\omega\cdot\xi}\hat{u}(\xi)/I(\xi)\, dS_\omega \quad \text{in } \cap_{\varepsilon>0}\widehat{\mathcal{S}}'_\varepsilon, \tag{3.35}$$

where $I(\xi)$ is as defined in Section 3.1. Let Γ be an open cone in $\boldsymbol{R}^n \setminus \{0\}$, and let $\{g^R(\xi)\}_{R\geq 2} \subset C^\infty(\boldsymbol{R}^n)$ be a family of symbols satisfying $\operatorname{supp} g^R \subset \Gamma$ and

$$\left|\partial_\xi^{\alpha+\tilde{\alpha}} g^R(\xi)\right| \leq C_{|\tilde{\alpha}|}(C/R)^{|\alpha|}\langle\xi\rangle^{-|\tilde{\alpha}|} \quad \text{if } \langle\xi\rangle \geq R|\alpha|.$$

First assume that $\overline{\Gamma} \cap \overline{S} = \emptyset$. Then there are $C_0 > 0$ and $c > 0$ such that

$$|e^{\omega\cdot\xi}| \leq C_0 e^{(1-c)\langle\xi\rangle} \quad \text{if } \omega \in S \text{ and } \xi \in \operatorname{supp} g^R.$$

It follows from Lemma 3.1. 3 and (3.35) that $g^R(D)u_S(x)$ is analytic, which gives

$$WF_A(u_S) \subset \{(x,\xi) \in T^*\boldsymbol{R}^n \setminus 0;\ \xi/|\xi| \in \overline{S}\}.$$

Next assume that $(x^0, \xi^0) \in T^*\boldsymbol{R}^n \setminus 0$ does not belong to $WF_A(u)$, $g^R(\xi) = 1$ if ξ belongs to a conic neighborhood of ξ^0 and $\langle\xi\rangle \geq R$ and that $g^R(D)u(x)$ is analytic at x^0 if $R \gg 1$. Note that

$$g^R(D)u_S(x) = \int_S K(\cdot - i\omega) * (g^R(D)u)(x)\, dS_\omega$$

in $\mathcal{F}_0$. It follows from Theorem 3.1. 6 that $K(\cdot + i\mathrm{Im}\ z) * (g^R(D)u)(\mathrm{Re}\ z)$ is an analytic function of z in a neighborhood of $\{x^0\} + iS^{n-1}$. Therefore, $g^R(D)u_S(x)$ is analytic at x^0, which implies that $(x^0, \xi^0) \notin WF_A(u_S)$. This proves the lemma. □

Theorem 3.3. 3 *Let Γ_j ($1 \le j \le J$) be closed cones in $\boldsymbol{R}^n \setminus \{0\}$ such that $\boldsymbol{R}^n \setminus \{0\} = \bigcup_{j=1}^J \Gamma_j$. We put $S_j = (\Gamma_j \setminus (\bigcup_{k=1}^{j-1} \Gamma_k)) \cap S^{n-1}$ ($1 \le j \le J$).* (i) *For $v \in \mathcal{F}_0$ we have*

$$WF_A(v_{S_j}) \subset WF_A(v) \cap (\boldsymbol{R}^n \times \Gamma_j) \quad (1 \le j \le J),$$
$$v = \sum_{j=1}^J v_{S_j}.$$

(ii) *Let $v \in \mathcal{A}'(\boldsymbol{R}^n)$, and assume that Γ_j ($1 \le j \le J$) are convex proper cones. We put $V_j(x, x_{n+1}) = \mathcal{H}(v_{S_j})(x, x_{n+1})$ for $x_{n+1} \ne 0$ and $1 \le j \le J$. Assume that $1 \le j \le J$ and that Γ is an open cone satisfying $\Gamma \subset\subset \Gamma_j^* \setminus \{0\}$. Then there is $\gamma > 0$ such that $V_j(x, x_{n+1})$ can be continued analytically from $\boldsymbol{R}^n \times (0, \infty)$ to a neighborhood of $(\boldsymbol{R}^n + i\Gamma_\gamma) \times [0, \infty)$, where $\Gamma_\gamma = \{y \in \Gamma;\ |y| < \gamma\}$. In particular, $V_j(x, +0)$ is defined and analytic in $\boldsymbol{R}^n + i\Gamma_\gamma$. Let X be a bounded open neighborhood of* supp v *in $\boldsymbol{R}^n$, and let $t_j(x) \in C(\boldsymbol{R}^n; \boldsymbol{R}^n)$ ($1 \le j \le J$) satisfy $t_j(x) \in \mathrm{int}\,(\Gamma_j^*)$ and $|t_j(x)| \ll 1$ for $x \in X$, and $t_j(x) = 0$ for $x \notin X$. Then we have*

$$v(\phi)(= \langle v, \phi\rangle) = 2\sum_{j=1}^J \int_{X_j} V_j(z, +0)\phi(z)\, dz_1 \wedge \cdots \wedge dz_n \tag{3.36}$$

for $\phi \in \mathcal{A}$, where $X_j = \{x + it_j(x) \in \boldsymbol{C}^n;\ x \in X\}$ ($1 \le j \le J$). (iii) *Let X be a bounded open subset of $\boldsymbol{R}^n$, and let $u \in \mathcal{B}(X)$.Assume that Γ_j ($1 \le j \le J$) are convex proper cones. Then there are an* int (Γ_j^*) *analytic function f_j at $\boldsymbol{R}^n$ ($1 \le j \le J$) such that*

$$u = \sum_{j=1}^J b_{\mathrm{int}\,(\Gamma_j^*)} f_j.$$

(iv) *Let X be an open subset of $\boldsymbol{R}^n$, and let $u \in \mathcal{B}(X)$. Assume that there is an open convex cone Γ in $\boldsymbol{R}^n \setminus \{0\}$ satisfying*

$$WF_A(u) \subset X \times (\Gamma^* \setminus \{0\}).$$

Then there is a Γ analytic function f at X such that $u = b_\Gamma f$.

Proof The assertion (i) easily follows from Lemmas 3.1. 5 and 3.3. 2 . (ii) Fix j so that $1 \le j \le J$. Since $\hat{v}(\xi) \in \mathcal{E}_0$, by (3.3) we have

$$\begin{aligned} V_j(x, x_{n+1}) &= (2\pi)^{-n}(\mathrm{sgn}\ x_{n+1})/2 \int \exp[ix \cdot \xi - |x_{n+1}|\langle\xi\rangle] \\ &\quad \times \Big(\int_{S_j} e^{\omega\cdot\xi}\, dS_\omega\Big)\hat{v}(\xi)/I(\xi)\, d\xi \end{aligned} \tag{3.37}$$

for $x_{n+1} \neq 0$. Let $\widetilde{\Gamma}$ be an open cone in $\boldsymbol{R}^n \setminus \{0\}$ such that $\Gamma \subset\subset \widetilde{\Gamma} \subset\subset \Gamma_j^* \setminus \{0\}$. Then we have $\Gamma^* \setminus \{0\} \supset\supset \widetilde{\Gamma}^* \setminus \{0\} \supset\supset \Gamma_j$ and there are $\gamma > 0$ and $c > 0$ such that

$$\begin{aligned} &\omega \cdot \xi \leq (1-\gamma)|\xi| \quad \text{if } \omega \in S_j \text{ and } \xi \notin \widetilde{\Gamma}^*, \\ &y \cdot \xi \geq c|y|\,|\xi| \quad \text{if } y \in \Gamma \text{ and } \xi \in \widetilde{\Gamma}^*. \end{aligned}$$

Lemma 3.1. 3 yields

$$\begin{aligned} &|\exp[iz\cdot\xi - |x_{n+1}|\langle\xi\rangle + \omega\cdot\xi]/I(\xi)| \\ &\leq \sup_{\omega\in S_j} \exp[(-|x_{n+1}|-1)\langle\xi\rangle + (\omega - \operatorname{Im} z)\cdot\xi] \\ &\leq \begin{cases} C\exp[-|x_{n+1}|\langle\xi\rangle - \gamma|\xi| + |\operatorname{Im} z|\,|\xi|] & \text{if } \xi \notin \widetilde{\Gamma}^*, \\ C\exp[-|x_{n+1}|\langle\xi\rangle - c|\operatorname{Im} z|\,|\xi|] & \text{if } \xi \in \widetilde{\Gamma}^* \end{cases} \end{aligned}$$

for $\omega \in S_j$, $\xi \in \boldsymbol{R}^n$ and $z \in \boldsymbol{R}^n + i\Gamma$. This, together with (3.37), implies that $V_j(x, x_{n+1})$ can be continued analytically to a neighborhood of $(\boldsymbol{R}^n + i\Gamma_\gamma) \times [0,\infty)$. It follows from the assertion (i) and Proposition 3.1. 2 that sing supp $v_{S_j} \subset$ supp v and $V_j(x, x_{n+1})$ can be continued analytically to a neighborhood of $(\boldsymbol{R}^n \setminus \operatorname{supp} v) \times [0,\infty)$. Therefore, we have

$$\begin{aligned} v(\phi) &= \lim_{x_{n+1}\to+0} 2\int_X \sum_{j=1}^J V_j(x, x_{n+1})\phi(x)\,dx \\ &= \lim_{x_{n+1}\to+0} 2\sum_{j=1}^J \int_{X_j} V_j(z, x_{n+1})\phi(z)\,dz_1\wedge\cdots\wedge dz_n \end{aligned}$$

for $\phi \in \mathcal{A}$, which gives (3.36). (iii) Choose $v \in \mathcal{A}'(\overline{X})$ so that $v|_X = u$. Let $V_j(z, x_{n+1})$ ($1 \leq j \leq J$) be as in the assertion (ii). If we take $f(z) = 2V_j(z, +0)$ ($1 \leq j \leq J$), then we can prove the assertion (iii). (iv) Now we choose closed convex cones Γ_j ($1 \leq j \leq J$) in $\boldsymbol{R}^n \setminus \{0\}$ so that $\boldsymbol{R}^n \setminus \{0\} = \bigcup_{j=1}^J \Gamma_j$, $\Gamma^* \setminus \{0\} \subset\subset \Gamma_1$ and $\Gamma^* \cap \Gamma_j = \emptyset$ ($2 \leq j \leq J$). Let $X_1 \subset\subset X$ be an open set, and let v_1 be an analytic functional in $\mathcal{A}'(\overline{X}_1)$ whose residue class in $\mathcal{B}(X_1)$ is $u|_{X_1}$. It follows from the assertions (ii) and (iii) that $f_j(z) \equiv 2\mathcal{H}(v_{1\,S_j})(z, +0)$ is an int (Γ_j^*) analytic at $\boldsymbol{R}^n$ ($1 \leq j \leq J$) and that $u|_{X_1} = \sum_{j=1}^J b_{\text{int }(\Gamma_j^*)} f_j$. Since $WF_A(v_1) \cap X_1 \times (\boldsymbol{R}^n \setminus \{0\}) \subset X_1 \times (\Gamma^* \setminus \{0\})$, the assertion (i) gives

$$WF_A(v_{1\,S_j}) \cap X_1 \times (\boldsymbol{R}^n \setminus \{0\}) = \emptyset \quad (2 \leq j \leq J).$$

This implies that $\mathcal{H}(v_1 s_j)(x, x_{n+1})$ ($2 \le j \le J$) can be continued analytically to a neighborhood of $X_1 \times [0, \infty)$. So $f_j(z)$ ($2 \le j \le J$) are analytic in a complex neighborhood of X_1. Putting $f = \sum_{j=1}^J f_j$, we can see that f is int (Γ_1^*) analytic at X_1 and that $u|_{X_1} = b_{\text{int }(\Gamma_1^*)} f$. This, together with the assertion (v) of Theorem 3.3. 1 , proves the assertion (iv). □

Corollary 3.3. 4 *Let X be an open subset of $\boldsymbol{R}$ which contains the origin, and let $u \in \mathcal{B}(X)$. If $0 \in \operatorname{supp} u \subset \{x \in X;\ x \ge 0\}$, then $(0, \pm 1) \in WF_A(u)$.*

Proof Assume that $0 \in \operatorname{supp} u \subset \{x \in X;\ x \ge 0\}$ and that $(0,1) \notin WF_A(u)$. Then, by Theorem 3.3. 3 there are an open neighborhood X_1 of 0, $\delta > 0$ and $f \in \mathcal{A}(X_1 - i(0, \delta))$ such that $u|_{X_1} = b_{\boldsymbol{R}_-} f$, where $\boldsymbol{R}_- = \{x \in \boldsymbol{R};\ x < 0\}$. Since $u|_{X_1 \cap \boldsymbol{R}_-} = 0$, the assertion (v) of Theorem 3.3. 1 yields $0 \notin \operatorname{supp} u$, which is a contradiction. So we have $(0,1) \in WF_A(u)$ and, similarly, $(0, -1) \in WF_A(u)$. □

Corollary 3.3. 5 (Theorem 9.3.7 in [Hr5]) *Let Γ_j ($1 \le j \le J$) be closed cones in $\boldsymbol{R}^n \setminus \{0\}$ such that $\boldsymbol{R}^n \setminus \{0\} = \bigcup_{j=1}^J \Gamma_j$. Let X be a bounded open subset of $\boldsymbol{R}^n$, and let $u \in \mathcal{B}(X)$. Then there are $u_j \in \mathcal{B}(X)$ ($1 \le j \le J$) such that $u = \sum_{j=1}^J u_j$ and $WF_A(u_j) \subset WF_A(u) \cap (X \times \Gamma_j)$ ($1 \le j \le J$). If $\{u_j'\}_{1 \le j \le J} \subset \mathcal{B}(X)$ satisfies $u = \sum_{j=1}^J u_j'$ and $WF_A(u_j') \subset WF_A(u) \cap (X \times \Gamma_j)$ ($1 \le j \le J$), then there are $u_{j,k} \in \mathcal{B}(X)$ ($1 \le j, k \le J$) such that $u_{j,k} = -u_{k,j}$ and $WF_A(u_{j,k}) \subset X \times (\Gamma_j \cap \Gamma_k)$ ($1 \le j, k \le J$) and $u_j' = u_j + \sum_{k=1}^J u_{j,k}$ ($1 \le j \le J$).*

Theorem 3.3. 6 *Let T be an open subset of $\boldsymbol{R}^m$, and let $u(t, \cdot) \in C(T; \mathcal{F}_0(\boldsymbol{R}^n))$. We choose open convex proper cones Γ_j ($1 \le j \le J$) in $\boldsymbol{R}^n \setminus \{0\}$ and $\{g_j^R(\xi)\} \subset C^\infty(\boldsymbol{R}^n)$ ($R \ge 2, 1 \le j \le J$) so that $\boldsymbol{R}^n \setminus \{0\} = \bigcup_{j=1}^J \Gamma_j$, $\operatorname{supp} g_j^R \cap \{|\xi| \ge 1\} \subset \Gamma_j$, $\sum_{j=1}^J g_j^R(\xi) = 1$ for $\xi \in \boldsymbol{R}^n$ and $|\partial_\xi^{\alpha+\gamma} g_j^R(\xi)| \le C_{|\gamma|}(C_*/R)^{|\alpha|}\langle\xi\rangle^{-|\gamma|}$ if $\langle\xi\rangle \ge R|\alpha|$. Put*

$$U_j^R(t, x, x_{n+1}) = (\operatorname{sgn} x_{n+1}) \mathcal{F}_\xi^{-1}\left[e^{-|x_{n+1}|\langle\xi\rangle} g_j^R(\xi) \hat{u}(t, \xi)\right](x)/2$$

for $x_{n+1} \ne 0$ and $1 \le j \le J$, where $\hat{u}(t, \xi) = \mathcal{F}[u(t, \cdot)](\xi)$. For every open set S with $S \Subset T$ we can define $u_S(t, x) \in \mathcal{F}_0(\boldsymbol{R}^{n+m})$ by

$$\langle u_S(t, x), \varphi(t, x)\rangle = \int_S \langle u(t, x), \varphi(t, x)\rangle_x \, dt$$

for $\varphi \in \mathcal{S}_\infty(\boldsymbol{R}^{n+m})$. (i) *By analytic continuation $U_j^R(t, x, x_{n+1})$ can be regarded as a continuous function in $T \times \mathcal{U}_j$ which is analytic in $\mathcal{U}_j$ for* *each* $t \in T$, *where* $\mathcal{U}_j$ *is a neighborhood of* $(\boldsymbol{R}^n + i\,\mathrm{int}(\Gamma_j^*)) \times [0,\infty)$ *in* $\boldsymbol{C}^n \times \boldsymbol{R}$ *and* $1 \le j \le J$. *In particular,* $U_j^R(t,x,+0)$ *belongs to* $C(T \times (\boldsymbol{R}^n + i\,\mathrm{int}(\Gamma_j^*)))$ *and is analytic in* $\boldsymbol{R}^n + i\,\mathrm{int}(\Gamma_j^*)$ *for each* $t \in T$, *where* $1 \le j \le J$. (ii) *If* T_k $(k=1,2)$ *and* S *are open subsets of* T, $S \subset T_k \Subset T$ $(k=1,2)$ *and* X *is a bounded open subset of* $\boldsymbol{R}^n$, *then*

$$u_{T_1}\Big|_{S\times X} = u_{T_2}\Big|_{S\times X} \quad \text{in } \mathcal{B}(S\times X), \tag{3.38}$$

i.e., u *uniquely determines an element* $\tilde{u}$ *in* $\mathcal{B}(T \times \boldsymbol{R}^n)$. *Moreover, for every open subset* $S \times X$ *of* $T \times \boldsymbol{R}^n$ *with* $S \times X \Subset T \times \boldsymbol{R}^n$ *and any neighborhood* $\mathcal{U}$ *of* $\partial(S\times X)$ *in* $\boldsymbol{R}^{n+1}$ *there is* $\gamma_0 > 0$ *such that the analytic functional*

$$v:\ \mathcal{A} \ni \varphi \longmapsto w(\varphi) - 2\sum_{j=1}^{J} \int_{S\times X} U_j^R(t, x+iy^j, +0)\varphi(t, x+iy^j)\,dtdx \in \boldsymbol{C},$$

satisfies $\operatorname{supp} v \subset \mathcal{U}$ *for* $y^j \in \Gamma_j^*$ $(1 \le j \le J)$ *with* $0 < |y^j| < \gamma_0$, *where* $w \in \mathcal{A}'(\overline{S} \times \overline{X})$ *satisfies* $w|_{S\times X} = \tilde{u}|_{S\times X}$ *and* $R \gg 1$.

Remark If $m = 0$, *i.e.*, $u(x) \in \mathcal{F}_0(\boldsymbol{R}^n)$, then we have

$$u|_X = 2\sum_{j=1}^{J} b_{\mathrm{int}(\Gamma_j^*)} U_j^R(\cdot, +0) \quad \text{in } \mathcal{B}(X)$$

for every bounded open subset X of $\boldsymbol{R}^n$

Proof (i) It is easy to see that

$$U_j^R(t,x,x_{n+1}) = (2\pi)^{-n}\langle \hat{u}(t,\xi), e^{ix\cdot\xi - x_{n+1}\langle\xi\rangle} g_j^R(\xi)\rangle_\xi/2$$

for $x_{n+1} > 0$. Since there is $c_j > 0$ satisfying $\mathrm{Im}\, z\cdot\xi \ge c_j|\mathrm{Im}\, z|\,|\xi|$ for $z \in \boldsymbol{R}^n + i\Gamma_j^*$ and $\xi \in \operatorname{supp} g_j^R \cap \{|\xi| \ge 1\}$, we have the assertion (i). (ii) Let $p(x,\xi) \in S^+(R_0, A)$. By Theorem 2.3. 3 and its remark we can define $p(x,D_x)u(t,x) \in C(T; \mathcal{F}_0(\boldsymbol{R}^n))$ if $R_0 \ge 2enA$. Moreover, applying the same argument as in the proof of Theorem 2.6. 1 we have

$$\begin{aligned} &\operatorname{supp}\,(p(x,D_x)u(t,x))_{T'} \\ &\subset \overline{T}' \times \mathrm{cl}(\{x \in \boldsymbol{R}^n;\ (x,\xi) \in \operatorname{supp} p \text{ for some } \xi\}) \end{aligned} \tag{3.39}$$

for every open set T' with $T' \Subset T$ if $R_0 \ge 2e\max\{n, 4\sqrt{n}\}A$. In fact, we can prove that $f_{k,\rho,\varepsilon}(t,x,s,\eta) \in C(\boldsymbol{R}^{2m+n}; \mathcal{S}(\boldsymbol{R}^n_\eta))$ and

$$\begin{aligned} &\langle (D_t, D_x)\rangle^k \exp[-\rho\langle (D_t,D_x)\rangle](p(x,D_x)u(t,x))_{T'} \\ &= (2\pi)^{-n}\int_{T'} \langle e^{-\varepsilon\langle\eta\rangle}\mathcal{F}[u(s,\cdot)](\eta), f_{k,\rho,\varepsilon}(t,x,s,\eta)\rangle_\eta\, ds \end{aligned}$$

if $0 < \rho, \varepsilon \le 1$, T' is open subset of T, $T' \subset\subset T$ and $R_0 \ge 2eA\max\{n, 4\sqrt{n}\}$, where $k = 0, 1$,

$$\begin{aligned} & f_{k,\rho,\varepsilon}(t,x,s,\eta) \\ &= (2\pi)^{-m-n} \int_{R^n \times R^n \times R^m} \exp[i(x-y)\cdot\xi + i(t-s)\cdot\tau + iy\cdot\eta + \varepsilon\langle\eta\rangle] \\ &\qquad \times p(y,\eta)\langle x-y\rangle_A^{-2M} \langle D_\xi\rangle_A^{2M}(\langle(\tau,\xi)\rangle^k e^{-\rho\langle(\tau,\xi)\rangle})\, dyd\xi d\tau, \end{aligned}$$

$\langle x\rangle_A = \langle Ax/(4+4\sqrt{2})\rangle$ and $M \ge [n/2]+1$. Then repetition of the same argument as in the proof of Theorem 2.6. 1 yields (3.39). Let T' and S be open subset of T such that $S \subset T' \subset\subset T$. Let X_j ($j = 1, 2$) and X be bounded open subsets of $\boldsymbol{R}^n$ such that $X \subset\subset X_1 \subset\subset X_2$, and let $a^R(x,\xi) \in S^0(R, A)$ ($R \ge 4$) satisfy $a^R(x,\xi) = 1$ for $x \in X_1$ and $\operatorname{supp} a^R \subset X_2 \times \boldsymbol{R}^n$. (3.39) yields $\operatorname{supp}\,(a^R(x,D_x)u(t,x))_{T'} \subset \overline{T}' \times X_2$ and $(a^R(x,D_x)u(t,x))_{T'}|_{S\times X} = u_{T'}|_{S\times X}$ in $\mathcal{B}(S\times X)$ if $R \gg 1$. Define

$$a^{R,R'}(x,\xi;y) = \sum_{k=1}^{\infty} \psi_k^{R'}(\xi) \sum_{|\beta|\le k-1} (iy)^\beta \partial_x^\beta a^R(x,\xi)/\beta!$$

for $x, y \in \boldsymbol{R}^n$ and $\xi \in \boldsymbol{R}^n$, where $R' \ge R$. Then we have

$$a^{R,R'}(x,\xi;y) \in S^{1,\delta(y)/R}(3R', 3\widehat{C} + 3AR'/R, 3AR'/R)$$

for each $y \in \boldsymbol{R}^n$, and

$$\begin{aligned} &\left|(\partial_{x_j} + i\partial_{y_j})\partial_\xi^{\alpha+\tilde{\alpha}} D_x^{\beta+\tilde{\beta}} a^{R,R'}(x,\xi;y)\right| \le C_{|\tilde{\alpha}|+|\tilde{\beta}|}(\widehat{C}/R' + A/R)^{|\alpha|} \\ &\qquad \times (A/R)^{|\beta|}\langle\xi\rangle^{|\beta|-|\tilde{\alpha}|} e^{(e\delta(y)/R - 1/(3R'))\langle\xi\rangle} \end{aligned}$$

if $1 \le j \le n$, $x, y \in \boldsymbol{R}^n$, $\xi \in \boldsymbol{R}^n$ and $\langle\xi\rangle \ge 3R'(|\alpha| + |\beta|)$, where $\delta(y) = \sqrt{n}A|y|$. Choose $R > 0$ so that $e\sqrt{n}A/R \le \min_{1\le j\le J} c_j/2$ and $R \gg 1$. Applying Stokes' formula we have

$$\begin{aligned} &\int_{T'} \langle a^R(x,D_x)u_\varepsilon(t,x), \varphi(t,x)\rangle_x\, dt \\ &= 2\sum_{j=1}^J \int_{T'} \langle a^R(x,D_x)U_j^R(t,x,\varepsilon), \varphi(t,x)\rangle_x\, dt \\ &= 2\sum_{j=1}^J \int_{T'} \Big\{ \int_{X_2} U_{j,1,\varepsilon}(t,x;y^j)\varphi(t,x+iy^j)\, dx \\ &\qquad + \int_0^1 \Big(\int_{X_2} U_{j,2,\varepsilon}(t,x;ry^j)\varphi(t,x+iry^j)\, dx \Big)\, dr \Big\}\, dt \end{aligned}$$

for $\varphi \in \mathcal{S}_\infty(\boldsymbol{R}^{m+n})$ and $\varepsilon > 0$, where $u_\varepsilon(t,x) = \exp[-\varepsilon\langle D_x\rangle]u(t,x)$, $y^j \in \Gamma_j^* \setminus \{0\}$ ($1 \le j \le J$) and

$$U_{j,1,\varepsilon}(t,x;y) = (2\pi)^{-n}\langle \mathcal{F}[u(t,\cdot)](\xi), \\ e^{i(x+iy)\cdot\xi-\varepsilon\langle\xi\rangle} g_j^R(\xi) a^{R,R'}(x,\xi;y)\rangle_\xi/2, \tag{3.40}$$

$$U_{j,2,\varepsilon}(t,x;y) = (2\pi)^{-n}\langle \mathcal{F}[u(t,\cdot)](\xi), e^{i(x+iy)\cdot\xi-\varepsilon\langle\xi\rangle} \\ \times g_j^R(\xi)\sum_{k=1}^n y_k(\partial_{x_k}+i\partial_{y_k})a^{R,R'}(x,\xi;y)\rangle_\xi/2 \tag{3.41}$$

for $y \in \Gamma_j^* \setminus \{0\}$. It is obvious that $U_{j,1,\varepsilon}(t,x;y) \rightrightarrows U_{j,1}(t,x;y)$ in $T' \times \boldsymbol{R}^n$ and $U_{j,2,\varepsilon}(t,x;ry) \rightrightarrows U_{j,2}(t,x;ry)$ in $T' \times \boldsymbol{R}^n \times [0,1]$ as $\varepsilon \downarrow 0$ for each $y \in \Gamma_j^* \setminus \{0\}$, where $1 \le j \le J$ and $U_{j,1}(t,x;y)$ and $U_{j,2}(t,x;y)$ are defined by (3.40) and (3.41) with $\varepsilon = 0$, respectively. So we have

$$\langle (a^R(x,D_x)u(t,x))_{T'}, \varphi(t,x)\rangle = \int_{T'} \langle a^R(x,D_x)u(t,x), \varphi(t,x)\rangle_x \, dt$$

$$= 2\sum_{j=1}^J \int_{T'} \Big\{ \int_{X_2} U_{j,1}(t,x;y^j)\varphi(t,x+iy^j)\,dx \\ + \int_0^1 \Big(\int_{X_2} U_{j,2}(t,x;ry^j)\varphi(t,x+iry^j)\,dx \Big)\,dr \Big\}\,dt$$

for $\varphi \in \mathcal{S}_\infty(\boldsymbol{R}^{m+n})$, where $y^j \in \Gamma_j^* \setminus \{0\}$ ($1 \le j \le J$). This implies that $(a^R(x,D_x)u(t,x))_{T'} \in \mathcal{A}'(\overline{T}' \times \overline{X}_2)$. Since $U_{j,1}(t,x;y) = U_j^R(t,x+iy,+0)$ and $U_{j,2}(t,x;y) = 0$ for $1 \le j \le J$, $(t,x) \in T' \times X$ and $y^j \in \Gamma_j^* \setminus \{0\}$, we have

$$\langle (a^R(x,D_x)u(t,x))_{T'}, \varphi(t,x)\rangle \\ -2\sum_{j=1}^J \int_{S\times X} U_j^R(t,x+iy^j,+0)\varphi(t,x+iy^j)\,dtdx$$

$$= 2\sum_{j=1}^J \Big[\int_{T'\times X_2\setminus S\times X} U_{j,1}(t,x;y^j)\varphi(t,x+iy^j)\,dtdx \\ + \int_0^1 \Big(\int_{T'\times(X_2\setminus X)} U_{j,2}(t,x;ry^j)\varphi(t,x+iry^j)\,dtdx \Big)\,dr \Big]$$

$$\equiv \langle \tilde{v}(t,x), \varphi(t,x)\rangle$$

for $\varphi \in \mathcal{S}_\infty(\boldsymbol{R}^{m+n})$, where $y^j \in \Gamma_j^* \setminus \{0\}$ ($1 \le j \le J$). Moreover, we have $\tilde{v} \in \mathcal{A}'(\mathcal{X}_\delta)$ and

$$\operatorname{supp} \tilde{v} \subset \{(t,x) \in \boldsymbol{R}^{m+n};\ |(t-s,x-y)| \le \delta \\ \text{for some } (s,y) \in \overline{T}' \times \overline{X}_2 \setminus S \times X\}, \tag{3.42}$$

where $\delta = \max_{1\le j\le J}|y^j|$ and $\mathcal{X}_\delta = \{(t,x) \in \overline{T}' \times \boldsymbol{C}^n;\ \mathrm{Re}\, x \in \overline{X}_2$ and $|\mathrm{Im}\, x| \le \delta\}$. By the same argument as in Theorem 3.3. 1 and (3.42) we can prove that

$$\varphi \mapsto 2\sum_{j=1}^{J}\int_{S\times X} U_j^R(t,x+iy^j,+0)\varphi(t,x+iy^j)\,dt dx$$

uniquely determines an element in $\mathcal{B}(S\times X)$ which is equal to $(a^R(x,D_x) \times u(t,x))_{T'}|_{S\times X}$. This gives (3.38). Then the assertion (ii) easily follows from (3.42). □

3.4 Operations on hyperfunctions

Using results in the preceding section one can define several operations on hyperfunctions and study their effects on the analytic wave front sets (see, *e.g.*, [Hr5], [KKK] and [Kn]). We will also use results obtained in Section 3.2 in order to define the operations on hyperfunctions.

Let X and Y be open subsets of $\boldsymbol{R}^n$ and $\boldsymbol{R}^m$, respectively, and let $h : X \to Y$ be a real analytic mapping. We denote

$$N_h := \{(h(x),\eta) \in T^*Y;\ {}^t h'(x)\eta = 0\},$$

where $h'(x)$ is the Jacobian matrix of h, *i.e.*, $h'(x) = ((\partial h_i/\partial x_j)(x))_{\substack{i\downarrow 1,\cdots,m\\ j\to 1,\cdots,n}}$ if $h(x) = {}^t(h_1(x),\cdots,h_m(x))$. Let $u \in \mathcal{B}(Y)$ satisfy $N_h \cap WF_A(u) = \emptyset$. For a fixed $x^0 \in X$ we can write

$$u = f_0 + \sum_{j=1}^{J} b_{G_j} f_j$$

in a neighborhood of $h(x^0)$, where f_0 is real analytic there, the G_j are open convex cones in $\boldsymbol{R}^m\setminus\{0\}$ and the f_j are G_j analytic near $h(x^0)$. From Theorem 3.3. 3 and Corollary 3.3. 4 we may assume that ${}^t h'(x^0)\eta \ne 0$ if $1 \le j \le J$ and $\eta \in G_j^* \setminus \{0\}$. Then one can show that $h'(x^0)^{-1}(G_j)$ is an open cone in $\boldsymbol{R}^n \setminus \{0\}$ and that

$$h'(x^0)^{-1}(G_j) = \mathrm{int}(({}^t h'(x^0)G_j^*)^*). \tag{3.43}$$

Moreover, one can show that $f_j \circ h$ is G_j' analytic near x^0 if G_j' is an open convex cone in $\boldsymbol{R}^n\setminus\{0\}$ and $G_j' \subset\subset h'(x^0)^{-1}(G_j)$. Therefore, we can define the pull-back $h^* : \mathcal{B}(Y) \to \mathcal{B}(X)$ by

$$h^*u = f_0\circ h + \sum_{j=1}^{J} b_{G_j'}(f_j\circ h)$$

in a neighborhood of x^0 if G'_j ($1 \le j \le J$) are open cones and $G'_j \Subset h'(x^0)^{-1}(G_j)$. This is well-defined by virtue of Corollary 3.3. 5 . It follows from (3.43), Theorems 3.3. 1 and 3.3. 3 and Corollary 3.3. 5 that

$$\text{supp } h^*u \subset h^{-1}(\text{supp } u), \tag{3.44}$$

$$\begin{aligned} &WF_A(h^*u) \subset h^*(WF_A(u)) \\ &\quad (= \{(x, {}^th'(x)\eta) \in T^*X \setminus 0;\ (h(x), \eta) \in WF_A(u)\}). \end{aligned} \tag{3.45}$$

This enables us to define $WF_A(u)$ ($\subset T^*X \setminus 0$) for $u \in \mathcal{B}(X)$ when X is a real analytic manifold. Let Z be an open subset of $\boldsymbol{R}^\ell$, and let $k : Y \to Z$ be a real analytic mapping. Then it easily follows from the definition that k^*u, $h^*(k^*u)$ and $(k \circ h)^*u$ are well-defined and $h^*(k^*u) = (k \circ h)^*u$ if $u \in \mathcal{B}(Z)$ and $WF_A(u) \cap N_{k \circ h} = \emptyset$. We note that the definition of h^*u here coincides with h^*u as defined in Section 1.4 when h is an analytic diffeomorphism. We omit the proof. Instead, we shall give an alternative definition of h^*u. Let $u \in \mathcal{F}_0(\boldsymbol{R}^m)$, and assume that $N_h \cap WF_A(u) = \emptyset$. Let X_j ($j = 1, 2$) be open subsets of X such that $X_1 \Subset X_2 \Subset X$. Choose $a^R(x, \eta) \in S^0(\boldsymbol{R}^n \times \boldsymbol{R}^m; R, A)$ ($R \ge 4$) so that $a^R(x, \eta) = 1$ for $x \in X_1$ and $\text{supp } a^R \subset X_2 \times \boldsymbol{R}^m$. Put $T(x, \eta) = h(x) \cdot \eta$. Then we can define $a^R_T(x, D_z)u$ by

$$a^R_T(x, D_z)u = \lim_{\varepsilon \downarrow 0} a^R_T(x, D_z)u_\varepsilon \quad \text{in } \mathcal{F}_0(\boldsymbol{R}^n) \tag{3.46}$$

if $R \gg 1$, where $u_\varepsilon(z) = e^{-\varepsilon\langle D \rangle}u(z)$. In fact, for any $x^0 \in X$ there are a bounded neighborhood $U(x^0)$ of x^0 in X, a bounded neighborhood $V(x^0)$ of $h(x^0)$ in Y and an open cone $\Gamma(x^0)$ in $\boldsymbol{R}^m \setminus \{0\}$ such that

$$\begin{aligned} &h(\overline{U(x^0)}) \Subset V(x^0), \\ &\{\eta \in \boldsymbol{R}^m;\ (y, \eta) \in WF_A(u) \text{ for some } y \in \overline{V(x^0)}\} \Subset \Gamma(x^0), \\ &\{\eta \in \boldsymbol{R}^m;\ {}^th'(x)\eta = 0 \text{ for some } x \in \overline{U(x^0)}\} \cap \overline{\Gamma(x^0)} = \emptyset. \end{aligned}$$

Since $\overline{X}_2$ is compact, there are $x^j \in X_2$ ($1 \le j \le N$) such that $\overline{X}_2 \subset \bigcup_{j=1}^N U(x^j)$. Choose $\Phi^R_j(x, \eta) \in S^0(\boldsymbol{R}^n \times \boldsymbol{R}^m; R, C_*, C(h, u))$ ($R \ge 4$, $1 \le j \le N$) so that $\text{supp } \Phi^R_j \subset U(x^j) \times \boldsymbol{R}^m$ and $\sum_{j=1}^N \Phi^R_j(x, \eta) = 1$ for $x \in X_2$, where $C_* > 0$ and $C(h, u)$ is a positive constant depending on h and u. Moreover, we choose open cones Γ^1_j ($1 \le j \le N$) in $\boldsymbol{R}^m \setminus \{0\}$ and $g^R_j(\eta) \in C^\infty(\boldsymbol{R}^m)$
($R \ge 2$, $1 \le j \le N$) so that

$$\{\eta \in \boldsymbol{R}^m;\ (y, \eta) \in WF_A(u) \text{ for some } y \in \overline{V(x^j)}\} \Subset \Gamma^1_j \Subset \Gamma(x^j),$$

$g_j^R(\eta) = 1$ for $\eta \in \Gamma_j^1$ with $\langle\eta\rangle \geq R$ and supp $g_j^R \subset \Gamma(x^j)$ ($1 \leq j \leq N$) and

$$\left|\partial_\eta^{\gamma+\tilde{\gamma}} g_j^R(\eta)\right| \leq C_{|\tilde{\gamma}|}\Big(C(\Gamma_j^1, \Gamma(x^j))/R\Big)^{|\gamma|} \langle\eta\rangle^{-|\tilde{\gamma}|}$$

if $\langle\eta\rangle \geq R|\gamma|$. Using a partition of unity we can show that $(1 - g_j^R(D))u$ is analytic in a neighborhood of $\overline{V(x^j)}$. Choose open subsets V_j of Y and $\chi_{j,k}(z) \in C_0^\infty(\boldsymbol{R}^m)$ ($1 \leq j \leq N$, $k \in \boldsymbol{N}$) so that $h(\overline{U(x^j)}) \subset\subset V_j \subset\subset V(x^j)$, $\chi_{j,k}(z) = 1$ in V_j, supp $\chi_{j,k} \subset V(x^j)$ and

$$\left|D_z^{\beta+\tilde{\beta}} \chi_{j,k}(z)\right| \leq C_{|\tilde{\beta}|}\Big(C(V_j, V(x^j))k\Big)^{|\beta|}$$

for $|\beta| \leq k$ ($1 \leq j \leq N$, $k \in \boldsymbol{N}$). Put, for $\varepsilon > 0$ and $\rho > 0$,

$$\begin{aligned} v_{1,\varepsilon}(x) &= a_{1T}^R(x, D_z)u_\varepsilon, \\ v_{2,j,\rho,\varepsilon}(x) &= \sum_{k=1}^\infty a_{2,j\,T}^R(x, D)\psi_k^R(D_z)\exp[(\rho - \varepsilon)\langle D_z\rangle] \\ &\quad \times(\chi_{j,k}(z)u_{j,\rho}(z)), \\ v_{3,j,\rho,\varepsilon}(x) &= \sum_{k=1}^\infty a_{2,j\,T}^R(x, D)\psi_k^R(D_z)\exp[(\rho - \varepsilon)\langle D_z\rangle] \\ &\quad \times((1 - \chi_{j,k}(z))u_{j,\rho}(z)), \end{aligned}$$

where

$$\begin{aligned} a_1^R(x, \eta) &= a^R(x, \eta)\sum_{j=1}^N \Phi_j^R(x, \eta)g_j^R(\eta), \\ a_{2,j}^R(x, \eta) &= a^R(x, \eta)\Phi_j^R(x, \eta), \\ u_{j,\rho}(z) &= e^{-\rho\langle D\rangle}(1 - g_j^R(D))u. \end{aligned}$$

Note that

$$a_T^R(x, D_z)u_\varepsilon = v_{1,\varepsilon} + \sum_{j=1}^N (v_{2,j,\rho,\varepsilon} + v_{3,j,\rho,\varepsilon}).$$

Replacing Ω by $\bigcup_{j=1}^N U(x^j)\times\Gamma(x^j)$ we can apply Theorem 2.3. 3 . So there is $R(h, u) > 0$ such that $a_{1T}^R(x, D_z)u_\varepsilon$ is well-defined for $\varepsilon \geq 0$ and $v_{1,\varepsilon} \to a_{1T}^R(x, D_z)u$ in $\mathcal{F}_0(\boldsymbol{R}^n)$ as $\varepsilon \downarrow 0$ if $R \geq R(h, u)$. The same argument as in the proof of Theorem 3.2. 4 yields, with $v_{2,j} \in \mathcal{S}'$ ($1 \leq j \leq N$), $v_{2,j,\rho,\varepsilon} + v_{3,j,\rho,\varepsilon} \to v_{2,j}$ in $\mathcal{S}'$ as $\varepsilon \downarrow 0$ if $0 < \rho \ll 1$ and $R \gg 1$, which proves (3.46). Theorem 2.6. 1 implies that supp $a_{1T}^R(x, D_z)u \subset X_2$. Moreover, we have supp $v_{2,j} \subset X_2$ by a usual calculus in $\mathcal{S}'$. Therefore, we have supp $a_T^R(x, D_z)u \subset X_2$. Let X_0 ($\subset\subset X_1$) be an open subset of X. Then, similarly we can show that $(a_T^R(x, D_z)u)|_{X_0}$ in $\mathcal{B}(X_0)$ does not depend on the choice of $a^R(x, \eta)$ if $R \gg 1$. This allows us to define $h^*u \in \mathcal{B}(X)$ by

$$h^*u|_{X_0} = (a_T^R(x, D_z)u)|_{X_0} \quad \text{in } \mathcal{B}(X_0)$$

for any open set $X_0 \subset\subset X_1$ if $u \in \mathcal{F}_0(\boldsymbol{R}^m)$ and $N_h \cap WF_A(u) = \emptyset$. Note that $a_T^R(x, D_z)u_\varepsilon$ is smooth in x for $\varepsilon > 0$ and

$$a_T^R(x, D_z)u_\varepsilon(x) = u_\varepsilon(h(x)) \quad \text{for } x \in X_1 \text{ and } \varepsilon > 0.$$

It follows from Theorem 3.2. 5 that

$$WF_A(a_{1T}^R(x, D_z)u) \cap X_1 \times \boldsymbol{R}^n \subset h^*(WF_A(u)).$$

Moreover, $v_{2,j}$ ($1 \le j \le N$) are analytic in X_1. So we have

$$WF_A(h^*u) \subset h^*(WF_A(u))$$

if $u \in \mathcal{F}_0(\boldsymbol{R}^m)$ and $N_h \cap WF_A(u) = \emptyset$.

Theorem 3.4. 1 *If $u \in \mathcal{F}_0(\boldsymbol{R}^m)$ and $N_h \cap WF_A(u) = \emptyset$, then we have $h^*u = h^*(u|_Y)$, that is, two definitions of h^*u given in this section are consistent.*

Proof Let $u \in \mathcal{A}'(\boldsymbol{R}^m)$ ($\subset \mathcal{F}_0(\boldsymbol{R}^m)$) satisfy $N_h \cap WF_A(u) = \emptyset$, and let $x^0 \in X$. Choose open neighborhoods X_j ($j = 1, 2$) of x^0 in X, a neighborhood V of $h(x^0)$ in Y and open cones Γ^j ($j = 1, 2$) in $\boldsymbol{R}^m \setminus \{0\}$ so that $X_1 \subset\subset X_2 \subset\subset X$, $\Gamma^1 \subset\subset \Gamma^2$, $h(X_2) \subset\subset V$, $WF_A(u) \cap V \times \boldsymbol{R}^m \subset\subset Y \times \Gamma^1$ and

$$\{\eta \in \boldsymbol{R}^m \setminus \{0\};\ {}^t h'(x)\eta = 0 \text{ for some } x \in \overline{X_2}\} \cap \overline{\Gamma^2} = \emptyset.$$

Moreover, we choose $g^R(\eta) \in C^\infty(\boldsymbol{R}^m)$ ($R \ge 2$) so that $g^R(\eta) = 1$ for $\eta \in \Gamma^1$ with $\langle \eta \rangle \ge R$, supp $g^R \subset \Gamma^2 \cap \{|\eta| \ge 1\}$ and $|\partial_\eta^{\alpha+\gamma} g^R(\eta)| \le C_{|\gamma|}(C(\Gamma^1, \Gamma^2)/R)^{|\alpha|}\langle \eta \rangle^{-|\gamma|}$ if $\langle \eta \rangle \ge R|\alpha|$. Let Γ_j ($1 \le j \le J$) be open convex cones in $\boldsymbol{R}^m \setminus \{0\}$ such that $\boldsymbol{R}^m \setminus \{0\} = \bigcup_{j=1}^J \Gamma_j$, and choose $g_j^R(\eta) \in C^\infty(\boldsymbol{R}^m)$ ($R \ge 2$, $1 \le j \le J$) so that supp $g_j^R \cap \{|\eta| \ge 1\} \subset \Gamma_j$, $\sum_{j=1}^J g_j^R(\eta) = 1$ for $\eta \in \boldsymbol{R}^m$ and $|\partial_\eta^{\alpha+\gamma} g_j^R(\eta)| \le C_{|\gamma|}(C_*/R)^{|\alpha|}\langle \eta \rangle^{-|\gamma|}$ if $\langle \eta \rangle \ge R|\alpha|$. We put

$$u_0 = (1 - g^R(D))u, \quad u_1 = g^R(D)u, \quad u_\varepsilon = e^{-\varepsilon\langle D \rangle}u,$$
$$u_{k,\varepsilon} = e^{-\varepsilon\langle D \rangle}u_k, \quad u_k^j = g_j^R(D)u_k, \quad u_{k,\varepsilon}^j = g_j^R(D)u_{k,\varepsilon},$$

where $k = 0, 1$ and $1 \le j \le J$. It is obvious that $u_\varepsilon = \sum_{\mu=0}^{1} \sum_{\nu=1}^{J} u_{\mu,\varepsilon}^{\nu}$, $u_{k,\varepsilon}^j \in \mathcal{S}'_{-\delta}$ if $\delta < \varepsilon$, $u_{k,\varepsilon}^j \to u_k^j$ in $\mathcal{F}_0(\boldsymbol{R}^m)$ as $\varepsilon \downarrow 0$, and that u_1^j and $u_{1,\varepsilon}^j$ (resp. u_0^j and $u_{0,\varepsilon}^j$) can be regarded as an analytic function in $\boldsymbol{R}^m + i\,\mathrm{int}((\Gamma_j \cap \Gamma^2)^*)$ (resp. $\boldsymbol{R}^m + i\,\mathrm{int}(\Gamma_j^*)$) and $u_{1,\varepsilon}^j(\cdot + i\zeta) \to u_1^j(\cdot + i\zeta)$ in $\mathcal{S}'_{-\delta}(\boldsymbol{R}^m)$ (resp. $u_{0,\varepsilon}^j(\cdot + i\zeta) \to u_0^j(\cdot + i\zeta)$ in $\mathcal{S}'_{\delta}(\boldsymbol{R}^m)$) as $\varepsilon \downarrow 0$ for $\zeta \in \mathrm{int}((\Gamma_j \cap \Gamma^2)^*)$ (resp. $\zeta \in \mathrm{int}(\Gamma_j^*)$) and $\delta < \inf\{\zeta \cdot \eta;\ \eta \in S^{m-1} \cap \Gamma_j \cap \Gamma^2\}$ (resp. $\delta < \inf\{\zeta \cdot \eta;\ \eta \in S^{m-1} \cap \Gamma_j\}$), where $k = 0, 1$ and $1 \le j \le J$. By assumption u_0 ($= \sum_{j=1}^{J} u_0^j$) is analytic in V if $R \gg 1$. Let G_j ($1 \le j \le J$) be open cones in $\boldsymbol{R}^m \setminus \{0\}$ such that $G_j \subset (\Gamma_j \cap \Gamma^2)^*$. Then u_1^j is G_j analytic at V ($1 \le j \le J$), and the same argument as in the proof of Theorem 3.3. 6 gives

$$u|_V = u_0|_V + \sum_{j=1}^{J} b_{G_j} u_1^j \quad \text{in } \mathcal{B}(V).$$

Let $a^R(x, \eta) \in S^0(\boldsymbol{R}^n \times \boldsymbol{R}^m; R, A)$ ($R \ge 4$) satisfy $a^R(x, \eta) = 1$ for $x \in X_1$ and $\mathrm{supp}\, a^R \subset X_2 \times \boldsymbol{R}^m$. Put, for $\varepsilon > 0$,

$$v_{k,\varepsilon} = a_T^R(x, D_z) u_{k,\varepsilon} \quad (k = 0, 1).$$

Recall that $v_{0,\varepsilon} \to v_0$ in $\mathcal{S}'$ and $v_{1,\varepsilon} \to a_T^R(x, D_z) u_1$ in $\mathcal{F}_0(\boldsymbol{R}^n)$ as $\varepsilon \downarrow 0$ for some $v_0 \in \mathcal{S}'$, where $R \gg 1$. It is easy to see that $v_0(x)$ is analytic in X_1 and

$$v_0(x) = u_0(h(x)) \quad \text{for } x \in X_1. \tag{3.47}$$

Put, for $1 \le j \le J$,

$$\widetilde{\Gamma}_j = \{{}^t h'(x)\eta;\ \eta \in \overline{\Gamma}_j \cap \overline{\Gamma}^2 \setminus \{0\} \text{ and } x \in \overline{X}_2\}.$$

By assumption there is $c > 0$ such that

$$|{}^t h'(x)\eta| \ge c|\eta| \quad \text{for } x \in X_2 \text{ and } \eta \in \Gamma^2.$$

This implies that the $\widetilde{\Gamma}_j$ are closed cones in $\boldsymbol{R}^n \setminus \{0\}$. We may assume that the convex hulls $\mathrm{ch}[\widetilde{\Gamma}_j]$ of the $\widetilde{\Gamma}_j$ are proper cones, modifying X_2 and $\{\Gamma_j\}$ if necessary. Let G'_j ($1 \le j \le J$) be open cones in $\boldsymbol{R}^n \setminus \{0\}$ such that $G'_j \subset\subset \widetilde{\Gamma}_j^*$. From Taylor's formula there are $\delta_j > 0$ ($1 \le j \le J$) and $c' > 0$ such that

$$\mathrm{Im}\, h(x + it\xi) \cdot \eta \ge c' t |\eta|$$

if $1 \le j \le J$, $x \in X_2$, $\eta \in \Gamma_j \cap \Gamma^2$, $\xi \in G_j' \cap S^{n-1}$ and $0 < t \le \delta_j$. Modifying G_j ($1 \le j \le J$) if necessary, we may assume that $\operatorname{Im} h(x+it\xi) \in G_j$ if $1 \le j \le J$, $x \in X_2$, $\xi \in G_j' \cap S^{n-1}$ and $0 < t \le \delta_j$. Define

$$a^{R,R'}(x,\eta;\xi) = \sum_{k=1}^{\infty} \psi_k^{R'}(\eta) \sum_{|\beta| \le k-1} (i\xi)^\beta \partial_x^\beta a^R(x,\eta)/\beta!$$

for $x \in \boldsymbol{R}^n$, $\eta \in \boldsymbol{R}^m$ and $\xi \in \boldsymbol{R}^n$, where $R' \ge R$. Then we have

$$a^{R,R'}(x,\eta;\xi) \in S^{1,\delta(\xi)/R}(\boldsymbol{R}^n \times \boldsymbol{R}^m; 3R', 3\widehat{C} + 3AR'/R, 3AR'/R)$$

for each $\xi \in \boldsymbol{R}^n$, and

$$\left|(\partial_{x_j} + i\partial_{\xi_j}) D_x^{\beta+\tilde{\beta}} \partial_\eta^{\gamma+\tilde{\gamma}} a^{R,R'}(x,\eta;\xi)\right| \le C_{|\tilde{\beta}|+|\tilde{\gamma}|}(A/R)^{|\beta|}$$
$$\times (\widehat{C}/R' + A/R)^{|\gamma|} \langle\eta\rangle^{|\beta|-|\tilde{\gamma}|} \exp[(e\delta(\xi)/R - 1/(3R'))\langle\eta\rangle]$$

if $1 \le j \le n$, $x \in \boldsymbol{R}^n$, $\eta \in \boldsymbol{R}^m$, $\xi \in \boldsymbol{R}^n$ and $\langle\eta\rangle \ge 3R'(|\beta| + |\gamma|)$, where $\delta(\xi) = \sqrt{n}A|\xi|$. Choose $R > 0$ so that $e\sqrt{n}A/R \le c'/2$. Applying Stokes' formula we have

$$\langle v_{1,\varepsilon}, \varphi\rangle = \sum_{j=1}^J \langle a_T^R(x, D_z) u_{1,\varepsilon}^j, \varphi\rangle = \sum_{j=1}^J \Big\{ \int_{X_2} V_{1,\varepsilon}^j(x;\xi^j)\varphi(x+i\xi^j)\,dx$$
$$+ \int_0^1 \Big(\int_{X_2} V_{2,\varepsilon}^j(x;t\xi^j)\varphi(x+it\xi^j)\,dx \Big)\,dt \Big\}$$

for $\varphi \in \mathcal{S}_\infty$, where $\xi^j \in G_j'$ ($1 \le j \le J$) satisfy $|\xi^j| \le \delta_j$, and for $\xi \in G_j'$ with $|\xi| \le \delta_j$

$$V_{1,\varepsilon}^j(x;\xi) = (2\pi)^{-m} \int e^{ih(x+i\xi)\cdot\eta} a^{R,R'}(x,\eta;\xi) \mathcal{F}[u_{1,\varepsilon}^j](\eta)\,d\eta, \quad (3.48)$$

$$V_{2,\varepsilon}^j(x;\xi) = (2\pi)^{-m} \int e^{ih(x+i\xi)\cdot\eta}\, i \sum_{k=1}^n \xi_k (\partial_{x_k} + i\partial_{\xi_k})$$
$$\times a^{R,R'}(x,\eta;\xi) \mathcal{F}[u_{1,\varepsilon}^j](\eta)\,d\eta. \quad (3.49)$$

It is obvious that $V_{1,\varepsilon}^j(x;\xi) \rightrightarrows V_1^j(x,\xi)$ in $\boldsymbol{R}_x^n$ and $V_{2,\varepsilon}^j(x;t\xi) \rightrightarrows V_2^j(x,t\xi)$ in $\boldsymbol{R}_x^n \times [0,1]$ as $\varepsilon \downarrow 0$ for $\xi \in G_j'$ with $|\xi| \le \delta_j$, where $1 \le j \le J$ and $V_1^j(x;\xi)$ and $V_2^j(x;\xi)$ are defined by (3.48) and (3.49) with $u_{1,\varepsilon}^j$ replaced by u_1^j, respectively. Let X_0 ($\subset\subset X_1$) be an open subset of X. Then we have $V_1^j(x;\xi) = u_1^j(h(x+i\xi))$ and $V_2^j(x;\xi) = 0$ for $1 \le j \le J$, $x \in X_0$ and $\xi \in G_j'$ with $|\xi| \le \delta_j$. Therefore, we have

$$v_1|_{X_0} = \sum_{j=1}^J b_{G_j'}(u_1^j \circ h) \quad \text{in } \mathcal{B}(X_0).$$

This, together with (3.47), proves the theorem. □

We can define the tensor product $u \otimes v \in \mathcal{F}_0(\boldsymbol{R}^{m+n})$ of $u \in \mathcal{F}_0(\boldsymbol{R}^n)$ and $v \in \mathcal{F}_0(\boldsymbol{R}^m)$ by

$$u \otimes v = \mathcal{F}_{\xi,\eta}^{-1}[\hat{u}(\xi) \otimes \hat{v}(\eta)](x, y),$$

where $\hat{u} \otimes \hat{v}$ is the tensor product of the distributions $\hat{u}$ and $\hat{v}$ (see, also, Section 1.4). Then we have the following

Proposition 3.4. 2 (i) *If $u \in \mathcal{F}_0(\boldsymbol{R}^n)$, $v \in \mathcal{F}_0(\boldsymbol{R}^m)$ and $\varepsilon > 0$, then*

$$\langle u(x) \otimes v(y), \varphi(x)\psi(y)\rangle = \langle u, \varphi\rangle\langle v, \psi\rangle \quad \textit{for } \varphi \in \mathcal{S}_\varepsilon(\boldsymbol{R}^n) \textit{ and } \psi \in \mathcal{S}_\varepsilon(\boldsymbol{R}^m).$$

(ii) *If $u \in \mathcal{F}_0(\boldsymbol{R}^n)$, $v \in \mathcal{F}_0(\boldsymbol{R}^m)$ and $\varepsilon > 0$, then*

$$\begin{aligned} &\langle u(x) \otimes v(y), \varphi(x, y)\rangle (= (2\pi)^{-m-n}\langle \hat{u}(\xi) \otimes \hat{v}(\eta), \hat{\varphi}(-\xi, -\eta)\rangle) \\ &= \langle u(x), \langle v(y), \varphi(x, y)\rangle_y\rangle_x = \langle v(y), \langle u(x), \varphi(x, y)\rangle_x\rangle_y \end{aligned}$$

for $\varphi(x, y) \in \mathcal{S}_\varepsilon(\boldsymbol{R}^{m+n})$. (iii) *If $u \in \mathcal{A}'(\boldsymbol{R}^n)$ and $v \in \mathcal{A}'(\boldsymbol{R}^m)$, then $u \otimes v \in \mathcal{A}'(\boldsymbol{R}^{m+n})$ and* supp $u \otimes v \subset$ supp $u \times$ supp v. (iv) *By the assertion* (iii) *we can define the tensor product $u \otimes v \in \mathcal{B}(X \times Y)$ of $u \in \mathcal{B}(X)$ and $v \in \mathcal{B}(Y)$. In other words,* supp $(u_1 \otimes v_1 - u_2 \otimes v_2) \cap X \times Y = \emptyset$ *if X and Y are bounded, $u_k \in \mathcal{A}'(\overline{X})$ and $v_k \in \mathcal{A}'(\overline{Y})$ ($k = 1, 2$),* supp $(u_1 - u_2) \subset \partial X$ *and* supp $(v_1 - v_2) \subset \partial Y$. *Let $(x^0, y^0) \in X \times Y$, and assume that*

$$u|_{X_1} = \sum_{j=1}^{J} b_{G_j} f_j, \quad v|_{Y_1} = \sum_{k=1}^{K} b_{G'_j} g_j,$$

where X_1 (resp. Y_1) is a neighborhood of x^0 (resp. y^0), the G_j (resp. the G'_k) are open convex cones in $\boldsymbol{R}^n \setminus \{0\}$ (resp. $\boldsymbol{R}^m \setminus \{0\}$) and the f_j (resp. the g_k) are G_j analytic at X_1 (resp. G'_k analytic at Y_1). Then we have

$$(u \otimes v)|_{X_1 \times Y_1} = \sum_{j=1}^{J}\sum_{k=1}^{K} b_{G_j \times G'_k} f_j \otimes g_k.$$

Moreover, we have

$$\begin{aligned} WF_A(u \otimes v) \quad \subset \quad & (WF_A(u) \times WF_A(v)) \cup (WF_A(u) \times (\text{supp } v \times \{0\})) \\ & \cup((\text{supp } u \times \{0\}) \times WF_A(v)), \end{aligned} \tag{3.50}$$

where we identify $A \times B$ with the set $\{(x, y, \xi, \eta) \in T^*(X \times Y);\ (x, \xi) \in A$ and $(y, \eta) \in B\}$ for $A \subset T^*X$ and $B \subset T^*Y$.

We omit the proof of Proposition 3.4. 2 since it is almost obvious. We can also prove directly from the definition of wave front sets (*i.e.*, Definition 3.1. 1) that (3.50) is valid for $u \in \mathcal{A}'(\boldsymbol{R}^n)$ and $v \in \mathcal{A}'(\boldsymbol{R}^m)$. The proof is a little long. So we also omit its proof.

Theorem 3.4. 3 *Let $u \in \mathcal{F}_0(\boldsymbol{R}^n)$ and $v \in \mathcal{F}_0(\boldsymbol{R}^m)$.* (i) *Let X and Y be bounded open subsets of $\boldsymbol{R}^n$ and $\boldsymbol{R}^m$, respectively. Then we have*

$$(u \otimes v)|_{X\times Y} = (u|_X) \otimes (v|_Y) \quad \text{in } \mathcal{B}(X \times Y).$$

(ii) supp $u \otimes v \subset$ supp $u \times$ supp v *and* (3.50) *is still valid for $u \in \mathcal{F}_0(\boldsymbol{R}^n)$ and $v \in \mathcal{F}_0(\boldsymbol{R}^m)$.*

Proof The assertion (ii) easily follows from the assertion (i). Let Γ_j ($1 \le j \le J$) and $\tilde{\Gamma}_k$ ($1 \le k \le K$) be open convex proper cones in $\boldsymbol{R}^n \setminus \{0\}$ and $\boldsymbol{R}^m \setminus \{0\}$, respectively, such that $\boldsymbol{R}^n \setminus \{0\} = \bigcup_{j=1}^J \Gamma_j$ and $\boldsymbol{R}^m \setminus \{0\} = \bigcup_{k=1}^K \tilde{\Gamma}_k$. Choose $\{g_j^R(\xi)\} \subset C^\infty(\boldsymbol{R}^n)$ ($R \ge 2, 1 \le j \le J$) and $\{\tilde{g}_k^R(\eta)\} \subset C^\infty(\boldsymbol{R}^m)$ ($R \ge 2$, $1 \le k \le K$) so that supp $g_j^R \cap \{|\xi| \ge 1\} \subset \Gamma_j$, supp $\tilde{g}_k^R \cap \{|\eta| \ge 1\} \subset \tilde{\Gamma}_k$, $\sum_{j=1}^J g_j^R(\xi) = 1$ for $\xi \in \boldsymbol{R}^n$, $\sum_{k=1}^K \tilde{g}_k^R(\eta) = 1$ for $\eta \in \boldsymbol{R}^m$, $|\partial_\xi^{\alpha+\tilde{\alpha}} g_j^R(\xi)| \le C_{|\tilde{\alpha}|}(C_*/R)^{|\alpha|}\langle\xi\rangle^{-|\tilde{\alpha}|}$ if $\langle\xi\rangle \ge R|\alpha|$, and $|\partial_\eta^{\gamma+\tilde{\gamma}} \tilde{g}_k^R(\eta)| \le C_{|\tilde{\gamma}|}(C_*/R)^{|\gamma|}\langle\eta\rangle^{-|\tilde{\gamma}|}$ if $\langle\eta\rangle \ge R|\gamma|$. We put $U_j^R(x, x_{n+1}) = \mathcal{H}(g_j^R(D)u)$ for $x_{n+1} \ne 0$ and $1 \le j \le J$ and $V_k^R(y, y_{m+1}) = \mathcal{H}(\tilde{g}_k^R(D)v)$ for $y_{m+1} \ne 0$ and $1 \le k \le K$. Let $a^R(x, y, \xi, \eta) \in S^0(\boldsymbol{R}^{n+m} \times \boldsymbol{R}^{n+m}; R, A)$ ($R \ge 4$) satisfy $a^R(x, y, \xi, \eta) = 1$ for $(x, y) \in X \times Y$ and supp $a^R \subset X_1 \times Y_1 \times \boldsymbol{R}^{n+m}$, where X_1 and Y_1 are bounded subsets of $\boldsymbol{R}^n$ and $\boldsymbol{R}^m$, respectively. Applying the same argument as in the proof of Theorem 3.3. 6 we have

$$(u \otimes v)|_{X\times Y} = \sum_{j=1}^J \sum_{k=1}^K (a^R(x, y, D_x, D_y)((g_j^R(D)u) \otimes (\tilde{g}_k^R(D)v))|_{X\times Y}$$

$$= \sum_{j=1}^J \sum_{k=1}^K b_{\mathrm{int}(\Gamma_j^*)\times\mathrm{int}(\tilde{\Gamma}_k^*)} f_{j,k}^R,$$

where $f_{j,k}^R(x, y) = 4U_j^R(x, +0)V_k^R(y, +0)$. On the other hand, Theorem 3.3. 6 (or its remark) yields

$$u|_X = 2\sum_{j=1}^J b_{\mathrm{int}(\Gamma_j^*)} U_j^R(\cdot, +0) \quad \text{in } \mathcal{B}(X),$$

$$v|_Y = 2\sum_{k=1}^K b_{\mathrm{int}(\tilde{\Gamma}_k^*)} V_k^R(\cdot, +0) \quad \text{in } \mathcal{B}(Y).$$

This, together with Proposition 3.4. 2 , proves the assertion (i). □

Let us define the push-forward h_*u. For $u \in \mathcal{A}'(\boldsymbol{R}^n)$ with supp $u \subset X$ we define $h_*u \in \mathcal{A}'(\boldsymbol{R}^m)$ by

$$h_*u(\varphi) = u(\varphi \circ h) \quad \text{for } \varphi \in \mathcal{A}(\omega),$$

where ω is a complex neighborhood of $h(\text{supp } u)$ in $\boldsymbol{C}^m$. Then it is obvious that supp $h_*u \subset h(\text{supp } u)$.

Theorem 3.4. 4 *Let $u \in \mathcal{A}'(\boldsymbol{R}^n)$ satisfy* supp $u \subset X$. *Then*

$$\begin{aligned} WF_A(h_*u) \subset \{&(y,\eta) \in T^*\boldsymbol{R}^m \setminus 0;\ y = h(x), \\ &(x, {}^th'(x)\eta) \in WF_A(u) \cup (\text{supp } u \times \{0\}) \textit{ for some } x \in X\}. \end{aligned}$$

Proof Let X_j ($j = 1, 2$) be open subsets of X such that supp $u \subset X_1 \subset\subset X_2 \subset\subset X$, and choose $\{\chi_{j,k}(x)\}_{j,k\in N} \subset C_0^\infty(X_2)$ so that $\chi_{j,k}(x) = 1$ for $x \in X_1$ and

$$\left|D^{\beta^1+\beta^2+\lambda}\chi_{j,k}(x)\right| \le C_{|\lambda|}(\widehat{C}_0 j)^{|\beta^1|}(\widehat{C}_0 k)^{|\beta^2|}$$

if $|\beta^1| \le 3j$ and $|\beta^2| \le 3k$, where $\widehat{C}_0$ is a positive constant depending on X_1 and X_2. Put $\Phi_j^R(x,\xi) = \sum_{k=1}^\infty \chi_{j,k}(x)\psi_k^R(\xi)$ for $j \in \boldsymbol{N}$ and $R > 0$. Then we have

$$\left|\Phi^{R(\alpha+\tilde{\alpha})}_{j(\beta^1+\beta^2+\tilde{\beta})}(x,\xi)\right| \le C_{|\tilde{\alpha}|+|\tilde{\beta}|}(\widehat{C}/R)^{|\alpha|}(\widehat{C}_0 j)^{|\beta^1|}(\widehat{C}_0/R)^{|\beta^2|}\langle\xi\rangle^{1+|\beta^2|-|\tilde{\alpha}|}$$

if $j \in \boldsymbol{N}$, $R > 0$, $\langle\xi\rangle \ge 3R|\alpha|$, $\langle\xi\rangle \ge R|\beta^2|$ and $|\beta^1| \le 3j$. Moreover, we have

$$\Phi_j^R(x,D)v = \sum_{k=1}^\infty \chi_{j,k}(x)\psi_k^R(D)v \quad \text{in } \mathcal{F}_0 \tag{3.51}$$

for $v \in \mathcal{F}_0$ if $R \gg 1$. In fact, we have

$$\left|\Phi^{R(\alpha+\tilde{\alpha})}_{j,\ell(\beta+\tilde{\beta})}(x,-\xi)\right| \le C_{|\tilde{\alpha}|+|\tilde{\beta}|}(\widehat{C}/R)^{|\alpha|}(\widehat{C}_0/R)^{|\beta|}\langle\xi\rangle^{2+|\beta|-|\tilde{\alpha}|}/(\ell R)$$

for $\ell \in \boldsymbol{N}$ if $\langle\xi\rangle \ge 3R|\alpha|$ and $\langle\xi\rangle \ge R|\beta|$, where $\Phi_{j,\ell}^R(x,\xi) = \sum_{k=\ell+1}^\infty \chi_{j,k}(x)\psi_k^R(\xi)$. From Theorem 2.3. 3 and its remark it follows that ${}^t\Phi_{j,\ell}^R(x,D)\varphi \to 0$ in $\mathcal{S}_{\varepsilon/2}$ as $\ell \to \infty$ for $\varphi \in \mathcal{S}_\varepsilon$ if $R > 2en\widehat{C}_0$ and $\varepsilon < 2/R$, which gives (3.51). Assume that $v \in \mathcal{F}_0$ and

$$|D^\beta\varphi(x)| \le C(\varphi)A(\varphi)^{|\beta|}|\beta|! \quad \text{for } x \in X_2.$$

We choose $\varepsilon > 0$ so that $\varepsilon \le 1/(3R)$. Since $e^{-\varepsilon\langle\xi\rangle}\hat{v}(\xi) \in \mathcal{S}'$ and

$$(\chi_{j,k}(x)\psi_k^R(D)v)(\varphi) = \langle e^{-\varepsilon\langle\xi\rangle}\hat{v}(\xi), e^{\varepsilon\langle\xi\rangle}\psi_k^R(\xi)\mathcal{F}^{-1}[\chi_{j,k}\varphi](\xi)\rangle,$$

there are $C > 0$ and $\nu_0 \in \boldsymbol{Z}_+$, which depend on v and ε (> 0), such that

$$\begin{aligned}&\left|(\chi_{j,k}(x)\psi_k^R(D)v)(\varphi)\right| \\ &\le C \max_{|\alpha|\le\nu_0} \sup_{\xi\in R^n} \left|\langle\xi\rangle^{\nu_0} D_\xi^\alpha\left\{e^{\varepsilon\langle\xi\rangle}\psi_k^R(\xi)\mathcal{F}^{-1}[\chi_{j,k}\varphi](\xi)\right\}\right|.\end{aligned}$$

Using

$$\left|L^\mu\left\{x^\alpha\chi_{j,k}(x)\varphi(x)\right\}\right| \le C_{|\alpha|}(\varphi)(\sqrt{n+1}(\widehat{C}_0 + A(\varphi))k/\langle\xi\rangle)^\mu$$

for $\mu \le k$, where $L = -\langle\xi\rangle^{-2}(\sum_{\nu=1}^n \xi_\nu D_{x_\nu} + 1)$, we have

$$\left|D_\xi^\alpha\mathcal{F}^{-1}[\chi_{j,k}\varphi](\xi)\right| \le C'_{|\alpha|}(\varphi)(\sqrt{n+1}(\widehat{C}_0 + A(\varphi))k/\langle\xi\rangle)^k.$$

Therefore, we have

$$\left|(\chi_{j,k}(x)\psi_k^R(D)v)(\varphi)\right| \le C_{\varepsilon,R}(v,\varphi)k^{\nu_0}(e\sqrt{n+1}(\widehat{C}_0 + A(\varphi))/R)^k,$$

which implies that $\sum_{k=1}^\infty(\chi_{j,k}(x)\psi_k^R(D)v)(\varphi)$ converges if $R \ge 2e\sqrt{n+1} \times(\widehat{C}_0 + A(\varphi))$. This, together with (3.51), yields $\Phi_j^R(x,D)v \in \mathcal{A}'(K)$ for $v \in \mathcal{F}_0$ if K is a compact complex neighborhood of $\overline{X}_2$ and $R \ge R(X_1, X_2, K)$, where $R(X_1, X_2, K)$ is a positive constant. Since $h(x)$ is analytic in X, we have, with positive constants $C(h)$ and $B(h)$,

$$|D^\beta h(x)| \le C(h)B(h)^{|\beta|}|\beta|! \quad \text{for } x \in X_2.$$

Then $h(z)$ is analytic in $\{z \in \boldsymbol{C}^n;\ \sum_{j=1}^n |z_j - x_j| < 1/B(h)$ for some $x \in X_2\}$ and Cauchy's estimates yield

$$\left|D_x^\beta e^{-ih(x)\cdot\eta}\right| \le e^{2C(h)|\eta|}(2nB(h))^{|\beta|}|\beta|! \quad \text{for } x \in X_2.$$

Therefore, we have

$$\begin{aligned}&\mathcal{F}[h_*(\Phi_j^R(x,D)u)](\eta) = (\Phi_j^R(x,D)u)_x(e^{-ih(x)\cdot\eta}) \\ &= \sum_{k=1}^\infty (2\pi)^{-n}\int e^{-ih(x)\cdot\eta+ix\cdot\xi}\chi_{j,k}(x)\psi_k^R(\xi)\hat{u}(\xi)\,d\xi dx\end{aligned}$$

for $R \gg 1$, where the right-hand side converges absolutely and locally uniformly in $\eta \in \boldsymbol{R}^m$. Put $S(x,\xi) = x \cdot \xi$ and

$$p(\eta, x, \xi) = \sum_{j,k=1}^{\infty} \psi_j^R(\eta)\chi_{j,k}(x)\psi_k^R(\xi),$$

Then $p(\eta, x, \xi) \in S^{1,1}(\boldsymbol{R}^m \times \boldsymbol{R}^n \times \boldsymbol{R}^n; 3R, 3\widehat{C}, 3\widehat{C}_0, 3\widehat{C})$ and

$$\sum_{j=1}^{\infty} \psi_j^R(D_y)h_*(\Phi_j^R(x,D)u) = p_{-T,S}(D_y, x, D_x)u \quad \text{in } \mathcal{F}_0(\boldsymbol{R}^m).$$

It is easy to see that $p^1_{-T,S}(D_y, x, D_x)u$ is analytic and

$$\begin{aligned} &WF_A(p^2_{-T,S}(D_y, x, D_x)u) \\ &\subset \{(h(x), \eta) \in T^*\boldsymbol{R}^m \setminus 0;\ x \in X_2 \text{ and } {}^t h'(x)\eta = 0\}, \end{aligned}$$

where $p^1(\eta, x, \xi) = \sum_{k=1}^{\infty} \psi_1^R(\eta)\chi_{1,k}(x)\psi_k^R(\xi)$ and $p^2(\eta, x, \xi) = \sum_{j=2}^{\infty} \psi_j^R(\eta)\chi_{j,1}(x)\psi_1^R(\xi)$. Applying Theorem 3.2. 4 to $p^0_{-T,S}(D_y, x, D_x)u$, we have

$$\begin{aligned} WF_A(p_{-T,S}(D_y, x, D_x)u) \subset \{&(y, \eta) \in T^*\boldsymbol{R}^m \setminus 0;\ y = h(x) \text{ and} \\ &(x, {}^t h'(x)\eta) \in WF_A(u) \cup (X_2 \times \{0\}) \text{ for some } x \in X\}, \end{aligned} \tag{3.52}$$

where $p^0(\eta, x, \xi) = p(\eta, x, \xi) - p^1(\eta, x, \xi) - p^2(\eta, x, \xi)$. Let X_0 be an open subset of X such that supp $u \subset\subset X_0 \subset\subset X_1$, and choose $\{\chi_k(x)\}_{k\in N} \subset C_0^\infty(X_1)$ so that $\chi_k(x) = 1$ for $x \in X_0$ and

$$|D^\beta \chi_k(x)| \le C(\widehat{C}_1 k)^{|\beta|} \quad \text{for } |\beta| \le k,$$

where $\widehat{C}_1$ depends on X_0 and X_1. Then we can write

$$\begin{aligned} (1 - \Phi_j^R(x,D))u \ &= \ \sum_{k=1}^{\infty} (1 - \Phi_j^R(x,D))\psi_k^R(D)e^{\rho\langle D\rangle} \\ &\qquad \times\{\chi_k u_\rho + (1-\chi_k)u_\rho\} \quad \text{in } \mathcal{F}_0 \end{aligned}$$

where $0 < \rho \le 1$. Since supp $(1-\chi_k) \cap$ supp $u = \emptyset$, supp $\chi_k \subset X_1$ and $1 - \Phi_j^R(x,\xi) = 0$ for $x \in X_1$, we can easily prove that

$$|D^\beta\{(1 - \Phi_j^R(x,D))u\}| \le C_R(u)(8Rj)^{|\beta|}$$

if $R \gg 1$ and $|\beta| \le j$ (see, *e.g.*, the proof of Lemma 2.6. 4). Moreover, we have supp $(1 - \Phi_j^R(x,D))u \subset X_2 \setminus X_1$. Therefore,

$$h_*((1 - \Phi_j^R(x,D))u) = \mathcal{F}_\eta^{-1}\Big[\int e^{-ih(x)\cdot\eta}(1 - \Phi_j^R(x,D))u(x)\,dx\Big]$$

and supp $h_*((1-\Phi_j^R(x,D))u) \subset h(X_2 \setminus X_1)$ for $R \gg 1$. Noting that

$$\sum_{j=1}^{\infty} \psi_j^R(D_y)h_*((1-\Phi_j^R(x,D))u)$$
$$= \sum_{j=1}^{\infty}(2\pi)^{-m}\int e^{iy\cdot\eta - ih(x)\cdot\eta}\psi_j^R(\eta)(1-\Phi_j^R(x,D))u(x)\,dx\,d\eta$$

in $\mathcal{S}'(\boldsymbol{R}^m)$, we can easily show that

$$WF_A\Big(\sum_{j=1}^{\infty}\psi_j^R(D_y)h_*((1-\Phi_j^R(x,D))u)\Big)$$
$$\subset \{(h(x),\eta) \in T^*\boldsymbol{R}^m \setminus 0;\ x \in X_2 \setminus X_1 \text{ and } {}^t h'(x)\eta = 0\}. \quad (3.53)$$

(3.52) and (3.53) prove the theorem. □

Let $u \in \mathcal{B}(X)$, and assume that the mapping $h|_{\text{supp } u} : \text{supp } u \ni x \mapsto h(x) \in Y$ is proper. Let us define $h_*u \in \mathcal{B}(Y)$. Choose a family of open subsets $\{Y_j\}$ of Y so that $Y_j \subset\subset Y$ and $Y = \cup Y_j$. Put $X_j = h^{-1}(Y_j)$ and $u_j = u|_{X_j} \in \mathcal{B}(X_j)$. We denote $\partial_X X_j = \partial X_j \cap X$ and $\text{cl}_X(X_j) = \overline{X}_j \cap X$. It is easy to see that $h(\partial_X X_j) \subset \partial Y_j$. Since supp $u \cap h^{-1}(\overline{Y}_j)$ is compact, supp $u \cap \text{cl}_X(X_j)$ is a compact subset of X. For each j we can choose a bounded open subsets $X_{j,0}$ of X_j and $\tilde{v}_j \in \mathcal{A}'(\overline{X}_{j,0})$ so that supp $u \cap X_j \subset X_{j,0}$ and $\tilde{v}_j|_{X_{j,0}} = u|_{X_{j,0}}$. We have

$$\text{supp } \tilde{v}_j \subset \partial X_{j,0} \cup (\text{supp } u \cap X_{j,0}) \subset \partial X_{j,0} \cup (\text{supp } u \cap \text{cl}_X(X_j)).$$

It follows from Theorem 1.3.4 that there is $v_j \in \mathcal{A}'(\text{supp } u \cap \text{cl}_X(X_j))$ satisfying supp $(\tilde{v}_j - v_j) \subset \partial X_{j,0}$. Then we have $v_j \in \mathcal{A}'(\overline{\Omega})$ and $v_j|_\Omega = u_j|_\Omega$ if Ω is a bounded open subset of X_j and supp $u \cap X_j \subset \Omega$. In fact, assume that Ω is a bounded open subset of X_j and supp $u \cap X_j \subset \Omega$. Then we have $u_j|_{\Omega \cap X_{j,0}} = [v_j]|_{\Omega\cap X_{j,0}}$, where $[v_j] = v_j|_\Omega$ in $\mathcal{B}(\Omega)$. On the other hand,

$$\text{supp } u_j|_\Omega = \text{supp } u \cap \Omega = \text{supp } u \cap X_j = \text{supp } u \cap X_{j,0}\ (\subset \Omega),$$
$$\text{supp } [v_j] \subset \text{supp } u \cap \text{cl}_X(X_j) \cap \Omega = \text{supp } u \cap \Omega = \text{supp } u \cap X_{j,0}.$$

Therefore, for every $x \in \Omega \setminus X_{j,0}$ there is an open neighborhood U of x in Ω such that $u_j|_U = [v_j]|_U = 0$. This yields $u_j|_\Omega = [v_j]$. If $v_j' \in \mathcal{A}'(\text{supp } u \cap \text{cl}_X(X_j))$ and $v_j'|_\Omega = u_j|_\Omega$ for any bounded open subset Ω of X_j with supp $u \cap X_j \subset \Omega$, then $v_j - v_j' \in \mathcal{A}'(\text{supp } u \cap \partial_X X_j)$. In fact, we can choose $R > 0$ so that supp $u \cap \text{cl}_X(X_j) \subset \{|x| < R\}$, and take

$\Omega = X_j \cap \{|x| < R\}$. Then we have $v_j - v_j' \in \mathcal{A}'(\partial\Omega \cap \text{supp } u \cap \text{cl}_X(X_j))$ and $v_j - v_j' \in \mathcal{A}'(\text{supp } u \cap \partial_X X_j)$ since $\partial\Omega \cap \text{supp } u \cap \text{cl}_X(X_j) \subset \text{supp } u \cap \partial_X X_j$. Now we put $w_j = (h_* v_j)|_{Y_j} \in \mathcal{B}(Y_j)$. Recall that $h(\text{supp } u \cap \partial_X X_j) \subset h(\partial_X X_j) \subset \partial Y_j$. This implies that $w_j \in \mathcal{B}(Y_j)$ is uniquely determined by u. So we define $h_* u \in \mathcal{B}(Y)$ by $h_* u = \{w_j\}$. In order to prove validity of this definition it suffices to verify that

$$w_j|_{Y_j \cap Y_k} = w_k|_{Y_j \cap Y_k}. \tag{3.54}$$

Note that $h^{-1}(Y_j \cap Y_k) = X_j \cap X_k$. By a repetition of the above argument with Y_j replaced by $Y_j \cap Y_k$ there is $v_{j,k} \in \mathcal{A}'(\text{supp } u \cap \text{cl}_X(X_j \cap X_k))$ such that $v_{j,k}|_\Omega = u|_\Omega$ for any bounded open subset Ω of $X_j \cap X_k$ with $\text{supp } u \cap X_j \cap X_k \subset \Omega$. Let Ω' be a bounded open subset of X_j such that $\text{supp } u \cap X_j \subset \Omega'$. Then we have $v_j|_{\Omega'} = u|_{\Omega'}$. This gives $\text{supp }(v_j - v_{j,k}) \cap (X_k \cap \Omega') = \emptyset$ and $v_j - v_{j,k} \in \mathcal{A}'(\text{supp } u \cap \text{cl}_X(X_j) \setminus (X_k \cap \Omega'))$. Since $\text{supp } u \cap \text{cl}_X(X_j) \setminus (X_k \cap \Omega') \subset (\text{supp } u \cap \text{cl}_X(X_j) \setminus X_k) \cup \partial_X X_j$, we have $\text{supp } h_*(v_j - v_{j,k}) \subset (\overline{Y}_j \setminus Y_k) \cup \partial Y_j$ and, therefore, $\text{supp } h_*(v_j - v_{j,k}) \cap (Y_j \cap Y_k) = \emptyset$. Similarly we have $\text{supp } h_*(v_k - v_{j,k}) \cap (Y_j \cap Y_k) = \emptyset$, which gives (3.54). It is easy to see that

$$\text{supp } h_* u \subset h(\text{supp } u), \tag{3.55}$$

$$\begin{aligned} WF_A(h_* u) \subset \{(y,\eta) \in T^*Y \setminus 0;\ & y = h(x) \text{ and } (x, {}^t h'(x)\eta) \\ & \in WF_A(u) \cup (\text{supp } u \times \{0\}) \text{ for some } x \in X\}. \end{aligned} \tag{3.56}$$

Let Z be an open subset of $\boldsymbol{R}^\ell$, and let $k : Y \to Z$ be a real analytic mapping. We can easily show that $h_* u$, $k_*(h_* u)$ and $(k \circ h)_* u$ are well-defined and $(k \circ h)_* u = k_*(h_* u)$ if $u \in \mathcal{B}(X)$ and the mapping $(k \circ h)|_{\text{supp } u}$: $\text{supp } u \to Z$ is proper.

Let X and T be real analytic manifolds, and let $\pi : X \times T \ni (x,t) \mapsto x \in X$. To define $\pi_* u$ for $u \in \mathcal{B}(X \times T)$ we must assign a volume element on T. Let $\mathcal{K}$ be an atlas for T, *i.e.*, let $\mathcal{K}$ be a family of homeomorphisms κ of open subsets T_κ of T onto open subsets $\widetilde{T}_\kappa$ of $\boldsymbol{R}^m$ such that $T = \bigcup_{\kappa \in \mathcal{K}} T_\kappa$ and the mapping

$$\kappa' \circ \kappa^{-1} :\ \kappa(T_\kappa \cap T_{\kappa'}) \to \kappa'(T_\kappa \cap T_{\kappa'})$$

is real analytic for every $\kappa, \kappa' \in \mathcal{K}$. We say that a collection $\omega \equiv \{\omega_\kappa\}_{\kappa \in \mathcal{K}}$ is a volume element with real analytic coefficients (a strictly positive real analytic density) on T if $\omega_\kappa \equiv \omega_\kappa(y)$ is a real analytic function in $\widetilde{T}_\kappa$ and $\omega_\kappa(y) > 0$ for every $\kappa \in \mathcal{K}$ and $y \in \widetilde{T}_\kappa$, and if

$$\omega_{\kappa_1}(y) = |\det(\kappa_2 \circ \kappa_1^{-1})'| \omega_{\kappa_2}(\kappa_2 \circ \kappa_1^{-1}(y))$$

for every $\kappa_1, \kappa_2 \in \mathcal{K}$ and $y \in \kappa_1(T_{\kappa_1} \cap T_{\kappa_2})$, where $(\kappa_2 \circ \kappa_1^{-1})'$ denotes the Jacobian matrix of $\kappa_2 \circ \kappa_1^{-1}$. Let $\omega \equiv \{\omega_\kappa\}_{\kappa \in \mathcal{K}}$ be a volume element with real analytic coefficients. For simplicity we assume that X is an open subset of $\boldsymbol{R}^n$. Let $u \in \mathcal{B}(X \times T)$, and assume that $\pi|_{\text{supp } u} : \text{supp } u \to X$ is proper. We may assume that there are closed subsets K_j of T and $\kappa_j \in \mathcal{K}$ ($1 \le j \le N$) such that $K_j \subset\subset T_{\kappa_j}$ ($1 \le j \le N$) and supp $u \subset \bigcup_{j=1}^N X \times K_j$, shrinking X if necessary. By flabbiness of the sheaf $\mathcal{B}_{X \times T}$ there are $u_j \in \mathcal{B}(X \times T)$ ($1 \le j \le N$) such that supp $u_j \subset X \times K_j$ and $u = \sum_{j=1}^N u_j$ (see, *e.g.*, Lemma 1.4.4 in [Kn]). We can identify u_j with an element $\tilde{u}_j(x,y)$ of $\mathcal{B}(X \times \widetilde{T}_{\kappa_j})$ in a natural way ($1 \le j \le N$). Put $v_j(x,y) = \omega_{\kappa_j}(y)\tilde{u}_j(x,y)$, and let $\pi_j : X \times \widetilde{T}_{\kappa_j} \ni (x,y) \mapsto x \in X$. Then we can define the integral of u on T with respect to the volume element ω by

$$\int_T u\omega = \sum_{j=1}^N \pi_{j*}(v_j)(x) \quad (\in \mathcal{B}(X)).$$

It is easy to see that the integral $\int_T u\omega$ is well-defined and that supp $\int_T u\omega \subset \pi(\text{supp } u)$ (see, *e.g.*, Lemma 4.2.3 in [Kn]). Since the sheaf $\mathcal{C}$ of microfunctions is a flabby sheaf (see Theorem 3.6. 1 below), we can also show that

$$\begin{aligned} WF_A(\textstyle\int_T u\omega) \;\subset\; & \{(x,\xi) \in T^*X \setminus 0; \\ & (x,t,\xi,0) \in WF_A(u) \text{ for some } t \in T\} \end{aligned} \tag{3.57}$$

if $u \in \mathcal{B}(X \times T)$ and $\pi|_{\text{supp } u}$ is proper.

Theorem 3.4. 5 *Let $u, v \in \mathcal{B}(X)$ satisfy $WF_A(u) \cap WF_A(v)' = \emptyset$, where $A' = \{(x,-\xi);\ (x,\xi) \in A\}$. Then we can define the product $uv \in \mathcal{B}(X)$ by $uv = \Delta^*(u \otimes v)$, where $\Delta : X \ni x \mapsto (x,x) \in X \times X$. Moreover, we have*

$$\begin{aligned} &\text{supp } uv \subset \text{supp } u \cap \text{supp } v, \\ &WF_A(uv) \subset \{(x,\xi+\eta); \\ &\qquad \text{“}(x,\xi) \in WF_A(u) \text{ and } (x,\eta) \in WF_A(v) \cup (\text{supp } v \times \{0\})\text{”} \\ &\qquad \text{or “}x \in \text{supp } u,\ \xi = 0 \text{ and } (x,\eta) \in WF_A(v)\text{”}\}. \end{aligned}$$

Proof Since $N_\Delta = \{(x,x,\xi,-\xi) \in T^*(X \times X);\ (x,\xi) \in T^*X\}$ and $N_\Delta \cap WF_A(u \otimes v) = \emptyset$, $\Delta^*(u \otimes v)$ is well-defined. (3.44), (3.45) and Proposition 3.4. 2 prove the latter part of the theorem. □

Let X_j be an open subset of $\boldsymbol{R}^{n_j}$ ($1 \le j \le 3$), and let $A \in \mathcal{B}(X_1 \times X_2)$ adn $B \in \mathcal{B}(X_2 \times X_3)$. We define the mapping Δ by $\Delta : X_1 \times X_2 \times X_3 \ni$

$(x,y,z) \mapsto (x,y,y,z) \in X_1 \times X_2 \times X_2 \times X_3$. Assume that $(y,z,-\eta,0) \notin WF_A(B)$ for any $z \in X_3$ if $(x,y,0,\eta) \in WF_A(A)$, and that the mapping $\pi : \Delta^{-1}(\text{supp } A \times \text{supp } B) \ni (x,y,z) \mapsto (x,z) \in X_1 \times X_2$ is proper. Then we have the following

Theorem 3.4. 6 *We can define $A \circ B \in \mathcal{B}(X_1 \times X_3)$ by $A \circ B := \pi_* \Delta^*(A \otimes B)$. Moreover, we have*

$$\begin{aligned} \text{supp } A \circ B \subset \text{supp } A \circ \text{supp } B := \{(x,z) \in X_1 \times X_3;\ (x,y) \in \text{supp } A \\ \textit{and } (y,z) \in \text{supp } B \textit{ for some } y \in X_2\}, \\ WF_A(A \circ B) \subset WF_A(A) \circ (WF_A(B)' \cup (\text{supp } B \times \{(0,0)\})) \\ \cup (\text{supp } A \times \{(0,0)\}) \circ WF_A(B)', \end{aligned}$$

where $T' = \{(y,z,-\eta,\zeta);\ (y,z,\eta,\zeta) \in T\}$ and $S \circ T = \{(x,z,\xi,\zeta);\ (x,y,\xi,\eta) \in S$ and $(y,z,\eta,\zeta) \in T$ for some $(y,\eta)\}$ for $S \subset T^*(X_1 \times X_2)$ and $T \subset T^*(X_2 \times X_3)$.

Proof Since $N_\Delta = \{(x,y,y,z,0,\eta,-\eta,0) \in T^*(X_1 \times X_2 \times X_2 \times X_3);\ x \in X_1,\ (y,\eta) \in T^*X_2$ and $z \in X_3\}$ and $N_\Delta \cap WF_A(A \otimes B) = \emptyset$, $\Delta^*(A \otimes B)$ is well-defined, $\text{supp } \Delta^*(A \otimes B) \subset \Delta^{-1}(\text{supp } A \times \text{supp } B)$ and $WF_A(\Delta^*(A \otimes B)) \subset \Delta^*(WF_A(A \otimes B))$ (see (3.44)). By assumption $\pi_*\Delta^*(A \otimes B)$ is well-defined. The latter part of the theorem easily follows from (3.55) and (3.56). □

Corollary 3.4. 7 *Let $A \in \mathcal{B}(X_1 \times X_2)$ and $u \in \mathcal{B}(X_2)$. Assume that $(y,-\eta) \notin WF_A(u)$ if $(x,y,0,\eta) \in WF_A(A)$, and that the mapping $\pi : \Delta^{-1}(\text{supp } A \times \text{supp } u) \ni (x,y) \mapsto x \in X_1$ is proper, where $\Delta : X_1 \times X_2 \ni (x,y) \mapsto (x,y,y) \in X_1 \times X_2 \times X_2$. Then $A \circ u := \pi_*\Delta^*(A \otimes u) \in \mathcal{B}(X_1)$ is well-defined and*

$$\begin{aligned} \text{supp } A \circ u &\subset \text{supp } A \circ \text{supp } u \\ &:= \{x \in X_1;\ (x,y) \in \text{supp } A \textit{ for some } y \in \text{supp } u\}, \\ WF_A(A \circ u) &\subset WF_A(A) \circ (WF_A(u)' \cup (\text{supp } u \times \{0\})), \end{aligned}$$

where $T' = \{(y,-\eta);\ (y,\eta) \in T\}$ and $S \circ T = \{(x,\xi);\ (x,y,\xi,\eta) \in S$ for some $(y,\eta) \in T\}$ for $S \subset T^*(X_1 \times X_2)$ and $T \subset T^*X_2$.

Theorem 3.4. 8 *Let X be an open subset of $\boldsymbol{R}^n$, and let $a(x,\xi) \in PS^+(X \times \boldsymbol{R}^n; R_0, C_a)$ $(\subset C^\infty(X; \cap_{\varepsilon>0} \widehat{\mathcal{S}}_{-\varepsilon}))$. Put*

$$K(x,y) = \mathcal{F}_\xi^{-1}[a(x,\xi)](y) \in C^\infty(X; \mathcal{E}_0(\boldsymbol{R}^n)),$$

which can be regard as an element in $\mathcal{B}(X \times \boldsymbol{R}^n)$. *Let* $h : X \times \boldsymbol{R}^n \ni (x, y) \mapsto (x, x - y) \in X \times \boldsymbol{R}^n$, *and put* $A(x, y)(\equiv K(x, x - y)) = h^*K \in \mathcal{B}(X \times \boldsymbol{R}^n)$. (i) *We have*

$$WF_A(A) \subset \{(x, x, \xi, -\xi) \in T^*(X \times \boldsymbol{R}^n);\ (x, \xi) \in T^*X \setminus 0\}.$$

(ii) *If* $u \in \mathcal{B}(X)$ *and* supp $u \subset\subset X$, *then* $A \circ u(\equiv \int K(x, x - y)u(y)\,dy = \pi_*(\Delta^*(A \otimes u)))$ $(\in \mathcal{B}(X))$ *is well-defined and* $A \circ u = a(x, D)u$, *where* $\Delta : X \times \boldsymbol{R}^n \ni (x, y) \mapsto (x, y, y) \in X \times \boldsymbol{R}^n \times \boldsymbol{R}^n$, $\pi : X \times \boldsymbol{R}^n \ni (x, y) \mapsto x \in X$ *and* $a(x, D)u$ *can be defined as in* Section 2.7.

Proof Let X_j $(j = 0, 1)$ be open subsets of X such that $X_0 \subset\subset X_1 \subset\subset X$. Choose $a^R(x, \xi) \in S^+(R, C_a')$ so that $a^R(x, \xi) = a(x, \xi)$ for $x \in X_0$ and supp $a^R \subset X_1 \times \boldsymbol{R}^n$. We put

$$K^R(x, y) = \mathcal{F}_\xi^{-1}[a^R(x, \xi)](y) \in C^\infty(\boldsymbol{R}^n; \mathcal{E}_0(\boldsymbol{R}_y^n)).$$

Then, $K^R(x, y) = b_T^R(x, y, D_y)\delta(y) \in \mathcal{E}_0(\boldsymbol{R}^{2n})$, where $T(x, y, \eta) = y \cdot \eta$ and $b^R(x, y, \eta) = a^R(x, \eta)$. It follows from Corollary 2.6. 2 that supp $K^R \subset \overline{X}_1 \times \boldsymbol{R}^n$ and

$$K^R(x, y)|_{X_0 \times R^n} = K(x, y)|_{X_0 \times R^n} \in \mathcal{B}(X_0 \times \boldsymbol{R}^n)$$

if $R \gg 1$. Put, for $\varepsilon > 0$,

$$K_\varepsilon^R(x, y) = \exp[-\varepsilon\langle D_y\rangle]K^R(x, y) \in \mathcal{S}(\boldsymbol{R}^{2n}).$$

Note that $K_\varepsilon^R \to K^R$ in $\mathcal{F}_0(\boldsymbol{R}^{2n})$ as $\varepsilon \downarrow 0$, and that

$$\mathcal{F}[K^R](\xi, \eta) = \int e^{-ix\cdot\xi}a^R(x, \eta)\,dx.$$

Theorem 3.2. 5 yields

$$WF_A(K) \subset \{(x, 0, 0, \eta) \in T^*(X \times \boldsymbol{R}^n);\ x \in X \text{ and } \eta \in \boldsymbol{R}^n \setminus \{0\}\}.$$

This, together with (3.45), proves the assertion (i). By Corollary 3.4. 7 $A \circ u$ is well-defined if $u \in \mathcal{B}(X)$ and supp $u \subset\subset X$. There is $C > 0$ such that

$$\left|K_\varepsilon^R(x, y)\right| \le C \tag{3.58}$$

for $(x, y) \in \boldsymbol{R}^{2n}$ with $|y| \ge 1$ and $0 < \varepsilon \le 1$ if $R \gg 1$, since

$$K_\varepsilon^R(x, y) = \sum_{k=1}^{\infty}(2\pi)^{-n}\int e^{iy\cdot\eta}L^k(e^{-\varepsilon\langle\eta\rangle}\psi_k^R(\eta)a^R(x, \eta))\,d\eta$$

for $|y| \geq 1$, where $L = |y|^{-2}\sum_{j=1}^n y_j D_{\eta_j}$. Applying Theorem 3.2. 5 and the same argument as used in Section 3.2 we can show that

$$\begin{aligned} WF_A(K^R) \subset \{(x,0,0,\eta) \in T^*\boldsymbol{R}^{2n}\setminus 0;\ x \in X_0\} \\ \cup\{(x,0,\xi,\eta) \in T^*\boldsymbol{R}^{2n}\setminus 0;\ x \in \overline{X}_1\setminus X_0\} \\ \cup\{(x,y,\xi,0) \in T^*\boldsymbol{R}^{2n}\setminus 0;\ x \in \overline{X}_1\setminus X_0\}. \end{aligned} \tag{3.59}$$

Let $\tilde{h} : \boldsymbol{R}^n \times \boldsymbol{R}^n \ni (x,y) \mapsto (x, x-y) \in \boldsymbol{R}^n \times \boldsymbol{R}^n$. Then we have

$$A(x,y)|_{X_0\times R^n} = (\tilde{h}^*K^R)|_{X_0\times R^n}\ (\in \mathcal{B}(X_0\times \boldsymbol{R}^n)).$$

Put $S(x,y,\xi,\eta) = x\cdot\xi + (x-y)\cdot\eta$ and $1(x,y,\xi,\eta) \equiv 1$. It follows from Corollary 2.6. 2 that

$$(\tilde{h}^*K^R)|_{X'\times Y'} = A^R(t,x,y)|_{X'\times Y'}$$

if $X'\times Y'$ is a bounded open subset of $\boldsymbol{R}^n\times\boldsymbol{R}^n$, where

$$A^R(x,y) = 1_S(x,y,D_w,D_z)K^R$$
$$\Big(= \lim_{\epsilon\downarrow 0}(2\pi)^{-2n}\int e^{ix\cdot\xi+i(x-y)\cdot\eta}\mathcal{F}[K_\epsilon^R](\xi,\eta)\,d\xi d\eta = \lim_{\epsilon\downarrow 0}K_\epsilon^R(x,x-y)\Big).$$

(3.44) gives supp A^R ($=$ supp $\tilde{h}^*K^R$) $\subset \overline{X}_1\times\boldsymbol{R}^n$. Note that $\mathcal{F}[A^R](\xi,\eta) = \mathcal{F}[K^R](\xi+\eta,-\eta)$. Now assume that $u \in \mathcal{B}(\boldsymbol{R}^n)$ and supp $u \subset\subset X_0$. By Theorem 1.3. 4 there is $v \in \mathcal{A}'(\boldsymbol{R}^n)$ such that $v|_Y = u|_Y$ for any bounded open subset Y of $\boldsymbol{R}^n$. From Theorem 3.4. 3 we have

$$\begin{aligned} &(A(x,y)\otimes u(w))|_{X_0\times R^n\times R^n} \\ &= (A^R|_{X_0\times R^n})\otimes u = (A^R\otimes v)|_{X_0\times R^n\times R^n}. \end{aligned}$$

Theorem 3.2. 5 and (3.59) give

$$\begin{aligned} WF_A(A^R) \subset \{(x,x,\eta,-\eta) \in T^*\boldsymbol{R}^{2n}\setminus 0;\ x \in X_0\} \\ \cup\{(x,x,\xi,\eta) \in T^*\boldsymbol{R}^{2n}\setminus 0;\ x \in \overline{X}_1\setminus X_0\} \\ \cup\{(x,y,\xi,0) \in T^*\boldsymbol{R}^{2n}\setminus 0;\ x \in \overline{X}_1\setminus X_0\}. \end{aligned}$$

Since $N_\Delta = \{(x,y,y,0,\eta,-\eta);\ x \in X,\ y \in \boldsymbol{R}^n$ and $\eta \in \boldsymbol{R}^n\setminus\{0\}\}$ and $N_\Delta \cap WF_A(A^R\otimes v) = \emptyset$, $\Delta^*(A^R\otimes v)$ ($\in \mathcal{B}(X\times\boldsymbol{R}^n)$) is well-defined and $\Delta^*(A\otimes u)|_{X_0\times R^n} = \Delta^*(A^R\otimes v)|_{X_0\times R^n}$. Put

$$\begin{aligned} F^R(x,y) &= \lim_{\epsilon\downarrow 0}F_\epsilon^R(x,y) \quad \text{in } \mathcal{F}_0(\boldsymbol{R}^{2n}), \\ F_\epsilon^R(x,y) &= (2\pi)^{-3n}\int e^{ix\cdot\xi+iy\cdot(\eta+\zeta)-\epsilon\langle\eta\rangle-\epsilon\langle\zeta\rangle} \\ &\qquad\times\mathcal{F}[K^R](\xi+\eta,-\eta)\hat{v}(\zeta)\,d\xi d\eta d\zeta \\ \Big(&= K_\epsilon^R(x,x-y)v_\epsilon(y)\Big), \end{aligned}$$

where $\varepsilon > 0$ and $v_\varepsilon(y) = e^{-\varepsilon\langle D\rangle}v$. Here we have applied the same argument as for (3.46). Moreover, we can show in a similar way that

$$\sup_{(\xi,\eta)\in R^{2n}} e^{-\delta\langle\xi\rangle-\delta\langle\eta\rangle}|\mathcal{F}[F^R - F^R_\varepsilon](\xi,\eta)| \to 0$$

as $\varepsilon \downarrow 0$ for any $\delta > 0$. Corollary 2.6. 2 yields

$$\Delta^*(A^R \otimes v)|_{X_0\times Y} = F^R|_{X_0\times Y}$$

for any bounded open subset Y of $\boldsymbol{R}^n$. Let $\tilde{\Delta} : \boldsymbol{R}^n \times \boldsymbol{R}^n \ni (x,y) \mapsto (x,y,y) \in \boldsymbol{R}^n \times \boldsymbol{R}^n \times \boldsymbol{R}^n$. Then we have

$$\mathrm{supp}\ F^R = \mathrm{supp}\ \tilde{\Delta}^*(A^R \otimes v) \subset \overline{X}_1 \times \mathrm{supp}\ v$$

and $F^R \in \mathcal{A}'(\boldsymbol{R}^{2n})$. In fact,

$$\langle F^R(x,y), \varphi(x,y)\rangle = \lim_{\varepsilon\downarrow 0} \int K^R_\varepsilon(x, x-y)v_\varepsilon(y)\varphi(x,y)\,dxdy$$

for $\varphi \in \mathcal{S}_\infty(\boldsymbol{R}^{2n})$. Since $v \in \mathcal{A}'(\boldsymbol{R}^n)$, (3.58) gives $|F^R_\varepsilon(x,y)| \le C$ if $y \notin V$ and $0 < \varepsilon \le 1$, where $C > 0$ and V is a neighborhood of $\overline{X}_1$. By Lebesgues' convergence theorem we have

$$\langle F^R(x,y), \varphi(x,y)\rangle = \lim_{\varepsilon\downarrow 0} \int_{\overline{X}_1\times V} F^R_\varepsilon(x,y)\varphi(x,y)\,dxdy$$

for $\varphi \in \mathcal{S}_\infty(\boldsymbol{R}^{2n})$. Applying the same argument as in the proof of the assertion (ii) of Theorem 3.3. 3 we have the same expression as (3.36), which implies that $F^R(x,y) \in \mathcal{A}'(\boldsymbol{R}^{2n})$. Recall that $\pi_* F^R$ ($\in \mathcal{A}'(\boldsymbol{R}^n)$) is defined by $(\pi_* F^R)(\varphi) = F^R(\varphi\circ\pi)$ for $\varphi \in \mathcal{A}(\boldsymbol{C}^n)$. By definition we have

$$(A\circ u)|_{X_0} \Big(= (\pi_*(\Delta^*(A\otimes u)))|_{X_0}\Big) = (\pi_* F^R)|_{X_0}.$$

Since

$$\mathcal{F}[\pi_* F^R](\xi) = F^R(e^{-ix\cdot\xi}) = \mathcal{F}[F^R](\xi, 0)$$

and

$$\begin{aligned}
\mathcal{F}[F^R](\xi,0) &= \lim_{\varepsilon\downarrow 0} \mathcal{F}[F^R_\varepsilon](\xi,0) \\
&= \lim_{\varepsilon\downarrow 0}(2\pi)^{-2n} \int e^{-ix\cdot\xi}\Big(\int e^{-ix\cdot\eta+iy\cdot(\eta+\zeta)-\varepsilon\langle\eta\rangle-\varepsilon\langle\zeta\rangle} \\
&\qquad\qquad \times a^R(x,-\eta)\hat{v}(\zeta)\,d\eta d\zeta\Big)\,dxdy \\
&= \lim_{\varepsilon\downarrow 0}(2\pi)^{-n} \int\Big(\int e^{-ix\cdot(\xi-\eta)-2\varepsilon\langle\eta\rangle} a^R(x,\eta)\hat{v}(\eta)\,d\eta\Big)\,dx,
\end{aligned}$$

we have

$$\begin{aligned} \pi_* F^R &= \lim_{\varepsilon \downarrow 0} (2\pi)^{-n} \int e^{ix\cdot\xi - 2\varepsilon\langle\xi\rangle} a^R(x,\xi)\hat{v}(\xi)\, d\xi \\ &= a^R(x,D)v \quad \text{in } \mathcal{F}_0(\boldsymbol{R}^n), \end{aligned}$$

which yields $A \circ u = a(x,D)u$ in $\mathcal{B}(X)$. □

3.5 Hyperfunctions supported by a half-space

Hyperfunctions with supports in a half-space have remarkable properties concerning their analytic wave front sets.

Theorem 3.5. 1 (Kashiwara-Kawai) *Let Ω ($\subset \boldsymbol{R}^n$) be an open neighborhood of the origin, and let $u \in \mathcal{B}(\Omega)$. If $0 \in \operatorname{supp} u \subset \{x \in \Omega;\ x_1 \ge 0\}$, then $(0, \pm e_1) \in WF_A(u)$, where $e_1 = (1, 0, \cdots, 0) \in \boldsymbol{R}^n$.*

Proof If $n = 1$, then Corollary 3.3. 4 proves the theorem. So we assume that $n \ge 2$, and that $0 \in \operatorname{supp} u \subset \{x_1 \ge 0\}$ and $(0, e_1) \notin WF_A(u)$. Moreover, we may assume that $u \in \mathcal{B}(\boldsymbol{R}^n)$ and $\operatorname{supp} u \subset\subset \boldsymbol{R}^n$. Let $\tau : \boldsymbol{R}^n \ni x \mapsto (x_1 - |x'|^2, x')$, where $x' = (x_2, \cdots, x_n)$. τ is called the Holmgren transformation. Since τ is an analytic diffeomorphism, it follows from (3.44) and (3.45) that $v \equiv \tau^* u$ ($\in \mathcal{B}(\boldsymbol{R}^n)$) satisfies $0 \in \operatorname{supp} v \subset \{x \in \boldsymbol{R}^n;\ x_1 \ge |x'|^2\}$ and

$$(0, e_1) \notin WF_A(v). \tag{3.60}$$

Let Γ_j ($1 \le j \le J$) be open convex proper cones in $\boldsymbol{R}^{n-1} \setminus \{0\}$ such that $\boldsymbol{R}^{n-1} \setminus \{0\} = \bigcup_{j=1}^J \Gamma_j$, and choose $\{g_j^R(\xi')\} \subset C^\infty(\boldsymbol{R}^{n-1})$ ($R \ge 2$, $1 \le j \le J$) so that $\operatorname{supp} g_j^R \cap \{|\xi'| \ge 1\} \subset \Gamma_j$, $\sum_{j=1}^J g_j^R(\xi') = 1$ for $\xi' \in \boldsymbol{R}^{n-1}$, and $|\partial_{\xi'}^{\alpha'+\gamma'} g_j^R(\xi')| \le C_{|\gamma'|}(C_*/R)^{|\alpha'|}\langle\xi'\rangle^{-|\gamma'|}$ if $\langle\xi'\rangle \ge R|\alpha'|$. Put

$$\begin{aligned} K_j^R(x) &= \mathcal{F}_\xi^{-1}\left[g_j^R(\xi')\right](x) \in \mathcal{S}'(\boldsymbol{R}^n), \\ \widetilde{K}_j^R(x') &= \mathcal{F}_{\xi'}^{-1}\left[g_j^R(\xi')\right](x') \in \mathcal{S}'(\boldsymbol{R}^{n-1}), \end{aligned}$$

where $1 \le j \le J$ and $\xi = (\xi_1, \xi')$. Then we have $K_j^R(x) = \delta(x_1) \otimes \widetilde{K}_j^R(x')$ and

$$\begin{aligned} &\operatorname{supp} K_j^R \subset \{x \in \boldsymbol{R}^n;\ x_1 = 0\}, \\ &WF_A(K_j^R) \subset \{(0,\xi) \in T^*\boldsymbol{R}^n \setminus 0;\ \xi' \in \Gamma_j \cup \{0\}\}, \end{aligned}$$

since $WF_A(\widetilde{K_j^R}) \subset \{0\} \times \Gamma_j$ ($\subset T^*\boldsymbol{R}^{n-1} \setminus 0$). Note that $\sum_{j=1}^J K_j^R(x) = \mathcal{F}^{-1}[1](x) = \delta(x)$ and the K_j^R can be regarded as elements of $\mathcal{B}(\boldsymbol{R}^n)$. Let $h : \boldsymbol{R}^n \times \boldsymbol{R}^n \ni (x,y) \mapsto x - y \in \boldsymbol{R}^n$, $\Delta : \boldsymbol{R}^n \times \boldsymbol{R}^n \ni (x,y) \mapsto (x,y,y) \in \boldsymbol{R}^n \times \boldsymbol{R}^n \times \boldsymbol{R}^n$ and $\pi : \boldsymbol{R}^n \times \boldsymbol{R}^n \ni (x,y) \mapsto x \in \boldsymbol{R}^n$. Then $\mathcal{K}_j^R(x,y) \equiv (h^*K_j^R)(x,y)$ ($\in \mathcal{B}(\boldsymbol{R}^n \times \boldsymbol{R}^n)$) is well-defined and

$$\begin{aligned}
&\mathrm{supp}\, \mathcal{K}_j^R \subset \{(x,y) \in \boldsymbol{R}^{2n};\ x_1 = y_1\},\\
&WF_A(\mathcal{K}_j^R) \subset \{(x,x,\xi,-\xi) \in T^*\boldsymbol{R}^{2n} \setminus 0;\ \xi' \in \Gamma_j \cup \{0\}\}.
\end{aligned}$$

Moreover, we have

$$\begin{aligned}
&\mathrm{supp}\, (\mathcal{K}_j^R \otimes v) \subset \mathrm{supp}\, \mathcal{K}_j^R \times \mathrm{supp}\, v,\\
&N_\Delta \cap WF_A(\mathcal{K}_j^R \otimes v) = \emptyset,
\end{aligned}$$

where $N_\Delta = \{(x,y,y,0,\xi,-\xi);\ x,y \in \boldsymbol{R}^n \text{ and } \xi \in \boldsymbol{R}^n\}$. Therefore, $\Delta^*(\mathcal{K}_j^R \otimes v)$ ($\in \mathcal{B}(\boldsymbol{R}^n \times \boldsymbol{R}^n)$) is well-defined and

$$\begin{aligned}
&\mathrm{supp}\, \Delta^*(\mathcal{K}_j^R \otimes v) \subset \{(x,y) \in \boldsymbol{R}^{2n};\ x_1 = y_1 \text{ and } y \in \mathrm{supp}\, v\},\\
&WF_A(\Delta^*(\mathcal{K}_j^R \otimes v)) \subset \{(x,x,\xi,-\xi+\eta) \in T^*\boldsymbol{R}^{2n} \setminus 0;\\
&\qquad \xi \neq 0,\ \xi' \in \Gamma_j \cup \{0\} \text{ and } (x,\eta) \in WF_A(v) \cup \mathrm{supp}\, v \times \{0\}\}\\
&\quad \cup \{(x,y,0,\eta) \in T^*\boldsymbol{R}^{2n} \setminus 0;\ x_1 = y_1 \text{ and } (y,\eta) \in WF_A(v)\}.
\end{aligned}$$

Since $\pi|_{\mathrm{supp}\, \Delta^*(\mathcal{K}_j^R \otimes v)}$ is proper, $v_j \equiv \pi_*(\Delta^*(\mathcal{K}_j^R \otimes v))$ ($\in \mathcal{B}(\boldsymbol{R}^n)$) is well-defined and

$$\begin{aligned}
&\mathrm{supp}\, v_j \subset \{x \in \boldsymbol{R}^n;\ x_1 \geq 0\},\\
&WF_A(v_j)\\
&\subset \{(x,\xi) \in T^*\boldsymbol{R}^n \setminus 0;\ \xi' \in \Gamma_j \cup \{0\} \text{ and } (x,\xi) \in WF_A(v)\}.
\end{aligned}$$

It follows from (3.60) that $\mathrm{ch}[\{\xi \in \boldsymbol{R}^n \setminus \{0\};\ \xi' \in \Gamma_j \cup \{0\} \text{ and } (0,\xi) \in WF_A(v)\}]$ is a proper convex cone. So there are a neighborhood U of the origin in $\boldsymbol{R}^n$ and an open convex cone G_j in $\boldsymbol{R}^n \setminus \{0\}$ such that

$$WF_A(v_j|_U) \subset U \times (G_j^* \setminus \{0\}).$$

Theorem 3.3. 3 (iv) implies that there is a G_j analytic function f_j at U satisfying $v_j|_U = b_{G_j} f_j$ in $\mathcal{B}(U)$. Moreover, Theorem 3.3. 1 (v) yields $v_j|_U = 0$. On the other hand, by Theorem 3.4. 8 we have $v \equiv \sum_{j=1}^J v_j$, which contradicts $0 \in \mathrm{supp}\, v$. Therefore, we have $(0,e_1) \in WF_A(u)$. Similarly we have $(0,-e_1) \in WF_A(u)$. □

Theorem 3.5. 1 , together with Theorem 3.2. 8 , proves the following Holmgren uniqueness theorem:

Theorem 3.5. 2 *Let $P(x,\xi)$ be a polynomial of ξ of degree m whose coefficients are real analytic functions of x defined in Ω. Assume that the principal part $P_m(x,\xi)$ of $P(x,\xi)$ satisfies $P_m(0,e_1) \neq 0$, where $e_1 = (1,0,\cdots,0) \in \boldsymbol{R}^n$. If $u \in \mathcal{B}(\Omega)$ satisfies $P(x,D)u = 0$ in $\mathcal{B}(\Omega)$ and* $\operatorname{supp} u \subset \{x_1 \geq 0\}$, *then $u = 0$ near the origin, i.e., $0 \notin \operatorname{supp} u$.*

Theorem 3.5. 3 (Kashiwara) *Let Ω ($\subset \boldsymbol{R}^n$) be an open neighborhood of the origin, and let $u \in \mathcal{B}(\Omega)$ satisfy* $\operatorname{supp} u \subset \{x \in \Omega;\, x_1 \geq 0\}$. *If there is $\xi^0 \in \boldsymbol{R}^n \setminus \{0\}$ such that $(0,\xi^0) \in WF_A(u)$, then $(0, te_1 + \xi^0) \in WF_A(u)$ for $t \in \boldsymbol{R}$ with $te_1 + \xi^0 \neq 0$, where $e_1 = (1,0,\cdots,0) \in \boldsymbol{R}^n$.*

Remark (i) The above theorem is called the watermelon-slicing theorem. (ii) We shall give the proof of the theorem, following [Hr5] (see, also, [Sj]). In doing so we shall implicitly prove that our definition of $WF_A(\cdot)$ is equivalent to the definition given by means of the FBI transformation. One can give a simpler proof if one use the boundary values of analytic functions. Although our proof here is a little long, it gives you a prototype of our calculus.

Proof We may assume that $u \in \mathcal{A}'$ and $\operatorname{supp} u \subset \{x_1 \geq 0\}$. Moreover, we assume that $\omega^0 \in S^{n-1}$ and $(0,\omega^0) \notin WF_A(u)$. In order to prove the theorem it suffices to show that $(0, te_1 + \omega^0) \notin WF_A(u)$ for any $t \in \boldsymbol{R}$. If ω^0 is parallel to e_1, then Theorem 3.5. 1 yields $0 \notin \operatorname{supp} u$. In particular, we have $(0,\xi) \notin WF_A(u)$ for $\xi \in \boldsymbol{R}^n \setminus \{0\}$. So we may assume that ω^0 is not parallel to e_1. Put

$$f_a(\xi;\lambda) = \langle u(y), \exp[-\lambda(iy\cdot\xi + ay^2)]\rangle_y$$

for $1 \leq a \leq 2$, $\lambda > 0$ and $\xi \in \boldsymbol{C}^n$. It is obvious that

$$\begin{aligned} &|f_a(ze_1+\omega;\lambda)| \\ &\leq C_\delta \sup_{\mathrm{Re}\, y_1 \geq -\delta,\, |\mathrm{Im}\, y| \leq \delta} |\exp[-\lambda(iy\cdot(ze_1+\omega) + ay^2)]| \\ &\leq \begin{cases} C'_\delta \exp[\lambda\{|\mathrm{Im}\, z|^2/4 + \delta(|\mathrm{Re}\, z| + 1 + a\delta)\}] & \text{if } \mathrm{Im}\, z \geq 0, \\ C'_\delta \exp[\lambda\delta\{|\mathrm{Im}\, z| + |\mathrm{Re}\, z| + 1 + a\delta)\}] & \text{if } \mathrm{Im}\, z < 0 \end{cases} \end{aligned}$$

for $z \in \boldsymbol{C}$, $\omega \in S^{n-1}$ and $\delta > 0$. By assumption there are a conic neighborhood Γ of ω^0 in $\boldsymbol{R}^n \setminus \{0\}$, a neighborhood U of 0 in $\boldsymbol{R}^n$ and $\{g^R(\xi)\}_{R \geq R_0} \subset C^\infty(\boldsymbol{R}^n)$ such that $g^R(\xi) = 1$ in $\Gamma \cap \{\langle\xi\rangle \geq R\}$, $|\partial_\xi^{\alpha+\gamma} g^R(\xi)| \leq C_{|\gamma|}(C/R)^{|\alpha|}$ if $\langle\xi\rangle \geq R|\alpha|$, and $g^R(D)u$ is analytic in U. We put

$$u_{1,\rho}^R = e^{-\rho\langle D\rangle} g^R(D)u, \qquad u_2^R = (1 - g^R(D))u,$$

where $\rho > 0$. Choose $\{\chi_k(x)\} \subset C_0^\infty(U)$ so that $\chi_k(x) = 1$ in U_0 and $|D^\beta \chi_k(x)| \le C(\hat{C}k)^{|\beta|}$ for $|\beta| \le k$, where U_0 is a neighborhood of 0. Applying the same argument as in the proof of Lemma 2.6. 4 , we have

$$\begin{aligned}&\left|e^{\rho\langle\xi\rangle}\psi_k^R(\xi)\mathcal{F}\left[\chi_k u_{1,\rho}^R\right](\xi)\right| \\ &\le C_R(u)k^{n+1}2^{-k}\langle\xi\rangle^{-n-1}\exp[(\rho-1/(3R))\langle\xi\rangle] \end{aligned} \tag{3.61}$$

if $R \ge R(u)$ and $0 < \rho \le 1$ where $R(u) \ge 1$. Let S_0 be an open subset of S^{n-1} such that $S_0 \subset\subset \Gamma \cap S^{n-1}$, and write

$$\begin{aligned} f_a(ite_1+\omega;\lambda) &= \langle u_{1,\rho}^R(y), e^{\rho\langle D_y\rangle}\exp[-\lambda(-ty_1+iy\cdot\omega+ay^2)]\rangle_y \\ &\quad +\langle \mathcal{F}[u_2^R](\xi), \mathcal{F}_y^{-1}(\exp[-\lambda(-ty_1+iy\cdot\omega+ay^2)])(\xi)\rangle_\xi \\ &=: F_{a,1}^R(t;\omega,\lambda)+F_{a,2}^R(t;\omega,\lambda), \\ F_{a,1}^R(t;\omega,\lambda) &= \sum_{k=1}^\infty \langle e^{\rho\langle\xi\rangle}\psi_k^R(\xi)\mathcal{F}[\chi_k u_{1,\rho}^R](\xi), \\ &\qquad \mathcal{F}_y^{-1}(\exp[-\lambda(-ty_1+iy\cdot\omega+ay^2)])(\xi)\rangle_\xi \\ &\quad +\sum_{k=1}^\infty \langle (1-\chi_k(y))u_{1,\rho}^R(u), e^{\rho\langle D_y\rangle}\psi_k^R(D_y)\exp[-\lambda(-ty_1+iy\cdot\omega+ay^2)])\rangle_y \\ &=: F_{a,1,1}^{R,\rho}(t;\omega,\lambda)+F_{a,1,2}^{R,\rho}(t;\omega,\lambda) \end{aligned}$$

for $\omega \in S_0$ and $t \in \boldsymbol{R}$. Noting that

$$\begin{aligned} &\mathcal{F}_y^{-1}(\exp[-\lambda(-ty_1+iy\cdot\omega+ay^2)])(\xi) \\ &= (4\pi\lambda a)^{-n/2}\exp[-(\xi-\lambda\omega-i\lambda te_1)^2/(4\lambda a)], \\ &|\mathcal{F}[u_2^R](\xi)| \le C_\delta \sup_{|\mathrm{Im}\, y|\le\delta} e^{\mathrm{Im}\, y\cdot\xi} \le C_\delta e^{\delta|\xi|} \quad \text{for } \delta>0 \end{aligned}$$

and $\mathcal{F}[u_2^R](\xi) = 0$ for $\xi \in \Gamma \cap \{\langle\xi\rangle \ge R\}$, we have, with some positive constant c,

$$\left|F_{a,2}^R(t;\omega,\lambda)\right| \le C_R\lambda^{-n/2}\exp[-(c-t^2)\lambda/(4a)]$$

for $\omega \in S_0$. Moreover, (3.61) yields

$$\left|F_{a,1,1}^{R,\rho}(t;\omega,\lambda)\right| \le C_R'\lambda^{-n/2}\exp[-(a/(2R)-t^2)\lambda/(4a)]$$

if $R \ge R(u)$ and $\rho \le 1/(6R)$. Indeed, we have

$$\begin{aligned} &-|\xi|/(6R)-(|\xi|^2+\lambda^2-2\lambda|\xi|-\lambda^2t^2)/(4\lambda a) \\ &\le -\lambda(1-t^2-(1-a/(3R))^2)/(4a) \le -(a/(2R)-t^2)\lambda/(4a) \end{aligned}$$

if $R \geq 2$. We can choose $\psi^R(\xi;\lambda) \in C^\infty(\boldsymbol{R}^n)$ ($\lambda > 0$) so that

$$\psi^R(\xi;\lambda) = \begin{cases} 1 & \text{if } \lambda/2 \leq |\xi| \leq 2\lambda \text{ and } \langle\xi\rangle \geq R, \\ 0 & \text{if } |\xi| \leq \lambda/4 \text{ or } |\xi| \geq 4\lambda, \end{cases}$$

$$\left|\partial_\xi^{\alpha+\gamma}\psi^R(\xi;\lambda)\right| \leq C_{|\gamma|}(C_*/R)^{|\alpha|}\langle\xi\rangle^{-|\gamma|} \quad \text{if } \langle\xi\rangle \geq R|\alpha|$$

(see Lemma 2.1. 3). Choose $c_1 > 0$ so that $U_1 \equiv \{y \in \boldsymbol{R}^n;\ |y| < c_1\} \subset\subset U_0$, and write

$$\begin{aligned}
&e^{\rho\langle D_y\rangle}\psi_k^R(D_y)\exp[-\lambda(-ty_1 + iy\cdot\omega + ay^2)] \\
&= \int e^{iy\cdot\xi+\rho\langle\xi\rangle}\psi_k^R(\xi)(1-\psi^R(\xi;\lambda)) \\
&\qquad \times(4\pi\lambda a)^{-n/2}\exp[-(\xi+\lambda\omega+i\lambda te_1)^2/(4\lambda a)]\,d\xi \\
&\quad +(2\pi)^{-n}\int_{w\notin U_1}\Big(\int \exp[i(y-w)\cdot\xi+\rho\langle\xi\rangle-\lambda(-tw_1+iw\cdot\omega+aw^2)] \\
&\qquad\qquad \times\psi_k^R(\xi)\psi^R(\xi;\lambda)\,d\xi\Big)\,dw \\
&\quad +(2\pi)^{-n}\int_{w\in U_1}\Big(\int \exp[i(y-w)\cdot\xi-\lambda(-tw_1+iw\cdot\omega+aw^2)] \\
&\qquad \times\sum_{|\alpha|=k}(k!/\alpha!)(w-y)^\alpha|y-w|^{-2k}D_\xi^\alpha\left(e^{\rho\langle\xi\rangle}\psi_k^R(\xi)\psi^R(\xi;\lambda)\right)d\xi\Big)\,dw \\
&=: \sum_{j=1}^{3} I_{a,k,j}^{R,\rho}(y;t,\omega,\lambda)
\end{aligned}$$

for $y \notin U_0$. Then it is obvious that

$$\left|I_{a,k,1}^{R,\rho}(y;t,\omega,\lambda)\right| \leq C_R k^{-2}\exp[-(1/12-t^2)\lambda/(4a)]$$

if $\rho \leq 1/48$ and $\lambda \geq 1$, and

$$\left|I_{a,k,2}^{R,\rho}(y;t,\omega,\lambda)\right| \leq C_R k^{-2}\exp[-ac_1^2\lambda/12]$$

if $\rho \leq ac_1^2/48$, $|t| \leq ac_1/4$ and $\lambda \geq 1$. Moreover, we have

$$\left|I_{a,k,3}^{R,\rho}(y;t,\omega,\lambda)\right| \leq C_R k^{-2}\exp[-(1/(6R)-t^2)\lambda/4]$$

if $y \notin U_0$, $\rho \leq 1/(12R)$ and $R \geq 2e\sqrt{n}(\hat{C}+C_*+3(1+\sqrt{2}))/c_0$, where $c_0 = \mathrm{dis}(U_1, \boldsymbol{R}^n \setminus U_0)$ (> 0). Therefore, there are positive constants $\hat{c} \equiv \hat{c}(u)$ and $\hat{a} \equiv \hat{a}(u)$ such that $\hat{a} \leq 1$ and

$$|f_a(ite_1+\omega;\lambda)| \leq C(u)\exp[-\hat{c}\lambda]$$

if $\omega \in S_0$, $\lambda \geq 1$, $t \in \boldsymbol{R}$, $|t| \leq \hat{a}$ and $1 \leq a \leq 2$. Put

$$g_a(z;\omega,\lambda) = \lambda^{-1} \log |f_a(ze_1 + \omega;\lambda)|$$

for $z \in \boldsymbol{C}$, $\omega \in S_0$, $\lambda \geq 1$ and $1 \leq a \leq 2$. Then $g_a(z;\omega,\lambda)$ is subharmonic in z, and for any $T > 0$ and any $\delta > 0$ there is $\lambda_{\delta,T} \geq 1$ such that

$$\begin{aligned} &g_a(z;\omega,\lambda) \leq (\text{Im } z)_+^2/4 + \delta \quad \text{when } |\text{Re } z| \leq T \text{ and } |\text{Im } z| \leq 1, \\ &g_a(it;\omega,\lambda) \leq -\hat{c} + \delta \quad \text{for } t \in \boldsymbol{R} \text{ with } |t| \leq \hat{a} \end{aligned}$$

if $\lambda \geq \lambda_{\delta,T}$. Let $t_0 \in \boldsymbol{R}$. Let us prove that $(0, t_0 e_1 + \omega^0) \notin WF_A(u)$. We may assume that $t_0 > 0$. Put

$$\begin{aligned} &D_\nu = \{t + i\tau \in \boldsymbol{C};\ 0 < t < 2t_0 \text{ and } -\hat{a} + \nu < \tau < \nu\}, \\ &G_{a,\nu}(z;\omega,\lambda) = g_a(z;\omega,\lambda) - \hat{c}_0 - \nu^2/4 \\ &\quad + \hat{c} \sinh(\pi(2t_0 - t)/\hat{a}) \sin(\pi(\nu - \tau)/\hat{a}) / \sinh(2\pi t_0/\hat{a}) \end{aligned}$$

for $0 < 2\nu \leq \hat{a}$ and $z = t + i\tau \in D_\nu$, where

$$\begin{aligned} \hat{c}_0 \quad = \quad & \min\{\hat{a}, 8\hat{c} \sinh(\pi t_0/\hat{a})/(\hat{a} \sinh(2\pi t_0/\hat{a}))\} \\ & \times \hat{c} \sinh(\pi t_0/\hat{a})/(4\hat{a} \sinh(2\pi t_0/\hat{a})). \end{aligned}$$

Then $G_{a,\nu}$ is subharmonic in z, and $G_{a,\nu}(z;\omega,\lambda) \leq 0$ if $z \in \partial D_\nu$, $\omega \in S_0$, $1 \leq a \leq 2$, $0 < 2\nu \leq \hat{a}$ and $\lambda \geq \lambda_{\hat{c}_0, 2t_0}$. This yields

$$g_a(t_0;\omega,\lambda) \leq \hat{c}_0 + \nu^2/4 - 2\nu \hat{c} \sinh(\pi t_0/\hat{a})/(\hat{a} \sinh(2\pi t_0/\hat{a}))$$

if $\omega \in S_0$, $1 \leq a \leq 2$, $0 < 2\nu \leq \hat{a}$ and $\lambda \geq \lambda_{\hat{c}_0, 2t_0}$. Therefore, we have

$$|f_a(t_0 e_1 + \omega;\lambda)| \leq \exp[-\hat{c}_0 \lambda] \tag{3.62}$$

if $\omega \in S_0$, $1 \leq a \leq 2$ and $\lambda \geq \lambda_{\hat{c}_0, 2t_0}$. For $\tau > 0$ we define M_τ : $\boldsymbol{R}^n \ni x \mapsto \tau x \in \boldsymbol{R}^n$. Put

$$u_\tau(x) = (M_{1/\tau}^* u)(x) \left(= u(x/\tau)\right)$$

for $\tau > 0$. Then we have

$$\begin{aligned} &\langle u_\tau(y), \exp[-iy \cdot \xi - \varepsilon|\xi|y^2] \rangle_y \\ &= \tau^n \langle u(y), \exp[-\tau i y \cdot \xi - \varepsilon \tau^2 |\xi| y^2] \rangle_y \end{aligned} \tag{3.63}$$

for $\varepsilon > 0$. Let S_1 be a neighborhood of ω^0 in S^{n-1} such that $S_1 \subset S_0$ and

$$2|t_0 e_1 + \omega^0|/3 \leq |t_0 e_1 + \omega| \leq 4|t_0 e_1 + \omega^0|/3 \quad \text{for } \omega \in S_1.$$

We put $\Gamma_1 = \{\lambda(t_0 e_1 + \omega);\ \lambda > 0$ and $\omega \in S_1\}$. It is obvious that $1 \le a \le 2$ if $\omega \in S_1$, $\lambda > 0$, $\varepsilon > 0$, $\lambda(t_0 e_1 + \omega) = \tau\xi$, $\lambda a = \varepsilon\tau^2|\xi|$ and $\tau = (2\varepsilon|t_0 e_1 + \omega^0|/3)^{-1}$. This, together with (3.62) and (3.63), gives

$$|\langle u_\tau(y), \exp[-iy\cdot\xi - \varepsilon|\xi|y^2]\rangle_y| \le C_{\varepsilon,t_0}(u)\exp[-\hat{c}_0'|\xi|/\varepsilon]$$

if $\xi \in \Gamma_1$, $\varepsilon > 0$ and $\tau = (2\varepsilon|t_0 e_1 + \omega^0|/3)^{-1}$, where $\hat{c}_0' = 9\hat{c}_0/(8|t_0 e_1 + \omega^0|^2)$. Let Γ_2 be a conic neighborhood of $t_0 e_1 + \omega^0$ satisfying $\Gamma_2 \subset\subset \Gamma_1$, and choose $\Phi^R(\xi, y, \eta) \subset S^{0,0}(R, A)$ so that $\Phi^R(\xi, y, \eta) = 1$ if $|y| \le \sqrt{\hat{c}_0'}/2$, $\eta \in \Gamma_2$ and $\langle\eta\rangle \ge R$ (≥ 2), and $\operatorname{supp}\Phi^R \subset \{(\xi, y, \eta) \in \boldsymbol{R}^n \times \boldsymbol{R}^n \times \boldsymbol{R}^n;\ |y| \le \sqrt{\hat{c}_0'}/2,\ \eta \in \Gamma_1$ and $\langle\eta\rangle \ge 1\}$. We assume that $0 < \varepsilon \le 1$ and $\tau = (2\varepsilon|t_0 e_1 + \omega^0|/3)^{-1}$. Put

$$\begin{aligned} &S_\varepsilon(y,\eta) = y\cdot\eta - i\varepsilon|\eta|y^2, \\ &v_\varepsilon(x) = 1'_{-S_\varepsilon}(D_x, y, D_y)u_\tau \\ &\Big(= \mathcal{F}_\xi^{-1}[\langle u_\tau(y), \exp[-iS_\varepsilon(y,\xi)]\rangle_y](x)\Big), \end{aligned}$$

where $1(\xi, y, \eta) \equiv 1$. Then it follows from Theorem 2.3. 3 that $\Phi^R_{S_\varepsilon}(D_x, y, D_z)v_\varepsilon$ is analytic if $R \ge R(S_\varepsilon)A$. Here $R(S_\varepsilon)$ depends only on $C(-S_\varepsilon)$ and $c_j(-S_\varepsilon)$ ($1 \le j \le 3$) with respect to S_ε (see the proof of Theorem 2.3. 3). Therefore, we can take $R(S_\varepsilon)$ to be independent of ε. Let $\Psi^R(\xi, y, \eta) \in S^{0,0}(R, A)$ and $\psi(\xi) \in C^\infty(\boldsymbol{R}^n)$ be symbols such that $\Psi^R(\xi, y, \eta) = 1$ if $|y| \le \sqrt{10\hat{c}_0'}$, $\Psi^R(\xi, y, \eta) = 0$ if $|y| \ge 4\sqrt{\hat{c}_0'}$, $\psi(\xi) = 1$ if $|\xi| \ge 2$, and $\psi(\xi) = 0$ if $|\xi| \le 1$. Put

$$\begin{aligned} &a^R(\xi, y, \eta) = \Psi^R(\xi, y, \eta)\psi(\xi), \\ &b^R_\varepsilon(\xi, y, \eta) = (1 - \Psi^R(\xi, y, \eta))\psi(\xi)e^{-\varepsilon|\xi|y^2}, \\ &\tilde{b}_\varepsilon(\xi, y) = (1 - \psi(\xi))e^{-\varepsilon|\xi|y^2}. \end{aligned}$$

It is obvious that $\tilde{b}_\varepsilon(D_x, y)u_\tau \in \mathcal{S}_\infty$ ($\subset \bigcap_\delta \mathcal{S}_\delta'$). Since

$$\begin{aligned} \left|\partial_\xi^\alpha D_y^\beta \partial_\eta^\gamma b^R_\varepsilon(\xi, y, \eta)\right| \ \le\ & C_{|\alpha|+|\gamma|}((A + 2(\sqrt{2} + \sqrt{3}))/R)^{|\beta|} \\ & \times\langle\eta\rangle^{|\beta|}\exp[-2\varepsilon\hat{c}_0'\langle\xi\rangle] \end{aligned}$$

if $\langle\eta\rangle \ge R|\beta|$, Theorem 2.3. 3 yields $b^R_\varepsilon(D_x, y, D_y)u_\tau \in \mathcal{S}'_{-\varepsilon\hat{c}_0'}$ if $R \ge 2en(A + 2(\sqrt{2} + \sqrt{3}))$. Therefore, by Theorem 2.3. 3 we see that

$$\Phi^R_{S_\varepsilon}(D_x, y, D_z)\Big(b^R_\varepsilon(D_x, y, D_y) + \tilde{b}_\varepsilon(D_x, y)\Big)u_\tau$$

is analytic if $R \geq R(S_\varepsilon)A$ and $R \geq 2en(A+2(\sqrt{2}+\sqrt{3}))$, modifying $R(S_\varepsilon)$ if necessary. This implies that $\Phi^R_{S_\varepsilon}(D_x,y,D_z)a^{R\prime}_{-S_\varepsilon}(D_x,y,D_y)u_\tau$ is analytic if $R \geq R(S_\varepsilon)A$ and $R \geq 2en(A+2(\sqrt{2}+\sqrt{3}))$. Now we apply Corollary 2.4. 5 to $\Phi^R_{S_\varepsilon}(D_x,y,D_z)a^{R\prime}_{-S_\varepsilon}(D_x,y,D_y)$. Let $Z(z;w,y)$ be the inverse function defined in Section 2.4 with $U(y,\zeta) = -S_\varepsilon(y,\zeta)$. Note that $Z(z;w,y)$ also depends on ε. We put

$$p^R(\xi,w,\zeta,y,\eta) = \Phi^R(\xi,w,\zeta)a^R(\zeta,y,\eta),$$
$$p^R_\kappa(\xi,w,\zeta,y,\eta;u) = \sum_{j=1}^\infty \psi_j^{\kappa R}(\eta) \sum_{|\alpha|\leq j-1} u^\alpha \partial_\zeta^\alpha p(\xi,w,\zeta,y,\eta)/\alpha!.$$

Then there are positive constants $R(S_\varepsilon,A)$, $\kappa_0(S_\varepsilon,A)$, $\kappa(S_\varepsilon,A)$ and $\delta(S_\varepsilon, A)$ such that

$$\Phi^R_{S_\varepsilon}(D_x,y,D_z)a^{R\prime}_{-S_\varepsilon}(D_x,y,D_y) = p^{R,\varepsilon}_\kappa(D_x,y,D_y) + Q^{R,\varepsilon} \quad \text{on } \mathcal{S}_\infty$$

and $Q^{R,\varepsilon}(\mathcal{F}_0) \subset \bigcup_{\delta>0} \mathcal{S}'_{-\delta}$ if

$$R \geq R(S_\varepsilon,A), \quad \kappa = \kappa_0(S_\varepsilon,A), \quad \varepsilon \leq \delta(S_\varepsilon,A)/R, \tag{3.64}$$

where $R' = \kappa(S_\varepsilon,A)R$ and

$$\begin{aligned} p^{R,\varepsilon}_\kappa(\xi,w,\eta) &= \sum_{j=1}^\infty \psi_j^{R'}(\eta) \sum_{|\gamma|<j} \frac{(-1)^n}{\gamma!}\Big[(-\partial_\zeta)^\gamma D_y^\gamma \\ &\times\Big\{p^R_\kappa(\xi,w,-\zeta+\eta,w+y,\eta;Z(-\zeta+\eta;w,w+y)+\zeta-\eta) \\ &\times \det\frac{\partial Z}{\partial z}(-\zeta+\eta;w,w+y)\Big\}\Big]_{y=0,\zeta=0}. \end{aligned}$$

From the proof of Corollary 2.4. 5 (and Theorem 2.4. 4) we see that the constants $R(S_\varepsilon,A)$, $\kappa_0(S_\varepsilon,A)$, $\kappa(S_\varepsilon,A)$ and $\delta(S_\varepsilon,A)$ depend only on $C(-S_\varepsilon)$, $c_1^\pm(-S_\varepsilon)$, $c_j(-S_\varepsilon)$ ($j=0,2,3$) and $c(-S_\varepsilon)$ with respect to S_ε. So we can assume that these constants are independent of ε. Moreover, we have

$$\begin{aligned} &\left|\partial_\xi^{\alpha+\tilde\alpha} D_y^{\beta^1+\beta^2+\tilde\beta} \partial_\eta^{\rho+\tilde\rho} p^{R,\varepsilon}_\kappa(\xi,y,\eta)\right| \leq C_{|\tilde\alpha|+|\tilde\beta|+|\tilde\rho|}(A'(A)/R)^{|\alpha|+|\beta^1|+|\beta^2|+|\rho|} \\ &\quad\times\langle\xi\rangle^{-|\tilde\alpha|+|\beta^1|}\langle\eta\rangle^{-|\tilde\rho|+|\beta^2|+2}\exp[(\nu+B(A)/R)\langle\eta\rangle/R] \end{aligned}$$

if (3.64) is valid, $\nu>0$, $\langle\xi\rangle \geq C'(A)R(|\alpha|+|\beta^1|)$, $\langle\eta\rangle \geq C'(A)R(|\beta^2|+|\rho|)$ and $\varepsilon \leq \delta''(A,\nu)$, where $A'(A)$, $B(A)$, $C'(A)$ and $\delta''(A,\nu)$ are positive constants independent of ε. If $\zeta \in \Gamma_2$, $\langle\zeta\rangle \geq R$, $|w| \leq \sqrt{\hat{c}_0'}/2$ and $|y| \leq \sqrt{10\hat{c}_0'}$, then $p_\kappa^R(\xi, w, \zeta, y, \eta; u) = 1$. Hence we have

$$p_\kappa^{R,\varepsilon}(\xi, w, \eta) = \sum_{j=1}^{\infty} \phi_j^{R'}(\eta) \sum_{|\gamma|=j} \frac{(-1)^n}{\gamma!} \Big[\partial_\eta^\gamma D_y^\gamma \det \frac{\partial Z}{\partial z}(\eta; w, w+y)\Big\}\Big]_{y=0}$$

if $\eta \in \Gamma_2$, $\langle\eta\rangle \geq R$ and $|w| \leq \sqrt{\hat{c}_0'}/2$. It follows from Lemma 2.4. 2 and its proof that

$$\begin{aligned} &\Big|D_w^\beta \partial_\eta^{\rho+\gamma}\Big[D_y^\gamma \det \frac{\partial Z}{\partial z}(\eta; w, w+y)\Big\}\Big]_{y=0}\Big| \\ &\leq C A_0^{|\beta|+|\rho|+2|\gamma|} |\beta|!|\rho|!|\gamma|!^2 \langle\eta\rangle^{-|\rho|-|\gamma|}, \\ &\Big|\det \frac{\partial Z}{\partial z}(\eta; w, w) - 1\Big| \leq 1/2 \end{aligned}$$

if $|w| \leq \sqrt{\hat{c}_0'}/2$ and $\varepsilon \ll 1$, where A_0 is a positive constant independent of ε. Lemma 2.2. 4 and its proof give

$$\begin{aligned} &\Big|D_w^\beta \partial_\eta^{\rho+\tilde{\rho}}\Big\{p_\kappa^{R,\varepsilon}(\xi, w, \eta) - (-1)^n \det \frac{\partial Z}{\partial z}(\eta; w, w)\Big\}\Big| \\ &\leq C_{R,|\tilde{\rho}|} A'^{|\beta|+|\rho|} |\beta|!|\rho|! \langle\eta\rangle^{-1-|\rho|-|\tilde{\rho}|} \end{aligned}$$

if $\eta \in \Gamma_2$, $\langle\eta\rangle \geq R$, $\langle\eta\rangle \geq 2R'|\rho|$, $|w| \leq \sqrt{\hat{c}_0'}/2$, $R'(= \kappa(S_\varepsilon, A)R) \geq \sqrt{n}A_0^2$, $\varepsilon \ll 1$ and $A' \geq 2A_0 + 3\widehat{C}$. Let Γ_j ($3 \leq j \leq 4$) be conic neighborhoods of $t_0 e_1 + \omega^0$ satisfying $\Gamma_4 \subset\subset \Gamma_3 \subset\subset \Gamma_2$, and choose $\Phi_1^R(\xi, y, \eta) \in S^{0,0}(R', A)$ such that $\Phi_1^R(\xi, y, \eta) = 1$ if $|y| \leq \sqrt{\hat{c}_0'}/4$, $\eta \in \Gamma_3$ and $\langle\eta\rangle \geq 2\max\{R, R'\}$, and supp $\Phi_1^R \subset \{(\xi, y, \eta) \in \boldsymbol{R}^n \times \boldsymbol{R}^n \times \Gamma_2;\ |y| \leq \sqrt{\hat{c}_0'}/2$ and $\langle\eta\rangle \geq R\}$. We put

$$\begin{aligned} p_1(\xi, y, \eta) &= p_\kappa^{R,\varepsilon}(\xi, y, \eta)\Phi_1^R(\xi, y, \eta), \\ p_2(\xi, y, \eta) &= p_\kappa^{R,\varepsilon}(\xi, y, \eta)(1 - \Phi_1^R(\xi, y, \eta)). \end{aligned}$$

Then we have $p_1(\xi, y, \eta) \in S^{0,0}(2R', 2A_0 + 3\widetilde{C} + A)$ if $R \gg 1$ and $\varepsilon \ll 1$. It follows from Corollary 2.6. 3 that

$$WF_A(p_2(D_x, y, D_y)u_\tau) \cap \{(y, \eta) \in \boldsymbol{R}^n \times \Gamma_4;\ |y| < \sqrt{\hat{c}_0'}/8\} = \emptyset$$

if $R \gg 1$ and $\varepsilon \ll 1$. This yields $(0, t_0 e_1 + \omega^0) \notin WF_A(p_1(D_x, y, D_y)u_\tau)$ if $R \gg 1$ and $\varepsilon \ll 1$. So, applying Theorem 3.2. 8 we have $(0, t_0 e_1 + \omega^0) \notin WF_A(u_\tau)$, *i.e.*, $(0, t_0 e_1 + \omega^0) \notin WF_A(u)$. This completes the proof. □

3.6 Microfunctions

Let $\mathcal{U}$ be an open subset of the cosphere bundle $S^*\boldsymbol{R}^n$ over $\boldsymbol{R}^n$, which is identified with $\boldsymbol{R}^n \times S^{n-1}$, and define

$$\mathcal{C}(\mathcal{U}) := \mathcal{B}(\boldsymbol{R}^n)/\{u \in \mathcal{B}(\boldsymbol{R}^n);\ WF_A(u) \cap \mathcal{U} = \emptyset\}.$$

Since $\mathcal{B}$ is a flabby sheaf (see Theorem 1.4. 8), we have

$$\mathcal{C}(\mathcal{U}) = \mathcal{B}(U)/\{u \in \mathcal{B}(U);\ WF_A(u) \cap \mathcal{U} = \emptyset\}.$$

if $\mathcal{U}$ is an open subset of $\boldsymbol{R}^n \times S^{n-1}$, U is an open subset of $\boldsymbol{R}^n$ and $\mathcal{U} \subset U \times S^{n-1}$. Elements of $\mathcal{C}(\mathcal{U})$ are called microfunctions on $\mathcal{U}$. We define the presheaf $\mathcal{C}$ by $\mathcal{U} \to \mathcal{C}(\mathcal{U})$ for every open subset $\mathcal{U}$ of $\boldsymbol{R}^n \times S^{n-1}$ (see Definition 1.4. 6). Let $\overline{\mathcal{C}}$ denote the sheaf associated with the presheaf $\mathcal{C}$. We shall show that $\overline{\mathcal{C}}$ can be identified with $\mathcal{C}$ (see Theorem 3.6. 1 below). For $f \in \overline{\mathcal{C}}(\mathcal{U})$ with an open subset $\mathcal{U}$ of $\boldsymbol{R}^n \times S^{n-1}$ we denote by supp f the support of f as a section of $\overline{\mathcal{C}}$, *i.e.*, the complement in $\mathcal{U}$ of the largest open subset of $\mathcal{U}$ where f is equal to 0.

Theorem 3.6. 1 (Kashiwara) (i) $\overline{\mathcal{C}}$ *is a flabby sheaf.* (ii) *If U is an open subset of $\boldsymbol{R}^n$, then we have $\overline{\mathcal{C}}(U \times S^{n-1}) = \mathcal{C}(U \times S^{n-1})$.* (iii) *$\overline{\mathcal{C}} = \mathcal{C}$, i.e., $\mathcal{C}$ is a sheaf.* (iv) *For each open subset U of $\boldsymbol{R}^n$ we can define the mapping* sp : $\mathcal{B}(U) \to \mathcal{C}(U \times S^{n-1})$ *such that the residue class in $\mathcal{C}(U \times S^{n-1})$ of $f \in \mathcal{B}(U)$ is equal to* sp(f). *Moreover,* sp *is surjective (an onto-mapping) and*

$$WF_A(f) \cap U \times S^{n-1} = \text{supp sp}(f) \quad \textit{for } f \in \mathcal{B}(U).$$

Proof Following [Kn] we shall give a proof of the theorem. Let Φ : $\mathcal{F}_0(\boldsymbol{R}^n) \to C(S^{n-1}_\omega; \mathcal{F}_0(\boldsymbol{R}^n))$ be the operator defined by

$$(\Phi u)(x, \omega) = \mathcal{F}_\xi^{-1}[e^{\langle\xi\rangle/2 + \omega\cdot\xi/2}\langle\xi\rangle^{n+2} I(\xi)^{-1}\hat{u}(\xi)](x)$$

for $u \in \mathcal{F}_0$ and $\omega \in S^{n-1}$, where $I(\xi)$ is as defined in Section 3.1. We denote by dS the surface element of S^{n-1}. Fix $\omega^0 \in S^{n-1}$. We may assume that $\omega^0 = (0, \cdots, 0, 1)$. We choose $S_0 = \{\omega \in S^{n-1};\ \omega_n > 1/2\}$ as a coordinate patch containing ω^0, and $\kappa : S_0 \ni \omega \mapsto \omega' = (\omega_1, \cdots, \omega_{n-1}) \in B_0$ as a local coordinate system, where $B_0 = \{y' = (y_1, \cdots, y_{n-1}) \in \boldsymbol{R}^{n-1};\ |y'| < \sqrt{3}/2\}$. Let $f(x, \omega) \in C(\overline{S}_0; \mathcal{F}_0(\boldsymbol{R}^n))$, and define $\tilde{f}(x, y') \in \mathcal{F}_0(\boldsymbol{R}^{2n-1})$ by

$$\langle \tilde{f}(x, y'), \varphi(x, y')\rangle = \int_{B_0} \langle f(x, \kappa^{-1}(y')), \varphi(x, y')\rangle_x \, dy'$$

for $\varphi \in \mathcal{S}_\infty(\boldsymbol{R}^{2n-1})$. We identify $f(x,\omega)$ with $\tilde{f}(x,y')$ in this local coordinate system κ. Moreover, $\tilde{f}(x,y')$ determines uniquely an element of $\mathcal{B}(\boldsymbol{R}^{2n-1})$, whose support is included in $\boldsymbol{R}^n \times \overline{B}_0$, and, therefore, one of $\mathcal{B}(\boldsymbol{R}^n \times S_0)$. We identify $f(x,\omega)$ with the corresponding element in $\mathcal{B}(\boldsymbol{R}^n \times S_0)$. Similarly, we can regard an element of $C(S^{n-1};\mathcal{F}_0(\boldsymbol{R}^n))$ as one of $\mathcal{B}(\boldsymbol{R}^n \times S^{n-1})$. In particular, we identify $(\Phi u)(x,\omega)$ with an element of $\mathcal{B}(\boldsymbol{R}^n \times S^{n-1})$ for $u \in \mathcal{F}_0$. Let $u \in \mathcal{F}_0$. Choose $\chi(\xi) \in C^\infty(\boldsymbol{R}^n)$ so that $\chi(\xi) = 1$ for $|\xi| \geq 2$ and $\chi(\xi) = 0$ for $|\xi| \leq 1$, and put

$$\begin{aligned}(\Phi_0 u)(x,\omega) &= \mathcal{F}_\xi^{-1}[e^{\langle\xi\rangle/2+\omega\cdot\xi/2}(1-\chi(\xi))\langle\xi\rangle^{n+2}I(\xi)^{-1}\hat{u}(\xi)](x),\\ (\Phi_1 u)(x,\omega) &= (\Phi u)(x,\omega) - (\Phi_0 u)(x,\omega).\end{aligned}$$

It is obvious that $WF_A(\Phi_0 u) = \emptyset$. Here we have regarded $\Phi_0 u$ as an element of $\mathcal{B}(\boldsymbol{R}^n \times S^{n-1})$. Let $\psi(\omega) \in C(S^{n-1})$ satisfy $\psi(\omega) = 1$ for $\omega \in S^{n-1}$ with $\omega_n \geq 2/3$ and $\psi(\omega) = 0$ for $\omega \notin S_0$, and put

$$\begin{aligned}&f_1(x,\omega) = \psi(\omega)(\Phi_1 u)(x,\omega) \ \in C(S_\omega^{n-1};\mathcal{F}_0(\boldsymbol{R}^n)),\\ &\tilde{f}_1(x,y') = \psi(\kappa^{-1}(y'))\mathcal{F}_\xi^{-1}[e^{iT(x,y',\xi)-ix\cdot\xi}a(\xi)\hat{u}(\xi)](x) \ \in \mathcal{F}_0(\boldsymbol{R}^{2n-1}),\end{aligned}$$

where $T(x,y',\xi) = x\cdot\xi + i(|\xi| - \kappa^{-1}(y')\cdot\xi)/2$ and

$$a(\xi) = e^{(\langle\xi\rangle+|\xi|)/2}\langle\xi\rangle^{n+2}\chi(\xi)/I(\xi).$$

Note that $f_1(x,\omega)$ is identified with $\tilde{f}_1(x,y')$ in the local coordinate system κ. By (3.6)–(3.8) we have

$$|\partial^\alpha a(\xi)| \leq CA^{|\alpha|}|\alpha|!\langle\xi\rangle^{3(n+1)/2-|\alpha|}$$

for $|\xi| \geq 2$, where C and A are some positive constants. Applying Theorem 3.2. 5 (or its proof) we have

$$\begin{aligned}&WF_A(\tilde{f}_1) \cap T^*(\boldsymbol{R}^n \times B_1) \subset\\ &\{(x,y',\xi,0) \in T^*(\boldsymbol{R}^n \times B_0)\setminus 0;\ \xi/|\xi| = \kappa^{-1}(y') \text{ and } (x,\xi) \in WF_A(u)\},\end{aligned}$$

where $B_1 = \{y' \in \boldsymbol{R}^{n-1};\ |y'| < \sqrt{5}/3\}$. This yields

$$\begin{aligned}&WF_A(\Phi u)(= WF_A(\Phi_1 u))\\ &\subset \{(x,\omega,\xi,0) \in T^*(\boldsymbol{R}^n \times S^{n-1})\setminus 0;\ \omega = \xi/|\xi| \text{ and } (x,\xi) \in WF_A(u)\}.\end{aligned}$$

Therefore, Φ induces a sheaf homomorphism $\widehat{\Phi} : \overline{\mathcal{C}} \to \mathcal{B}/\mathcal{A}$. Put

$$H(x,\omega) = \mathcal{F}_\xi^{-1}[e^{\omega\cdot\xi/2-\langle\xi\rangle/2}\langle\xi\rangle^{-n-2}](x) \ \in C(\boldsymbol{R}^n \times S^{n-1}).$$

Similarly, we have

$$\begin{aligned}&WF_A(H)\\&\subset \{(0,\omega,\xi,0)\in T^*(\boldsymbol{R}^n\times S^{n-1})\setminus 0;\ \omega=\xi/|\xi| \text{ and } \xi\in\boldsymbol{R}^n\setminus\{0\}\}.\end{aligned}$$

Here we have regarded $H(x,\omega)$ as an element of $\mathcal{B}(\boldsymbol{R}^n\times S^{n-1})$ (or $C(S^{n-1}_\omega;\ \mathcal{F}_0(\boldsymbol{R}^n))$). We put $G(x,z,\omega)=H(x-z,\omega)\in C(\boldsymbol{R}^n\times\boldsymbol{R}^n\times S^{n-1})$. Then we have

$$\begin{aligned}WF_A(G)\subset\{(x,x,\omega,\xi,-\xi,0)\in T^*(\boldsymbol{R}^n\times\boldsymbol{R}^n\times S^{n-1})\setminus 0;\\ \omega=\xi/|\xi| \text{ and } \xi\in\boldsymbol{R}^n\setminus\{0\}\}.\end{aligned}$$

Let $\Delta:\boldsymbol{R}^n\times\boldsymbol{R}^n\times S^{n-1}\ni(x,z,\omega)\mapsto(x,z,\omega,z,\omega)\in\boldsymbol{R}^n\times\boldsymbol{R}^n\times S^{n-1}\times\boldsymbol{R}^n\times S^{n-1}$ and $\pi:\boldsymbol{R}^n\times\boldsymbol{R}^n\times S^{n-1}\ni(x,z,\omega)\mapsto(x,\omega)\in\boldsymbol{R}^n\times S^{n-1}$. If $f(x,\omega)\in\mathcal{B}(\boldsymbol{R}^n\times S^{n-1})$, then we can define $\Delta^*(G\otimes f)\in\mathcal{B}(\boldsymbol{R}^n\times\boldsymbol{R}^n\times S^{n-1})$ since $N_\Delta\cap WF_A(G\otimes f)=\emptyset$ (see Section 3.4). Moreover, we have

$$\begin{aligned}&\operatorname{supp}\Delta^*(G\otimes f)\subset\{(x,z,\omega)\in\boldsymbol{R}^n\times\boldsymbol{R}^n\times S^{n-1};\ (z,\omega)\in\operatorname{supp} f\},\\&WF_A(\Delta^*(G\otimes f))\subset\{(x,x,\omega,\xi,-\xi+\zeta,\eta')\in T^*(\boldsymbol{R}^n\times\boldsymbol{R}^n\times S^{n-1})\setminus 0;\\&\quad\omega=\xi/|\xi|,\ \xi\in\boldsymbol{R}^n\setminus\{0\} \text{ and } (x,\omega,\zeta,\eta')\in WF_A(f)\cup\operatorname{supp} f\times\{0\}\}\\&\cup\{(x,z,\omega,0,\zeta,\eta')\in T^*(\boldsymbol{R}^n\times\boldsymbol{R}^n\times S^{n-1})\setminus 0;\ (z,\omega,\zeta,\eta')\in WF_A(f)\}.\end{aligned}$$

Assume that $f(x,\omega)\in\mathcal{B}(\boldsymbol{R}^n\times S^{n-1})$ and $\operatorname{supp} f$ is compact. Then we can define $\pi_*(\Delta^*(G\otimes f))\in\mathcal{B}(\boldsymbol{R}^n\times S^{n-1})$ and we have

$$\begin{aligned}&\operatorname{supp}\pi_*(\Delta^*(G\otimes f))\\&\subset\{(x,\omega)\in\boldsymbol{R}^n\times S^{n-1};\ (z,\omega)\in\operatorname{supp} f \text{ for some } z\in\boldsymbol{R}^n\}\end{aligned}$$

(see Section 3.4). Although S^{n-1} is a manifold, we can show without using flabbiness of the sheaf of microfunctions that

$$\begin{aligned}WF_A(\pi_*(\Delta^*(G\otimes f)))\subset\{(x,\omega,\xi,\eta')\in WF_A(f);\ \omega=\xi/|\xi|\\ \text{and } \xi\in\boldsymbol{R}^n\setminus\{0\}\}\cup\{(x,\omega,0,\eta')\in T^*(\boldsymbol{R}^n\times S^{n-1})\setminus 0;\\ (z,\omega,0,\eta')\in WF_A(f) \text{ for some } z\in\boldsymbol{R}^n\}.\end{aligned}$$

Finally we can define

$$(\Psi f)(x)=\int_{S^{n-1}}\pi_*(\Delta^*(G\otimes f))(x,\omega)\,dS_\omega.$$

Since we have used flabbiness of the sheaf of microfunctions to prove (3.57), we can not apply (3.57) to estimate $WF_A(\Psi f)$. By flabbiness of the sheaf $\mathcal{B}_{R^n\times S^{n-1}}$ we may assume without loss of generality that $\operatorname{supp} f \subset \{(x,\omega)\in \boldsymbol{R}^n\times S^{n-1};\ \omega_n > 2/3\} \subset \boldsymbol{R}^n \times S_0$. $f(x,\omega)$ can be identified with an element $\tilde{f}(x,y')$ of $\mathcal{B}(\boldsymbol{R}^{2n-1})$ in a natural way, *i.e.*, $\operatorname{supp}\tilde{f}\subset \boldsymbol{R}^n\times B_1$ and $\tilde{\kappa}^*(\tilde{f}|_{R^n\times B_0}) = f|_{R^n\times S_0}$, where $\tilde{\kappa} : \boldsymbol{R}^n\times S_0 \ni (x,\omega)\mapsto (x,\kappa(\omega))\in \boldsymbol{R}^n\times B_0$. Then we have

$$\Psi f = \pi_{0*}(\tilde{\pi}_*(\widetilde{F})) \tag{3.65}$$

where $\pi_0 : \boldsymbol{R}^n\times\boldsymbol{R}^{n-1}\ni (x,y')\mapsto x\in\boldsymbol{R}^n$, $\tilde{\pi} : \boldsymbol{R}^n\times\boldsymbol{R}^n\times\boldsymbol{R}^{n-1}\ni (x,z,y')\mapsto (x,y')\in \boldsymbol{R}^n\times\boldsymbol{R}^{n-1}$, $\widetilde{\Delta} : \boldsymbol{R}^n\times\boldsymbol{R}^n\times\boldsymbol{R}^{n-1}\ni (x,z,y')\mapsto (x,z,y',z,y')\in \boldsymbol{R}^n\times\boldsymbol{R}^n\times\boldsymbol{R}^{n-1}\times\boldsymbol{R}^n\times\boldsymbol{R}^{n-1}$,

$$\widetilde{F}(x,z,y') = (1-|y'|^2)^{-1/2}\widetilde{\Delta}^*(\widetilde{G}\otimes\tilde{f})(x,z,y')$$

and $\widetilde{G}(x,z,y') = G(x,z,\kappa^{-1}(y'))$. This gives

$$WF_A(\Psi f)\subset\{(x,\xi)\in T^*\boldsymbol{R}^n\setminus 0;\ (x,\xi/|\xi|)\in \operatorname{sing\,supp} f\},$$

using flabbiness of the sheaf $\mathcal{B}/\mathcal{A}$. Therefore, Ψ induces a sheaf homomorphism $\widehat{\Psi} : \mathcal{B}/\mathcal{A}\to\overline{\mathcal{C}}$. Next we assume that $f(x,\omega)\in C(S^{n-1};\mathcal{A}'(K))$, *i.e.*, $f(\cdot,\omega)\in\mathcal{A}'(K)$ for $\omega\in S^{n-1}$ and $\langle f(\cdot,\omega),\varphi\rangle\in C(S^{n-1}_\omega)$ for every $\varphi\in\mathcal{A}(\boldsymbol{C}^n)$, and that $f(\cdot,\omega)=0$ in $\mathcal{F}_0$ if $\omega_n\le 3/4$, where K is a compact subset of $\boldsymbol{R}^n$. Then we can regard $f(x,\omega)$ as an element of $\mathcal{B}(\boldsymbol{R}^n\times S^{n-1})$ and, therefore, Ψf is given by (3.65). Note that $\tilde{f}(x,y')\in C(\overline{B}_0;\mathcal{A}'(K))$ ($\subset\mathcal{B}(\boldsymbol{R}^{2n-1})$) and $\operatorname{supp}\tilde{f}\subset K\times B_1$. Choose $\chi(y')\in C_0^\infty(B_0)$ so that $\chi(y')=1$ in B_1. Let X_0 and X_1 be open subsets of $\boldsymbol{R}^n$ such that $X_0\Subset X_1\Subset\boldsymbol{R}^n$. We choose symbols $a^R(x,\xi)\in S^0(R,A)$ ($R\ge 2$) so that $a^R(x,\xi)=1$ for $x\in X_0$ and $\operatorname{supp} a^R\subset X_1\times\boldsymbol{R}^n$. Put

$$\begin{aligned}\widetilde{G}^R(x,z,y') &= (2\pi)^{-n}\chi(y')\int \exp[i(x-z)\cdot\xi+\kappa^{-1}(y')\cdot\xi/2-\langle\xi\rangle/2]\\ &\qquad\times a^R(x,\xi)\langle\xi\rangle^{-n-2}\,d\xi\ \in C(\boldsymbol{R}^n\times\boldsymbol{R}^n\times\boldsymbol{R}^{n-1}).\end{aligned}$$

Then it is obvious that $\operatorname{supp}\widetilde{G}^R\subset X_1\times\boldsymbol{R}^n\times B_0$ and $|\widetilde{G}^R(x,z,y')|\le C$ for $(x,z,y')\in\boldsymbol{R}^n\times\boldsymbol{R}^n\times\boldsymbol{R}^{n-1}$, where C is a positive constant. We have also $\widetilde{G}^R(x,z,y')=\widetilde{G}(x,z,y')$ for $(x,z,y')\in X_0\times\boldsymbol{R}^n\times B_1$ and

$$\begin{aligned}&\mathcal{F}[\widetilde{G}^R](\xi,\zeta,\eta')\\ &= \hat{a}^R(\xi+\zeta,-\zeta)\langle\zeta\rangle^{-n-2}\mathcal{F}_{y'}(\chi(y')\exp[-\kappa^{-1}(y')\cdot\zeta/2-\langle\zeta\rangle/2])(\eta')\ \in L^1,\end{aligned}$$

where $\hat{a}^R(\xi,\zeta)=\mathcal{F}_x[a^R(x,\zeta)](\xi)$. Theorem 3.4. 3 yields

$$(\widetilde{G}\otimes\tilde{f})\Big|_{X_0\times R^n\times B_1\times R^{2n-1}} = (\widetilde{G}^R\otimes\tilde{f})\Big|_{X_0\times R^n\times B_1\times R^{2n-1}}.$$

For $\varepsilon > 0$ we put

$$F_\varepsilon^R(x,z,y') = \tilde{G}^R(x,z,y')\tilde{f}_\varepsilon(z,y'),$$

where $\tilde{f}_\varepsilon(x,y') = \exp[-\varepsilon\langle D_x\rangle]\tilde{f}(x,y') \in C(B_0;\mathcal{S})$. Then we have

$$\begin{aligned} &F_\varepsilon^R(x,z,y') \\ &= (2\pi)^{-2n}\int \exp[ix\cdot\zeta - iz\cdot(\zeta-\tilde{\zeta}) + \kappa^{-1}(y')\cdot\zeta/2 - \langle\zeta\rangle/2 - \varepsilon\langle\tilde{\zeta}\rangle] \\ &\qquad \times a^R(x,\zeta)\langle\zeta\rangle^{-n-2}\mathcal{F}_x[\tilde{f}(x,y')](\tilde{\zeta})\, d\zeta d\tilde{\zeta}. \end{aligned} \tag{3.66}$$

Applying the same argument as in the proof of Theorem 3.4.8 we can show that there is $F^R(x,z,y') \in \mathcal{F}_0(\boldsymbol{R}^{3n-1})$ satisfying

$$\begin{aligned} &\sup_{(\xi,\zeta,\eta')\in R^{3n-1}} \left|e^{-\delta(\langle\xi\rangle+\langle\zeta\rangle+\langle\eta'\rangle)}\mathcal{F}[F^R - F_\varepsilon^R](\xi,\zeta,\eta')\right| \\ &\qquad \longrightarrow 0 \quad \text{as } \varepsilon\downarrow 0 \quad \text{for any } \delta>0. \end{aligned} \tag{3.67}$$

Moreover, we have

$$F^R\Big|_{X_0\times Y\times B_1} = \Delta^*(\tilde{G}\otimes\tilde{f})\Big|_{X_0\times Y\times B_1}$$

for any bounded open subset Y of $\boldsymbol{R}^n$, $\operatorname{supp} F^R \subset X_1\times K\times B_1$ and $F^R \in \mathcal{A}'(\boldsymbol{R}^{3n-1})$. By definition we have

$$(\tilde{\pi}_*\tilde{F}^R)(\varphi) = \tilde{F}^R(\varphi\circ\tilde{\pi}) \quad \text{for } \varphi\in\mathcal{A}(\boldsymbol{C}^{2n-1}),$$

where $\tilde{F}^R(x,z,y') = (1-|y'|^2)^{-1/2}F^R(x,z,y')$. Since $\operatorname{supp}\tilde{F} \subset \boldsymbol{R}^n\times K\times B_1$, we have

$$\tilde{\pi}_*\tilde{F}\Big|_{X_0\times R^{n-1}} = \tilde{\pi}_*\tilde{F}^R\Big|_{X_0\times R^{n-1}}. \tag{3.68}$$

This, together with (3.66) and (3.67), yields

$$\begin{aligned} &\mathcal{F}[\tilde{\pi}_*\tilde{F}^R](\xi,\eta') = \mathcal{F}[\tilde{F}^R](\xi,0,\eta') = \lim_{\varepsilon\downarrow 0}\mathcal{F}[\tilde{F}_\varepsilon^R](\xi,0,\eta') \\ &= \lim_{\varepsilon\downarrow 0}(2\pi)^{-n}\int \exp[-ix\cdot\xi - iy'\cdot\eta' + iT(x,\zeta;y') - \varepsilon\langle\zeta\rangle] \\ &\qquad \times a^R(x,\zeta)\langle\zeta\rangle^{-n-2}(1-|y'|^2)^{-1/2}\mathcal{F}_z[\tilde{f}(z,y')](\zeta)\, d\zeta dx dy', \end{aligned}$$

where $T(x,\zeta;y') = x\cdot\zeta + i(\langle\zeta\rangle - \kappa^{-1}(y')\cdot\zeta)/2$. Therefore, we have

$$(\tilde{\pi}_*\tilde{F}^R)(x,y') = (1-|y'|^2)^{-1/2}a_T^R(x,D_z)\langle D_z\rangle^{-n-2}\tilde{f}(z,y') \tag{3.69}$$

$\in C(\boldsymbol{R}^{n-1}_{y'};\mathcal{E}_0(\boldsymbol{R}^n))$ ($\subset \mathcal{F}_0(\boldsymbol{R}^{2n-1})$). Here we have regarded y' as a parameter in the definition of $a^R_T(x, D_z)$. (3.65), (3.68) and (3.69) give

$$\Psi f\Big|_{X_0} = \pi_{0*}(\tilde{\pi}_*\widetilde{F}^R)\Big|_{X_0}$$
$$= \Big(\int_{S^{n-1}} \big(a^R_{T'}(x, D_z)\langle D_z\rangle^{-n-2} f\big)(x,\omega)\, dS_\omega\Big)\Big|_{X_0},$$

where $T'(x,\zeta;\omega) = T(x,\zeta;\kappa(\omega))$. Therefore, we have

$$\Psi g\Big|_{X_0} = \Big(\int_{S^{n-1}} \big(a^R_{T'}(x, D_z)\langle D_z\rangle^{-n-2} g\big)(x,\omega)\, dS_\omega\Big)\Big|_{X_0} \tag{3.70}$$

for $g(x,\omega) \in C(S^{n-1}_\omega;\mathcal{A}'(K))$. The right-hand side of (3.70) is well-defined and belongs to $\mathcal{B}(X_0)$ even if $g(x,\omega) \in C(S^{n-1}_\omega;\mathcal{F}_0(\boldsymbol{R}^n))$. Moreover, with repetition of the above argument, we have

$$(x,\omega) \notin WF_A\Big(\Big(\int_{S^{n-1}} \big(a^R_{T'}(x, D_z)\langle D_z\rangle^{-n-2} g\big)(x,\omega)\, dS_\omega\Big)\Big|_{X_0}\Big) \tag{3.71}$$

if $(x,\omega) \notin \operatorname{sing\,supp} g$. Let $u \in \mathcal{A}'(\boldsymbol{R}^n)$. Then we have $(\Phi u)(x,\omega) \in C(S^{n-1}_\omega;\mathcal{E}_0(\boldsymbol{R}^n))$. Choose $f(x,\omega) \in C(S^{n-1}_\omega;\mathcal{A}'(\overline{X}_0))$ so that $f(x,\omega)|X_0 = (\Phi u)(x,\omega)|X_0$ in $\mathcal{B}(X_0)$ for $\omega \in S^{n-1}$. By (3.70) and (3.71) we see that $\Psi f - \int_{S^{n-1}}\big(a^R_{T'}(x,D_z)\langle D_z\rangle^{-n-2}(\Phi u)\big)(x,\omega)\, dS_\omega$ is analytic in X_0. On the other hand the definition of $I(\xi)$ and Corollary 2.6.2 yield

$$\Big(\int_{S^{n-1}} \big(a^R_{T'}(x, D_z)\langle D_z\rangle^{-n-2}(\Phi u)\big)(x,\omega)\, dS_\omega\Big)\Big|_{X_0}$$
$$= (2\pi)^{-n}\Big(\int_{S^{n-1}} \Big(\lim_{\varepsilon\downarrow 0}\int e^{ix\cdot\zeta+\omega\cdot\zeta-\varepsilon\langle\zeta\rangle} a^R(x,\zeta) I(\zeta)^{-1}\hat{u}(\zeta)\, d\zeta\Big)\, dS_\omega\Big)\Big|_{X_0}$$
$$= (a^R(x,D)u)\Big|_{X_0} = u\Big|_{X_0}.$$

Therefore, we have $WF_A(\Psi f - u) \cap X_0 \times \boldsymbol{R}^n = \emptyset$, which implies that $\widehat{\Psi}\circ\widehat{\Phi} = id : \overline{\mathcal{C}} \ni u \mapsto u \in \overline{\mathcal{C}}$. Now we can prove the assertion (i). Let $\mathcal{U}$ be an open subset of $\boldsymbol{R}^n \times S^{n-1}$, and let $f \in \overline{\mathcal{C}}(\mathcal{U})$. From Theorem 1.5.2 there is $g \in \mathcal{B}/\mathcal{A}(\boldsymbol{R}^n \times S^{n-1})$ such that $g|_{\mathcal{U}} = \widehat{\Phi} f$. Putting $F = \widehat{\Psi} g$ ($\in \overline{\mathcal{C}}(\boldsymbol{R}^n \times S^{n-1})$) we have

$$F|_{\mathcal{U}} = \widehat{\Psi}(g|_{\mathcal{U}}) = \widehat{\Psi}(\widehat{\Phi} f) = f,$$

which proves the assertion (i). Define the operator $\Upsilon : \mathcal{F}_0(\boldsymbol{R}^n) \to C(S^{n-1}_\omega;\mathcal{F}_0(\boldsymbol{R}^n))$ by

$$(\Upsilon u)(x,\omega) = \mathcal{F}_\xi^{-1}[e^{\omega\cdot\xi} I(\xi)^{-1}\hat{u}(\xi)](x)$$

for $u \in \mathcal{F}_0$ and $\omega \in S^{n-1}$. Similarly, we can show that Υ induces a sheaf homomorphism $\widehat{\Upsilon} : \overline{\mathcal{C}} \to \mathcal{B}/\mathcal{A}$. Let U be an open subset of $\boldsymbol{R}^n$, and let $f \in \overline{\mathcal{C}}(U \times S^{n-1})$. It follows from Theorem 1.5.2 that there is $g \in \mathcal{B}(\boldsymbol{R}^n \times S^{n-1})$ such that the residue class in $\mathcal{B}(U \times S^{n-1})/\mathcal{A}(U \times S^{n-1})$ of $g|_{U\times S^{n-1}}$ is equal to $\widehat{\Upsilon} f$. We put

$$F(x) = \int_{S^{n-1}} g(x,\omega)\, dS_\omega \in \mathcal{B}(\boldsymbol{R}^n).$$

Repeating the above argument we can show that the residue class in $\mathcal{C}(U \times S^{n-1})$ of F is equal to f. This proves the assertion (ii). The assertions (iii) and (iv) easily follows from the assertions (i) and (ii). □

For a real analytic manifold X we can define the sheaf $\mathcal{C}_X$ on the cosphere bundle $S^*(X)$ of microfunctions in a similar way. Theorem 3.6. 1 is still valid for $\mathcal{C}_X$.

Let Ω be an open conic subset of $\boldsymbol{R}^n \times (\boldsymbol{R}^m \setminus \{0\})$, and let $T(x,\eta) \in \mathcal{P}(\Omega; A_0, A_0, c_0(T), 0, c_2(T), c_3(T))$. We assume that $T(x,\eta)$ is positively homogeneous of degree 1 in η. We may assume that $T(x,\eta)$ is defined and analytic in $\overline{\Omega} \cap \{|\eta| > 0\}$, and that $T(y,0) = 0$. We say that $p(x,\eta) \in PS^+_{loc}(\Omega)$ if for any open conic subset $\Gamma \subset\subset \Omega$ there are $A_\Gamma > 0$ and $R_\Gamma \geq 1$ satisfying $p(x,\eta) \in PS^+(\Gamma; R_\Gamma, A_\Gamma)$. Now assume that $p(x,\eta) \in PS^+_{loc}(\Omega)$. We put

$$\begin{aligned}
\widetilde{\mathcal{N}}(x,\xi) &= \{\lambda\eta \in \boldsymbol{R}^m;\ (x,\eta) \in \overline{\Omega},\ \mathrm{Im}\, T(x,\eta) = 0, \\
&\qquad \xi = \nabla_x T(x,\eta) \text{ and } \lambda > 0\}, \\
\mathcal{N}_0(x) &= \{\eta \in \boldsymbol{R}^m \setminus \{0\};\ (x,\eta) \in \overline{\Omega},\ \mathrm{Im}\, T(x,\eta) = 0 \\
&\qquad \text{and } \nabla_x T(x,\eta) = 0\}, \\
\Gamma(x) &= \{(\nabla_\eta T(x,\eta),\eta);\ \eta \in \mathcal{N}_0(x)\}
\end{aligned}$$

for $(x,\xi) \in \boldsymbol{R}^n \times S^{n-1}$. Let $(x^0,\xi^0) \in \boldsymbol{R}^n \times S^{n-1}$ satisfy $(x^0,\eta) \in \Omega$ for some η, and assume that

$$\{x^0\} \times \widetilde{\mathcal{N}}(x^0,\xi^0) \subset\subset \Omega. \tag{3.72}$$

Let $u \in \mathcal{F}_0(\boldsymbol{R}^m)$ satisfy $WF_A(u) \cap \Gamma(x^0) = \emptyset$. It is easy to see that for any conic neighborhood γ of $\Gamma(x^0)$ in $\boldsymbol{R}^m \times (\boldsymbol{R}^m \setminus \{0\})$ there is a neighborhood U of x^0 satisfying $\Gamma(x) \subset\subset \gamma$ for $x \in \overline{U}$. Therefore, there is a neighborhood U of x^0 satisfying $WF_A(u) \cap \Gamma(x) = \emptyset$ for $x \in \overline{U}$. We choose an open conic subset Γ of $\Omega \cap U \times \boldsymbol{R}^m$ and $\Phi^R(\xi,y,\eta) \in S^{0,0}(\boldsymbol{R}^n \times \boldsymbol{R}^n \times \boldsymbol{R}^m; R, A_1)$ ($R \geq 1$) so that $\{x^0\} \times \widetilde{\mathcal{N}}(x^0,\xi^0) \subset\subset \Gamma \subset\subset \Omega$ and $\Phi^R(\xi,y,\eta) = 0$ for $(y,\eta) \notin \Gamma$. Put $\tilde{p}^R(\xi,y,\eta) = \Phi^R(\xi,y,\eta)p(y,\eta)$.

Then we have $\tilde{p}^R(\xi, y, \eta) \in S^+(\boldsymbol{R}^n \times \boldsymbol{R}^n \times \boldsymbol{R}^m; R, A)$ if $R \geq \max\{1, R_\Gamma\}$ and $A \geq A_1 + A_\Gamma$. Applying Theorem 3.2. 4 with Ω' replaced by Γ, we see that there is $R(T,u) > 0$ such that $\tilde{p}_T^R(D_x, y, D_z)u$ is well-defined and in $\mathcal{F}_0(\boldsymbol{R}^n)$ if $R \geq R(T,u)A$. Moreover, there is $R(T,u,x^0,\xi^0) > 0$ such that $R(T,u,x^0,\xi^0) \geq R(T,u)$ and $(x^0,\xi^0) \notin WF_A(\tilde{p}_T^R(D_x,y,D_z)u)$ if $R \geq R(T,u,x^0,\xi^0)A$ and $WF_A(u) \cap \{(\nabla_\eta T(x^0,\eta),\eta);\ \eta \in \widetilde{\mathcal{N}}(x^0,\xi^0)$ and $(\xi^0,x^0,\eta) \in ECS(\Phi^R)\} = \emptyset$. In particular, we choose an open conic subset Γ_1 of Γ and $\Phi^R(\xi,y,\eta)$ so that $\{x^0\} \times \widetilde{\mathcal{N}}(x^0,\xi^0) \subset\subset \Gamma_1 \subset\subset \Gamma$, $\Phi^R(\xi,y,\eta) = 1$ for $(y,\eta) \in \Gamma_1$ and $\Phi^R(\xi,y,\eta) = 0$ for $(y,\eta) \notin \Gamma$. Then $\tilde{p}_T^R(D_x,y,D_z)u$ determines an element of $\mathcal{C}_{(x^0,\xi^0)}$, which does not depend on the choice of Φ^R and $R \gg 1$. Here $\mathcal{C}_{(x^0,\xi^0)}$ denotes the stalk of $\mathcal{C}$ at (x^0,ξ^0) (see Section 1.5). Therefore, we have the following

Theorem 3.6. 2 *Assume that* (3.72) *is valid. Let $\mathcal{U}$ be an open neighborhood of $\mathcal{U}(x^0,\xi^0) \equiv \{(\nabla_\eta T(x^0,\eta),\eta);\ \eta \in (\widetilde{\mathcal{N}}(x^0,\xi^0) \cup \mathcal{N}_0(x^0)) \cap S^{m-1}\}$ in $\boldsymbol{R}^m \times S^{m-1}$. Then $p_T(x,D_z)u \in \mathcal{C}_{(x^0,\xi^0)}$ is well-defined for $u \in \mathcal{C}(\mathcal{U})$ with* supp $u \cap \Gamma(x^0) = \emptyset$. *Moreover, $p_T(x,D_z)u = 0$ (in $\mathcal{C}_{(x^0,\xi^0)}$) if $u \in \mathcal{C}(\mathcal{U})$ and* supp $u \cap \mathcal{U}(x^0,\xi^0) = \emptyset$.

Next let us define ${}^r p_T(x,D_z)$ for microfunctions. Let $(z^0,\eta^0) \in \boldsymbol{R}^m \times S^{m-1}$ satisfy $(y,\eta^0) \in \Omega$ for some y. We assume that

$$\nabla_y T(y,\eta^0) \neq 0 \quad \text{if } y \in \mathcal{Y}(z^0,\eta^0), \tag{3.73}$$
$$\mathcal{Y}(z^0,\eta^0) \times \{\eta^0\} \subset\subset \Omega, \tag{3.74}$$

where

$$\mathcal{Y}(z^0,\eta^0) = \{y \in \boldsymbol{R}^n;\ (y,\eta^0) \in \overline{\Omega},\ z^0 = -\nabla_\eta T(y,\eta^0) \text{ and } \operatorname{Im} T(y,\eta^0) = 0\}.$$

Choose an open conic subset Γ of Ω and $\Psi^R(\eta,y,\xi) \in S^{0,0}(\boldsymbol{R}^m \times \boldsymbol{R}^n \times \boldsymbol{R}^n; R, A_1)$ ($R \geq 1$) so that $\mathcal{Y}(z^0,\eta^0) \times \{\eta^0\} \subset\subset \Gamma \subset\subset \Omega$ and $\Psi^R(\eta,y,\xi) = 0$ for $(y,\eta) \notin \Gamma$. We put $q^R(\eta,y,\xi) = \Psi^R(\eta,y,\xi)p(y,\eta)$. Then we have $q^R(\eta,y,\xi) \in S^+(\boldsymbol{R}^m \times \boldsymbol{R}^n \times \boldsymbol{R}^n; R, A)$ if $R \geq \max\{1, R_\Gamma\}$ and $A \geq A_1 + A_\Gamma$. Applying Proposition 3.2. 3 with S, T and Ω replaced by T, $y \cdot \xi$ and Γ, respectively, we have

$$(z^0,\eta^0) \notin WF_A(q_T^{R\prime}(D_z,y,D_x)v$$

if $v \in \mathcal{F}_0(\boldsymbol{R}^n)$,

$$WF_A(v) \cap \{(y, -\nabla_y T(y,\eta^0));\ y \in \mathcal{Y}(z^0,\eta^0) \text{ and } (\eta^0, y, -\nabla_y T(y,\eta^0)) \in ECS(\Psi^R)\} = \emptyset$$

and $R \geq R(T, v, z^0, \eta^0)$, where $R(T, v, z^0, \eta^0)$ is positive constant depending on T, v, z^0 and η^0. In particular, we choose an open conic subset Γ_1 of Γ and $\Psi^R(\eta, y, \xi)$ so that $\mathcal{Y}(z^0, \eta^0) \times \{\eta^0\} \subset\subset \Gamma_1 \subset\subset \Gamma$, $\Psi^R(\eta, y, \xi) = 1$ for $(y, \eta) \in \Gamma_1$ with $|\eta| \geq R$ and $\Psi^R(\eta, y, \xi) = 0$ for $(y, \eta) \notin \Gamma$. Then $q_T^{R\prime}(D_z, y, D_x)v$ determines an element of $\mathcal{C}_{(z^0, \eta^0)}$ for each $v \in \mathcal{F}_0(\boldsymbol{R}^n)$, which does not depend on the choice of Ψ^R and $R \gg 1$. Therefore, we can define ${}^r p_T(x, D_z) : \mathcal{C}(\mathcal{V}) \to \mathcal{C}_{(z^0, \eta^0)}$ by ${}^r p_T(x, D_z) = q_T^{R\prime}(D_z, y, D_x)$, where $\mathcal{V}$ is an open neighborhood of $\mathcal{V}(z^0, \eta^0) \equiv \{(y, -\lambda \nabla_y T(y, \eta^0));$ $y \in \mathcal{Y}(z^0, \eta^0)$ and $\lambda > 0\} \cap \boldsymbol{R}^n \times S^{n-1}$ in $\boldsymbol{R}^n \times S^{n-1}$.

Theorem 3.6. 3 *Assume that* (3.73) *and* (3.74) *are valid. Let $\mathcal{V}$ be an open neighborhood of $\mathcal{V}(z^0, \eta^0)$. Then the linear operator (mapping) ${}^r p_T(x, D_z) : \mathcal{C}(\mathcal{V}) \to \mathcal{C}_{(z^0, \eta^0)}$ is well-defined, and ${}^r p_T(x, D_z)v = 0$ (in $\mathcal{C}_{(z^0, \eta^0)}$) if $v \in \mathcal{C}(\mathcal{V})$ and* supp $v \cap \mathcal{V}(z^0, \eta^0) = \emptyset$.

We can easily extend the results given in Sections 2.7 and 2.8 for microfunctions. Let Ω be an open conic subset of $T^*\boldsymbol{R}^n \setminus 0$. If $p(x, \xi) \in PS^+_{loc}(\Omega)$, then we can define a sheaf homomorphism $p(x, D) : \mathcal{C}|_{\Omega_0} \to \mathcal{C}|_{\Omega_0}$, where $\Omega_0 = \Omega \cap \boldsymbol{R}^n \times S^{n-1}$ and $\mathcal{C}|_{\Omega_0}$ denotes the sheaf on Ω_0 defined by associating $\mathcal{C}(\mathcal{U})$ to every open subset $\mathcal{U}$ of Ω_0 (see Theorem 3.6. 2). Applying Theorem 3.2. 5 instead of Theorem 3.2. 4 we can also define $p(x, D)$ on $\mathcal{C}|_{\Omega_0}$ as follows: Let X and X_1 be open subsets of $\boldsymbol{R}^n$, and let γ and γ_1 be open cones in $\boldsymbol{R}^n \setminus \{0\}$. We assume that $X \times \gamma \subset\subset X_1 \times \gamma_1 \subset\subset \Omega$, and choose $\varphi^R(x, \xi) \in S^0(R, A)$ and $g^R(\xi) \in S^0(R, A/2)$ ($R \geq 4$) so that $\varphi^R(x, \xi) = 1$ for $x \in X$, supp $\varphi^R \subset X_1 \times \boldsymbol{R}^n$, $g^R(\xi) = 1$ for $\xi \in \gamma$ with $|\xi| \geq R$ and supp $g^R \subset \gamma_1 \cap \{|\xi| \geq R/2\}$. Put $\Phi^R(x, \xi) = \varphi^R(x, \xi) g^R(\xi)$. Then we have $\Phi^R(x, \xi) \in S^0(R, A)$ and

$$\left|\Phi^{R(\alpha+\tilde{\alpha})}_{(\beta)}(x, \xi)\right| = \left|\partial_\xi^{\alpha+\tilde{\alpha}} D_x^\beta g^R(\xi)\right| \leq C_{|\tilde{\alpha}|}(A/R)^{|\alpha|} A^{|\beta|} |\beta|!$$

if $x \in X$ and $\langle \xi \rangle \geq R|\alpha|$. Applying Theorem 3.2. 5 to $\tilde{p}^R(x, \xi) \equiv \Phi^R(x, \xi) \times p(x, \xi)$ we can show that

$$WF_A(\tilde{p}^R(x, D)u) \cap X \times \boldsymbol{R}^n \subset WF_A(u) \cap X \times \gamma_1$$

and $\tilde{p}^R(x, D)u$ determines an element of $\mathcal{C}(X \times (\gamma \cap S^{n-1}))$ for each $u \in \mathcal{F}_0$ if $R \gg 1$, which does not depend on the choice of Φ^R and $R \gg 1$. So we can define $p(x, D) : \mathcal{C}(X \times (\gamma \cap S^{n-1})) \to \mathcal{C}(X \times (\gamma \cap S^{n-1}))$, and $p(x, D) : \mathcal{C}|_{\Omega_0} \to \mathcal{C}|_{\Omega_0}$. It is easy to see that this definition of $p(x, D)$ coincides with the previous one. Therefore, we can prove the following theorems in the same way as in Sections 2.7 and 2.8.

Theorem 3.6. 4 *Let $p_j(x,\xi) \in PS^+(\Omega; R_0, A, B)$ ($j = 1,2$). We fix $R_1 \geq R_0 + enAB$ and put*

$$p(x,\xi) = \sum_{j=0}^{\infty} \phi_j^{R_1}(\xi) \sum_{|\gamma|=j} \frac{1}{\gamma!} p_1^{(\gamma)}(x,\xi) p_{2(\gamma)}(x,\xi).$$

Then we have

$$p(x,\xi) \in PS^+(\Omega; 2R_1, A', B)$$

and $p_1(x,D)(p_2(x,D)u) = p(x,D)u$ in $\mathcal{C}(\mathcal{U})$ for $u \in \mathcal{C}(\mathcal{U})$, where $A' = A + 3\widehat{C}$ and $\mathcal{U}$ is an open subset of Ω_0.

Theorem 3.6. 5 *Let $p(x,\xi) \in PS^m(\Omega; R, A)$. Assume that there are positive constants C and c satisfying*

$$|p(x,\xi)| \geq c\langle\xi\rangle^m \quad \text{if } (x,\xi) \in \Omega \text{ and } |\xi| \geq C.$$

Then there are positive constants R' and A' and a symbol $q(x,\xi) \in PS^{-m}(\Omega; R', A')$ such that

$$q(x,D)p(x,D)u = p(x,D)q(x,D)u = u \quad \text{for } u \in \mathcal{C}(\mathcal{U}),$$

where $\mathcal{U}$ is an open subset of Ω_0 ($= \Omega \cap \boldsymbol{R}^n \times S^{n-1}$), i.e., $p(x,D) : \mathcal{C}|_{\Omega_0} \to \mathcal{C}|_{\Omega_0}$ is a sheaf isomorphism and $p(x,D)^{-1} = q(x,D)$ in $\mathcal{C}|_{\Omega_0}$.

3.7 Formal analytic symbols

The symbol of the product of two pseudodifferential operators with pseudo-analytic symbols, for example, is also pseudo-analytic if one ignores analytic regularizers. However, pseudo-analytic symbols are not always transformed to pseudo-analytic symbols by canonical transformations. On the other hand we could not construct the symbols of products of pseudodifferential operators in the framework of analytic symbols (see, *e.g.*, Theorem 3.6. 4). By these reasons we shall introduce some classes of formal analytic symbols.

Definition 3.7. 1 Let Γ be an open conic subset of $\boldsymbol{R}^n \times (\boldsymbol{R}^{n'} \setminus \{0\})$, and let $\{a_j(x,\eta)\}_{j\in Z_+} \subset C^\infty(\Gamma)$. We say that $a(x,\eta) \equiv \{a_j(x,\eta)\}_{j\in Z_+} \in FS^{m,\delta}(\Gamma; C_0, A)$ if $a(x,\eta)$ satisfies

$$\left|a_{j(\beta)}^{(\alpha)}(x,\eta)\right| \leq CC_0^j A^{|\alpha|+|\beta|} j!|\alpha|!|\beta|!\langle\eta\rangle^{m-j-|\alpha|} e^{\delta\langle\eta\rangle}$$

for $j \in \boldsymbol{Z}_+$ and $(x,\eta) \in \Gamma$ with $|\eta| \ge C$, where C is a positive constant. Symbols in $FS^{m,\delta}(\Gamma; C_0, A)$ are called formal analytic symbols. We also write $FS^m(\Gamma; C_0, A) \equiv FS^{m,0}(\Gamma; C_0, A)$. We define

$$FS^+(\Gamma; C_0, A) := \bigcap_{\delta>0} FS^{m,\delta}(\Gamma; C_0, A). \tag{3.75}$$

Here the right-hand side of (3.75) does not depend on the choice of $m \in \boldsymbol{R}$. For $a(x,\eta) = \{a_j(x,\eta)\} \in FS^{m,\delta}(\Gamma; C_0, A)$ we also write $a(x,\eta) = \sum_{j=0}^\infty a_j(x,\eta)$ formally.

Let Γ be an open conic subset of $\boldsymbol{R}^n \times (\boldsymbol{R}^{n'} \setminus \{0\})$, and let $T(x,\eta) \in \mathcal{P}(\Gamma; A_0, A_0, c_0(T), 0, c_2(T), c_3(T))$. We assume that $T(x,\eta)$ is positively homogeneous of degree 1 in η. Let (x^0,ξ^0) be a point in $\boldsymbol{R}^n \times S^{n-1}$ such that $(x^0,\eta) \in \Gamma$ for some η, and assume that (3.72) is valid with Ω replaced by Γ. Let $a(x,\eta) = \sum_{j=0}^\infty a_j(x,\eta) \in FS^+(\Gamma; C_0, A)$. It follows from Lemma 2.2.4 that

$$\tilde{a}(x,\eta) \equiv \sum_{j=0}^\infty \phi_j^R(\eta) a_j(x,\eta) \in PS^+(\Gamma; 2R, 2A + 3\widehat{C}, A)$$

if $R \ge C_0$. Moreover, we have

$$\Big|\tilde{a}^{(\alpha)}_{(\beta)}(x,\eta) - \sum_{j=0}^{N-1} a^{(\alpha)}_{j(\beta)}(x,\eta)\Big| \le C_{\delta,\alpha}(C_0/(2R))^N A^{|\beta|}|\beta|! e^{\delta\langle\eta\rangle}$$

if $\delta > 0$, $N \in \boldsymbol{N}$, $(x,\eta) \in \Gamma$, $\langle\eta\rangle \ge 3RN$ and $R \ge C_0$. Let $\mathcal{U}$ be an open neighborhood in $\boldsymbol{R}^{n'} \times S^{n'-1}$ of $\mathcal{U}(x^0,\xi^0)$ which is defined as in Section 3.6 with Ω replaced by Γ. It follows from Theorem 3.6.2 that $\tilde{a}_T(x,D_z): \mathcal{C}(\mathcal{U}) \to \mathcal{C}_{(x^0,\xi^0)}$ is well-defined and $\tilde{a}_T(x,D_z)u = 0$ in $\mathcal{C}_{(x^0,\xi^0)}$ if $u \in \mathcal{C}(\mathcal{U})$ and $\operatorname{supp} u \cap \mathcal{U}(x^0,\xi^0) = \emptyset$. Now assume that a symbol $b(x,\eta)$ ($\in C^\infty(\Gamma)$) satisfies

$$\Big|b^{(\alpha)}_{(\beta)}(x,\eta) - \sum_{j=0}^{N-1} a^{(\alpha)}_{j(\beta)}(x,\eta)\Big| \le C'_{\delta,\alpha}(C_1/R_1)^N A_1^{|\beta|}|\beta|! e^{\delta\langle\eta\rangle}$$

if $\delta > 0$, $N \in \boldsymbol{N}$, $(x,\eta) \in \Gamma$ and $\langle\eta\rangle \ge R_1 N$, where the $C'_{\delta,\alpha}$, C_1, R_1 and A_1 are some positive constants. Then, putting $c(x,\eta) = \tilde{a}(x,\eta) - b(x,\eta)$, we have

$$\Big|c^{(\alpha)}_{(\beta)}(x,\eta)\Big| \le e(C_{\delta,\alpha} + C'_{\delta,\alpha}) A_2^{|\beta|}|\beta|! \exp[(\delta - 1/R_2)\langle\eta\rangle]$$

for $(x,\eta) \in \Gamma$ if $\delta > 0$, $R \ge eC_0/2$ and $R_1 \ge eC_1$, where $A_2 = \max\{A, A_1\}$ and $R_2 = \max\{3R, R_1\}$. Theorem 2.3.3 yields $\tilde{a}_T(x,D_z)u = b_T(x,D_z)u$

for $u \in \mathcal{C}(\mathcal{U})$ if $R \geq eC_0/2$ and $R_1 \geq eC_1$. This implies that $a_T(x, D_z) : \mathcal{C}(\mathcal{U}) \to \mathcal{C}_{(x^0,\xi^0)}$ can be defined by $a_T(x, D_z) = \tilde{a}_T(x, D_z)$. We write $a(x, D) = a_T(x, D_z)$ if $n' = n$ and $T(x, \eta) = x \cdot \eta$. Then $a(x, D) : \mathcal{C}(\Gamma_0) \to \mathcal{C}(\Gamma_0)$, where $\Gamma_0 = \Gamma \cap \boldsymbol{R}^n \times S^{n-1}$. Similarly, we can define ${}^r a_T(x, D_z)$. In fact, let (z^0, η^0) be a point in $\boldsymbol{R}^{n'} \times S^{n'-1}$ such that $(y, \eta^0) \in \Gamma$ for some y. We assume that (3.73) and (3.74) are satisfied with Ω replaced by Γ. Let $\mathcal{V}$ be an open neighborhood in $\boldsymbol{R}^n \times S^{n-1}$ of $\mathcal{V}(z^0, \eta^0)$ which is defined as in Section 3.6 with Ω replaced by Γ. Then, by Theorem 3.6.3 ${}^r\tilde{a}_T(x, D_z) : \mathcal{C}(\mathcal{V}) \to \mathcal{C}_{(z^0,\eta^0)}$ is well-defined and ${}^r\tilde{a}_T(x, D_z)v = 0$ in $\mathcal{C}_{(z^0,\eta^0)}$ if $v \in \mathcal{C}(\mathcal{V})$ and $\operatorname{supp} v \cap \mathcal{V}(z^0, \eta^0) = \emptyset$. Therefore, we can define ${}^r a_T(x, D_z) : \mathcal{C}(\mathcal{V}) \to \mathcal{C}_{(z^0,\eta^0)}$ by ${}^r a_T(x, D_z) = {}^r\tilde{a}_T(x, D_z)$.

Theorem 3.7.2 *Let Γ be an open conic subset of $\boldsymbol{R}^n \times (\boldsymbol{R}^n \setminus \{0\})$, and let $a(x, \xi) \equiv \sum_{j=0}^\infty a_j(x, \xi)$ and $b(x, \xi) \equiv \sum_{j=0}^\infty b_j(x, \xi)$ be in $FS^+(\Gamma; C_0, A)$. We put*

$$c_j(x, \xi) = \sum_{|\gamma|+\mu+\nu=j} a_\mu^{(\gamma)}(x, \xi) b_{\nu(\gamma)}(x, \xi)/\gamma!.$$

Then we have $c(x, \xi) \equiv \sum_{j=0}^\infty c_j(x, \xi) \in FS^+(\Gamma; C_0', 2A)$ with $C_0' = \max \{C_0, 4nA^2\}$. Moreover, we have $c(x, D) = a(x, D)b(x, D)$ on $\mathcal{C}(\Gamma_0)$. If $a(x, \xi) \in FS^{m_1}(\Gamma; C_0, A)$ and $b(x, \xi) \in FS^{m_2}(\Gamma; C_0, A)$, then $c(x, \xi) \in FS^{m_1+m_2}(\Gamma; C_0', 2A)$.

Proof It is easy to see that $c(x, \xi) \in FS^+(\Gamma; C_0', 2A)$. Put, for $R \geq C_0$,

$$\tilde{a}(x, \xi) = \sum_{j=0}^\infty \phi_j^R(\xi) a_j(x, \xi), \quad \tilde{b}(x, \xi) = \sum_{j=0}^\infty \phi_j^R(\xi) b_j(x, \xi).$$

From Theorem 3.6.4 we have

$$\tilde{a}(x, D)\tilde{b}(x, D)\ (\equiv a(x, D)b(x, D)) = \tilde{c}(x, D) \quad \text{on } \mathcal{C}(\Gamma_0),$$

where $R_1 \geq 2R + enA(2A + 3\widehat{C})$ and

$$\tilde{c}(x, D) = \sum_{j=0}^\infty \phi_j^{R_1}(\xi) \sum_{|\gamma|=j} \tilde{a}^{(\gamma)}(x, \xi)\tilde{b}_{(\gamma)}(x, \xi)/\gamma!.$$

A simple calculation yields

$$\tilde{c}(x, \xi) = \sum_{j=0}^\infty \phi_j^{R_2}(\xi) c_j(x, \xi) + I_1(x, \xi) + I_2(x, \xi),$$

$$I_1(x, \xi) = \sum_{j=0}^\infty \sum_{|\gamma|+\mu+\nu=j} \left(\phi_{|\gamma|}^{R_1}(\xi)\phi_\mu^R(\xi)\phi_\nu^R(\xi) - \phi_j^{R_2}(\xi)\right)$$

$$\times a_\mu^{(\gamma)}(x,\xi) b_{\nu(\gamma)}(x,\xi)/\gamma!,$$

$$I_2(x,\xi) = \sum_{\gamma>0} \sum_{\rho} \sum_{\mu=1}^{\infty} \sum_{\nu=0}^{\infty} \phi_{|\gamma|+|\rho|}^{R_1}(\xi) \phi_\mu^{R(\gamma)}(\xi) \phi_\nu^R(\xi) \times a_\mu^{(\rho)}(x,\xi) b_{\nu(\gamma+\rho)}(x,\xi)/(\gamma!\rho!),$$

where $R_2 \geq 3R_1/2$. Moreover, we have

$$\left| I_{1(\beta)}^{(\alpha)}(x,\xi) \right| \leq C_{\delta,\alpha}(2A)^{|\beta|}|\beta|! \exp[(\delta - 1/(3R_2))\langle\xi\rangle]$$

if $\delta > 0$ and $R \geq 3eC_0'$, since $2Rj/3 \leq \langle\xi\rangle \leq 3R_2 j$ when $|\gamma| + \mu + \nu = j$ and $\phi_{|\gamma|}^{R_1}(\xi)\phi_\mu^R(\xi)\phi_\nu^R(\xi) - \phi_j^{R_2}(\xi) \neq 0$. Similarly, we have

$$\left| I_{2(\beta)}^{(\alpha)}(x,\xi) \right| \leq C_{\delta,\alpha}(2A)^{|\beta|}|\beta|! \exp[(\delta - 1/(3R))\langle\xi\rangle]$$

if $R \geq \max\{eC_0/2, 8nA\widehat{C}\}$ and $R_1 \geq 8nA^2$, since $2R\mu \leq \langle\xi\rangle \leq 3R\mu$, $\langle\xi\rangle \geq 2R_1 j$, $\langle\xi\rangle \geq 2R\nu$ and $j \leq 2\mu$ when $\gamma > 0$ and $\phi_j^{R_1}(\xi)\phi_\mu^{R(\gamma)}(\xi)\phi_\nu^R(\xi) \neq 0$. This yields $a(x,D)b(x,D) = c(x,D)$ on $\mathcal{C}(\Gamma_0)$. The remaining part of the theorem is obvious. □

Chapter 4

Microlocal uniqueness

A microlocal version of the Holmgren uniqueness theorem will be given in Section 4.2. Carleman type estimates prove microlocal uniqueness. As applications of the results in Section 4.2, we shall give theorems on propagation of analytic singularities for microhyperbolic operators in Section 4.3. In Section 4.5 we shall give several results on analytic hypoellipticity. Metivier-Okaji's result and Grigis-Sjöstrand's result will be proved in the framework of hyperfunctions. The lemmas and tools given in Sections 4.1 and 4.4 are very useful to our microlocal analysis.

4.1 Preliminary lemmas

In this section we shall give a series of Lemmas used in this chapter.

Lemma 4.1.1 *Let Γ be an open conic subset of $\boldsymbol{R}^n \times (\boldsymbol{R}^n \setminus \{0\})$ satisfying $\Gamma \subset\subset \boldsymbol{R}^n \times (\boldsymbol{R}^n \setminus \{0\})$, and assume that $u \in \mathcal{F}_0$ and $WF_A(u) \cap \overline{\Gamma}_\varepsilon = \emptyset$ for some $\varepsilon > 0$, where $\Gamma_\varepsilon = \{(x,\xi) \in \boldsymbol{R}^n \times (\boldsymbol{R}^n \setminus \{0\});\ |(x,\xi/|\xi|) - (y,\eta/|\eta|)| < \varepsilon$ for some $(y,\eta) \in \Gamma\}$. Moreover, we assume that a symbol $p(\xi,y,\eta)$ satisfies* supp $p \subset \boldsymbol{R}^n \times \Gamma$ *and*

$$\left|\partial_\xi^\alpha D_y^{\beta+\tilde{\beta}} \partial_\eta^{\gamma+\tilde{\gamma}} p(\xi,y,\eta)\right| \\ \le C_{|\alpha|+|\tilde{\beta}|+|\tilde{\gamma}|} (A/R_0)^{|\beta|+|\gamma|} \langle\xi\rangle^{|\beta|} \exp[\delta_1\langle\xi\rangle + \delta_2\langle\eta\rangle]$$

if $\langle\xi\rangle \ge R_0|\beta|$ and $\langle\eta\rangle \ge R_0|\gamma|$. Then there are positive constants $R_0(\varepsilon)$, $\delta(\varepsilon,u)$ and $\delta_j(\varepsilon,u)$ ($j=1,2$) such that $p(D_x,y,D_y)u \in \mathcal{S}'_{-\delta}$ and, more precisely,

$$|\mathcal{F}[p(D_x,y,D_y)u](\xi)| \le C_{R_0,\delta}(u,p) e^{-\delta\langle\xi\rangle} \tag{4.1}$$

if $R_0 \geq R_0(\varepsilon)A$, $2\delta_1 + (\delta_2)_+ < 1/R_0$, $\delta \leq \min\{1/(2R_0), \delta(\varepsilon, u)\}$ *and* $\delta_j \leq \delta_j(\varepsilon, u)$ ($j = 1, 2$), *where* $C_{R_0,\delta}(u, p)$ *is a positive constant depending on* R_0, δ, u *and* p.

Proof Choose open conic subsets $\Gamma_j \equiv X_j \times \gamma_j$ ($1 \leq j \leq N$) of $\boldsymbol{R}^n \times (\boldsymbol{R}^n \setminus \{0\})$ so that $\Gamma \subset\subset \bigcup_{j=1}^N \Gamma_j \subset \Gamma_{\varepsilon/2}$. Fix j so that $1 \leq j \leq N$. Since $WF_A(u) \cap \overline{X^j_{\varepsilon/3}} \times \overline{\gamma^j_{\varepsilon/3}} = \emptyset$, it follows from Proposition 3.1.2 (ii) that for any $y \in \overline{X^j_{\varepsilon/3}}$ there are a symbol $g_y(\xi)$, a neighborhood U_y of y and $R_y > 0$ such that $|\partial_\xi^\alpha g_y(\xi)| \leq C_{|\alpha|}$, $g_y(\xi) = 1$ in $\gamma^j_{\varepsilon/3} \cap \{\langle\xi\rangle \geq R_y\}$ and $g_y(D)u$ is analytic in U_y, where $X^j_\varepsilon = \{x \in \boldsymbol{R}^n;\ |x - y| < \varepsilon$ for some $y \in X^j\}$ and $\gamma^j_\varepsilon = \{\xi \in \boldsymbol{R}^n \setminus \{0\};\ |\xi/|\xi| - \eta/|\eta|| < \varepsilon$ for some $\eta \in \gamma^j\}$. Applying Proposition 3.1.2 (i) and Proposition 2.2.3 , we see that there are a symbol $\tilde{g}_j^R(\xi)$ ($R \geq 4$) and $R_1 > 0$ such that $\tilde{g}_j^R(\xi) = 1$ in $\gamma^j_{\varepsilon/6} \cap \{\langle\xi\rangle \geq R/2\}$, supp $\tilde{g}_j^R \subset \gamma^j_{\varepsilon/3}$,

$$\left|\partial_\xi^{\alpha+\tilde{\alpha}} \tilde{g}_j^R(\xi)\right| \leq C_{|\tilde{\alpha}|}(C/(\varepsilon R))^{|\alpha|}\langle\xi\rangle^{-|\tilde{\alpha}|} \tag{4.2}$$

if $\langle\xi\rangle \geq R|\alpha|$ and $\tilde{g}_j^R(D)u$ is analytic in $X^j_{\varepsilon/3}$ for $R \geq R_1$, since $\overline{X^j_{\varepsilon/3}}$ is compact. Therefore, the application of Proposition 3.1.2 (i) shows that there is a symbol $g_j^R(\xi)$ ($R \geq 4$) such that $g_j^R(\xi) = 1$ in $\gamma^j_{\varepsilon/12} \cap \{\langle\xi\rangle \geq R/2\}$, supp $g_j^R \subset \gamma^j_{\varepsilon/6}$, (4.2) with $\tilde{g}_j^R$ replaced by g_j^R is satisfied for some positive number C independent of ε and $g_j^R(D)u$ is analytic in $X^j_{\varepsilon/6}$ for $R \geq 436e\sqrt{n}C/\varepsilon^2$. By Proposition 2.2.3 there are symbols $\Phi_j^R(\xi, y, \eta) \in S^{0,0}(R, A(\varepsilon))$ ($1 \leq j \leq N$, $R \geq 4$) such that supp $\Phi_j^R \subset \boldsymbol{R}^n \times X^j_{\varepsilon/12} \times \gamma^j_{\varepsilon/12} \cap \{\langle\eta\rangle \geq R/2\}$ and $\sum_{k=1}^N \Phi_k^R(\xi, y, \eta) = 1$ for $(\xi, y, \eta) \in \boldsymbol{R}^n \times \Gamma$ with $\langle\eta\rangle \geq R$. Put

$$p_j^R(\xi, y, \eta) = \Phi_j^R(\xi, y, \eta)p(\xi, y, \eta) \quad (1 \leq j \leq N).$$

We note that $p_j^R(\xi, y, \eta)g_j^R(\eta) = p_j^R(\xi, y, \eta)$. It follows from Theorem 2.6.7 and its remark that there are positive constants $R_0(\varepsilon)$, $\delta'(\varepsilon, u)$ and $\delta'_k(\varepsilon, u)$ ($k = 1, 2$) such that (4.1) is valid if $R_0 \geq R_0(\varepsilon)A$, $R \geq \max\{R_0, R_0(\varepsilon)A(\varepsilon), 436e\sqrt{n}C/\varepsilon^2\}$, $2\delta_1 + (\delta_2)_+ < 1/R$, $\delta \leq \min\{1/(2R), \delta'(\varepsilon, u)\}$ and $\delta_k \leq \delta'_k(\varepsilon, u)$ ($k = 1, 2$). This proves the lemma. □

Lemma 4.1.2 *Let* K *be a compact subset of* $\boldsymbol{R}^n$, *and let* $p(\xi, y, \eta) \in C^\infty(\boldsymbol{R}^n \times \boldsymbol{R}^n \times \boldsymbol{R}^n)$ *be a symbol satisfying* supp $p \subset \boldsymbol{R}^n \times K \times \boldsymbol{R}^n$ *and*

$$\left|D_y^\beta \partial_\eta^\gamma p(\xi, y, \eta)\right| \leq C_{|\beta|+|\gamma|}\langle\xi\rangle^m \exp[\delta\langle\eta\rangle],$$

where $m, \delta \in \boldsymbol{R}$. *Then we have* $p(D_x, y, D_y)u \in H^\infty$ *and*

$$\|D^\alpha p(D_x, y, D_y)u\| \le C_{K,\delta,\varepsilon,m+|\alpha|}(u)$$

for $u \in \mathcal{S}_\varepsilon{}'$ *if* $\delta + \varepsilon < 0$, *where* $\|f\|$ *denotes the* L^2*-norm of* f, *i.e.,* $\|f\| = (\int |f(x)|^2\, dx)^{1/2}$. *Here* H^s ($\equiv H^s(\boldsymbol{R}^n)$) *denotes the Sobolev space of order* s *and* $H^\infty \equiv \bigcap_s H^s$. *In particular,* $p(D_x, y, D_y)$ *maps continuously* $\mathcal{S}_\varepsilon{}'$ *to* H^s *if* $\delta + \varepsilon < 0$.

Proof Let $u \in \mathcal{S}_\varepsilon{}'$, and assume that $\delta + \varepsilon < 0$. It is easy to see that

$$\begin{aligned}&\mathcal{F}[p(D_x, y, D_y)u](\xi)\\ &= (2\pi)^{-n} \int e^{-iy\cdot\xi} \langle e^{-\varepsilon\langle\eta\rangle}\hat{u}(\eta), e^{iy\cdot\eta+\varepsilon\langle\eta\rangle} p(\xi, y, \eta)\rangle_\eta\, dy,\end{aligned}$$

since $e^{-\varepsilon\langle\eta\rangle}\hat{u}(\eta) \in \mathcal{S}'$ and

$$\left|D_y^\beta \partial_\eta^\gamma (e^{iy\cdot\eta+\varepsilon\langle\eta\rangle} p(\xi, y, \eta))\right| \le C_{\beta,\gamma,K,\delta,\varepsilon,N} \langle\xi\rangle^m \langle\eta\rangle^{-N}$$

($N = 0, 1, 2, \cdots$). Integration by part gives

$$|\xi^\alpha \mathcal{F}[p(D_x, y, D_y)u](\xi)| \le C'_{K,\delta,\varepsilon,m+|\alpha|}(u)\langle\xi\rangle^{-(n+1)/2},$$

which proves the lemma. □

The following lemma is a simple consequence of Lemma 2.1. 7 .

Lemma 4.1. 3 *Let* Ω *be an open conic subset of* $\boldsymbol{R}^n \times (\boldsymbol{R}^n \setminus \{0\})$, *and let a symbol* $\Lambda(x, \xi) \in C^\infty(\Omega)$ *satisfy*

$$\left|\Lambda^{(\alpha)}_{(\beta)}(x, \xi)\right| \le C_0 A_0^{|\alpha|+|\beta|} |\alpha|! |\beta|! \langle\xi\rangle^{1-|\alpha|} \tag{4.3}$$

for $(x, \xi) \in \Omega$ *with* $|\xi| \ge 1$. *Then we have*

$$\left|\partial_\xi^\alpha D_x^\beta e^{t\Lambda(x,\xi)}\right| \le e^{t\mathrm{Re}\ \Lambda(x,\xi)} (6A_0/\rho)^{|\alpha|+|\beta|} |\alpha|! |\beta|! \langle\xi\rangle^{-|\alpha|} e^{\rho C_0 |t| \langle\xi\rangle}$$

if $(x, \xi) \in \Omega$, $|\xi| \ge 1$, $t \in \boldsymbol{R}$ *and* $0 < \rho \le 2^{-4}$.

From Theorems 2.3. 3 and 2.4. 6 and the remark of Theorem 2.4. 6 we have the following

Lemma 4.1. 4 *Let* $p(\xi, w, \zeta, y, \eta) \in C^\infty(\boldsymbol{R}^n \times \boldsymbol{R}^n \times \boldsymbol{R}^n \times \boldsymbol{R}^n \times \boldsymbol{R}^n)$ *be a symbol satisfying* (P–2) *and* (P–4), *where the conditions* (P–2) *and* (P–4) *are as in* Section 2.4. *We assume that* $p(\xi, w, \zeta, y, \eta) = 0$ *if* $|w -$

$y| \le \varepsilon_0$ and $|\zeta - \eta| \le \varepsilon_0'|\eta|$, where $\varepsilon_0, \varepsilon_0' > 0$. Then there are a symbol $p(\xi, y, \eta)$, positive constants $A_1' \equiv A_1'(A_1, \varepsilon_0)$ and $R(A_1, \varepsilon_0, \varepsilon_0')$, which are polynomials of A_1 of degree 1, and $c_0 > 0$ such that

$$p(D_x, w, D_w, y, D_y) = p(D_x, y, D_y) \quad \text{on } \mathcal{S}_\infty,$$
$$\left|\partial_\xi^\alpha D_y^{\beta+\tilde\beta}\partial_\eta^\gamma p(\xi, y, \eta)\right| \le C_{|\alpha|+|\tilde\beta|+|\gamma|, \varepsilon_0, \varepsilon_0', R_0}((R_0/(50en) + A_1')/R_0)^{|\beta|}$$
$$\times \langle\xi\rangle^{m_1 - |\alpha| + |\beta|}\exp[\delta_1\langle\xi\rangle - c_0\langle\eta\rangle/R_0]$$

if $R_0 \ge R(A_1, \varepsilon_0, \varepsilon_0')$, $\langle\xi\rangle \ge R_0|\beta|$, $\delta_2 \le 1/R_0$ and $\max\{4(\delta_2)_+ + \delta_3, 4\delta_2 + 2|\delta_2| + 2\delta_3\} \le c_0/R_0$. Moreover, $p(D_x, w, D_w, y, D_y)$ maps continuously $\mathcal{S}_{\varepsilon_2}$ and $\mathcal{S}_{-\varepsilon_2}'$ to $\mathcal{S}_{\varepsilon_1}$ and $\mathcal{S}_{-\varepsilon_1}'$, respectively, if $\varepsilon_2 = -c_0/R_0 + 2(\delta_1 + \varepsilon_1)_+$, $\delta_1 + \varepsilon_1 \le 1/R_0$, $R_0 \ge \max\{R(A_1, \varepsilon_0, \varepsilon_0'), 4enA_1'\}$, $\delta_2 \le 1/R_0$ and $\max\{4(\delta_2)_+ + \delta_3, 4\delta_2 + 2|\delta_2| + 2\delta_3\} \le c_0/R_0$.

Lemma 4.1. 5 *Let Ω_j ($1 \le j \le 4$) be open conic subsets of $\boldsymbol{R}^n \times (\boldsymbol{R}^n \setminus \{0\})$ satisfying $\Omega_j \subset\subset \Omega_{j+1}$ ($1 \le j \le 3$), and let $p(\xi, y, \eta) \in S^{m_1, m_2}(R_0, A)$ and $\varphi(\xi, y, \eta) \in S^{0,0}(R_0, A)$. We assume that*

$$\left|\partial_\xi^{\alpha+\tilde\alpha} D_y^\beta \partial_\eta^{\gamma+\tilde\gamma} p(\xi, y, \eta)\right|$$
$$\le C_{|\tilde\alpha|+|\tilde\gamma|}(A/R_0)^{|\alpha|+|\gamma|}B^{|\beta|}|\beta|!\langle\xi\rangle^{m_1 - |\tilde\alpha|}\langle\eta\rangle^{m_2 - |\tilde\gamma|}$$

if $(y, \eta) \in \Omega_2$, $|\eta| \ge R_0$, $\langle\xi\rangle \ge R_0|\alpha|$ and $\langle\eta\rangle \ge R_0|\gamma|$, and that $\varphi(\xi, y, \eta) = 1$ for $(\xi, y, \eta) \in \boldsymbol{R}^n \times \Omega_2$ with $|\eta| \ge R_0$ and $\operatorname{supp} \varphi \subset \boldsymbol{R}^n \times \Omega_3$. Then, for any $\varepsilon \in (0, 1]$ there are positive constants $A_1 \equiv A_1(A, \varepsilon, \{\Omega_j\})$, $A_2 \equiv A_2(A, \{\Omega_j\})$ and $R(A, \varepsilon, \{\Omega_j\})$, which are polynomials of A of degree 1, symbols $q_\varepsilon(\xi, y, \eta) \in \bigcap_{\delta > 0} S^{m_1, m_2, 0, \delta}(4R_0, A_1 + 4\varepsilon/\delta)$ and $r_\varepsilon(\xi, y, \eta)$, and $c_0 > 0$ such that $\operatorname{supp} q_\varepsilon \subset \boldsymbol{R}^n \times (\Omega_4 \setminus \Omega_1)$ and

$$\left|\partial_\xi^\alpha D_y^{\beta+\tilde\beta}\partial_\eta^\gamma r_\varepsilon(\xi, y, \eta)\right| \le C_{|\alpha|+|\tilde\beta|+|\gamma|, \varepsilon, R_0}(\{\Omega_j\})((R_0/(4en) + A_2)/R_0)^{|\beta|}$$
$$\times \langle\xi\rangle^{(m_1)_+ - |\alpha| + |\beta|}\exp[-c_0\langle\eta\rangle/R_0] \quad \text{if } \langle\xi\rangle \ge R_0|\beta|,$$
$$[p(D_x, y, D_y), \varphi(D_x, y, D_y)]\,(\equiv p(D_x, y, D_y)\varphi(D_x, y, D_y)$$
$$-\varphi(D_x, y, D_y)p(D_x, y, D_y)) = q_\varepsilon(D_x, y, D_y) + r_\varepsilon(D_x, y, D_y) \quad \text{on } \mathcal{S}_\infty$$

if $R_0 \ge R(A, \varepsilon, \{\Omega_j\})$. Moreover, $r_\varepsilon(D_x, y, D_y)$ maps continuously $\mathcal{S}_{\varepsilon_2}$ and $\mathcal{S}_{-\varepsilon_2}'$ to $\mathcal{S}_{\varepsilon_1}$ and $\mathcal{S}_{-\varepsilon_1}'$, respectively, if $0 < \varepsilon \le 1$, $\varepsilon_2 = -c_0/R_0 + 2(\varepsilon_1)_+$, $\varepsilon_1 \le 1/R_0$ and $R_0 \ge \max\{R(A, \varepsilon, \{\Omega_j\}), 4enA_2\}$.

Proof Let $0 < \varepsilon \le 1$. It follows from Theorem 2.4. 6 that there are positive constants $A_1' \equiv A_1'(A, \varepsilon)$, $A_2' \equiv A_2'(A)$ and $R(A, \varepsilon)$, which are polynomials of A of degree 1, symbols

$$p_{0,\varepsilon}^j(\xi, y, \eta) \in \bigcap_{\delta > 0} S^{m_1, m_2, 0, \delta}(4R_0, 4A, 6A + 4\varepsilon/\delta, A_1')$$

and $r^j_{0,\varepsilon}(\xi,y,\eta)$ ($j=1,2$) and $c_0>0$ such that

$$\begin{aligned}&\left|\partial_\xi^\alpha D_y^{\beta+\tilde\beta}\partial_\eta^\gamma r^j_{0,\varepsilon}(\xi,y,\eta)\right|\le C_{|\alpha|+|\tilde\beta|+|\gamma|,\varepsilon,R_0}((R_0/(4en)+A_2')/R_0)^{|\beta|}\\&\qquad\times\langle\xi\rangle^{m_1-|\alpha|+|\beta|}\exp[-c_0\langle\eta\rangle/R_0]\quad\text{if } j=1,2 \text{ and } \langle\xi\rangle\ge R_0|\beta|,\\&p(D_x,y,D_y)\varphi(D_x,y,D_y)=p^1_{0,\varepsilon}(D_x,y,D_y)+r^1_{0,\varepsilon}(D_x,y,D_y)\quad\text{on } S_\infty,\\&\varphi(D_x,y,D_y)p(D_x,y,D_y)=p^2_{0,\varepsilon}(D_x,y,D_y)+r^2_{0,\varepsilon}(D_x,y,D_y)\quad\text{on } S_\infty\end{aligned}$$

if $R_0\ge R(A,\varepsilon)$. Choose open conic subsets $\tilde\Omega_1$ and $\tilde\Omega_4$ of $\boldsymbol{R}^n\times(\boldsymbol{R}^n\setminus\{0\})$ so that $\Omega_1\subset\subset\tilde\Omega_1\subset\subset\Omega_2$ and $\Omega_3\subset\subset\tilde\Omega_4\subset\subset\Omega_4$. We also choose $\psi(\eta)\in C^\infty(\boldsymbol{R}^n)$ so that $\psi(\eta)=1$ for $|\eta|\ge 2R_0$ and $\psi(\eta)=0$ for $|\eta|\le R_0$. Applying Theorem 2.4. 6 (ii) to $p(\xi,w,\zeta)\varphi(\zeta,y,\eta)\psi(\eta)$ and $\varphi(\xi,w,\zeta)p(\zeta,y,\eta)\psi(\eta)$ we can show that there are positive constants $A_2\equiv A_2(A,\tilde\Omega_1,\Omega_2,\Omega_3,\tilde\Omega_4)$ and $R_1(A,\varepsilon,\tilde\Omega_1,\Omega_2,\Omega_3,\tilde\Omega_4)$, which are polynomials of A of degree 1, such that

$$\begin{aligned}&\left|\partial_\xi^\alpha D_y^{\beta+\tilde\beta}\partial_\eta^\gamma\left\{p^j_{0,\varepsilon}(\xi,y,\eta)-p(\xi,y,\eta)\varphi(\xi,y,\eta)\right\}\right|\\&\le C_{|\alpha|+|\tilde\beta|+|\gamma|,\varepsilon,R_0}(\tilde\Omega_1,\Omega_2,\Omega_3,\tilde\Omega_4)((R_0/(4en)+A_2)/R_0)^{|\beta|}\\&\qquad\times\langle\xi\rangle^{(m_1)_+-|\alpha|+|\beta|}\exp[-c_0\langle\eta\rangle/R_0]\quad(j=1,2)\end{aligned}$$

if $(y,\eta)\in\tilde\Omega_1$ or $(y,\eta)\notin\tilde\Omega_4$ and if $R_0\ge R_1(A,\varepsilon,\tilde\Omega_1,\Omega_2,\Omega_3,\tilde\Omega_4)$ and $\langle\xi\rangle\ge R_0|\beta|$, modifying c_0 if necessary. Indeed, $p(\xi,w,\zeta)\varphi(\zeta,y,\eta)\psi(\eta)$ and $\varphi(\xi,w,\zeta)p(\zeta,y,\eta)\psi(\eta)$ are analytic in y if $(w,\eta)\in\tilde\Omega_1$ or $(w,\eta)\notin\tilde\Omega_4$ and if $|w-y|\ll 1$ and $|\zeta-\eta|/|\eta|\ll 1$. Choose $\Phi(\xi,y,\eta)\in S^{0,0}(R_0,C(\Omega_1,\tilde\Omega_1,\tilde\Omega_4,\Omega_4))$ so that $\Phi(\xi,y,\eta)=1$ for $(\xi,y,\eta)\in\boldsymbol{R}^n\times(\tilde\Omega_4\setminus\tilde\Omega_1)$ with $|\eta|\ge R_0$ and $\operatorname{supp}\Phi\subset\boldsymbol{R}^n\times(\Omega_4\setminus\Omega_1)$. If we put

$$\begin{aligned}q_\varepsilon(\xi,y,\eta)&=\Phi(\xi,y,\eta)(p^1_{0,\varepsilon}(\xi,y,\eta)-p^2_{0,\varepsilon}(\xi,y,\eta)),\\r_\varepsilon(\xi,y,\eta)&=r^1_{0,\varepsilon}(\xi,y,\eta)-r^2_{0,\varepsilon}(\xi,y,\eta)\\&\quad+(1-\Phi(\xi,y,\eta))(p^1_{0,\varepsilon}(\xi,y,\eta)-p^2_{0,\varepsilon}(\xi,y,\eta)),\end{aligned}$$

then this choice of q_ε and r_ε proves the first part of the lemma. The second part easily follows from Theorem 2.3. 3 . □

Let Ω and $\Gamma_j\equiv X_j\times\gamma_j$ ($0\le j\le 3$) be open conic subsets of $\boldsymbol{R}^n\times(\boldsymbol{R}^n\setminus\{0\})$ satisfying $\Gamma_0\subset\subset\Gamma_1\subset\subset\Gamma_2\subset\subset\Gamma_3\subset\subset\Omega$. Let $\Lambda(x,\xi)\in C^\infty(\Omega)$ be a symbol satisfying (4.3), and choose $\Psi_j(\xi,y,\eta)\in S^{0,0}(R_0,A)$ and $\chi_j(\xi)\in S^0(R_0,A)$ ($R\ge 4$, $j=1,2$) so that $\Psi_j(\xi,y,\eta)=1$ for $y\in X_j$, $\operatorname{supp}\Psi_j\subset\boldsymbol{R}^n\times X_{j+1}\times\boldsymbol{R}^n$, $\chi_j(\xi)=1$ for $\xi\in\gamma_j$ with $|\xi|\ge R_0$ and $\operatorname{supp}\chi_j\subset\gamma_{j+1}\cap\{|\xi|\ge R_0/2\}$ ($j=1,2$). Put $\psi_j(\xi,y,\eta)=\Psi_j(\xi,y,\eta)\chi_j(\eta)$ ($j=1,2$). We write

$$Op(a(\xi,y,\eta))=a(D_x,y,D_y).$$

Lemma 4.1. 6 *There are positive constants* $A(\Lambda, A)$, $R_0(\Lambda, A)$, $C(\Lambda, A)$, $C'(\Lambda, A)$, $\tau(\Lambda, A)$, $\delta(\Lambda, A)$ *and* C_1 *and* $q(\xi, y, \eta; t) \in S^{0,0}(C(\Lambda, A)R_0, A(\Lambda, A))$ *such that*

$$\Big\{Op\Big(e^{t\Lambda(y,\xi)}\psi_1(\eta, y, \xi)\Big)Op\Big(e^{-t\Lambda(y,\eta)}\psi_2(\xi, y, \eta)\Big)q(D_x, y, D_y) - 1\Big\}$$
$$\times\varphi(D_x, y, D_y)$$

maps continuously $\mathcal{S}_{-\delta}$ (resp. $\mathcal{S}_\delta{}'$) *to* $\mathcal{S}_\delta$ (resp. $\mathcal{S}'_{-\delta}$) *if*

$$\begin{cases} R_0 \ge R_0(\Lambda, A), \qquad 0 \le t \le \tau(\Lambda, A)/R_0, \\ \varphi(\xi, y, \eta) \in S^{m_1, m_2}(R', B), \qquad \operatorname{supp}\varphi \subset \boldsymbol{R}^n \times \Gamma_0, \\ R' \ge C'(\Lambda, A)R_0, \qquad R' \ge R(B), \\ \delta \le \min\{\delta(\Lambda, A)/R_0, C_1/R'\}, \end{cases} \tag{4.4}$$

where $R(B)$ *is a polynomial of* B *of degree* 1.

Remark (i) From the proof of the lemma we see that

$$\varphi(D_x, y, D_y)\Big\{Op\Big(e^{t\Lambda(y,\xi)}\psi_1(\eta, y, \xi)\Big)Op\Big(e^{-t\Lambda(y,\eta)}\psi_2(\xi, y, \eta)\Big)$$
$$\times q(D_x, y, D_y) - 1\Big\}$$

maps continuously $\mathcal{S}_{-\delta}$ (resp. $\mathcal{S}_\delta{}'$) to $\mathcal{S}_\delta$ (resp. $\mathcal{S}'_{-\delta}$) if (4.4) is satisfied.
(ii) In the condition (4.4) we can drop the condition $R' \ge C'(\Lambda, A)R_0$, modifying $R_0(\Lambda, A)$, $R(B)$ and $\delta(\Lambda, A)$ if necessary. In fact, if $R' \le C'(\Lambda, A)R_0$, then we replace R' and B by $\widetilde{R}' \equiv C'(\Lambda, A)R_0$ and $\widetilde{B} \equiv C'(\Lambda, A)R_0B/R'$, respectively. Since $R(\widetilde{B}) = C'(\Lambda, A)R_0BK_1/R' + K_2$ with some $K_j > 0$ ($j = 1, 2$), we have $\widetilde{R}' \ge R(\widetilde{B})$ if $R_0 \ge 2K_2/C'(\Lambda, A)$ and $R' \ge 2BK_1$. Moreover, we have $C_1/\widetilde{R}' = C_1/(C'(\Lambda, A)R_0) \ge \delta(\Lambda, A)/R_0$ if $\delta(\Lambda, A) \le C_1/C'(\Lambda, A)$.

Proof It follows from Corollary 2.4. 5 (iii) with $p(\xi, w, \zeta, y, \eta) = \psi_1(\zeta, w, \xi)\psi_2(\zeta, y, \eta)$, $S(y, \xi) = -y \cdot \xi - it\Lambda(y, \xi)$, $T(y, \eta) = y \cdot \eta + it\Lambda(y, \eta)$ and $U(y, \zeta) = -y \cdot \zeta$ that there are $R_1(\Lambda, A)$, $\kappa_0(\Lambda, A)$, $\kappa(\Lambda, A)$, $\tau_1(\Lambda, A)$ and $\delta_1(\Lambda, A)$ such that

$$Op\Big(e^{t\Lambda(y,\xi)}\psi_1(\eta, y, \xi)\Big)Op\Big(e^{-t\Lambda(y,\eta)}\psi_2(\xi, y, \eta)\Big)$$
$$= Op\Big(e^{t(\Lambda(y,\xi)-\Lambda(y,\eta))}p_\kappa^R(\xi, y, \eta; t)\Big) + Q_1$$

on $\mathcal{S}_\infty$, and Q_1 maps continuously $\mathcal{S}_{-\delta}$ (resp. $\mathcal{S}_\delta{}'$) to $\mathcal{S}_\delta$ (resp. $\mathcal{S}'_{-\delta}$) if

$$\begin{cases} R_0 \ge R_1(\Lambda, A), \quad \kappa = \kappa_0(\Lambda, A), \quad R = \kappa(\Lambda, A)R_0, \\ 0 \le t \le \tau_1(\Lambda, A)/R_0, \quad \delta \le \delta_1(\Lambda, A), \end{cases} \tag{4.5}$$

where

$$p_\kappa^R(\xi, w, \eta; t) = \sum_{j=1}^{\infty} \psi_j^R(\eta) \sum_{|\gamma|<j} \Big\{ \partial_\zeta^\gamma D_y^\gamma$$
$$\times p_\kappa(\xi, w, \zeta+\eta, \eta; it\tilde{\nabla}_y \Lambda(w, w+y, \eta)) \Big\}_{y=0,\, \zeta=0} / \gamma!,$$
$$p_\kappa(\xi, w, \zeta, \eta; u) = \sum_{j=1}^{\infty} \psi_j^{\kappa R_0}(\eta) \sum_{|\alpha|<j} u^\alpha \partial_\zeta^\alpha \psi_1(\zeta, w, \xi) \chi_2(\eta)/\alpha!.$$

Moreover, there are positive constants $C_1(\Lambda, A)$, $A_1(\Lambda, A)$, $B_1(\Lambda, A)$ and $t_1(\Lambda, A, \varepsilon)$ ($\varepsilon > 0$) such that

$$p_\kappa^R(\xi, y, \eta; t) \in S^{0,2,0,\delta(\varepsilon)}(R_1, A_1(\Lambda, A))$$

if $\varepsilon > 0$, (4.5) is valid and $0 \le t \le t_1(\Lambda, A, \varepsilon)$, where $R_1 = C_1(\Lambda, A) R_0$ and $\delta(\varepsilon) = \varepsilon/R_0 + B_1(\Lambda, A)/R_0^2$. Note that

$$\operatorname{supp} p_\kappa^R \subset \{(\xi, y, \eta) \in \gamma_2 \times X_2 \times \gamma_3;\ |\xi| \ge R_0/2 \text{ and } |\eta| \ge R_0/2\},$$
$$p_\kappa^R(\xi, y, \eta; t) = \chi_1(\xi)\chi_2(\eta) \quad \text{for } y \in X_1.$$

Next let us apply Theorem 2.5. 3 to $Op\Big(e^{t(\Lambda(y,\xi)-\Lambda(y,\eta))} p_\kappa^R(\xi, y, \eta; t)\Big)$ with $p(\xi, w, \zeta, y, \eta) = p_\kappa^R(\zeta, y, \eta; t)$, $S(w, \xi) = -w \cdot \xi$, $T(w, \eta) = w \cdot \eta$ and $U(y, \eta) = -y \cdot \eta - it\Lambda(y, \eta)$. From Lemma 2.5. 2 (or its proof) there are positive constants $t_2(\Lambda)$, $C_0(\Lambda)$, $A_0(\Lambda)$ and ε_0, a complex neighborhood $\tilde{X}_2$ of $\overline{X}_2$ in $\boldsymbol{C}^n$ and an analytic function $Z(z; \eta, \zeta; t)$ defined in $D \equiv \{(z, \eta, \zeta, t) \in \tilde{X}_2 \times \gamma_3 \times \gamma_3 \times [0, t_2(\Lambda)];\ |\zeta - \eta| \le \varepsilon_0 |\eta|,\ |\zeta| \ge 1 \text{ and } |\eta| \ge 1\}$ such that $y = Z(z; \eta, \zeta; t)$ satisfies

$$\begin{aligned} &y + it\tilde{\nabla}_\eta \Lambda(y, \eta, \zeta) = z, \\ &\Big| D_z^\beta \partial_\eta^\rho D_\zeta^\gamma \partial_t^j Z(z; \eta, \zeta; t) \Big| \\ &\le C_0(\Lambda) A_0(\Lambda)^{|\beta|+|\rho|+|\gamma|+j} |\beta|! |\rho|! |\gamma|! j! \langle \eta \rangle^{-|\gamma|-|\rho|} \end{aligned} \tag{4.6}$$

if $(z, \eta, \zeta, t) \in D$. Let $g(\xi, \eta) \in S^{0,0}(8R_1, C_*)$ be a symbol such that $g(\xi, \eta) = 1$ if $|\xi| \le 4|\eta|$ and $|\xi|^2 + 2|\eta|^2 \ge 8$, and $g(\xi, \eta) = 0$ if $|\xi| \ge 5|\eta|$ or $|\xi|^2 + 2|\eta|^2 \le 4$. We put

$$p(\zeta, y, \eta; u; t) = \sum_{j=1}^{\infty} \psi_j^{R_1}(\eta) \sum_{|\beta|<j} u^\beta \partial_y^\beta p_\kappa^R(\zeta, y, \eta; t)/\beta!,$$
$$p(\xi, w, \eta; t) = \sum_{j=1}^{\infty} g(\xi, \eta) \psi_j^{8R_1}(\eta) \sum_{|\gamma|<j} \Big[(-\partial_\zeta)^\gamma D_y^\gamma$$

$$\times\Big\{p(\zeta, \text{Re}\ Z(y,\eta,\zeta,w;t),\eta; i\,\text{Im}\ Z(y,\eta,\zeta,w;t);t)$$
$$\times(-1)^n \det\frac{\partial Z}{\partial z}(y,\eta,\zeta,w;t)\Big\}\Big]_{y=0,\,\zeta=\eta}\Big/\gamma!,$$

where $R = \kappa(\Lambda, A)R_0$, $\kappa = \kappa_0(\Lambda, A)$ and

$$Z(z,\eta,\zeta,w;t) = Z(-z+w;\eta,\zeta;t).$$

It follows from Theorem 2.5. 3 that there are positive constants $R_2(\Lambda, A)$, $\tau_2(\Lambda, A)$, $\varepsilon(\Lambda, A)$ and $\delta_2(\Lambda, A)$ such that

$$Op\Big(e^{t(\Lambda(y,\xi)-\Lambda(y,\eta))}p_\kappa^R(\xi,y,\eta;t)\Big) = p(D_x,y,D_y;t) + Q_2 \quad \text{on } \mathcal{S}_\infty,$$

and Q_2 maps continuously $\mathcal{S}_{-\delta}$ (resp. $\mathcal{S}_\delta{}'$) to $\mathcal{S}_\delta$ (resp. $\mathcal{S}'_{-\delta}$) if (4.5) is valid and

$$\begin{cases} R_0 \ge R_2(\Lambda, A), \qquad \delta \le \delta_2(\Lambda, A)/R_0, \\ 0 \le t \le \min\{t_1(\Lambda, A, \varepsilon(\Lambda, A)), \tau_2(\Lambda, A)/R_0\}. \end{cases} \tag{4.7}$$

Moreover, there are positive constants $A_2(\Lambda, A)$, $B_2(\Lambda, A)$ and $t_2(\Lambda, A)$ such that

$$\Big|\partial_\xi^{\alpha+\tilde\alpha} D_y^{\beta+\tilde\beta}\partial_\eta^{\gamma+\tilde\gamma}p(\xi,y,\eta;t)\Big| \le C_{|\tilde\alpha|+|\tilde\beta|+|\tilde\gamma|}(A_2(\Lambda, A)/R_1)^{|\alpha|+|\beta|+|\gamma|}\langle\xi\rangle^{-|\tilde\alpha|}$$
$$\times\langle\eta\rangle^{4-|\tilde\gamma|+|\beta|}\exp[(\varepsilon/R_0 + B_1(\Lambda, A)/R_0^2 + \varepsilon'/R_1 + B_2(\Lambda, A)/R_1^2)\langle\eta\rangle]$$

if $\varepsilon, \varepsilon' > 0$, (4.5) and (4.7) are valid, $0 \le t \le \min\{t_1(\Lambda, A, \varepsilon), \varepsilon' t_2(\Lambda, A)\}$, $\langle\xi\rangle \ge 8R_1|\alpha|$, $\langle\eta\rangle \ge 24R_1|\gamma|$ and $\langle\eta\rangle \ge 8R_1(|\beta|+|\gamma|)$. Assume that (4.5) and (4.7) are valid. Let $g_1(\xi,\eta) \in S^{0,0}(8R_1, C_*)$ be a symbol such that $g_1(\xi,\eta) = 1$ if $|\xi| \ge |\eta|/2$ and $|\xi|^2+|\eta|^2 \ge 2$, and $g_1(\xi,\eta) = 0$ if $|\xi| \le |\eta|/4$ or $|\xi|^2+|\eta|^2 \le 1$, and put

$$p_1(\xi,y,\eta;t) = g_1(\xi,\eta)p(\xi,y,\eta;t),$$
$$p_2(\xi,y,\eta;t) = (1-g_1(\xi,\eta))p(\xi,y,\eta;t).$$

By Lemma 4.1. 4 with $p(\xi,w,\zeta,y,\eta) = p_2(\zeta,y,\eta;t)$ there are positive constants $R_3(\Lambda, A)$, $\delta_3(\Lambda, A)$, $\varepsilon_1(\Lambda, A)$ and $\varepsilon_1'(\Lambda, A)$ such that $p_2(D_x,y,D_y;t)$ maps continuously $\mathcal{S}_{-\delta}$ (resp. $\mathcal{S}_\delta{}'$) to $\mathcal{S}_\delta$ (resp. $\mathcal{S}'_{-\delta}$) if

$$\begin{cases} R_0 \ge R_3(\Lambda, A), \qquad \delta \le \delta_3(\Lambda, A)/R_0, \\ 0 \le t \le \min\{t_1(\Lambda, A, \varepsilon_1(\Lambda, A)), \varepsilon_1'(\Lambda, A)t_2(\Lambda, A)\}. \end{cases}$$

It is obvious that $p_1(\xi,y,\eta;t) \in S^{0,4,0,\delta(\varepsilon,\varepsilon')}(48R_1, 192A_2(\Lambda,A)+6C_*)$ if $\varepsilon,\varepsilon'>0$ and $0\le t\le \min\{t_1(\Lambda,A,\varepsilon),\varepsilon' t_2(\Lambda,A)\}$, where

$$\delta(\varepsilon,\varepsilon')=\varepsilon/R_0+B_1(\Lambda,A)/R_0^2+\varepsilon'/R_1+B_2(\Lambda,A)/R_1^2.$$

Choose an open subset $\widetilde{X}_1$ of $\boldsymbol{R}^n$ and $t(\widetilde{X}_1,X_1)>0$ so that $X_0 \subset\subset \widetilde{X}_1 \subset\subset X_1$ and $\mathrm{Re}\, Z(0,\eta,\eta,w;t)\in X_1$ if $0\le t\le t(\widetilde{X}_1,X_1)$, $w\in\widetilde{X}_1$ and $|\eta|\ge 1$. Then we have

$$p_1(\xi,w,\eta;t)=g(\xi,\eta)g_1(\xi,\eta)p(w,\eta;t)$$

if $0\le t\le t(\widetilde{X}_1,X_1)$, $w\in\widetilde{X}_1$, $\eta\in\gamma_1$ and $|\eta|\ge R_0$, where

$$\begin{aligned}&p(w,\eta;t)\\&=\sum_{j=1}^{\infty}\psi_j^{8R_1}(\eta)\sum_{|\gamma|<j}(-1)^n\Big[(-\partial_\zeta)^\gamma D_y^\gamma \det\frac{\partial Z}{\partial z}(y,\eta,\zeta,w;t)\Big]_{y=0,\,\zeta=\eta}/\gamma!.\end{aligned}$$

From (4.6) and Lemma 2.2.4 we have $p(x,\xi;t)\in PS^0(\Gamma_1;16R_1,6A_0(\Lambda)+3\widehat{C})$ if $8R_1\ge 18nA_0(\Lambda)^2$. Let $\widetilde{\Gamma}_0\equiv\widetilde{X}_0\times\tilde{\gamma}_0$ be an open conic subset of $\boldsymbol{R}^n\times(\boldsymbol{R}^n\setminus\{0\})$ satisfying $\Gamma_0\subset\subset\widetilde{\Gamma}_0\subset\subset\widetilde{X}_1\times\gamma_1$, and choose $\psi(\xi,y,\eta)\in S^{0,0}(16R_1,A_1)$ so that $\psi(\xi,y,\eta)=1$ for $(\xi,y,\eta)\in\boldsymbol{R}^n\times\widetilde{\Gamma}_0$ with $|\eta|\ge 2R_0$ and $\mathrm{supp}\,\psi\subset\boldsymbol{R}^n\times\widetilde{X}_1\times\gamma_1\cap\{|\eta|\ge R_0\}$. We put

$$\begin{aligned}\tilde{p}_0(\xi,y,\eta;t)&=\psi(\xi,y,\eta)p(y,\eta;t),\\ \tilde{p}_1(\xi,y,\eta;t)&=(g(\xi,\eta)g_1(\xi,\eta)-1)\tilde{p}_0(\xi,y,\eta;t),\\ \tilde{p}_2(\xi,y,\eta;t)&=(1-\psi(\xi,y,\eta))p_1(\xi,y,\eta;t).\end{aligned}$$

Note that $p_1(\xi,y,\eta;t)=\sum_{j=0}^2\tilde{p}_j(\xi,y,\eta;t)$. From Lemma 4.1.4 we can see that $\tilde{p}_1(D_x,y,D_y;t)$ maps continuously $\mathcal{S}_{-\delta}$ (resp. $\mathcal{S}_\delta{}'$) to $\mathcal{S}_\delta$ (resp. $\mathcal{S}'_{-\delta}$) if $R_0\ge R_4(\Lambda,A,A_1)$ and $\delta\le\delta_4(\Lambda,A)/R_0$, where $R_4(\Lambda,A,A_1)$ and $\delta_4(\Lambda,A)$ are some positive constants. Let $\varphi(\xi,y,\eta)\in S^{m_1,m_2}(R',B)$ be a symbol satisfying $\mathrm{supp}\,\varphi\subset\boldsymbol{R}^n\times\Gamma_0$. It follows from Lemma 4.1.4 (or its variant) that there are positive constants $R_5(\Lambda,A,A_1,\Gamma_0,\widetilde{\Gamma}_0)$, $R'(B,\Gamma_0,\widetilde{\Gamma}_0)$, $\delta_5(\Lambda,A)$, $\varepsilon_2(\Lambda,A)$ and $\varepsilon_2'(\Lambda,A)$ such that $R'(B,\Gamma_0,\widetilde{\Gamma}_0)$ is a polynomial of B of degree 1, and $\tilde{p}_2(D_x,y,D_y;t)\varphi(D_x,y,D_y)$ maps continuously $\mathcal{S}_{-\delta}$ (resp. $\mathcal{S}_\delta{}'$) to $\mathcal{S}_\delta$ (resp. $\mathcal{S}'_{-\delta}$) if $R_0\ge R_5(\Lambda,A,A_1,\Gamma_0,\widetilde{\Gamma}_0)$, $R'\ge R'(B,\Gamma_0,\widetilde{\Gamma}_0)$, $0\le t\le\min\{t_1(\Lambda,A,\varepsilon_2(\Lambda,A)),\varepsilon_2'(\Lambda,A)t_2(\Lambda,A)\}$ and $\delta\le\delta_5(\Lambda,A)/R_0$. Note that

$$\begin{aligned}&\tilde{p}_0(\xi,y,\eta;t)\in S^{0,0}(16R_1,A'(\Lambda,A_1)),\\ &\tilde{p}_0(\xi,y,\eta;t)=p(y,\eta;t)\quad\text{for } (\xi,y,\eta)\in\boldsymbol{R}^n\times\widetilde{\Gamma}_0 \text{ with } |\eta|\ge 2R_0,\\ &|p(x,\xi;t)|\ge 1/2\quad\text{if } (x,\xi)\in\widetilde{\Gamma}_0 \text{ and } 0\le t\ll 1,\end{aligned}$$

where $A'(\Lambda, A_1)$ is a positive constant. Applying the same argument as in Theorem 3.2.7 we can construct a right microlocal parametrix of $\tilde{p}_0(D_x, y, D_y; t)$, *i.e.*, there are positive constants $R_6(\Lambda, A, A_1, \Gamma_0, \widetilde{\Gamma}_0)$, $A(\Lambda, A_1, \Gamma_0, \widetilde{\Gamma}_0)$, $R_2'(B, \Gamma_0, \widetilde{\Gamma}_0)$, $\delta_6(\Lambda, A)$ and C_1 and a symbol $q(\xi, y, \eta; t) \in S^{0,0}(32R_1, A(\Lambda, A_1, \Gamma_0, \widetilde{\Gamma}_0))$ such that $R_2'(B, \Gamma_0, \widetilde{\Gamma}_0)$ is a polynomial of B of degree 1, and

$$(\tilde{p}_0(D_x, y, D_y; t)q(D_x, y, D_y; t) - 1)\varphi(D_x, y, D_y)$$

maps continuously $\mathcal{S}_{-\delta}$ (resp. $\mathcal{S}_\delta'$) to $\mathcal{S}_\delta$ (resp. $\mathcal{S}_{-\delta}'$) if $R_0 \geq R_6(\Lambda, A, A_1, \Gamma_0, \widetilde{\Gamma}_0)$, $R' \geq 96R_1$, $R' \geq R_2'(B, \Gamma_0, \widetilde{\Gamma}_0)$, $\delta \leq \min\{\delta_6(\Lambda, A)/R_0, C_1/R'\}$ and $0 \leq t \ll 1$. This completes the proof. □

Let $p(x, \xi) \equiv \sum_{j=0}^\infty p_j(x, \xi) \in FS^m(\Omega; C_0, A_1)$, and choose open conic subsets $\Gamma_j \equiv X_j \times \gamma_j$ ($j = 4, 5$) of $\boldsymbol{R}^n \times (\boldsymbol{R}^n \setminus \{0\})$ and a symbol $\Phi^R(\xi, y, \eta) \in S^{0,0}(R, A)$ ($R \geq 4$) so that $\Gamma_3 \subset\subset \Gamma_4 \subset\subset \Gamma_5 \subset\subset \Omega$, $\Phi^R(\xi, y, \eta) = 1$ for $(\xi, y, \eta) \in \Gamma_4$ with $|\eta| \geq R$ and $\operatorname{supp} \Phi^R \subset \boldsymbol{R}^n \times \Gamma_5 \cap \{|\eta| \geq R/2\}$. We put

$$\tilde{p}(x, \xi) = \sum_{j=0}^\infty \phi_j^{R_0/2}(\xi) p_j(x, \xi),$$
$$\tilde{p}(\xi, y, \eta) = \Phi^{R_0}(\xi, y, \eta)\tilde{p}(y, \eta).$$

Recall that $\tilde{p}(x, \xi) \in PS^m(\Omega; R_0, A_1 + 3\widehat{C}, A_1)$ and $\tilde{p}(\xi, y, \eta) \in S^{0,m}(R_0, A, A_1 + A + 3\widehat{C}, A_1 + A)$ if $R_0 \geq 2C_0$.

Lemma 4.1.7 *Put*

$$p(\xi, y, \eta; t\Lambda) = \sum_{j=2}^\infty \psi_j^{8R_0}(\eta) \times \sum_{|\gamma|<j} \Big[\partial_\zeta^\gamma \Big\{\psi_1(\xi, y, \eta + \zeta)\tilde{p}_{(\gamma)}(y + it\widehat{\nabla}_\eta \Lambda(y, \eta, \zeta), \eta)\Big\}\Big]_{\zeta=0} / \gamma!$$

for $0 \leq t \ll 1$, *where*

$$\widehat{\nabla}_\eta \Lambda(y, \eta, \zeta) \, (= \widetilde{\nabla}_\eta \Lambda(y, \eta, \eta + \zeta)) = \int_0^1 \nabla_\eta \Lambda(y, \eta + \theta\zeta) \, d\theta.$$

Then there are positive constants $R_0(\Lambda, A_1, A)$, $\tau(\Lambda)$ *and* $\delta(\Lambda)$ *such that*

$$Op\Big(e^{-t\Lambda(y,\eta)}\psi_1(\xi, y, \eta)\Big)\tilde{p}(D_x, y, D_y) = Op\Big(e^{-t\Lambda(y,\eta)}p(\xi, y, \eta; t\Lambda)\Big) + Q \quad \text{on } \mathcal{S}_\infty,$$

and Q maps continuously $\mathcal{S}_{-\delta}$ (resp. $\mathcal{S}_\delta{}'$) to $\mathcal{S}_\delta$ (resp. $\mathcal{S}'_{-\delta}$) if

$$R_0 \geq R_0(\Lambda, A_1, A), \quad 0 \leq t \leq \tau(\Lambda)/R_0, \quad \delta \leq \delta(\Lambda)/R_0. \tag{4.8}$$

Moreover, we have $p(\xi, y, \eta; t\Lambda) \in S^{0,m+1}(2R_0, A'(\Lambda, A_1, A))$ if (4.8) is valid, where $A'(\Lambda, A_1, A)$ is a positive constant.

Remark The first part of the lemma is still valid for $p(x,\xi) \in FS^+(\Omega; C_0, A_1)$. Then we have $p(\xi, y, \eta; t\Lambda) \in S^+(2R_0, A'(\Lambda, A_1, A))$.

Proof Applying Corollary 2.5. 4 with $p(\xi, w, \zeta, y, \eta) = \psi_1(\xi, w, \zeta)\tilde{p}(\zeta, y, \eta)$, $S(w,\xi) = -w\cdot\xi$, $T(w,\eta) = w\cdot\eta + it\Lambda(w,\eta)$ and $U(y,\eta) = -y\cdot\eta$, we can easily prove the lemma. □

Applying Lemma 2.4. 2 with $U(y,\zeta) = -y\cdot\zeta - it\Lambda(y,\zeta)$ we can see that there are $t_0(\Lambda) > 0$ and a symbol $N(x,y,\xi;t)$ defined in $\tilde{\Omega}$ such that $\eta = N(x,y,\xi;t)$ is the solution of

$$\eta + it\hat{\nabla}_x\Lambda(x,y,\eta) = \xi,$$

where $\varepsilon_0 \ll 1$, $\tilde{\Omega} = \{(x,y,\xi,t) \in \boldsymbol{R}^n \times \boldsymbol{R}^n \times \boldsymbol{R}^n \times [0, t_0(\Lambda)];\ (x,\xi) \in \Omega,\ |y| \leq \varepsilon_0 \text{ and } |\xi| \geq 1\}$ and

$$\hat{\nabla}_x\Lambda(x,y,\eta)\ (= \tilde{\nabla}_x\Lambda(x, x+y, \eta)) = \int_0^1 \nabla_x\Lambda(x+\theta y, \eta)\, d\theta,$$

modifying Ω if necessary. Put

$$p_{t\Lambda}(x,\xi) = \sum_{j=0}^{\infty} p^j_{t\Lambda}(x,\xi) \quad \text{as a formal analytic symbol,}$$

$$\begin{aligned} p^j_{t\Lambda}(x,\xi) = \sum_{|\gamma|+|\lambda|+k=j} \Big[&\partial_\xi^\gamma \partial_\eta^\lambda D_y^\gamma \\ &\times\Big\{p_{k(\lambda)}(x + it\hat{\nabla}_\eta\Lambda(x, N(x,y,\xi;t), \eta), N(x,y,\xi;t)) \\ &\times \det \frac{\partial N}{\partial \xi}(x,y,\xi;t)\Big\}\Big]_{y=0,\,\eta=0}/(\gamma!\lambda!) \end{aligned}$$

for $(x,\xi) \in \Omega$ and $t \in [0, t_0(\Lambda)]$, modifying $t_0(\Lambda)$ if necessary. Using Cauchy's estimates or induction we can see that there are positive constants $C_0(\Lambda, C_0, A_1)$ and $A_1(\Lambda, A_1)$ such that

$$p_{t\Lambda}(x,\xi) \in FS^m(\Omega; C_0(\Lambda, C_0, A_1), A_1(\Lambda, A_1))$$

for $t \in [0, t_0(\Lambda)]$. A simple calculation gives

$$p^0_{t\Lambda}(x,\xi) = p_0(z(x,\xi;t), \zeta(x,\xi;t)) \det \mathcal{N}(x,\xi;t), \tag{4.9}$$

$$\begin{aligned}
&p^1_{t\Lambda}(x,\xi) + (i/2)\sum_{j=1}^n (\partial^2/\partial x_j \partial\xi_j) p^0_{t\Lambda}(x,\xi) \ \Big(= (p_{t\Lambda})^1_s(x,\xi)\Big) \\
&= p^1_s(z(x,\xi;t), \zeta(x,\xi;t)) \det \mathcal{N}(x,\xi;t) \\
&\quad +(i/2) \sum_{j,k=1}^n (\partial_{x_j} p_0)(z(x,\xi;t), \zeta(x,\xi;t)) \\
&\qquad \times \partial_{\xi_k}\{(\delta_{j,k} + it(\partial^2\Lambda/\partial x_k \partial\eta_j)(x, \zeta(x,\xi;t)) \det \mathcal{N}(x,\xi;t)\},
\end{aligned} \tag{4.10}$$

where $z(x,\xi;t) = x + it\nabla_\eta \Lambda(x, N(x,0,\xi;t))$, $\zeta(x,\xi;t) = N(x,0,\xi;t)$, $\mathcal{N}(x,\xi;\, t) = (\partial N/\partial\xi)(x,0,\xi;t)$ and

$$p^1_s(x,\xi) = p_1(x,\xi) + (i/2)\sum_{j=1}^n (\partial^2 p_0/\partial x_j \partial\xi_j)(x,\xi).$$

Let $\tilde\Gamma_0$ be an open conic subset of $\boldsymbol{R}^n \times (\boldsymbol{R}^n \setminus \{0\})$ satisfying $\Gamma_0 \subset\subset \tilde\Gamma_0 \subset\subset \Gamma_1$, and choose $\psi(\xi,y,\eta) \in S^{0,0}(R_0, A)$ so that $\psi(\xi,y,\eta) = 1$ for $(\xi,y,\eta) \in \boldsymbol{R}^n \times \tilde\Gamma_0$ with $|\eta| \ge 2R_0$ and $\operatorname{supp}\psi \subset \boldsymbol{R}^n \times \Gamma_1 \cap \{|\eta| \ge R_0\}$. We put

$$\tilde{p}^R_{t\Lambda}(\xi,y,\eta) = \psi(\xi,y,\eta)\sum_{j=0}^\infty \phi^R_j(\eta) p^j_{t\Lambda}(y,\eta)$$

for $R \ge C_0(\Lambda, C_0, A_1)$.

Lemma 4.1. 8 *There are positive constants $R_0(\Lambda, A_1, A)$, $\tau(\Lambda, A_1, A)$, $\delta(\Lambda,\, A_1, A)$, $\kappa(\Lambda, A_1, A)$, $R'(\Lambda, A_1, A)$ and c_0 such that*

$$\begin{aligned}
&Op\Big(e^{-t\Lambda(y,\eta)}\psi_1(\xi,y,\eta)\Big)\tilde p(D_x,y,D_y) Op\Big(e^{t\Lambda(y,\xi)}\psi_2(\eta,y,\xi)\Big)\varphi(D_x,y,D_y) \\
&= \tilde p^R_{t\Lambda}(D_x,y,D_y)\varphi(D_x,y,D_y) + Q \quad \text{on } \mathcal{S}_\infty,
\end{aligned}$$

and Q maps continuously $\mathcal{S}_{-\delta}$ (resp. $\mathcal{S}_\delta{}'$) to $\mathcal{S}_\delta$ (resp. $\mathcal{S}'_{-\delta}$) if $R_0 \ge R_0(\Lambda, A_1, A)$, $R = \kappa(\Lambda, A_1, A)R_0$, $0 \le t \le \tau(\Lambda, A_1, A)/R_0$, $\varphi(\xi,y,\eta) \in S^{m_1,m_2,\delta',\delta'}(R', B)$, $\operatorname{supp}\varphi \subset \boldsymbol{R}^n \times \Gamma_0$, $R' \ge R'(\Lambda, A_1, A)R_0$, $R' \ge R'(B)$ and $\max\{\delta, \delta'\} \le \min\{\delta(\Lambda, A_1, A)/R_0, c_0/R'\}$, where $R'(B)$ is a polynomial of B of degree 1.

Remark (i) The lemma is also valid for $p(x,\xi) \in FS^+(\Omega; C_0, A_1)$. (ii) In the conditions of the lemma we can drop the condition $R' \ge R'(\Lambda, A_1, A)R_0$ (see the remark of Lemma 4.1. 6).

Proof It follows from Lemma 4.1.7 that

$$\begin{aligned} P_{t\Lambda} &\equiv Op\Big(e^{-t\Lambda(y,\eta)}\psi_1(\xi,y,\eta)\Big)\tilde{p}(D_x,y,D_y)Op\Big(e^{t\Lambda(y,\xi)}\psi_2(\eta,y,\xi)\Big) \\ &= Op\Big(e^{-t\Lambda(y,\eta)}p(\xi,y,\eta;t\Lambda)\Big)Op\Big(e^{t\Lambda(y,\xi)}\psi_2(\eta,y,\xi)\Big) + Q_1 \qquad (4.11) \end{aligned}$$

on $\mathcal{S}_\infty$, and Q_1 maps continuously $\mathcal{S}_{-\delta}$ (resp. $\mathcal{S}_\delta{}'$) to $\mathcal{S}_\delta$ (resp. $\mathcal{S}'_{-\delta}$) if (4.8) is satisfied, modifying $\tau(\Lambda)$ and $\delta(\Lambda)$ if necessary. We assume that (4.8) is valid. Let us apply Corollary 2.4.5 (iii) to the first term on the right-hand side of (4.11) with $p(\xi,w,\zeta,y,\eta) = p(\xi,w,\zeta;t\Lambda)\psi_2(\eta,y,\zeta)$, $S(y,\xi) = -y\cdot\xi$, $T(y,\eta) = y\cdot\eta$ and $U(y,\zeta) = -y\cdot\zeta - it\Lambda(y,\zeta)$. Put

$$\begin{aligned} &p_\kappa(\xi,w,\zeta,y,\eta;u) \\ &= \sum_{j=1}^{\infty}\psi_j^{\kappa R_0}(\eta)\sum_{|\alpha|<j} u^\alpha\partial_\zeta^\alpha\big(p(\xi,w,\zeta;t\Lambda)\psi_2(\eta,y,\zeta)\big)/\alpha!, \\ &p_\kappa^R(\xi,w,\eta) = \sum_{j=1}^{\infty}\psi_j^R(\eta)\sum_{|\gamma|<j}\Big[\partial_\zeta^\gamma D_y^\gamma \\ &\quad\times\Big\{p_\kappa(\xi,w,\eta+\zeta,w+y,\eta;N(w,y,\eta+\zeta;t)-\eta-\zeta) \\ &\quad\times\det\frac{\partial N}{\partial\eta}(w,y,\eta+\zeta;t)\Big\}\Big]_{y=0,\,\zeta=0}\Big/\gamma!. \end{aligned}$$

Then there are positive constants $R(\Lambda,A_1,A)$, $\kappa_0(\Lambda,A_1,A)$, $\kappa(\Lambda,A_1,A)$, $\delta(\Lambda,A_1,A)$ and $\tau(\Lambda,A_1,A)$ such that

$$P_{t\Lambda} = p_\kappa^R(D_x,y,D_y) + Q_2 \quad \text{on } \mathcal{S}_\infty,$$

and Q_2 maps continuously $\mathcal{S}_{-\delta}$ (resp. $\mathcal{S}_\delta{}'$) to $\mathcal{S}_\delta$ (resp. $\mathcal{S}'_{-\delta}$) if $\delta \le \delta(\Lambda,A_1,A)/R_0$ and

$$\begin{cases} R_0 \ge R(\Lambda,A_1,A), & \kappa = \kappa_0(\Lambda,A_1,A), \\ R = \kappa(\Lambda,A_1,A)R_0, & 0 \le t \le \tau(\Lambda,A_1,A)/R_0. \end{cases} \qquad (4.12)$$

Now assume that (4.12) is valid. Moreover, there are positive constants $C(\Lambda,A_1,A)$, $A_1(\Lambda,A_1,A)$, $B(\Lambda,A_1,A)$ and $t(\Lambda,A_1,A,\varepsilon)$ ($\varepsilon > 0$) such that

$$p_\kappa^R(\xi,y,\eta) \in S^{0,m+3,0,\delta(\varepsilon)}(C(\Lambda,A_1,A)R_0, A_1(\Lambda,A_1,A))$$

if $\varepsilon > 0$, $\delta(\varepsilon) = \varepsilon/R_0 + B(\Lambda,A_1,A)/R_0^2$ and $0 \le t \le t(\Lambda,A_1,A,\varepsilon)$. Let $\varphi(\xi,y,\eta) \in S^{m_1,m_2,\delta',\delta'}(R',B)$ satisfy $\operatorname{supp}\varphi \subset \boldsymbol{R}^n \times \Gamma_0$. Then, by Lemma 4.1.4 and Theorem 2.3.3 we see that the operator

$$\Big\{Op\Big(p_\kappa^R(\xi,y,\eta)(1-\psi(\xi,y,\eta))\Big) + Q_2\Big\}\varphi(D_x,y,D_y)$$

maps continuously $\mathcal{S}_{-\delta}$ (resp. $\mathcal{S}_\delta{}'$) to $\mathcal{S}_\delta$ (resp. $\mathcal{S}_{-\delta}'$) if $R' \geq R'(B)$ and $\max\{\delta,\delta'\} \leq \min\{\delta(\Lambda,A_1,A)/R_0, c_0/R'\}$, where $R'(B)$ is a polynomial of B of degree 1 and c_0 is a positive constant, modifying $R(\Lambda,A_1,A)$, $\tau(\Lambda,A_1,A)$ and $\delta(\Lambda,A_1,A)$ if necessary. We have

$$\begin{aligned} p(\xi,y,\eta;t\Lambda) = &\sum_\gamma \sum_{k=0}^\infty \tilde{\phi}_{|\gamma|}^{8R_0}(\eta)\phi_k^{R_0/2}(\eta) \\ &\times \Big\{\partial_\zeta^\gamma p_{k(\gamma)}(y+it\widehat{\nabla}_\eta\Lambda(y,\eta,\zeta),\eta)\Big\}_{\zeta=0}/\gamma! \end{aligned}$$

for $(\xi,y,\eta) \in \boldsymbol{R}^n \times \Gamma_1$ with $|\eta| \geq R_0$, where $\tilde{\phi}_0^{8R_0}(\eta) = \phi_1^{8R_0}(\eta)$ and $\tilde{\phi}_j^{8R_0}(\eta) = \phi_j^{8R_0}(\eta)$ for $j \geq 1$. Moreover, we have

$$\begin{aligned} p_\kappa(\xi,w,\zeta,y,\eta;u) = &\sum_\gamma \sum_{k=0}^\infty \tilde{\phi}_{|\gamma|}^{8R_0}(\zeta)\phi_k^{R_0/2}(\zeta) \\ &\times \Big\{\partial_z^\gamma p_{k(\gamma)}(w+it\widehat{\nabla}_\eta\Lambda(w,\zeta+u,z),\zeta+u)\Big\}_{z=0}/\gamma! \\ &+q_\kappa(w,\zeta,\eta;u) \end{aligned}$$

if $(w,\zeta),(y,\zeta) \in \Gamma_1$, $|\zeta| \geq R_0$ and $|u|/|\zeta| \ll 1$, where

$$\begin{aligned} q_\kappa(w,\zeta,\eta;u) = &\sum_{\alpha,\gamma}\sum_{k=0}^\infty u^\alpha/(\alpha!\gamma!)\Big(\phi_{|\alpha|}^{\kappa R_0}(\eta)-1\Big) \\ &\times \tilde{\phi}_{|\gamma|}^{8R_0}(\zeta)\phi_k^{R_0/2}(\zeta)\Big\{\partial_\zeta^\alpha\partial_z^\gamma p_{k(\gamma)}(w+it\widehat{\nabla}_\eta\Lambda(w,\zeta,z),\zeta)\Big\}_{z=0} \\ &+ \sum_{\substack{\alpha^1,\alpha^2,\alpha^3,\gamma \\ \alpha^1+\alpha^2>0}}\sum_{k=0}^\infty u^{\alpha^1+\alpha^2+\alpha^3}/(\alpha^1!\alpha^2!\alpha^3!\gamma!)\phi_{|\alpha^1|+|\alpha^2|+|\alpha^3|}^{\kappa R_0}(\eta) \\ &\times \tilde{\phi}_{|\gamma|}^{8R_0(\alpha^1)}(\zeta)\phi_k^{R_0/2(\alpha^2)}(\zeta)\Big\{\partial_\zeta^{\alpha^3}\partial_z^\gamma p_{k(\gamma)}(w+it\widehat{\nabla}_\eta\Lambda(w,\zeta,z),\zeta)\Big\}_{z=0} \\ \equiv\ &q_\kappa^1(w,\zeta,\eta;u)+q_\kappa^2(w,\zeta,\eta;u). \end{aligned}$$

Then there are positive constants $A(\Lambda,A_1)$, $R_0(\Lambda,C_0,A_1)$ and $c(\Lambda,A_1)$ such that

$$\begin{aligned} \Big|\partial_\zeta^{\alpha+\tilde{\alpha}}D_w^\beta\partial_\eta^\gamma\partial_u^\lambda q_\kappa(w,\zeta,\eta;u)\Big| \leq\ &C_{|\tilde{\alpha}|+|\gamma|}(A(\Lambda,A_1)/R_0)^{|\alpha|} \\ &\times A(\Lambda,A_1)^{|\beta|+|\lambda|}|\beta|!|\lambda|!\langle\zeta\rangle^{m-|\lambda|-|\tilde{\alpha}|-|\gamma|}\exp[-\langle\zeta\rangle/(6\kappa R_0)] \end{aligned}$$

if $(w,\zeta),(y,\zeta) \in \Gamma_1$ $|\zeta| \geq R_0$, $|\eta-\zeta| \leq |\zeta|/2$, $|u| \leq c(\Lambda,A_1)\langle\zeta\rangle$, $R_0 \geq R_0(\Lambda,C_0,A_1)$ and $\langle\zeta\rangle \geq 24R_0|\alpha|$, modifying $\kappa_0(\Lambda,A_1,A)$ if necessary. In fact, for example, we have $\langle\eta\rangle \geq 2\kappa R_0(|\rho^1|+|\rho^2|+|\rho^3|)$, $\langle\zeta\rangle \geq$

$16R_0 \max\{|\delta|, 1\}$, $\langle\zeta\rangle \geq R_0 k$, $\langle\zeta\rangle \leq 24R_0 \max\{|\delta|, 1\}$ for $|\alpha^1| + |\rho^1| > 0$, $\langle\zeta\rangle \leq 3R_0 k/2$ for $|\alpha^2| + |\rho^2| > 0$, $|\alpha^1| + |\rho^1| \leq 2\max\{|\delta|, 1\}$ and $|\alpha^2| + |\rho^2| \leq k/8$ if $\phi^{\kappa R_0}_{|\rho^1|+|\rho^2|+|\rho^3|}(\eta)\tilde{\phi}^{8R_0(\alpha^1+\rho^1)}_{|\delta|}(\zeta)\phi^{R_0/2(\alpha^2+\rho^2)}_k(\zeta) \neq 0$, $|\eta - \zeta| \leq |\zeta|/2$, $\langle\zeta\rangle \geq 24R_0(|\alpha^1| + |\alpha^2|)$ and $\kappa \geq 18$. Using Cauchy's estimates for $p_{k(\gamma)}(w + it\widehat{\nabla}_\eta\Lambda(w, \zeta, z), \zeta)$, we have

$$
\begin{aligned}
&\left|\partial_\zeta^{\alpha+\tilde{\alpha}} D_w^\beta \partial_\eta^\gamma \partial_u^\lambda q_\kappa^2(w, \zeta, \eta; u)\right| \\
&\leq \sum_{\rho,\delta} \sum_{\substack{\rho^1+\rho^2+\rho^3=\rho+\lambda \\ \rho^1+\rho^2>0}} \sum_{k=0}^{\infty} \sum_{\alpha^1+\alpha^2+\alpha^3=\alpha} C_{|\tilde{\alpha}|+|\gamma|}\langle\zeta\rangle^{m-|\tilde{\alpha}|-|\gamma|}(\rho+\lambda)! \\
&\quad \times \alpha!|u^\rho|/(\rho!\rho^1!\rho^2!\rho^3!\alpha^1!\alpha^2!\alpha^3!\delta!)(\widehat{C}/(8R_0))^{|\alpha^1|+|\rho^1|} \\
&\quad \times (2\widehat{C}/R_0)^{|\alpha^2|+|\rho^2|} C_0^k A_1(\Lambda, A_1)^{|\alpha^3|+|\rho^3|+2|\delta|+|\beta|} k! \\
&\quad \times |\alpha^3|!|\rho^3|!|\delta|!^2|\beta|!\langle\zeta\rangle^{-k-|\alpha^3|-|\rho^3|-|\delta|}\chi_{\rho^1,\rho^2,\rho^3,\delta,\alpha^1,k,\alpha^2}(\zeta, \eta)
\end{aligned}
$$

if $(w, \zeta), (y, \zeta) \in \Gamma_1$, $|\zeta| \geq R_0$, $|\eta - \zeta| \leq |\zeta|/2$, $|u|/|\zeta| \ll 1$, $\kappa \geq 18$ and $\langle\zeta\rangle \geq 24R_0|\alpha|$, where $A_1(\Lambda, A_1) > 0$ and $\chi_{\rho^1,\rho^2,\rho^3,\delta,\alpha^1,k,\alpha^2}(\zeta, \eta)$ is the characteristic function of the set $\{(\zeta, \eta) \in \boldsymbol{R}^n \times \boldsymbol{R}^n;\ \langle\eta\rangle \geq 2\kappa R_0(|\rho^1| + |\rho^2| + |\rho^3|),\ \langle\zeta\rangle \geq R_0 \max\{16|\delta|, k\},\ \langle\zeta\rangle \leq 24R_0|\delta|$ when $|\alpha^1| + |\rho^1| > 0$, and $\langle\zeta\rangle \leq 3R_0 k/2$ when $|\alpha^2| + |\rho^2| > 0\}$. Note that

$$
\begin{aligned}
&\sum_{\alpha^1+\alpha^2+\alpha^3=\alpha} \alpha!/(\alpha^1!\alpha^2!\alpha^3!)(\widehat{C}/(8R_0))^{|\alpha^1|}(2\widehat{C}/R_0)^{|\alpha^2|} \\
&\qquad \times A_1(\Lambda, A_1)^{|\alpha^3|}|\alpha^3|!\langle\zeta\rangle^{-|\alpha^3|}\chi_{\rho^1,\rho^2,\rho^3,\delta,\alpha^1,k,\alpha^2}(\zeta, \eta) \\
&\leq \{(17\widehat{C} + A_1(\Lambda, A_1)/3)/(8R_0)\}^{|\alpha|}, \\
&j^\ell \leq 2^{j+\ell-1}\ell!, \\
&\sum_{\substack{\rho^1+\rho^2+\rho^3=\rho+\lambda \\ \rho^1+\rho^2>0}} (\rho+\lambda)!/(\rho^1!\rho^2!\rho^3!)(\widehat{C}/(8R_0))^{|\rho^1|}(2\widehat{C}/R_0)^{|\rho^2|} \\
&\qquad \times A_1(\Lambda, A_1)^{|\rho^3|}|\rho^3|!\langle\zeta\rangle^{-|\rho^3|}\chi_{\rho^1,\rho^2,\rho^3,\delta,\alpha^1,k,\alpha^2}(\zeta, \eta) \\
&\leq (12\widehat{C} + A_1(\Lambda, A_1))^{|\rho|+|\lambda|}(2e)^{|\delta|+k}(|\rho| + |\lambda|)!\langle\zeta\rangle^{-|\rho|-|\lambda|} \\
&\qquad \times \exp[-\langle\zeta\rangle/(24R_0)]/4, \\
&\sum_{k=0}^{\infty}(2eC_0/R_0)^k k!/k^k \leq C, \\
&\sum_{\rho,\delta} |u^\rho|/(\rho!\delta!)(12\widehat{C} + A_1(\Lambda, A_1))^{|\rho|+|\lambda|}(2eA_1(\Lambda, A_1)^2)^{|\delta|} \\
&\qquad \times (|\rho| + |\lambda|)!|\delta|!^2\langle\zeta\rangle^{-|\rho|-|\lambda|-|\delta|}\chi_{\rho^1,\rho^2,\rho^3,\delta,\alpha^1,k,\alpha^2}(\zeta, \eta)
\end{aligned}
$$

$$\leq \sum_{\mu=0}^{\infty}(2\sqrt{n}(12\widehat{C}+A_1(\Lambda,A_1))|u|/\langle\zeta\rangle)^{\mu}\sum_{\nu=0}^{\infty}\nu!/\nu^{\nu}(enA_1(\Lambda,A_1)^2/(8R_0))^{\nu}$$
$$\times(24\widehat{C}+2A_1(\Lambda,A_1))^{|\lambda|}|\lambda|!\langle\zeta\rangle^{-|\lambda|}$$
$$\leq C(24\widehat{C}+2A_1(\Lambda,A_1))^{|\lambda|}|\lambda|!\langle\zeta\rangle^{-|\lambda|}$$

if $\langle\zeta\rangle \geq 24R_0|\alpha|$, $R_0 \geq 2eC_0$, $R_0 \geq enA_1(\Lambda,A_1)^2/8$ and $|u| \leq \langle\zeta\rangle/(4\sqrt{n} \times(12\widehat{C}+A_1(\Lambda,A_1)))$. This gives

$$\left|\partial_\zeta^{\alpha+\tilde{\alpha}}D_w^{\beta}\partial_\eta^{\gamma}\partial_u^{\lambda}q_\kappa^2(w,\zeta,\eta;u)\right| \leq C_{|\tilde{\alpha}|+|\gamma|}\{(17\widehat{C}+A_1(\Lambda,A_1)/3)/(8R_0)\}^{|\alpha|}$$
$$\times A_1(\Lambda,A_1)^{|\beta|}(24\widehat{C}+2A_1(\Lambda,A_1))^{|\lambda|}|\beta|!|\lambda|!\langle\zeta\rangle^{m-|\lambda|-|\tilde{\alpha}|-|\gamma|}$$
$$\times\exp[-\langle\zeta\rangle/(24R_0)]$$

if $(w,\zeta),(y,\zeta)\in\Gamma_1$, $|\zeta|\geq R_0$, $|\eta-\zeta|\leq|\zeta|/2$, $\langle\zeta\rangle\geq 24R_0|\alpha|$, $R_0 \geq \max\{2eC_0, enA_1(\Lambda,A_1)^2/8\}$ and $|u|\leq\langle\zeta\rangle/(4\sqrt{n}(12\widehat{C}+A_1(\Lambda,A_1)))$. Similarly, we can deal with $q_\kappa^1(w,\zeta,\eta;u)$. A simple calculation yields

$$p_\kappa^R(\xi,w,\eta)\psi(\xi,w,\eta)=\tilde{p}_{t\Lambda}^R(\xi,w,\eta)+\sum_{\nu=1}^{3}q_\nu(\xi,w,\eta),$$

where

$$q_1(\xi,w,\eta)=\psi(\xi,w,\eta)\sum_{j=0}^{\infty}\sum_{|\gamma|+|\lambda|+k=j}\left(\phi_{|\lambda|}^R(\eta)\tilde{\phi}_{|\gamma|}^{8R_0}(\eta)\phi_k^{R_0/2}(\eta)-\phi_j^R(\eta)\right)$$
$$\times\Big[\partial_\eta^\lambda D_y^\lambda\partial_\zeta^\gamma\Big\{p_{k(\gamma)}(w+it\widehat{\nabla}_\eta\Lambda(w,N(w,y,\eta;t),\zeta),N(w,y,\eta;t))$$
$$\times\det\frac{\partial N}{\partial\eta}(w,y,\eta;t)\Big\}\Big]_{y=0,\,\zeta=0}/(\gamma!\lambda!),$$

$$q_2(\xi,w,\eta)=\psi(\xi,w,\eta)\sum_{\substack{\lambda^1,\lambda^2,\lambda^3,\gamma\\ \lambda^1+\lambda^2>0}}\sum_{k=0}^{\infty}\phi_{|\lambda^1|+|\lambda^2|+|\lambda^3|}^R(\eta)\tilde{\phi}_{|\gamma|}^{8R_0(\lambda^1)}(\eta)$$
$$\times\phi_k^{R_0/2(\lambda^2)}(\eta)\Big[\partial_\eta^{\lambda^3}D_y^{\lambda^1+\lambda^2+\lambda^3}\partial_\zeta^\gamma\Big\{p_{k(\gamma)}(w+it\widehat{\nabla}_\eta\Lambda(w,N(w,y,\eta;t),\zeta),$$
$$N(w,y,\eta;t))\det\frac{\partial N}{\partial\eta}(w,y,\eta;t)\Big\}\Big]_{y=0,\,\zeta=0}/(\gamma!\lambda^1!\lambda^2!\lambda^3!),$$

$$q_3(\xi,w,\eta)=\psi(\xi,w,\eta)\sum_\lambda\phi_{|\lambda|}^R(\eta)\Big[\partial_\zeta^\lambda D_y^\lambda\Big\{q_\kappa(w,\eta+\zeta,\eta;$$
$$N(w,y,\eta+\zeta;t)-\eta-\zeta)\det\frac{\partial N}{\partial\eta}(w,y,\eta+\zeta;t)\Big\}\Big]_{y=0,\,\zeta=0}/\lambda!,$$

modifying $\kappa(\Lambda, A_1, A)$ so that $R \geq C_0(\Lambda, C_0, A_1)$ if necessary. Similarly, we can show that

$$\left|\partial_\xi^\alpha D_w^{\beta^1+\beta^2+\tilde{\beta}} \partial_\eta^\gamma q_j(\xi, w, \eta)\right| \leq C_{|\alpha|+|\gamma|}(A(\Lambda, A_1, A)/R_0)^{|\beta^1|+|\beta^2|} \times \langle\xi\rangle^{-|\alpha|+|\beta^1|} \langle\eta\rangle^{m-|\gamma|+|\beta^2|} \exp[-\langle\eta\rangle/(3R)]$$

if $1 \leq j \leq 3$, $\langle\xi\rangle \geq R_0|\beta^1|$ and $\langle\eta\rangle \geq R_0|\beta^2|$, where $A(\Lambda, A_1, A)$ is a positive constant, modifying $\kappa(\Lambda, A_1, A)$ and $R(\Lambda, A_1, A)$ if necessary. By Theorem 2.3. 3 the operator $\sum_{\nu=1}^3 q_\nu(D_x, y, D_y)\varphi(D_x, y, D_y)$ maps continuously $\mathcal{S}_{-\delta}$ (resp. $\mathcal{S}_\delta{}'$) to $\mathcal{S}_\delta$ (resp. $\mathcal{S}'_{-\delta}$) if $R' \geq R'(B)$ and $\max\{\delta, \delta'\} \leq \min\{\delta(\Lambda, A_1, A)/R_0, c_0/R'\}$, modifying $R(\Lambda, A_1, A)$, $\delta(\Lambda, A_1, A)$, $R'(B)$ and c_0 if necessary. This proves the lemma. □

Lemma 4.1. 9 *Let $a_j(\xi, y, \eta)$ ($j \in \boldsymbol{N}$) be symbols such that*

$$\left|\partial_\xi^\alpha D_y^\beta \partial_\eta^\gamma a_j(\xi, y, \eta)\right| \leq C_{|\alpha|+|\gamma|}(A/R)^{|\beta|}\langle\xi\rangle^{m_1+|\beta|}\langle\eta\rangle^{m_2} \times \exp[\delta_1\langle\xi\rangle + \delta_2\langle\eta\rangle]$$

if $j \in \boldsymbol{N}$ and $\langle\xi\rangle \geq R|\beta|$, and for any compact subset K of $\boldsymbol{R}^n$ and any $k \in \boldsymbol{N}$ there are $C_{K,k,j} > 0$ ($j \in \boldsymbol{N}$) satisfying $C_{K,k,j} \to 0$ as $j \to \infty$ and

$$\left|\partial_\xi^\alpha D_y^\beta \partial_\eta^\gamma a_j(\xi, y, \eta)\right| \leq C_{K,|\alpha|+|\gamma|,j}(A/R)^{|\beta|}\langle\xi\rangle^{m_1+|\beta|}\langle\eta\rangle^{m_2} \times \exp[\delta_1\langle\xi\rangle + \delta_2\langle\eta\rangle]$$

for $j \in \boldsymbol{N}$ and $(\xi, y, \eta) \in K \times \boldsymbol{R}^n \times K$ if $\langle\xi\rangle \geq R|\beta|$. Then we have

$$a_j(D_x, y, D_y)u \longrightarrow 0 \quad \text{in } \mathcal{S}_{\varepsilon_1} \text{ (resp. in } \mathcal{S}'_{-\varepsilon_1}) \text{ as } j \to \infty$$

if $u \in \mathcal{S}_{\varepsilon_2}$ (resp. in $\mathcal{S}'_{-\varepsilon_2}$), $\varepsilon_2 = \delta_2 + 2(\delta_1 + \varepsilon_1)_+$, $\delta_1 + \varepsilon_1 \leq 1/R$ and $R \geq 2enA$.

Proof Put $K_M = \{\xi \in \boldsymbol{R}^n;\ \langle\xi\rangle \leq M\}$ for $M > 0$, and

$$\widehat{C}_{k,j} = \inf_{M>0} \max\{C_{K_M,k,j}, C_k/M\}.$$

Then we can easily show that $\widehat{C}_{k,j} \to 0$ as $j \to \infty$ and that

$$\left|\partial_\xi^\alpha D_y^\beta \partial_\eta^\gamma a_j(\xi, y, \eta)\right| \leq \widehat{C}_{|\alpha|+|\gamma|,j}(A/R)^{|\beta|}\langle\xi\rangle^{m_1+1+|\beta|}\langle\eta\rangle^{m_2+1} \times \exp[\delta_1\langle\xi\rangle + \delta_2\langle\eta\rangle]$$

if $j \in \boldsymbol{N}$ and $\langle\xi\rangle \geq R|\beta|$. Therefore, the lemma follows from Theorem 2.3. 3 (and its remark). □

4.2 General results

Carleman type estimates played an important role in the studies on uniqueness theorems and microlocal uniqueness theorems in the framework of distributions. Microlocal uniqueness theorems gave results on propagation of singularities and hypoellipticity (see [KW2]). This is still true in the framework of hyperfunctions. We shall show that it also suffices to derive microlocal *a priori* estimates (Carleman type estimates) in order to prove microlocal uniqueness theorems for hyperfunctions.

Let $(x^0,\xi^0) \in \boldsymbol{R}^n \times S^{n-1}$, and let Ω be a conic neighborhood of (x^0,ξ^0) in $T^*\boldsymbol{R}^n \setminus 0$ ($\simeq \boldsymbol{R}^n \times (\boldsymbol{R}^n \setminus \{0\})$). For vector-valued functions and matrix-valued symbol (and operators) we use the same notations as in the scalar case. Let $L(x,\xi) \equiv (L_{i,j}(x,\xi))$ be an $N \times N$ matrix-valued symbol in $FS^m(\Omega; C_0, A)$, *i.e.*, $L_{i,j}(x,\xi) \equiv \sum_{k=0}^{\infty} L_{i,j,k}(x,\xi) \in FS^m(\Omega; C_0, A)$. We assume that $u = {}^t(u_1,\cdots,u_N) \in \mathcal{C}(\Omega_0)$ satisfies

(A–1) $L(x,D)u = 0$ in $\mathcal{C}(\Omega_0)$,

where $\Omega_0 = \Omega \cap \boldsymbol{R}^n \times S^{n-1}$. Here we have represented $\mathcal{C}(\Omega_0)^N$ simply as $\mathcal{C}(\Omega_0)$. Under the above assumption we study conditions which give "$u = 0$ near (x^0,ξ^0) (in $\mathcal{C}(\Omega_0)$)." Let S be a closed conic subset of $\boldsymbol{R}^n \times (\boldsymbol{R}^n \setminus \{0\})$ satisfying $(x^0,\xi^0) \in S$, and assume that

(A–2) $\operatorname{supp} u \equiv \bigcup_{j=1}^N \operatorname{supp} u_j \subset S_0 \equiv S \cap \boldsymbol{R}^n \times S^{n-1}$.

Let $\{\Lambda_j(x,\xi;t,\delta)\}_{j\in \boldsymbol{N}}$ ($\subset C^\infty(\Omega \times [0,t_j] \times [0,\delta_j])$) be a family of symbols satisfying the following conditions, where the t_j and the δ_j are positive constants less than or equal to 1:

(Λ–1) There are positive constants $A_{0,j} \equiv A(\Lambda_j)$ and $C_j \equiv C(\Lambda_j)$ ($j \in \boldsymbol{N}$) such that

$$\left|\partial_\xi^\alpha D_x^\beta \Lambda_j(x,\xi;t,\delta)\right| \le C_j A_{0,j}^{|\alpha|+|\beta|} |\alpha|!|\beta|!\langle\xi\rangle^{1-|\alpha|}$$

for $(x,\xi) \in \Omega$ with $|\xi| \ge 1$, $t \in [0,t_j]$ and $\delta \in [0,\delta_j]$.

(Λ–2) There are $\hat{c}_j \equiv \hat{c}(\Lambda_j) > 0$ ($j \in \boldsymbol{N}$) such that

$$\operatorname{Re} \Lambda_j(x,\xi;t,\delta) \ge \hat{c}_j\langle\xi\rangle - C_{j,\delta}$$

for $(x,\xi) \in S \cap \Omega$ with $|\xi| \ge 1$, $t \in [0,t_j]$ and $\delta \in (0,\delta_j]$, where the $C_{j,\delta}$ are constants depending on j and δ.

(Λ–3) There is a conic neighborhood $\mathcal{U}^0$ of (x^0,ξ^0) satisfying $\mathcal{U}^0 \subset\subset \Omega$ such that for any conic neighborhood $\mathcal{U}$ of (x^0,ξ^0) with $\mathcal{U} \subset\subset \mathcal{U}^0$ there are $j_0 \in \boldsymbol{N}$ and $c_j^0 \equiv c^0(\Lambda_j) > 0$ ($j \geq j_0$) satisfying

$$\mathrm{Re}\ \Lambda_j(x,\xi;t,\delta) \geq c_j^0\langle\xi\rangle$$

for $(x,\xi) \in (\mathcal{U}^0 \setminus \mathcal{U}) \cap S$ with $|\xi| \geq 1$, $t \in [0,t_j]$ and $\delta \in [0,\delta_j]$.

(Λ–4) There are $c_j \equiv c(\Lambda_j) > 0$ ($j \in \boldsymbol{N}$) such that

$$\mathrm{Re}\ \Lambda_j(x^0,\lambda\xi^0;t,0) \leq -c_j\lambda \quad \text{for } \lambda \geq 1 \text{ and } t \in [0,t_j].$$

Choose a symbol $\varphi(x,\xi) \in AS^0(\Omega;A)$ so that $\varphi(x,\xi)$ is positively homogeneous of degree 0, Re $\varphi(x^0,\xi^0) = 0$ and Re $\varphi(x,\xi) > 0$ for $(x,\xi) \in S \cap \Omega_0 \setminus \{(x^0,\xi^0)\}$. Let $\psi(\xi)$ and $\lambda(\xi)$ be symbols in $S_{1,0}^0$ and $S_{1,0}^1$, respectively, such that $\psi(\xi) = 1$ for $|\xi| \geq 1$ and $\psi(\xi) = 0$ for $|\xi| \leq 1/2$, $C^{-1}\langle\xi\rangle \leq \lambda\langle\xi\rangle \leq C\langle\xi\rangle$ and

$$|\partial^\alpha \lambda(\xi)| \leq C_0 A_0^{|\alpha|}|\alpha|!\langle\xi\rangle^{1-|\alpha|}$$

for $(x,\xi) \in \Omega$ with some $x \in \boldsymbol{R}^n$ and $|\xi| \geq 1$. Here $S_{\rho,\nu}^m$ denotes the usual symbol class, *i.e.*, we say that $a(x,\xi) \in S_{\rho,\nu}^m$ if

$$\left|a_{(\beta)}^{(\alpha)}(x,\xi)\right| \leq C_{\alpha,\beta}\langle\xi\rangle^{m-\rho|\alpha|+\nu|\beta|} \quad \text{for } (x,\xi) \in \boldsymbol{R}^n \times \boldsymbol{R}^n.$$

If we put, for $j \in \boldsymbol{N}$,

$$\Lambda_j(x,\xi;t,\delta) = \{(\varphi(x,\xi) - 1/j)\lambda(\xi)^{1-\delta} + (\varphi(x,\xi) + 1/j)\lambda(\xi)/2\}\psi(\xi),$$

then $\{\Lambda_j(x,\xi;t,\delta)\}_{j\in N}$ satisfies the above conditions (Λ–1)–(Λ–4) with $t_j = 1$ and $\delta_j = 1$.

It follows from Lemma 2.4.2 with $U(y,\zeta) = -y\cdot\zeta - it\Lambda_j(y,\zeta;t,\delta)$ that for each $j \in \boldsymbol{N}$ there is a symbol $N^j(x,y,\xi)$ ($\equiv N^j(x,y,\xi;t,\delta)$) defined in $\widetilde{\Omega}_j \equiv \widetilde{\Omega}(\Lambda_j)$ such that $\eta = N^j(x,y,\xi)$ is the solution of

$$\eta + it\widehat{\nabla}_x\Lambda_j(x,y,\eta;t,\delta) = \xi,$$

where $\varepsilon_0 \ll 1$, $\widetilde{\Omega}_j = \{(x,y,\xi,t,\delta) \in \boldsymbol{R}^n \times \boldsymbol{R}^n \times \boldsymbol{R}^n \times [0,t_j] \times [0,\delta_j];$ $(x,\xi) \in \Omega,\ |y| \leq \varepsilon_0$ and $|\xi| \geq 1\}$ and

$$\widehat{\nabla}_x\Lambda_j(x,y,\eta;t,\delta) = \int_0^1 \nabla_x\Lambda_j(x+\theta y,\eta;t,\delta)\,d\theta,$$

modifying t_j and Ω if necessary. We note that $N^j(x,y,\xi;t,\delta) \in C^\infty(\widetilde{\Omega}_j)$ and

$$\left|\partial_\xi^\alpha D_x^\beta D_y^\lambda N^j(x,y,\xi;t,\delta)\right| \leq C_{j,\alpha,\beta,\lambda}\langle\xi\rangle^{1-|\alpha|}$$

for $(x,y,\xi,t,\delta)\in\widetilde{\Omega}_j$. Let $L(x,\xi;t\Lambda_k)$ ($\equiv(L_{i,j}(x,\xi;t\Lambda_k))$) be a symbol in $S^m_{1,0}$ satisfying

$$L(x,\xi;t\Lambda_j)\sim\sum_{k=0}^{\infty}L^k_{t\Lambda_j}(x,\xi) \tag{4.13}$$

in $S^m_{1,0}(\Omega)$ uniformly in $\delta\in[0,\delta_j]$, where $0\le t\le t_j$, $L_\ell(x,\xi)=(L_{i,j,\ell}(x,\xi))$ and

$$\begin{aligned}L^k_{t\Lambda_j}(x,\xi)=&\sum_{|\gamma|+|\lambda|+\ell=k}\Big[\partial_\xi^\gamma\partial_\eta^\lambda D_y^\gamma\\&\times\Big\{L_{\ell(\lambda)}\Big(x+it\widehat{\nabla}_\xi\Lambda_j(x,N^j(x,y,\xi;t,\delta),\eta;t,\delta),N^j(x,y,\xi;t,\delta)\Big)\\&\times\det\frac{\partial N^j}{\partial\xi}(x,y,\xi;t,\delta)\Big\}\Big]_{y=0,\,\eta=0}\Big/(\gamma!\lambda!),\\ \widehat{\nabla}_\xi\Lambda_j(x,\xi,\eta;t,\delta)=&\int_0^1\nabla_\xi\Lambda_j(x,\xi+\theta\eta;t,\delta)\,d\theta.\end{aligned}$$

Here (4.13) means that

$$\Big|\partial_\xi^\alpha D_x^\beta\Big(L(x,\xi;t\Lambda_j)-\sum_{k=0}^{N-1}L^k_{t\Lambda_j}(x,\xi)\Big)\Big|\le C_{N,\alpha,\beta}\langle\xi\rangle^{m-N-|\alpha|}$$

for $(x,\xi)\in\Omega$ with $|\xi|\ge1$, $N\in\boldsymbol{N}$ and $\delta\in[0,\delta_j]$, where $|M|$ denotes the matrix norm of M for an $N\times N$ matrix M. Here and after the constants do not depend on δ if not stated. Note that $L(x,\xi;t\Lambda_j)$ also depends on t and δ. We assume that

(ME) there are $j_0\in\boldsymbol{N}$ and $\chi(x,\xi)\in S^0_{1,0}$ such that "$\chi(x,\xi)$ is positively homogeneous of degree 0 for $|\xi|\ge1$, $\chi(x,\xi)=1$ near (x^0,ξ^0), and for any $j\ge j_0$ there is $t_j^0\in(0,t_j]$ such that for any $t\in(0,t_j^0]$ there are $\delta_0\,(\equiv\delta_0(j,t))\in(0,\delta_j]$, $\ell_k\in\boldsymbol{R}$ ($1\le k\le3$) and $C>0$ satisfying

$$\begin{aligned}\|\langle D\rangle^{\ell_1}v\|\ \le\ &C\Big\{\|\langle D\rangle^{\ell_2}L(x,D;t\Lambda_j)v\|+\|\langle D\rangle^{\ell_1-1}v\|\\&+\|\langle D\rangle^{\ell_3}(1-\chi(x,D))v\|\Big\}\end{aligned}$$

if $v\equiv{}^t(v_1,\cdots,v_N)\in C_0^\infty(\boldsymbol{R}^n)$ and $0<\delta\le\delta_0$, where $\|v\|$ denotes the L^2-norm of a vector-valued function v.

Theorem 4.2.1 *Assume that* (A–1), (A–2) *and* (ME) *are satisfied. Then,* $u=0$ *near* (x^0,ξ^0), *i.e.,* $(x^0,\xi^0)\notin\operatorname{supp}u$.

Proof Let $\Gamma \equiv X \times \gamma$ and $\Gamma^j \equiv X_j \times \gamma^j$ ($0 \le j \le 11$) be conic neighborhoods of (x^0, ξ^0) satisfying $\Gamma^{j+1} \subset\subset \Gamma^j \subset\subset \Gamma \subset\subset \Omega$ ($0 \le j \le 10$), and choose $\Phi^R(\xi, y, \eta) \in S^{0,0}(R, A)$ ($R \ge 4$) so that $\Phi^R(\xi, y, \eta) = 1$ for $(\xi, y, \eta) \in \boldsymbol{R}^n \times \Gamma^0$ with $|\eta| \ge R$ and supp $\Phi^R \subset \boldsymbol{R}^n \times \Gamma \cap \{|\eta| \ge R/2\}$, modifying A if necessary. We put

$$\widetilde{L}(x, \xi) = \sum_{j=0}^{\infty} \phi_j^{R_0/2}(\xi) L_j(x, \xi) \quad \text{for } R_0 \ge 2C_0,$$

$$\widetilde{L}(\xi, y, \eta) = \Phi^{R_0}(\xi, y, \eta)\widetilde{L}(y, \eta).$$

We may assume that $\widetilde{L}(x, \xi) \in PS^m(\Omega; R_0, A)$ and $\widetilde{L}(\xi, y, \eta) \in S^{0,m}(R_0, A)$. Let $v \equiv {}^t(v_1, \cdot, v_N) \in \mathcal{A}'(\overline{X}_1)$ satisfy

$$\mathrm{sp}([v])|_{\Gamma_0^1} = u|_{\Gamma_0^1} \quad \text{in } \mathcal{C}(\Gamma_0^1),$$

where $[v]$ denotes the residue class in $\mathcal{B}(X_1)$ of v and $\Gamma_0^1 = \Gamma^1 \cap \boldsymbol{R}^n \times S^{n-1}$. It is obvious that $WF_A(v) \cap \Gamma^1 \subset S$. We have $WF_A(f) \cap \Gamma^1 = \emptyset$ if $R_0 \gg 1$, where $f = \widetilde{L}(D_x, y, D_y)v \in \mathcal{E}_0$ ($\subset \mathcal{F}_0$), since $\mathrm{sp}(f) = L(x, D)u$ in $\mathcal{C}(\Gamma_0^1)$. In order to prove the theorem it suffices to show that $(x^0, \xi^0) \notin WF_A(v)$. We may assume that $\Gamma^8 \subset\subset \mathcal{U}^0 \subset\subset \Gamma^7$ and that $\chi(x, \xi) = 1$ for $(x, \xi) \in \Gamma^6$ with $|\xi| \ge 1$, where $\mathcal{U}^0$ is as in the condition (Λ–3) and $\chi(y, \eta)$ is the symbol in the condition (ME). Choose $\varphi(\xi, y, \eta) \in S^{0,0}(R, A_1)$ so that $\varphi(\xi, y, \eta) = 1$ for $(\xi, y, \eta) \in \boldsymbol{R}^n \times \Gamma^9$ with $|\eta| \ge 2R_0$ and supp $\varphi \subset \{(\xi, y, \eta) \in \boldsymbol{R}^n \times \Gamma^8;\ \chi(y, \eta) = 1\}$. It follows from Lemma 4.1.1 that there are $R(v) > 0$ and $\delta(v) > 0$ such that

$$\varphi(D_x, y, D_y) f \in \mathcal{S}'_{-\varepsilon} \tag{4.14}$$

if $R_0 \ge R(v)A_1$ and $\varepsilon \le \min\{1/(2R_0), \delta(v)\}$. Now assume that $R_0 \ge R(v)A_1$. Choose $\Psi_1^R(\xi, y, \eta) \in S^{0,0}(R, A_1)$ and $\chi_1^R(\xi) \in S^0(R, A_1)$ so that $\Psi_1^R(\xi, y, \eta) = 1$ for $y \in X_7$, supp $\Psi_1^R \subset \boldsymbol{R}^n \times X_6 \times \boldsymbol{R}^n$, $\chi_1^R(\xi) = 1$ for $\xi \in \gamma^7$ with $|\xi| \ge R$ and supp $\chi_1^R \subset \gamma^6 \cap \{|\xi| \ge R/2\}$. Put $\psi^{1,R}(\xi, y, \eta) = \Psi_1^R(\xi, y, \eta)\chi_1^R(\eta)$. By the assumption ($\Lambda$–3) on $\{\Lambda_j(x, \xi; t, \delta)\}_{j \in N}$ we can find $j_1 \in \boldsymbol{N}$, $\varepsilon_0 > 0$ and a conic neighborhood $\mathcal{U}$ of S so that $j_1 \ge j_0$ and

$$\mathrm{Re}\, \Lambda_{j_1}(x, \xi; t, \delta) \ge \varepsilon_0 \langle \xi \rangle \tag{4.15}$$

for $(x, \xi) \in \mathcal{U} \cap (\mathcal{U}^0 \setminus \Gamma^{11})$ with $|\xi| \ge 1$, $t \in [0, t_{j_1}]$ and $\delta \in [0, \delta_{j_1}]$, where j_0 is as in (ME). Now we write $\Lambda(x, \xi; t, \delta) = \Lambda_{j_1}(x, \xi; t, \delta)$ and $\delta_0 = \delta_{j_1}$. Note that $Op(e^{-t\Lambda(y,\eta;t,\delta)}\psi^{1,R_0}(\xi, y, \eta)) = \psi_T^{1,R_0}(D_x, y, D_z)$ if $T(y, \eta) = y \cdot \eta + it\Lambda(y, \eta; t, \delta)$. It follows from Lemmas 4.1.2 and 4.1.3 that there are $C_0' > 0$ and $R_0(v) > 0$ such that

$$\left\| D^\alpha Op\left(e^{-t\Lambda(y,\eta;t,\delta)}\psi^{1,R_0}(\xi, y, \eta)\right)\varphi(D_x, y, D_y) f \right\| \le C_\alpha \tag{4.16}$$

for $\alpha \in \boldsymbol{Z}_+^n$, $0 \le t \le C_0'/R_0$ and $\delta \in [0, \delta_0]$ if $R_0 \ge R_0(v)$. Here the C_α also depend on R_0. Applying Lemma 4.1.5, for $0 < \varepsilon \le 1$ we can find positive constants $A_1'(\varepsilon)$, $A_1''(\varepsilon)$, $R_1(\varepsilon)$ and $R_2(\varepsilon)$, which also depend on the choice of Γ^j ($9 \le j \le 11$), a symbol $q_\varepsilon(\xi, y, \eta) \in \bigcap_{\nu > 0} S^{0,m,0,\nu}(4R_0, A_1'(\varepsilon)\max\{A, A_1\} + A_1''(\varepsilon) + 4\varepsilon/\nu)$ and $c_0 > 0$ such that $\operatorname{supp} q_\varepsilon \subset \boldsymbol{R}^n \times (\Gamma^8 \setminus \Gamma^{10})$ and

$$r_\varepsilon \equiv [\widetilde{L}(D_x, y, D_y), \varphi(D_x, y, D_y)] - q_\varepsilon(D_x, y, D_y)$$

maps continuously $\mathcal{F}_0$ to $\mathcal{S}'_{-c_0/R_0}$ if $R_0 \ge R_1(\varepsilon)\max\{A, A_1\} + R_2(\varepsilon)$. It is obvious that

$$\left\| D^\alpha Op\left(e^{-t\Lambda(y,\eta;t,\delta)}\psi^{1,R_0}(\xi, y, \eta)\right) r_\varepsilon v \right\| \le C_{\varepsilon,\alpha} \tag{4.17}$$

for $\alpha \in \boldsymbol{Z}_+^n$, $0 \le t \le C_0'/R_0$ and $\delta \in [0, \delta_0]$ if $R_0 \ge R_1(\varepsilon)\max\{A, A_1\} + R_2(\varepsilon)$, modifying C_0' if necessary. Let $\mathcal{U}_j$ ($1 \le j \le 3$) be conic neighborhoods of $S \cap \overline{\Gamma}^8$ satisfying $\mathcal{U}_1 \subset\subset \mathcal{U}_2 \subset\subset \mathcal{U}_3 \subset\subset \mathcal{U} \cap \mathcal{U}^0$, and let $\widetilde{\Gamma}^8$ and $\widetilde{\Gamma}^{10}$ be conic neighborhoods of (x^0, ξ^0) satisfying $\Gamma^{11} \subset\subset \widetilde{\Gamma}^{10} \subset\subset \Gamma^{10} \subset\subset \Gamma^8 \subset\subset \widetilde{\Gamma}^8 \subset\subset \mathcal{U}^0$. Choose $\psi^{j,R}(\xi, y, \eta) \in S^{0,0}(R, A_1)$ ($j = 2, 3$) so that $\psi^{2,R}(\xi, y, \eta) = 1$ for $(\xi, y, \eta) \in \boldsymbol{R}^n \times \mathcal{U}_1$ with $|\eta| \ge R$, $\operatorname{supp} \psi^{2,R} \subset \boldsymbol{R}^n \times \mathcal{U}_2$, $\psi^{3,R}(\xi, y, \eta) = 1$ for $(\xi, y, \eta) \in \boldsymbol{R}^n \times (\mathcal{U}_3 \cap (\widetilde{\Gamma}^8 \setminus \widetilde{\Gamma}^{10}))$ with $|\eta| \ge R/2$ and $\operatorname{supp} \psi^{3,R} \subset \boldsymbol{R}^n \times (\mathcal{U} \cap (\mathcal{U}^0 \setminus \Gamma^{11}))$. We put

$$\begin{aligned} q_{\varepsilon,1}(\xi, y, \eta) &= (1 - \psi^{2,R_0}(\xi, y, \eta)) q_\varepsilon(\xi, y, \eta), \\ q_{\varepsilon,2}(\xi, y, \eta) &= \psi^{2,R_0}(\xi, y, \eta) q_\varepsilon(\xi, y, \eta). \end{aligned}$$

Since $\operatorname{supp} q_{\varepsilon,1} \cap \{|\eta| \ge R_0\} \subset \boldsymbol{R}^n \times (\Gamma^8 \setminus \mathcal{U}_1)$ and $WF_A(v) \cap (\overline{\Gamma}^8 \setminus \mathcal{U}_1) = \emptyset$, it follows from Lemma 4.1.1 that there are $R'(v) > 0$ and $R''(v) > 0$ such that $q_{\varepsilon,1}(D_x, y, D_y)v \in \mathcal{S}'_{-1/(8R_0)}$ if

$$\begin{cases} R_0 \ge R'(v)(A_1'(\varepsilon)\max\{A, A_1\} + A_1''(\varepsilon) + 4A_1), \\ R_0 \ge R''(v), \qquad 0 < \varepsilon \le 1/(32R'(v)). \end{cases} \tag{4.18}$$

Indeed, if (4.18) is valid and $\nu = 1/(8R_0)$, then we have

$$2R_0 \ge R'(v)(A_1'(\varepsilon)\max\{A, A_1\} + A_1''(\varepsilon) + 4\varepsilon/\nu + 4A_1).$$

By Lemmas 4.1.2 and 4.1.3 we have

$$\left\| D^\alpha Op\left(e^{-t\Lambda(y,\eta;t,\delta)}\psi^{1,R_0}(\xi, y, \eta)\right) q_{\varepsilon,1}(D_x, y, D_y)v \right\| \le C_{\varepsilon,\alpha} \tag{4.19}$$

for $\alpha \in Z_+^n$, $0 \le t \le C_0'/R_0$ and $\delta \in [0,\delta_0]$ if (4.18) is valid, modifying C_0' if necessary. Put

$$\begin{aligned} e_1(\xi,y,\eta;t,\delta) &= e^{-t\Lambda(y,\eta;t,\delta)}\psi^{1,R_0}(\xi,y,\eta)\psi^{3,R_0}(\xi,y,\eta) \\ e_2(\xi,y,\eta;t,\delta) &= e^{-t\Lambda(y,\eta;t,\delta)}\psi^{1,R_0}(\xi,y,\eta)(1-\psi^{3,R_0}(\xi,y,\eta)). \end{aligned}$$

From (4.15) and Theorem 2.3. 3 it follows that there is $R(A_1) > 0$ such that $e_1(D_x,y,D_y;t,\delta)$ maps continuously $\mathcal{S}_{-t\varepsilon_0}$ to $\mathcal{S}$ (uniformly in $\delta \in [0,\delta_0]$) if $t \in [0,t_{j_1}]$, $\delta \in [0,\delta_0]$ and $R_0 \ge R(A_1)$. By Theorem 2.3. 3 we also see that $q_{\varepsilon,2}(D_x,y,D_y)$ maps continuously $\mathcal{S}_{-a/(4R_0)}$ to $\mathcal{S}_{-a/R_0}$ if $0 < a \le 1/2$ and

$$\begin{cases} R_0 \ge en(A_1'(\varepsilon)\max\{A,A_1\} + A_1''(\varepsilon) + 4A_1), \\ 0 < \varepsilon \le a/(16en). \end{cases} \tag{4.20}$$

Therefore, we have

$$\left\| D^a e_1(D_x,y,D_y;t,\delta)q_{\varepsilon,2}(D_x,y,D_y)v \right\| \le C_{\varepsilon,\tau,\alpha} \tag{4.21}$$

if $\alpha \in Z_+^n$, $0 < \tau \le 1/(2\varepsilon_0)$, $\tau/R_0 \le t \le t_{j_1}$, $\delta \in [0,\delta_0]$ and (4.20) is valid with $a = \tau\varepsilon_0$, where the $C_{\varepsilon,\tau,\alpha}$ also depend on R_0. Applying Lemma 4.1. 4 we can see that there are positive constants C_j ($1 \le j \le 3$), τ_0 and c_0 such that $e_2(D_x,y,D_y;t,\delta)q_{\varepsilon,2}(D_x,y,D_y)$ maps continuously $\mathcal{S}_{-c_0/R_0}$ to $\mathcal{S}$ (uniformly in $\delta \in [0,\delta_0]$) and, therefore,

$$\left\| D^\alpha e_2(D_x,y,D_y;t,\delta)q_{\varepsilon,2}(D_x,y,D_y)v \right\| \le C_{\varepsilon,\alpha} \tag{4.22}$$

if $\alpha \in Z_+^n$, $0 \le t \le \tau_0/R_0$, $\delta \in [0,\delta_0]$, $R_0 \ge C_1(A_1'(\varepsilon)\max\{A,A_1\} + A_1''(\varepsilon) + 4A_1)$, $R_0 \ge C_2(A(\Lambda)+A_1)+C_3$ and $0 < \varepsilon \le c_0/(32C_1)$. We take

$$\varepsilon = \min\{1/(32R'(v)), \tau\varepsilon_0/(16en), c_0/(32C_1)\}$$

for τ with $0 < \tau \le \min\{C_0', 1/(2\varepsilon_0), \tau_0\}$. (4.16), (4.17), (4.19), (4.21) and (4.22) yield

$$\begin{aligned} &\left\| D^\alpha Op\Big(e^{-t\Lambda(y,\eta;t,\delta)}\psi^{1,R_0}(\xi,y,\eta)\Big)\widetilde{L}(D_x,y,D_y)\varphi(D_x,y,D_y)v \right\| \\ &\le C_{\alpha,\tau,R_0}(v) \end{aligned} \tag{4.23}$$

if $\alpha \in Z_+^n$, $\delta \in [0,\delta_0]$ and

$$\begin{cases} t = \tau/R_0, \quad 0 < \tau \le \min\{C_0', 1/(2\varepsilon_0), \tau_0\}, \\ R_0 \ge R(v,\Lambda,\tau)\max\{A,A_1\} + R'(v,\Lambda,\tau), \end{cases} \tag{4.24}$$

where $R(v,\Lambda,\tau)$ and $R'(v,\Lambda,\tau)$ are some positive constants. Now assume that (4.24) is satisfied. We choose $\psi^{j,R}(\xi,y,\eta)\in S^{0,0}(R,A_1)$ ($j=4,5$) so that $\psi^{j,R}(\xi,y,\eta)=1$ for $(\xi,y,\eta)\in \boldsymbol{R}^n\times\Gamma^{13-2j}$ with $|\eta|\geq R$ and supp $\psi^{j,R}\subset \boldsymbol{R}^n\times\Gamma^{12-2j}\cap\{|\eta|\geq R/2\}$ ($j=4,5$). It follows from Lemma 4.1.6 (and its remark) that there are positive constants $A(\Lambda,A_1)$, $R(\Lambda,A_1)$, $C(\Lambda,A_1)$, $\tau(\Lambda,A_1)$ and $\delta(\Lambda,A_1)$ and a symbol $q(\xi,y,\eta;t,\delta)\in S^{0,0}(C(\Lambda,A_1)R_0,A(\Lambda,A_1))$ such that

$$\Big\{Op\Big(e^{t\Lambda(y,\xi;t,\delta)}\psi^{4,R_0}(\eta,y,\xi)\Big)Op\Big(e^{-t\Lambda(y,\eta;t,\delta)}\psi^{5,R_0}(\xi,y,\eta)\Big)\\ \times q(D_x,y,D_y;t,\delta)-1\Big\}\varphi(D_x,y,D_y)$$

maps continuously $\mathcal{S}_{-\varepsilon}$ to $\mathcal{S}_\varepsilon$ (uniformly in $\delta\in[0,\delta_0]$) if $R_0\geq R(\Lambda,A_1)$, $0<\tau\leq\tau(\Lambda,A_1)$ and $\varepsilon\leq\delta(\Lambda,A_1)/R_0$. We put

$$\begin{aligned} L_{t\Lambda}&=Op\Big(e^{-t\Lambda(y,\eta;t,\delta)}\psi^{1,R_0}(\xi,y,\eta)\Big)\widetilde{L}(D_x,y,D_y)\\ &\qquad\times Op\Big(e^{t\Lambda(y,\xi;t,\delta)}\psi^{4,R_0}(\eta,y,\xi)\Big),\\ v_\delta&\ (\equiv v_\delta(x;t))\\ &=Op\Big(e^{-t\Lambda(y,\eta;t,\delta)}\psi^{5,R_0}(\xi,y,\eta)\Big)q(D_x,y,D_y;t,\delta)\varphi(D_x,y,D_y)v. \end{aligned}$$

Then, by (4.23) we have

$$\Big\|D^\alpha L_{t\Lambda}v_\delta\Big\|\leq C_{\alpha,\tau,R_0}(v) \tag{4.25}$$

if $\alpha\in\boldsymbol{Z}_+^n$, $\delta\in[0,\delta_0]$, $R_0\geq R(\Lambda,A_1)$ and $0<\tau\leq\tau(\Lambda,A_1)$. On the other hand, we have

$$v_\delta\in C^\infty(\boldsymbol{R}^n)\quad\text{for }\delta\in(0,\delta_0], \tag{4.26}$$

modifying $R(v,\Lambda,\tau)$ and τ_0 in (4.24) if necessary. Indeed, from (Λ–2) there are $\hat{c}\equiv\hat{c}(\Lambda)>0$, $C_\delta>0$ ($\delta\in(0,\delta_0]$) and a conic neighborhood $\mathcal{U}_4$ of $S\cap\Gamma^1$ such that

$$\operatorname{Re}\Lambda(x,\xi;t,\delta)\geq\hat{c}\langle\xi\rangle-C_\delta$$

for $(x,\xi)\in\mathcal{U}_4$ with $|\xi|\geq 1$ and $\delta\in(0,\delta_0]$. Choose $\psi^{6,R}(\xi,y,\eta)\in S^{0,0}(R,A_1)$ so that $\psi^{6,R}(\xi,y,\eta)=1$ for $(\xi,y,\eta)\in\boldsymbol{R}^n\times\widetilde{\mathcal{U}}_4$ with $|\eta|\geq R$ and supp $\psi^{6,R}\subset\mathcal{U}_4$, where $\widetilde{\mathcal{U}}_4$ is an open conic set satisfying $S\cap\Gamma^2\subset\subset\widetilde{\mathcal{U}}_4\subset\subset\mathcal{U}_4$. We put

$$\begin{aligned} v_\delta^1&=Op\Big(e^{-t\Lambda(y,\eta;t,\delta)}\psi^{5,R_0}(\xi,y,\eta)\psi^{6,R_0}(\xi,y,\eta)\Big)\\ &\qquad\times q(D_x,y,D_y;t,\delta)\varphi(D_x,y,D_y)v,\\ v_\delta^2&=v_\delta-v_\delta^1. \end{aligned}$$

Since $q(D_x, y, D_y; t, \delta)\varphi(D_x, y, D_y)v \in \mathcal{E}_0$ if $R_0 \geq 2en \max\{A_1, A(\Lambda, A_1) \times C(\Lambda, A_1)^{-1}\}$, Lemma 4.1. 2 yields $v_\delta^1 \in H^\infty$ if $R_0 \geq 2en \max\{A_1, A(\Lambda, A_1)C(\Lambda, A_1)^{-1}\}$ and $\delta \in (0, \delta_0]$. Note that $\left(\overline{\Gamma}^2 \setminus \widetilde{\mathcal{U}}_4\right) \cap S = \emptyset$. From Proposition 3.2. 3 (or its proof) we have

$$WF_A(q(D_x, y, D_y; t, \delta)\varphi(D_x, y, D_y)v) \cap \left(\overline{\Gamma}^2 \setminus \widetilde{\mathcal{U}}_4\right) = \emptyset$$

for $\delta \in [0, \delta_0]$, modifying $R(v, \Lambda, \tau)$ in (4.24) if necessary. Therefore, by Lemma 4.1. 1 we have $v_\delta^2 \in C^\infty(\boldsymbol{R}^n)$ for $\delta \in [0, \delta_0]$, modifying $R(v, \Lambda, \tau)$ and τ_0 if necessary. This proves (4.26). Let $\widetilde{\mathcal{U}}^0$ and $\widetilde{\Gamma}^7$ be conic neighborhoods of (x^0, ξ^0) satisfying $\mathcal{U}^0 \subset\subset \widetilde{\mathcal{U}}^0 \subset\subset \widetilde{\Gamma}^7 \subset\subset \Gamma^7$, and choose $\varphi_0^R(\xi, y, \eta) \in S^{0,0}(R, A_1)$ and $\psi(\xi, y, \eta) \in S^{0,0}(R_0, A_1)$ so that $\varphi_0^R(\xi, y, \eta) = 1$ for $(\xi, y, \eta) \in \boldsymbol{R}^n \times \widetilde{\Gamma}^8$ with $|\eta| \geq R$, $\psi(\xi, y, \eta) = 1$ for $(\xi, y, \eta) \in \boldsymbol{R}^n \times \widetilde{\Gamma}^7$ with $|\eta| \geq 2R_0$, supp $\varphi_0^R \subset \boldsymbol{R}^n \times \mathcal{U}^0 \cap \{|\eta| \geq R/2\}$ and supp $\psi \subset \boldsymbol{R}^n \times \Gamma^7 \cap \{|\eta| \geq R_0\}$. We put

$$\begin{aligned} w_\delta &= Op\left(e^{-t\Lambda(y,\eta;t,\delta)}\psi^{5,R_0}(\xi, y, \eta)\varphi_0^{R_0}(\xi, y, \eta)\right) \\ &\quad \times q(D_x, y, D_y; t, \delta)\varphi(D_x, y, D_y)v, \\ \tilde{v}_\delta &= v_\delta - w_\delta. \end{aligned}$$

Applying Lemma 4.1. 4 twice and Theorem 2.3. 3 , we see that $\tilde{v}_\delta \in H^\infty$ and

$$\left\| D^\alpha L_{t\Lambda} \tilde{v}_\delta \right\| \leq C_{\alpha,\tau,R}(v) \tag{4.27}$$

if $\alpha \in \boldsymbol{Z}_+^n$ and $\delta \in [0, \delta_0]$, modifying $R(v, \Lambda, \tau)$ and τ_0 if necessary. Therefore, Lemma 4.1. 8 , together with (4.25) and (4.27), yields

$$\left\| D^\alpha \widetilde{L}_{t\Lambda}^R(D_x, y, D_y) w_\delta \right\| \leq C_{\alpha,\tau,R_0}(v)$$

if $\alpha \in \boldsymbol{Z}_+^n$ and $\delta \in [0, \delta_0]$ and

$$R_0 \geq R_0(\Lambda, A, A_1), \quad R = \kappa(\Lambda, A, A_1)R_0, \quad 0 < \tau \leq \tau(\Lambda, A, A_1), \tag{4.28}$$

where $R_0(\Lambda, A, A_1)$, $\kappa(\Lambda, A, A_1)$ and $\tau(\Lambda, A, A_1)$ are some positive constants and

$$\widetilde{L}_{t\Lambda}^R(\xi, y, \eta) = \psi(\xi, y, \eta) \sum_{k=0}^\infty \phi_k^R(\eta) L_{t\Lambda}^k(y, \eta).$$

Note that $w_\delta \in C^\infty(\boldsymbol{R}^n)$ for $\delta \in (0, \delta_0]$. We assume that (4.28) is valid. Let $\Phi(x, \xi) \in S^0(R_0, A_1)$ be a symbol such that $\Phi(x, \xi) = 1$ for $(x, \xi) \in \widetilde{\mathcal{U}}^0$ with $|\xi| \geq R_0$ and supp $\Phi \subset \widetilde{\Gamma}^7$, and put

$$w_\delta^1 = \Phi(x, D) w_\delta, \qquad w_\delta^2 = w_\delta - w_\delta^1.$$

Note that $w_\delta^1 \in C_0^\infty(\boldsymbol{R}^n)$ for $\delta \in (0, \delta_0]$. Applying Corollary 2.4. 7 we have

$$\left\|D^\alpha w_\delta^2\right\| \leq C_{\alpha,\tau,R_0}(v) \tag{4.29}$$

if $\alpha \in \boldsymbol{Z}_+^n$ and $\delta \in [0, \delta_0]$, modifying τ_0 and $R(\Lambda, A, A_1)$ if necessary. Choose $\widetilde{\Phi}(x,\xi) \in S^0(R_0, A_1)$ so that $\widetilde{\Phi}(x,\xi) = 1$ for $(x,\xi) \in \Gamma^7$ with $|\xi| \geq R_0$ and supp $\widetilde{\Phi} \subset \Gamma^6$. Similarly, we have

$$\left\|D^\alpha\left(w_\delta^1 - \widetilde{\Phi}(x,D)w_\delta^1\right)\right\| \leq C_{\alpha,\tau,R_0}(v).$$

Moreover, by the standard calculus in $S_{1,0}^m$, we have

$$\begin{aligned}
&\left\|\langle D\rangle^{\ell_2}\left(\widetilde{L}_{t\Lambda}^R(D_x,y,D_y) - L(x,D;t\Lambda)\right)w_\delta^1\right\| \leq C_{\ell_2,\tau,R_0}(v),\\
&\left\|\langle D\rangle^{\ell_2}L(x,D;t\Lambda)w_\delta^1\right\| \leq C_{\ell_2,\tau,R_0}(v),\\
&\left\|\langle D\rangle^{\ell_3}(1-\chi(x,D))w_\delta^1\right\| \leq \left\|\langle D\rangle^{\ell_3}(1-\chi(x,D))\widetilde{\Phi}(x,D)w_\delta^1\right\| + C_{\ell_3,\tau,R_0}(v)\\
&\leq C_{\ell_3,\ell_1}\left\|\langle D\rangle^{\ell_1-1}w_\delta^1\right\| + C_{\ell_3,\tau,R_0}(v)
\end{aligned}$$

for $\delta \in [0, \delta_0]$. On the other hand, we have

$$\begin{aligned}
\left\|\langle D\rangle^{\ell_1-1}w_\delta^1\right\| &\leq \varepsilon\left\|\langle D\rangle^{\ell_1}w_\delta^1\right\| + C_\varepsilon\left\|e^{-\langle D\rangle}w_\delta^1\right\|\\
&\leq \varepsilon\left\|\langle D\rangle^{\ell_1}w_\delta^1\right\| + C_{\varepsilon,\tau,R_0}(v)
\end{aligned}$$

for $\varepsilon > 0$, where C_ε and $C_{\varepsilon,\tau,R_0}(v)$ are constants depending on ε. Therefore, from the assumption (ME) and (4.29) we have

$$\left\|\langle D\rangle^{\ell_1}w_\delta\right\| \leq C(v) \tag{4.30}$$

for $\delta \in (0, \delta_0]$, modifying δ_0 if necessary. Here we have fixed R_0 and τ (and $t = \tau/R_0$) so that R_0 and τ satisfy the conditions in this proof and $0 < t \leq t_{j_1}^0$, where t_j^0 is as in (ME). It follows from (4.30), Lemma 4.1. 9 and the construction of $q(\xi, y, \eta; t, \delta)$ that $w_\delta \to w_0$ weakly in H^{ℓ_1} as $\delta \downarrow 0$. In particular, we have $w_0 \in H^{\ell_1}$. By (Λ–4) there are $c > 0$ and a conic neighborhood $\mathcal{V}_0$ of (x^0, ξ^0) such that $\mathcal{V}_0 \subset\subset \Gamma^9$ and

$$\text{Re } \Lambda(x,\xi;t,0) \leq -c|\xi| \quad \text{for } (x,\xi) \in \mathcal{V}_0 \text{ with } |\xi| \geq 1.$$

Let $\mathcal{V}_j$ ($1 \leq j \leq 3$) be conic neighborhoods of (x^0, ξ^0) satisfying $\mathcal{V}_j \subset\subset \mathcal{V}_{j-1}$ ($1 \leq j \leq 3$), and choose $\varphi_j(\xi, y, \eta) \in S^{0,0}(R_0, A_1)$ ($j = 1, 2$) so that $\varphi_j(\xi, y, \eta) = 1$ for $(\xi, y, \eta) \in \boldsymbol{R}^n \times \mathcal{V}_{2j-1}$ with $|\eta| \geq R_0/2$ and supp $\varphi_j \subset \boldsymbol{R}^n \times V_{2j-2}$ ($j = 1, 2$). Then it follows from the construction of $q(\xi, y, \eta; t, 0)$ in Lemma 4.1. 6 and its remark and Lemma 4.1. 4 that

$$\varphi_2(D_x, y, D_y) Op\Big(e^{t\Lambda(y,\xi;t,0)} \varphi_1(\eta, y, \xi)\Big) w_0 - \varphi_2(D_x, y, D_y) v$$

is analytic, modifying R_0 if necessary. On the other hand, by Theorem 2.3. 3 we have

$$\varphi_2(D_x, y, D_y) Op\Big(e^{t\Lambda(y,\xi;t,0)} \varphi_1(\eta, y, \xi)\Big) w_0 \in \mathcal{S}'_{-ct/2},$$

since $e^{t\Lambda(y,\xi;t,0)} \varphi_1(\eta, y, \xi) \in S^{0,0,-ct/2,0}(R_0, A_1')$ with some $A_1' > 0$. This implies that $\varphi_2(D_x, y, D_y) v$ is analytic. Thus, by Lemma 3.2. 6 we have $(x^0, \xi^0) \notin WF_A(v)$, modifying R_0 if necessary, which proves the theorem. □

4.3 Microhyperbolic operators

In this section we shall first give the definition of microhyperbolicity, which was given by Kashiwara and Kawai [KK]. Using generalized Hamilton flows (or generalized bicharacteristics), we shall give a theorem on propagation of singularities (see Theorem 4.3. 8 below).

Let Ω be an open subset of $\boldsymbol{R}^N$, and let $f(z)$ ($\not\equiv 0$) be a real analytic function defined in Ω.

Definition 4.3. 1 Let $z^0 \in \Omega$ and $\vartheta \in T_{z^0}\Omega \simeq \boldsymbol{R}^N$. (i) We say that $f(z)$ is microhyperbolic at z^0 with respect to ϑ if there are a neighborhood U of z^0 in Ω and $t_0 > 0$ such that

$$f(z - it\vartheta) \neq 0 \quad \text{for } z \in U \text{ and } t \in (0, t_0].$$

(ii) We define the localization polynomial $f_{z^0}(\xi)$ of $f(z)$ at z^0 by

$$f(z^0 + t\xi) = t^\mu (f_{z^0}(\xi) + o(1)) \quad \text{as } t \to 0,$$
$$f_{z^0}(\xi) \not\equiv 0 \quad \text{in } \xi \in T_{z^0}\Omega.$$

We call the number μ the multiplicity of z^0 relative to f.

Remark (i) One can easily prove that $f(z)$ is microhyperbolic at z^0 with respect to ϑ if and only if there are a neighborhood U of z^0 in Ω, $\mu \in \boldsymbol{Z}_+$ and positive constants c and t_0 such that

$$\Big|\sum_{j=0}^{\mu} (-it\vartheta)^j f(z)/j!\Big| \geq ct^\mu \quad \text{for } z \in U \text{ and } t \in (0, t_0],$$

where ϑ is regarded as a vector field $\vartheta = \vartheta \cdot (\partial/\partial z)$. Therefore, the definition of microhyperbolicity can be extended to the case where $f(z) \in C^\infty(\Omega)$ by the above equivalence (see [Wk2] and [KW1]). (ii) When $f(z)$ is a polynomial of z, the localization polynomial $f_{z^0}(\xi)$ was defined by Hörmander [Hr2] and Atiyah, Bott and Gårding [ABG] in the study of singularities of solutions of partial differential operators with constant coefficients.

The following lemma easily follows from the definition of mirohyperbolicity (see [Hr3], [IP] and Lemma 8.7.2 in [Hr5]).

Lemma 4.3. 2 *If $f(z)$ is microhyperbolic at z^0 ($\in \Omega$) with respect to ϑ ($\in T_{z^0}\Omega$), then $f_{z^0}(\xi)$ is a homogeneous polynomial of degree μ and hyperbolic with respect to ϑ, i.e.,*

$$f_{z^0}(\xi - it\vartheta) \neq 0 \quad \text{for } \xi \in T_{z^0}\Omega \text{ and } t > 0,$$

where μ is the multiplicity of z^0 relative to f.

Let $z^0 \in \Omega$ and $\vartheta \in T_{z^0}\Omega$, and assume that f is microhyperbolic at z^0 with respect to ϑ. By Lemma 4.3. 2 we can define $\Gamma(f_{z^0}, \vartheta)$ as the connected component of the set $\{\xi \in T_{z^0}\Omega;\ f_{z^0}(\xi) \neq 0\}$ which contains ϑ. It is well-known that $\Gamma(f_{z^0}, \vartheta)$ is an open convex cone and that

$$f_{z^0}(\xi - it\eta) \neq 0 \quad \text{for } \xi \in T_{z^0}\Omega,\ \eta \in \Gamma(f_{z^0}, \vartheta) \text{ and } t > 0$$

(see, *e.g.*, [ABG] and [Hr1]).

Proposition 4.3. 3 (i) *For any compact subset M of $\Gamma(f_{z^0}, \vartheta)$ there are a neighborhood U of z^0 in Ω, $t_0 > 0$ and $c_0 > 0$ such that*

$$|f(z - it\eta)| \geq c_0 t^\mu \quad \text{for } z \in U,\ \eta \in M \text{ and } 0 \leq t \leq t_0,$$

where μ is the multiplicity of z^0 relative to f. (ii) *For any compact subset M of $\Gamma(f_{z^0}, \vartheta)$ there is a neighborhood U of z^0 in Ω such that f is microhyperbolic at z with respect to η if $z \in U$ and $\eta \in M$. Moreover, the family $\{\Gamma(f_z, \vartheta)\}_{z \in U}$ is inner semi-continuous at z^0, i.e., for any compact subset M of $\Gamma(f_{z^0}, \vartheta)$ there is a neighborhood U_0 ($\subset U$) of z^0 in Ω such that $M \subset \Gamma(f_z, \vartheta)$ for $z \in U_0$, where U is a neighborhood of z^0.*

Remark (i) The proposition was essentially proved in [ABG] (see, also, [Hr5]). (ii) Proposition 4.3. 3 is still valid, with some modifications, in the case where $f(z) \in C^\infty(\Omega)$ (see [Wk1] and [Wk5]).

Proof (i) We may assume that M is a compact subset of $\Gamma(f_{z^0},\vartheta)$ containing ϑ. By assumption there are a neighborhood U_0 of z^0 in Ω and $s_0>0$ such that $f(z-is\vartheta)\neq 0$ for $z\in U_0$ and $s\in(0,s_0]$. Put

$$F(t,\eta,\xi,r,s)=f(z^0+r\xi-it\eta-is\vartheta)$$

for $t\in\boldsymbol{C}$ with $|t|\le t_0$ and $\mathrm{Re}\,t\ge 0$, $\eta\in M$, $\xi\in\boldsymbol{R}^N$ with $|\xi|\le 1$, $0<r\le r_0$ and $0\le s\le s_1$. Here we have chosen t_0, r_0 and s_1 so that $z^0+r\xi+(\mathrm{Im}\,t)\eta\in U_0$, $s_1\le s_0$ and $F(t,\eta,\xi,r,s)$ is well-defined and analytic in t for $t\in\boldsymbol{C}$ with $|t|\le t_0$ and $\mathrm{Re}\,t\ge 0$, $\eta\in M$, $\xi\in\boldsymbol{R}^N$ with $|\xi|\le 1$, $0<r\le r_0$ and $0\le s\le s_1$. It is obvious that

$$F(t,\eta,\xi,r,s)\neq 0\quad\text{if } \mathrm{Re}\,t=0 \text{ and } s>0. \tag{4.31}$$

Moreover, we have

$$F(t,\eta,\xi,r,s)\neq 0\quad\text{if } |t|=t_0,\ t_0\ll 1 \text{ and } r/t_0,\,s/t_0\ll 1. \tag{4.32}$$

Indeed, there is $c>0$ such that

$$|f_{z^0}(\eta+i(r/t)\xi+(s/t)\vartheta)|\ge c\quad\text{if } |t|=t_0 \text{ and } r/t_0,\,s/t_0\ll 1.$$

On the other hand, we have

$$|(i/t)^\mu F(t,\eta,\xi,r,s)-f_{z^0}(\eta+i(r/t)\xi+(s/t)\vartheta)|\ge c/2$$

if $|t|=t_0$, $t_0\ll 1$ and $r,s\le t_0$, where μ is the multiplicity at z^0 relative to p. This yields (4.32). Now assume that there are $t_1\in\boldsymbol{C}$, $\eta^1\in M$ and $\xi^1\in\boldsymbol{R}^N$ such that $|t_1|<t_0$, $\mathrm{Re}\,t_1>0$, $|\xi^1|\le 1$ and $F(t_1,\eta^1,\xi^1,r_0,0)=0$. It follows form Rouché's theorem that for a sufficiently small $s>0$ there is $t_2\in\boldsymbol{C}$ satisfying $|t_2|<t_0$, $\mathrm{Re}\,t_2>0$ and $F(t_2,\eta^1,\xi^1,r_0,s)=0$. By Rouché's theorem there is a continuous function $t(\theta)$ defined on $[0,\theta_0)$ ($\subset[0,1]$) such that

$$F(t(\theta),(1-\theta)\eta^1+\theta\vartheta,\xi^1,r_0,s)=0,\quad t(0)=t_2.$$

Since (4.31) and (4.32) imply that $|t(\theta)|<t_0$ and $\mathrm{Re}\,t(\theta)>0$, $t(\theta)$ can be defined on $[0,1]$. This is a contradiction to $F(t,\vartheta,\xi^1,r_0,s)\neq 0$ for $\mathrm{Re}\,t\ge 0$. Therefore, we have

$$f(z^0+r_0\xi-it\eta)\neq 0 \tag{4.33}$$

if $t\in\boldsymbol{C}$, $|t|<t_0$, $\mathrm{Re}\,t>0$, $\eta\in M$, $\xi\in\boldsymbol{R}^N$ and $|\xi|\le 1$. It follows from the Weierstrass preparation theorem that there are a neighborhood U of z^0

in Ω and real analytic functions $a_j(z,\eta)$ ($1 \le j \le \mu$) defined near $U \times M$ and an analytic function $e(t,z,\eta)$ defined near $\{(t,z,\eta) \in \boldsymbol{C} \times U \times M;\ |t| < t_0\}$ such that $a_j(z^0,\eta) = 0$ for $\eta \in M$ and $1 \le j \le \mu$, $e(t,z,\eta) \ne 0$ and $f(z+t\eta) = e(t,z,\eta)g(t,z,\eta)$ for $(t,z,\eta) \in \boldsymbol{C} \times U \times M$ with $|t| < t_0$, where

$$g(t,z,\eta) = t^\mu + a_1(z,\eta)t^{\mu-1} + \cdots + a_\mu(z,\eta),$$

modifying t_0 if necessary. Moreover, (4.33) implies that $g(t,z,\eta)$ is a hyperbolic polynomial in t. This proves the assertion (i). (ii) The assertion (i) and the definition of microhyperbolicity imply the first part of the assertion (ii). Let $z \in \Omega$. Then it is obvious that $M \subset \Gamma(f_z,\vartheta)$ if M is a connected subset of $\boldsymbol{R}^N$, $\vartheta \in M$ and f is microhyperbolic at z with respect to any $\eta \in M$. In fact, we have $\Gamma(f_z,\vartheta) = \Gamma(f_z,\eta)$ ($\ni \eta$) for $\eta \in M$. Therefore, the second part of the assertion (ii) is also obvious. □

Now we assume that Ω is an open conic subset of $\boldsymbol{R}^n \times (\boldsymbol{R}^n \setminus \{0\})$. Let $p(x,\xi) \equiv \sum_{j=0}^{\infty} p_{m-j}(x,\xi) \in FS^m(\Omega;C_0,A)$, and assume that $p_m(x,\xi)$ is positively homogeneous of degree m in ξ. We impose the following condition (P) on p in this section:

(P) There is a continuous vector field $\vartheta : \Omega \ni z \mapsto \vartheta(z) \in T_z\Omega$ such that p_m is microhyperbolic with respect to $\vartheta(z)$ at each $z \in \Omega$.

Following [Wk3] we define generalized Hamilton flows.

Definition 4.3. 4 For $z^0 \in \Omega$ Lipschitz continuous curves $\{z(s)\}_{\pm s \in [0,a_\pm)}$ in Ω are called generalized semi-bicharacteristics of p starting from z^0 in the positive and the negative directions (with respect to ϑ), respectively, if

$$\begin{cases} (d/ds)z(s) \in \Gamma(p_{m\,z(s)}, \vartheta(z(s)))^\sigma \cap \{\delta z;\ |\delta z| = 1\} \\ \qquad\qquad\qquad\qquad \text{for } a.e.\ s \text{ with } \pm s \in [0,a_\pm), \\ z(0) = z^0, \end{cases}$$

where $a_\pm > 0$, σ denotes the canonical symplectic form on $T^*\boldsymbol{R}^n$ ($\simeq \boldsymbol{R}^n \times \boldsymbol{R}^n$), *i.e.*, $\sigma((\delta x,\delta\xi),(\delta y,\delta\eta)) = \delta y \cdot \delta\xi - \delta x \cdot \delta\eta$ for $(\delta x,\delta\xi),(\delta y,\delta\eta) \in \boldsymbol{R}^{2n} \equiv \boldsymbol{R}^n \times \boldsymbol{R}^n$, and

$$\Gamma^\sigma := \{\delta z \in T_z\Omega;\ \sigma(\delta w,\delta z) \ge 0 \text{ for any } \delta w \in \Gamma\}$$

for $z \in \Omega$ and $\Gamma \subset T_z\Omega$. We define the generalized Hamilton flow $K^+(z^0;\Omega,\vartheta)$ of p starting from z^0 in the positive direction (with respect to ϑ) by the union of all generalized semi-bicharacteristics of p starting from z^0 in the positive direction (with respect to ϑ). Similarly, we define $K^-(z^0;\Omega,\vartheta)$.

For simplicity we assume that p_m is microhyperbolic at each $z \in \Omega$ with respect to some $\vartheta \in \boldsymbol{R}^{2n}$, *i.e.*, $\vartheta(z) \equiv \vartheta$. Put $\Gamma_z = \Gamma(p_{m\,z}, \vartheta)$. If $\vartheta = 0$, then $\Gamma_z^\sigma = \{0\}$ and $K^{\pm}(z;\Omega,\vartheta) = \{z\}$ for $z \in \Omega$. We assume that $\vartheta \neq 0$. Then we can assume without loss of generality that $\vartheta = (0, e_1) \in \boldsymbol{R}^{2n}$, where $e_1 = (1, 0, \cdots, 0) \in \boldsymbol{R}^n$. Let $z^0 \in \Omega$, and K be a compact neighborhood of z^0 in Ω. For any $h > 0$ and $z \in K$ there is a compact set $M(z,h)$ in Γ_z such that $\vartheta \in \mathrm{int}(M(z,h))$ and

$$\begin{aligned}
&\Gamma_z^\sigma \subset M(z,h)^\sigma \subset \\
&\left(\Gamma_z^\sigma\right)_h \equiv \{\delta z;\ \delta z = 0 \text{ or } |\delta z/|\delta z| - \delta z^1/|\delta z^1|| < h \text{ for some } \delta z^1 \in \Gamma_z^\sigma\}.
\end{aligned}$$

Here we take $M(z,h)^\sigma = \{0\}$ if $\Gamma_z^\sigma = \{0\}$. By Proposition 4.3. 3 , for each $h > 0$ and $z \in K$ there is $r(z,h) > 0$ such that $r(z,h) < h$ and

$$M(z,h) \subset \Gamma_{z^1} \quad \text{for } z^1 \in U(z,h) \equiv \{w;\ |w - z| < r(z,h)\}.$$

Since K is compact, there are a finite number of $z^{h,j} \in K$ ($1 \le j \le N(h)$), such that $K \subset \bigcup_{j=1}^{N(h)} U'(z^{h,j}, h)$, where $U'(z,h) = \{z^1;\ |z^1 - z| < r(z,h)/2\}$. Put $M_1 = \bigcap_{j=1}^{N(1)} M(z^{1,j}, 1)$. Then it is obvious that $\vartheta \in \mathrm{int}(M_1)$ and $M_1 \subset \Gamma_z$ for any $z \in K$. Thus there is $\varepsilon > 0$ such that

$$\delta x_1 \ge \varepsilon |(\delta x', \delta\xi)| \quad \text{for } (\delta x_1, \delta x', \delta \xi) \in M_1^\sigma.$$

We may assume that $M_1 \subset M(z,h)$. Choose $\tau > 0$ so that $A_\tau^\pm \equiv \{z^0\} \pm \{(x,\xi) \in M_1^\sigma;\ x_1 \le \tau\} \Subset K$, and put $K_\tau^\pm = \{(x,\xi) \in K;\ \pm(x_1 - x_1^0) \le \tau\}$. In this situation, the polygonal line connecting z^0 with $\tilde{z}^1$, $\tilde{z}^1$ with $\tilde{z}^2$, $\cdots$ and $\tilde{z}^{\nu-1}$ with $\tilde{z}^\nu$ is called an h-approximate semi-bicharacteristics of p starting from z^0 in the positive direction if $j_\ell \in \{1, 2, \cdots, N(h)\}$ ($0 \le \ell \le \nu$), $\tilde{z}^\ell \in A_\tau^+ \cap U'(z^{h,j_\ell}, h)$ ($0 \le \ell \le \nu$) and $\tilde{z}^{\ell+1} \in \{\tilde{z}^\ell\} + M(z^{h,j_\ell}, h)^\sigma \cap \{\delta z;\ |\delta z| \le \rho(h)\}$ ($0 \le \ell \le \nu - 1$), where $\tilde{z}^0 = z^0$ and $\rho(h) = \min_{1 \le j \le N(h)} r(z^{h,j}, h)/2$. Similarly, we define h-approximate semi-bicharacteristics of p starting from z^0 in the negative direction. We denote by $K^{\pm}(z^0; K_\tau^\pm, \vartheta; h)$ the unions of all h-approximate semi-bicharacteristics of p starting from z^0 in the positive and the negative directions, respectively.

Proposition 4.3. 5 (i) *We have*

$$\begin{aligned}
&\bigcap_{h>0} K^+(z^0; K_\tau^+, \vartheta, h) = \bigcap_{h>0} \overline{K^+(z^0; K_\tau^+, \vartheta, h)} \\
&= K^+(z^0; \Omega, \vartheta) \cap K_\tau^+.
\end{aligned}$$

Moreover, for any neighborhood $\mathcal{U}$ of $K^+(z^0;\Omega,\vartheta)\cap K_\tau^+$ there is $h>0$ such that

$$\mathcal{U}\supset K^+(z^0;K_\tau^+,\vartheta,h') \quad for \quad 0<h'\le h.$$

(ii) *If $\{h_j\}\subset(0,1]$, $h_j\to 0$ as $j\to\infty$ and $\{L_j\}_{j\in N}$ is a family of polygonal lines in K_τ^+ such that $L_j\equiv\{z^j(s)\}_{s\in[0,a_j]}$ ($j\in N$) are h_j-approximate semi-bicharacteristics of p starting from z^0 in the positive direction and s represents the arc length from z^0 to $z^j(s)$, then we can find a subsequence $\{h_{j_k}\}$ of $\{h_j\}$ and a generalized semi-bicharacteristics $\gamma\equiv\{z(s)\}_{s\in[0,a]}$ of p starting from z^0 in the positive direction so that $a_{j_k}\to a$ and $\sup_{s\in[0,a_0]}|z^{j_k}(s)-z(s)|\to 0$ as $k\to\infty$, where $a_0=\sup_{j\in N}a_j$ ($\le\tau(1+\varepsilon^{-2})^{1/2}$) and we have defined $z^j(s)=z^j(a_j)$ for $j\in N$ and $s\ge a_j$, and $z(s)=z(a)$ for $s\ge a$.*

Remark We can also replace the + sign by the − sign in the assertion of the proposition.

Applying the same argument as in the proof of the existence theorem for ordinary differential equations, one can easily prove Proposition 4.3. 5 (see
§3 of [Wk3]). For further properties of $K^\pm(z^0;\Omega,\vartheta)$ we refer to [Wk3], [Wk1] and [Wk4].

First we obtain the following microlocal version of Holmgren's uniqueness theorem as an immediate consequence of Theorem 4.2. 1 .

Theorem 4.3. 6 *Let $z^0\equiv(x^0,\xi^0)\in\Omega_0\equiv\Omega\cap\mathbf{R}^n\times S^{n-1}$, and assume that $\varphi(z)\in C^2(\Omega)$ is real-valued and positively homogeneous of degree 0 in ξ, $\varphi(z^0)=0$ and that p_m is microhyperbolic with respect to $-H_\varphi(z^0)$ at z^0, where $H_\varphi(z)$ denotes the Hamilton vector field of φ, i.e.,*

$$H_\varphi(z):=\sum_{j=1}^n\{(\partial\varphi/\partial\xi_j)(z)(\partial/\partial x_j)-(\partial\varphi/\partial x_j)(z)(\partial/\partial\xi_j)\}\in T_z\Omega$$

for $z=(x,\xi)\in\Omega$. Then $z^0\notin\operatorname{supp}u$ if $u\in C(\Omega_0)$, $z^0\notin\operatorname{supp}p(x,D)u$ and $\operatorname{supp}u\cap\{z\in\mathcal{U};\ \varphi(z)<0\}=\emptyset$ for a neighborhood $\mathcal{U}$ of z^0 in Ω_0.

Proof If $H_\varphi(z^0)=0$, then p is elliptic at z^0, *i.e.*, $p_m(z^0)\ne 0$. By Theorem 3.6. 5 Theorem 4.3. 6 is valid if $H_\varphi(z^0)=0$. So we assume that $H_\varphi(z^0)\ne 0$. We take $S=\{z\in\Omega;\ \varphi(z)\ge 0\}$. By assumption there is $\kappa>0$ such that $\Phi_\kappa(z)>\varphi(z)$ for $z\in\Omega_0\setminus\{z^0\}$, where

$$\begin{aligned}\Phi_\kappa(x,\xi) &= (x-x^0)\cdot\nabla_x\varphi(z^0)+(\xi/|\xi|-\xi^0)\cdot\nabla_\xi\varphi(z^0)\\ &\quad+\kappa(|x-x^0|^2+|\xi/|\xi|-\xi^0|^2),\end{aligned}$$

modifying Ω if necessary. Put, for $j \in \boldsymbol{N}$,

$$\begin{aligned}&\Lambda_j(z;t,\delta)\ (\equiv \Lambda_j(z;\delta))\\&= \{(\Phi_\kappa(z)-1/j)|\xi|^{1-\delta}+(\Phi_\kappa(z)+1/j)|\xi|/2\}\psi(\xi),\end{aligned}$$

where $\psi(\xi)\in S^0_{1,0}$ satisfies $\psi(\xi)=1$ for $|\xi|\ge 1$ and $\psi(\xi)=0$ for $|\xi|\le 1/2$. Then $\{\Lambda_j\}_{j\in N}$ satisfies the conditions (Λ–1)–(Λ–4) in Section 4.2. Let us take $p(x,\xi)$ as $L(x,\xi)$ in Section 4.2 (and in Theorem 4.2. 1). Then, using the notations in Section 4.2 with L replaced by p, we have

$$\left|\partial_\xi^\alpha D_x^\beta\left(p(x,\xi;t\Lambda_j)-p^0_{t\Lambda_j}(x,\xi)\right)\right|\le C_{\alpha,\beta}\langle\xi\rangle^{m-1-|\alpha|}$$

for $(x,\xi)\in\Omega$ with $|\xi|\ge 1$ and $\delta\in[0,1]$, and

$$\begin{aligned}p^0_{t\Lambda_j}(x,\xi) &= p_m(x+it\nabla_\eta\Lambda_j(x,N^j(x,0,\xi;t,\delta);t,\delta),\\&\qquad N^j(x,0,\xi;t,\delta))\det\frac{\partial N^j}{\partial\eta}(x,0,\xi;t,\delta),\end{aligned}$$

where $j\in\boldsymbol{N}$ and $0\le t\le t_j$. Since $\xi^0\cdot\nabla_\xi\varphi(z^0)=0$ and

$$\begin{aligned}N^j(x,0,\xi;t,\delta)&=\xi-it\nabla_x\varphi(z^0)(|\xi|^{1-\delta}+|\xi|/2)\\&\quad+O(t|x-x^0||\xi|+t^2|\xi|),\\ \nabla_\eta\Lambda_j(x,N^j(x,0,\xi;t,\delta);t,\delta)&=\nabla_\xi\varphi(z^0)(|\xi|^{-\delta}+1/2)\\&\quad+O(j^{-1}+|x-x^0|+|\xi/|\xi|-\xi^0|+t)\end{aligned}$$

as $x\to x^0$ and $t\downarrow 0$ for $|\xi|\ge 2$, it follows from Proposition 4.3. 3 that there are $j_0\in\boldsymbol{N}$, a conic neighborhood $\mathcal{U}$ of z^0 in Ω, $t_0>0$ and $c(t)>0$ ($t\in(0,t_0]$) such that

$$\left|p^0_{t\Lambda_j}(x,\xi)\right|\ge c(t)|\xi|^m$$

if $j\ge j_0$, $(x,\xi)\in\mathcal{U}$, $|\xi|\ge 2$, $0<t\le t_0$ and $\delta\in[0,1]$. By the ellipticity of $p^0_{t\Lambda_j}$ one can easily see that p satisfies (ME) with $\ell_1=\ell_3=m$ and $\ell_2=0$. Therefore, Theorem 4.2. 1 yields Theorem 4.3. 6 . □

Applying the method of sweeping out in [Jh], we can obtain the following theorem from Theorem 4.3. 6 (see the proof of Proposition 5.2 in [KW1]).

Theorem 4.3. 7 *Assume that the condition* (P) *is satisfied. Let $z^0\equiv(x^0,\xi^0)\in\Omega_0$, and let M be compact subset of $\Gamma(p_{m\,z^0},\vartheta(z^0))$. We assume that $p_m(z^0)=0$. Then there is $\tau_0>0$ such that*

$$\begin{aligned}\operatorname{supp} u\ \cap\ \{(x,\xi/|\xi|)\in\Omega_0;\ (x,\xi)\in\{z^0\}-M^\sigma\\ \text{and } \sigma((x-x^0,\xi/|\xi|-\xi^0),\vartheta(z^0))=\tau\}\ne\emptyset\end{aligned}$$

for $0\le\tau\le\tau_0$ if $u\in\mathcal{C}(\Omega_0)$, $p(x,D)u=0$ in $\mathcal{C}(\Omega_0)$ and $z^0\in\operatorname{supp} u$.

Remark If $\Gamma(p_{m\,z^0}, \vartheta(z^0))^\sigma$ contains $r_0 \equiv \sum_{j=1}^n \xi_j^0(\partial/\partial\xi_j)$ or $-r_0$, then the theorem becomes trivial.

Finally, combining Theorem 4.3. 7 with Proposition 4.3. 5 we have the following

Theorem 4.3. 8 *Assume that the condition* (P) *is satisfied.* (i) *If* $u \in \mathcal{C}(\Omega_0)$, $p(x,D)u = 0$ *in* $\mathcal{C}(\Omega_0)$ *and* $z^0 \in \operatorname{supp} u$, *then there are* $a \in (-\infty, 0) \cup \{-\infty\}$ *and a generalized semi-bicharacteristics* $\{z(s)\}_{s\in(a,0]}$ $(\subset \Omega)$ *of* p *starting from* z^0 *in the negative direction such that* $(x(s), |\xi(s)|^{-1} \times \xi(s)) \in \operatorname{supp} u$ *for* $s \in (a, 0]$ *and*

$$\lim_{s\to a+0} z(s) \in \partial\Omega \quad \text{if } a > -\infty,$$

where the parameter s of the curve is chosen so that $-s$ *coincides with the arc length from* z^0 *to* $z(s)$ *and* $z(s) = (x(s), \xi(s))$. (ii) *Assume that there is a real-valued function* $t(z) \in C^1(\Omega)$, *which is called a time function for* p *in* Ω, *such that* $t(x,\xi)$ *is positively homogeneous of degree* 0 *in* ξ *and* p_m *is microhyperbolic with respect to* $-H_t(z)$ *at every* $z \in \Omega$. *Let* $z^0 \in \Omega_0$, *and let* $t_0 \in \boldsymbol{R}$ *satisfy* $t_0 \le t(z^0)$. *We assume that*

$$K^-(z^0;\Omega,-H_t) \cap \{z \in \Omega;\ t(z) \ge t_0\} \subset\subset \Omega.$$

Then $z^0 \notin \operatorname{supp} u$ *if* $u \in \mathcal{C}(\Omega_0)$,

$$\begin{aligned}&\operatorname{supp} p(x,D)u\\ &\cap\{(x,\xi/|\xi|) \in \Omega_0;\ (x,\xi) \in K^-(z^0;\Omega,-H_t) \text{ and } t(x,\xi) \ge t_0\} = \emptyset\end{aligned}$$

and

$$\operatorname{supp} u \cap \{(x,\xi/|\xi|) \in \Omega_0;\ (x,\xi) \in K^-(z^0;\Omega,-H_t) \text{ and } t(x,\xi) = t_0\} = \emptyset.$$

We shall give several remarks concerning Theorem 4.3. 8 . We first note that for systems of pseudodifferential operators we have the corresponding results (see [KW1]). As for the hyperebolic Cauchy problem in the space of hyperfunctions we can also prove its well-posedness (Bony-Schapira's result) from Theorem 4.3. 8 and well-posedness of the hyperbolic Cauchy problem in some Gevrey class. In fact, the fundamental solutions can be constructed as ultradistributions (see, *e.g.*, [Wk3]). Theorem 4.3. 8 gives the estimates of their analytic wave front sets. If we apply Theorem 4.3. 8 (i) to the case where $K^-(z^0;\Omega,\vartheta) = \{z^0\}$, then we have the following

Corollary 4.3. 9 *Assume that the condition* (P) *is satisfied. If* $z^0 \in \Omega_0$, $K^-(z^0;\Omega,\vartheta) = \{z^0\}$, $u \in \mathcal{C}(\Omega_0)$ *and* $z^0 \notin \operatorname{supp} p(x,D)u$, *then* $z^0 \notin \operatorname{supp} u$.

Remark If $n=2$, $m=1$ and $p_1(x,\xi) = \xi_1 + ix_1^{2k}\xi_2$, then $K^-(z;\boldsymbol{R}^2 \times (\boldsymbol{R}^2 \setminus \{0\}),(0,0,1,0)) = \{z^0\}$. Therefore, the corollary implies that $p(x,D)$ is analytic hypoelliptic (see Definition 4.5. 1 below).

Let us consider a simple example to see a difference between Theorems 4.3. 7 and 4.3. 8 . Let $n=2$, $m=1$ and $p_1(x,\xi) = x_1\xi_1$. We put $z^0 = (0,0,0,1)$ and $\vartheta = (-1,0,1,0)$. Then we have $\Gamma(p_{1\,z^0},\vartheta)^\sigma = \{(s,0,t,0);\ s,t \geq 0\}$ and $K^\pm(z^0;\boldsymbol{R}^2 \times (\boldsymbol{R}^2 \setminus \{0\}),\vartheta) = \{(x_1,0,\xi_1,1);$ "$\pm x_1 \geq 0$ and $\xi_1 = 0$" or "$x_1 = 0$ and $\pm\xi_1 \geq 0$"$\}$. So Theorem 4.3. 8 gives better results than Theorem 4.3. 7 .

4.4 Canonical transformation

In microlocal analysis we often reduce problems for general pseudodifferential operators to those for model operators (pseudodifferential operators of canonical forms). Homogeneous canonical transformation at the symbol level can be realized by conjugation with the corresponding Fourier integral operators (see Propositions 4.4. 2 and 4.4. 3 and Theorem 4.4. 4 below). This is one of basic ideas in microlocal analysis. Following [Hr4] we shall give a brief introduction to canonical transformations. Let us begin with the definition of homogeneous canonical transformations. Let Ω_1 and Ω_2 be conic subsets of $T^*\boldsymbol{R}^n \setminus 0$ ($\simeq \boldsymbol{R}^n \times (\boldsymbol{R}^n \setminus \{0\})$), and let $\chi : \Omega_1 \to \Omega_2$ be a real analytic mapping.

Definition 4.4. 1 We say that χ is a homogeneous canonical transformation if $\chi(y,\lambda\eta) = (x,\lambda\xi)$ for $(y,\eta) \in \Omega_1$ with $(x,\xi) = \chi(y,\eta)$ and $\lambda > 0$ and $\chi^*\sigma = \sigma$, where $\sigma = \sum_{j=1}^n d\xi_j \wedge dx_j$.

Let $\chi : \Omega_1 \to \Omega_2$ be a homogeneous canonical transformation, and write $\chi(y,\eta) = (x(y,\eta),\xi(y,\eta))$. Then the $x_j(y,\eta)$ and $\xi_j(y,\eta)$ are real analytic and positively homogeneous of degree 0 and 1 in η, respectively. Moreover, we have

$$\{x_j,x_k\} = \{\xi_j,\xi_k\} = 0, \quad \{\xi_j,x_k\} = \delta_{j,k} \quad (\,1 \leq j,k \leq n),$$

where $\{a,b\} = \sum_{j=1}^n \{(\partial a/\partial\eta_j)(y,\eta)(\partial b/\partial y_j)(y,\eta) - (\partial a/\partial y_j)(y,\eta) \times(\partial b/\partial\eta_j)(y,\eta)\}$. Conversely, if $x(y,\eta)$ and $\xi(y,\eta)$, which are real analytic, satisfy the above conditions, then $\chi(y,\eta) = (x(y,\eta),\xi(y,\eta))$ is a homogeneous canonical transformation. To construct homogeneous canonical transformations the following proposition, called the homogeneous Darboux theorem, plays a key role.

Proposition 4.4. 2 *Let A and B be subsets of $\{1,2,\cdots,n\}$, and let $w^0 \in T^*\boldsymbol{R}^n \setminus 0$. Assume that real analytic functions $q_j(y,\eta)$ ($j \in A$) and $p_k(y,\eta)$ ($k \in B$) defined in a conic neighborhood of w^0 are given and that they satisfy the following:* (i) *The q_j and the p_k are positively homogeneous of degree* 0 *and* 1, *respectively.* (ii) $\{q_j, q_{j'}\} = \{p_k, p_{k'}\} = 0$ *and* $\{p_k, q_j\} = \delta_{j,k}$ *if $j, j' \in A$ and $k, k' \in B$.* (iii) *The Hamilton vector fields H_{q_j} ($j \in A$) and H_{p_k} ($k \in B$) and the radial vector field $r \equiv \sum_{j=1}^n \eta_j(\partial/\partial\eta_j)$ are linearly independent at w^0. Let a_j and b_k ($1 \le j,k \le n$) be real numbers such that*

$$q_j(w^0) = a_j \quad (j \in A), \qquad p_k(w^0) = b_k \quad (k \in B), \tag{4.34}$$

and assume that there is $k_0 \in \{1,2,\cdots,n\} \setminus A$ satisfying $b_{k_0} \neq 0$. Then one can find real analytic functions $q_j(y,\eta)$ ($j \in \{1,2,\cdots,n\} \setminus A$) and $p_k(y,\eta)$ ($k \in \{1,2,\cdots,n\} \setminus B$) defined in a conic neighborhood of w^0 so that (i), (ii) *and* (4.34) *remain valid with A and B replaced by $\{1,2,\cdots,n\}$ and* (iii) *remains valid with A and B replaced by $\{1,2,\cdots,n\} \setminus \{k_0\}$ and $\{1,2,\cdots,n\}$, respectively.*

Proposition 4.4. 2 is a restatement of Theorem 21.1.9 in [Hr6] in the framework of the space of real analytic functions, and the proof is completely parallel.

Let $x^0 \in \boldsymbol{R}^n$ and $\eta^0 \in \boldsymbol{R}^n \setminus \{0\}$, and let $S(x,\eta)$ be a real analytic function defined in a conic neighborhood of (x^0,η^0) in $\boldsymbol{R}^n \times (\boldsymbol{R}^n \setminus \{0\})$ such that $S(x,\eta)$ is positively homogeneous of degree 1 in η and $\det(\partial^2 S/\partial x\partial\eta)(x^0,\eta^0) \neq 0$. Then, by the implicit function theorem the equation $y = (\partial S/\partial\eta)(x,\eta)$ defines a real analytic function $x(y,\eta)$ defined in a conic neighborhood of (y^0,η^0) such that $y = (\partial S/\partial\eta)(x(y,\eta),\eta)$ and $x = x((\partial S/\partial\eta)(x,\eta),\eta)$, where $y^0 = (\partial S/\partial\eta)(x^0,\eta^0)$. Define

$$\begin{aligned} \chi(y,\eta) &= (x(y,\eta), \xi(y,\eta)), \\ \xi(y,\eta) &= (\partial S/\partial x)(x(y,\eta),\eta). \end{aligned}$$

Then χ defines a homogeneous canonical transformation from a conic neighborhood of (y^0,η^0) to a conic neighborhood of (x^0,ξ^0), where $\xi^0 = \xi(y^0,\eta^0)$ ($= (\partial S/\partial x)(x^0,\eta^0)$). S is called a generating function of χ.

Proposition 4.4. 3 *If χ is a homogeneous canonical transformation from a conic neighborhood of $(y^0,\eta^0) \in T^*\boldsymbol{R}^n \setminus 0$ to a conic neighborhood of $(x^0,\xi^0) \in T^*\boldsymbol{R}^n \setminus 0$, then one can choose local coordinates in $\boldsymbol{R}^n$ at y^0 so that χ has a generating function in a conic neighborhood of (y^0,η^0).*

For the proof of Proposition 4.4. 3 we refer to [Hr4]. Let $w^0 \equiv (y^0, \eta^0)$, $z^0 \equiv (x^0, \xi^0) \in T^*\boldsymbol{R}^n \setminus 0$, and let χ be a homogeneous canonical transformation from a conic neighborhood $\Omega^1 \equiv X_1 \times \gamma_1$ of w^0 onto a conic neighborhood $\Omega^2 \equiv X_2 \times \gamma_2$ of z^0. Assume that $z^0 = \chi(w^0)$ and that χ has a generating function $S(x,\eta)$ which is real analytic and positively homogeneous of degree 1 in η and satisfies $\det(\partial^2 S/\partial x \partial\eta)(x,\eta) \neq 0$ in $X_2 \times \gamma_1$. Then we may assume that $S(x,\eta) \in \mathcal{P}(X_2 \times \gamma_1; A_0, A_0, c_0(S), 0, c_2(S), c_3(S))$. Let $a(x,\eta) \equiv \sum_{j=0}^\infty a_j(x,\eta)$ and $b(x,\eta) \equiv \sum_{j=0}^\infty b_j(x,\eta)$ be formal analytic symbols in $FS^+(X_2 \times \gamma_1; C_0, A)$, and let $p(x,\xi) \equiv \sum_{j=0}^\infty p_j(x,\xi) \in FS^+(\Omega^2; C_0, A)$. Put

$$\widetilde{S}(x,\eta) = S(x,\eta) - \eta \cdot \nabla S(x^0,\eta^0) - S_0(x - x^0) \cdot \eta,$$

where $S_0 = (\partial^2 S/\partial\eta\partial x)(x^0,\eta^0)$. Note that $\det S_0 \neq 0$. It follows from Lemma 2.3. 2 that for any $\varepsilon > 0$ there is a conic neighborhood Ω of (x^0,η^0) in $X_2 \times \gamma_1$ such that

$$\langle\eta\rangle^{-1}|\nabla_x \widetilde{S}(x,\eta)| + |\nabla_\eta \widetilde{S}(x,\eta)| + |(\partial^2\widetilde{S}/\partial\eta\partial x)(x,\eta)| < \varepsilon$$

for $(x,\eta) \in \Omega$ with $|\eta| \geq 1$. Noting that $S(x,\eta)$ is real-valued and shrinking Ω^1 and Ω^2, we can apply the same arguments as in the proofs of Lemmas 4.1. 6 –4.1. 8 and obtain the following

Theorem 4.4. 4 (i) *Put*

$$q_j(x,\eta) = \sum_{|\gamma|+k+\ell=j} \Big[D_y^\gamma\big\{p_k^{(\gamma)}(x, \widehat{\nabla}_x S(x,y,\eta))a_\ell(x+y,\eta)\big\}\Big]_{y=0}/\gamma!$$

for $j \in \boldsymbol{Z}_+$ and $(x,\eta) \in \Omega$ with $|\eta| \geq C$, where Ω is a conic neighborhood of (x^0,η^0), $C > 0$ and $\widehat{\nabla}_x S(x,y,\eta) = \int_0^1 \nabla_x S(x+\theta y,\eta)\,d\theta$. Then we have, with positive constants $\widetilde{C}_0(C_0,A,S)$ and $\widetilde{A}(A,S)$,

$$q(x,\eta) \equiv \sum_{j=0}^\infty q_j(x,\eta) \in FS^+(\Omega; \widetilde{C}_0(C_0,A,S), \widetilde{A}(A,S))$$

and

$$p(x,D)a_S(x,D_z)u = q_S(x,D_z)u \quad \text{in } \mathcal{C}(\Omega_0^2)$$

for $u \in \mathcal{C}(\Omega_0^1)$, where $\Omega_0^j = \Omega^j \cap \boldsymbol{R}^n \times S^{n-1}$ ($j = 1,2$), modifying Ω^1 and Ω^2 if necessary. (ii) *Put*

$$\begin{aligned} c_j(y,\eta) \quad = \quad & \sum_{|\gamma|+k+\ell=j} \Big[\partial_\zeta^\gamma D_y^\gamma\{b_k(X(y,\eta,\zeta),\eta+\zeta) \\ & \times a_\ell(X(y,\eta,\zeta),\eta)|\det(\partial X/\partial y)(y,\eta,\zeta)|\}\Big]_{\zeta=0}/\gamma! \end{aligned}$$

for $j \in Z_+$ *and* $(y,\eta) \in \Omega^1$ *with* $|\eta| \geq C$, *where* $x = X(y,\eta,\zeta)$ *satisfies* $y = \widehat{\nabla}_\eta S(x,\eta,\zeta)$ ($\equiv \int_0^1 \nabla_\eta S(x,\eta+\theta\zeta)\,d\theta$), *modifying* Ω^1 *and* C *if necessary. Then we have*

$$c(y,\eta) \equiv \sum_{j=0}^{\infty} c_j(y,\eta) \in FS^+(\Omega^1;\widetilde{C}_0(C_0,A,S),\widetilde{A}(A,S))$$

and

$$^r b_{-S}(x,D_z)a_S(x,D_z)u = c(y,D)u \quad \text{in } \mathcal{C}(\Omega_0^1)$$

for $u \in \mathcal{C}(\Omega_0^1)$. (iii) *Put*

$$\begin{aligned} d_j(x,\xi) \;=\; & \sum_{|\gamma|+k+\ell=j} \Big[\partial_\xi^\gamma D_y^\gamma \{a_k(x,N(x,y,\xi)) \\ & \times b_\ell(x+y,N(x,y,\xi))|\det(\partial N/\partial\xi)(x,y,\xi)|\}\Big]_{y=0}/\gamma! \end{aligned}$$

for $j \in Z_+$ *and* $(x,\xi) \in \Omega^2$ *with* $|\xi| \geq C$, *where* $\eta = N(x,y,\xi)$ *satisfies* $\xi = \widehat{\nabla}_x S(x,y,\eta)$, *modifying* Ω^2 *and* C *if necessary. Then we have*

$$d(x,\xi) \equiv \sum_{j=0}^{\infty} d_j(x,\xi) \in FS^+(\Omega^2;\widetilde{C}_0(C_0,A,S),\widetilde{A}(A,S))$$

and

$$a_S(x,D_z)\,^r b_{-S}(x,D_z)v = d(x,D)v \quad \text{in } \mathcal{C}(\Omega_0^2)$$

for $v \in \mathcal{C}(\Omega_0^2)$. (iv) *If* $a(x,\eta) \in FS^{m_1}(\Omega;C_0,A)$, $b(x,\eta) \in FS^{m_2}(\Omega;C_0,A)$ *and there are* $c > 0$ *and* $C_1 > 0$ *such that*

$$|a_0(x,\eta)| \geq c|\eta|^{m_1} \quad \text{if } (x,\eta) \in \Omega \text{ and } |\eta| \geq C_1, \tag{4.35}$$
$$|b_0(x,\eta)| \geq c|\eta|^{m_2} \quad \text{if } (x,\eta) \in \Omega \text{ and } |\eta| \geq C_1, \tag{4.36}$$

then $a_S(x,D_z) : \mathcal{C}(\Omega_0^1) \to \mathcal{C}(\Omega_0^2)$, $^r b_{-S}(x,D_z) : \mathcal{C}(\Omega_0^2) \to \mathcal{C}(\Omega_0^1)$ *and*

$$\begin{aligned} &\text{supp } a_S(x,D_z)u = \{(x,\xi/|\xi|) \in \Omega_0^2;\ (x,\xi) \in \chi(\text{supp } u)\}, \\ &\text{supp } ^r b_{-S}(x,D_z)v = \{(y,\eta/|\eta|) \in \Omega_0^1;\ (y,\eta) \in \chi^{-1}(\text{supp } v)\} \end{aligned}$$

for $u \in \mathcal{C}(\Omega_0^1)$ *and* $v \in \mathcal{C}(\Omega_0^2)$. *Moreover, if* $a(x,\eta) \in FS^{m_1}(\Omega;C_0,A)$ (*resp.* $b(x,\eta) \in FS^{m_2}(\Omega;C_0,A)$) *satisfies* (4.35) (*resp.* (4.36)), *then we can find* $b(x,\eta) \in FS^{-m_1}(\Omega;C_0,A)$ (*resp.* $a(x,\eta) \in FS^{-m_2}(\Omega;C_0,A)$) *so that* $a_S(x,D_z)\,^r b_{-S}(x,D_z) = id$ *on* $\mathcal{C}(\Omega_0^2)$ *and* $^r b_{-S}(x,D_z)a_S(x,D_z) = id$

on $\mathcal{C}(\Omega_0^1)$ modifying Ω^1 and Ω^2 if necessary. (v) *Put*

$$\tilde{p}_j(y,\eta) = \sum_{|\gamma|+|\lambda|+k+\ell+\mu=j} \Big[\partial_\zeta^\gamma D_y^\gamma D_w^\lambda \times\Big\{b_k(X(y,\eta,\zeta),\eta+\zeta)p_\mu^{(\lambda)}(X(y,\eta,\zeta),\hat{\nabla}_x S(X(y,\eta,\zeta),w,\eta)) \times a_\ell(X(y,\eta,\zeta)+w,\eta)|\det(\partial X/\partial y)(y,\eta,\zeta)|\Big\}\Big]_{w=0,\,\zeta=0}/(\gamma!\lambda!)$$

for $j \in Z_+$ and $(y,\eta) \in \Omega^1$ with $|\eta| \geq C$. Then we have

$$\tilde{p}(y,\eta) \equiv \sum_{j=0}^\infty \tilde{p}_j(y,\eta) \in FS^+(\Omega^1;\tilde{C}_0(C_0,A,S),\tilde{A}(A,S))$$

and

$$^r b_{-S}(x,D_z)p(x,D)a_S(x,D_z)u = \tilde{p}(y,D)u \quad \text{in } \mathcal{C}(\Omega_0^1)$$

for $u \in \mathcal{C}(\Omega_0^1)$. Moreover, we have

$$\begin{aligned}
\tilde{p}_0(y,\eta) &= b_0(x(y,\eta),\eta)a_0(x(y,\eta),\eta)|\det(\partial^2 S/\partial x\partial\eta)(x(y,\eta),\eta)|^{-1} \\
&\quad \times p_0(\chi(y,\eta)), \\
\tilde{p}_1(y,\eta) &= b_0(x(y,\eta),\eta)a_0(x(y,\eta),\eta)|\det(\partial^2 S/\partial x\partial\eta)(x(y,\eta),\eta)|^{-1} \\
&\quad \times\Big\{p_1(\chi(y,\eta)) + \frac{i}{2}\sum_{k=1}^n\Big(\big(\partial^2 p_0/\partial\xi_k\partial x_k\big)(\chi(y,\eta)) \\
&\qquad - \big(\partial^2/\partial\eta_k\partial y_k\big)p_0(\chi(y,\eta))\Big)\Big\} \\
&\quad + \sum_{|\alpha|+|\beta|\leq 1} c_{\alpha,\beta}(y,\eta)p_{0(\beta)}^{(\alpha)}(\chi(y,\eta)),
\end{aligned}$$

where the $c_{\alpha,\beta}(y,\eta)$ are analytic symbols and $(x(y,\eta),\xi(y,\eta)) = \chi(y,\eta)$.

Remark We call $a_S(x,D_z)$ and $^r b_{-S}(x,D_z)$ Fourier integral operators corresponding to χ and χ^{-1}, respectively.

From Theorem 4.4. 4 and Proposition 4.4. 3 , for example, it follows that a pseudodifferential operator with constant multiple characteristics of order m is microlocally equivalent to $D_1^m + q(x,D)$, where $q(x,D)$ is a pseudodifferential operator of order $m-1$. In the framework of microfunctions (or hyperfunctions) we can conjugate operators by invertible pseudodifferential operators of infinte order. So, finally we see that our operator is microlocally equivalent to D_1^m (see §5.2 of [SKK]).

4.5 Hypoellipticity

In this section we shall prove several results on analytic hypoellipticity, applying Theorem 4.2. 1 . Let us begin with the definition of analytic hypoellipticity.

Definition 4.5. 1 (i) Let Ω be an open conic subset of $T^*\boldsymbol{R}^n \setminus 0$, and let $p(x,\xi) \in PS^+_{loc}(\Omega)$ (or $p(x,\xi) \in FS^+(\Omega; C_0, A)$). For $z^0 = (x^0, \xi^0) \in \Omega$ we say that $p(x,D)$ is analytic microhypoelliptic at z^0 if there is an open neighborhood $\mathcal{U}$ of $(x^0, \xi^0/|\xi^0|)$ in $\boldsymbol{R}^n \times S^{n-1}$ satisfying supp $u =$ supp $p(x,D)u$ for any $u \in \mathcal{C}(\mathcal{U})$, *i.e.*, the sheaf homomorphism $p(x,D)$: $\mathcal{C}|_{\mathcal{U}} \to \mathcal{C}|_{\mathcal{U}}$ is injective. We also say that $p(x,D)$ is analytic microhypoelliptic in a conic set Γ ($\subset \Omega$) if $p(x,D)$ is analytic microhypoelliptic at each $z \in \Gamma$. (ii) Let X be an open subset of $\boldsymbol{R}^n$, and let $p(x,\xi) \in PS^+_{loc}(X)$ (or $p(x,\xi) \in FS^+(X \times (\boldsymbol{R}^n \setminus \{0\}); C_0, A)$). For $x^0 \in X$ we say that $p(x,D)$ is analytic microhypoelliptic at x^0 if $p(x,D)$ is analytic microhypoelliptic in $\{x^0\} \times (\boldsymbol{R}^n \setminus \{0\})$. Moreover, we say that $p(x,D)$ is analytic hypoelliptic at x^0 if there is an open neighborhood U of x^0 in $\boldsymbol{R}^n$ satisfying supp $u =$ supp $p(x,D)u$ for any $u \in \mathcal{B}(U)/\mathcal{A}(U)$, *i.e.*, the sheaf homomorphism $p(x,D) : \mathcal{B}_U/\mathcal{A}_U \to \mathcal{B}_U/\mathcal{A}_U$ is injective. We also say that $p(x,D)$ is analytic hypoelliptic (resp. microhypoelliptic) in a subset Y ($\subset X$) if $p(x,D)$ is analytic hypoelliptic (resp. microhypoelliptic) at each $x \in Y$.

Remark By definition, $p(x,D)$ is analytic hypoelliptic at x^0 if $p(x,D)$ is analytic microhypoelliptic at x^0.

We first consider a class of operators studied by Metivier [Mt] and Okaji [Ok]. Let $(x^0,\xi^0) \in \boldsymbol{R}^n \times S^{n-1}$, and let Ω be an open conic neighborhood of (x^0,ξ^0). Let $p(x,\xi) \equiv \sum_{j=0}^{\infty} p_{m-j}(x,\xi) \in FS^m(\Omega; C_0, A)$, and assume that $p_{m-j}(x,\xi)$ is positively homogeneous of degree $m-j$ ($j \in \boldsymbol{Z}_+$). We impose on $p(x,\xi)$ the following condition:

(M–O–1) $1 \le r \le n-1$ and $\Sigma \equiv \{(x,\xi) \in \Omega;\ p_m(x,\xi) = 0\}$ is a real analytic symplectic submanifold of codimension $2r$ of Ω, *i.e.*, there is a homogeneous canonical transformation $\chi : \widetilde{\Omega} \ni (y,\eta) \overset{\sim}{\mapsto} \chi(y,\eta) \in \Omega$ such that $\chi(0,\eta^0) = (x^0,\xi^0)$ and $\Sigma = \chi(S)$, where $\eta^0 = (0,\cdots,0,1) \in \boldsymbol{R}^n$, $\widetilde{\Omega}$ is an open conic neighborhood of $(0,\eta^0)$ and $S = \{(y,\eta) \in \widetilde{\Omega};\ y_1 = \cdots = y_r = 0 \text{ and } \eta_1 = \cdots = \eta_r = 0\}$ (see Theorem 21.2.4 in [Hr6]).

Under the above choice of χ we put $\Sigma_j = \chi(S_j)$ ($j = 1,2$), where $S_1 = \{(y,\eta) \in \widetilde{\Omega};\ \eta_1 = \cdots = \eta_r = 0\}$ and $S_2 = \{(y,\eta) \in \widetilde{\Omega};\ y_1 = \cdots = y_r = 0\}$. Define

$$d_{\Sigma_j}(x,\xi) = \text{dis}(\{(x,\xi/|\xi|)\}, \Sigma_j) \quad (j = 1,2).$$

Moreover, we assume that

(M–O–2) there are $\mu > 0$, $M \in \boldsymbol{Z}_+$ and $C > 0$ such that

$$|p_{m-j}(x,\xi)|/|\xi|^{m-j} \le C(d_{\Sigma_1}(x,\xi) + d_{\Sigma_2}(x,\xi)^\mu)^{M-j(\mu+1)/\mu}$$

for $(x,\xi) \in \Omega$ and $j(\mu+1) \le \mu M$, and

$$|p_m(x,\xi)|/|\xi|^m \ge C^{-1}(d_{\Sigma_1}(x,\xi) + d_{\Sigma_2}(x,\xi)^\mu)^M \tag{4.37}$$

for $(x,\xi) \in \Omega$.

By Proposition 4.4. 3 we may assume that χ has a generating function. Then we can choose Fourier integral operators F_1 and F_2 corresponding to χ and χ^{-1}, respectively, so that $F_2F_1 = id$ on $\mathcal{C}(\tilde{\Omega}_0)$ and $F_1F_2 = id$ on $\mathcal{C}(\Omega_0)$, where $\Omega_0 = \Omega \cap \boldsymbol{R}^n \times S^{n-1}$ and $\tilde{\Omega}_0 = \tilde{\Omega} \cap \boldsymbol{R}^n \times S^{n-1}$ (see Theorem 4.4. 4). It follows from Theorem 4.4. 4 that there is a formal analytic symbol $\tilde{p}(y,\eta) \equiv \sum_{j=0}^\infty \tilde{p}_{m-j}(y,\eta) \in FS^m(\tilde{\Omega};\tilde{C}_0,\tilde{A})$ satisfying $F_2p(x,D)F_1u = \tilde{p}(y,D)u$ in $\mathcal{C}(\tilde{\Omega}_0)$. The $\tilde{p}_{m-j}(y,\eta)$ are explicitly given by Theorem 4.4. 4 (v). Therefore, the conditions (M–O–1) and (M–O–2) imply that $\tilde{p}_{m-j}(y,\eta)$ ($j \le \mu M/(\mu+1)$) can be written in the form

$$\tilde{p}_{m-j}(y,\eta) = \sum_{\mu|\alpha'|+|\beta'|=\mu M-j(\mu+1)} a^j_{\alpha',\beta'}(y,\eta) y'^{\beta'}\eta'^{\alpha'},$$

where $\alpha' = (\alpha_1,\cdots,\alpha_r)$, $\beta' = (\beta_1,\cdots,\beta_r)$, $y' = (y_1,\cdots,y_r)$, $\eta' = (\eta_1, \cdots,\eta_r)$ and $a^j_{\alpha',\beta'}(y,\eta) \in AS^{m-j-|\alpha'|}(\tilde{\Omega};\tilde{A})$ is positively homogeneous of degree $m-j-|\alpha'|$. Define

$$L(y',\eta';\eta_n) = \sum_{\substack{j\le\mu M/(\mu+1)\\ \mu|\alpha'|+|\beta'|=\mu M-j(\mu+1)}} a^j_{\alpha',\beta'}(0,\eta^0) y'^{\beta'}\eta'^{\alpha'}\eta_n^{M-j-|\alpha'|}.$$

We also assume that

(M–O–3) the equation $L(y',D_{y'};1)v(y') = 0$ in $\mathcal{S}(\boldsymbol{R}^r)$ has no non-trivial solution.

For the meaning of the above condition we refer to Theorem 2.1 in [PR].

Theorem 4.5. 2 *Under the conditions* (M–O–1)–(M–O–3) $p(x,D)$ *is analytic microhypoelliptic at* (x^0,ξ^0).

Proof By Theorem 4.4. 4 it suffices to prove that $\tilde{p}(y,D)$ is analytic microhypoelliptic at $(0,\eta^0)$. Thus, we may assume that (M–O–1)–(M–O–3) are satisfied with $\chi = id$ and $(x^0,\xi^0) = (0,\eta^0)$. If $(x^1,\xi^1) \in \Omega \setminus S$, then $p(x,\xi)$ is elliptic at (x^1,ξ^1), *i.e.*, $p_m(x^1,\xi^1) \neq 0$. Therefore, by Theorem 3.6. 5 , we have $(x^1,\xi^1) \notin \operatorname{supp} u$ if $u \in \mathcal{C}(\Omega_0)$ and $(x^1,\xi^1) \notin \operatorname{supp} p(x,D)u$. Now assume that (x^1,ξ^1) is a point in S sufficiently close to $(0,\eta^0)$. Put

$$\begin{aligned}\varphi(x'',\xi'') &= \sum_{k=r+1}^{n} (x_k - x_k^1)^2 + \sum_{k=r+1}^{n-1} (\xi_k/\xi_n - \xi_k^1/\xi_n^1)^2, \\ \Lambda_j(x,\xi;\delta) &\equiv \Lambda_j(x'',\xi'';\delta) \\ &= \Big\{(\varphi(x'',\xi'') - 1/j)\xi_n^{1-\delta} + (\varphi(x'',\xi'') + 1/j)\xi_n/2\Big\}\psi(\xi),\end{aligned}$$

where $x'' = (x_{r+1},\cdots,x_n) \in \boldsymbol{R}^{n-r}$, $\xi'' = (\xi_{r+1},\cdots,\xi_n) \in \boldsymbol{R}^{n-r}$, $j \in \boldsymbol{N}$, $0 \le \delta \le 1$ and $\psi(\xi) \in S^0_{1,0}$ satisfies $\psi(\xi) = 1$ for $|\xi| \ge 1$ and $\psi(\xi) = 0$ for $|\xi| \le 1/2$. Let us apply Theorem 4.2. 1 with (x^0,ξ^0) and $L(x,\xi)$ replaced by (x^1,ξ^1) and $p(x,\xi)$, respectively. Let $\eta = N^j(x,y,\xi;t,\delta)$ be the solution of

$$\eta + it\widehat{\nabla}_x\Lambda_j(x,y,\eta;\delta) = \xi$$

for $(x,\xi),(y,\xi) \in \Omega$, $t \in [0,t_0]$ and $\delta \in [0,1]$, where $t_0 \ll 1$. Note that $N^j_k(x,y,\xi;t,\delta) = \xi_k$ ($1 \le k \le r$), $|N^j(x,y,\xi;t,\delta) - \xi| = O(t)$ as $t \to 0$ and that $N^j_k(x,y,\xi;t,\delta)$ ($r+1 \le k \le n$) do not depend on x' and ξ'. Let $p(x,\xi;t\Lambda_j)$ be a symbol in $S^m_{1,0}$ satisfying

$$p(x,\xi;t\Lambda_j) \sim \sum_{k=0}^{\infty} p^k_{t\Lambda_j}(x,\xi)$$

in $S^m_{1,0}(\Omega)$ uniformly in $\delta \in [0,1]$, where $t \in [0,t_0]$ and

$$\begin{aligned}p^k_{t\Lambda_j}(x,\xi) = &\sum_{|\gamma|+|\lambda|+\ell=k} \Big[\partial_\xi^\gamma \partial_\eta^\lambda D_y^\gamma \\ &\times\Big\{p_{m-\ell(\lambda)}\Big(x + it\widehat{\nabla}_\xi\Lambda_j(x,N^j(x,y,\xi;t,\delta),\eta;\delta), N^j(x,y,\xi;t,\delta)\Big) \\ &\times \det(\partial N^j/\partial\xi)(x,y,\xi;t,\delta)\Big\}\Big]_{y=0,\,\eta=0}/(\gamma!\lambda!).\end{aligned}$$

For a fixed $j \in \boldsymbol{N}$ we put

$$P(x,\xi) = \xi_n^{-m+M} p(x,\xi;t\Lambda_j).$$

Since $\widehat{\nabla}_{\xi_k}\Lambda_j(x,\xi,\eta;\delta) = 0$ ($1 \le k \le r$) and $\det(\partial N^j/\partial\xi)(x,y,\xi;t,\delta) = 1 + O(t)$ as $t \to 0$, we can write

$$P(x,\xi) \equiv L(x',\xi';\xi_n)$$

$$- \sum_{\substack{k \le \mu M/(\mu+1) \\ \mu|\alpha'|+|\beta'|=\mu M - k(\mu+1)}} \Big(a^k_{\alpha',\beta'}(x, \xi/\xi_n) - a^k_{\alpha',\beta'}(0, \eta^0) + t b^k_{\alpha',\beta'}(x, \xi/\xi_n; t) \Big) x'^{\beta'} \xi'^{\alpha'} \xi_n^{M-k-|\alpha'|}$$

mod $S^{M/(\mu+1)-1}_{1,0}$, where the $b^k_{\alpha',\beta'}(x,\xi;t) \in S^{m-k-|\alpha'|}_{1,0}$ are positively homogeneous of degree $m-k$ in ξ. By (4.37) $L^0(x',\xi') \equiv \sum_{\mu|\alpha'|+|\beta'|=\mu M} a^0_{\alpha',\beta'}(0,\eta^0) x'^{\beta'} \xi'^{\alpha'}$ satisfies, with some $C > 0$,

$$C^{-1}(\langle \xi' \rangle + \langle x' \rangle^{\mu})^M \le |L^0(x',\xi')| \le C(\langle \xi' \rangle + \langle x' \rangle^{\mu})^M$$

for $(x',\xi') \in \boldsymbol{R}^r \times \boldsymbol{R}^r$ with $|\xi'| \ge 1$. Hence, we can apply Theorem B.1.1 and show that there are $\nu > 0$ and $C > 0$ such that

$$\begin{aligned} \|\langle D \rangle^{M/(\mu+1)} v\| \;\; &\le \;\; C(\|\langle D \rangle^{-m+M} p(x,D;t\Lambda_j)v\| + \|u\|_{-1} \\ &\quad + \|\langle D \rangle^M (1 - \Psi_\nu(x,D))v\|_M) \end{aligned}$$

for $v \in C_0^\infty(\boldsymbol{R}^n)$ if $j \in \boldsymbol{N}$, $t \in [0,t_0]$ and $\delta \in [0,1]$, modifying t_0 if necessary, where $\Psi_\nu(x,\xi) = \varphi(x/\nu)\psi(\xi_1,\cdots,\xi_{n-1},\nu\xi_n)$, $\varphi(x) \in C_0^\infty(\boldsymbol{R}^n)$, $\psi(\xi) \in S^0_{1,0}$, $\varphi(x) = 1$ for $|x| \le 1$, $\varphi(x) = 0$ for $|x| \ge 2$, $\psi(\xi) = 1$ if $|(\xi_1,\cdots,\xi_{n-1})| \le \xi_n$ and $\xi_n \ge 1$, and $\psi(\xi) = 0$ if $|(\xi_1,\cdots,\xi_{n-1})| \ge 2\xi_n$ or $\xi_n \le 1/2$. Since (ME) is valid for p, from Theorem 4.2.1 we have $(x^1,\xi^1/|\xi^1|) \notin \operatorname{supp} u$ if (x^1,ξ^1) is sufficiently close to $(0,\eta^0)$, $u \in \mathcal{C}(\Omega_0)$ and $(x^1,\xi^1/|\xi^1|) \notin \operatorname{supp} p(x,D)u$. This proves the theorem. □

Next we consider some examples of analytic hypoelliptic operators with double characteristics. Let $\xi^0 = (0,\cdots,0,1) \in \boldsymbol{R}^n$, and let Ω be a conic neighborhood of $(0,\xi^0)$ in $\boldsymbol{R}^n \times (\boldsymbol{R}^n \setminus \{0\})$. Let $p(x,\xi) \equiv \sum_{j=0}^\infty p_{2-j}(x,\xi)$ be a formal analytic symbol in $FS^2(\Omega; C_0, A)$ such that $p_2(x,\xi) = \xi_1^2 + a(x,\xi_2,\cdots,\xi_n)$ and $a(x,\xi_2,\cdots,\xi_n)$ is positively homogeneous of degree 2. Put

$$S = \{(x,\xi) \in \boldsymbol{R}^n \times (\boldsymbol{R}^n \setminus \{0\});\; x' = 0 \text{ and } \xi' = 0\},$$

where $1 \le r \le n-1$, $x' = (x_1,\cdots,x_r) \in \boldsymbol{R}^r$ and $\xi' = (\xi_1,\cdots,\xi_r) \in \boldsymbol{R}^r$. We impose the following conditions on p:

(H–1) $p(x,D)$ is analytic microhypoelliptic in $\Omega \setminus S$.

(H–2) There are a conic neighborhood Γ of $(0,\xi^0)$ in $\boldsymbol{R}^r \times \boldsymbol{C}^{n-r} \times (\boldsymbol{R}^r \times \boldsymbol{C}^{n-r} \setminus \{0\})$, $\varepsilon > 0$, $C > 0$ and $q_j(x,\xi) \in C^\infty(\Gamma)$ ($1 \le j \le 2J$) such that the $q_j(x,\xi)$ are analytic in $(x'',\xi'') = (x_{r+1},\cdots,x_n,\xi_{r+1},\cdots,\xi_n)$ and

$$\left|q^{(\alpha)}_{j(\beta)}(x,\xi)\right| \le C_{\alpha,\beta}\langle\xi\rangle^{1-|\alpha|},$$

$$(1-\varepsilon)\mathrm{Re}\ p_2(x,\xi) + \mathrm{Re}\ p_1^s(x,\xi) - \sum_{j=1}^{2J}|q_j(x,\xi)|^2$$

$$+\mathrm{Re}\ \sum_{j=1}^{J}\{q_{2j-1},q_{2j}\}(x,\xi) \ge -C$$

for $(x,\xi)\in\Gamma$, where $p_1^s(x,\xi) = p_1(x,\xi) + (i/2)\sum_{j=1}^n(\partial^2 p_2/\partial x_j\partial\xi_j)(x,\xi)$.
(H–3) There are a neighborhood $\mathcal{U}$ of $(0,\xi^0)$ in $\boldsymbol{R}^r\times\boldsymbol{C}^{n-r}\times(\boldsymbol{R}^r\times\boldsymbol{C}^{n-r}\setminus\{0\})$ and $C>0$ such that

$$\left|p^{(\alpha)}_{2(\beta)}(x,\xi)\right|^2 \le C\,\mathrm{Re}\ p_2(x,\xi)$$

if $|\alpha|+|\beta| = 1$ and $(x,\xi)\in\mathcal{U}$.

Theorem 4.5. 3 *Under the conditions* (H–1)–(H–3) $p(x,D)$ *is analytic microhypoelliptic at* $(0,\xi^0)$.

Remark Let $n=3$, and let $P(x,\xi) = \xi_1^2+\xi_2^2+\alpha(x_1,x_2)b(x)\xi_3^2$ be an analytic symbols. Assume that $\alpha(x_1,x_2) > 0$ for $(x_1,x_2)\ne(0,0)$ and $b(x)>0$. Then, by the theorem $p(x,D)$ is analytic microhypoelliptic.

Proof By (H–1), $(x^1,\xi^1/|\xi^1|)\notin \mathrm{supp}\ u$ if $(x^1,\xi^1)\notin\Omega\setminus S$, $u\in\mathcal{C}(\Omega_0)$ and $(x^1,\xi^1/|\xi^1|)\notin\mathrm{supp}\ p(x,D)u$, where $\Omega_0 = \Omega\cap\boldsymbol{R}^n\times S^{n-1}$. We use the same notations as in the proof of Theorem 4.5. 2 . Fix $j\in\boldsymbol{N}$. From (4.9) and (4.10) we have

$$p^0_{t\Lambda_j}(x,\xi) = p_2(x',z'',\xi',\zeta'')d(x'',\xi'';t,\delta),$$

$$p^1_{t\Lambda_j}(x,\xi) = \Big(p_1^s(x',z'',\xi',\zeta'') - (i/2)\sum_{j=1}^n\Big(\partial^2/\partial x_j\partial\xi_j\Big)p_2(x',z'',\xi',\zeta'')$$

$$+\sum_{|\alpha|+|\beta|\le 1} c_{\alpha,\beta}(x,\xi;t,\delta)p^{(\alpha)}_{2(\beta)}(x',z'',\xi',\zeta'')\Big)d(x',z'',\xi',\zeta'')$$

for $(x,\xi)\in\Omega$ and $0\le t\ll 1$, where $\zeta_k\equiv\zeta_k(x'',\xi'';t,\delta) = N^j_k(x'',0,\xi'';t,\delta)$ ($r+1\le k\le n$), *i.e.*, $\eta'' = \zeta''(x'',\xi'';t,\delta)$ is the solution of $\eta'' + it\nabla_x\Lambda_j(x'',\eta'';\delta) = \xi''$, $z''\equiv z''(x'',\xi'';t,\delta) = x''+it(\nabla_{\xi''}\Lambda_j)(x'',\zeta'';\delta)$, $d(x'',\xi'';t,\delta) = \det(\partial N^j/\partial\xi)(x,0,\xi;t,\delta)$ and $c_{\alpha,\beta}(x,\xi;t,\delta)\in S^{-1+|\alpha|}_{1,0}$, modifying Ω if necessary. Choose $e(x,\xi;t,\delta) \in S^0_{1,0}$ so that $e(x,\xi;t,\delta) = d(x'',\xi'';t,\delta)^{-1}$ in $\Omega \cap \{|\xi| \geq 1\}$, and put

$$P(x,D;t,\delta) = e(x,D;t,\delta)p(x,D;t\Lambda_j).$$

Then, a simple calculation gives

$$\begin{aligned} P(x,\xi;t,\delta) &\equiv p_2(x',z'',\xi',\zeta'') + p_1^s(x',z'',\xi',\zeta'') \\ &-(i/2)\sum_{j=1}^{n}\left(\partial^2/\partial x_j\partial\xi_j\right)p_2(x',z'',\xi',\zeta'') \\ &+ \sum_{|\alpha|+|\beta|\leq 1} \tilde{c}_{\alpha,\beta}(x,\xi;t,\delta)p_{2(\beta)}^{(\alpha)}(x',z'',\xi',\zeta'') \end{aligned}$$

mod $S^0_{1,0}$ in Ω, where $z'' = z''(x'',\xi'';t,\delta)$, $\zeta'' = \zeta''(x'',\xi'';t,\delta)$ and $\tilde{c}_{\alpha,\beta}(x,\xi;t,\delta) \in S^{-1+|\alpha|}_{1,0}$. Putting $R(x,\eta) = x\cdot\eta + it\Lambda_j(x,\eta;\delta)$, we have

$$\begin{aligned} &(\nabla_x R)(x,\xi',\zeta''(x'',\xi'';t,\delta)) = \xi, \\ &(\nabla_\eta R)(x,\xi',\zeta''(x'',\xi'';t,\delta)) = (x',z''(x'',\xi'';t,\delta)). \end{aligned}$$

Therefore, we have

$$\{f,g\}(x',z'',\xi',\zeta'') = \{\tilde{f},\tilde{g}\}(x,\xi) \tag{4.38}$$

if $f,g \in C^\infty(\Gamma)$, $f(x,\xi)$ and $g(x,\xi)$ are analytic in (x'',ξ'') and $0 \leq t \ll 1$, where $z'' = z''(x'',\xi'';t,\delta)$, $\zeta'' = \zeta''(x'',\xi'';t,\delta)$, $\tilde{f}(x,\xi) = f(x',z'',\xi',\zeta'')$ and $\tilde{g}(x,\xi) = g(x',z'',\xi',\zeta'')$. Indeed, (4.38) can be verified by the same argument as for (real) canonical transformations. Write

$$\begin{aligned} &P(x,\xi;t,\delta) \equiv P_2(x,\xi;t,\delta) + P_1(x,\xi;t,\delta) \quad \text{mod } S^0_{1,0}, \\ &P_2(x,\xi;t,\delta) = p_2(x',z'',\xi',\zeta''), \\ &P_1(x,\xi;t,\delta) = p_1^s(x',z'',\xi',\zeta'') - (i/2)\sum_{j=1}^{n}\left(\partial^2/\partial x_j\partial\xi_j\right)P_2(x,\xi;t,\delta) \\ &\qquad + \sum_{|\alpha|+|\beta|\leq 1} \tilde{c}_{\alpha,\beta}(x,\xi;t,\delta)p_{2(\beta)}^{(\alpha)}(x',z'',\xi',\zeta'') \end{aligned}$$

in Ω, where $z'' = z''(x'',\xi'';t,\delta)$ and $\zeta'' = \zeta''(x'',\xi'';t,\delta)$. Let us apply Corollary B.2. 2 to $P(x,D;t,\delta)$. Note that

$$\begin{aligned} &P_1^s(x,\xi;t,\delta) \left(\equiv P_1(x,\xi;t,\delta) + (i/2)\sum_{j=1}^{n}\left(\partial^2/\partial x_j\partial\xi_j\right)P_2(x,\xi;t,\delta)\right) \\ &= p_1^s(x',z'',\xi',\zeta'') + \sum_{|\alpha|+|\beta|\leq 1} \tilde{c}_{\alpha,\beta}(x,\xi;t,\delta)p_{2(\beta)}^{(\alpha)}(x',z'',\xi',\zeta'') \end{aligned}$$

in Ω. The condition (H–3) implies that for any $\nu > 0$ there is $C_\nu > 0$ satisfying

$$\Big| \sum_{|\alpha|+|\beta|\le 1} \tilde{c}_{\alpha,\beta}(x,\xi;t,\delta) p_{2(\beta)}^{(\alpha)}(x',z'',\xi',\zeta'')\Big| \le \nu \mathrm{Re}\ P_2(x,\xi;t,\delta) + C_\nu \tag{4.39}$$

in Ω. Thus, it follows from the condition (H–2) and (4.38) and (4.39) that $q_j^1(x,\xi;t,\delta) \equiv \mathrm{Re}\ q_j(x',z'',\xi',\zeta'') \in S^1_{1,0}$ and $q_j^2(x,\xi;t,\delta) \equiv \mathrm{Im}\ q_j(x',z'', \xi',\zeta'') \in S^1_{1,0}$ ($1 \le j \le 2J$), and

$$(1-\varepsilon/2)\mathrm{Re}\ P_2(x,\xi;t,\delta) + \mathrm{Re}\ P_1^s(x,\xi;t,\delta) - \sum_{k=1}^{2}\sum_{j=1}^{2J} q_j^k(x,\xi;t,\delta)^2$$
$$+\sum_{k=1}^{2}\sum_{j=1}^{J}(-1)^{k-1}\{q_{2j-1}^k, q_{2j}^k\}(x,\xi;t,\delta) \ge -C$$

in Ω with $0 \le t \ll 1$, where $z'' = z''(x'',\xi'';t,\delta)$ and $\zeta'' = \zeta''(x'',\xi'';t,\delta)$, modifying Ω if necessary. From Corollary B.2. 2 there are positive constants ν and C such that

$$\|v\| \le C(\|p(x,D;t\Lambda_j)v\| + \|v\|_{-1} + \|(1-\Psi_\nu(x,D))v\|_2)$$

for $v \in C_0^\infty(\boldsymbol{R}^n)$ if $j \in \boldsymbol{N}$, $t \in [0,t_0]$ and $\delta \in [0,1]$, where t_0 is a positive constant. Since (ME) is valid for p, from Theorem 4.2. 1 we have $(x^1,\xi^1/|\xi^1|) \notin \mathrm{supp}\ u$ if (x^1,ξ^1) is sufficiently close to $(0,\xi^0)$, $u \in \mathcal{C}(\Omega_0)$ and $(x^1,\xi^1/|\xi^1|) \notin \mathrm{supp}\ p(x,D)u$. This proves the theorem. □

Grigis and Sjöstrand [GS] gave interesting examples of second-order partial differential operators which are analytic hypoelliptic. The characteristic sets of these operators are not symplectic. We shall give a slight generalization of their result. Let $\xi^0 = (0,0,1) \in \boldsymbol{R}^3$, and let Ω be an open conic neighborhood of $(0,\xi^0)$ in $T^*\boldsymbol{R}^3 \setminus 0$. Let $a(x')$ be a real-valued real analytic function in U' satisfying $a(0) = 0$, where $x = (x',x_3) = (x_1,x_2,x_3) \in \boldsymbol{R}^3$ and $U' = \{x' \in \boldsymbol{R}^2;\ (x',x_3,\xi) \in \Omega$ for some x_3 and $\xi\}$. Put

$$f(x',\xi) = \xi_2 + a(x')\xi_3, \quad g(x') = \frac{\partial a}{\partial x_1}(x')\ (= \{\xi_1, f(x',\xi)\}/\xi_3).$$

Let $a_j(x,\xi)$ ($j = 1,2$) be symbols in $AS^0(\Omega;A)$ which are real-valued and positively homogeneous of degree 0 in ξ for $|\xi| \ge 1$, and put

$$\begin{aligned} p(x,D) &= D_1^2 + a_1(x,D)(D_1 f(x',D) + f(x',D)D_1) \\ &\quad + a_2(x,D)f(x',D)^2 + b(x,D), \end{aligned}$$

where $b(x,\xi) \equiv \sum_{j=0}^{\infty} b_{1-j}(x,\xi) \in FS^1(\Omega; C_0, A)$ and $b_{1-j}(x,\xi)$ is positively homogeneous of degree $1-j$ ($j \in \boldsymbol{Z}_+$). Then we have $p(x,\xi) \equiv \sum_{j=0}^{\infty} p_{2-j}(x,\xi) \in FS^2(\Omega; C_0, A)$ and

$$\begin{aligned}
p_2(x,\xi) &= \xi_1^2 + 2a_1(x,\xi)\xi_1 f(x',\xi) + a_2(x,\xi) f(x',\xi)^2, \\
p_1(x,\xi) &= b_1(x,\xi) - i\Big\{a_1(x,\xi)g(x')\xi_3 + a_2(x,\xi)\frac{\partial a}{\partial x_2}(x')\xi_3 \\
&\quad +2\sum_{k=1}^{2}\Big(\xi_1 \frac{\partial a_1}{\partial \xi_k}(x,\xi) + f(x',\xi)\frac{\partial a_2}{\partial \xi_k}(x,\xi)\Big)\frac{\partial a}{\partial x_k}(x')\xi_3\Big\}.
\end{aligned}$$

Moreover, we have

$$\begin{aligned}
p_1^s(x,\xi) &\Big(\equiv p_1(x,\xi) + (i/2)\sum_{k=1}^{3}\Big(\partial^2 p_2/\partial x_k \partial \xi_k\Big)(x,\xi)\Big) \\
&= b_1(x,\xi) + i\beta_1(x,\xi)\xi_1 + i\beta_2(x,\xi) f(x',\xi),
\end{aligned} \tag{4.40}$$

where $\beta_j(x,\xi) \in AS^0(\Omega; A)$ ($j = 1,2$) are positively homogeneous of degree 0 in ξ for $|\xi| \ge 1$, modifying A if necessary. We now assume the following conditions:

(G–S–1) There is $c > 0$ such that

$$\zeta_1^2 + 2a_1(x,\xi)\zeta_1\zeta_2 + a_2(x,\xi)\zeta_2^2 \ge c(\zeta_1^2 + \zeta_2^2)$$

for $(x,\xi) \in \Omega$ with $|\xi| \ge 1$ and $(\zeta_1,\zeta_2) \in \boldsymbol{R}^2$.

(G–S–2) $g(x') = 0$ if and only if $x' = 0$.

(G–S–3) If $(x,\xi) \in \Omega$, $|\xi| = 1$, $\xi_1 = 0$, $f(x',\xi) = 0$ and $x' \ne 0$, then

$$-(a_2(x,\xi) - a_1(x,\xi)^2)^{-1/2}|g(x')|^{-1} b_1(x,\xi/\xi_3) \notin \{1,3,5,\cdots\}.$$

(G–S–4) There are real-valued real analytic functions $A(x')$ and $B(x')$ defined in a neighborhood U_0' of 0 in $\boldsymbol{R}^2$ satisfying

$$(\partial/\partial x_1)(A(x')g(x')) + (\partial/\partial x_2)(B(x')g(x')) = g(x') \quad \text{in } U_0'.$$

(G–S–5) There are positive constants ε, C and δ, a conic neighborhood $\mathcal{U}$ of $(0,\xi^0)$ in $T^*\boldsymbol{R}^3 \setminus 0$ and $q_j(x,\xi) \in S_{1,0}^1$ ($1 \le j \le 2J$) such that the $q_j(x,\xi)$ are real-valued and

$$\begin{aligned}
&(1-\varepsilon)p_2(x,\xi) + \mathrm{Re}\, b_{1,0}(x+z,\xi+\zeta) + g(x')\mathrm{Re}\, b_{1,1}(x,\xi) \\
&-\sum_{j=1}^{2J} q_j(x,\xi)^2 + \sum_{j=1}^{J}\{q_{2j-1}, q_{2j}\}(x,\xi) \ge -C
\end{aligned}$$

for $(x,\xi) \in \mathcal{U}$, $z \in \boldsymbol{C}^3$ with $|z| \le \delta$ and $\zeta \in \boldsymbol{C}^3$ with $|\zeta| \le \delta|\xi|$, where $b_{1,0}(x,\xi), b_{1,1}(x,\xi) \in AS^1(\Omega; A)$, $\tilde{\beta}_k(x,\xi) \in AS^0(\Omega; A)$ ($k = 1,2$) and

$$b_1(x,\xi) = b_{1,0}(x,\xi) + b_{1,1}(x,\xi)g(x') + \tilde{\beta}_1(x,\xi)\xi_1 + \tilde{\beta}_2(x,\xi)f(x',\xi).$$

Theorem 4.5. 4 *Under the conditions* (G–S–1)–(G–S–5) $p(x,D)$ *is analytic microhypoelliptic at* $(0,\xi^0)$.

Remark (i) If $p(x,D) = D_1^2 + f(x,D)^2$, then one can directly prove the above theorem from Grigis-Sjöstrand's results, using canonical transformations. (ii) Let $p(x,D) = D_1^2 + f(x',D)^2$, where $f(x',\xi) = \xi_2 + (x_1^3 + x_1x_2^2)\xi_3$. Then, by the theorem $p(x,D)$ is analytic microhypoelliptic at $(0,\xi^0)$.

Proof It is obvious that $p(x,D)$ is analytic microhypoelliptic at (x^1,ξ^1) if $(x^1,\xi^1) \in \Omega$ and "$\xi_1^1 \neq 0$ or $f(x^{1\prime},\xi^1) \neq 0$." Let (x^1,ξ^1) be a point in Ω satisfying $\xi_1^1 = f(x^{1\prime},\xi^1) = 0$ and $x^{1\prime} \neq 0$. We note that $\xi_3^1 \neq 0$ and $g(x^{1\prime}) \neq 0$. Let $N(x',\xi_1)$ be a function defined in a neighborhood of $(x^{1\prime},0)$ in $\boldsymbol{R}^2 \times \boldsymbol{R}$ such that $t = N(x',\xi_1)$ satisfies

$$\xi_1 = \int_0^t g(x_1, x_2 + s)\, ds, \quad N(x^{1\prime},0) = 0.$$

Then we can write $N(x,\xi_1) = h(x',\xi_1)\xi_1$, where $h(x',\xi_1)$ is analytic. Moreover, we have $h(x^{1\prime},0) = 1/g(x^{1\prime})$. If we put

$$\eta_1 = N(x',\xi_1/\xi_3)\xi_3, \quad y_1 = f(x',\xi)/\xi_3, \quad \eta_3 = \xi_3, \tag{4.41}$$

then we have

$$\{\eta_1, y_1\} = 1, \quad \{\eta_1,\eta_3\} = 0, \quad \{y_1,\eta_3\} = 0.$$

It follows from Proposition 4.4. 2 that there are conic neighborhoods Ω^1 and Ω^2 of $(0,\xi^0)$ and (x^1,ξ^1) in $T^*\boldsymbol{R}^3\backslash 0$, respectively, and a homogeneous canonical transformation $\chi : \Omega^1 \xrightarrow{\sim} \Omega^2$ such that $\chi(0,\xi^0) = (x^1,\xi^1/\xi_3^1)$ and (4.41) is satisfied if $(x,\xi) = \chi(y,\eta)$. By Proposition 4.4. 3 we may assume that χ has a generating function. Then we can choose Fourier integral operators F_1 and F_2 corresponding to χ and χ^{-1}, respectively, so that $F_2F_1 = id$ on $\mathcal{C}(\Omega_0^1)$ and $F_1F_2 = id$ on $\mathcal{C}(\Omega_0^2)$, where $\Omega_0^j = \Omega^j \cap \boldsymbol{R}^3 \times S^2$ ($j = 1,2$). It follows from Theorem 4.4. 4 that there is a formal analytic symbol $\tilde{p}(y,\eta) \equiv \sum_{j=1}^{\infty} \tilde{p}_{2-j}(y,\eta) \in FS^2(\Omega^1; \tilde{C}_0, \tilde{A})$ satisfying $F_2p(x,D)F_1 = \tilde{p}(y,D)$ in $\mathcal{C}(\Omega_0^1)$ and

$$\tilde{p}_2(y,\eta) = e(y,\eta)^2\eta_1^2 + 2a_1(\chi(y,\eta))e(y,\eta)y_1\eta_1\eta_3$$

$$+a_2(\chi(y,\eta))y_1^2\eta_3^2,$$

$$\tilde{p}_1^s(y,\eta)\left(\equiv \tilde{p}_1(y,\eta)+(i/2)\sum_{k=1}^{3}\left(\partial^2\tilde{p}_2/\partial y_k\partial\eta_k\right)(y,\eta)\right)$$
$$= b_1(\chi(y,\eta))+c_1(y,\eta)\eta_1+c_2(y,\eta)y_1\eta_3,$$

where $e(y,\eta)\in AS^0(\Omega^1;\widetilde{A})$ satisfies $\xi_1=e(y,\eta)\eta_1$ if $(x,\xi)=\chi(y,\eta)$ and $c_j(y,\eta)\in AS^0(\Omega^1;\widetilde{A})$ ($j=1,2$). Let us apply Theorem 4.5. 2 to $\tilde{p}(y,D)$ at $(0,\xi^0)$. Note that $e(0,\xi^0)=1/h(x^{1\prime},0)=g(x^{1\prime})$. Put

$$L(y_1,\eta_1;\eta_3)=g(x^{1\prime})^2\eta_1^2+2a_1(x^1,\xi^1)g(x^{1\prime})y_1\eta_1\eta_3$$
$$+a_2(x^1,\xi^1)y_1^2\eta_3^2+\left(b_1(x^1,\xi^1/\xi_3^1)-ia_1(x^1,\xi^1)g(x^{1\prime})\right)\eta_3.$$

Then it is easy to see that the equation $L(y_1,D_1;1)v(y_1)=0$ in $\mathcal{S}(\boldsymbol{R})$ has no non-trivial solution if (G–S–3) is satisfied. Indeed, if one makes the change of variable

$$y_1=\left(a_2(x^1,\xi^1)-a_1(x^1,\xi^1)^2\right)^{-1/4}|g(x^{1\prime})|^{1/2}\tau,$$

then the equation $L(y_1,D_1;1)v(y_1)=0$ is equivalent to

$$\left((\partial/\partial\tau)^2-\tau^2-(a_2(x^1,\xi^1)-a_1(x^1,\xi^1)^2)^{-1/2}|g(x^{1\prime})|^{-1}b_1(x^1,\xi^1/\xi_3^1)\right)$$
$$\times w(\tau)=0,$$

where $w(\tau)=\exp[ia_1(x^1,\xi^1)y_1^2/(2g(x^{1\prime}))]v(y_1)$. Therefore, Theorem 4.5. 2 implies that $p(x,D)$ is analytic microhypoelliptic at (x^1,ξ^1) (see, also, Theorem 4.5. 3 , [Tr1] and [Tf]). Put $S=\{(x,\xi)\in T^*\boldsymbol{R}^3\setminus 0;\ x_1=x_2=0$ and $\xi_1=\xi_2=0\}$. We have proved that $p(x,D)$ is analytic microhypoelliptic in $\Omega\setminus S$. Let (x^1,ξ^0) be a point in $\mathcal{U}_1\cap S$, where $\mathcal{U}_1$ is a conic neighborhood of $(0,\xi^0)$, $\mathcal{U}_1\subset\subset\mathcal{U}_0\equiv\mathcal{U}\cap(U_0'\times\boldsymbol{R}\times\boldsymbol{R}^3)$ and U_0' and $\mathcal{U}$ are as in the conditions (G–S–4) and (G–S–5), respectively. We can assume without loss of generality that $x^1=0$. Put

$$\begin{aligned}\tilde{\varphi}(x,\xi) &= (x_3+X(x'))^2\\ &\quad +2(x_3+X(x'))(A(x')\xi_1+B(x')f(x',\xi))/\xi_3,\\ X(x') &= -\int_0^{x_1}B(x')g(x')\,dx_1\\ &\quad +\int_0^{x_2}\{A(0,x_2)g(0,x_2)-a(0,x_2)\}\,dx_2,\end{aligned}$$

where $A(x')$ and $B(x')$ are as in the condition (G–S–4). Then it follows from (G–S–4) that

$$\begin{aligned}\tilde{\varphi}(x,\xi) &= x_3^2+Y(x)+\{f,x_3^2+Y\}\xi_1/\{\xi_1,f\}\\ &\quad +\{\xi_1,x_3^2+Y\}f/\{f,\xi_1\}\end{aligned}$$

if $x' \neq 0$, where $Y(x) = X(x')^2 + 2x_3X(x')$. Note that

$$\tilde{\varphi}(x,\xi) = \tilde{\varphi}(0,0,x_3,0,0,\xi_3) = x_3^2 \quad \text{for } (x,\xi) \in S \cap \mathcal{U}_0.$$

Define

$$\widetilde{\Lambda}(x,\xi;\nu,\delta) = (\tilde{\varphi}(x,\xi) - \nu)\xi_3^{1-\delta} + (\tilde{\varphi}(x,\xi) + \nu)\xi_3/2,$$

where $(x,\xi) \in \mathcal{U}_0$, $|\delta| \leq 1$ and $|\nu| \leq 1$. Let us consider the Cauchy problem

$$\begin{cases} (\partial S/\partial t)(x,\eta;t,\nu,\delta) - i\widetilde{\Lambda}((\partial S/\partial \eta)(x,\eta;t,\nu,\delta),\eta;\nu,\delta) = 0, \\ S(x,\eta;0,\nu,\delta) = x \cdot \eta. \end{cases} \tag{4.42}$$

Under the change of variables

$$s = |\eta|^{-\delta}, \quad \omega \equiv (\omega_1,\omega_2) = (\eta_1/|\eta|, \eta_2/|\eta|),$$

(4.42) is transformed into

$$\begin{cases} (\partial \Sigma/\partial t)(x,s,\omega;t,\nu,\delta) \\ \quad -i\Lambda^0(s,\omega;t,\nu,\delta;\Sigma,\partial\Sigma/\partial s,\partial\Sigma/\partial\omega) = 0, \\ \Sigma(x,s,\omega;0,\nu,\delta) = x_1\omega_1 + x_2\omega_2 + x_3\sqrt{1-\omega_1^2-\omega_2^2}, \end{cases} \tag{4.43}$$

where $\Sigma(x,s,\omega;t,\nu,\delta) = s^{1/\delta}S(x,\eta;t,\nu\delta)$ and

$$\begin{aligned}
&\Lambda^0(s,\omega;t,\nu,\delta;\Sigma,\partial\Sigma/\partial s,\partial\Sigma/\partial\omega) \\
&\Big(\equiv s^{1/\delta}\widetilde{\Lambda}(\sigma_1,\sigma_2,\sigma_3,s^{-1/\delta}\omega,s^{-1/\delta}\sqrt{1-\omega_1^2-\omega_2^2};\nu,\delta)\Big) \\
&= (\tilde{\varphi}(\sigma_1,\sigma_2,\sigma_3,\omega,\sqrt{1-\omega_1^2-\omega_2^2}) - \nu)s(1-\omega_1^2-\omega_2^2)^{(1-\delta)/2} \\
&\quad +(\tilde{\varphi}(\sigma_1,\sigma_2,\sigma_3,\omega,\sqrt{1-\omega_1^2-\omega_2^2}) + \nu)\sqrt{1-\omega_1^2-\omega_2^2}/2, \\
&\sigma_1\,(\equiv(\partial S/\partial\eta_1)(x,\eta;t,\nu,\delta)) \\
&= \omega_1\Sigma - s\omega_1\delta(\partial\Sigma/\partial s) + (1-\omega_1^2)(\partial\Sigma/\partial\omega_1) - \omega_1\omega_2(\partial\Sigma/\partial\omega_2), \\
&\sigma_2\,(\equiv(\partial S/\partial\eta_2)(x,\eta;t,\nu,\delta)) \\
&= \omega_2\Sigma - s\omega_2\delta(\partial\Sigma/\partial s) - \omega_1\omega_2(\partial\Sigma/\partial\omega_1) + (1-\omega_2^2)(\partial\Sigma/\partial\omega_2), \\
&\sigma_3\,(\equiv(\partial S/\partial\eta_3)(x,\eta;t,\nu,\delta)) \\
&= \sqrt{1-\omega_1^2-\omega_2^2}(\Sigma - s\delta(\partial\Sigma/\partial s) - \omega_1(\partial\Sigma/\partial\omega_1) - \omega_2(\partial\Sigma/\partial\omega_2)).
\end{aligned}$$

We can apply the Cauchy-Kowalevsky theorem to (4.43), and we see that there is an analytic function $\Sigma(x,s,\omega;t,\nu,\delta)$ defined near $(x,s,\omega,t,\nu,\delta) =$

$(0,0,0,0,0,0)$ which satisfies (4.43). Therefore, there are a conic neighborhood $\mathcal{U}_2$ of $(0,\xi^0)$ in $T^*\boldsymbol{R}^3\setminus 0$ and positive constants t_0, ν_0, δ_0 and R such that $\mathcal{U}_2 \subset\subset \mathcal{U}_1$, $R\ge 1$ and

$$S_R(x,\eta;t,\nu,\delta)\equiv|\eta|\Sigma(x,(R|\eta|)^{-\delta},\eta_1/|\eta|,\eta_2/|\eta|;t,\nu,\delta)$$

is analytic in $(x,\eta)\in\mathcal{U}_2$ and satisfies

$$\begin{cases} (\partial S_R/\partial t)(x,\eta;t,\nu,\delta)-i\widetilde{\Lambda}((\partial S_R/\partial\eta)(x,\eta;t,\nu,\delta),R\eta;\nu,\delta)/R=0,\\ S_R(x,\eta;0,\nu,\delta)=x\cdot\eta \end{cases}$$

if $|\eta|\ge 1/2$, $t\in[0,t_0]$, $0\le\nu\le\nu_0$ and $\delta\in[0,\delta_0]$. We can write

$$S_R(x,\eta;t,\nu,\delta)=x\cdot\eta+it\Lambda(x,\eta;t,\nu,\delta)$$

and $\Lambda(x,\xi;t,\nu,\delta)$ satisfies

$$\left|\partial_\xi^\alpha D_x^\beta\Lambda(x,\xi;t,\nu,\delta)\right|\le C_0A_0^{|\alpha|+|\beta|}|\alpha|!|\beta|!\langle\xi\rangle^{1-|\alpha|}$$

if $(x,\xi)\in\mathcal{U}_2$, $|\xi|\ge 1/2$, $t\in[0,t_0]$, $0\le\nu\le\nu_0$ and $\delta\in[0,\delta_0]$, where C_0 and A_0 are positive constants. Note that

$$\begin{aligned}\Lambda(x,\xi;0,\nu,\delta)&=-i(\partial S_R/\partial t)(x,\xi;0,\nu,\delta)=\widetilde{\Lambda}(x,R\xi;\nu,\delta)/R\\ &=(\tilde{\varphi}(x,\xi)-\nu)R^{-\delta}\xi_3^{1-\delta}+(\tilde{\varphi}(x,\xi)+\nu)\xi_3/2,\end{aligned}$$

and that there is $C>0$ such that

$$|\Lambda(x,\xi;t,\nu,\delta)-\widetilde{\Lambda}(x,R\xi;\nu,\delta)/R|\le Ct|\xi|$$

if $(x,\xi)\in\mathcal{U}_2$, $|\xi|\ge 1/2$, $t\in[0,t_0]$, $0\le\nu\le\nu_0$ and $\delta\in[0,\delta_0]$. We put

$$\Lambda_j(x,\xi;t,\delta)=\Lambda(x,\xi;t,\nu_0/j,\delta)\quad(j\in\boldsymbol{N}).$$

We may assume that $\Lambda_j(x,\xi;t,\delta)\in C^\infty(\mathcal{U}_2\times[0,t_j]\times[0,\delta_0])$, where $0<t_j\le t_0$. Then it is easy to see that $\{\Lambda_j(x,\xi;t,\delta)\}$ satisfies the conditions (Λ–1)–(Λ–4) in Section 4.2 if $t_j\le\nu_0/(6Cj)$ ($j\in\boldsymbol{N}$) and (x^0,ξ^0), Ω and δ_j are replaced by $(0,\xi^0)$, $\mathcal{U}_2$ and δ_0, respectively. Let $\eta=N^j(x,y,\xi;t,\delta)$ be the solution of

$$\eta+it\widehat{\nabla}_x\Lambda_j(x,y,\eta;t,\delta)=\xi$$

for $(x,\xi),(y,\xi)\in\mathcal{U}_2$, $t\in[0,t_j]$ and $\delta\in[0,\delta_0]$, modifying t_j if necessary. Let $p(x,\xi;t\Lambda_j)$ be a symbol in $S^2_{1,0}$ satisfying

$$p(x,\xi;t\Lambda_j)\sim\sum_{k=0}^\infty p^k_{t\Lambda_j}(x,\xi)$$

in $S^2_{1,0}(\mathcal{U}_2)$ uniformly in $\delta \in [0, \delta_0]$, where $t \in [0, t_j]$ and

$$p^k_{t\Lambda_j}(x,\xi) = \sum_{|\gamma|+|\lambda|+\ell=k} \Big[\partial^\gamma_\xi \partial^\lambda_\eta D^\gamma_y \times \Big\{p_{2-\ell(\lambda)}(x + it\widehat{\nabla}_\xi \Lambda_j(x, N^j(x,y,\xi;t,\delta),\eta;t,\delta), N^j(\cdots)) \times \det(\partial N^j/\partial\xi)(x,y,\xi;t,\delta)\Big\}\Big]_{y=0,\,\eta=0} / (\gamma!\lambda!).$$

Fix $j \in \boldsymbol{N}$. It follows from (4.9) and (4.10) that

$$\begin{aligned} p^0_{t\Lambda_j}(x,\xi) &= \{\zeta_1^2 + 2a_1(z,\zeta)\zeta_1 f(z',\zeta) + a_2(z,\zeta) f(z',\zeta)^2\} \\ &\qquad \times \det \mathcal{N}(x,\xi;t,\delta), \\ (p_{t\Lambda_j})^1_s(x,\xi) &\Big(\equiv p^1_{t\Lambda_j}(x,\xi) + (i/2)\sum_{k=1}^{3}(\partial^2/\partial x_k\partial\xi_k) p^0_{t\Lambda_j}(x,\xi)\Big) \\ &= p^1_s(z,\zeta)\det\mathcal{N}(x,\xi;t,\delta) + c_1(x,\xi;t,\delta)\zeta_1 + c_2(x,\xi;t,\delta) f(z',\zeta), \end{aligned}$$

where $\zeta \equiv \zeta(x,\xi;t,\delta) = N^j(x,0,\xi;t,\delta)$, $z \equiv z(x,\xi;t,\delta) = x + it(\nabla_\xi\Lambda_j)(x, \zeta(x,\xi;t,\delta);t,\delta)$, $\mathcal{N}(x,\xi;t,\delta) = (\partial\zeta/\partial\xi)(x,\xi;t,\delta)$, $c_k(x,\xi;t,\delta) \in S^0_{1,0}(\mathcal{U}_2)$ uniformly in $\delta \in [0,\delta_0]$ and $t \in [0,t_j]$ ($k = 1, 2$), and $p^1_s(x,\xi) = p_1(x,\xi) + (i/2)\sum_{k=1}^3(\partial^2 p_2/\partial x_k\partial\xi_k)(x,\xi)$. A simple calculation yields

$$\begin{aligned} (\partial/\partial t)\zeta(x,\xi;t,\delta) &= -i(\nabla_x\widetilde{\Lambda})(z(x,\xi;t,\delta), R\zeta(x,\xi;t,\delta);\nu_0/j,\delta)/R, \\ (\partial/\partial t)z(x,\xi;t,\delta) &= i(\nabla_\xi\widetilde{\Lambda})(z(x,\xi;t,\delta), R\zeta(x,\xi;t,\delta);\nu_0/j,\delta). \end{aligned}$$

Indeed, for example, we have

$$(\partial/\partial t)z(x,\xi;t,\delta) = \Big[\nabla_\eta\{(\partial S_R/\partial t)(x,\eta;t,\nu_0/j,\delta)\} + (\partial^2 S_R/\partial\eta\partial\eta)(x,\eta;t,\nu_0/j,\delta)(\partial\zeta/\partial t)(x,\xi;t,\delta)\Big]_{\eta=\zeta(x,\xi;t,\delta)}.$$

Put $\widehat{\Lambda}(x,\xi) = \widetilde{\Lambda}(x, R\xi;\nu_0/j,\delta)/R$. Let $F(y,\eta)$ be an analytic function. Then we have

$$\begin{aligned} &(\partial/\partial t)^k F(z(x,\xi;t,\delta),\zeta(x,\xi;t,\delta)) \\ &= \Big[(iH_{\widehat{\Lambda}}(y,\eta))^k F(y,\eta)\Big]_{y=z(x,\xi;t,\delta),\,\eta=\zeta(x,\xi;t,\delta)}. \end{aligned}$$

Therefore, we have

$$F(z(x,\xi;t,\delta),\zeta(x,\xi;t,\delta)) = \sum_{k=0}^{\infty} t^k (iH_{\widehat{\Lambda}}(x,\xi))^k F(x,\xi)/k!$$

if $0 \le t \ll 1$. Since

$$\begin{aligned} H_{\widehat{\Lambda}}(x,\xi)\xi_1 = \Big(&\xi_1(\partial/\partial x_1)(A(x')(x_3+X(x'))) \\ &+f(x',\xi)(\partial/\partial x_1)(B(x')(x_3+X(x')))\Big)(2R^{-\delta}\xi_3^{-\delta}+1), \\ H_{\widehat{\Lambda}}(x,\xi)f(x',\xi) = (&\xi_1\{A(x')(x_3+X(x')),f(x',\xi)\} \\ &+f(x',\xi)\{B(x')(x_3+X(x')),f(x',\xi)\})(2R^{-\delta}\xi_3^{-\delta}+1), \end{aligned}$$

we have

$$\begin{aligned} &\zeta_1(x,\xi;t,\delta) \\ &= (1+t\alpha_{1,1}(x,\xi;t,\delta))\xi_1 + t\alpha_{1,2}(x,\xi;t,\delta)f(x',\xi), \qquad (4.44) \\ &f(z'(x,\xi;t,\delta),\zeta(x,\xi;t,\delta)) \\ &= t\alpha_{2,1}(x,\xi;t,\delta)\xi_1 + (1+t\alpha_{2,2}(x,\xi;t,\delta))f(x',\xi), \qquad (4.45) \end{aligned}$$

where $\alpha_{i,k}(x,\xi;t,\delta) \in S^0_{1,0}(\mathcal{U}_2)$ uniformly in $\delta \in [0,\delta_0]$ and $t \in [0,t_j]$, modifying t_j and $\mathcal{U}_2$ if necessary. Put

$$P(x,D;t,\delta) = e(x,D;t,\delta)p(x,D;t\Lambda_j),$$

where $e(x,\xi;t,\delta) \in S^0_{1,0}$ and $e(x,\xi;t,\delta) = 1/\det \mathcal{N}(x,\xi;t,\delta)$ for $(x,\xi) \in \mathcal{U}_2$ with $|\xi| \ge 1$. A simple calculation yields

$$P(x,\xi;t,\delta) \equiv P_2(x,\xi;t,\delta) + P_1(x,\xi;t,\delta)$$

mod $S^0_{1,0}(\mathcal{U}_2)$ uniformly in $\delta \in [0,\delta_0]$ for $t \in [0,t_j]$,

$$P_2(x,\xi;t,\delta) = \zeta_1^2 + 2a_1(z,\zeta)\zeta_1 f(z',\zeta) + a_2(z,\zeta)f(z',\zeta)^2, \qquad (4.46)$$

$$\begin{aligned} &P_1^s(x,\xi;t,\delta)\Big(\equiv P_1(x,\xi;t,\delta) + (i/2)\sum_{k=1}^{3}(\partial^2/\partial x_k\partial\xi_k)P_2(x,\xi;t,\delta)\Big) \\ &= p_s^1(z,\zeta) + \tilde{c}_1(x,\xi;t,\delta)\zeta_1 + \tilde{c}_2(x,\xi;t,\delta)f(z',\zeta), \qquad (4.47) \end{aligned}$$

where $\zeta \equiv \zeta(x,\xi;t,\delta)$, $z \equiv z(x,\xi;t,\delta)$ and the $\tilde{c}_k(x,\xi;t,\delta)$ have the same properties as the $c_k(x,\xi;t,\delta)$. It follows from the condition (G–S–1) and (4.44)–(4.46) that

$$\mathrm{Re}\ P_2(x,\xi;t,\delta) \ge (1-\varepsilon/4)p_2(x,\xi) \qquad (4.48)$$

if $(x,\xi) \in \mathcal{U}_2$, $|\xi| \ge 1$, $0 \le t \ll 1$ and $\delta \in [0,\delta_0]$, where ε is the constant in the condition (G–S–5). The Jacobi identity gives

$$\begin{aligned} H_{\widehat{\Lambda}}(x,\xi)(g(x')\xi_3) &= -\{\xi_1,\{f,\widehat{\Lambda}\}\} - \{f,\{\widehat{\Lambda},\xi_1\}\} \\ &= d_1(x,\xi;t,\delta)\xi_1 + d_2(x,\xi;t,\delta)f(x',\xi) + d_3(x,\xi;t,\delta)g(x')\xi_3, \end{aligned}$$

where the $d_k(x,\xi;t,\delta)$ have the same properties as the $c_k(x,\xi;t,\delta)$. Hence we have

$$\begin{aligned} g(z'(x,\xi;t,\delta)) &= t\tilde{d}_1(x,\xi;t,\delta)\xi_1/\xi_3 + t\tilde{d}_2(x,\xi;t,\delta)f(x',\xi)/\xi_3 \\ &\quad +(1+t)\tilde{d}_3(x,\xi;t,\delta)g(x'), \end{aligned}$$

where the $\tilde{d}_k(x,\xi;t,\delta)$ have the same properties as the $c_k(x,\xi;t,\delta)$. This, together with (4.40), (4.44), (4.45) and (4.47), yields

$$\begin{aligned} P_1^s(x,\xi;t,\delta) &= b_{1,0}(z(x,\xi;t,\delta),\zeta(x,\xi;t,\delta)) \\ &\quad +(b_{1,1}(x,\xi)+te_1(x,\xi;t,\delta)\xi_3)g(x') \\ &\quad +e_2(x,\xi;t,\delta)\xi_1 + e_3(x,\xi;t,\delta)f(x',\xi), \qquad (4.49) \end{aligned}$$

where the $e_k(x,\xi;t,\delta)$ have the same properties as the $c_k(x,\xi;t,\delta)$. Note that

$$\begin{aligned} & t\,\mathrm{Re}\ e_1(x,\xi;t,\delta)g(x')\xi_3 \\ & = \{\sqrt{t}\xi_1, \sqrt{t}\,\mathrm{Re}\ e_1(x,\xi;t,\delta)f(x',\xi)\} - t\,\mathrm{Re}\ (\partial e_1/\partial x_1)(x,\xi;t,\delta)f(x',\xi) \end{aligned}$$

and there is $C>0$ such that

$$\begin{aligned} & t(\xi_1^2 + (\mathrm{Re}\ e_1(x,\xi;t,\delta))^2 f(x',\xi)^2) + t\,\mathrm{Re}\ (\partial e_1/\partial x_1)(x,\xi;t,\delta)f(x',\xi) \\ & -\mathrm{Re}\ e_2(x,\xi;t,\delta)\xi_1 - \mathrm{Re}\ e_3(x,\xi;t,\delta)f(x',\xi) \le \varepsilon p_2(x,\xi)/2 + C \end{aligned}$$

if $(x,\xi)\in\mathcal{U}_2$ and $0\le t\ll 1$. Therefore, by the condition (G–S–5), (4.48) and (4.49) we have

$$\begin{aligned} & (1-\varepsilon/4)\,\mathrm{Re}\ P_2(x,\xi;t,\delta) + \mathrm{Re}\ P_1^s(x,\xi;t,\delta) \\ & -\sum_{j=1}^{2J+2} q_j(x,\xi)^2 + \sum_{j=1}^{J+1}\{q_{2j-1},q_{2j}\}(x,\xi) \ge -C' \end{aligned}$$

if $(x,\xi)\in\mathcal{U}_2$ and $0\le t\ll 1$, where $q_{2J+1}(x,\xi) = -\sqrt{t}\xi_1$ and $q_{2J+2}(x,\xi) = \sqrt{t}\,\mathrm{Re}\ e_1(x,\xi;t,\delta)f(x',\xi)$. From Corollary B.2. 2 there are positive constants ν and C such that

$$\|v\| \le C(\|p(x,D;t\Lambda_j)v\| + \|v\|_{-1} + \|(1-\Psi_\nu(x,D))v\|_2)$$

for $v\in C_0^\infty(\boldsymbol{R}^3)$ if $j\in\boldsymbol{N}$, $t\in[0,t_j]$ and $\delta\in[0,\delta_0]$, modifying t_j if necessary. Since (ME) is valid for p and $x^0=0$, Theorem 4.2. 1 implies that $(0,\xi^0)\notin\mathrm{supp}\ u$ if $u\in\mathcal{C}(\Omega_0)$ and $(0,\xi^0)\notin\mathrm{supp}\ p(x,D)u$, where $\Omega_0=\Omega\cap\boldsymbol{R}^3\times S^2$. This proves the theorem. □

Chapter 5

Local solvability

Applying the same arguments as in the framework of C^∞ and distributions, we shall prove in Section 5.2 that $p(x, D)$ (resp. ${}^t p(x, D)$) satisfies certain energy estimates if $p(x, D)$ is analytic hypoelliptic (resp. locally solvable) in the space of hyperfunctions. Conversely, under a little strenghen estimates we shall prove local solvability in the space of hyperfunctions in Section 5.3. As applications of the results in Sections 5.2 and 5.3, we shall prove local solvability of microhyperbolic operators and some second order operators of a special form in Section 5.4. The so-called Mizohata operators $D_1 + ix_1^k D_n$ will be also treated.

5.1 Preliminaries

In this section we shall give several fundamental results and theorems on locally convex spaces in the form of propositions, which will be used in Section 5.2. For terminology we refer to Schaefer [Sr] and Grothendieck [Gk]. We shall also give some lemmas used in this chapter.

Proposition 5.1.1 *Let $\mathcal{X}$ and $\mathcal{Y}$ be a Fréchet space and a metrizable locally convex space, respectively, and let $\varphi : \mathcal{X} \times \mathcal{Y} \to \boldsymbol{C}$ be a bilinear form. Then φ is continuous if φ is separately continuous.*

For the proof of the proposition we refer to [Sr] and [Gk]. Let $\{\mathcal{X}_j\}$ and $\{u_j\}$ be sequences of locally convex spaces and linear mappings, respectively, such that $u_j : \mathcal{X}_j \to \mathcal{X}_{j+1}$ and u_j is one-to-one ($j = 1, 2, \cdots$). We assume that u_j is weakly compact, *i.e.*, there is a neighborhood V of 0 in $\mathcal{X}_j$ such that $u_j(V)$ is a relatively weakly compact subset of $\mathcal{X}_{j+1}$ ($j \in \boldsymbol{N}$). If the $\mathcal{X}_j$ are reflexive locally convex spaces and $\mathcal{X}_j \subset \mathcal{X}_{j+1}$ ($j \in \boldsymbol{N}$), then $\{\mathcal{X}_j\}$ and $\{u_j\}$ satisfy the above requirement, where u_j :

$\mathcal{X}_j \to \mathcal{X}_{j+1}$ is the inclusion mapping ($j \in \boldsymbol{N}$). We define the locally convex space $\mathcal{X}$ as $\mathcal{X} = \varinjlim \mathcal{X}_j$, *i.e.*, $\mathcal{X}$ is the injective limit of weakly compact sequence $\{\mathcal{X}_j\}$. By definition there are mappings $v_j : \mathcal{X}_j \to \mathcal{X}$ ($j \in \boldsymbol{N}$) such that $v_{j+1} \circ u_j = v_j$ ($j \in \boldsymbol{N}$). Note that, for a convex subset V of $\mathcal{X}$, V is a neighborhood of 0 in $\mathcal{X}$ if and only if $v_j^{-1}(V)$ is a neighborhood of 0 in $\mathcal{X}_j$.

Proposition 5.1. 2 (Komatsu [Km]) *$\mathcal{X}$ is a Hausdorff, complete, reflexive and bornologic* (DF) *space. In particular, $\mathcal{X}$ is the (strong) dual space of Fréchet space, and for any bounded subset X of $\mathcal{X}$ there is $j \in \boldsymbol{N}$ such that $v_j^{-1}(X)$ is a bounded subset of $\mathcal{X}_j$.*

We shall use the following closed graph theorem in Section 5.2 (see, *e.g.*, [Sr]).

Proposition 5.1. 3 (Robertson-Robertson) *Let E and F be a barrelled locally convex space and a B-complete locally convex space, respectively, and let $T : E \to F$ be a closed linear operator whose domain $D(T)$ is equal to E. Then T is continuous.*

Proposition 5.1. 4 *The strong dual of a reflexive Fréchet space is barrelled, bornologic and B-complete. In particular, the injective limit $\mathcal{X}$ of the weakly compact sequence of locally convex spaces is barrelled and B-complete.*

For the proof of the above proposition we refer to [Sr].

Lemma 5.1. 5 *Let $p(\xi, y, \eta) \in S^+(R_0, A)$, and let Γ be a conic subset of $\boldsymbol{R}^n \times (\boldsymbol{R}^n \setminus \{0\})$. We assume that there is $\varepsilon > 0$ such that $p(\xi, y, \eta) = 0$ if $(y, \eta) \in \Gamma_\varepsilon$ and $\langle \eta \rangle \geq R_0/2$, where $\Gamma_\varepsilon = \{(x, \xi) \in \boldsymbol{R}^n \times (\boldsymbol{R}^n \setminus \{0\})$; $|(x, \xi/|\xi|) - (y, \eta/|\eta|)| < \varepsilon$ for some $(y, \eta) \in \Gamma\}$. Then we can find $R(\varepsilon) > 0$ so that*

$$WF_A(p(D_x, y, D_y)u) \cap \Gamma = \emptyset$$

if $u \in \mathcal{F}_0$ and $R_0 \geq R(\varepsilon)A$.

Proof We may assume that $\varepsilon \leq 1$. Fix $(x^0, \xi^0) \in \Gamma$ with $|\xi^0| = 1$, and choose a symbol $g^R(\xi)$ so that

$$\left|\partial_\xi^{\alpha+\gamma} g^R(\xi)\right| \leq C_{|\gamma|}(C/(\varepsilon R))^{|\alpha|}\langle \xi \rangle^{-|\gamma|} \quad \text{if } \langle \xi \rangle \geq R|\alpha|,$$

$\operatorname{supp} g^R \subset \{\xi \in \boldsymbol{R}^n;\ |\xi/|\xi| - \xi^0| < \varepsilon/2$ and $\langle \xi \rangle \geq R\}$ and $g^R(\xi) = 1$ if $|\xi/|\xi| - \xi^0| \leq \varepsilon/4$ and $\langle \xi \rangle \geq 2R$, where C is independent of ε and $R \geq R_0$. We put

$$q^R(\xi, y, \eta) \equiv g^R(\xi)p(\xi, y, \eta) \in S^+(R, RA/R_0 + C/\varepsilon).$$

We see that $(y,\eta) \in \Gamma_\varepsilon$ and $\langle\eta\rangle \geq R_0/2$ if $|y-x^0| \leq \varepsilon/4$, $|\xi-\eta| \leq \varepsilon|\eta|/8$, $|\xi/|\xi| - \xi^0| \leq \varepsilon/2$ and $\langle\xi\rangle \geq R$. Therefore, we have $q^R(\xi,y,\eta) = 0$ if $|y-x^0| \leq \varepsilon/4$ and $|\xi-\eta| \leq \varepsilon|\eta|/8$. From Corollary 2.6. 3 there are $R_j(\varepsilon) > 0$ ($j = 1,2$) such that $q^R(D_x,y,D_y)u$ is analytic at x^0 for $u \in \mathcal{F}_0$ if $R \geq R_1(\varepsilon) + R_2(\varepsilon)(RA/R_0 + C/\varepsilon)$. In particular, $(x^0,\xi^0) \notin WF_A(p(D_x,y,D_y)u)$ for $u \in \mathcal{F}_0$ if $R_0 \geq 2R_2(\varepsilon)A$ and $R \geq \max\{R_0, 2R_1(\varepsilon) + 2R_2(\varepsilon)C/\varepsilon\}$, which proves the lemma. □

We define

$$L^2_\varepsilon := \{f \in \mathcal{S}'_{-\varepsilon};\ e^{\varepsilon\langle\xi\rangle}\hat{f}(\xi) \in L^2(\boldsymbol{R}^n)\}$$

and introduce the norm $\|f\|_{L^2_\varepsilon} := (2\pi)^{-n/2}\|e^{\varepsilon\langle\xi\rangle}\hat{f}(\xi)\|$ to L^2_ε, where $\|\cdot\|$ denotes the L^2-norm.

Lemma 5.1. 6 *Let $p(\xi,y,\eta)$ be a symbol satisfying*

$$\begin{aligned}\left|\partial_\xi^\alpha D_y^{\beta+\tilde{\beta}}\partial_\eta^\gamma p(\xi,y,\eta)\right| \;&\leq\; C_{|\alpha|+|\tilde{\beta}|+|\gamma|}(A/R_0)^{|\beta|}\langle\xi\rangle^{-|\alpha|+|\beta|} \\ &\quad\times\langle\eta\rangle^{-|\gamma|}\exp[\delta_1\langle\xi\rangle - \delta_2\langle\eta\rangle]\end{aligned}$$

if $\langle\xi\rangle \geq R_0|\beta|$. Then $p(D_x,y,D_y)$ maps continuously $L^2_{\varepsilon_2}$ to $L^2_{\varepsilon_1}$ if $R_0 \geq 25e\sqrt{n}A$, $2(\varepsilon_1+\delta_1)_+ < \varepsilon_2+\delta_2$ and $3(\varepsilon_1+\delta_1) + 2(\varepsilon_2+\delta_2)_- < 1/R_0$.

Proof Choose a symbol $g(\xi,\eta)$ so that $|\partial_\xi^\alpha\partial_\eta^\gamma g(\xi,\eta)| \leq C_{|\alpha|+|\gamma|}\langle\xi\rangle^{-|\alpha|} \times\langle\eta\rangle^{-|\gamma|}$, $g(\xi,\eta) = 1$ if $|\xi| \leq 3|\eta|/2$ or $|\xi| \leq 1$, and $g(\xi,\eta) = 0$ if $|\xi| \geq 2|\eta|$ and $|\xi| \geq 2$. We put

$$p_1(\xi,y,\eta) = g(\xi,\eta)p(\xi,y,\eta), \quad p_2(\xi,y,\eta) = (1-g(\xi,\eta))p(\xi,y,\eta).$$

Then $p_1(\xi,y,\eta)$ satisfies

$$\left|\partial_\xi^\alpha D_y^\beta\partial_\eta^\gamma\{\exp[\varepsilon_1\langle\xi\rangle - \varepsilon_2\langle\eta\rangle]p_1(\xi,y,\eta)\}\right| \leq C_{|\alpha|+|\beta|+|\gamma|}\langle\xi\rangle^{-|\alpha|}\langle\eta\rangle^{-|\gamma|}$$

if $2(\varepsilon_1+\delta_1)_+ < \varepsilon_2+\delta_2$. This implies that $p_1(D_x,y,D_y) : L^2_{\varepsilon_2} \to L^2_{\varepsilon_1}$ is continuous if $2(\varepsilon_1+\delta_1)_+ < \varepsilon_2+\delta_2$. Since

$$\begin{aligned}&\left|\partial_\xi^\alpha D_y^\beta\partial_\eta^\gamma\{\exp[-\delta\langle\xi\rangle + \delta_2\langle\eta\rangle]p_2(\xi,y,\eta)\}\right| \\ &\leq C_{|\alpha|+|\beta|+|\gamma|,\delta}\langle\xi\rangle^{-|\alpha|}\langle\eta\rangle^{-|\gamma|}\exp[-(\delta-\delta_1)\langle\xi\rangle/2]\end{aligned}$$

if $\delta > \delta_1$, $p_2(D_x,y,D_y)v$ belongs to $\mathcal{S}_{-\delta}$ if $v \in \mathcal{S}_\infty$ and $\delta > \delta_1$. So $\sum_{j=1}^\infty \psi_j^{R_0}(D)p_2(D_x,y,D_y)v$ converges to $p_2(D_x,y,D_y)v$ in $\mathcal{S}_{-\delta}$ if $v \in \mathcal{S}_\infty$

and $\delta > \delta_1$. Using oscillatory integrals and Lebesgue's convergence theorem, we have

$$\psi_j^{R_0}(D)p_2(D_x, y, D_y)v = (2\pi)^{-2n} \int e^{ix\cdot\xi - iy\cdot(\xi-\eta)} \times \langle x-y\rangle^{-2M} \langle D_\xi\rangle^{2M} \Big\{\psi_j^{R_0}(\xi) K^j p_2(\xi, y, \eta)\Big\} \hat{v}(\eta)\, d\eta dy d\xi,$$

where $M = [n/2] + 1$ and $K = |\xi - \eta|^{-2} \sum_{k=1}^n (\xi_k - \eta_k) D_{y_k}$. This implies that $\psi_j^{R_0}(\xi)p_2(\xi, y, \eta) = a_j(D_x, y, D_y)$ on $\mathcal{S}_\infty$, where $a_j(\xi, y, \eta) = \psi_j^{R_0}(\xi) K^j p_2(\xi, y, \eta)$. Putting $a(\xi, y, \eta) = \sum_{j=1}^\infty a_j(\xi, y, \eta)$, we have $p_2(D_x, y, D_y) = a(D_x, y, D_y)$ on $\mathcal{S}_\infty$. On the other hand, a simple calculation yields

$$\Big|\partial_\xi^\alpha D_y^\beta \partial_\eta^\gamma \{\exp[\varepsilon_1\langle\xi\rangle - \varepsilon_2\langle\eta\rangle] a(\xi, y, \eta)\}\Big| \le C_{|\alpha|+|\beta|+|\gamma|} \exp[(\delta_1 - 1/(3R_0) + \varepsilon_1 + 2(\varepsilon_2 + \delta_2)_-/3)\langle\xi\rangle]$$

if $R_0 \ge 25e\sqrt{n}A$, since

$$\Big|\partial_\xi^\alpha D_y^\beta \partial_\eta^\gamma a_j(\xi, y, \eta)\}\Big| \le C_{|\alpha|+|\beta|+|\gamma|} \langle\xi\rangle^{-|\alpha|}\langle\eta\rangle^{-|\gamma|} \times (24e\sqrt{n}A/R_0)^j (\langle\xi\rangle/|\xi|)^j \exp[(\delta_1 - 1/(3R_0))\langle\xi\rangle - \delta_2\langle\eta\rangle].$$

Therefore, $a(D_x, y, D_y)$ maps continuously $L^2_{\varepsilon_2}$ to $L^2_{\varepsilon_1}$ if $R_0 \ge 25e\sqrt{n}A$ and $3(\varepsilon_1 + \delta_1) + 2(\varepsilon_2 + \delta_2)_- < 1/R_0$, which proves the lemma. □

The following lemma is a variant of Lemma 4.1. 4 .

Lemma 5.1. 7 *Let $p(\xi, w, \zeta, y, \eta) \in C^\infty(\boldsymbol{R}^n \times \boldsymbol{R}^n \times \boldsymbol{R}^n \times \boldsymbol{R}^n \times \boldsymbol{R}^n)$ be a symbol satisfying* (P–2) *and* (P–4), *where the conditions* (P–2) *and* (P–4) *are as in* Section 2.4. *We assume that $p(\xi, w, \zeta, y, \eta) = 0$ if $|w - y| \le \varepsilon_0$, $|\xi - \eta| \le \varepsilon_0'|\eta|$ and $|\zeta - \eta| \le \varepsilon_0'|\eta|$, where $\varepsilon_0, \varepsilon_0' > 0$. Then there are symbols $p_j(\xi, y, \eta)$ ($j = 1, 2$), positive constants $A_1' \equiv A_1'(A_1, \varepsilon_0, \varepsilon_0')$ and $R(A_1, \varepsilon_0, \varepsilon_0')$, which are polynomials of A_1 of degree 1, and $c > 0$ such that*

$$p(D_x, w, D_w, y, D_y) = p_1(D_x, y, D_y) + {}^t p_2(D_x, y, D_y) \quad \text{on } \mathcal{S}_\infty,$$
$$\Big|\partial_\xi^\alpha D_y^{\beta+\tilde\beta} \partial_\eta^\gamma p_j(\xi, y, \eta)\Big| \le C_{|\alpha|+|\tilde\beta|+|\gamma|, \varepsilon_0, \varepsilon_0', R_0} ((R_0/(50en) + A_1')/R_0)^{|\beta|} \times \langle\xi\rangle^{m_1 - |\alpha| + |\beta|} \exp[\delta_{2j-1}\langle\xi\rangle - c_0\langle\eta\rangle/R_0] \quad (j = 1, 2)$$

if $R_0 \ge R(A_1, \varepsilon_0, \varepsilon_0')$, $\langle\xi\rangle \ge R_0|\beta|$, $\delta_2 \le 1/R_0$ and $\max\{4(\delta_2)_+ + \delta_{5-2j}, 4\delta_2 + 2|\delta_2| + 2\delta_{5-2j}\} \le c_0/R_0$ ($j = 1, 2$).

Proof By Lemma 2.1.3 we can choose a symbol $g(\xi,\eta)\in S^{0,0}(R_0,C(\varepsilon_0'))$ so that $g(\xi,\eta)=0$ if $|\xi-\eta|\ge\varepsilon_0'|\eta|/3$, and $g(\xi,\eta)=1$ if $|\xi-\eta|\le\varepsilon_0'|\eta|/6$ and $\langle\eta\rangle\ge R_0$. Put

$$\begin{aligned}
p_1(\xi,w,\zeta,y,\eta)&=p(\xi,w,\zeta,y,\eta)g(\xi,\zeta),\\
p_2(\xi,w,\zeta,y,\eta)&=p(-\eta,y,-\zeta,w,-\xi)(1-g(-\eta,-\zeta)),\\
p_{2,1}(\xi,w,\zeta,y,\eta)&=p_2(\xi,w,\zeta,y,\eta)\phi_1^{R_0}(\eta),\\
p_{2,2}(\xi,w,\zeta,y,\eta)&=p_2(\xi,w,\zeta,y,\eta)(1-\phi_1^{R_0}(\eta)).
\end{aligned}$$

Note that

$$p(D_x,w,D_w,y,D_y)=p_1(D_x,w,D_w,y,D_y)+{}^tp_2(D_x,w,D_w,y,D_y).$$

We may assume that $\varepsilon_0'\le 3$. Since $p_1(\xi,w,\zeta,y,\eta)$ satisfies (P–2) and (P–4) with A_1 replaced by $A_1+C(\varepsilon_0')$ and $p_1(\xi,w,\zeta,y,\eta)=0$ if $|w-y|\le\varepsilon_0$ and $|\zeta-\eta|\le\varepsilon_0'|\eta|/3$, we can apply Lemma 4.1.4 to $p_1(\xi,w,\zeta,y,\eta)$. Since $p_{2,1}(\xi,w,\zeta,y,\eta)=0$ if $|\zeta-\eta|\le\varepsilon_0'|\zeta|/6$, we can also apply Lemma 4.1.4 to $p_{2,1}(\xi,w,\zeta,y,\eta)$. Therefore, applying Theorem 2.4.6 (i) to $p_{2,2}(\xi,w,\zeta,y,\eta)$ we can prove the lemma. □

Lemma 5.1.8 *Let $f(t)$ be a continuous function on $[0,\infty)$ such that $f(t)\ge 0$ ($t\in[0,\infty)$), and for any $\varepsilon>0$ there is $T_\varepsilon>0$ satisfying $f(t)\le\varepsilon t$ for $t\ge T_\varepsilon$. Then there is $F(t)\in\mathcal{A}(\boldsymbol{C}\setminus(-\infty,0])$ satisfying the following:* (i) *$F(t)\ge\max_{0\le s\le t}f(s)$ for $t>0$.* (ii) *For any $\varepsilon>0$ there is $\widehat{T}_\varepsilon>0$ such that $F(t)\le\varepsilon t$ for $t\ge\widehat{T}_\varepsilon$.* (iii) *For any $\varepsilon>0$ there is $\widehat{T}_\varepsilon>0$ such that $t/(1+\log t)\le\varepsilon F(t)$ for $t\ge\widehat{T}_\varepsilon$.* (iv) *$0<F'(t)\le F(t)/t$ for $t>0$.* (v) *There is $C>0$ such that $F(t)/t\le CF'(t)$ for $t\ge C$.* (vi) *$F''(t)<0$ for $t>0$.* (vii) *For any $\varepsilon>0$ there is $\widehat{T}_\varepsilon>0$ such that $t^2|F''(t)|\le\varepsilon F(t)$ for $t\ge\widehat{T}_\varepsilon$.* (viii) *There is $C>0$ such that*

$$|F^{(k)}(t)|\le C(2/t)^k k!F(t)\quad \text{for } t>0 \text{ and } k\in\boldsymbol{Z}_+.$$

Proof By assumption we can choose $T(j)\in[0,\infty)$ ($j\in\boldsymbol{N}$) so that $T(j)>jT(j-1)$ ($j=2,3,\cdots$) and

$$\max_{0\le s\le t}f(s)\le t/j\quad \text{for } t\ge T(j)$$

($j\in\boldsymbol{N}$). Put

$$a_1=\max_{0\le t\le T(1)}f(t),\quad a_j=a_{j-1}+T(j)/(j(j-1))\ (j=2,3,\cdots),$$

and define

$$g_0(t) = \begin{cases} t + a_1 & \text{if } 0 \le t \le T(2), \\ t/j + a_j & \text{if } j \ge 2 \text{ and } T(j) \le t \le T(j+1). \end{cases}$$

Then it is obvious that $g_0(t) \ge f(t)$ for $t \ge 0$, $g_0(t)$ is continuous and monotone increasing on $[0, \infty)$. Moreover, for $\varepsilon > 0$ we have $g_0(t) \le \varepsilon t$ if $j > 1/\varepsilon$ and $t \ge \max\{T(j), a_j/(\varepsilon - 1/j)\}$. Put $g(s) = 2g_0(e^s)$. Then $g(s)$ is a monotone increasing continuous function defined on $\boldsymbol{R}$ and satisfies $g(s) \ge 2f(e^s)$ ($s \in \boldsymbol{R}$) and $\lim_{s \to -\infty} g(s) = 2a_1$. Moreover, for any $\varepsilon > 0$ there is S_ε such that $g(s) \le \varepsilon e^s$ for $s \ge S_\varepsilon$. Define

$$G(s) = E * g(s) = \int_{-\infty}^{\infty} E(s - \tau) g(\tau)\, d\tau$$

for $s \in \boldsymbol{C}$, where $E(s) = (2\pi)^{-1/2} \exp[-s^2/2]$. It is obvious that $G(s)$ is entire analytic and $G(s) \ge G(s')$ if $s \ge s'$. We have

$$G(s) \ge \int_s^{\infty} E(s - \tau) g(s)\, d\tau = g_0(e^s) \quad (s \in \boldsymbol{R}).$$

Moreover, for $\varepsilon > 0$

$$\begin{aligned} G(s) &\le \int_{-\infty}^{S_{\varepsilon'}} E(s - \tau) g(S_{\varepsilon'})\, d\tau + \varepsilon' \int_{S_{\varepsilon'}}^{\infty} E(s - \tau) e^\tau\, d\tau \\ &\le \varepsilon' \exp[S_{\varepsilon'}] + \varepsilon' e^{s + 1/2} = \varepsilon e^s \end{aligned}$$

if $s \ge S_{\varepsilon'}$, where $\varepsilon' = \varepsilon/(1 + e^{1/2})$. Finally, we put $F(t) = G(\log t)$ for $t \in \boldsymbol{C} \setminus (-\infty, 0]$. Here $\log t = \log|t| + i \arg t$ and $|\arg t| < \pi$ for $t \in \boldsymbol{C} \setminus (-\infty, 0]$. It is obvious that $F \in \mathcal{A}(\boldsymbol{C} \setminus (-\infty, 0])$ and F has the properties (i) and (ii) of the lemma. Let $j \in \boldsymbol{N}$ satisfy $j \ge 2$ and $T(j) \ge 1$. Then we have

$$t/(F(t)(1 + \log t)) \le j/(1 + \log T(j))$$

if $T(j) \le t < T(j+1)$. Since $T(2) > 0$ and $T(j) > j!T(2)/2$, the Stirling formula proves the assertion (iii). In order to prove that F has the property (iv), it suffices to show that $0 < G'(s) \le G(s)$ for $s \in \boldsymbol{R}$. Since

$$g'(s) = \begin{cases} 2e^s & \text{if } s < \log T(2), \\ 2e^s/j & \text{if } j \ge 2 \text{ and } \log T(j) < s < \log T(j+1), \end{cases}$$

we have

$$0 < G'(s) = \int_{-\infty}^{\infty} E(\tau)g'(s-\tau)\,d\tau \le \int_{-\infty}^{\infty} E(\tau)g(s-\tau)\,d\tau = G(s)$$

for $s \in \boldsymbol{R}$. Note that

$$\begin{aligned} a_j &= a_1 + \sum_{k=2}^{j} T(k)/(k(k-1)) \\ &\le \begin{cases} a_1 + T(j)/(j(j-1)) + T(j-1)(1-1/(j-1)) & \text{if } j \ge 3, \\ a_1 + T(2)/2 & \text{if } j = 2. \end{cases} \end{aligned}$$

Therefore, we have

$$\begin{aligned} &g(s)/g'(s) \le 1 + ja_j/T(j) \le 1 + j(a_1 + T(j)/j)/T(j) \\ &\le 2 + a_1/T(1) \quad \text{if } j \ge 2 \text{ and } \log T(j) < s < \log T(j+1). \end{aligned}$$

Here we have used the condition that $T(j) > jT(j-1)$. This yields

$$\begin{aligned} &G(s) \le 2\int_{\log T(2)}^{\infty} E(s-\tau)g(\tau)\,d\tau \\ &\le 2(2+a_1/T(1))\int_{-\infty}^{\infty} E(s-\tau)g'(\tau)\,d\tau = 2(2+a_1/T(1))G'(s) \end{aligned}$$

for $s \ge \log T(2)$. So we have

$$F(t)/t \le 2(2+a_1/T(1))F'(t) \quad \text{for } t \ge T(2),$$

which implies that F has the property (v). Since $\widetilde{E}(\tau) \equiv e^{-\tau}E(\tau) \in \mathcal{S}(\boldsymbol{R})$, $\tilde{g}(\tau) \equiv e^{-\tau}g'(\tau) \in \mathcal{S}'$ and $G'(s) = e^s\widetilde{E} * \tilde{g}(s)$, we have

$$G''(s) = G'(s) + e^s\widetilde{E} * \tilde{g}'(s).$$

On the other hand, we have

$$\begin{aligned} &\tilde{g}'(\tau) = -\tilde{g}(\tau) + e^{-\tau}g''(\tau), \\ &g''(\tau) = g'(\tau) - \sum_{j=2}^{\infty} 2T(j)\delta(\tau - \log T(j))/(j(j-1)). \end{aligned}$$

Therefore, we have

$$\begin{aligned} &G''(s) = \langle e^{-\tau}g''(\tau), e^{\tau}E(s-\tau)\rangle_\tau \\ &= G'(s) - 2(2\pi)^{-1/2}\sum_{j=2}^{\infty} T(j)\exp[-(s-\log T(j))^2/2]/(j(j-1)), \end{aligned}$$

which gives

$$F''(t) = (G''(\log t) - G'(\log t))/t^2 < 0 \quad \text{for } t > 0.$$

This proves the assertion (vi). We can write

$$tF''(t) = -2(2\pi)^{-1/2} \sum_{j=2}^{\infty} h(T(j)/t)/(j(j-1)) \quad \text{for } t > 0,$$

where

$$h(X) = X \exp[-(\log X)^2/2] \quad (X > 0).$$

Note that $h(X)$ is monotone increasing on $(0, e]$ and monotone decreasing on $[e, \infty)$. Since $F(t) \geq t/k$ if $k \geq 2$ and $T(k) \leq t < T(k+1)$, we have

$$\begin{aligned} &t^2|F''(t)|/F(t) \\ &\leq 2(2\pi)^{-1/2}\Big\{\sum_{j=2}^{k-1} kh(T(j)/T(k))/(j(j-1)) + h(T(k)/t)/(k-1) \\ &\quad +h(T(k+1)/t)/(k+1) + \sum_{j=k+2}^{\infty} kh(T(j)/T(k+1))/(j(j-1))\Big\} \end{aligned}$$

if $k \geq 2$ and $T(k) \leq t < T(k+1)$. Noting that $h(X) \leq h(e) = \sqrt{e}$, $h(T(j)/T(k)) \leq h(1/k) \leq k^{-2}$ if $2 \leq j < k$ and $k \geq e^2$ and that $h(T(j)/T(k+1)) \leq h(j) \leq j^{-1}$ if $j \geq k+2 \geq e^4$, we can prove the assertion (vii). Let $s_0 \in \boldsymbol{R}$ and $s \in \boldsymbol{C}$. We have

$$\begin{aligned} |G(s)| &\leq (2\pi)^{-1/2} \int_{-\infty}^{\infty} \exp[-\tau^2/2 + (\text{Im } s)^2/2] g(\text{Re } s - \tau)\, d\tau \\ &= \exp[(\text{Im } s)^2/2] G(\text{Re } s). \end{aligned} \tag{5.1}$$

If Re $s \leq s_0$, then $g(\text{Re } s - \tau) \leq g(s_0 - \tau)$. If Re $s > s_0$, then

$$g(\text{Re } s - \tau) \leq \exp[\text{Re } s - s_0] g(s_0 - \tau). \tag{5.2}$$

Indeed, $g_0(t) \leq t/j + a_j \leq (t/t_0) g_0(t_0)$ if $j \geq 2$, $T(j) \leq t_0 \leq T(j+1)$ and $t \geq t_0$. We have also $g_0(t) \leq (t/t_0) g_0(t_0)$ if $0 < t_0 \leq T(2)$ and $t \geq t_0$. Taking $t_0 = \exp[s_0 - \tau]$ and $t = \exp[\text{Re } s - \tau]$ we have (5.2). Hence (5.1) yields

$$|G(s)| \leq \exp[(\text{Im } s)^2/2] \max\{1, \exp[\text{Re } s - s_0]\} G(s_0),$$

and for $t_0 > 0$ and $t \in \boldsymbol{C} \setminus (-\infty, 0]$

$$|F(t)| \leq \exp[\pi^2/2] \max\{1, |t|/t_0\} F(t_0).$$

So Cauchy's estimates prove that F has the property (viii). □

Definition 5.1. 9 We say that a symbol $\omega(\xi) \in C^\infty(\boldsymbol{R}^n)$ belongs to $\mathcal{W}$ if there is a real analytic function $F(t)$ defined near $[1,\infty)$ satisfying the following conditions: (0) $\omega(\xi) = F(\langle\xi\rangle)$. (i) $F(t) \geq t/(1+\log t)$ for $t \geq 1$. (ii) For any $\varepsilon > 0$ there is $T_\varepsilon \geq 1$ such that $F(t) \leq \varepsilon t$ for $t \geq T_\varepsilon$. (iii) $0 < F'(t) \leq F(t)/t$ for $t \geq 1$. (iv) There is $C \geq 1$ such that $F(t)/t \leq CF'(t)$ for $t \geq C$. (v) $F''(t) < 0$ for $t \geq 1$. (vi) For any $\varepsilon > 0$ there is $T_\varepsilon \geq 1$ such that $t^2|F''(t)| \leq \varepsilon F(t)$ for $t \geq T_\varepsilon$. (vii) There is $C > 0$ such that $|F^{(k)}(t)| \leq C(2/t)^k k! F(t)$ for $t \geq 1$ and $k \in \boldsymbol{Z}_+$.

Lemma 5.1. 10 (i) *If $\omega(\xi) \in \mathcal{W}$, then*

$$|\partial^\alpha \omega(\xi)| \leq 10^{|\alpha|} C|\alpha|! \langle\xi\rangle^{-|\alpha|}\omega(\xi) \quad \textit{for } \xi \in \boldsymbol{R}^n.$$

(ii) *If $\omega(\xi) \in \mathcal{W}$, then $\omega(\xi+\eta) \leq \omega(\xi) + \omega(\eta)$.* (iii) *For any $u \in \mathcal{A}'$ ($\equiv \mathcal{A}'(\boldsymbol{R}^n)$) there is $\omega(\xi) \in \mathcal{W}$ such that $|\hat{u}(\xi)| \leq e^{\omega(\xi)}$ for $\xi \in \boldsymbol{R}^n$.* (iv) *Assume that $u \in \bigcup_{\varepsilon>0} L^2_{-\varepsilon}$. If there is $\omega(\xi) \in \mathcal{W}$ satisfying $|\hat{u}(\xi)| \leq e^{\omega(\xi)}$ for $\xi \in \boldsymbol{R}^n$, then $u \in \bigcap_{\varepsilon>0} L^2_{-\varepsilon}$ ($\subset \mathcal{F}_0$).*

Proof (i) By induction we can show that, with some constant $C > 0$,

$$|\partial^\alpha F^{(k)}(\langle\xi\rangle)| \leq CA^{|\alpha|}2^k(k+|\alpha|)!\langle\xi\rangle^{-k-|\alpha|}\omega(\xi)$$

if $A \geq (2+\sqrt{3})(1+\sqrt{2})$ and $\omega(\xi) = F(\langle\xi\rangle)$. (ii) Let $\omega(\xi) = F(\langle\xi\rangle)$. Then we have

$$F(t+s) = F(t) + sF'(t+\theta s) \leq F(t) + sF'(s) \leq F(t) + F(s)$$

for $t \geq s \geq 1$ with some $\theta \in (0,1)$, since $F''(t) < 0$ for $t \geq 1$. Therefore, we have

$$\omega(\xi+\eta) = F(\langle\xi+\eta\rangle) \leq F(\langle\xi\rangle + \langle\eta\rangle) \leq \omega(\xi) + \omega(\eta).$$

(iii) Let $u \in \mathcal{A}'$, and put

$$w(t) = \max_{\xi\in R^n, |\xi|=t} |\hat{u}(\xi)| + 1,$$
$$f(t) = \max\{2t/(1+\log(1+t)), \log w(t)\}$$

for $t \geq 0$. Since for any $\varepsilon > 0$ there is $C_\varepsilon \equiv C_\varepsilon(u) > 0$ satisfying $w(t) \leq C_\varepsilon e^{\varepsilon t/2} + 1$ for $t \geq 0$, $f(t)$ satisfies the conditions in Lemma 5.1. 8 . Note that $f(t) \geq t/(1+\log t)$ for $t \geq 1$. Therefore, from Lemma 5.1. 8 there is $\omega(\xi) \in \mathcal{W}$ satisfying $\omega(\xi) \geq \max_{|\eta|\leq\langle\xi\rangle} \log(1+|\hat{u}(\eta)|)$, which proves the assertion (iii). (iv) The assertion (iv) is obvious. □

Lemma 5.1. 11 *Let $\omega(\xi) \in \mathcal{W}$, and let $p(\xi, y, \eta)$ be a symbol satisfying*

$$\begin{aligned}\left|\partial_\xi^\alpha D_y^{\beta+\tilde{\beta}} \partial_\eta^\gamma p(\xi, y, \eta)\right| \;&\le\; C_{|\alpha|+|\tilde{\beta}|+|\gamma|}(A/R_0)^{|\beta|}\langle\xi\rangle^{-|\alpha|+|\beta|} \\ &\quad \times\langle\eta\rangle^{-|\gamma|}\exp[\delta_1\omega(\xi) - \delta_2\omega(\eta)]\end{aligned}$$

if $\langle\xi\rangle \ge R_0|\beta|$. Then we have

$$\|\exp[\varepsilon_1\omega(D)]p(D_x, y, D_y)u\| \le C_{\varepsilon_1,\varepsilon_2}\|\exp[\varepsilon_2\omega(D)]u\| \tag{5.3}$$

if $u \in \mathcal{S}_\infty$, $R_0 \ge 25e\sqrt{n}A$ and $2(\varepsilon_1 + \delta_1)_+ < \varepsilon_2 + \delta_2$.

Proof We can apply the same argument as in the proof of Lemma 5.1. 6 . We shall use the same notations as in the proof of Lemma 5.1. 6 . Noting that $\omega(\xi) \le 2\omega(\eta)$ if $|\xi| \le 2|\eta|$, we have

$$\left|\partial_\xi^\alpha D_y^\beta \partial_\eta^\gamma\{\exp[\varepsilon_1\omega(\xi) - \varepsilon_2\omega(\eta)]p_1(\xi, y, \eta)\}\right| \le C_{|\alpha|+|\beta|+|\gamma|}\langle\xi\rangle^{-|\alpha|}\langle\eta\rangle^{-|\gamma|}$$

if $2(\varepsilon_1 + \delta_1)_+ < \varepsilon_2 + \delta_2$. This yields (5.3) with $p(D_x, y, D_y)$ replaced by $p_1(D_x, y, D_y)$ if $u \in \mathcal{S}_\infty$ and $2(\varepsilon_1 + \delta_1)_+ < \varepsilon_2 + \delta_2$. Repetition of the argument in the proof of Lemma 5.1. 6 shows that

$$\begin{aligned}&\left|\partial_\xi^\alpha D_y^\beta \partial_\eta^\gamma\{\exp[\varepsilon_1\omega(\xi) - \varepsilon_2\omega(\eta)]a(\xi, y, \eta)\}\right| \\ &\le C_{|\alpha|+|\beta|+|\gamma|,\varepsilon}\exp[(\varepsilon - 1/(3R_0))\langle\xi\rangle]\end{aligned}$$

if $R_0 \ge 25e\sqrt{n}A$ and $\varepsilon > 0$, since $\omega(\eta) \le \omega(\xi) \le \varepsilon\langle\xi\rangle + C_\varepsilon$ when $g(\xi, \eta) \ne 1$. This proves the lemma. □

5.2 Necessary conditions on local solvability and hypoellipticity

Let us begin with the definition of local solvability. Let Ω be an open conic subset of $T^*\boldsymbol{R}^n \setminus 0$, and let $p(x, \xi) \in PS^+_{loc}(\Omega)$ (or $p(x, \xi) \in FS^+(\Omega; C_0, A)$).

Definition 5.2. 1 (i) For $z^0 = (x^0, \xi^0) \in \Omega$ we say that $p(x, D)$ is microlocally solvable at z^0 if there is a neighborhood $\mathcal{U}$ of $(x^0, \xi^0/|\xi^0|)$ in $\Omega \cap \boldsymbol{R}^n \times S^{n-1}$ such that for any $f \in \mathcal{C}(\mathcal{U})$ there is $u \in \mathcal{C}(\mathcal{U})$ satisfying $p(x, D)u = f$ in $\mathcal{C}(\mathcal{U})$, *i.e.*, $p(x, D) : \mathcal{C}(\mathcal{U}) \to \mathcal{C}(\mathcal{U})$ is surjective. (ii) Let X be an open subset of $\boldsymbol{R}^n$, and assume that $\Omega = X \times (\boldsymbol{R}^n \setminus \{0\})$. For $x^0 \in X$ we say that $p(x, D)$ is locally solvable at x^0 modulo analytic functions if there is a neighborhood U of x^0 in X such that for any

$f \in \mathcal{B}(U)$ there is $u \in \mathcal{B}(U)$ satisfying $p(x,D)u - f \in \mathcal{A}(U)$, *i.e.*, $p(x,D) : \mathcal{B}(U)/\mathcal{A}(U) \to \mathcal{B}(U)/\mathcal{A}(U)$ is surjective. Assume that $p(x,\xi)$ is a polynomial of ξ whose coefficients are real analytic functions of x defined in X. Then, for $x^0 \in X$ we say that $p(x,D)$ is locally solvable at x^0 if there is a neighborhood U of x^0 in X such that for any $f \in \mathcal{B}(U)$ there is $u \in \mathcal{B}(U)$ satisfying $p(x,D)u = f$ in $\mathcal{B}(U)$, *i.e.*, $p(x,D) : \mathcal{B}(U) \to \mathcal{B}(U)$ is surjective.

Remark Let $\Omega = X \times (\boldsymbol{R}^n \setminus \{0\})$ and $x^0 \in X$. It is obvious that $p(x,D)$ is microlocally solvable at (x^0,ξ) if $p(x,D)$ is locally solvable at x^0 modulo analytic functions and $\xi \in \boldsymbol{R}^n \setminus \{0\}$. Moreover, if p(x,D) is a differential operator and locally solvable at x^0, then $p(x,D)$ is locally solvable at x^0 modulo analytic functions.

Applying the same argument as in the framework of distributions we can obtain necessary conditions on local solvability and hypoellipticity. Let $z^0 = (x^0,\xi^0) \in \Omega \cap (\boldsymbol{R}^n \times S^{n-1})$, and let Γ and Γ_0 be conic neighborhoods of z^0 satisfying $\Gamma_0 \subset\subset \Gamma \subset\subset \Omega$. Choose $\Phi^R(\xi,y,\eta) \in S^{0,0}(R,C_*,C(\Gamma_0,\Gamma),C(\Gamma_0,\Gamma))$ ($R \geq 4$) so that supp $\Phi^R \subset \boldsymbol{R}^n \times \Gamma$ and $\Phi^R(\xi,y,\eta) = 1$ for $(\xi,y,\eta) \in \boldsymbol{R}^n \times \Gamma_0$ with $|\eta| \geq R$, and put $\tilde{p}^R(\xi,y,\eta) = \Phi^R(\xi,y,\eta)p(y,\eta)$. In the case where $p(x,\xi) \equiv \sum_{j=0}^{\infty} p_j(x,\xi) \in FS^+(\Omega;C_0, A)$ we define $\tilde{p}^R(\xi,y,\eta)$ replacing $p(y,\eta)$ by $\tilde{p}(x,\xi) = \sum_{j=0}^{\infty} \phi_j^{R/2}(\xi)p_j(x, \xi)$. We may assume that $\tilde{p}^R(\xi,y,\eta) \in S^{0,0,0,\delta}(R,A)$ for any $\delta > 0$, where A is a positive constant depending on $C(\Gamma_0,\Gamma)$ and p. It follows from Lemma 5.1.6 that $e^{\varepsilon\langle D\rangle}\, {}^t\tilde{p}^R(D_x,y,D_y)v \in L^2$ if $v \in \mathcal{S}_\infty$, $R \geq 25e\sqrt{n}A$ and $\varepsilon < 1/(3R)$.

Theorem 5.2.2 *Assume that $p(x,D)$ is microlocally solvable at z^0. Then there are conic neighborhoods Γ_j ($j = 1,2$) of z^0 and $R_0 \geq 1$ such that $\Gamma_2 \subset\subset \Gamma_1 \subset\subset \Gamma_0$, and for any $R,R' \geq R_0$ and any $\varepsilon \in (0,\varepsilon(R,R'))$ there are positive constants δ and C satisfying*

$$\begin{aligned}|\langle f,v\rangle| \;\leq\; & C\|e^{-\delta\langle D\rangle}f\|\Big(\|e^{\varepsilon\langle D\rangle}\,{}^t\tilde{p}^R(D_x,y,D_y)v\| \\ & +\|e^{\varepsilon\langle D\rangle}(1-{}^t\Psi^{R'}(D_x,y,D_y))v\| + \|v\|\Big) \qquad (5.4)\end{aligned}$$

for $f,v \in \mathcal{S}_\infty$, where $\varepsilon(R,R') = \min\{1/(3R),1/(3R')\}$. Here $\Psi^R(\xi,y,\eta)$ is a symbol in $S^{0,0}(R,C_*,C(\Gamma_2,\Gamma_1),C(\Gamma_2,\Gamma_1))$ satisfying supp $\Psi^R \subset \boldsymbol{R}^n$ *$\times\Gamma_1$ and $\Psi(\xi,y,\eta) = 1$ for $(\xi,y,\eta) \in \boldsymbol{R}^n \times \Gamma_2$ with $|\eta| \geq R$. Moreover, for any $R,R' \geq R_0$ and any $\varepsilon \in (0,\varepsilon(R,R'))$ there are positive constants δ and C satisfying*

$$\|e^{\delta\langle D\rangle}v\| \;\leq\; C\Big(\|e^{\varepsilon\langle D\rangle}\,{}^t\tilde{p}^R(D_x,y,D_y)v\|$$

$$+\|e^{\varepsilon\langle D\rangle}(1-{}^t\Psi^{R'}(D_x,y,D_y))v\|+\|v\|\Big) \quad (5.5)$$

for $v \in \mathcal{S}_\infty$.

Proof Let $\mathcal{X} = \lim\limits_{\longleftarrow} L^2_{-1/j}$ ($\subset \mathcal{F}_0$). Then $\mathcal{X}$ is a Fréchet space whose topology is defined by the seminorms $\|e^{-\langle D\rangle/j}f\|$ ($f \in \mathcal{X}$, $j \in \boldsymbol{N}$). By assumption, there are $R_0 > 0$ and a conic neighborhood γ ($\subset\subset \Gamma_0$) of z^0 such that for any $f \in \mathcal{X}$ there is $u \in \mathcal{A}'$ ($\subset \mathcal{X}$) satisfying $WF_A(\tilde{p}^R(D_x,y,D_y)u - f) \cap \gamma = \emptyset$ for $R \geq R_0$. Indeed, assume that $WF_A(\tilde{p}^{R_1}(D_x,y,D_y)u-f)\cap\gamma = \emptyset$ for $R_1 \geq R(f)$, where $R(f)$ is a positive constant depending on f. Since $\tilde{p}^{R_1}(\xi,y,\eta)-\tilde{p}^R(\xi,y,\eta) = 0$ for $(\xi,y,\eta) \in \boldsymbol{R}^n \times \Gamma_0$ with $|\eta| \geq \max(R,R_1)$, it follows from Lemma 5.1.5 that there is $R_0 > 0$ such that $WF_A(\tilde{p}^{R_1}(D_x,y,D_y)w - \tilde{p}^R(D_x,y,D_y)w) \cap \gamma = \emptyset$ if $w \in \mathcal{F}_0$ and $R_1, R \geq R_0$. This yields $WF_A(\tilde{p}^R(D_x,y,D_y)u - f) \cap \gamma = \emptyset$ for $R \geq R_0$. Choose conic neighborhoods Γ_j ($j = 1,2$) of z^0 so that $\Gamma_2 \subset\subset \Gamma_1 \subset\subset \gamma$. Fix $R, R' \geq R_0$ and $\varepsilon \in (0,\varepsilon(R,R'))$, and let $\mathcal{Y}$ be a normed space which coincides with $\mathcal{S}_\infty$ as a set and has the norm

$$\|v\|_{\mathcal{Y}} \equiv \|e^{\varepsilon\langle D\rangle}\,{}^t\tilde{p}^R(D_x,y,D_y)v\| + \|e^{\varepsilon\langle D\rangle}(1-{}^t\Psi^{R'}(D_x,y,D_y))v\| + \|v\|),$$

where $v \in \mathcal{Y}$. We define a bilinear form $\varphi : \mathcal{X} \times \mathcal{Y} \to \boldsymbol{C}$ by $\varphi(f,v) = \langle f,v\rangle$ for $(f,v) \in \mathcal{X} \times \mathcal{Y}$. It is obvious that for a fixed $v \in \mathcal{Y}$ $\varphi(f,v)$ is continuous in $f \in \mathcal{X}$. Fix $f \in \mathcal{X}$. Then there is $u \in \mathcal{X}$ such that $WF_A(\tilde{p}^R(D_x,y,D_y)u - f) \cap \gamma = \emptyset$. It follows from Lemma 4.1.1 that $\Psi^{R'}(D_x,y,D_y)(\tilde{p}^R(D_x,y,D_y)u - f) \in L^2$, modifying R_0 if necessary. Since

$$\begin{aligned}\langle f,v\rangle \;=\; & -\langle\Psi^{R'}(D_x,y,D_y)(\tilde{p}^R(D_x,y,D_y)u-f),v\rangle \\ & -\langle\tilde{p}^R(D_x,y,D_y)u-f,(1-{}^t\Psi^{R'}(D_x,y,D_y))v\rangle \\ & +\langle u,{}^t\tilde{p}^R(D_x,y,D_y)v\rangle,\end{aligned}$$

we have

$$\begin{aligned}&|\langle f,v\rangle| \leq \|\Psi^{R'}(D_x,y,D_y)(\tilde{p}^R(D_x,y,D_y)u-f)\|\,\|v\| \\ &\quad +\|e^{-\varepsilon\langle D\rangle}(\tilde{p}^R(D_x,y,D_y)u-f)\|\,\|e^{\varepsilon\langle D\rangle}(1-{}^t\Psi^{R'}(D_x,y,D_y))v\| \\ &\quad +\|e^{-\varepsilon\langle D\rangle}u\|\,\|e^{\varepsilon\langle D\rangle}\,{}^t\tilde{p}^R(D_x,y,D_y)v\|\end{aligned}$$

for $v \in \mathcal{Y}$. This implies that φ is separately continuous. By Proposition 5.1.1 φ is continuous. Therefore, there are $j \in \boldsymbol{N}$ and $C > 0$ such that

$$|\langle f,v\rangle| \leq C\|e^{-\langle D\rangle/j}f\|\,\|v\|_{\mathcal{Y}} \quad \text{for } f \in \mathcal{X} \text{ and } v \in \mathcal{Y},$$

which gives (5.4). If we put $f = e^{2\delta\langle D\rangle}\overline{v}$ in (5.4), we obtain (5.5). □

Theorem 5.2. 3 *Assume that $p(x,D)$ is analytic microhypoelliptic at z^0. Then there are conic neighborhoods Γ_j ($j=1,2$) of z^0 and $R_0 \geq 1$ such that $\Gamma_2 \subset\subset \Gamma_1 \subset\subset \Gamma_0$, and for any $R, R' \geq R_0$ and any $\varepsilon \in (0,\varepsilon(R,R'))$ there are positive constants δ and C satisfying*

$$\begin{aligned} \|e^{\delta\langle D\rangle}v\| \;&\leq\; C\Big(\|e^{\varepsilon\langle D\rangle}\tilde{p}^R(D_x,y,D_y)v\| \\ &\quad +\|e^{\varepsilon\langle D\rangle}(1-\Psi^{R'}(D_x,y,D_y))v\| + \|v\|\Big) \end{aligned} \tag{5.6}$$

for $v \in \mathcal{S}_\infty$, where $\varepsilon(R,R') = \min\{1/(3R), 1/(3R')\}$ and $\Psi^R(\xi,y,\eta)$ is as in Theorem 5.2. 2 .

Proof By assumption there is a conic neighborhood γ ($\subset\subset \Gamma_0$) of z^0 such that

$$\mathrm{supp}\; p(x,D)u \cap \gamma_0 = \mathrm{supp}\; u \cap \gamma_0 \quad \text{for } u \in \mathcal{C}(\gamma_0),$$

where $\gamma_0 = \gamma \cap \boldsymbol{R}^n \times S^{n-1}$. Fix $R, R' \geq R_0$ and $\varepsilon_0 \in (0,\varepsilon(R,R'))$, where $R_0 \geq 25e\sqrt{n}\max\{A, C(\Gamma_2,\Gamma_1)\}$ and R_0 is determined later. Define $\mathcal{X} = \varinjlim L^2_{1/j}$ and $\mathcal{Y} = \varinjlim \mathcal{Y}_j$, where

$$\begin{aligned} \mathcal{Y}_j \;&=\; \Big\{u \in \mathcal{F}_0;\ |u|_j \equiv \Big(\|\exp[\varepsilon_0\langle D\rangle/j]\tilde{p}^R(D_x,y,D_y)u\|^2 \\ &\quad +\|\exp[\varepsilon_0\langle D\rangle/j](1-\Psi^{R'}(D_x,y,D_y))u\|^2 + \|u\|^2\Big)^{1/2} < \infty\Big\} \end{aligned}$$

and $\mathcal{Y}_j$ is a Hilbert space with the norm $|\cdot|_j$. We note that $\mathcal{S}_\infty \subset \mathcal{Y}_j$ if $R_0 \geq 25e\sqrt{n}\max\{A, C(\Gamma_2,\Gamma_1)\}$. Let us prove that $\mathcal{X} = \mathcal{Y}$ as a set. It follows from Lemma 5.1. 6 that

$$|u|_k \leq C\|e^{\langle D\rangle/j}u\| \quad \text{if } 2\varepsilon_0/k < 1/j \text{ and } j \gg 1, \tag{5.7}$$

i.e., $L^2_{1/j} \subset \mathcal{Y}_k$ if $2\varepsilon_0/k < 1/j$ and $j \gg 1$. This implies that $\mathcal{X} \subset \mathcal{Y}$. Let $u \in \mathcal{Y}_j$. Then, $u \in L^2$, $\tilde{p}^R(D_x,y,D_y)u$ is analytic and $(1-\Psi^{R'}(D_x,y,D_y))u \in L^2_{\varepsilon_0/j}$. From Lemma 5.1. 5 there is $R(\gamma,\Gamma_0) > 0$ such that

$$WF_A(\tilde{p}^{R_1}(D_x,y,D_y)u) \cap \gamma = \emptyset \quad \text{if } R, R_1 \geq R(\gamma,\Gamma_0).$$

We take $R_0 \geq R(\gamma,\Gamma_0)$. Then, $p(x,D)\mathrm{sp}(u) = 0$ in $\mathcal{C}(\gamma_0)$, where $\mathrm{sp}(u)$ denotes the element in $\mathcal{C}(\gamma_0)$ corresponding to u ($\in L^2$). So we have $WF_A(u) \cap \gamma = \emptyset$. Choose conic neighborhoods Γ_j ($j=1,2$) of z^0 so that $\Gamma_2 \subset\subset \Gamma_1 \subset\subset \gamma$. By Lemma 4.1. 1 there are positive constants $R(\Gamma_2,\Gamma_1,\gamma)$ and $\delta(u,\Gamma_1,\gamma)$ such that $\Psi^{R'}(D_x,y,D_y)u \in L^2_\delta$ if $R' \geq R(\Gamma_2,\Gamma_1,\gamma)$ and

$\delta \leq \min\{1/(2R'), \delta(u, \Gamma_1, \gamma)\}$. Therefore, taking $R_0 \geq R(\Gamma_2, \Gamma_1, \gamma)$, we have $u \in \mathcal{X}$, which proves $\mathcal{X} = \mathcal{Y}$. Next we shall prove that the mapping $\iota : \mathcal{X} \ni u \mapsto u \in \mathcal{Y}$ is continuous. Since $\mathcal{X}$ is bornologic (see Proposition 5.1. 2), it suffices to show that B is bounded in $\mathcal{Y}$ if B is bounded in $\mathcal{X}$. Let B be a bounded subset of $\mathcal{X}$. Then, by Proposition 5.1. 2 there are $j \in \boldsymbol{N}$ and $b > 0$ such that $B \subset \{u \in L^2_{1/j};\ \|e^{\langle D\rangle/j}u\| < b\}$. So it follows from (5.7) that B is bounded in $\mathcal{Y}_k$ if $2\varepsilon_0/k < 1/j$ and $j \gg 1$, which implies that B is bounded in $\mathcal{Y}$. Therefore, $\iota^{-1} : \mathcal{Y} \ni u \mapsto u \in \mathcal{X}$ is a closed operator. By Proposition 5.1. 3 ι^{-1} is continuous, *i.e.*, for any $j \in \boldsymbol{N}$ there are $k \in \boldsymbol{N}$ and $C > 0$ such that $\|e^{\langle D\rangle/k}u\| \leq C|u|_j$ for $u \in \mathcal{Y}_j$. □

Similarly, we have the following theorems when $\Omega = X \times (\boldsymbol{R}^n \setminus \{0\})$, where X is an open subset of $\boldsymbol{R}^n$.

Theorem 5.2. 4 *Assume that U and U_0 are neighborhoods of x^0 satisfying $U_0 \subset\subset U \subset\subset X$, $\Omega = X \times (\boldsymbol{R}^n \setminus \{0\})$, $\Gamma = U \times (\boldsymbol{R}^n \setminus \{0\})$ and $\Gamma_0 = U_0 \times (\boldsymbol{R}^n \setminus \{0\})$. Moreover, we assume that $p(x, D)$ is locally solvable at x^0 modulo analytic functions. Then there are neighborhoods U_j ($j = 1, 2$) of x^0 and $R_0 \geq 1$ such that $U_2 \subset\subset U_1 \subset\subset U_0$, and for any $R, R' \geq R_0$ and any $\varepsilon \in (0, \varepsilon(R, R'))$ there are positive constants δ and C satisfying* (5.4) *for $f, v \in \mathcal{S}_\infty$, where $\Psi^R(\xi, y, \eta) \in S^{0,0}(R, C_*, C(U_2, U_1), C_*)$ satisfies* supp $\Psi^R \subset \boldsymbol{R}^n \times U_1 \times \boldsymbol{R}^n$ *and $\Psi(\xi, y, \eta) = 1$ for $(\xi, y, \eta) \in \boldsymbol{R}^n \times U_2 \times \boldsymbol{R}^n$. Moreover, for any $R, R' \geq R_0$ and any $\varepsilon \in (0, \varepsilon(R, R'))$ there are positive constants δ and C satisfying* (5.5) *for $v \in \mathcal{S}_\infty$.*

Theorem 5.2. 5 *Assume that U and U_0 are neighborhoods of x^0 satisfying $U_0 \subset\subset U \subset\subset X$, $\Omega = X \times (\boldsymbol{R}^n \setminus \{0\})$, $\Gamma = U \times (\boldsymbol{R}^n \setminus \{0\})$ and $\Gamma_0 = U_0 \times (\boldsymbol{R}^n \setminus \{0\})$. Moreover, we assume that $p(x, D)$ is analytic hypoelliptic at x^0. Then there are neighborhoods U_j ($j = 1, 2$) of x^0 and $R_0 \geq 1$ such that $U_2 \subset\subset U_1 \subset\subset U_0$, and for any $R, R' \geq R_0$ and any $\varepsilon \in (0, \varepsilon(R, R'))$ there are positive constants δ and C satisfying* (5.6) *for $v \in \mathcal{S}_\infty$, where $\Psi^R(\xi, y, \eta)$ is as in* Theorem 5.2. 4 .

5.3 Sufficient conditions on local solvability

In the framework of distributions necessary conditions on local solvability as obtained in the same way as in Theorems 5.2. 2 and 5.2. 4 are also sufficient conditions. However, we can not prove sufficiency of the conditions in Theorems 5.2. 2 and 5.2. 4 in the framework of hyperfunctions, applying the same argument as for distributions. We need to strengthen conditions in order to prove local solvability.

Let Ω be an open conic subset of $T^*\boldsymbol{R}^n \setminus 0$, and let $p(x,\xi) \in PS^+_{loc}(\Omega)$ (or $p(x,\xi) \in FS^+(\Omega; C_0, A)$). Let $z^0 = (x^0, \xi^0) \in \Omega \cap (\boldsymbol{R}^n \times S^{n-1})$. We choose conic neighborhoods Γ and Γ_0 and $\Phi^R(\xi, y, \eta) \in S^{0,0}(R, C_*, C(\Gamma_0, \Gamma), C(\Gamma_0, \Gamma))$ ($R \geq 4$) so that $\Gamma_0 \subset\subset \Gamma \subset\subset \Omega$, supp $\Phi^R \subset \boldsymbol{R}^n \times \Gamma \cap \{|\eta| \geq R/2\}$ and $\Phi^R(\xi, y, \eta) = 1$ for $(\xi, y, \eta) \in \boldsymbol{R}^n \times \Gamma_0$ with $|\eta| \geq R$. Put $\tilde{p}^R(\xi, y, \eta) = \Phi^R(\xi, y, \eta)p(y, \eta)$, with obvious modification in the case where $p(x,\xi) \in FS^+(\Omega; C_0, A)$, as in Section 5.2. We may assume that $\tilde{p}^R(\xi, y, \eta) \in S^{0,0,0,\delta}(R, A)$ for any $\delta > 0$, where A is a positive constant depending on $C(\Gamma_0, \Gamma)$ and p. Recall that ${}^t\tilde{p}^R(D_x, y, D_y)v \in L^2_\varepsilon$ if $v \in \mathcal{S}_\infty$, $R \geq 25e\sqrt{n}A$ and $\varepsilon < 1/(3R)$.

Theorem 5.3. 1 *Let Γ_j ($j = 1, 2$) be conic neighborhoods of z^0 such that $\Gamma_2 \subset\subset \Gamma_1 \subset\subset \Gamma_0$, and let $\Psi^R(\xi, y, \eta)$ be a symbol in $S^{0,0}(R, C_*, C(\Gamma_2, \Gamma_1), C(\Gamma_2, \Gamma_1))$ satisfying* supp $\Psi^R \subset \boldsymbol{R}^n \times \Gamma_1$ *and $\Psi^R(\xi, y, \eta) = 1$ for $(\xi, y, \eta) \in \boldsymbol{R}^n \times \Gamma_2$ with $|\eta| \geq R$. Assume that for any $\omega(\xi) \in \mathcal{W}$ there is $R_0 \geq 4$ such that for any $R \geq R_0$ there are $\nu(\xi) \in \mathcal{W}$ and $C > 0$ satisfying*

$$\|e^{\omega(D)}v\| \leq C\Big(\|e^{\nu(D)}\,{}^t\tilde{p}^R(D_x, y, D_y)v\| + \|e^{\nu(D)}(1 - {}^t\Psi^R(D_x, y, D_y))v\| + \|v\|\Big) \tag{5.8}$$

for $v \in \mathcal{S}_\infty$. Then $p(x, D)$ is microlocally solvable at z^0.

Proof Let $f \in \mathcal{A}'$. Then, by Lemma 5.1. 10 there is $\omega(\xi) \in \mathcal{W}$ such that $f \in L^2(-\omega)$, where

$$L^2(-\omega) := \{u \in \mathcal{F}_0;\ e^{-\omega(\xi)}\hat{u}(\xi) \in L^2\}.$$

From the assumptions of the lemma there is $R_0 \geq 4$ such that for each fixed $R \geq R_0$ there are $\nu(\xi) \in \mathcal{W}$ and $C > 0$ satisfying

$$\|e^{\omega(D)}v\| \leq C\Big(\|e^{\nu(D)}\,{}^t\tilde{p}^R(D_x, y, D_y)v\| + \|e^{\nu(D)}(1 - {}^t\Psi^R(D_x, y, D_y))v\| + \|e^{-\langle D\rangle}v\|\Big) \tag{5.9}$$

for $v \in \mathcal{S}_\infty$, since $\|v\| \leq a\|e^{\omega(D)}v\| + C_a\|e^{-\langle D\rangle}v\|$ for $a > 0$ and $v \in \mathcal{S}_\infty$. Put

$$\mathcal{X} = L^2(\nu) \times L^2(\nu) \times L^2_{-1},$$
$$\mathcal{Y} = \Big\{\Big({}^t\tilde{p}^R(D_x, y, D_y)v, (1 - {}^t\Psi^R(D_x, y, D_y))v, v\Big) \in \mathcal{X};\ v \in \mathcal{S}_\infty\Big\}.$$

Let F be a linear functional on $\mathcal{Y}$ defined by $F(w) = \langle f, v_3\rangle$ for $w = (v_1, v_2, v_3) \in \mathcal{Y}$. Then, by (5.9) we have

$$|\langle f, v\rangle| \leq C\|e^{-\omega(D)}f\|\,\|({}^t\tilde{p}^R(D_x, y, D_y)v, (1 - {}^t\Psi^R(D_x, y, D_y))v, v)\|_{\mathcal{X}}$$

for $v \in \mathcal{S}_\infty$, where $\|(v_1, v_2, v_3)\|_{\mathcal{X}} = \|e^{\nu(D)} v_1\| + \|e^{\nu(D)} v_2\| + \|e^{-\langle D\rangle} v_3\|$. By the Hahn-Banach theorem there is $\widetilde{F} \equiv (u, -\varphi, -\psi) \in \mathcal{X}' \equiv L^2(-\nu) \times L^2(-\nu) \times L^2_1$ such that $\widetilde{F}|_{\mathcal{Y}} = F$, *i.e.*,

$$\langle f, v\rangle = \langle u, {}^t\tilde{p}^R(D_x, y, D_y)v\rangle - \langle \varphi, (1 - {}^t\Psi^R(D_x, y, D_y))v\rangle - \langle \psi, v\rangle$$

for $v \in \mathcal{S}_\infty$. This implies that

$$\tilde{p}^R(D_x, y, D_y)u = f + (1 - \Psi^R(D_x, y, D_y))\varphi + \psi \quad \text{in } \mathcal{F}_0,$$

which gives

$$WF_A(\tilde{p}^R(D_x, y, D_y)u - f) = WF_A((1 - \Psi^R(D_x, y, D_y))\varphi).$$

Choose a conic neighborhood Γ_3 of z^0 so that $\Gamma_3 \subset\subset \Gamma_2$. It follows from Lemma 5.1. 5 that we can find $R(\Gamma_3, \Gamma_2, \Gamma_1) > 0$ satisfying $WF_A((1 - \Psi^R(D_x, y, D_y))\varphi) \cap \Gamma_3 = \emptyset$ for $R \geq R(\Gamma_3, \Gamma_2, \Gamma_1)$. Therefore, we have

$$WF_A(\tilde{p}^R(D_x, y, D_y)u - f) \cap \Gamma_3 = \emptyset \quad \text{if } R \geq R(\Gamma_3, \Gamma_2, \Gamma_1).$$

Applying Lemma 5.1. 5 again, we see that there is $R(\Gamma_3, \Gamma_0, \Gamma) > 0$ such that

$$WF_A(\tilde{p}^{R_1}(D_x, y, D_y)u - \tilde{p}^R(D_x, y, D_y)u) \cap \Gamma_3 = \emptyset$$

if $R, R' \geq R(\Gamma_3, \Gamma_0, \Gamma)$. So we have $p(x, D)\mathrm{sp}(u) = \mathrm{sp}(f)$ in $\mathcal{C}(\mathcal{U})$, where $\mathcal{U} = \Gamma_3 \cap (\boldsymbol{R}^n \times S^{n-1})$ and $\mathrm{sp}(u)$ denotes the element in $\mathcal{C}(\mathcal{U})$ corresponding to $u \in \mathcal{F}_0$. This proves the theorem. □

Similarly, we have the following theorem when $\Omega = X \times (\boldsymbol{R}^n \setminus \{0\})$, where X is an open subset of $\boldsymbol{R}^n$.

Theorem 5.3. 2 *Assume that U and U_0 are neighborhoods of x^0 satisfying $U_0 \subset\subset U \subset\subset X$, $\Omega = X \times (\boldsymbol{R}^n \setminus \{0\})$, $\Gamma = U \times (\boldsymbol{R}^n \setminus \{0\})$ and $\Gamma_0 = U_0 \times (\boldsymbol{R}^n \setminus \{0\})$. Let U_j ($j = 1, 2$) be neighborhoods of x^0 such that $U_2 \subset \subset U_1 \subset\subset U_0$, and let $\Psi^R(\xi, y, \eta)$ be a symbol in $S^{0,0}(R, C_*, C(U_2, U_1), C_*)$ satisfying* supp $\Psi^R \subset \boldsymbol{R}^n \times U_1 \times \boldsymbol{R}^n$ *and $\Psi^R(\xi, y, \eta) = 1$ for $(\xi, y, \eta) \in \boldsymbol{R}^n \times U_2 \times \boldsymbol{R}^n$. Assume that for any $\omega(\xi) \in \mathcal{W}$ there is $R_0 \geq 4$ such that for any $R \geq R_0$ there are $\nu(\xi) \in \mathcal{W}$ and $C > 0$ satisfying* (5.8) *for $v \in \mathcal{S}_\infty$. Then $p(x, D)$ is locally solvable at x^0 modulo analytic functions.*

Theorem 5.3. 3 *Let $p(x, D)$ be a differential operator in X whose coefficients belong to $\mathcal{A}(X)$. Assume that the hypothese of* Theorem 5.3. 2 *are fulfilled, and that $p^0(x^0, \xi) \not\equiv 0$ in ξ, where $p^0(x, \xi)$ denotes the principal symbol of $p(x, D)$. Then $p(x, D)$ is locally solvable at x^0.*

Proof We may assume that $p^0(x, e_1) \neq 0$ for $x \in U_2$, where $e_1 = (1, 0, \cdots, 0) \in \boldsymbol{R}^n$. Let $g \in \mathcal{B}(U_2)$, and choose $f \in \mathcal{A}'(\overline{U}_2)$ so that $g = f|_{U_2}$ in $\mathcal{B}(U_2)$. Then there are $\omega(\xi) \in \mathcal{W}$ and $R_0 \geq 4$ such that $f \in L^2(-\omega)$, and for each fixed $R \geq R_0$ there are $\nu(\xi) \in \mathcal{W}$ and $C > 0$ satisfying (5.9) for $v \in \mathcal{S}_\infty$. It follows from the proof of Theorem 5.3. 2 (see the proof of Theorem 5.3. 1) that there is $(u, \varphi, \psi) \in L^2(-\nu) \times L^2(-\nu) \times L^2_1$ such that

$$\tilde{p}^R(D_x, y, D_y)u = f + (1 - \Psi^R(D_x, y, D_y))\varphi + \psi \quad \text{in } \mathcal{F}_0.$$

$\psi(x)$ can be continued analytically to the set $\{x + iy \in \boldsymbol{C}^n; x, y \in \boldsymbol{R}^n$ and $|y| < 1\}$. Let U_3 be a neighborhood of x^0 such that $U_3 \subset\subset U_2$. Applying the same argument as in the proof of Corollary 2.6. 3 we can write

$$1 - \Psi^R(D_x, y, D_y) = \sum_{j=0}^{2} q_j(x, D) \quad \text{on } \mathcal{S}_\infty,$$

where the $q_j(x, \xi)$ satisfy

$$\begin{aligned}
&|q_{0(\beta)}^{(\alpha)}(x, \xi)| \leq C_{|\alpha|,R}(4R + 1)^{|\beta|}|\beta|! \exp[-\langle\xi\rangle/R], \\
&|q_1^{(\alpha)}(x, \xi)| \leq C_{|\alpha|,R} \exp[-\langle\xi\rangle/R], \\
&|q_{1(\beta)}^{(\alpha)}(x, \xi)| \leq C_{|\alpha|,R}(R + 1)^{|\beta|}|\beta|! \exp[-\langle\xi\rangle/R] \quad \text{for } x \in U_3
\end{aligned}$$

and $\operatorname{supp} q_2(x, D)w \cap U_3 = \emptyset$ for $w \in \mathcal{F}_0$ if $R \geq R(U_3, U_2, U_1)$, where $R(U_3, U_2, U_1)$ is some positive constant. Assume that $R \geq R(U_3, U_2, U_1)$, and let $w \in \mathcal{F}_0$. Then we can easily show that $((1 - \Psi^R(D_x, y, D_y))w)|_{U_3}$ can be continued analytically from U_3 to the complex δ-neighborhood $(U_3)_\delta$ of U_3 in $\boldsymbol{C}^n$ if $\delta \leq 1/((4R+1)\sqrt{n})$. Indeed, $((1-\Psi^R(D_x, y, D_y))w)|_{U_3}$ is an analytic function and

$$\begin{aligned}
&(1 - \Psi^R(D_x, y, D_y))w(x) \\
&= (2\pi)^{-n}\langle e^{-\rho\langle\xi\rangle}\hat{w}(\xi), e^{ix\cdot\xi + \rho\langle\xi\rangle}(q_0(x, \xi) + q_1(x, \xi))\rangle
\end{aligned}$$

if $0 < \rho < 1/R$ and $x \in U_3$. By Theorem 1.3. 3 we can choose $v \in \mathcal{A}'(\overline{U}_1)$ so that $\operatorname{supp}(u - v) \cap U_1 = \emptyset$. It easily follows from the proof of Corollary 2.6. 6 , Theorem 2.6. 5 , Lemma 2.6. 4 and Lemma 1.2. 7 (iii) that there are positive constants $\delta_0 \equiv \delta_0(U_3, U_1)$ and $R(A, U_3, U_1)$, which are independent of $u - v$, such that $(\tilde{p}^R(D_x, y, D_y)(u - v))|_{U_3}$ can be continued analytically from U_3 to $(U_3)_\delta$ if $R \geq R(A, U_3, U_1)$ and $\delta \leq \min\{\delta_0, 1/((4R+1)\sqrt{n})\}$. Let $\Phi^R(x, \xi)$ be a symbol in $S^0(R, C_*, C(U_0, U))$

($R \geq 4$) such that supp $\Phi^R \subset U \times \boldsymbol{R}^n$ and $\Phi^R(x,\xi) = 1$ for $(x,\xi) \in U_0 \times \boldsymbol{R}^n$, and put $\tilde{p}(x,\xi) = \Phi^R(x,\xi)p(x,\xi)$. Applying Corollary 2.4.7 and the same argument as for $1 - \Psi^R(D_x, y, D_y)$, we can show that $(\tilde{p}^R(D_x, y, D_y)v - \tilde{p}^R(x, D)v)|_{U_3})$ can be continued analytically from U_3 to $(U_3)_\delta$ if $R \geq R(U_3, U_1, U_0, U)$ and $\delta \leq 1/((4R + C(U_1, U_0))\sqrt{n})$, where $R(U_3, U_1, U_0, U)$ and $C(U_1, U_0)$ are some positive constants independent of v. On the other hand, from Theorem 2.7.1 we have

$$(\tilde{p}^R(x,D)v)|_{U_1} = (p(x,D)v))|_{U_1} \quad \text{in } \mathcal{B}(U_1)$$

if $R \geq R(U_1, U_0, U)$, where $R(U_1, U_0, U)$ is a positive constant. Since

$$\begin{aligned}(p(x,D)v - g)|_{U_3} &= (\tilde{p}^R(x,D)v - \tilde{p}^R(D_x, y, D_y)v)|_{U_3} \\ &+ (\tilde{p}^R(D_x, y, D_y)(v - u))|_{U_3} + ((1 - \Psi^R(D_x, y, D_y))\varphi)|_{U_3} + \psi|_{U_3},\end{aligned}$$

there are positive constants $R(\{U_j\}, U)$ and $\delta \equiv \delta(\{U_j\}, R)$ ($R \geq R(\{U_j\}, U)$) such that $(p(x,D)v - g)|_{U_3}$ can be continued analytically from U_3 to $(U_3)_\delta$ if $R \geq R(\{U_j\}, U)$. Assume that $R \geq R(\{U_j\}, U)$. Note that v depends on R. The Cauchy-Kowalevsky theorem shows that there are a neighborhood V of x^0, which is independent of g, and $w \in \mathcal{A}(V)$ satisfying $p(x,D)w = (p(x,D)v - g)|_V$ in V (see, *e.g.*, [Wk4]). Thus we have $(p(x,D)(v - w))|_V = g|_V$ in $\mathcal{B}(V)$, which proves the theorem. □

Assume that $\Gamma = U \times \gamma$, $\Gamma_0 = U_0 \times \gamma_0$ and $p(x,\xi) \equiv \sum_{j=0}^\infty p_j(x,\xi) \in FS^m(\Omega; C_0, A)$. Let $\Omega_0 \equiv X_0 \times \sigma_0$ ($\subset T^*\boldsymbol{R}^n \setminus 0$) be a conic neighborhood of z^0 satisfying $\Omega_0 \subset\subset \Gamma_0$, and fix $\omega(\xi) \in \mathcal{W}$. Let $\Lambda(x,\xi) \in C^\infty(\Omega)$ be an analytic symbol such that

$$\left|\Lambda^{(\alpha)}_{(\beta)}(x,\xi)\right| \leq C_1 A_1^{|\alpha|+|\beta|}|\alpha|!|\beta|!\langle\xi\rangle^{-|\alpha|}\omega(\xi) \quad \text{for } (x,\xi) \in \Omega, \tag{5.10}$$

$$\Lambda(x,\xi) \geq \omega(\xi) \quad \text{for } (x,\xi) \in \Omega_0. \tag{5.11}$$

Recall that for any $\varepsilon > 0$ there is $T_\varepsilon > 0$ satisfying $\omega(\xi) \leq \varepsilon\langle\xi\rangle$ for $|\xi| \geq T_\varepsilon$. Applying Lemma 2.4.2 we see that there are $T \equiv T(\Lambda) > 0$ and a symbol $N(x,y,\xi) \in C^\infty(\widetilde{\Omega}_T)$ such that $\eta = N(x,y,\xi)$ is the solution of $\eta + i\widehat{\nabla}_x\Lambda(x,y,\eta) = \xi$, where $0 < \varepsilon_0 \ll 1$ and $\widetilde{\Omega}_T = \{(x,y,\xi) \in \boldsymbol{R}^n \times \boldsymbol{R}^n \times \boldsymbol{R}^n;\ (x,\xi) \in \Omega,\ |y| \leq \varepsilon_0 \text{ and } |\xi| \geq T\}$. Choose $\tilde{\psi}(x,\xi) \in S^0_{1,0}$ so that supp $\tilde{\psi} \subset \Gamma \cap \{|\xi| \geq T(\Lambda)\}$ and $\tilde{\psi}(x,\xi) = 1$ for $(x,\xi) \in \Gamma_0$ with $|\xi| \geq 2T(\Lambda)$. For a fixed $\ell \in \boldsymbol{Z}_+$ we put

$$p_{\Lambda,\ell}(x,\xi) = \sum_{j=0}^{\ell} \tilde{\psi}(x,\xi)p^j_\Lambda(x,\xi),$$

where

$$\begin{aligned} p_\Lambda^j(x,\xi) = & \sum_{|\gamma|+|\lambda|+k=j} \Big[\partial_\xi^\gamma \partial_\eta^\lambda D_y^\gamma \\ & \times\Big\{p_{k(\lambda)}(x+i\widehat{\nabla}_\eta\Lambda(x,N(x,y,\xi),\eta),N(x,y,\xi)) \\ & \times \det \frac{\partial N}{\partial \xi}(x,y,\xi)\Big\}\Big]\Big|_{y=0,\,\eta=0}/(\gamma!\lambda!) \end{aligned} \tag{5.12}$$

for $(x,\xi) \in \Omega$ with $|\xi| \ge T(\Lambda)$. Note that

$$p_\Lambda(x,\xi) \equiv \sum_{j=0}^{\infty} p_\Lambda^j(x,\xi) \in FS^m(\Omega; C_0(\Lambda, C_0, A), A_0(\Lambda, A))$$

if the $p_\Lambda^j(x,\xi)$ are defined suitably for $|\xi| \le T(\Lambda)$, where $C_0(\Lambda, C_0, A)$ and $A_0(\Lambda, A)$ are some positive constants.

Theorem 5.3.4 *Let $\Gamma_j \equiv U_j \times \gamma_j$ ($1 \le j \le 4$) be conic neighborhoods of z^0 satisfying $\Gamma_4 \subset\subset \Gamma_3 \subset\subset \Omega_0 \subset\subset \Gamma_2 \subset\subset \Gamma_1 \subset\subset \Gamma_0$, and let $\tilde{\psi}_0(x,\xi)$ be a symbol in $S^0_{1,0}$ such that* supp $\tilde{\psi}_0 \subset \check{\Gamma}_1 \equiv \{(x,\xi);\ (x,-\xi) \in \Gamma_1\}$ *and $\tilde{\psi}_0(x,\xi) = 1$ for $(x,-\xi) \in \Gamma_2$ with $|\xi| \ge 1$. Assume that there are $s \in \boldsymbol{R}$ and $C > 0$ such that*

$$\begin{aligned} \|w\| \ \le\ & C\Big(\|\langle D\rangle^{\ell-m}\,{}^t p_{\Lambda,\ell}(x,D)w\| \\ & +\|\langle D\rangle^s(1-\tilde{\psi}_0(x,D))w\| + \|\langle D\rangle^{-1}w\|\Big) \end{aligned} \tag{5.13}$$

for $w \in C_0^\infty(X_0)$. Let $\Psi(\xi,y,\eta)$ be a symbol in $S^{0,0}(R, C_*, C(\Gamma_4,\Gamma_3), C(\Gamma_4, \Gamma_3))$ satisfying supp $\Psi^R \subset \boldsymbol{R}^n \times \Gamma_3$ *and $\Psi^R(\xi,y,\eta) = 1$ for $(\xi,y,\eta) \in \boldsymbol{R}^n \times \Gamma_4$ with $|\eta| \ge R$. Then there is $R_0 \ge 4$ such that for any $R \ge R_0$ there is $C > 0$ satisfying*

$$\begin{aligned} \|e^{\omega(D)}v\| \ \le\ & C\Big(\|\exp[5C_1\omega(D)]\,{}^t\tilde{p}^R(D_x,y,D_y)v\| \\ & +\|\exp[11C_1\omega(D)](1-{}^t\Psi^R(D_x,y,D_y)v\| + \|v\|\Big) \end{aligned}$$

for $v \in \mathcal{S}_\infty$. In particular, if for any $\omega(\xi) \in \mathcal{W}$ one can choose $\Lambda(x,\xi) \in C^\infty(\Omega)$ so that $\Lambda(x,\xi)$ satisfies (5.10) and (5.11) and (5.13) is valid, then $p(x,D)$ is microlocally solvable at z^0

Remark When $\Omega = X \times (\boldsymbol{R}^n \setminus \{0\})$, $\Omega_0 = X_0 \times (\boldsymbol{R}^n \setminus \{0\})$, $\Gamma_j = U_j \times (\boldsymbol{R}^n \setminus \{0\})$ ($1 \le j \le 4$) in the above theorem, the theorem gives a sufficient condition that $p(x,D)$ is locally solvable at x^0 modulo analytic functions.

Proof Put

$$\tilde{p}_\Lambda^R(\xi, y, \eta) = \Phi^R(\xi, y, \eta) \sum_{j=0}^{\infty} \phi_j^{R/2}(\eta) p_\Lambda^j(y, \eta)$$

for $R > C_0(\Lambda, C_0, A)$. Then we have $\tilde{p}_\Lambda^R(\xi, y, \eta) \in S^{0,m}(R, A(\Lambda, A, \Gamma_0, \Gamma))$ with some positive constant $A(\Lambda, A, \Gamma_0, \Gamma)$. Choose $\chi(x) \in C_0^\infty(X_0)$ and $\chi_1(x) \in C_0^\infty(\boldsymbol{R}^n)$ so that $\chi(x) = 1$ for $x \in X_1$ and $\chi_1(x) = 1$ for $x \in U_1$, where X_1 is a neighborhood of x^0 satisfying $U_3 \Subset X_1$. Since $\sigma((\langle D\rangle^{\ell-m} \times {}^t p_{\Lambda,\ell}(x, D) - \chi_1(x)\langle D\rangle^{\ell-m}\, {}^t\tilde{p}_\Lambda^R(D_x, y, D_y))\tilde{\psi}_0(x, D)) \in S_{1,0}^{-1}$, we have

$$\begin{aligned}\|w\| \;\le\; & C\Big(\|\chi_1(x)\langle D\rangle^{\ell-m}\, {}^t\tilde{p}_\Lambda^R(D_x, y, D_y)w\| \\ & + \|\langle D\rangle^{\ell(s)}(1 - \tilde{\psi}_0(x, D))w\| \\ & + \|\langle D\rangle^{\ell(s)}(1 - \chi(x))w\| + \|\langle D\rangle^{-1}w\|\Big)\end{aligned}$$

for $w \in \mathcal{S}$, where $\ell(s) = \max\{\ell, s\}$. Let $\Gamma_{0,j} \equiv U_{0,j} \times \gamma_{0,j}$ ($j = 1, 2$) be conic neighborhoods of z^0 satisfying $\Gamma_1 \Subset \Gamma_{0,2} \Subset \Gamma_{0,1} \Subset \Gamma_0$, and choose symbols $\Psi_j^R(\xi, y, \eta) \in S^{0,0}(R, C_*, C(U_{0,j+1}, U_{0,j}), C(U_{0,j+1}, U_{0,j}))$ and $g_j^R(\xi) \in S^0(R, C(\gamma_{0,j+1}, \gamma_{0,j}))$ ($j = 0, 1$) so that $\operatorname{supp} \Psi_j^R \subset \boldsymbol{R}^n \times U_{0,j} \times \boldsymbol{R}^n$. $\operatorname{supp} g_j^R \subset \gamma_{0,j} \cap \{|\xi| \ge T(\Lambda)\}$, $\Psi_j^R(\xi, y, \eta) = 1$ for $(\xi, y, \eta) \in \boldsymbol{R}^n \times U_{0,j+1} \times \boldsymbol{R}^n$, and $g_j^R(\xi) = 1$ for $\xi \in \gamma_{0,j+1}$ with $|\xi| \ge 2T(\Lambda)$, where $U_{0,0} = U_0$ and $\gamma_{0,0} = \gamma_0$. We put $\psi_{j,R}(\xi, y, \eta) = \Psi_j^R(\xi, y, \eta) g_j^R(\eta)$ ($j = 0, 1$). Let $\Gamma_{2,j} \equiv U_{2,j} \times \gamma_{2,j}$ ($1 \le j \le 3$) and $\Omega_1 \equiv X_1 \times \sigma_1$ be conic neighborhoods of z^0 satisfying $\Gamma_3 \Subset \Gamma_{2,3} \Subset \Gamma_{2,2} \Subset \Gamma_{2,1} \Subset \Omega_1 \Subset \Omega_0$. Similarly, we choose symbols $\psi_{j,R}(\xi, y, \eta) \in S^{0,0}(R, C_*, C(\Gamma_{2,2j-3}, \Gamma_{2,2j-4}), C(\Gamma_{2,2j-3}, \Gamma_{2,2j-4}))$ ($j = 2, 3$) so that $\operatorname{supp} \psi_{j,R} \subset \Gamma_{2,2j-4} \cap \{|\eta| \ge T(\Lambda)\}$ and $\psi_{j,R}(\xi, y, \eta) = 1$ for $(\xi, y, \eta) \in \boldsymbol{R}^n \times \Gamma_{2,2j-3}$ with $|\eta| \ge 2T(\Lambda)$, where $\Gamma_{2,0} = \Omega_1$. Noting that $\sigma((1 - \tilde{\psi}_0(x, D))\, {}^t\psi_{2,R}(D_x, y, D_y))$ and $\sigma((1 - \chi(x))\, {}^t\psi_{2,R}(D_x, y, D_y))$ belong to $S^{-\infty}$, we have

$$\begin{aligned}\|w\| \;\le\; & C\Big(\|\chi_1(x)\langle D\rangle^{\ell-m}\, {}^t\tilde{p}_\Lambda^R(D_x, y, D_y)\, {}^t\psi_{2,R}(D_x, y, D_y)w\| \\ & + \|\langle D\rangle^{\ell(s)}(1 - {}^t\psi_{2,R}(D_x, y, D_y))w\| + \|\langle D\rangle^{-1}w\|\Big)\end{aligned}$$

for $w \in \mathcal{S}$. It follows from Lemma 4.1.8 and its proof that there are positive constants $R(\Lambda, A, \{\Gamma_{j,k}\})$, $\kappa \equiv \kappa(\Lambda, A, \{\Gamma_{j,k}\})$ (≥ 1) and $\delta_0 \equiv \delta(\Lambda, A, \{\Gamma_{j,k}\})$ such that Q_1 maps continuously $\mathcal{S}_{-\delta}$ to $\mathcal{S}_\delta$ if $R \ge R(\Lambda, A, \{\Gamma_{j,k}\})$, $R' = \kappa R$ and $\delta \le \delta_0/R$, modifying $T(\Lambda)$ if necessary, where

$$\begin{aligned}Q_1 = \psi_{2,R}(D_x, y, D_y)\Big\{ & Op(e^{-\Lambda(y,\eta)}\psi_{1,R}(\xi, y, \eta))\tilde{p}^R(D_x, y, D_y) \\ & \times Op(e^{\Lambda(y,\xi)}\psi_{0,R}(\eta, y, \xi)) - \tilde{p}_\Lambda^{R'}(D_x, y, D_y)\Big\}.\end{aligned}$$

Assume that $R \geq R(\Lambda, A, \{\Gamma_{j,k}\})$. Since $\|\chi_1(x)\langle D\rangle^{\ell-m}\,{}^tQ_1 w\| \leq C\|\langle D\rangle^{-1} \times w\|$ for $w \in \mathcal{S}$, we have

$$\begin{aligned}\|w\| \;\leq\; & C\Big(\|\chi_1(x)\langle D\rangle^{\ell-m}\,{}^tOp(e^{\Lambda(y,\xi)}\psi_{0,R}(\eta,y,\xi))\,{}^t\tilde{p}^R(D_x,y,D_y)\\ & \times{}^tOp(e^{-\Lambda(y,\eta)}\psi_{1,R}(\xi,y,\eta))\,{}^t\psi_{2,R}(D_x,y,D_y)w\|\\ & +\|\langle D\rangle^{\ell(s)}(1-{}^t\psi_{2,R}(D_x,y,D_y))w\|+\|e^{-\langle D\rangle}w\|\Big)\end{aligned}$$

for $w \in \mathcal{S}$. From Lemma 4.1. 6 (or its proof) there are positive constants $R(\Lambda, \{\Gamma_{j,k}\})$, $C \equiv C(\Lambda, \{\Gamma_{j,k}\})$, $\widetilde{A} \equiv A(\Lambda, \{\Gamma_{j,k}\})$ and $\delta_0 \equiv \delta(\Lambda, \{\Gamma_{j,k}\})$ and a symbol $q(\xi, y, \eta) \in S^{0,0}(CR, \widetilde{A})$ such that Q_2 maps continuously $\mathcal{S}_{-\delta}$ to $\mathcal{S}_\delta$ if $R \geq R(\Lambda, \{\Gamma_{j,k}\})$ and $\delta \leq \delta_0/R$, modifying $T(\Lambda)$ if necessary, where

$$\begin{aligned}Q_2 \;=\; & \Psi^R(D_x,y,D_y)\Big\{q(D_x,y,D_y)Op(e^{\Lambda(y,\xi)}\psi_{3,R}(\eta,y,\xi))\\ & \times Op(e^{-\Lambda(y,\eta)}\psi_{1,R}(\xi,y,\eta))-1\Big\}.\end{aligned}$$

We assume that $R \geq R(\Lambda, \{\Gamma_{j,k}\})$. For $v \in \mathcal{S}_\infty$ we put

$$w = {}^tOp(e^{\Lambda(y,\xi)}\psi_{3,R}(\eta,y,\xi))\,{}^tq(D_x,y,D_y)\,{}^t\Psi^R(D_x,y,D_y)v.$$

It follows from Lemmas 5.1. 7 and 5.1. 6 that there are $C > 0$ and $\delta_0 > 0$ such that

$$\Big\|e^{\delta\langle D\rangle}(1-{}^t\psi_{2,R}(D_x,y,D_y))\,{}^tOp(e^{\Lambda(y,\xi)}\psi_{3,R}(\eta,y,\xi))u\Big\| \leq C\|u\|$$

if $u \in \mathcal{S}$ and $\delta \leq \delta_0/R$, modifying $R(\Lambda, \{\Gamma_{j,k}\})$ if necessary. This, together with Theorem 2.3. 3 , yields

$$\begin{aligned}\|w\| \;\leq\; & C\Big(\Big\|\chi_1(x)\langle D\rangle^{\ell-m}\,{}^tOp(e^{\Lambda(y,\xi)}\psi_{0,R}(\eta,y,\xi))\,{}^t\tilde{p}^R(D_x,y,D_y)\\ & \times{}^tOp(e^{-\Lambda(y,\eta)}\psi_{1,R}(\xi,y,\eta))w\Big\|+\|v\|\Big).\end{aligned}$$

Since Q_2 is a regularizer, we have

$$\begin{aligned}\|w\| \;\leq\; & C\Big(\Big\|\langle D\rangle^{\ell-m}\,{}^tOp(e^{\Lambda(y,\xi)}\psi_{0,R}(\eta,y,\xi))\,{}^t\tilde{p}^R(D_x,y,D_y)\\ & \times{}^t\Psi^R(D_x,y,D_y)v\|+\|v\|\Big).\end{aligned} \tag{5.14}$$

Moreover, we have

$$\begin{aligned}\Big\|e^{\omega(D)}\,{}^t\Psi^R(D_x,y,D_y)v\Big\| \leq & \Big\|{}^t\psi_{3,R}(D_x,y,D_y)e^{\omega(D)}\\ & \times{}^tOp(e^{-\Lambda(y,\eta)}\psi_{1,R}(\xi,y,\eta))w\Big\|+C\|v\|.\end{aligned} \tag{5.15}$$

By Lemma 4.1.4 we have

$$\begin{aligned}
&\left\| {}^t\psi_{3,R}(D_x, y, D_y) e^{\omega(D)}\, {}^t Op(e^{-\Lambda(y,\eta)}\psi_{1,R}(\xi, y, \eta)) w \right\| \\
&\le \left\| {}^t Op(e^{-\Lambda(y,\eta)+\omega(\eta)}\psi_{2,R}(\xi, y, \eta)) w \right\| + C\|v\| \\
&\le C(\|w\| + \|v\|), \qquad (5.16)
\end{aligned}$$

since $-\Lambda(y,\eta) + \omega(\eta) \le 0$ if $\psi_{2,R}(\xi, y, \eta) \ne 0$. We can easily prove that

$$\begin{aligned}
\left| \{e^{\Lambda(x,\xi)}\}^{(\alpha)}_{(\beta)} \right| \;&\le\; e^{\mathrm{Re}\,\Lambda(x,\xi)} \tilde{A}_1^{|\alpha|+|\beta|} (|\alpha| + |\beta|)! \\
&\quad \times \langle\xi\rangle^{-|\alpha|} \sum_{k=0}^{|\alpha|+|\beta|} (\rho\omega(\xi))^k / k!
\end{aligned}$$

if $\rho > 0$ and $\tilde{A}_1 \ge \max\{2, 8C_1/\rho\} A_1$. So Lemma 5.1.11 gives

$$\left\| \langle D\rangle^{\ell-m}\, {}^t Op(e^{\Lambda(y,\xi)}\psi_{0,R}(\eta, y, \xi)) u \right\| \le C\| \exp[5C_1\omega(D)] u\|, \quad (5.17)$$

$$\| \exp[5C_1\omega(D)]\, {}^t\tilde{p}^R(D_x, y, D_y) u\| \le C\| \exp[11C_1\omega(D)] u\| \quad (5.18)$$

if $R \gg 1$. (5.14)–(5.18) prove the first part of the theorem. The second part of the theorem easily follows from Theorem 5.3.1. □

Next we shall give a sufficient condition that (5.13) holds. For simplicity we assume that $m \in N$. A simple calculation yields

$$\|{}^t p_{\Lambda,m}(x, D) w\|^2 = \mathrm{Re}\,(q(x, D) w, w)_{L^2},$$

where $(f, g)_{L^2} = \int_{R^n} f(x)\overline{g(x)}\, dx$ and

$$\begin{aligned}
q(x,\xi) \;&\equiv\; |p_{\Lambda,1}(x,\xi)|^2 + \frac{i}{2}\{p_{\Lambda,1}(x,-\xi), \overline{p_{\Lambda,1}(x,-\xi)}\} \\
&\quad + \frac{i}{2}\sum_{j=1}^n \Big(p_{\Lambda,1}(x,-\xi)(\partial^2/\partial x_j \partial\xi_j)\overline{p_{\Lambda,1}(x,-\xi)} \\
&\qquad - \overline{p_{\Lambda,1}(x,-\xi)}(\partial^2/\partial x_j \partial\xi_j) p_{\Lambda,1}(x,-\xi) \Big)
\end{aligned}$$

mod $S^{2m-2}_{1,0}$. By definition we have

$$q(x,\xi) \equiv |p^0_\Lambda(x,-\xi) + (p^1_\Lambda)_s(x,-\xi)|^2 + \frac{i}{2}\{p^0_\Lambda(x,-\xi), \overline{p^0_\Lambda(x,-\xi)}\}$$

in $\check{\Gamma}_0$ ($\equiv \{(x,\xi);\ (x,-\xi) \in \Gamma_0\}$) mod $S^{2m-2}_{1,0}$, where

$$(p^1_\Lambda)_s(x,\xi) = p^1_\Lambda(x,\xi) + \frac{i}{2}\sum_{j=1}^n (\partial^2 p^0_\Lambda/\partial x_j \partial\xi_j)(x,\xi).$$

Therefore, the Fefferman-Phong inequality gives the following

Theorem 5.3. 5 *Assume that $m \in \boldsymbol{N}$ and that there are positive constants c, C and δ satisfying*

$$|p_\Lambda^0(x,\xi) + (p_\Lambda^1)_s(x,\xi)|^2 + \frac{i}{2}\{\overline{p_\Lambda^0(x,\xi)}, p_\Lambda^0(x,\xi)\} \geq c\langle\xi\rangle^{2m-2+\delta} \quad (5.19)$$

for $(x,\xi) \in \Gamma_1$ with $|\xi| \geq C$, where Γ_1 is a conic neighborhood of z^0 as in Theorem 5.3. 4 . *Then the inequality* (5.13) *holds with $\ell = m$. In particular, if for any $\omega(\xi) \in \mathcal{W}$ one can choose $\Lambda(x,\xi) \in C^\infty(\Omega)$ so that $\Lambda(x,\xi)$ satisfies* (5.10) *and* (5.11) *and* (5.19) *is valid, then $p(x,D)$ is microlocally solvable at z^0.*

Remark If $\Lambda(x,\xi)$ does not depend on x, *i.e.*, $\Lambda(x,\xi) \equiv \Lambda(\xi)$, then $N(x,y,\xi) \equiv \xi$ and

$$p_\Lambda^0(x,\xi) = p_0(x + i\nabla_\xi\Lambda(\xi), \xi),$$
$$(p_\Lambda^1)_s(x,\xi) = p_1^s(x + i\nabla_\xi\Lambda(\xi), \xi),$$

where $p_1^s(x,\xi) = p_1(x,\xi) + (i/2)\sum_{j=1}^n (\partial^2 p_0/\partial x_j\partial\xi_j)(x,\xi)$.

In order to show local solvability we need to derive local estimates instead of microlocal ones in Theorems 5.3. 2 and 5.3. 5 . Let us consider how to glue microlocal estimates. Assume that $\Omega = X \times (\boldsymbol{R}^n \setminus \{0\})$, $\Gamma = U \times (\boldsymbol{R}^n \setminus \{0\})$ and $\Gamma_0 = U_0 \times (\boldsymbol{R}^n \setminus \{0\})$. Let $\omega(\xi) \in \mathcal{W}$, and choose open conic subsets γ^j, σ_0^j and γ_k^j ($1 \leq j \leq N$, $0 \leq k \leq 3$) of $\boldsymbol{R}^n \setminus \{0\}$ and symbols $\Lambda_j(x,\xi) \in C^\infty(X \times \gamma^j)$ ($1 \leq j \leq N$) so that $\gamma_3^j \subset\subset \sigma_0^j \subset\subset \gamma_2^j \subset\subset \gamma_1^j \subset\subset \gamma_0^j \subset\subset \gamma^j$, $\boldsymbol{R}^n \setminus \{0\} = \bigcup_{j=1}^N \gamma_3^j$,

$$\left|\Lambda_{j(\beta)}^{(\alpha)}(x,\xi)\right| \leq C_1 A_1^{|\alpha|+|\beta|}|\alpha|!|\beta|!\langle\xi\rangle^{-|\alpha|}\omega(\xi) \quad \text{for } (x,\xi) \in X \times \gamma^j,$$
$$\Lambda_j(x,\xi) \geq \omega(\xi) \quad \text{for } (x,\xi) \in X \times \sigma_0^j,$$
$$\Lambda_j(x,\xi) \leq 0 \quad \text{for } (x,\xi) \in X \times (\gamma^j \setminus \gamma_2^j).$$

By Lemma 2.4. 2 there are $T \equiv T(\{\Lambda_j\}) > 0$ and symbols $N^j(x,y,\xi) \in C^\infty(\tilde{\Omega}_T^j)$ ($1 \leq j \leq N$) such that $\eta = N^j(x,y,\xi)$ is the solution of $\eta + i\hat{\nabla}_x\Lambda_j(x,y,\xi) = \xi$, where $0 < \varepsilon_0 \ll 1$ and $\tilde{\Omega}_T^j = \{(x,y,\xi) \in X \times \boldsymbol{R}^n \times \gamma^j;\ |y| \leq \varepsilon_0$ and $|\xi| \geq T\}$. Choose $\tilde{\psi}^j(x,\xi) \in S_{1,0}^0$ ($1 \leq j \leq N$) so that $\operatorname{supp} \tilde{\psi}^j \subset U \times \gamma^j \cap \{|\xi| \geq T(\{\Lambda_j\})\}$ and $\tilde{\psi}^j(x,\xi) = 1$ for $(x,\xi) \in U_0 \times \gamma_0^j$ with $|\xi| \geq 2T(\{\Lambda_j\})$. For $\ell \in \boldsymbol{Z}_+$ we put

$$p_{\Lambda_j,\ell}(x,\xi) = \sum_{\mu=0}^{\ell} \tilde{\psi}^j(x,\xi)p_{\Lambda_j}^\mu(x,\xi),$$

where the $p^{\mu}_{\Lambda_j}(x,\xi)$ are defined with Λ replaced by Λ_j in (5.12). Let U_k ($k=1,2$) be neighborhoods of x^0 satisfying $U_2 \subset\subset U_1 \subset\subset U_0$, and let $\tilde{\psi}^j_0(x,\xi)$ ($1 \le j \le N$) be symbols in $S^0_{1,0}$ such that supp $\tilde{\psi}^j_0 \subset U_1 \times \check{\gamma}^j_1$ and $\tilde{\psi}^j_0 = 1$ for $(x,-\xi) \in U_2 \times \gamma^j_2$ with $|\xi| \ge 1$, where $\check{\gamma}^j_1 = \{\xi;\ -\xi \in \gamma^j_1\}$. We impose the following condition for each $\omega(\xi) \in \mathcal{W}$:

(ω) There are $\ell \in \boldsymbol{Z}_+$, $s \in \boldsymbol{R}$ and $C > 0$ such that

$$\begin{aligned}\|w\| \ \le\ & C\Big(\Big\|\langle D\rangle^{\ell-m}\,{}^t p_{\Lambda_j,\ell}(x,D)w\Big\| \\ & +\Big\|\langle D\rangle^s(1-\tilde{\psi}^j_0(x,D))w\Big\| + \|\langle D\rangle^{-1}w\|\Big)\end{aligned}$$

for $w \in C^\infty_0(X_0)$ and $1 \le j \le N$, where X_0 is an open neighborhood of x^0 satisfying $X_0 \subset\subset U_2$.

Theorem 5.3.6 *Let U_k ($k=3,4$) be neighborhoods of x^0 satisfying $U_4 \subset\subset U_3 \subset\subset X_0$, and let $\omega(\xi) \in \mathcal{W}$. We assume that ($\omega$) is satisfied. Let $\Psi^R(\xi,y,\eta)$ be a symbol in $S^{0,0}(R,C_*,C(U_4,U_3),C_*)$ satisfying* supp $\Psi^R \subset$ *$\boldsymbol{R}^n \times U_3 \times \boldsymbol{R}^n$ and $\Psi^R(\xi,y,\eta)=1$ for $(\xi,y,\eta) \in \boldsymbol{R}^n \times U_4 \times \boldsymbol{R}^n$. Then there is $R_0 \ge 4$ such that for any $R \ge R_0$ there is $C>0$ satisfying*

$$\begin{aligned}\|e^{\omega(D)}v\| \ \le\ & C\Big(\|\exp[5C_1\omega(D)]\,{}^t\tilde{p}^R(D_x,y,D_y)v\| \\ & +\|\exp[23C_1\omega(D)](1-{}^t\Psi^R(D_x,y,D_y)v\| + \|v\|\Big)\end{aligned}$$

for $v \in \mathcal{S}_\infty$.

Proof Let $\Gamma^j_{0,\mu} \equiv U_{0,\mu} \times \gamma^j_{0,\mu}$ and $\Gamma^j_{1,\nu} \equiv U_{1,\nu} \times \gamma^j_{1,\nu}$ ($1 \le j \le N$, $1 \le \mu \le 3$ and $1 \le \nu \le 4$) be open conic subsets of $\boldsymbol{R}^n \times (\boldsymbol{R}^n \setminus \{0\})$ satisfying $\Gamma^j_2 \equiv U_2 \times \gamma^j_2 \subset\subset \Gamma^j_{1,4} \subset\subset \Gamma^j_{1,3} \subset\subset \Gamma^j_{1,2} \subset\subset \Gamma^j_{1,1} \subset \subset \Gamma^j_1 \equiv U_1 \times \gamma^j_1 \subset\subset \Gamma^j_{0,3} \subset\subset \Gamma^j_{0,2} \subset\subset \Gamma^j_{0,1} \subset\subset \Gamma^j_0 \equiv U_0 \times \gamma^j_0$, and choose symbols $\Psi^R_\mu(\xi,y,\eta) \in S^{0,0}(R,C_*,C(U_{0,\mu+2},U_{0,\mu+1}),C_*)$, $g^{j,R}_\mu(\xi) \in S^0(R,C(\gamma^j_{0,\mu+2},\gamma^j_{0,\mu+1}))$, $\Psi^R_\nu(\xi,y,\eta) \in S^{0,0}(R,C_*,C(U_{1,2\nu-2},U_{1,2\nu-3}),C_*)$ and $g^{j,R}_\nu(\xi) \in S^0(R,C(\gamma^j_{1,2\nu-2},\ \gamma^j_{1,2\nu-3}))$ ($1 \le j \le N$, $\mu = 0,1$ and $\nu = 2,3$) so that supp $\Psi^R_\mu \subset \boldsymbol{R}^n \times U_{0,\mu+1} \times \boldsymbol{R}^n$, supp $g^{j,R}_\mu \subset \gamma^j_{0,\mu+1} \cap \{|\xi| \ge T(\{\Lambda_j\})\}$, supp $\Psi^R_\nu \subset \boldsymbol{R}^n \times U_{1,2\nu-3} \times \boldsymbol{R}^n$, supp $g^{j,R}_\nu \subset \gamma^j_{1,2\nu-3} \cap \{|\xi| \ge T(\{\Lambda_j\})\}$, $\Psi^R_\mu(\xi,y,\eta) = 1$ for $(\xi,y,\eta) \in \boldsymbol{R}^n \times U_{0,\mu+2} \times \boldsymbol{R}^n$, $g^{j,R}_\mu(\xi) = 1$ for $\xi \in \gamma^j_{0,\mu+2}$ with $|\xi| \ge 2T(\{\Lambda_j\})$, $\Psi^R_\nu(\xi,y,\eta) = 1$ for $(\xi,y,\eta) \in \boldsymbol{R}^n \times U_{1,2\nu-2} \times \boldsymbol{R}^n$ and $g^{j,R}_\nu(\xi) = 1$ for $\xi \in \gamma^j_{1,2\nu-2}$ with $|\xi| \ge 2T(\{\Lambda_j\})$. We put $\psi^{j,R}_k(\xi,y,\eta) = \Psi^R_k(\xi,y,\eta) g^{j,R}_k(\eta)$ ($1 \le j \le N$ and $0 \le k \le 3$). From Lemma 4.1.6 there are positive constants $R(\{\Lambda_j\},\{\Gamma^j_{i,k}\})$, $C \equiv$

$C(\{\Lambda_j\},\{\Gamma^j_{i,k}\})$, $\tilde{A} \equiv A(\{\Lambda_j\},\{\Gamma^j_{i,k}\})$ and $\delta_0 \equiv \delta(\{\Lambda_j\},\{\Gamma^j_{i,k}\})$ and symbols $q^j(\xi,y,\eta) \in S^{0,0}(CR,\tilde{A})$ ($1 \le j \le N$) such that Q^j maps continuously $\mathcal{S}_{-\delta}$ to $\mathcal{S}_\delta$ if $R \ge R(\{\Lambda_j\},\{\Gamma^j_{i,k}\})$ and $\delta \le \delta_0/R$, modifying $T(\{\Lambda_j\})$ if necessary, where

$$\begin{aligned} Q^j &= \psi_3^{j,R}(D_x,y,D_y)\Big\{q^j(D_x,y,D_y)Op(e^{\Lambda_j(y,\xi)}\psi_2^{j,R}(\eta,y,\xi)) \\ &\quad \times Op(e^{-\Lambda_j(y,\eta)}\psi_1^{j,R}(\xi,y,\eta)) - 1\Big\}. \end{aligned}$$

We assume that $R \ge R(\{\Lambda_j\},\{\Gamma^j_{i,k}\})$. Then it follows from the proof of Theorem 5.3. 4 and (5.14), (5.17) and (5.18) that

$$\begin{aligned} &\Big\| {}^tOp\Big(e^{\Lambda_j(y,\xi)}\psi_2^{j,R}(\eta,y,\xi)\Big)\, {}^tq^j(D_x,y,D_y)\, {}^t\psi_3^{j,R}(D_x,y,D_y)v\Big\| \\ &\le C\Big(\Big\|\langle D\rangle^{\ell-m}\, {}^tOp\Big(e^{\Lambda_j(y,\xi)}\psi_0^{j,R}(\eta,y,\xi)\Big)\, {}^t\tilde{p}^R(D_x,y,D_y)v\Big\| \\ &\qquad + \|\exp[11C_1\omega(D)](1-{}^t\psi_3^{j,R}(D_x,y,D_y))v\| + \|v\|\Big) \end{aligned} \tag{5.20}$$

for $v \in \mathcal{S}_\infty$. Let $\gamma^j_{1,k}$ ($1 \le j \le N$ and $k = 5,6$) be open conic subsets of $\boldsymbol{R}^n \setminus \{0\}$ satisfying $\gamma^j_2 \subset\subset \gamma^j_{1,6} \subset\subset \gamma^j_{1,5} \subset\subset \gamma^j_{1,4}$, and choose $g^{j,R}(\xi) \in S^0(R, C(\gamma^j_{1,6},\gamma^j_{1,5}))$ so that supp $g^{j,R} \subset \gamma^j_{1,5}$ and $g^{j,R}(\xi) = 1$ for $\xi \in \gamma^j_{1,6}$ with $|\xi| \ge R$. By Corollary 2.4. 7 there are positive constants $A' \equiv A'(A,\{\gamma^j_{1,k}\})$ and $R(A,\{\gamma^j_{1,k}\})$ and symbols $q^j_k(x,\xi)$ and $r^j_k(x,\xi)$ ($1 \le j \le N$ and $0 \le k \le 2$) such that $q^j_\mu(x,\xi) \in S^m(4R,A')$ for $\mu = 1,2$,

$$\begin{aligned} &g^{j,R}(D)\tilde{p}^R(D_x,y,D_y) = q^j_1(x,D) + r^j_1(x,D) \quad \text{on } \mathcal{S}_\infty, \\ &\tilde{p}^R(D_x,y,D_y)g^{j,R}(D) = q^j_2(x,D) + r^j_2(x,D) \quad \text{on } \mathcal{S}_\infty, \\ &\Big| r^{j(\alpha)}_{k(\beta)}(x,\xi)\Big| \le C_{|\alpha|,R}(4R+1)^{|\beta|}|\beta|!\langle\xi\rangle^m \exp[-\langle\xi\rangle/R] \end{aligned}$$

if $k = 1,2$ and $R \ge R(A,\{\gamma^j_{1,k}\})$, supp $q^j_0 \cap U_0 \times (\boldsymbol{R}^n \cap \{|\xi| \ge R\}) \subset U_0 \times (\gamma^j_{1,5} \setminus \gamma^j_{1,6})$ and

$$\begin{aligned} &q^j_1(x,\xi) - q^j_2(x,\xi) = q^j_0(x,\xi) + r^j_0(x,\xi), \\ &\Big| q^{j(\alpha+\tilde{\alpha})}_{0(\beta)}(x,\xi)\Big| \le C_{|\tilde{\alpha}|}(A'/R)^{|\alpha|}A'^{|\beta|}|\beta|!\langle\xi\rangle^{m-|\tilde{\alpha}|} \quad \text{for } \langle\xi\rangle \ge 6R|\alpha|, \\ &\Big| r^{j(\alpha)}_{0(\beta)}(x,\xi)\Big| \le C_{|\alpha|,R}(A'+R)^{|\beta|}|\beta|!\langle\xi\rangle^{m-|\alpha|} \exp[-\langle\xi\rangle/R] \end{aligned}$$

if $x \in U_0$ and $R \ge R(A,\{\gamma^j_{1,k}\})$. Assume that $R \ge R(A,\{\gamma^j_{1,k}\})$. Then we have

$${}^tOp\Big(e^{\Lambda_j(y,\xi)}\psi_0^{j,R}(\eta,y,\xi)\Big)\, {}^t\tilde{p}^R(D_x,y,D_y)({}^tg^{j,R}(D)v)$$

$$
\begin{aligned}
&= {}^tOp\Big(g^{j,R}(\xi)e^{\Lambda_j(y,\xi)}\psi_0^{j,R}(\eta,y,\xi)\Big)\,{}^t\tilde{p}^R(D_x,y,D_y)v \\
&\quad +{}^tOp\Big((r_1^j(x,\xi)-r_2^j(x,\xi))e^{\Lambda_j(y,\xi)}\psi_0^{j,R}(\eta,y,\xi)\Big)v \\
&\quad +{}^tOp\Big(e^{\Lambda_j(y,\xi)}\psi_0^{j,R}(\eta,y,\xi)\Big)({}^tq_1^j(x,D)-{}^tq_2^j(x,D))v, \\
&\equiv I_1^jv+I_2^jv+I_3^jv,
\end{aligned} \tag{5.21}
$$

where $1 \le j \le N$ and $Op(a(x,\xi,y,\eta)) = a(x,D_x,y,D_y)$. Lemma 5.1. 11 yields

$$
\|\langle D\rangle^{\ell-m}I_1^jv\| \le C\|\exp[5C_1\omega(D)]\,{}^t\tilde{p}^R(D_x,y,D_y)v\| \tag{5.22}
$$

if $v \in \mathcal{S}_\infty$ and $R \ge R(\{\Lambda_j\},U_{0,1},U_{0,0})$, where $R(\{\Lambda_j\},U_{0,1},U_{0,0})$ is some positive constant. Since $(r_1^j(x,\xi)-r_2^j(x,\xi))\exp[\Lambda_j(y,\xi)] \in S^{-\infty}$ as a multiple symbol, we have

$$
\|\langle D\rangle^{\ell-m}I_2^jv\| \le C\|v\| \tag{5.23}
$$

Let $\widetilde{U}_0$ be an open subset of $\boldsymbol{R}^n$ satisfying $U_{0,1} \subset\subset \widetilde{U}_0 \subset\subset U_0$, and choose a symbol $\Psi_{0,0}^R(\xi,y,\eta) \in S^{0,0}(R,C_*,C(\widetilde{U}_0,U_0),C_*)$ so that supp $\Psi_{0,0}^R \subset \boldsymbol{R}^n \times U_0 \times \boldsymbol{R}^n$ and $\Psi_{0,0}^R(\xi,y,\eta)=1$ for $(\xi,y,\eta) \in \boldsymbol{R}^n \times \widetilde{U}_0 \times \boldsymbol{R}^n$. We put

$$
\begin{aligned}
q_{0,0}^{j,R}(\xi,y,\eta) &= (1-\Psi_{0,0}^R(\xi,y,\eta))(q_1^j(y,\eta)-q_2^j(y,\eta)), \\
q_{0,1}^{j,R}(\xi,y,\eta) &= \Psi_{0,0}^R(\xi,y,\eta)q_0^j(y,\eta), \\
q_{0,2}^{j,R}(\xi,y,\eta) &= \Psi_{0,0}^R(\xi,y,\eta)r_0^j(y,\eta).
\end{aligned}
$$

By Lemma 4.1. 4 we have

$$
\Big\|\langle D\rangle^{\ell-m}\,{}^tOp\Big(e^{\Lambda_j(y,\xi)}\psi_0^{j,R}(\eta,y,\xi)\Big)\,{}^tq_{0,0}^{j,R}(D_x,y,D_y)v\Big\| \le C\|v\| \tag{5.24}
$$

if $v \in \mathcal{S}_\infty$ and $R \ge R(\{\Lambda_j\},A,\{\Gamma_{i,k}^j\},\widetilde{U}_0)$, where $R(\{\Lambda_j\},A,\{\Gamma_{i,k}^j\},\widetilde{U}_0)$ is some positive constant. Assume that $R \ge R(\{\Lambda_j\},A,\{\Gamma_{i,k}^j\},\widetilde{U}_0)$. It is obvious that

$$
\Big\|\langle D\rangle^{\ell-m}\,{}^tOp\Big(e^{\Lambda_j(y,\xi)}\psi_0^{j,R}(\eta,y,\xi)\Big)\,{}^tq_{0,2}^{j,R}(D_x,y,D_y)v\Big\| \le C\|v\| \tag{5.25}
$$

for $v \in \mathcal{S}_\infty$. Since supp $q_{0,1}^{j,R} \cap \{|\eta| \ge R\} \subset \boldsymbol{R}^n \times U_0 \times (\gamma_{1,5}^j \setminus \gamma_{1,6}^j)$ and $\Lambda_j(x,\xi) \le 0$ for $(x,\xi) \in X \times (\gamma^j \setminus \gamma_2^j)$, we have

$$
\Big\|\langle D\rangle^{\ell-m}\,{}^tOp\Big(e^{\Lambda_j(y,\xi)}\psi_0^{j,R}(\eta,y,\xi)\Big)\,{}^tq_{0,1}^{j,R}(D_x,y,D_y)v\Big\| \le C\|\langle D\rangle^{\ell}v\| \tag{5.26}
$$

for $v \in \mathcal{S}_\infty$. Lemma 4.1. 4 shows that

$$\begin{aligned}&\left\|\exp[11C_1\omega(D)]({}^t\psi_3^{j,R}(D_x,y,D_y) - {}^t\Psi_3^R(D_x,y,D_y))\,{}^tg^{j,R}(D)v\right\| \\ &\le C\|v\| \qquad (5.27)\end{aligned}$$

for $v \in \mathcal{S}_\infty$, with modification of $R(\{\Lambda_J\}, A, \{\Gamma_{i,k}^j\}, \widetilde{U}_0)$ if necessary. (5.20)–(5.27) give

$$\begin{aligned}&\left\|{}^tOp\Big(e^{\Lambda_j(y,\xi)}\psi_2^{j,R}(\eta,y,\xi)\Big)\,{}^tq^j(D_x,y,D_y)\,{}^t\psi_3^{j,R}(D_x,y,D_y)\,{}^tg^{j,R}(D)v\right\| \\ &\le C\Big(\|\exp[5C_1\omega(D)]\,{}^t\tilde{p}^R(D_x,y,D_y)v\| \\ &\quad +\|\exp[11C_1\omega(D)](1 - {}^t\Psi_3^R(D_x,y,D_y))\,{}^tg^{j,R}(D)v\| \\ &\quad +\|\langle D\rangle^\ell v\|\Big) \qquad (5.28)\end{aligned}$$

for $v \in \mathcal{S}_\infty$. Choose symbols $\chi_j^R(\xi) \in S^0(R, C(\{\gamma_3^j\}))$ ($R \ge 4$, $1 \le j \le N$) so that supp $\chi_j^R \subset \gamma_3^j$, $0 \le \chi_j^R \le 1$, and $\sum_{j=1}^N \chi_j^R(\xi) = 1$ if $|\xi| \ge R$ (see Proposition 2.2. 3). Since the Q^j are regularizers and $\exp[\omega(\xi) - \Lambda_j(x,\xi)]\chi_j^R(\xi) \le 1$, applying Lemmas 4.1. 4 and 5.1. 11 we have

$$\begin{aligned}&\|e^{\omega(D)}\,{}^t\chi_j^R(D)\,{}^t\Psi_3^R(D_x,y,D_y)v\| - C\|v\| \\ &\le \|e^{\omega(D)}\,{}^t\chi_j^R(D)\,{}^t\psi_3^{j,R}(D_x,y,D_y)\,{}^tg^{j,R}(D)v\| - C'\|v\| \\ &\le \Big\|e^{\omega(D)}\,{}^t\chi_j^R(D)\,{}^tOp\Big(e^{-\Lambda_j(y,\eta)}\psi_1^{j,R}(\xi,y,\eta)\Big)\,{}^tOp\Big(e^{\Lambda_j(y,\xi)}\psi_2^{j,R}(\eta,y,\xi)\Big) \\ &\quad \times{}^tq^j(D_x,y,D_y)\,{}^t\psi_3^{j,R}(D_x,y,D_y)\,{}^tg^{j,R}(D)v\Big\| \\ &\le C\Big\|{}^tOp\Big(e^{\Lambda_j(y,\xi)}\psi_2^{j,R}(\eta,y,\xi)\Big) \\ &\qquad \times{}^tq^j(D_x,y,D_y)\,{}^t\psi_3^{j,R}(D_x,y,D_y)\,{}^tg^{j,R}(D)v\Big\|, \\ &\|\exp[11C_1\omega(D)](1 - {}^t\Psi_3^R(D_x,y,D_y))\,{}^tg^{j,R}(D)v\| \\ &\le C(\|\exp[23C_1\omega(D)](1 - {}^t\Psi^R(D_x,y,D_y))v\| + \|v\|)\end{aligned}$$

for $v \in \mathcal{S}_\infty$. This, together with (5.28), proves the lemma. □

5.4 Some examples

In this section we shall give several examples as applications of the results in Sections 5.2 and 5.3. Let Ω be an open conic subset of $\boldsymbol{R}^n \times (\boldsymbol{R}^n \setminus \{0\})$, and let $p(x,\xi) \equiv \sum_{j=0}^\infty p_j(x,\xi) \in FS^m(\Omega; C_0, A)$. Let $z^0 = (x^0,\xi^0) \in \Omega \cap \boldsymbol{R}^n \times S^{n-1}$, and assume that $p_0(x,\xi)$ is positively homogeneous of degree m in ξ and microhyperbolic with respect to $\vartheta = (0, e_1) \in \boldsymbol{R}^{2n}$ at z^0, where $e_1 = (1, 0, \cdots, 0) \in \boldsymbol{R}^n$. Then we have the following

Theorem 5.4. 1 *Under the above assumptions $p(x, D)$ is microlocally solvable at z^0.*

Proof Let $\omega(\xi) \in \mathcal{W}$, and put

$$\Lambda(x, \xi) = ((x_1 - x_1^0)/\delta + 2)\omega(\xi),$$

where $\delta > 0$. Let $\eta = N(\xi)$ be the solution of $\eta + i(\omega(\eta)/\delta)e_1 = \xi$. Then there is $C_\delta > 0$ such that

$$|N(\xi) - (\xi - i(\omega(\xi)/\delta)e_1)| + |\xi||(\nabla_\xi \omega)(N(\xi))| \leq \omega(\xi)$$

if $|\xi| \geq C_\delta$. Therefore, there are a conic neighborhood Γ of z^0 in Ω, a compact subset M of $\Gamma((p_0)_{z^0}, \vartheta)$, $\delta > 0$ and $C > 0$ such that

$$-(\delta/\omega(\xi))(|\xi| \text{Re } (\nabla_\xi \Lambda)(x, N(\xi)), \text{Im } N(\xi)) \in M$$

for $(x, \xi) \in \Gamma$ with $|\xi| \geq C$ and $|x_1 - x_1^0| < \delta$. It follows from Proposition 4.3. 3 that there is $c > 0$ such that

$$\begin{aligned}
|p_\Lambda^0(x, \xi)| &= |p_0(x + i(\nabla_\xi \Lambda)(x, N(\xi)), N(\xi))| \, |\det(\partial N/\partial \xi)(\xi)| \\
&= |\xi|^m |p_0(x + i(\nabla_\xi \Lambda)(x, N(\xi)), |\xi|^{-1} N(\xi))| \, |\det(\partial N/\partial \xi)(\xi)| \\
&\geq c|\xi|^m (\omega(\xi)/(\delta|\xi|))^\mu \geq c|\xi|^m (\delta(1 + \log\langle\xi\rangle))^{-\mu} \geq |\xi|^{m-1/4}
\end{aligned}$$

for $(x, \xi) \in \Gamma$ with $|\xi| \geq C$ and $|x_1 - x_1^0| < \delta$, modifying Γ and C if necessary, where μ is the multiplicity of z^0 relative to p_0. Therefore, applying Theorem 5.3. 4 or Theorem 5.3. 5 we can prove the theorem. □

Next we consider some examples of operators with double characteristics. Some of them were studied by Funakoshi [Fn]. Let X be an open neighborhood of the origin in $\boldsymbol{R}^n$, and assume that $\Omega = X \times (\boldsymbol{R}^n \setminus \{0\})$, $p(x, \xi) \equiv \sum_{j=0}^{\infty} p_j(x, \xi) \in FS^2(\Omega; C_0, A)$ and

$$p_0(x, \xi) = \xi_1^2 + \cdots + \xi_{n-1}^2 + \alpha x_n^k \xi_n^2,$$

where $\alpha \in \boldsymbol{C} \setminus \{0\}$ and $k \in \boldsymbol{N}$. Then we have the following

Theorem 5.4. 2 *Under the above assumptions $p(x, D)$ is locally solvable at the origin of $\boldsymbol{R}^n$ modulo analytic functions.*

Proof We shall apply Theorem 5.3.5 and its remark. Let $\omega(\xi) \in \mathcal{W}$, and put $\Lambda(x,\xi) \equiv \Lambda(\xi) = \omega(\xi)$. Then we have

$$p_\Lambda^0(x,\xi) = \xi_1^2 + \cdots + \xi_{n-1}^2 + \alpha(x_n + i\partial_{\xi_n}\omega(\xi))^k \xi_n^2,$$
$$(p_\Lambda^1)_s(x,\xi) = p_1(x + i\nabla_\xi \omega(\xi), \xi) + \alpha k i (x_n + i\partial_{\xi_n}\omega(\xi))^{k-1}\xi_n.$$

A straightforward calculation yields

$$\begin{aligned}
&(i/2)\{\overline{p_\Lambda^0(x,\xi)}, p_\Lambda^0(x,\xi)\} \\
&= k|\alpha|^2 \xi_n^2 (x_n^2 + |\partial_{\xi_n}\omega(\xi)|^2)^{k-1}(2\xi_n \partial_{\xi_n}\omega(\xi) + k\xi_n^2 \partial_{\xi_n}^2 \omega(\xi)) \\
&= k|\alpha|^2 \xi_n^{2k} \langle\xi\rangle^{-2k+2} F'(\langle\xi\rangle)^{2k-2} \\
&\quad \times \{F'(\langle\xi\rangle)\xi_n^2 \langle\xi\rangle^{-1}(2 + k(1 - \xi_n^2/\langle\xi\rangle^2)) + kF''(\langle\xi\rangle)\xi_n^4/\langle\xi\rangle^2\},
\end{aligned}$$

where $\omega(\xi) = F(\langle\xi\rangle)$. Since $t^2|F''(t)|/F(t) \to 0$ as $t \to \infty$ and $tF'(t) \geq C^{-1}F(t) \geq C^{-1}t/(1+\log t)$ for $t \geq C$ with some $C > 0$, there is $c > 0$ such that

$$(i/2)\{\overline{p_\Lambda^0(x,\xi)}, p_\Lambda^0(x,\xi)\} \geq c\langle\xi\rangle^3 (\log\langle\xi\rangle)^{-2k+1}$$

if $|\xi_n| \geq \langle\xi\rangle/2$ and $\langle\xi\rangle \geq C$. On the other hand, we have, with some $C > 0$,

$$|p_\Lambda^0(x,\xi) + (p_\Lambda^1)_s(x,\xi)|^2 \geq \langle\xi\rangle^4/5$$

if $|x_n| < |\alpha|^{-1/k}$, $|\xi_n| \leq \langle\xi\rangle/2$ and $\langle\xi\rangle \geq C$. Therefore, Theorem 5.3.5 proves the theorem. □

Let Ω be an open conic neighborhood of $z^0 \equiv (0, e_n) \in \boldsymbol{R}^n \times (\boldsymbol{R}^n \setminus \{0\})$, where $e_n = (0,\cdots,0,1) \in \boldsymbol{R}^n$, and $p(x,\xi) \equiv \sum_{j=0}^\infty p_j(x,\xi) \in FS^1(\Omega; C_0, A)$. We assume that $p_0(x,\xi)$ is positively homogeneous of degree 1 in ξ, $p_0(z^0) = 0$ and that $p(x,\xi)$ is of principal type at z^0, *i.e.*, $H_{p_0}(z^0)$ and $\partial/\partial\xi_n$ are linearly independent. We may assume that $p_0(x,\xi) = \xi_1 + ia(x,\xi')$, where $\xi' = (\xi_2,\cdots,\xi_n)$ (see Theorem 21.3.6 of [Hr6]). It is known that there is $A(x,\xi) \equiv \sum_{j=0}^\infty A_j(x,\xi) \in FS^0(\Omega; C_0', A')$ such that $A(x,D)$ is elliptic and $A(x,D)p(x,D) = p_0(x,D)A(x,D)$ in $\mathcal{C}(\Omega \cap \boldsymbol{R}^n \times S^{n-1})$, modifying Ω if necessary (see Theorem 2.1.2 of [SKK]). If there is $k \in \boldsymbol{N}$ satisfying $(\partial/\partial x_1)^j a(x,\xi') \equiv 0$ for $j < k$ and $\pm(\partial^k a/\partial x_1^k)(0, e_n') > 0$, then we may assume that $a(x,\xi') = \pm x_1^k \xi_n$ (see Theorems 21.3.5 and 21.3.6 of [Hr6]). Let us consider the simple cases where $p(x,\xi) = \xi_1 \pm i x_1^k \xi_n$.

Theorem 5.4.3 *Let $p(x,\xi) = \xi_1 + i x_1^k \xi_n$, and assume that k is even. Then $p(x,D)$ is analytic microhypoelliptic and microlocally solvable at $(0, \pm e_n)$.*

Remark $p(x, D)$ has a twosided fundamental solution (see §26.3 of [Hr7]). This also proves Theorem 5.4. 3 .

Proof We proved in Corollary 4.3. 9 that $p(x, D)$ is analytic microhypoelliptic at $(0, \pm e_n)$. Since $p(x, \xi)$ is microhyperbolic with respect to $(0, \mp e_1)$ at $(0, \pm e_n)$, by Theorem 5.4. 1 $p(x, D)$ is microlocally solvable. □

Theorem 5.4. 4 *Let $p(x, \xi) = \xi_1 + ix_1^k \xi_n$, and assume that k is odd.* (i) *$p(x, D)$ is analytic microhypoelliptic at $(0, e_n)$ and microlocally solvable at $(0, -e_n)$.* (ii) *$p(x, D)$ is not microlocally solvable at $(0, e_n)$ and is not analytic microhypoelliptic at $(0, -e_n)$.*

Remark The argument as in §26.3 of [Hr7] may be more suitable to deal with $D_1 + ix_1^k D_n$ (see, also, [KKK]). However, we believe that our argument here is applicable to a wider class of operators (see, also, [Hr7]).

Proof (i) Applying Theorem 4.5. 2 with $L(x_1, \xi_1; \xi_n) = \xi_1 + ix_1^k \xi_n$ to $p(x, \xi)$, we can prove that $p(x, D)$ is analytic microhypoelliptic at $(0, e_n)$. For $\omega(\xi) \in \mathcal{W}$ we take $\Lambda(x, \xi) \equiv \Lambda(\xi_n) = 2F(1 + |\xi_n|)$, where $\omega(\xi) = F(\langle\xi\rangle)$. Then we have

$$\omega(\xi) \leq F(2(1 + |\xi_n|)) \leq \Lambda(\xi_n) \leq 4\omega(\xi),$$
$$\left|\partial_{\xi_n}^k \Lambda(\xi_n)\right| \leq C4^k k! \langle\xi\rangle^{-k} \omega(\xi)$$

for $\xi \in \gamma \equiv \{\xi \in \boldsymbol{R}^n \setminus \{0\}; -\xi_n > |\xi|/2\}$. On the other hand, we have $p_\Lambda^0(x, \xi) = p(x, \xi)$ and $p_\Lambda^j(x, \xi) = 0$ for $j \geq 1$. Let $\tilde{\psi}(\xi)$ be a symbol in $S_{1,0}^0$ such that $0 \leq \tilde{\psi} \leq 1$, supp $\tilde{\psi} \subset \gamma$ and $\tilde{\psi}(\xi) = 1$ if $-\xi_n > 2|\xi|/3$ and $|\xi| \geq 1$, and put $\tilde{p}(x, \xi) = p(x, \xi)\tilde{\psi}(\xi)$. Then a simple calculation yields

$$\begin{aligned}
&-\text{Re } ({}^t\tilde{p}(x, D)w, D_1 w)_{L^2} = \text{Re } ({}^t\tilde{\psi}(D)(D_1 + iX(x_1)^k D_n)w, D_1 w)_{L^2} \\
&\geq \|D_1 w\|^2 + \text{Re } (r(x_1, D)\,{}^t\tilde{\psi}_1(D)w, {}^t\tilde{\psi}_0(D)w)_{L^2} \\
&\quad -C(\|\langle D\rangle^2(1 - {}^t\tilde{\psi}_0)w\|^2 + \|w\|^2),
\end{aligned} \tag{5.29}$$

for $w \in C_0^\infty(U)$, where $U = \{x \in \boldsymbol{R}^n; |x| < 1\}$, $X(t) \in C_0^\infty(\boldsymbol{R})$ satisfies $X(t) = t$ for $|t| \leq 1$ and $X(t) \geq 0$, $\tilde{\psi}_0(\xi) \in S_{1,0}^0$ is a symbol satisfying supp $\tilde{\psi}_0 \subset \{\xi \in \boldsymbol{R}^n; -\xi_n > 2|\xi|/3\}$ and $\tilde{\psi}_0(\xi) = 1$ for $\xi \in \boldsymbol{R}^n$ with $-\xi_n > 3|\xi|/4$ and $|\xi| \geq 1$, and $r(x_1, \xi) = kX(x_1)^{k-1}\xi_n \tilde{\psi}(-\xi)$ (≥ 0). From the sharp Gårding inequality or the Fefferman-Phong inequality it follows that, with some C,

$$\text{Re } (r(x_1, D)\,{}^t\tilde{\psi}_0(D)w, {}^t\tilde{\psi}_0(D)w)_{L^2} \geq -C\|w\|^2 \tag{5.30}$$

for $w \in C_0^\infty(U)$. On the other hand, by Poincaré's inequality we have

$$\|w\| \le (2\delta/\pi)\|D_1 w\| \quad \text{if } w \in C_0^\infty(U(\delta)),$$

where $U(\delta) = \{x \in U;\ |x_1| \le \delta\}$. This, together with (5.29) and (5.30), yields

$$\|w\| \le C(\|{}^t\tilde{p}(x,D)w\| + \|\langle D\rangle^2(1 - {}^t\tilde{\psi}_0(D))w\|)$$

for $w \in C_0^\infty(U(\delta))$ if $\delta \ll 1$. Therefore, Theorem 5.3.4 implies that $p(x,D)$ is microlocally solvable at $(0,-e_n)$. (ii) We shall prove that $p(x,D)$ is not microlocally solvable at $(0,e_n)$, using Theorem 5.2.2. Similarly, we can prove that $p(x,D)$ is not analytic microhypoelliptic at $(0,-e_n)$, using Theorem 5.2.3 instead of Theorem 5.2.2. It suffices to show that $q(x,D) \equiv D_1 - ix_1^k D_n$ is not microlocally solvable at $(0,-e_n)$. Put

$$\begin{aligned}\varphi_\tau(x,\xi) &= -x\cdot\xi + \tau x_n + i\tau\Phi(x) \quad (\tau \ge 1),\\ \Phi(x) &= x_1^{k+1}/(k+1) + x_2^2/2 + \cdots + x_{n-1}^2/2\\ &\qquad + (x_n + ix_1^{k+1}/(k+1))^2/2.\end{aligned}$$

We note that $q(x,D)\exp[i\tau x_n - \tau\Phi(x)] = 0$. Let $\chi^R(x,\xi) \in S^0(R,A)$ be a symbol such that $0 \le \chi^R \le 1$, $\operatorname{supp}\chi^R \subset \{(x,\xi) \in \boldsymbol{R}^n \times \boldsymbol{R}^n;\ |x| < 1/2\}$ and $\chi^R(x,\xi) = 1$ if $|x| \le 1/4$. We define

$$v_\tau(x) = (2\pi)^{-n}\int\Big(\int \exp[ix\cdot\xi + i\varphi_\tau(y,\xi)]\chi^R(y,\xi)\,dy\Big)\,d\xi.$$

Then we have

$$\hat{v}_\tau(\xi) = \int \exp[i\varphi_\tau(y,\xi)]\chi^R(y,\xi)\,dy.$$

Put

$$K = \langle\xi\rangle^{-2}\Big(\sum_{j=1}^n \xi_j D_{y_j} + 1\Big).$$

Then we have

$$\begin{aligned}|\hat{v}_\tau(\xi)| &= \Big|\int \exp[-i\xi\cdot y]K^{\ell_0}(e^{i\tau y_n - \tau\Phi(y)}\chi^R(y,\xi))\,dy\Big|\\ &\le C\tau^{\ell_0}\langle\xi\rangle^{-\ell_0},\end{aligned}$$

where $\tau \ge 1$ and $\ell_0 = [n/2]+1$. Therefore, we have

$$\|v_\tau\| \le C\tau^{\ell_0} \quad \text{for } \tau \ge 1. \tag{5.31}$$

We see that

$$\begin{aligned}&\text{Re } i\varphi_\tau(y,\xi)\\&= -\tau(y_1^{k+1}(1-y_1^{k+1}/(2(k+1)))/(k+1)+|y'|^2/2) \ge -1/4,\\&|\text{Im } i\varphi_\tau(y,\xi)| = |(\xi-\tau e_n)\cdot y+\tau y_1^{k+1}y_n/(k+1)| \le 9/16,\\&\text{Re } \exp[i\varphi_\tau(y,\xi)] \ge e^{-1/4}/2\end{aligned}$$

if $\tau \ge 1$, $|\xi-\tau e_n| \le 1$ and $|y| < \tau^{-1/2}/2$, where $y' = (y_2,\cdots,y_n)$. Moreover, we have

$$\begin{aligned}&|\text{Im } i\varphi_\tau(y,\xi)| \le \pi/2 \quad \text{when } |y| \le \tau^{-1/(k+2)},\\&|\exp[i\varphi_\tau(y,\xi)]| \le \exp[-c(k)\tau^{1/(k+2)}] \quad \text{when } \tau^{-1/(k+2)} < |y| \le 1,\end{aligned}$$

if $\tau \ge 1$ and $|\xi-\tau e_n| \le 1$, where $c(k) = 2^{-(k+3)/2}/(k+1)$. Let $\delta > 0$. Since

$$\begin{aligned}\text{Re } e^{\delta\langle\xi\rangle}\hat{v}_\tau(\xi) \ge &\int_{|y|<\tau^{-1/2}/4} e^{-1/4+\delta(\tau-1)}/2\,dy\\&-\left|\int_{\tau^{-1/(k+2)}<|y|<1/2} \exp[-c(k)\tau^{1/(k+2)}+\delta(\tau+2)]\,dy\right|\end{aligned}$$

if $\tau \ge 1$ and $|\xi-\tau e_n| \le 1$, there are $c_0 > 0$ and $T \ge 1$ such that

$$\|e^{\delta\langle D\rangle}v_\tau(x)\| = (2\pi)^{-n/2}\|e^{\delta\langle\xi\rangle}\hat{v}_\tau(\xi)\| \ge c_0\tau^{-n/2}e^{\delta\tau} \tag{5.32}$$

if $0 < \delta \le 1$ and $\tau \ge T$. Let $\psi^R(x,\xi) \in S^0(R,A)$ be a symbol satisfying

$$\text{supp } \psi^R \cap \{\langle\xi\rangle \ge R\} \subset \{(x,\xi);\ |\xi/|\xi|-e_n|^2+|x|^2 \ge c \text{ and } |x| < 1/2\},$$

where c is a positive constant. Put

$$w_\tau(\xi) = \int \exp[i\varphi_\tau(y,\xi)]\psi^R(y,\xi)\,dy.$$

We shall prove that there is $C(A,c) > 0$ such that

$$|w_\tau(\xi)| \le C_R\,e^{-\langle\xi\rangle/R} \tag{5.33}$$

if $R \ge eC(A,c)$. It is easy to see that

$$\begin{aligned}|\nabla_y\varphi_\tau(y,\xi)|^2 &= (\xi_1+\tau y_1^k y_n)^2+\tau^2y_1^{2k}(1-y_1^{k+1}/(k+1))^2\\&\quad+\xi_2^2+\cdots+\xi_{n-1}^2+\tau^2|y'|^2+(\xi_n-\tau+\tau y_1^{k+1}/(k+1))^2\\&\ge |\xi-\tau e_n|^2/2+\tau^2(y_1^{2k}/3+y_2^2+\cdots+y_{n-1}^2+y_n^2/2) \ge c_1(|\xi|^2+\tau^2)\end{aligned}$$

if $(y,\xi) \in \text{supp}\ \psi^R$ and $\langle\xi\rangle \geq R$, where $c_1 \equiv c_1(c,k)$ is a positive constant. Here we have used the facts that $(a+b)^2 \geq a^2/2 - 2b^2$ for $a,b \in \boldsymbol{R}$ and $|\xi - \tau e_n|^2 \geq (1-c/4)(|\xi|-\tau)^2 + c(|\xi|^2+\tau^2)/4$ if $|\xi/|\xi| - e_n|^2 \geq c/2$. Moreover, we have

$$\left|\varphi_{\tau(\beta)}^{(\alpha)}(y,\xi)\right| \leq C(|\xi|^2+\tau^2)^{(1-|\alpha|)/2}$$

for $|y| \leq 1/2$ and $|\alpha|+|\beta| \geq 1$. Let L be a differential operator defined by

$${}^tL = |\nabla_y\varphi_\tau(y,\xi)|^{-2}\sum_{j=1}^n \overline{(\partial\varphi_\tau/\partial y_j)(y,\xi)}D_{y_j}.$$

From Lemma 2.1. 5 there is $C(A,c) > 0$ such that

$$|L^j\psi^R(y,\xi)| \leq C(C(A,c)/R)^j$$

if $j \geq 1$ and $Rj \leq \langle\xi\rangle$ $(\leq R(j+1))$. This proves (5.33). Since $x_1^k e^{i(x-y)\cdot\xi} = \sum_{j=0}^k \binom{k}{j} y_1^{k-j} D_{\xi_1}^j e^{i(x-y)\cdot\xi}$ and $D_{x_j}e^{i(x-y)\cdot\xi} = -D_{y_j}e^{i(x-y)\cdot\xi} = \xi_j e^{i(x-y)\cdot\xi}$, we have

$$\begin{aligned}
q(x,D)v_\tau(x) &= (2\pi)^{-n}\int\Big(\int e^{i(x-y)\cdot\xi}\Big\{D_{y_1} \\
&\quad -i\sum_{j=0}^k\binom{k}{j}y_1^{k-j}(-D_{\xi_1})^jD_{y_n}\Big\}(\exp[i\tau y_n - \tau\Phi(y)]\chi^R(y,\xi))\,dy\Big)\,d\xi \\
&= (2\pi)^{-n}\int\Big(\int \exp[ix\cdot\xi + i\varphi_\tau(y,\xi)] \\
&\quad \times\Big\{D_{y_1} - iy_1^kD_{y_n} - i\sum_{j=1}^k\binom{k}{j}y_1^{k-j}\xi_n(-D_{\xi_1})^j\Big\}\chi^R(y,\xi))\,dy\Big)\,d\xi.
\end{aligned}$$

By (5.33) we have

$$\|e^{\varepsilon\langle D\rangle}q(x,D)v_\tau(x)\| \leq C_{\varepsilon,R} \tag{5.34}$$

if $R \geq R_0$ and $\varepsilon < 1/R$, where R_0 is some positive constant. We note that ${}^tq(x,D) = -q(x,D)$. Let $\psi^R(\xi,x,\eta,y)$ be a symbol such that $\text{supp}\ \psi^R \subset \{(\xi,x,\eta,y) \in \boldsymbol{R}^n \times \boldsymbol{R}^n \times \boldsymbol{R}^n \times \boldsymbol{R}^n;\ |x| \geq a+b \text{ and } |y| \leq a\}$ and

$$\left|D_x^\beta\partial_\eta^{\alpha+\gamma}D_y^\lambda\psi^R(\xi,x,\eta,y)\right| \leq C_{|\beta|+|\gamma|+|\lambda|}(A/R)^{|\alpha|}\langle x\rangle^k\langle\xi\rangle$$

if $\langle\eta\rangle \geq R|\alpha|$, where $a,b > 0$, and put

$$w_\tau(\xi) = \int \exp[-ix\cdot(\xi-\eta) + i\varphi_\tau(y,\eta)]\psi^R(\xi,x,\eta,y)\,dyd\eta dx,$$

where the integral is defined as an oscillatory integral. Then we have, with some $C > 0$,

$$|w_\tau(\xi)| \le C\langle\xi\rangle e^{-\langle\xi\rangle/R} \quad \text{if } R \ge 8e\sqrt{n}A/b. \tag{5.35}$$

Indeed,

$$\begin{aligned} w_\tau(\xi) &= \int \exp[-ix\cdot(\xi-\eta) + i\varphi_\tau(y,\eta)] \\ &\qquad \times K_1^{n+1} L_1^{k+n+1+j} \psi^R(\xi,x,\eta,y)\, dy d\eta dx \end{aligned}$$

if $\langle\eta\rangle \ge Rj$, where

$$L_1 = |x-y|^{-2}\sum_{\mu=1}^{n}(y_\mu - x_\mu)D_{\eta_\mu},$$

$$K_1 = \langle\xi-\eta\rangle^{-2}\Big(\sum_{\mu=1}^{n}(\xi_\mu-\eta_\mu)D_{x_\mu} + 1\Big).$$

This gives (5.35). By (5.35) we have

$$\|e^{\varepsilon\langle\xi\rangle} w_\tau(\xi)\| \le C_\varepsilon \quad \text{if } R \ge 8e\sqrt{n}A/b \text{ and } \varepsilon < 1/R. \tag{5.36}$$

Let $\Phi^R(\xi,y,\eta) \in S^{0,0}(R,A)$ be a symbol satisfying supp $\Phi^R \subset \{(\xi,y,\eta) \in \boldsymbol{R}^n \times \boldsymbol{R}^n \times \boldsymbol{R}^n;\ |y| \le 2\}$ and $\Phi^R(\xi,y,\eta) = 1$ for $|y| \le 1$, and put $\tilde{q}^R(\xi,y,\eta) = q(y,\eta)\Phi^R(\xi,y,\eta)$. (5.34) and (5.36) yield

$$\|e^{\varepsilon\langle D\rangle}\, {}^t\tilde{q}^R(D_x,y,D_y)v_\tau(x)\| \le C_\varepsilon \tag{5.37}$$

if $R \ge R_0$ and $\varepsilon < 1/R$, modifying R_0 if necessary. Let $\Gamma_j \equiv U_j \times \gamma_j$ ($1 \le j \le 5$) be conic neighborhoods of $(0, e_n)$ satisfying $\Gamma_5 \subset\subset \Gamma_4 \subset\subset \Gamma_3 \subset\subset \Gamma_2 \subset\subset \Gamma_1$, and choose $\Psi^R(\xi,y,\eta) \in S^{0,0}(R, C(\Gamma_2,\Gamma_1), C(\Gamma_2,\Gamma_1), C_*)$ so that supp $\Psi^R \subset \{(\xi,y,\eta);\ (y,\xi) \in \Gamma_1 \text{ and } \eta \in \boldsymbol{R}^n\}$ and $\Psi^R(\xi,y,\eta) = 1$ if $(y,\xi) \in \Gamma_2$ and $|\xi| \ge R$. Moreover, we choose $\Phi_0^R(x,\xi) \in S^0(R, C_*, C(U_5, U_4))$ and $g_j^R(\xi) \in S^0(R, C(\gamma_{2j+3},\gamma_{2j+2}))$ ($j = 0, 1$) so that supp $\Phi_0^R \subset U_4 \times \boldsymbol{R}^n$, $\Phi_0^R(x,\xi) = 1$ for $x \in U_5$, supp $g_j^R \subset \gamma_{2j+2} \cap \{|\xi| \ge R\}$ and $g_j^R(\xi) = 1$ for $\xi \in \gamma_{2j+3}$ with $\langle\xi\rangle \ge 2R$. Write

$$\mathcal{F}[e^{\varepsilon\langle D\rangle}(1-\Psi^R(D_x,y,D_y))v_\tau](\xi) = I_1(\xi) + I_2(\xi) + I_3(\xi),$$

where

$$I_1(\xi) \;=\; (2\pi)^{-n}\int \exp[\varepsilon\langle\xi\rangle - ix\cdot(\xi-\eta) + i\varphi_\tau(y,\eta)](1-\Psi^R(\xi,x,\eta))$$

$$\begin{aligned}
&\times(1-\Phi_0^R(y,\eta)g_1^R(\eta))\chi^R(y,\eta)\,dyd\eta dx,\\
I_2(\xi) \;=\;& (2\pi)^{-n}\int \exp[\varepsilon\langle\xi\rangle - ix\cdot(\xi-\eta)+i\varphi_\tau(y,\eta)](1-\Psi^R(\xi,x,\eta))\\
&\times g_0^R(\xi)\Phi_0^R(y,\eta)g_1^R(\eta))\chi^R(y,\eta)\,dyd\eta dx,\\
I_3(\xi) \;=\;& (2\pi)^{-n}\int \exp[\varepsilon\langle\xi\rangle - ix\cdot(\xi-\eta)+i\varphi_\tau(y,\eta)](1-\Psi^R(\xi,x,\eta))\\
&\times(1-g_0^R(\xi))\Phi_0^R(y,\eta)g_1^R(\eta))\chi^R(y,\eta)\,dyd\eta dx.
\end{aligned}$$

Since $(1-\Phi_0^R(y,\eta)g_1^R(\eta))\chi^R(y,\eta)=0$ for $(y,\eta)\in\Gamma_5$ with $\langle\eta\rangle\geq 2R$, the same argument as for (5.34) yields $\|I_1(\xi)\|\leq C_{\varepsilon,R}$ for $R\geq R_0$ and $\varepsilon<1/R$, modifying R_0 if necessary. Since $x\notin U_2$ if $(1-\Psi^R(\xi,x,\eta))g_0^R(\xi)\neq 0$, it follows from (5.35) that $\|I_2(\xi)\|\leq C_\varepsilon$ for $R\geq R_0$ and $\varepsilon<1/R$, modifying R_0 if necessary. Since $\xi\notin\gamma_3$ and $\eta\in\gamma_4$ if $(1-g_0^R(\xi))g_1^R(\eta)\neq 0$ and $\langle\xi\rangle\geq 2R$, integration by parts with respect to x gives $\|I_3(\xi)\|\leq C_\varepsilon$ for $R\geq R_0$ and $\varepsilon<1/R$ in a similar way, modifying R_0 if necessary. Therefore, we have

$$\|e^{\varepsilon\langle D\rangle}(1-\Psi^R(D_x,y,D_y))v_\tau\|\leq C_{\varepsilon,R} \tag{5.38}$$

for $R\geq R_0$ and $\varepsilon<1/R$. Theorem 5.2. 2 , together with (5.31), (5.32), (5.37) and (5.38), proves that $q(x,D)$ is not microlocally solvable at $(0,-e_n)$. □

We remark that Trépreau proved that the condition (Ψ) is necessary and sufficient for operators of principal type to be microlocally solvable (see [Tp]).

Appendix A

Proofs of product formulae

A.1 Proof of Theorem 2.4. 4

We shall prove Theorem 2.4. 4 in this section. In doing so, we assume that the hypotheses of Section 2.4 are satisfied and use the notations there. Let $g_1^R(\xi,\zeta,\eta) \in C^\infty(\boldsymbol{R}^{n'} \times \boldsymbol{R}^n \times \boldsymbol{R}^{n''})$ ($R \geq 3$) be a symbol such that $0 \leq g_1^R(\xi,\zeta,\eta) \leq 1$ and

$$g_1^R(\xi,\zeta,\eta) = \begin{cases} 1 & \text{if } |\xi| \geq (4c_2(U)|\zeta| + 6c_2(T)|\eta|)/c_3(S) \\ & \text{or } |\xi|^2 + |\zeta|^2 + |\eta|^2 \leq 1, \\ 0 & \text{if } |\xi| \leq (2c_2(U)|\zeta| + 3c_2(T)|\eta|)/c_3(S) \\ & \text{and } |\xi|^2 + |\zeta|^2 + |\eta|^2 \geq 4, \end{cases}$$

$$\left|\partial_\xi^{\alpha+\tilde{\alpha}} \partial_\zeta^{\gamma+\tilde{\gamma}} \partial_\eta^{\rho+\tilde{\rho}} g_1^R(\xi,\zeta,\eta)\right| \leq C_{|\tilde{\alpha}|+|\tilde{\gamma}|+|\tilde{\rho}|}(\widehat{C}(S,T,U)/R)^{|\alpha|+|\gamma|+|\rho|}$$
$$\times\langle\xi\rangle^{-|\tilde{\alpha}|-|\tilde{\gamma}|-|\tilde{\rho}|} \quad \text{if } \langle\xi\rangle \geq R|\alpha|,\ \langle\zeta\rangle \geq R|\gamma| \text{ and } \langle\eta\rangle \geq R|\rho|,$$

where the C_j also depend on S, T and U. Similarly, we choose symbols $g_j^R(\zeta,\eta) \in C^\infty(\boldsymbol{R}^n \times \boldsymbol{R}^{n''})$ ($R \geq 3$) ($j = 2, 3$) so that $0 \leq g_j^R(\zeta,\eta) \leq 1$ ($j = 2, 3$) and

$$g_2^R(\zeta,\eta) = \begin{cases} 1 & \text{if } |\zeta| \geq 8c_2(T)|\eta|/c_3(U) \text{ and } |\zeta| \geq 1, \\ 0 & \text{if } |\zeta| \leq 4c_2(T)|\eta|/c_3(U) \text{ or } |\zeta| \leq 1/2, \end{cases}$$

$$g_3^R(\zeta,\eta) = \begin{cases} 1 & \text{if } |\eta| \geq 8c_2(U)|\zeta|/c_3(T) \text{ and } |\eta| \geq 1, \\ 0 & \text{if } |\eta| \leq 4c_2(U)|\zeta|/c_3(T) \text{ or } |\eta| \leq 1/2, \end{cases}$$

$$\left|\partial_\zeta^{\gamma+\tilde{\gamma}} \partial_\eta^{\rho+\tilde{\rho}} g_2^R(\zeta,\eta)\right| \leq C_{|\tilde{\gamma}|+|\tilde{\rho}|}(\widehat{C}(T,U)/R)^{|\gamma|+|\rho|}\langle\zeta\rangle^{-|\tilde{\gamma}|-|\tilde{\rho}|}$$
$$\text{if } \langle\zeta\rangle \geq R|\gamma| \text{ and } \langle\eta\rangle \geq R|\rho|,$$

$$\left|\partial_\zeta^{\gamma+\tilde{\gamma}} \partial_\eta^{\rho+\tilde{\rho}} g_3^R(\zeta,\eta)\right| \leq C_{|\tilde{\gamma}|+|\tilde{\rho}|}(\widehat{C}(T,U)/R)^{|\gamma|+|\rho|}\langle\eta\rangle^{-|\tilde{\gamma}|-|\tilde{\rho}|}$$
$$\text{if } \langle\zeta\rangle \geq R|\gamma| \text{ and } \langle\eta\rangle \geq R|\rho|.$$

We assume that $R_0 \geq R(T, U, A_1, \varepsilon_0)$ and $\delta_2 + c_1^+(U) + c_1^-(U) \leq 1/(3R_0)$, where $R(T, U, A_1, \varepsilon_0)$ is the quantity in Lemma 2.4.1 . Write, for $u \in \mathcal{S}_\infty(\boldsymbol{R}^{n''})$,

$$\begin{aligned}
&\mathcal{F}_x[p_{S,-U,U,T}(D_x, w, D_w, y, D_z)u(x)](\xi) \\
&\equiv \widehat{P}u(\xi) + (2\pi)^{-n-n''} \sum_{j=1}^{5} Q_j \hat{u}(\xi),
\end{aligned}$$

where

$$\begin{aligned}
Q_1\hat{u}(\xi) &= \int E(\xi, w, \zeta, y, \eta) \\
&\quad \times L_1^N L^M \{g_1^{R'}(\xi, \zeta, \eta) p(\xi, w, \zeta, y, \eta) \hat{u}(\eta)\}\, d\eta dy d\zeta dw, \\
Q_2\hat{u}(\xi) &= \int E(\xi, w, \zeta, y, \eta) \sum_{j=1}^{\infty} \psi_j^{R_1}(\zeta) \\
&\quad \times L^M K_1^j \{(1 - g_1^{R'}(\xi, \zeta, \eta)) g_2^{R'}(\zeta, \eta) \chi_j^{\varepsilon_0}(w - y) \\
&\quad \times p(\xi, w, \zeta, y, \eta) \hat{u}(\eta)\}\, d\eta dy d\zeta dw, \\
Q_3\hat{u}(\xi) &= \int E(\xi, w, \zeta, y, \eta) \sum_{j=1}^{\infty} (1 - \chi_j^{\varepsilon_0}(w - y)) \\
&\quad \times L^M \widetilde{L}_1^{j+n+1} \{\psi_j^{R_1}(\zeta)(1 - g_1^{R'}(\xi, \zeta, \eta)) g_2^{R'}(\zeta, \eta) \\
&\quad \times p(\xi, w, \zeta, y, \eta) \hat{u}(\eta)\}\, d\eta dy d\zeta dw, \\
Q_4\hat{u}(\xi) &= \int E(\xi, w, \zeta, y, \eta) L_1^N L^M \{(1 - g_1^{R'}(\xi, \zeta, \eta))(1 - g_2^{R'}(\zeta, \eta)) \\
&\quad \times g_3^{R'}(\zeta, \eta) p(\xi, w, \zeta, y, \eta) \hat{u}(\eta)\}\, d\eta dy d\zeta dw, \\
Q_5\hat{u}(\xi) &= \int E(\xi, w, \zeta, y, \eta) \sum_{j=1}^{\infty} L_1^N L^M \{(1 - g_1^{R'}(\xi, \zeta, \eta)) \\
&\quad \times (1 - g_2^{R'}(\zeta, \eta))(1 - g_3^{R'}(\zeta, \eta))(1 - \chi_j^{\varepsilon_0}(w - y)) \psi_j^R(\eta) \\
&\quad \times p(\xi, w, \zeta, y, \eta) \hat{u}(\eta)\}\, d\eta dy d\zeta dw, \\
\widehat{P}u(\xi) &= (2\pi)^{-n-n''} \int E(\xi, w, \zeta, y, \eta) \sum_{j=1}^{\infty} L^M \{(1 - g_1^{R'}(\xi, \zeta, \eta)) \\
&\quad \times (1 - g_2^{R'}(\zeta, \eta))(1 - g_3^{R'}(\zeta, \eta)) \chi_j^{\varepsilon_0}(w - y) \psi_j^R(\eta) \\
&\quad \times p(\xi, w, \zeta, y, \eta) \hat{u}(\eta)\}\, d\eta dy d\zeta dw, \\
E(\xi, w, \zeta, y, \eta) &= \exp[iS(w, \xi) - iU(w, \zeta) + iU(y, \zeta) + iT(y, \eta)],
\end{aligned}$$

$M, N \geq n + 1$, $R' \geq 3$ and $R_1, R > 0$. Here $\{\psi_j^R(\zeta)\}$ and $\{\psi_j^R(\eta)\}$ are the families of symbols defined in Section 2.2. If $n \neq n''$, then the $\psi_j^R(\zeta)$

and the $\psi_j^R(\eta)$ are different. However, we do not use different notations to represent these symbols, as there is no confusion.

Lemma A.1. 1 *Q_1 maps continuously $\hat{\mathcal{S}}_{\varepsilon_2}(\boldsymbol{R}^{n''})$ and $\hat{\mathcal{S}}_{-\varepsilon_2}'(\boldsymbol{R}^{n''})$ to $\hat{\mathcal{S}}_{\varepsilon_1}(\boldsymbol{R}^{n'})$ and $\hat{\mathcal{S}}_{-\varepsilon_1}'(\boldsymbol{R}^{n'})$, respectively, if $R_0 \geq R(S,U,A_1)$, $R' \geq R'(S,U)$, $R' \geq R_0$ and*

$$\max\{\delta_1 + c_1(S) + \varepsilon_1, \delta_2 + c_1^+(U) + c_1^-(U), \delta_3 + c_1(T) - \varepsilon_2\} \leq \delta(S,T,U)/R',$$

where $R(S,U,A_1)$, $R'(S,U)$ and $\delta(S,T,U)$ are positive constants.

Proof We can write

$$\begin{aligned} Q_1\hat{u}(\xi) = &\int E(\xi,w,\zeta,y,\eta) \\ &\times L_1^N L^M K_2^j\{g_1^{R'}(\xi,\zeta,\eta)p(\xi,w,\zeta,y,\eta)\hat{u}(\eta)\}\, d\eta dy d\zeta dw \end{aligned}$$

for $R'j \leq \langle\xi\rangle \leq R'(j+1)$ if $R' \geq R_0$, where K_2 is a differential operator defined by

$${}^tK_2 = |\nabla_w S(w,\xi) - \nabla_w U(w,\zeta)|^{-2} \sum_{k=1}^n \left(\overline{\partial_{w_k} S(w,\xi)} - \overline{\partial_{w_k} U(w,\zeta)}\right) D_{w_k}.$$

It follows from Lemma 2.1. 5 that

$$\begin{aligned} \left|\partial_\xi^\alpha \partial_\zeta^\gamma \partial_\eta^\rho K_2^j p(\xi,w,\zeta,y,\eta)\right| &\leq C_{|\alpha|+|\gamma|+|\rho|}(S,U,p) \\ &\times \Gamma(S,U,A_1,R_0,R')^j \langle\xi\rangle^{m_1-|\alpha|}\langle\zeta\rangle^{m_2-|\gamma|}\langle\eta\rangle^{m_3-|\rho|} \\ &\times \exp[\delta_1\langle\xi\rangle + \delta_2\langle\zeta\rangle + \delta_3\langle\eta\rangle] \end{aligned}$$

if $(\xi,\zeta,\eta) \in \operatorname{supp} g_1^{R'}$, $\langle\xi\rangle \geq R'j$ and $R' \geq R_0$, where $j \in \boldsymbol{Z}_+$ and

$$\begin{aligned} \Gamma(S,U,A_1,R_0,R') &= 2^{10} n A_0 c_3(S)^{-2} C(S,U) \\ &\times \max\{A_1/R_0, 2^9 \cdot 15 n A_0^3 c_3(S)^{-2} C(S,U)^2/R'\}, \\ C(S,U) &= C(S) + c_3(S)C(U)/(2c_2(U)). \end{aligned}$$

Therefore, we have

$$\begin{aligned} &\left|\langle\xi\rangle^k D_\xi^\alpha\left\{e^{\varepsilon_1\langle\xi\rangle} Q_1\hat{u}(\xi)\right\}\right| \leq C_{|\alpha|,\varepsilon_1,\varepsilon_2,M,N}(S,T,U,p) \\ &\quad\times \int_{\Omega_\xi} \exp[c_1(S)|\xi| + (c_1^+(U) + c_1^-(U))|\zeta| + c_1(T)|\eta|] \\ &\quad\times \langle\xi\rangle^k \langle y-w\rangle^{-N}\langle y\rangle^{-M}\langle w\rangle^{|\alpha|}\Gamma(S,U,A_1,R_0,R')^j\langle\xi\rangle^{m_1}\langle\zeta\rangle^{m_2} \\ &\quad\times \langle\eta\rangle^{m_3}\exp[(\delta_1+\varepsilon_1)\langle\xi\rangle + \delta_2\langle\zeta\rangle + (\delta_3-\varepsilon_2)\langle\eta\rangle] \\ &\quad\times \left|e^{\varepsilon_2\langle\eta\rangle}\hat{u}(\eta)\right|_{S,M} d\eta dy d\zeta dw \\ &\leq C'_{|\alpha|,\varepsilon_1,\varepsilon_2,M,N}(S,T,U,p)\langle\xi\rangle^{k+m_1+(m_2)_+ + (m_3)_+ + n + n''} \\ &\quad\times \exp[(\delta(\varepsilon_1,\varepsilon_2,S,T,U) - 1/R')\langle\xi\rangle]\left|e^{\varepsilon_2\langle\eta\rangle}\hat{u}(\eta)\right|_{S,M} \end{aligned}$$

if $R'j \le \langle\xi\rangle \le R'(j+1)$, $R' \ge R_0$ and

$$\begin{cases} R_0 \ge 2^{10} enA_0A_1C(S,U)/c_3(S)^2, \\ R' \ge 2^{19}\cdot 15en^2A_0^4C(S,U)^3/c_3(S)^4, \end{cases} \tag{A.1}$$

where $j \in \boldsymbol{Z}_+$, $M, N \ge n+1+|\alpha|$, $\Omega_\xi = \{(w,\zeta,y,\eta);\ 2c_2(U)|\zeta| + 3c_2(T)|\eta| \le c_3(S)|\xi| \text{ or } |\xi|^2+|\zeta|^2+|\eta|^2 \le 4\}$ and $\delta(\varepsilon_1,\varepsilon_2,S,T,U) = \delta_1 + c_1(S) + \varepsilon_1 + c_3(S)\max\{(\delta_2 + c_1^+(U) + c_1^-(U))/(2c_2(U)), (\delta_3 + c_1(T) - \varepsilon_2)/(3c_2(T)), 0\}$. This proves that Q_1 maps continuously $\widehat{\mathcal{S}}_{\varepsilon_2}$ to $\widehat{\mathcal{S}}_{-\varepsilon_1}$ if $R' \ge R_0$, R_0 and R' satisfy (A.1), and $\delta(\varepsilon_1,\varepsilon_2,S,T,U) < 1/R'$. Define ${}^tQ_1$ by $\langle {}^tQ_1 v, \varphi\rangle = \langle v, Q_1\varphi\rangle$ for $v \in \widehat{\mathcal{S}}_\infty(\boldsymbol{R}^{n'})$ and $\varphi \in \widehat{\mathcal{S}}_\infty(\boldsymbol{R}^{n''})$. For $v \in \widehat{\mathcal{S}}_\infty(\boldsymbol{R}^{n'})$ we can represent

$$\begin{aligned} {}^tQ_1 v(\eta) = \sum_{j=1}^{\infty} \int & E(\xi,w,\zeta,y,\eta) \\ & \times L_1^N L_2^M \Big[\psi_j^{R'}(\xi) K_2^j \{g_1^{R'}(\xi,\zeta,\eta) p(\xi,w,\zeta,y,\eta) v(\xi)\}\Big]\, d\xi dw d\zeta dy, \end{aligned}$$

where $M, N \ge n+1$ and L_2 is a differential operator defined by

$${}^tL_2 = (1 + |\nabla_\xi S(w,\xi)|^2)^{-1}\Big(\sum_{j=1}^{n'} \overline{\partial_{\xi_j} S(w,\xi)} D_{\xi_j} + 1\Big).$$

Thus we can show that

$$\begin{aligned} & \Big|\langle\eta\rangle^k D_\eta^\rho \Big\{e^{-\varepsilon_2\langle\eta\rangle}\, {}^tQ_1 v(\eta)\Big\}\Big| \le C_{|\rho|,k,\varepsilon_1,\varepsilon_2,M,N,R'}(S,T,U,p) \\ & \times \exp[\{\delta_3 + c_1(T) - \varepsilon_2 - 3c_2(T)((3R')^{-1} - \delta_1 - c_1(S) - \varepsilon_1)/c_3(S)\}\langle\eta\rangle] \\ & \times \Big|e^{-\varepsilon_1\langle\xi\rangle} v(\xi)\Big|_{S,M} \end{aligned}$$

if $M, N \ge n+1+|\rho|$, $R' \ge R_0$, R_0 and R' satisfy (A.1) with the right-hand sides multiplied by 2, and

$$1/(3R') \ge \delta_1 + c_1(S) + \varepsilon_1 + c_3(S)(\delta_2 + c_1^+(U) + c_1^-(U))_+/(2c_2(U)).$$

This proves the lemma. □

Lemma A.1.2 *Q_k ($k = 2,3$) map continuously $\widehat{\mathcal{S}}_{\varepsilon_2}(\boldsymbol{R}^{n''})$ and $\widehat{\mathcal{S}}_{-\varepsilon_2}{}'(\boldsymbol{R}^{n''})$ to $\widehat{\mathcal{S}}_{\varepsilon_1}(\boldsymbol{R}^{n'})$ and $\widehat{\mathcal{S}}_{-\varepsilon_1}{}'(\boldsymbol{R}^{n'})$, respectively, if $R_0 \ge R(T,U,A_1,\varepsilon_0)$, $R' \ge R'(S,T,U,\varepsilon_0)$, $R_1 \ge R_1(T,U,\varepsilon_0)$, $R' \ge R_0$, $R_1 \ge R'$ and*

$$\max\{\delta_1 + c_1(S) + \varepsilon_1, \delta_2 + c_1^+(U) + c_1^-(U), \delta_3 + c_1(T) - \varepsilon_2\} \le \delta(S,T,U)/R_1,$$

where $R(T,U,A_1,\varepsilon_0)$, $R'(S,T,U,\varepsilon_0)$, $R_1(T,U,\varepsilon_0)$ and $\delta(S,T,U)$ are positive constants.

Proof It follows from Lemma 2.1. 5 that

$$\begin{aligned}&\left|\partial_\xi^\alpha \partial_\eta^\rho K_1^j\Big\{(1-g_1^{R'}(\xi,\zeta,\eta))g_2^{R'}(\zeta,\eta)\chi_j^{\varepsilon_0}(w-y)p(\xi,w,\zeta,y,\eta)\Big\}\right|\\ &\le C_{|\alpha|+|\rho|,\varepsilon_0,R_1}(S,T,U,p)\Gamma(T,U,A_1,\varepsilon_0,R_0,R_1)^j\langle\zeta\rangle^{(m_1)_+ + m_2+(m_3)_+}\\ &\quad\times\exp[\delta_1\langle\xi\rangle+\delta_2\langle\zeta\rangle+\delta_3\langle\eta\rangle] \qquad\qquad (A.2)\end{aligned}$$

if $\langle\zeta\rangle \ge 2R_1(j-1)$ and $R_1 \ge R_0$, where $j \in \boldsymbol{N}$ and

$$\begin{aligned}&\Gamma(T,U,A_1,\varepsilon_0,R_0,R_1)=2^8 nA_0c_3(U)^{-2}C(T,U)\\ &\quad\times\max\{C_*/(\varepsilon_0R_1)+A_1/R_0,\,2^7\cdot 15nA_0^3c_3(U)^{-2}C(T,U)^2/R_1\},\\ &C(T,U)=C(U)+c_3(U)C(T)/(2c_2(T))\end{aligned}$$

(see, also, (2.41)). We have also

$$\begin{aligned}&\left|\partial_\xi^\alpha \partial_\eta^\rho \widetilde{L}_1^{j+k+n+1}\Big\{\psi_j^{R_1}(\zeta)(1-g_1^{R'}(\xi,\zeta,\eta))g_2^{R'}(\zeta,\eta)p(\xi,w,\zeta,y,\eta)\Big\}\right|\\ &\le C_{|\alpha|+|\rho|,k,\varepsilon_0,R_1}(S,T,U,p)\langle y-w\rangle^{-k-n-1}\\ &\quad\times\Gamma(S,T,U,A_1,\varepsilon_0,R_0,R',R_1)^j\\ &\quad\times\langle\xi\rangle^{m_1-|\alpha|}\langle\zeta\rangle^{m_2}\langle\eta\rangle^{m_3-|\rho|}\exp[\delta_1\langle\xi\rangle+\delta_2\langle\zeta\rangle+\delta_3\langle\eta\rangle]\end{aligned}$$

if $|w-y| \ge \varepsilon_0/2$ and $R_1 \ge \max\{R_0, R'\}$, where

$$\begin{aligned}&\Gamma(S,T,U,A_1,\varepsilon_0,R_0,R',R_1)=2^7n^{3/2}A_0^2\varepsilon_0^{-1}c_0(U)^{-2}C(U)\\ &\quad\times\max\{\widehat{C}/R_1+(\widehat{C}(S,T,U)+\widehat{C}(T,U)/R'+A_1/R_0,\\ &\qquad 2^5\cdot 15n^2A_0^5c_0(U)^{-2}C(U)^2/R_1\}\end{aligned}$$

(see (2.43)). Since, for $M \ge n+1$,

$$\begin{aligned}{}^tQ_2v(\eta)&=\int\sum_{j=1}^\infty E(\xi,w,\zeta,y,\eta)\psi_j^{R_1}(\zeta)L_2^MK_1^j\Big\{(1-g_1^{R'}(\xi,\zeta,\eta))\\ &\quad\times g_2^{R'}(\zeta,\eta)\chi_j^{\varepsilon_0}(w-y)p(\xi,w,\zeta,y,\eta)v(\xi)\Big\}\,d\xi dw d\zeta dy,\\ {}^tQ_3v(\eta)&=\int\sum_{j=1}^\infty E(\xi,w,\zeta,y,\eta)(1-\chi_j^{\varepsilon_0}(w-y))L_2^M\widetilde{L}_1^{j+M}\\ &\quad\times\Big\{\psi_j^{R_1}(\zeta)(1-g_1^{R'}(\xi,\zeta,\eta))g_2^{R'}(\zeta,\eta)p(\xi,w,\zeta,y,\eta)v(\xi)\Big\}\,d\xi dw d\zeta dy,\end{aligned}$$

we can prove the lemma, applying the same argument as in the proof of Lemma A.1. 1 . □

Lemma A.1. 3 *Q_4 maps continuously $\widehat{\mathcal{S}}_{\varepsilon_2}(\boldsymbol{R}^{n''})$ and $\widehat{\mathcal{S}}_{-\varepsilon_2}{}'(\boldsymbol{R}^{n''})$ to $\widehat{\mathcal{S}}_{\varepsilon_1}(\boldsymbol{R}^{n'})$ and $\widehat{\mathcal{S}}_{-\varepsilon_1}{}'(\boldsymbol{R}^{n'})$, respectively, if $R_0 \ge R(T,U,A_1)$ and*

$$\max\{\delta_1+c_1(S)+\varepsilon_1, \delta_2+c_1^+(U)+c_1^-(U), \delta_3+c_1(T)-\varepsilon_2\} \le \delta(S,T,U)/R_0,$$

where $R(T,U,A_1)$ and $\delta(S,T,U)$ are positive constants.

Proof We can represent

$$\begin{aligned} Q_4\hat{u}(\xi) = \int E(\xi,w,\zeta,y,\eta) \sum_{j=1}^{\infty} L_1^N L^M \Big\{ \psi_j^{R_2}(\eta)(1-g_1^{R'}(\xi,\zeta,\eta)) \\ \times(1-g_2^{R'}(\zeta,\eta))g_3^{R'}(\zeta,\eta)(K_1^j p(\xi,w,\zeta,y,\eta))\hat{u}(\eta)\Big\}\, d\eta dy d\zeta dw, \end{aligned}$$

using an oscillatory integral, where $M, N \ge n+1$ and $R_2 > 0$. Similarly, the estimates

$$\begin{aligned} \Big|\partial_\xi^\alpha \partial_\zeta^\gamma \partial_\eta^\rho \Big\{ \psi_j^{R_2}(\eta)(1-g_1^{R'}(\xi,\zeta,\eta))(1-g_2^{R'}(\zeta,\eta)) \\ \times g_3^{R'}(\zeta,\eta) K_1^j p(\xi,w,\zeta,y,\eta)\Big\}\Big| \\ \le C_{|\alpha|+|\gamma|+|\rho|,R_2}(S,T,U,p)\Gamma(T,U,R_0,R_2)^j \langle\eta\rangle^{(m_1)_+ +(m_2)_+ + m_3} \\ \times \exp[\delta_1\langle\xi\rangle + \delta_2\langle\zeta\rangle + \delta_3\langle\eta\rangle] \end{aligned}$$

hold if $R_2 \ge R_0$, where $j \in \boldsymbol{N}$ and

$$\begin{aligned} \Gamma(T,U,R_0,R_2) &= 2^6 n A_0 c_3(T)^{-2} C'(T,U) \\ &\quad \times \max\{A_1/R_0, 2^3\cdot 15 n A_0^3 c_3(T)^{-2} C'(T,U)^2/R_2\}, \\ C'(T,U) &= 4C(T) + c_3(T)C(U)/c_2(U) \end{aligned}$$

(see, also, (A.2)). ${}^tQ_4$ is also represented as

$$\begin{aligned} {}^tQ_4 v(\eta) = \int E(\xi,w,\zeta,y,\eta) L_2^M L_1^M \Big\{(1-g_1^{R'}(\xi,\zeta,\eta)) \\ \times(1-g_2^{R'}(\zeta,\eta))g_3^{R'}(\zeta,\eta)(K_1^j p(\xi,w,\zeta,y,\eta))v(\xi)\Big\}\, d\xi dw d\zeta dy \end{aligned}$$

if $R_2 j \le \langle\eta\rangle \le R_2(j+1)$ and $R_2 \ge R_0$, where $j \in \boldsymbol{Z}_+$ and $M \ge n+1$. Therefore, we can prove the lemma. □

Lemma A.1. 4 *Q_5 maps continuously $\widehat{\mathcal{S}}_{\varepsilon_2}(\boldsymbol{R}^{n''})$ and $\widehat{\mathcal{S}}_{-\varepsilon_2}{}'(\boldsymbol{R}^{n''})$ to $\widehat{\mathcal{S}}_{\varepsilon_1}(\boldsymbol{R}^{n'})$ and $\widehat{\mathcal{S}}_{-\varepsilon_1}{}'(\boldsymbol{R}^{n'})$, respectively, if $R_0 \ge R(U,A_1,\varepsilon_0)$, $R' \ge R'(S,T,U,\varepsilon_0)$, $R \ge R(T,U,\varepsilon_0)$, $R \ge \kappa_1(T,U)\max\{R_0,R'\}$ and $\max\{\delta_1+c_1(S)+\varepsilon_1, \delta_2+c_1^+(U)+c_1^-(U), \delta_3+c_1(T)-\varepsilon_2\} \le \delta(S,T,U)/R$, where $R(U,A_1,\varepsilon_0)$, $R'(S,T,U,\varepsilon_0)$, $R(T,U,\varepsilon_0)$, $\kappa_1(T,U)$ and $\delta(S,T,U)$ are positive constants.*

Proof We can represent

$$Q_5\hat{u}(\xi) = \int \sum_{j=1}^{\infty} E(\xi, w, \zeta, y, \eta) L^M \widetilde{L}_1^{j+M} \Big\{ (1 - g_1^{R'}(\xi, \zeta, \eta))$$
$$\times (1 - g_2^{R'}(\zeta, \eta))(1 - g_3^{R'}(\zeta, \eta))(1 - \chi_j^{\varepsilon_0}(w - y))\psi_j^R(\eta)$$
$$\times p(\xi, w, \zeta, y, \eta))\hat{u}(\eta) \Big\}\, d\eta dy d\zeta dw,$$

$${}^tQ_5 v(\eta) = \int \sum_{j=1}^{\infty} E(\xi, w, \zeta, y, \eta) L_2^M \widetilde{L}_1^{j+M} \Big\{ (1 - g_1^{R'}(\xi, \zeta, \eta))$$
$$\times (1 - g_2^{R'}(\zeta, \eta))(1 - g_3^{R'}(\zeta, \eta))(1 - \chi_j^{\varepsilon_0}(w - y))\psi_j^R(\eta)$$
$$\times p(\xi, w, \zeta, y, \eta))v(\xi) \Big\}\, d\xi dw d\zeta dy,$$

where $M \geq n + 1$. Moreover, we have

$$\Big| \partial_\xi^\alpha \partial_\eta^\rho \widetilde{L}_1^{j+M} \Big\{ (1 - g_1^{R'}(\xi, \zeta, \eta))(1 - g_2^{R'}(\zeta, \eta))(1 - g_3^{R'}(\zeta, \eta))$$
$$\times (1 - \chi_j^{\varepsilon_0}(w - y))\psi_j^R(\eta) p(\xi, w, \zeta, y, \eta) \Big\} \Big|$$
$$\leq C_{|\alpha|+|\rho|, M, \varepsilon_0, R}(S, T, U, p) \langle w - y \rangle^{-M} \Gamma(S, T, U, A_1, \varepsilon_0, R_0, R', R)^j$$
$$\times \langle \xi \rangle^{m_1 - |\alpha|} \langle \zeta \rangle^{m_2} \langle \eta \rangle^{m_3 - |\rho|} \exp[\delta_1 \langle \xi \rangle + \delta_2 \langle \zeta \rangle + \delta_3 \langle \eta \rangle]$$

if $R \geq 2^4 c_2(U) \max\{R_0, R'\}/c_3(T)$, where

$$\Gamma(S, T, U, A_1, \varepsilon_0, R_0, R', R) = 2^7 n^{3/2} A_0^2 \varepsilon_0^{-1} c_3(U)^{-2} C(U)$$
$$\times \max\{(\widehat{C}(S, T, U) + 2\widehat{C}(T, U))/R' + A_1/R_0,$$
$$2^9 \cdot 15 n A_0^5 c_2(U) c_3(T)^{-1} c_0(U)^{-2} C(U)^2 / R\}.$$

This proves the lemma. □

We can write

$$\widehat{P}u(\xi) = (2\pi)^{-n-n''} \lim_{\nu \downarrow 0} \int e^{-\nu |w|^2} E(\xi, w, \zeta, y, \eta)(1 - g_1^{R'}(\xi, \zeta, \eta))$$
$$\times (1 - g_2^{R'}(\zeta, \eta))(1 - g_3^{R'}(\zeta, \eta)) \sum_{j=1}^{\infty} \chi_j^{\varepsilon_0}(w - y)\psi_j^R(\eta)$$
$$\times p(\xi, w, \zeta, y, \eta)\hat{u}(\eta)\, d\eta dy d\zeta dw.$$

Put

$$\tilde{p}(\xi, w, \eta) \quad = \quad (2\pi)^{-n} \int e^{-iU(w,\zeta) + iU(y,\zeta) - iT(w,\eta) + iT(y,\eta)}$$

$$\times \sum_{j=1}^{\infty} \chi_j^{\varepsilon_0}(w-y)\psi_j^R(\eta) f^{R'}(\xi,w,\zeta,y,\eta)\, dyd\zeta,$$

$$\begin{aligned} f^{R'}(\xi,w,\zeta,y,\eta) &= (1-g_1^{R'}(\xi,\zeta,\eta))(1-g_2^{R'}(\zeta,\eta))(1-g_3^{R'}(\zeta,\eta)) \\ &\quad \times p(\xi,w,\zeta,y,\eta). \end{aligned}$$

Then,

$$\begin{aligned} \widehat{P}u(\xi) &= (2\pi)^{-n''} \lim_{\nu \downarrow 0} \int e^{-\nu|w|^2} e^{iS(w,\xi)+iT(w,\eta)} \tilde{p}(\xi,w,\eta)\hat{u}(\eta)\, d\eta dw \\ &= \mathcal{F}_x[\tilde{p}_{S,T}(D_x,y,D_z)u(x)](\xi). \end{aligned}$$

Write

$$\begin{aligned} &-U(w,\zeta)+U(y,\zeta)-T(w,\eta)+T(y,\eta) = (y-w)\cdot z(w,\zeta,y,\eta), \\ &z(w,\zeta,y,\eta) = {}^tU_0\zeta + {}^tT_0\eta + \tilde{\nabla}_y\tilde{U}(w,y,\zeta) + \tilde{\nabla}_y\tilde{T}(w,y,\eta). \end{aligned}$$

Then $z(w,\zeta,y,\eta)$ can be regarded as an analytic function of ζ in $\mathcal{C}_{w,1/(\sqrt{n}A_0)}$ for $y, w \in \boldsymbol{R}^n$ with $|w-y| \le \varepsilon_0$, $\eta \in \boldsymbol{R}^{n''}$ with $(y,\eta) \in \Omega_1'$. Recall that $\delta \equiv \delta(U)$ and $c(U) \le \min\{\gamma_1\delta, \gamma_2\}/|U_0^{-1}|$, where $\delta(U)$ is the quantity in Lemma 2.4. 2 . Let $0 < \delta' \le \delta$, and assume that

$$c(U) \le \min\{\gamma_1\delta', \gamma_2\}/|U_0^{-1}|.$$

Now we assume that $y, w \in \boldsymbol{R}^n$, $(y,\eta) \in \Omega_1'$, $|\eta| \ge 1$ and $|y-w| \le \varepsilon_0$. Put

$$\begin{aligned} &\hat{\zeta}(\equiv \hat{\zeta}(w,\zeta,y,\eta)) \\ &= \zeta + \Psi_w(\zeta;\delta)^t U_0^{-1} \mathrm{Re}\, \tilde{\nabla}_y\tilde{U}(w,y,\zeta) + {}^tU_0^{-1}\mathrm{Re}\, \tilde{\nabla}_y\tilde{T}(w,y,\eta) \end{aligned}$$

for $\zeta \in \boldsymbol{C}^n$. Then we have the following

Lemma A.1. 5 *Assume that $\hat{\zeta} \in \Gamma_w$, $|\hat{\zeta}| \ge 1$ and $|\eta| \le 8c_2(U)|\hat{\zeta}|/c_3(T)$. Then, $|\hat{\zeta}-\zeta| < \delta'\langle\hat{\zeta}\rangle/2$, $\zeta \in \mathcal{C}_{w,\delta'/2}$ and $\hat{\zeta} = {}^tU_0^{-1}\mathrm{Re}\, z(w,\zeta,y,\eta) - {}^tU_0^{-1}\,{}^tT_0\eta$ if*

$$c(T) \le c_3(T)\delta'/(2^8 c_2(U)|U_0^{-1}|), \tag{A.3}$$

modifying γ_1 if necessary.

Proof Assume that (A.3) is valid. Now suppose that $\zeta \notin \mathcal{C}_{w,5\delta/2}$. Then

$$|\hat{\zeta}-\zeta| \le |U_0^{-1}|c(T)\langle\eta\rangle < 16c(T)c_2(U)|U_0^{-1}|\langle\hat{\zeta}\rangle/c_3(T) \le \delta\langle\hat{\zeta}\rangle/16.$$

This is a contradiction. So we have $\zeta \in \mathcal{C}_{w,5\delta/2}$. There is $\tilde{\zeta} \in \Gamma_w$ such that $|\tilde{\zeta}| \geq 1$, $|\zeta - \tilde{\zeta}| < 5\delta\langle\tilde{\zeta}\rangle/2$ and $|\zeta - \tilde{\zeta}| \leq |\hat{\zeta} - \zeta|$. Since $5\delta/2 \leq (2\sqrt{n}A_0)^{-1}$, (2.46) yields

$$|\hat{\zeta} - \zeta| < |U_0^{-1}|\{c(U)\langle\tilde{\zeta}\rangle + (c(U) + 10n^{3/2}C(\tilde{U})A_0^3\delta)|\zeta - \tilde{\zeta}|\} \\ +\delta'\langle\hat{\zeta}\rangle/16.$$

Noting that $|U_0^{-1}|(c(U) + 10n^{3/2}C(\tilde{U})A_0^3\delta) \leq 1/2$, we have

$$|\hat{\zeta} - \zeta| < 2c(U)|U_0^{-1}|\langle\tilde{\zeta}\rangle + \delta'\langle\hat{\zeta}\rangle/8.$$

Since $5\delta/2 \leq 1/2$, we have $\langle\tilde{\zeta}\rangle < 2\langle\hat{\zeta}\rangle + 2|\hat{\zeta} - \zeta|$. Therefore, $|\hat{\zeta} - \zeta| < \delta'\langle\hat{\zeta}\rangle/2$ if $c(U) \leq 2^{-5}\delta'/|U_0^{-1}|$, *i.e.*, $\gamma_1 \leq 2^{-5}$. This proves the lemma. □

Assume that (A.3) is valid. Let $\kappa > 0$, and put

$$f_\kappa^{R'}(\xi, w, \zeta, y, \eta; u) = \sum_{j=1}^{\infty} \psi_j^{\kappa R_0}(\eta) \sum_{|\alpha| \leq j-1} u^\alpha \partial_\zeta^\alpha f^{R'}(\xi, w, \zeta, y, \eta)/\alpha!$$

for $(\xi, w, \zeta, y, \eta, u) \in \boldsymbol{R}^{n'} \times \boldsymbol{R}^n \times \boldsymbol{R}^n \times \boldsymbol{R}^n \times \boldsymbol{R}^{n''} \times \boldsymbol{C}^n$. For simplicity we assume that $R' \geq R_0$. Applying the same argument as in the proof of Lemma 2.4. 3 we have the following

Lemma A.1. 6 *$f_\kappa^{R'}(\xi, w, \zeta, y, \eta; u)$ is analytic in u, $\operatorname{supp} f_\kappa^{R'}(\cdot; u) \subset \operatorname{supp} f^{R'}(\cdot)$ for $u \in \boldsymbol{C}^n$ and*

$$f_\kappa^{R'}(\xi, w, \zeta, y, \eta; 0) = f^{R'}(\xi, w, \zeta, y, \eta),$$

$$\begin{aligned} &\left|\partial_\xi^{\alpha+\tilde{\alpha}} D_w^{\beta^1+\beta^2+\tilde{\beta}} \partial_\zeta^{\gamma+\tilde{\gamma}} D_y^{\lambda^1+\lambda^2+\tilde{\lambda}} \partial_\eta^{\rho+\tilde{\rho}} \partial_u^{\delta+\tilde{\delta}} f_\kappa^{R'}(\xi, w, \zeta, y, \eta; u)\right| \\ &\leq C_{|\tilde{\alpha}|+|\tilde{\beta}|+|\tilde{\gamma}|+|\tilde{\delta}|+|\tilde{\lambda}|,|\tilde{\rho}|,\kappa} (\widehat{C}(S,T,U)/R' + A_1/R_0)^{|\alpha|} \\ &\quad \times (A_1/R_0)^{|\beta^1|+|\beta^2|+|\lambda^1|+|\lambda^2|} (A_1(S,T,U,A_1,R_0/R')/R_0)^{|\gamma|+|\delta|} \\ &\quad \times ((\widehat{C}/\kappa + A_1(S,T,U,A_1,R_0/R'))/R_0)^{|\rho|} \\ &\quad \times \langle\xi\rangle^{m_1-|\tilde{\alpha}|+|\beta^1|} \langle\zeta\rangle^{m_2-|\tilde{\gamma}|-|\tilde{\delta}|+|\beta^2|+|\lambda^1|} \langle\eta\rangle^{m_3-|\tilde{\rho}|+|\lambda^2|+1} \\ &\quad \times \exp[\delta_1\langle\xi\rangle + \delta_2\langle\zeta\rangle + \delta_3\langle\eta\rangle \\ &\qquad + \sqrt{n}A_1(S,T,U,A_1,R_0/R')\langle u\rangle/R_0], \end{aligned} \tag{A.4}$$

$$\begin{aligned} &\left|(\partial_{\zeta_j} - \partial_{u_j})\partial_\xi^{\alpha+\tilde{\alpha}} D_w^{\beta^1+\beta^2+\tilde{\beta}} \partial_\zeta^{\gamma+\tilde{\gamma}} D_y^{\lambda^1+\lambda^2+\tilde{\lambda}} \partial_\eta^{\rho+\tilde{\rho}} \partial_u^{\delta+\tilde{\delta}} f_\kappa^{R'}(\xi, w, \zeta, y, \eta; u)\right| \\ &\leq C_{|\tilde{\alpha}|+|\tilde{\beta}|+|\tilde{\gamma}|+|\tilde{\delta}|+|\tilde{\lambda}|,|\tilde{\rho}|}((\widehat{C}(S,T,U)/R' + A_1/R_0)^{|\alpha|} \end{aligned}$$

$$\times(A_1/R_0)^{|\beta^1|+|\beta^2|+|\lambda^1|+|\lambda^2|}(A_1(S,T,U,A_1,R_0/R')/R_0)^{|\gamma|}$$
$$\times((\widehat{C}/\kappa + A_1(S,T,U,A_1,R_0/R'))/R_0)^{|\rho|}$$
$$\times(eA_1(S,T,U,A_1,R_0/R')/R_0)^{|\delta|}$$
$$\times\langle\xi\rangle^{m_1-|\tilde\alpha|+|\beta^1|}\langle\zeta\rangle^{m_2-1-|\tilde\gamma|-|\tilde\delta|+|\beta^2|+|\lambda^1|}\langle\eta\rangle^{m_3-|\tilde\rho|+|\lambda^2|}$$
$$\times\exp[\delta_1\langle\xi\rangle+\delta_2\langle\zeta\rangle+(\delta_3-1/(3\kappa R_0))\langle\eta\rangle$$
$$+e\sqrt{n}A_1(S,T,U,A_1,R_0/R')\langle u\rangle/R_0], \quad (A.5)$$

if $R'\langle\eta\rangle \le \kappa R_0\langle\zeta\rangle$, $|w-y|\le\varepsilon_0$, $\langle\xi\rangle \ge R'(|\alpha|+|\beta^1|)$, $\langle\zeta\rangle \ge 2R'(|\gamma|+|\beta^2|+|\lambda^1|)$ *and* $\langle\eta\rangle \ge \max\{1,3\kappa\}R'(|\rho|+|\lambda^2|)$, *where* $A_1(S,T,U,A_1,R_0/R') = (\widehat{C}(S,T,U)+2\widehat{C}(T,U))R_0/R' + A_1$. *Moreover, we have*

$$\Big|\partial_\xi^{\alpha+\tilde\alpha}D_w^{\beta^1+\beta^2+\tilde\beta}\partial_\zeta^{\gamma+\tilde\gamma}D_y^{\lambda^1+\lambda^2+\tilde\lambda}\partial_\eta^{\rho+\tilde\rho}\partial_u^{\delta+\tilde\delta}\Big\{f_\kappa^{R'}(\xi,w,\zeta+u,y,\eta;-u)$$
$$-f^{R'}(\xi,w,\zeta,y,\eta)\Big\}\Big|$$
$$\le C'_{|\tilde\alpha|+|\tilde\beta|+|\tilde\gamma|+|\tilde\delta|+|\tilde\lambda|,|\tilde\rho|}((\widehat{C}(S,T,U)/R'+A_1/R_0)^{|\alpha|}$$
$$\times(A_1/R_0)^{|\beta^1|+|\lambda^2|}(3A_1/(2R_0))^{|\beta^2|+|\lambda^1|}$$
$$\times(A_1(S,T,U,A_1,R_0/R')/R_0)^{|\gamma|}$$
$$\times((\widehat{C}/\kappa+A_1(S,T,U,A_1,R_0/R'))/R_0)^{|\rho|}$$
$$\times((1+e)A_1(S,T,U,A_1,R_0/R')/R_0)^{|\delta|}$$
$$\times\langle\xi\rangle^{m_1-|\tilde\alpha|+|\beta^1|}\langle\zeta\rangle^{m_2-1-|\tilde\gamma|-|\tilde\delta|+|\beta^2|+|\lambda^1|}\langle\eta\rangle^{m_3-|\tilde\rho|+|\lambda^2|}$$
$$\times\exp[\delta_1\langle\xi\rangle+(\delta_2+|\delta_2|/2)\langle\zeta\rangle+(\delta_3-1/(3\kappa R_0))\langle\eta\rangle$$
$$+e\sqrt{n}A_1(S,T,U,A_1,R_0/R')\langle u\rangle/R_0], \qquad (A.6)$$

if $u\in\boldsymbol{R}^n$, $|u|\le\langle\zeta\rangle/2$, $2R'\langle\eta\rangle\le\kappa R_0\langle\zeta\rangle$, $|w-y|\le\varepsilon_0$, $\langle\xi\rangle\ge R'(|\alpha|+|\beta^1|)$, $\langle\zeta\rangle\ge 4R'(|\gamma|+|\beta^2|+|\lambda^1|+|\delta|)$ *and* $\langle\eta\rangle\ge\max\{1,3\kappa\}R'(|\rho|+|\lambda^2|)$.

Remark For the proof of Theorem 2.4. 4 it is sufficient to use more rough estimates than (A.5) and (A.6).

Proof One can prove (A.4) and (A.5) in the same way as in the proof of Lemma 2.4. 3 . (A.6) follows from (A.5) and the identity that

$$f_\kappa^{R'}(\xi,w,\zeta+u,y,\eta;-u)-f^{R'}(\xi,w,\zeta,y,\eta)$$
$$=\sum_{k=1}^n\int_0^1\Big((\partial_{\zeta_k}-\partial_{u_k})f_\kappa^{R'}\Big)(\xi,w,\zeta+\theta u,y,\eta;-\theta u)u_k\,d\theta,$$
$$\partial_u^{\delta+\tilde\delta}f_\kappa^{R'}(\xi,w,\zeta+u,y,\eta;-u)=\Big((\partial_\zeta-\partial_u)^{\delta+\tilde\delta}f_\kappa^{R'}\Big)(\xi,w,\zeta+u,y,\eta;-u).$$

□

$\tilde{p}(\xi, w, \zeta)$ can be written as

$$\tilde{p}(\xi, w, \zeta) = \tilde{p}_\kappa(\xi, w, \zeta) + q_\kappa(\xi, w, \zeta),$$

where

$$\tilde{p}_\kappa(\xi, w, \zeta) = (2\pi)^{-n} \int e^{i(y-w)\cdot z(w,\zeta,y,\eta)} \sum_{j=1}^{\infty} \chi_j^{\varepsilon_0}(w-y)$$
$$\times \psi_j^R(\eta) f_\kappa^{R'}(\xi, w, \hat{\zeta}(w, \zeta, y, \eta), y, \eta; \zeta - \hat{\zeta}(w, \zeta, y, \eta))\, dy d\zeta,$$
$$q_\kappa(\xi, w, \eta) = (2\pi)^{-n} \int e^{-iU(w,\zeta)+iU(y,\zeta)-iT(w,\eta)+iT(y,\eta)}$$
$$\times \sum_{j=1}^{\infty} \chi_j^{\varepsilon_0}(w-y) \psi_j^R(\eta) F_\kappa^{R'}(\xi, w, \zeta, y, \eta)\, dy d\zeta,$$
$$F_\kappa^{R'}(\xi, w, \zeta, y, \eta)$$
$$= f^{R'}(\xi, w, \zeta, y, \eta) - f_\kappa^{R'}(\xi, w, \hat{\zeta}(w, \zeta, y, \eta), y, \eta; \zeta - \hat{\zeta}(w, \zeta, y, \eta)).$$

Lemma A.1. 5 implies that $\zeta \in C_{w,\delta'/2}$,

$$\begin{cases} |\hat{\zeta}(w, \zeta, y, \eta)| \geq 1, \quad |\eta| \leq 8c_2(U)|\hat{\zeta}(w, \zeta, y, \eta)|/c_3(T), \\ \hat{\zeta}(w, \zeta, y, \eta) = {}^tU_0^{-1}\mathrm{Re}\ z(w, \zeta, y, \eta) - {}^tU_0^{-1}\,{}^tT_0\eta \in \Gamma_w \end{cases} \tag{A.7}$$

if $\zeta \in \boldsymbol{R}^n$ and $f_\kappa^{R'}(\xi, w, \hat{\zeta}(w, \zeta, y, \eta), y, \eta; \zeta - \hat{\zeta}(w, \zeta, y, \eta)) \neq 0$. Let us first consider $q_\kappa(\xi, w, \eta)$. To simplify the proof we use the following lemma instead of Theorem 2.3. 3 , although the results can be improved if we use Theorem 2.3. 3 .

Lemma A.1. 7 *Let $a(\xi, y, \eta) \in C^\infty(\boldsymbol{R}^{n'} \times \boldsymbol{R}^n \times \boldsymbol{R}^{n''})$ satisfy* supp $a \subset \{(\xi, y, \eta) \in \boldsymbol{R}^{n'} \times \Omega';\ (y, \xi) \in \Omega,\ |\xi| \geq 1$ *and* $|\eta| \geq 1\}$ *and*

$$\left|\partial_\xi^\alpha \partial_\eta^\gamma a(\xi, y, \eta)\right| \leq C_{|\alpha|+|\gamma|}(a) \exp[\hat{\delta}_1\langle\xi\rangle + \hat{\delta}_2\langle\eta\rangle].$$

Then we can define

$$a_{S,T}(D_x, y, D_z)u$$
$$= (2\pi)^{-n''} \mathcal{F}_\xi^{-1}\Big[\int \Big(\int e^{iS(y,\xi)+iT(y,\eta)} a(\xi, y, \eta)\hat{u}(\eta)\, d\eta\Big)\, dy\Big](x)$$

for $u \in \mathcal{S}_\infty(\boldsymbol{R}^{n''})$. Moreover, $a_{S,T}(D_x, y, D_z)$ can be extended to continuous linear operators from $\mathcal{S}_{\varepsilon_2}(\boldsymbol{R}^{n''})$ to $\mathcal{S}_{\varepsilon_1}(\boldsymbol{R}^{n'})$ and from $\mathcal{S}_{-\varepsilon_2}{}'(\boldsymbol{R}^{n''})$ to $\mathcal{S}_{-\varepsilon_1}{}'(\boldsymbol{R}^{n'})$, respectively, if $\hat{\delta}_1 + c_1(S) + \varepsilon_1 < 0$ and $\hat{\delta}_2 + c_1(T) - \varepsilon_2 < 0$.

Remark In the above lemma it suffices to assume that $S(y,\xi) \in C^\infty(\Omega)$ and $T(y,\eta) \in C^\infty(\Omega')$ satisfy ($\mathcal{P}$–1), ($\mathcal{P}$–2) and

$$|S^{(\alpha)}(y,\xi)| \le C_{|\alpha|}(S)\langle y\rangle \quad \text{in } \Omega \cap \{|\xi| \ge 1\} \text{ for } |\alpha| \ge 1,$$
$$|T^{(\gamma)}(y,\eta)| \le C_{|\gamma|}(T)\langle y\rangle \quad \text{in } \Omega' \cap \{|\eta| \ge 1\} \text{ for } |\gamma| \ge 1,$$

as assumptions on S and T.

Proof We have

$$\mathcal{F}_x[a_{S,T}(D_x,y,D_z)u](\xi) = \int e^{iS(y,\xi)+iT(y,\eta)} L^M(a(\xi,y,\eta)\hat{u}(\eta))\, d\eta dy$$

for $u \in \mathcal{S}_\infty(\boldsymbol{R}^{n'})$, where $M \ge n+1$. Since

$$\begin{aligned}&\left|\partial_\xi^\alpha L^M(a(\xi,y,\eta)\hat{u}(\eta))\right| \\ &\le C_{|\alpha|,M}(T,p)\langle y\rangle^{-M} \exp[\hat{\delta}_1\langle\xi\rangle + \hat{\delta}_2\langle\eta\rangle] \sup_{|\gamma|\le M} |\partial_\eta^\gamma \hat{u}(\eta)|,\end{aligned}$$

we have

$$\begin{aligned}&\left|\langle\xi\rangle^k D_\xi^\alpha\left\{e^{\varepsilon_1\langle\xi\rangle}\mathcal{F}_x[a_{S,T}(D_x,y,D_z)u](\xi)\right\}\right| \\ &\le C_{|\alpha|,\varepsilon_1,\varepsilon_2}(S,T,p)|u|_{\mathcal{S}_{\varepsilon_2},|\alpha|+n+n''+2} \\ &\quad \times\langle\xi\rangle^k \exp[(\hat{\delta}_1 + c_1(S)+\varepsilon_1)\langle\xi\rangle]\end{aligned}$$

if $\hat{\delta}_2+c_1(T)-\varepsilon_2 \le 0$. This implies that $a_{S,T}(D_x,y,D_z)$ can be extended to continuous linear operators from $\mathcal{S}_{\varepsilon_2}(\boldsymbol{R}^{n''})$ to $\mathcal{S}_{\varepsilon_1}(\boldsymbol{R}^{n'})$ if $\hat{\delta}_1+c_1(S)+\varepsilon_1 < 0$ and $\hat{\delta}_2+c_1(T)-\varepsilon_2 \le 0$. Applying the same argument to the transposed operator ${}^t a_{S,T}(D_x,y,D_z)$, we can prove the lemma. □

Lemma A.1. 8 *Assume that* $|w-y| \le \varepsilon_0$. *Then we have*

$$\left\{\begin{array}{l}\zeta \in \mathcal{C}_{w,\delta'/2},\ |\hat{\zeta}-\zeta| < \delta'\langle\hat{\zeta}\rangle/2,\ 5|\zeta|/6 \le |\hat{\zeta}| \le 5|\zeta|/4, \\ 10\langle\zeta\rangle/11 \le \langle\hat{\zeta}\rangle \le 10\langle\zeta\rangle/9,\end{array}\right. \tag{A.8}$$

$$\left\{\begin{array}{l}|\xi| \le (5c_2(U)|\zeta| + 6c_2(T)|\eta|)/c_3(S), \\ |\zeta| \le 48c_2(T)|\eta|/(5c_3(U)),\ |\eta| \le 10c_2(U)|\zeta|/c_3(T)\end{array}\right. \tag{A.9}$$

if $F_\kappa^{R'}(\xi,w,\zeta,y,\eta) \ne 0$, *where* $\hat{\zeta} = \hat{\zeta}(w,\zeta,y,\eta)$. *Moreover, we have*

$$\begin{aligned}\left|\partial_\xi^\alpha \partial_\eta^\rho F_\kappa^{R'}(\xi,w,\zeta,y,\eta)\right| &\le C_{|\alpha|+|\rho|}\langle\xi\rangle^{m_1-|\alpha|}\langle\zeta\rangle^{m_2}\langle\eta\rangle^{m_3-|\rho|} \\ &\times \exp[\delta_1\langle\xi\rangle + (\delta_2+|\delta_2|/2+5e\delta'\sqrt{n}A_1(S,T,U,A_1,R_0/R')/(9R_0))\langle\zeta\rangle \\ &\qquad +(\delta_3-(3\kappa R_0)^{-1})\langle\eta\rangle]\end{aligned} \tag{A.10}$$

if

$$\kappa \geq 40c_2(U)R'/(c_3(T)R_0), \tag{A.11}$$

where $A_1(S,T,U,A_1,R_0/R')$ *is the constant in* Lemma A.1. 6 *and the* C_j *also depend on* S, T, U *and* p.

Proof (A.8) and (A.9) easily follows from (A.7) and Lemma A.1. 5 if $f_\kappa^{R'}(\xi,w,\hat{\zeta},y,\eta;\zeta-\hat{\zeta}) \neq 0$, since $\delta' \leq 1/5$. Assume that $f^{R'}(\xi,w,\zeta,y,\eta) \neq 0$. Then (A.9) is valid, $\zeta \in \Gamma_w$ and $|\zeta| \geq 1$. Moreover, we have

$$\begin{aligned}|\hat{\zeta}-\zeta| &\leq |U_0^{-1}|(c(U)\langle\zeta\rangle + 2c(T)|\eta|)\\ &\leq (c(U)|U_0^{-1}| + \delta'/16)\langle\zeta\rangle < \delta'\langle\zeta\rangle/8,\end{aligned}$$

using (A.3). This gives (A.8). Since

$$\left|\partial_\eta^\rho(\hat{\zeta}(w,\zeta,y,\eta)-\zeta)\right| \leq C_{|\rho|}(T)\langle\eta\rangle^{1-|\rho|}$$

for $(y,\eta) \in \Omega_1'$ with $|\eta| \geq 1$ and $|\rho| \geq 1$, (A.6), (A.8) and (A.9) yield (A.10). □

We assume that (A.11) is valid. Then we have the following

Lemma A.1. 9 $q_\kappa(\xi,w,\eta)$ *satisfies the estimates*

$$\begin{aligned}&\left|\partial_\xi^\alpha\partial_\eta^\rho q_\kappa(\xi,w,\eta)\right| \leq C_{|\alpha|+|\rho|,\kappa,\varepsilon_0}\langle\xi\rangle^{m_1}\langle\eta\rangle^{m_2+m_3+n+2}\\ &\quad\times\exp\Big[(\delta_1-\nu_1(S,T,U)/(\kappa R_0))\langle\xi\rangle + (\delta_3(T,U,\delta_2,\delta_3)-1/(9\kappa R_0))\langle\eta\rangle\Big]\end{aligned}$$

if

$$\left\{\begin{array}{l}\delta_2 + |\delta_2|/2 + c_1^+(U) + c_1^-(U) \leq 5c_3(U)/(2^5\cdot 3^3 c_2(T)\kappa R_0),\\ R' \geq (\widehat{C}(S,U) + 2\widehat{C}(T,U))R_0/A_1\\ \delta' \leq c_3(U)/(2^4\cdot 6e\sqrt{n}A_1c_2(T)\kappa),\end{array}\right. \tag{A.12}$$

where $\nu_1(S,T,U) = (4\cdot 3^3 c_2(T)(8c_2(U)/c_3(U)+1))^{-1}c_3(S)$ *and* $\delta_3(T,U,\delta_2,\delta_3) = \delta_3 + c_1(T) + c_1^-(T) - c_3(T)(\delta_2/2 + c_1^+(U) + c_1^-(U))_-/(10c_2(U))$. *Moreover,* $q_{\kappa\, S,T}(D_x,y,D_z)$ *maps continuously* $\mathcal{S}_{\varepsilon_2}(\boldsymbol{R}^{n''})$ *and* $\mathcal{S}_{-\varepsilon_2}'(\boldsymbol{R}^{n''})$ *to* $\mathcal{S}_{\varepsilon_1}(\boldsymbol{R}^{n'})$ *and* $\mathcal{S}_{-\varepsilon_1}'(\boldsymbol{R}^{n'})$, *respectively, if* (A.12) *is satisfied and*

$$\begin{aligned}&\delta_1 + c_1(S) + \varepsilon_1 \leq \nu_1(S,T,U)/(\kappa R_0),\\ &\delta_3(T,U,\delta_2,\delta_3) + c_1(T) - \varepsilon_2 \leq 1/(18\kappa R_0).\end{aligned}$$

Remark In the above lemma $(c_1(T) + c_1^-(T))$ and $(c_1^+(U) + c_1^-(U))$ can be replaced by $\varepsilon_0\hat{c}_1(\tilde{T})$ and $\varepsilon_0\hat{c}_1(\tilde{U})$, respectively, where $\hat{c}_1(\tilde{T})$ and $\hat{c}_1(\tilde{U})$ are constants satisfying $|\mathrm{Im}\ \nabla_y\tilde{T}(y,\eta)| \le \hat{c}_1(\tilde{T})|\eta| + \widehat{C}_1(\tilde{T})$ and $|\mathrm{Im}\ \nabla_y\tilde{U}(y,\zeta)| \le \hat{c}_1(\tilde{U})|\zeta| + \widehat{C}_1(\tilde{U})$ with some constants $\widehat{C}_1(\tilde{T})$ and $\widehat{C}_1(\tilde{U})$.

Proof It follows from Lemma A.1. 8 that

$$\Big|\partial_\xi^\alpha \partial_\eta^\rho \Big\{ e^{-iU(w,\zeta)+iU(y,\zeta)-iT(w,\eta)+iT(y,\eta)} \times \sum_{j=1}^{\infty} \chi_j^{\varepsilon_0}(w-y)\psi_j^R(\eta) F_\kappa^{R'}(\xi,w,\zeta,y,\eta)\Big\}\Big|$$
$$\le C_{|\alpha|+|\rho|,\kappa,\varepsilon_0}\langle\xi\rangle^{m_1-|\alpha|}\langle\eta\rangle^{m_2+m_3+1} \times \exp\Big[\delta_1\langle\xi\rangle + (\delta_2 + |\delta_2|/2 + 5e\delta'\sqrt{n}A_1(S,T,U,A_1,R_0/R')/(9R_0)$$
$$+ c_1^+(U) + c_1^-(U))\langle\zeta\rangle + (\delta_3 + c_1(T) + c_1^-(T) - (3\kappa R_0)^{-1})\langle\eta\rangle\Big],$$

since $T(y,\eta) - T(w,\eta) = (y-w)\cdot({}^tT_0\eta + \tilde{\nabla}_y\tilde{T}(w,y,\eta))$. If $F_\kappa^{R'}(\xi,w,\zeta,y,\eta) \ne 0$, then $|\eta|/(18\kappa R_0) \ge \nu_1(S,T,U)|\xi|/(\kappa R_0)$, $|\eta|/(6\kappa R_0) \ge 5c_3(U)|\zeta| \times (2^5\cdot 3^2 c_2(T)\kappa R_0)^{-1}$ and $\langle\zeta\rangle \ge c_3(T)|\eta|/(10c_2(U))$. This proves the first part of the lemma. The second part follows from Lemma A.1. 7 . □

Next consider $\tilde{p}_\kappa(\xi,w,\eta)$. Fix $y, w \in \boldsymbol{R}^n$ so that $|w-y| \le \varepsilon_0$, and define the map $\tilde{\mathcal{Z}}_\delta$ by

$$\tilde{\mathcal{Z}}_\delta : \boldsymbol{R}^n \ni \zeta \mapsto \zeta + {}^tU_0^{-1}\Psi_w(\zeta;\delta)\mathrm{Re}\ \tilde{\nabla}_y\tilde{U}(w,y,\zeta) \in \boldsymbol{R}^n.$$

Then, applying the same argument as in the proof of Lemma 2.4. 2 we can show that $\tilde{\mathcal{Z}}_\delta$ is a diffeomorphism on $\boldsymbol{R}^n$ and

$$\Big\{\tilde{Z}_\delta(z;w,y) \in \boldsymbol{R}^n;\ |z-\tilde{z}| < \delta\langle\tilde{z}\rangle/8 \text{ for some } \tilde{z} \in \widehat{\Gamma}_{w,y,\delta}\Big\} \subset \mathcal{C}_{w,\delta/2} \cap \boldsymbol{R}^n,$$

where $\tilde{Z}_\delta(z;w,y)$ is the inverse function (map) of $\tilde{\mathcal{Z}}_\delta$ and $\widehat{\Gamma}_{w,y,\delta} = \tilde{\mathcal{Z}}_\delta(\{\zeta \in \Gamma_w;\ |\zeta| \ge 1\})$, modifying δ and γ_j ($j = 1,2$) if necessary. Assume that $(y,\eta) \in \Omega_1'$, and put

$$\Omega_{w,y,\eta,t} = \Big\{\mathrm{Re}\ z(w,\zeta,y,\eta) + i(1-t)\mathrm{Im}\ z(w,\zeta,y,\eta) \in \boldsymbol{C}^n;\ \zeta \in \mathcal{C}_{w,\delta'/2} \cap \boldsymbol{R}^n\Big\} \quad (t \in [0,1]),$$

$$\Omega_{w,y,\eta} = \bigcup_{t\in[0,1]} \Omega_{w,y,\eta,t}.$$

Note that $\mathrm{Re}\ \Omega_{w,y,\eta,t}$ ($\equiv \{\mathrm{Re}\ z;\ z \in \Omega_{w,y,\eta,t}\}$) $= \Omega_{w,y,\eta,1}$ for $t \in [0,1]$.

Lemma A.1. 10 *Assume that* $y, w \in \boldsymbol{R}^n$, $|w-y| \le \varepsilon_0$, $(y,\eta) \in \Omega_1'$, $|\eta| \ge 1$ *and*

$$c(T) \le \gamma_3(T,U)\delta'/|U_0^{-1}|, \tag{A.13}$$

where $\gamma_3(T,U)$ *is a positive constant.* (i) *If* $\hat\zeta \in \Gamma_w$, $|\hat\zeta| \ge 1$, $|\hat\zeta| \ge c_3(T)|\eta| /(8c_2(U))$ *and*

$$\gamma_3(T,U) \le c_3(T)/(2^8 c_2(U)), \tag{A.14}$$

then ${}^tU_0\hat\zeta + {}^tT_0\eta \in \Omega_{w,y,\eta,1}$, *modifying* γ_1 *if necessary.* (ii) *If* $z \in \Omega_{w,y,\eta}$, $\hat\zeta \equiv {}^tU_0^{-1}\mathrm{Re}\ z - {}^tU_0^{-1}\,{}^tT_0\eta \in \Gamma_w$, $|\hat\zeta| \ge 1$, $|\hat\zeta| \ge c_3(T)|\eta|/(8c_2(U))$ *and*

$$\gamma_3(T,U) \le 11c_3(T)/(2^9 \cdot 3^2 c_2(U)), \tag{A.15}$$

then $\zeta^1 \equiv {}^tU_0^{-1}z - {}^tU_0^{-1}\tilde\nabla_y T(w,y,\eta) \in \tilde{\mathcal{C}}_{w,y,\delta'}$.

Proof (i) Put $\zeta = \tilde Z_\delta(\hat\zeta - {}^tU_0^{-1}\mathrm{Re}\ \tilde\nabla_y \tilde T(w,y,\eta); w,y)$. Then we have

$$\hat\zeta = \zeta + {}^tU_0^{-1}\Psi_w(\zeta;\delta)\mathrm{Re}\ \tilde\nabla_y\tilde U(w,y,\zeta) + {}^tU_0^{-1}\mathrm{Re}\ \tilde\nabla_y\tilde T(w,y,\eta).$$

Applying the same argument as in the proof of Lemma A.1. 5 , we have

$$|\hat\zeta - \zeta| < \delta'\langle\hat\zeta\rangle/2 \tag{A.16}$$

if (A.14) is valid and $\gamma_1 \le 2^{-5}$. This implies that $\zeta \in \mathcal{C}_{w,\delta'/2} \cap \boldsymbol{R}^n$ and $\hat\zeta = {}^tU_0^{-1}\mathrm{Re}\ z(w,\zeta,y,\eta) - {}^tU_0^{-1}\,{}^tT_0\eta$, *i.e.*, ${}^tU_0\hat\zeta + {}^tT_0\eta = \mathrm{Re}\ z(w,\zeta,y,\eta)$, which proves the assertion (i). (ii) There are $t \in [0,1]$ and $\zeta \in \mathcal{C}_{w,\delta'/2}$ such that

$$z = \mathrm{Re}\ z(w,\zeta,y,\eta) + i(1-t)\mathrm{Im}\ z(w,\zeta,y,\eta).$$

It is obvious that

$$\hat\zeta - \zeta = {}^tU_0^{-1}\mathrm{Re}\ (\tilde\nabla_y\tilde U(w,y,\zeta) + \tilde\nabla_y\tilde T(w,y,\eta)).$$

Putting $\tilde z = \hat\zeta + {}^tU_0^{-1}\tilde\nabla_y\tilde U(w,y,\hat\zeta)$ ($\in \tilde\Gamma_{w,y,\delta}$), we have

$$\begin{aligned}\zeta^1 - \tilde z = {} & i(1-t){}^tU_0^{-1}\mathrm{Im}\ \tilde\nabla_y\tilde U(w,y,\zeta) - {}^tU_0^{-1}\tilde\nabla_y\tilde U(w,y,\hat\zeta) \\ & - {}^tU_0^{-1}(\mathrm{Re}\ \tilde\nabla_y\tilde T(w,y,\eta) + it\,\mathrm{Im}\ \tilde\nabla_y\tilde T(w,y,\eta)).\end{aligned} \tag{A.17}$$

(A.16), which is still valid in this case, gives

$$|\zeta - \hat\zeta| < \langle\hat\zeta\rangle/10, \quad \langle\zeta\rangle < 11\langle\hat\zeta\rangle/10. \tag{A.18}$$

Moreover, we have

$$|\tilde z - \hat\zeta| < \langle\hat\zeta\rangle/12, \quad \langle\hat\zeta\rangle < 12\langle\tilde z\rangle/11.$$

Therefore, (A.17), together with (2.46), yields

$$\begin{aligned}|\zeta^1 - \tilde{z}| &\leq 12|U_0^{-1}|\{c(U)(1+\delta'/2) + n^{3/2}C(\tilde{U})A_0^3\delta'^2 \\ &\qquad +2^4 c_2(U)c(T)/c_3(T)\}\langle\tilde{z}\rangle/11 \\ &\leq \delta'\langle\tilde{z}\rangle/8,\end{aligned}$$

i.e., $\zeta^1 \in \tilde{\mathcal{C}}_{w,y,\delta'}$, if (A.15) is valid. □

We assume that (A.13) and (A.15) are satisfied. Put

$$\begin{aligned}G_\kappa^{R',R}(\xi,w,y,\eta;z) &= \sum_{j=1}^{\infty} \chi_j^{\varepsilon_0}(w-y)\psi_j^R(\eta) \\ &\times f_\kappa^{R'}(\xi, w, {}^tU_0^{-1}\mathrm{Re}\ z - {}^tU_0^{-1}\,{}^tT_0\eta, y, \eta; Z(z,\eta,w,y) \\ &\qquad - {}^tU_0^{-1}\mathrm{Re}\ z + {}^tU_0^{-1}\,{}^tT_0\eta) \\ &\times \det \frac{\partial Z}{\partial z}(z,\eta,w,y),\end{aligned}$$

where

$$Z(z,\eta,w,y) = Z({}^tU_0^{-1}z - {}^tU_0^{-1}\tilde{\nabla}_y T(w,y,\eta); w, y).$$

By Lemma A.1. 5 , $\zeta \in \mathcal{C}_{w,\delta'/2}$ if $f_\kappa^{R'}(\xi, w, \hat{\zeta}(w,\zeta,y,\eta), y, \eta; u) \neq 0$. We can write

$$\begin{aligned}\tilde{p}_\kappa(\xi,w,\eta) = (2\pi)^{-n} \int \Big(\int_{\Omega_{w,y,\eta,0}} & e^{i(y-w)\cdot z} \\ &\times G_\kappa^{R',R}(\xi,w,y,\eta;z)\, dz_1 \wedge \cdots \wedge dz_n \Big)\, dy.\end{aligned}$$

Lemma A.1. 10 implies that Re $z \in \Omega_{w,y,\eta,1}$ if $G_\kappa^{R',R}(\xi,w,y,\eta;z) \neq 0$. Therefore, applying Stokes' formula we have

$$\begin{aligned}\tilde{p}_\kappa(\xi,w,\eta) &= (2\pi)^{-n} \int \Big(\int_{\Omega_{w,y,\eta,1}} e^{i(y-w)\cdot\zeta} \\ &\qquad \times G_\kappa^{R',R}(\xi,w,y,\eta;\zeta)\, d\zeta_1 \wedge \cdots \wedge d\zeta_n \Big)\, dy + q_\kappa^1(\xi,w,\eta), \\ &= (2\pi)^{-n} \int \Big(\int_{R^n} e^{i(y-w)\cdot\zeta} G_\kappa^{R',R}(\xi,w,y,\eta;\zeta) \\ &\qquad \times \mathrm{sgn}(\det U_0)\, d\zeta \Big)\, dy + q_\kappa^1(\xi,w,\eta),\end{aligned}$$

where

$$\begin{aligned}q_\kappa^1(\xi,w,\eta) = \pm(2\pi)^{-n} \int \Big(\int_{\Omega_{w,y,\eta}} & d_z \Big(e^{i(y-w)\cdot z} \\ &\times G_\kappa^{R',R}(\xi,w,y,\eta;z)\, dz_1 \wedge \cdots \wedge dz_n \Big)\Big)\, dy,\end{aligned}$$

d_z denotes the exterior differential with respect to z and "$\pm$" depends on the orientation of the chain $\Omega_{w,y,\eta}$. If $G_\kappa^{R',R}(\xi, w, y, \eta; z) \neq 0$ and $z \in \Omega_{w,y,\eta}$, then by Lemma A.1. 10 we have ${}^tU_0^{-1}z - {}^tU_0^{-1}\tilde{\nabla}_y T(w, y, \eta) \in \tilde{C}_{w,y,\delta'}$. This implies that $Z(z, \eta, w, y)$ is analytic with respect to z in a neighborhood of $\{z \in \Omega_{w,y,\eta};\ G_\kappa^{R',R}(\xi, w, y, \eta; z) \neq 0\}$.

Lemma A.1. 11 *Put*

$$F_\ell(\zeta, t; w, y, \eta) = \begin{pmatrix} {}^te_\ell + {}^te_\ell\,{}^tU_0^{-1}\dfrac{\partial(\mathrm{Re}\ \tilde{\nabla}_y\tilde{U})}{\partial\zeta} & 0 \\ I + {}^tU_0^{-1}\dfrac{\partial(\mathrm{Re}\ \tilde{\nabla}_y\tilde{U})}{\partial\zeta} + i(1-t)\,{}^tU_0^{-1}\dfrac{\partial(\mathrm{Im}\ \tilde{\nabla}_y\tilde{U})}{\partial\zeta} & -{}^tU_0^{-1}\mathrm{Im}\ (\tilde{\nabla}_y\tilde{U} + \tilde{\nabla}_y\tilde{T}) \end{pmatrix}$$

$$(\ 1 \le \ell \le n\ ),$$

$$\hat{z}(\zeta, t; w, y, \eta) = \mathrm{Re}\ z(w, \zeta, y, \eta) + i(1-t)\mathrm{Im}\ z(w, \zeta, y, \eta),$$

$$\zeta^1(\zeta, t; w, y, \eta) = \zeta + {}^tU_0^{-1}\tilde{\nabla}_y\tilde{U} - it\,{}^tU_0^{-1}\mathrm{Im}\ (\tilde{\nabla}_y\tilde{U} + \tilde{\nabla}_y\tilde{T})$$

for $\zeta \in C_{w,\delta'/2} \cap \boldsymbol{R}^n$, $y, w \in \boldsymbol{R}^n$ *with* $|w - y| \le \varepsilon_0$ *and* $\eta \in \boldsymbol{R}^{n''}$ *with* $(y, \eta) \in \Omega_1'$, *where* $\tilde{\nabla}_y\tilde{U} = \tilde{\nabla}_y\tilde{U}(w, y, \zeta)$ *and* $\tilde{\nabla}_y\tilde{T} = \tilde{\nabla}_y\tilde{T}(w, y, \eta)$. *Then we have*

$$\begin{aligned} q_\kappa^1(\xi, w, \eta) = (2\pi)^{-n} \int \Big(\int_0^1 \Big(\int_{C_{w,\delta'/2}\cap R^n} &\exp[iy \cdot \hat{z}(\zeta, t; w, w+y, \eta)] \\ \times \sum_{j=1}^\infty \chi_j^{\varepsilon_0}(-y)\psi_j^R(\eta) \det \frac{\partial Z}{\partial z}&(\zeta^1(\zeta, t; w, w+y, \eta); w, w+y) \\ \times \sum_{\ell=1}^n ((\partial_{\zeta_\ell} - \partial_{u_\ell}) f_\kappa^{R'})(\xi, w, &\hat{\zeta}(w, \zeta, w+y, \eta), w+y, \eta; \\ Z(\zeta^1(\zeta, t; w, w+y, \eta); w, w+y) &- \hat{\zeta}(w, \zeta, w+y, \eta)) \\ \times \det F_\ell(\zeta, t; w, w+y, \eta)\, \mathrm{sgn}(\det U_0)\, d\zeta \Big)\, dt \Big)\, dy, \end{aligned}$$

where

$$\hat{\zeta}(w, \zeta, y, \eta) = \zeta + {}^tU_0^{-1}\mathrm{Re}\ (\tilde{\nabla}_y\tilde{U}(w, y, \zeta) + \tilde{\nabla}_y\tilde{T}(w, y, \eta)).$$

Remark We note that

$$\begin{aligned} \hat{z}(\zeta, t; w, y, \eta) &= {}^tU_0^{-1}\zeta + \tilde{\nabla}_y\tilde{U}(w, y, \zeta) + \tilde{\nabla}_y T(w, y, \eta) \\ &\quad -it\ \mathrm{Im}\ (\tilde{\nabla}_y\tilde{U}(w, y, \zeta) + \tilde{\nabla}_y\tilde{T}(w, y, \eta)), \\ \hat{\zeta}(w, \zeta, y, \eta) &= {}^tU_0^{-1}\mathrm{Re}\ \hat{z}(\zeta, t; w, y) - {}^tU_0^{-1}\,{}^tT_0\eta, \\ \zeta^1(\zeta, t; w, y, \eta) &= {}^tU_0^{-1}\hat{z}(\zeta, t; w, y) - {}^tU_0^{-1}\tilde{\nabla}_y T(w, y, \eta). \end{aligned}$$

Proof Write $U_0^{-1} = (a_{k\ell})$. Then we have

$$
\begin{aligned}
& d_z\Big(e^{i(y-w)\cdot z} G_\kappa^{R',R}(\xi, w, y, \eta; z)\, dz_1 \wedge \cdots \wedge dz_n\Big) \\
&= e^{i(y-w)\cdot z} \sum_{j=1}^{\infty} \chi_j^{\varepsilon_0}(w-y)\psi_j^R(\eta) \\
&\quad \times \sum_{k,\ell=1}^{n} ((\partial_{\zeta_\ell} - \partial_{u_\ell}) f_\kappa^{R'})(\xi, w, {}^tU_0^{-1}(\mathrm{Re}\, z - {}^tT_0\eta), y, \eta; \\
&\qquad\qquad Z(z, \eta, w, y) - {}^tU_0^{-1}(\mathrm{Re}\, z - {}^tT_0\eta)) \\
&\quad \times \det \frac{\partial Z}{\partial z}(z, \eta, w, y) a_{k\ell}\, d(\mathrm{Re}\, z_k) \wedge dz_1 \wedge \cdots \wedge dz_n.
\end{aligned}
$$

Fix $w, y \in \boldsymbol{R}^n$ and $\eta \in \boldsymbol{R}^{n''}$ so that $(y, \eta) \in \Omega_1'$ and $|w - y| \le \varepsilon_0$. If $\zeta \in \mathcal{C}_{w,\delta'/2} \cap \boldsymbol{R}^n$, $t \in [0,1]$ and $z = \hat{z}(\zeta, t; w, y, \eta)$ $(\in \Omega_{w,y,\eta,t})$, then

$$
\begin{aligned}
& \sum_{k=1}^{n} a_{k\ell}\, d(\mathrm{Re}\, z_k) \wedge dz_1 \wedge \cdots \wedge dz_n \\
&= \det U_0 \cdot \det F_\ell(\zeta, t; w, y, \zeta)\, d\zeta_1 \wedge \cdots \wedge d\zeta_n \wedge dt.
\end{aligned}
$$

In fact,

$$
\begin{aligned}
& {}^t\Big(\sum_{k=1}^{n} a_{k\ell}\, d(\mathrm{Re}\, z_k), dz_1, \cdots, dz_n)\Big) \\
&= \begin{pmatrix} 1 & 0 \\ 0 & {}^tU_0 \end{pmatrix} F_\ell(\zeta, t; w, y, \zeta)\, {}^t(d\zeta_1, \cdots, d\zeta_n, dt).
\end{aligned}
$$

On the other hand,

$$
\begin{aligned}
& Z(\hat{z}(\zeta, t; w, y, \eta), \eta, w, y) = Z(\zeta^1(\zeta, t; w, y, \eta); w, y), \\
& \det \frac{\partial Z}{\partial z}(\hat{z}(\zeta, t; w, y, \eta), \eta, w, y) \\
&= (\det U_0)^{-1} \det \frac{\partial Z}{\partial z}(\zeta^1(\zeta, t; w, y, \eta); w, y).
\end{aligned}
$$

This proves the lemma. □

Lemma A.1. 12 *$q_\kappa^1(\xi, w, \eta)$ satisfies the estimates*

$$
\begin{aligned}
& \Big|\partial_\xi^\alpha \partial_\eta^\rho q_\kappa^1(\xi, w, \eta)\Big| \le C_{|\alpha|+|\rho|,\kappa,\varepsilon_0} \langle\xi\rangle^{m_1} \langle\eta\rangle^{m_2+m_3+1} \\
& \times \exp\Big[\Big(\delta_1 - \nu_1'(S, T, U)/(\kappa R_0)\Big)\langle\xi\rangle + \Big(\delta_3'(T, U, \delta_2, \delta_3) - 1/(9\kappa R_0)\Big)\langle\eta\rangle\Big]
\end{aligned}
$$

if

$$\begin{cases} \delta_2 + 11(c_1^+(U) + c_1^-(U))/10 \le c_3(U)/(2^4 \cdot 3^2 c_2(T)\kappa R_0), \\ R' \ge (\widehat{C}(S,T,U) + 2\widehat{C}(T,U))R_0/A_1, \\ \delta' \le c_3(U)/(39enA_1C_0(U)c_2(T)\kappa), \end{cases} \tag{A.19}$$

where $\nu_1'(S,T,U) = (36c_2(T)(16c_2(U)/c_3(U)+3))^{-1}c_3(S)$ *and* $\delta_3'(T,U,\delta_2,\delta_3) = \delta_3 + c_1(T) + c_1^-(T) - c_3(T)(\delta_2 + 11(c_1^+(U) + c_1^-(U))/10)_-/(8c_2(U))$. *Moreover,* $q^1_{\kappa\, S,T}(D_x, y, D_z)$ *maps continuously* $\mathcal{S}_{\varepsilon_2}(\boldsymbol{R}^{n''})$ *and* $\mathcal{S}_{-\varepsilon_2}'(\boldsymbol{R}^{n''})$ *to* $\mathcal{S}_{\varepsilon_1}(\boldsymbol{R}^{n'})$ *and* $\mathcal{S}_{-\varepsilon_1}'(\boldsymbol{R}^{n'})$*, respectively, if* (A.19) *is satisfied and*

$$\begin{aligned} &\delta_1 + c_1(S) + \varepsilon_1 \le \nu_1'(S,T,U)/(\kappa R_0), \\ &\delta_3'(T,U,\delta_2,\delta_3) + c_1(T) - \varepsilon_2 \le 1/(18\kappa R_0). \end{aligned}$$

Proof Assume that $\zeta \in \mathcal{C}_{w,\delta'/2} \cap \boldsymbol{R}^n$, $t \in [0,1]$, $(\xi, w, \hat{\zeta}(w,\zeta,w+y,\eta), w+y,\eta) \in \operatorname{supp} f^{R'}$ and $|y| \le \varepsilon_0$. Then, $\hat{\zeta}(w,\zeta,w+y,\eta) \in \Gamma_w$, $|\hat{\zeta}(w,\zeta,w+y,\eta)| \ge 1$ and $R'\langle\eta\rangle \le \kappa R_0\langle\hat{\zeta}\rangle$. Write $\hat{\zeta} = \hat{\zeta}(w,\zeta,w+y,\eta)$, and put

$$\begin{aligned} &z = \hat{z}(\zeta, t; w, w+y, \eta), \\ &\tilde{z} = \hat{\zeta} + {}^tU_0^{-1}\tilde{\nabla}_y\tilde{U}(w, w+y, \hat{\zeta}) \ (\in \tilde{\Gamma}_{w,w+y,\delta}). \end{aligned}$$

Then, $\zeta^1(\zeta, t; w, w+y, \eta) = {}^tU_0^{-1}z - {}^tU_0^{-1}\tilde{\nabla}_yT(w, w+y, \eta)$. We showed in the proof of Lemma A.1. 10 that

$$|\zeta^1(\zeta, t; w, w+y, \eta) - \tilde{z}| < \delta'\langle\tilde{z}\rangle/8.$$

We have also

$$|\tilde{z} - \hat{\zeta}| \le c(U)|U_0^{-1}|\langle\hat{\zeta}\rangle \le \langle\hat{\zeta}\rangle/12, \quad \langle\tilde{z}\rangle \le 13\langle\hat{\zeta}\rangle/12.$$

So we have

$$|\zeta^1(\zeta, t; w, w+y, \eta) - \tilde{z}| < 13\delta'\langle\hat{\zeta}\rangle/96 \le 13\delta' c_2(T)\langle\eta\rangle/(6c_3(U)).$$

Since $Z(\tilde{z}; w, w+y, \eta) = \hat{\zeta}$, (2.50) yields

$$\begin{aligned} &|Z(\zeta^1(\zeta, t; w, w+y, \eta); w, w+y) - \hat{\zeta}| \\ &\le \sqrt{n}C_0(U)|\zeta^1(\zeta, t; w, w+y, \eta) - \tilde{z}| \\ &\le 13\sqrt{n}\delta' C_0(U)c_2(T)\langle\eta\rangle/(6c_3(U)). \end{aligned}$$

Using (2.50) and (A.18), we have

$$\begin{aligned}
&\left|\partial_\eta^\rho(\partial_z^\alpha Z)(\zeta^1(\zeta,t;w,w+y,\eta);w,w+y)\right| \le C_{|\alpha|+|\rho|} \text{ if } |\alpha|+|\rho|\ge 1,\\
&\left|\partial_\eta^\rho(y\cdot\hat{z}(\zeta,t;w,w+y,\eta))\right| + \left|\partial_\eta^\rho\hat{\zeta}(w,\zeta,w+y,\eta)\right| \le C_{|\rho|}\langle\eta\rangle^{1-|\rho|},\\
&|\mathrm{Im}\ y\cdot\hat{z}(\zeta,t;w,w+y,\eta)|\\
&\le 11(c_1^+(U)+c_1^-(U))\langle\hat{\zeta}\rangle/10 + (c_1(T)+c_1^-(T))\langle\eta\rangle,\\
&\left|\partial_\eta^\rho \det F_\ell(\zeta,t;w,w+y,\eta)\right| \le C_{|\rho|}\langle\eta\rangle^{1-|\rho|}.
\end{aligned}$$

Therefore, (A.5) and Lemma A.1. 11 prove the first part of the lemma. The second part follows from Lemma A.1. 7 . □

Using Taylor's formula, we can write

$$\begin{aligned}
&\tilde{p}_\kappa(\xi,w,\eta) - q_\kappa^1(\xi,w,\eta)\\
&= (2\pi)^{-n}\lim_{\nu\downarrow 0}\sum_{j=1}^{\infty}\int e^{iy\cdot\zeta}\chi(\nu y)\sum_{|\gamma|<j}\left[(-\partial_\zeta)^\gamma D_y^\gamma G_{\kappa,j}^{R',R}(\xi,w,\zeta,y,\eta)\right]_{y=0}/\gamma!\\
&\qquad\qquad \times d\zeta dy\\
&+(2\pi)^{-n}\sum_{j=1}^{\infty}\int e^{iy\cdot\zeta}\sum_{|\gamma|=j}\langle y\rangle^{-2M}\langle D_\zeta\rangle^{2M}\\
&\times\Big(\int_0^1 j(1-\theta)^{j-1}\big((-\partial_\zeta)^\gamma D_y^\gamma G_{\kappa,j}^{R',R}\big)(\xi,w,\zeta,\theta y,\eta)\,d\theta\Big)/\gamma!\,d\zeta dy\\
&\equiv p_\kappa^1(\xi,w,\eta) + q_\kappa^2(\xi,w,\eta),
\end{aligned}$$

where $M \ge [n/2]+1$, $\chi(y) \in C_0^\infty(\boldsymbol{R}^n)$ satisfies $\chi(y)=1$ near 0,

$$\begin{aligned}
G_{\kappa,j}^{R',R}(\xi,w,\zeta,y,\eta) &= \chi_j^{\varepsilon_0}(-y)\psi_j^R(\eta)\\
&\times f_\kappa^{R'}(\xi,w,v(\zeta,\eta),w+y,\eta;Z(\zeta,\eta,w,w+y)-v(\zeta,\eta))\\
&\times|\det U_0|^{-1}\det\frac{\partial Z}{\partial z}(v(\zeta,\eta)-{}^tU_0^{-1}\tilde{\nabla}_y\tilde{T}(w,w+y,\eta);w,w+y),
\end{aligned}$$

for $(\xi,w,\zeta,y,\eta)\in\boldsymbol{R}^{n'}\times\boldsymbol{R}^n\times\boldsymbol{R}^n\times\boldsymbol{R}^n\times\boldsymbol{R}^{n''}$ and $v(\zeta,\eta)$ is defined by (2.63). Since

$$(2\pi)^{-n}\lim_{\nu\downarrow 0}\int e^{iy\cdot\zeta}\chi(\nu y)(-\partial_\zeta)^\gamma f(\zeta)\,d\zeta dy = ((-\partial_\zeta)^\gamma f)(0)$$

for $f\in C_0^\infty(\boldsymbol{R}^n)$, we have

$$p_\kappa^1(\xi,w,\eta) = \sum_{j=1}^{\infty}\sum_{|\gamma|<j}\left[(-\partial_\zeta)^\gamma D_y^\gamma G_{\kappa,j}^{R',R}(\xi,w,\zeta,y,\eta)\right]_{y=0,\,\zeta=0}/\gamma!.$$

Lemma A.1. 13 *Let $j \in \boldsymbol{N}$ and $(\xi, w, \zeta, y, \eta) \in \operatorname{supp} G^{R',R}_{\kappa,j}$. Then, for $|\alpha|+|\beta|+|\beta'|+|\gamma|+|\lambda|+|\lambda'|+|\rho| \ge 1$*

$$\begin{aligned}
&\Big|D_w^\beta \partial_\zeta^\gamma D_y^\lambda \partial_\eta^\rho \Big(D_w^{\beta'} D_y^{\lambda'} \partial_z^\alpha Z\Big)({}^tU_0^{-1}\zeta - {}^tU_0^{-1}\widetilde{\nabla}_y T(w, w+y, \eta); w, w+y)\Big| \\
&\le C_0(T,U) A_0'(T,U)^{|\beta|+|\lambda|+|\rho|} (\sqrt{n}|U_0^{-1}| A_0(T,U))^{|\gamma|} \\
&\quad \times A_0(T,U)^{|\alpha|} (2A_0(U)))^{|\beta'|+|\lambda'|} \\
&\quad \times (|\alpha|+|\beta|+|\beta'|+|\gamma|+|\lambda|+|\lambda'|+|\rho|)! \langle\eta\rangle^{1-|\alpha|-|\gamma|-|\rho|},
\end{aligned}$$

where

$$\begin{aligned}
A_0(T,U) &= 2^5 c_2(U) A_0(U)/c_3(T), \\
A_0'(T,U) &= \max\{2A_0, 2A_0(U) + 2^5 n C(T) A_0^2 A_0(T,U)|U_0^{-1}|\}, \\
C_0(T,U) &= 24 c_2(T) C_0(U)/(c_3(U) A_0(U))
\end{aligned}$$

and $A_0(U)$ is the constant in Lemma 2.4. 2 , *modifying δ' if necessary.*

Proof It follows from Lemma A.1. 10 (or its proof) that

$$\begin{aligned}
&\hat{\zeta} \equiv {}^tU_0^{-1}\zeta - {}^tU_0^{-1}\,{}^tT_0\eta \in \Gamma_w, \quad |\hat{\zeta}| \ge 1, \\
&\zeta^1 \equiv {}^tU_0^{-1} - {}^tU_0^{-1}\widetilde{\nabla}_y T(w, w+y, \eta) \in \widetilde{\mathcal{C}}_{w,w+y,\delta'}, \\
&\zeta = {}^tU_0\hat{\zeta} + {}^tT_0\eta \in \Omega_{w,w+y,\eta,1}, \\
&c_3(T)|\eta|/(8c_2(U)) \le |\hat{\zeta}| \le 8c_2(T)|\eta|/c_3(U), \quad |\eta| \ge 1, \\
&|\zeta^1 - \tilde{z}| < \delta'\langle\tilde{z}\rangle/8, \quad Z(\tilde{z}; w, w+y) = \hat{\zeta},
\end{aligned}$$

where $\tilde{z} = \hat{\zeta} + {}^tU_0^{-1}\widetilde{\nabla}_y \widetilde{U}(w, w+y, \hat{\zeta})$ ($\in \widetilde{\Gamma}_{w,w+y,\delta}$). We also showed in the proof of Lemma A.1. 12 that

$$|Z(\zeta^1; w, w+y) - \hat{\zeta}| \le 13\sqrt{n}\delta' C_0(U)\langle\hat{\zeta}\rangle/96. \tag{A.20}$$

So, taking $\delta' \le 48/(13\sqrt{n}C_0(U))$, we have

$$c_3(T)\langle\eta\rangle/(2^5 c_2(U)) \le \langle Z(\zeta^1; w, w+y)\rangle \le 24 c_2(T)\langle\eta\rangle/c_3(U).$$

By Lemma 2.4. 2 we have

$$\begin{aligned}
&\Big|D_w^\beta \partial_\zeta^\gamma D_y^\lambda \partial_z^\alpha Z({}^tU_0^{-1}\zeta - z; w, w+y)\Big| \\
&\le 24 c_2(T) c_3(U)^{-1} A_0(U)^{-1} C_0(U) (\sqrt{n}|U_0^{-1}| A_0(T,U))^{|\gamma|} \\
&\quad \times (2A_0(U))^{|\beta|+|\lambda|} A_0(T,U)^{|\alpha|} (|\alpha|+|\beta|+|\gamma|+|\lambda|)! \langle\eta\rangle^{1-|\alpha|-|\gamma|}
\end{aligned}$$

if $|\alpha|+|\beta|+|\gamma|+|\lambda| \ge 1$. Applying the same argument as for Lemma 2.1. 6 we can prove the lemma (by induction on $|\beta|+|\lambda|+|\rho|$). □

Lemma A.1. 14 (i) *The estimates*

$$\begin{aligned}
&\left|\partial_\xi^\alpha \partial_\zeta^{\gamma+\tilde\gamma} D_y^\beta \partial_\eta^\rho G_{\kappa,j}^{R',R}(\xi,w,\zeta,y,\eta)\right| \\
&\le C_{|\alpha|+|\tilde\gamma|+|\rho|,\kappa,R_0,R',R,\varepsilon_0}(S,T,U,p) \\
&\quad \times (A_1'(S,T,U,A_1,R_0/R',R_0/R)/R_0)^{|\gamma|} \\
&\quad \times (jB_1'(S,T,U,A_1,R_0/R',R/R_0,\varepsilon_0))^{|\beta|} \\
&\quad \times \langle\xi\rangle^{m_1-|\alpha|}\langle\eta\rangle^{m_2+m_3-|\tilde\gamma|-|\rho|+1} \\
&\quad \times \exp[\delta_1\langle\xi\rangle + (\tilde\delta_3(T,U,\delta_2,\delta_3) \\
&\qquad\qquad +\delta'\nu'(S,T,U,A_1,R_0/R')/R_0)\langle\eta\rangle]
\end{aligned} \tag{A.21}$$

hold for $j \in \boldsymbol{N}$ *if* $|\gamma| \le j$, $|\beta| \le j$ *and*

$$R \ge \max\{32c_2(U)/c_3(T), 3\kappa, 1\}R' \ (\ge R_0), \tag{A.22}$$

where

$$\begin{aligned}
&A_1'(S,T,U,A_1,R_0/R',R_0/R) \\
&= \max\{12\sqrt{n}|U_0^{-1}|A_0(T,U)R_0/R, 4\sqrt{n}|U_0^{-1}|A_1(S,T,U,A_1,R_0/R'), \\
&\qquad 2^6\cdot 3n|U_0^{-1}|C_0(T,U)A_0(T,U)A_1(S,T,U,A_1,R_0/R')\} \\
&\quad +6\sqrt{n}|U_0^{-1}|A_0(T,U)R_0/R, \\
&B_1'(S,T,U,A_1,R_0/R',R/R_0,\varepsilon_0) \\
&= \max\{12A_0'(T,U), 2(3A_1R/R_0 + C_*/\varepsilon_0), \\
&\qquad 2^6\cdot 3^2\sqrt{n}C_0(T,U)A_0'(T,U)A_1(S,T,U,A_1,R_0/R')R/R_0\}, \\
&\quad +6A_0'(T,U), \\
&\nu'(S,T,U,A_1,R_0/R') \\
&= 13nC_0(U)c_2(T)A_1(S,T,U,A_1,R_0/R')/(12c_3(U)), \\
&\tilde\delta_3(T,U,\delta_2,\delta_3) = \delta_3 + 8(\delta_2)_+c_2(T)/c_3(U) - (\delta_2)_-c_3(T)/(8c_2(U)).
\end{aligned}$$

(ii) $q_\kappa^2(\xi,w,\eta)$ *satisfies the estimates*

$$\begin{aligned}
\left|\partial_\xi^\alpha\partial_\eta^\rho q_\kappa^2(\xi,w,\eta)\right| &\le C_{|\alpha|+|\rho|,\kappa,R_0,R',R,\varepsilon_0}(S,T,U,p) \\
&\quad\times \exp[(\delta_1 - \tilde\nu_1(S,T,U)/R))\langle\xi\rangle \\
&\qquad +(\tilde\delta_3(T,U,\delta_2,\delta_3) - 1/(6R))\langle\eta\rangle]
\end{aligned} \tag{A.23}$$

if (A.22) *is satisfied and*

$$\begin{aligned}
R_0 &\ge 2e^2\sqrt{n}A_1'(S,T,U,A_1,R_0/R',R_0/R) \\
&\qquad \times B_1'(S,T,U,A_1,R_0/R',R/R_0,\varepsilon_0),
\end{aligned} \tag{A.24}$$

$$\delta' \le R_0/(12R\nu'(S,T,U,A_1,R_0/R')) \tag{A.25}$$

where

$$\tilde{\nu}_1'(S,T,U) = c_3(S)/(24c_2(T)(2^4c_2(U)/c_3(U)+3)).$$

(iii) $q^2_{\kappa\,S,T}(D_x, y, D_z)$ *maps continuously* $\mathcal{S}_{\varepsilon_2}(\boldsymbol{R}^{n''})$ *and* $\mathcal{S}_{-\varepsilon_2}{}'(\boldsymbol{R}^{n''})$ *to* $\mathcal{S}_{\varepsilon_1}(\boldsymbol{R}^{n'})$ *and* $\mathcal{S}_{-\varepsilon_1}{}'(\boldsymbol{R}^{n'})$, *respectively, if* (A.22), (A.24) *and* (A.25) *are satisfied and*

$$\delta_1 + c_1(S) + \varepsilon_1 \le \tilde{\nu}_1(S,T,U)/R,$$
$$\tilde{\delta}_3(T,U,\delta_2,\delta_3) + c_1(T) - \varepsilon_2 \le 1/(12R).$$

Proof It follows from Lemma A.1. 6 that

$$\begin{aligned}
&\Big|\partial_\xi^\alpha \partial_\zeta^{\gamma+\tilde{\gamma}} D_y^\beta \partial_\eta^\rho \partial_u^{\delta+\tilde{\delta}}\Big\{\chi_j^{\varepsilon_0}(-y)\psi_j^R(\eta) \\
&\quad \times f_\kappa^{R'}(\xi, w, v(\zeta,\eta), w+y, \eta; u - v(\zeta,\eta))\Big\}\Big|_{u=Z(\zeta,\eta,w,w+y)} \\
&\le C_{|\alpha|+|\tilde{\gamma}|+|\rho|+|\tilde{\delta}|,\kappa,R_0,R',\varepsilon_0}(S,T,U,p)(3A_1 jR/R_0 + C_* j/\varepsilon_0)^{|\beta|} \\
&\quad \times (2\sqrt{n}|U_0^{-1}|A_1(S,T,U,A_1,R_0/R')/R_0)^{|\gamma|} \\
&\quad \times (A_1(S,T,U,A_1,R_0/R')/R_0)^{|\delta|} \\
&\quad \times \langle\xi\rangle^{m_1-|\alpha|}\langle v(\zeta,\eta)\rangle^{m_2-|\tilde{\gamma}|-|\tilde{\delta}|}\langle\eta\rangle^{m_3-|\rho|+1} \\
&\quad \times \exp[\delta_1\langle\xi\rangle + \delta_2\langle v(\zeta,\eta)\rangle + (\delta_3 + \delta'\nu'(S,T,U,A_1,R_0/R')/R_0)\langle\eta\rangle]
\end{aligned}$$

if $|\beta|, |\gamma| \le j$, and (A.22) is valid. In fact, $1 \le |\eta| \le 8c_2(U)|v(\zeta,\eta)|/c_3(T)$, and $2R(j-1) \le \langle\eta\rangle \le 3Rj$ if $f_\kappa^{R'}(\xi, w, v(\zeta,\eta), w+y, \eta; u - v(\zeta,\eta)) \ne 0$ and $\psi_j^R(\eta) \ne 0$. Moreover, (A.20) yields

$$\begin{aligned}
&|Z(\zeta,\eta,w,w+y) - v(\zeta,\eta)| \\
&\le 13\sqrt{n}\delta' C_0(U)c_2(T)|\eta|/(12c_3(U)) + 13\sqrt{n}\delta' C_0(U)/16 \qquad \text{(A.26)}
\end{aligned}$$

if $f_\kappa^{R'}(\xi, w, v(\zeta,\eta), w+y, \eta; u-v(\zeta,\eta)) \ne 0$. On the other hand, by Lemma A.1. 13 we have

$$\begin{aligned}
&\Big|\partial_\zeta^{\gamma+\tilde{\gamma}} D_y^\beta \partial_\eta^\rho \det\frac{\partial Z}{\partial z}(v(\zeta,\eta) - {}^tU_0^{-1}\tilde{\nabla}_y\tilde{T}(w, w+y, \eta); w, w+y)\Big| \\
&\le C_{|\tilde{\gamma}|+|\rho|}(T,U)(6A_0'(T,U))^{|\beta|}(6\sqrt{n}|U_0^{-1}|A_0(T,U))^{|\gamma|} \\
&\quad \times |\beta|!|\gamma|!\langle\eta\rangle^{-|\gamma|-|\tilde{\gamma}|-|\rho|}
\end{aligned}$$

if $G_\kappa^{R',R}(\xi, w, \zeta, y, \eta) \ne 0$. Since $|\beta|! \le j^{|\beta|}$ and $|\gamma|!\langle\eta\rangle^{-|\gamma|} \le R^{-|\gamma|}$ if $j \ge 2$, $|\beta|, |\gamma| \le j$ and $\psi_j^R(\eta) \ne 0$, (A.21) easily follows from Lemmas 2.1. 6 and A.1. 13 . Since $1 \le e^j \exp[-\langle\eta\rangle/(3R)]$ and

$$|\xi| \le 2(2^4c_2(U)/c_3(U)+3)c_2(T)|\eta|/c_3(S)$$

if $G^{R',R}_{\kappa,j}(\xi,w,\zeta,y,\eta) \neq 0$, (A.23) also easily follows from (A.21). Lemma A.1.7 proves the assertion (iii) by virtue of (A.23). □

If $v(0,\eta) \in \Gamma_w$, $(w,\eta) \in \Omega_1'$, $|v(0,\eta)| \geq 1$ and $|\eta| \geq 1$, then

$$c_3(T)|\eta|/(4c_2(U)) \leq |v(0,\eta)| \leq 4c_2(T)|\eta|/c_3(U). \tag{A.27}$$

In fact, ${}^tU_0 v(0,\eta) = {}^tT_0\eta$ and

$$(1-\sqrt{2}/4)c_3(T)|\eta| \leq |{}^tT_0\eta| \leq 5c_2(T)|\eta|/(2\sqrt{2}),$$
$$(1-\sqrt{2}/4)c_3(U)|v(0,\eta)| \leq |{}^tU_0 v(0,\eta)| \leq 5c_2(U)|v(0,\eta)|/(2\sqrt{2}).$$

(A.27) implies that

$$\begin{aligned}(-\partial_\zeta)^\gamma D_y^\gamma G^{R',R}_{\kappa,j}(\xi,w,\zeta,y,\eta)\Big|_{y=0,\,\zeta=0} &= \psi_j^R(\eta)\\ \times(-\partial_\zeta)^\gamma D_y^\gamma\Big\{g^{R'}_\kappa(\xi,w,v(\zeta,\eta),w+y,\eta;Z(\zeta,\eta,w,w+y)-v(\zeta,\eta))\\ \times(\det U_0)^{-1}\det\frac{\partial Z}{\partial z}(v(\zeta,\eta)-{}^tU_0^{-1}\widetilde{\nabla}_y\widetilde{T}(w,w+y,\eta);\\ w,w+y)\Big\}\Big|_{y=0,\,\zeta=0},\end{aligned}$$

where

$$\begin{aligned}g^{R'}_\kappa(\xi,w,\zeta,y,\eta;u) = \sum_{j=1}^\infty \psi_j^{\kappa R_0}(\eta)\sum_{|\alpha|\leq j-1} u^\alpha\\ \times\partial_\zeta^\alpha\{(1-g_1^{R'}(\xi,\zeta,\eta))p(\xi,w,\zeta,y,\eta)\}/\alpha!\end{aligned}$$

Since $g_1^{R'}(\xi,\zeta,\eta)=0$ if $|\xi| \leq 3c_2(T)|\eta|/c_3(S)$ and $|\xi|^2+|\eta|^2 \geq 4$, we have

$$p_\kappa(\xi,w,\eta) = \tilde{g}^R(\xi,\eta)p^1_\kappa(\xi,w,\eta).$$

Lemma A.1.15 (i) *There are positive constants $C_{j,\kappa}(S,T,U,p)$ ($j \in Z_+$), $\widetilde{A}_1 \equiv \widetilde{A}_1(S,T,U,A_1,R_0/R')$, $\widetilde{A}_2 \equiv \widetilde{A}_2(S,T,U,A_1,R_0/R',\kappa)$, $\widetilde{A}_3 \equiv \widetilde{A}_3(S,T,U,A_1,R_0/R',R_0/R,\kappa)$, $\widetilde{B}_1 \equiv \widetilde{B}_1(S,T,U,A_1,R_0/R')$ and $\widetilde{B}_2 \equiv \widetilde{B}_2(S,T,U,A_1,R_0/R',R_0/R,\kappa)$ such that*

$$\begin{aligned}\left|\partial_\xi^{\alpha+\tilde\alpha}D_w^{\beta+\tilde\beta}\partial_\eta^{\rho+\tilde\rho}p^1_\kappa(\xi,w,\eta)\right| &\leq C_{|\tilde\alpha|+|\tilde\beta|+|\tilde\rho|,\kappa}(S,T,U,p)\\ &\times(\widetilde{A}_1/R_0)^{|\alpha|}(\widetilde{A}_2/R_0)^{|\beta|}(\widetilde{A}_3/R_0)^{|\rho|}\\ &\times\langle\xi\rangle^{m_1-|\tilde\alpha|}\langle\eta\rangle^{m_2+m_3-|\tilde\rho|+|\beta|+2}\\ &\times\exp[\delta_1\langle\xi\rangle+(\tilde\delta(T,U,\delta_2,\delta_3)+\delta'\widetilde{B}_1/R_0+\widetilde{B}_2/R_0^2)\langle\eta\rangle]\end{aligned}\tag{A.28}$$

if $\langle\xi\rangle \geq R'|\alpha|$, $\langle\eta\rangle \geq 2\tilde{\kappa}(T,U,\kappa)R'(|\beta|+|\rho|)$, $\langle\eta\rangle \geq 3R|\rho|$ *and*

$$R \geq \tilde{\kappa}(T,U,\kappa)R', \tag{A.29}$$

where

$$\tilde{\delta}(T,U,\delta_2,\delta_3) = 4c_2(T)(\delta_2)_+/c_3(U) - c_3(T)(\delta_2)_-/(4c_2(U)) + \delta_3,$$
$$\tilde{\kappa}(T,U,\kappa) = \max\{2^4c_2(U)/c_3(T), 1, 3\kappa\}$$

(ii) *There are positive constants* $C_{j,\kappa}(S,T,U,p)$ ($j \in \boldsymbol{Z}_+$), $A \equiv A(S,T,U,A_1,R_0/R,\kappa)$, $B_1 \equiv B_1(T,U,A_1)$ *and* $B_2 \equiv B_2(T,U,A_1,R_0/R,\kappa)$ *such that*

$$\begin{aligned}&\left|\partial_\xi^{\alpha+\tilde{\alpha}}D_w^{\beta+\tilde{\beta}}\partial_\eta^{\rho+\tilde{\rho}}p_\kappa(\xi,w,\eta)\right| \leq C_{|\tilde{\alpha}|+|\tilde{\beta}|+|\tilde{\rho}|,\kappa}(S,T,U,p)\\ &\quad\times(A/R_0)^{|\alpha|+|\beta|+|\rho|}\langle\xi\rangle^{m_1-|\tilde{\alpha}|}\langle\eta\rangle^{m_2+m_3-|\tilde{\rho}|+|\tilde{\beta}|+2}\\ &\quad\times\exp[\delta_1\langle\xi\rangle+(\tilde{\delta}(T,U,\delta_2,\delta_3)+\delta' B_1/R_0+B_2/R_0^2)\langle\eta\rangle]\end{aligned}$$

if $\langle\xi\rangle \geq R|\alpha|$, $\langle\eta\rangle \geq 2\tilde{\kappa}(T,U,\kappa)R_0(|\beta|+|\rho|)$, $\langle\eta\rangle \geq 3R|\rho|$ *and* $R \geq \tilde{\kappa}(T,U,\kappa)R_0$. (iii) *Put* $q_\kappa^3(\xi,w,\eta) = (1-\tilde{g}^R(\xi,\eta))p_\kappa^1(\xi,w,\eta)$. *Then* $q^3_{\kappa\,S,T}(D_x,y,D_z)$ *maps* $\mathcal{S}_{\varepsilon_2}(\boldsymbol{R}^{n''})$ *and* $\mathcal{S}_{-\varepsilon_2}{}'(\boldsymbol{R}^{n''})$ *to* $\mathcal{S}_{\varepsilon_1}(\boldsymbol{R}^{n'})$ *and* $\mathcal{S}_{-\varepsilon_1}{}'(\boldsymbol{R}^{n'})$, *respectively, if* $R_0 \geq R_0(S,T,U,A_1)$, $R' \geq R'(S,T,U,\kappa)$, $R \geq \tilde{\kappa}(T,U,\kappa)R'$ *and*

$$\begin{aligned}&\max\{\delta_1+c_1(S)+\varepsilon_1, \tilde{\delta}(T,U,\delta_2,\delta_3)+\delta'\widetilde{B}_1/R_0+\widetilde{B}_2/R_0^2+c_1(T)-\varepsilon_2\}\\ &\leq \delta(S,T,U,\kappa)/R',\end{aligned}$$

where $R_0(S,T,U,A_1)$, $R'(S,T,U,\kappa)$ *and* $\delta(S,T,U,\kappa)$ *are positive constants.*

Proof (i) Let us prove the assertion (i), applying the same argument as in the proof of Lemma A.1. 14 . First we have, in the same way as in Lemma A.1. 6 ,

$$\begin{aligned}&\left|\partial_\xi^{\alpha+\tilde{\alpha}}D_w^{\beta+\tilde{\beta}}\partial_\zeta^\gamma D_y^\lambda\partial_\eta^{\rho+\tilde{\rho}}\partial_u^{\delta+\tilde{\delta}}\left\{\psi_j^R(\eta)g_\kappa^{R'}(\xi,w,\zeta,y,\eta;u)\right\}\right|\\ &\leq C_{|\tilde{\alpha}|+|\tilde{\beta}|+|\tilde{\rho}|+|\tilde{\delta}|,\kappa}(\widetilde{A}_1/R_0)^{|\alpha|+|\gamma|+|\delta|}(A_1/R_0)^{|\beta|+|\lambda|}\\ &\quad\times((\widehat{C}/\kappa+\widetilde{A}_1)/R_0+\widehat{C}/R)^{|\rho|}\langle\xi\rangle^{m_1-|\tilde{\alpha}|}\langle\zeta\rangle^{m_2-|\tilde{\delta}|+|\tilde{\beta}|}\langle\eta\rangle^{m_3-|\tilde{\rho}|+|\lambda|+1}\\ &\quad\times\exp[\delta_1\langle\xi\rangle+\delta_2\langle\zeta\rangle+\delta_3\langle\eta\rangle+\sqrt{n}\widetilde{A}_1\langle u\rangle/R_0]\end{aligned}$$

if $R'\langle\eta\rangle \leq \kappa R_0\langle\zeta\rangle$, $|w-y| \leq \varepsilon_0$, $\langle\xi\rangle \geq R'|\alpha|$, $\langle\zeta\rangle \geq 2R'(|\beta|+|\gamma|)$, $\langle\eta\rangle \geq \max\{1,3\kappa\}R'(|\lambda|+|\rho|)$ and $\langle\eta\rangle \geq 3R|\rho|$, where

$$\widetilde{A}_1 \equiv \widetilde{A}_1(S,T,U,A_1,R_0/R') = \widehat{C}(S,T,U)R_0/R' + A_1.$$

This gives

$$\Big|\partial_\xi^{\alpha+\tilde\alpha} D_w^{\beta+\tilde\beta}\partial_\zeta^\gamma D_y^\lambda \partial_\eta^{\rho+\tilde\rho}\partial_u^{\delta+\tilde\delta}\Big\{\psi_j^R(\eta)$$
$$g_\kappa^{R'}(\xi,w,v(\zeta,\eta),w+y,\eta;u-v(\zeta,\eta))\Big\}\Big|_{u=Z(\zeta,\eta,w,w+y)}$$
$$\le C'_{|\tilde\alpha|+|\tilde\beta|+|\tilde\rho|+|\tilde\delta|,\kappa}(\widetilde A_1/R_0)^{|\alpha|+|\delta|}(2A_1/R_0)^{|\beta|}(A_1/R_0)^{|\lambda|}$$
$$\times(2\sqrt{n}|U_0^{-1}|\widetilde A_1/R_0)^{|\gamma|}((\widehat C/\kappa+(2\sqrt{n}|U_0^{-1}||T_0|+1)\widetilde A_1)/R_0+\widehat C/R)^{|\rho|}$$
$$\times\langle\xi\rangle^{m_1-|\tilde\alpha|}\langle v(\zeta,\eta)\rangle^{m_2-|\tilde\delta|+|\beta|}\langle\eta\rangle^{m_3-|\tilde\rho|+|\lambda|+1}$$
$$\times\exp[\delta_1\langle\xi\rangle+\delta_2\langle v(\zeta,\eta)\rangle+\delta_3\langle\eta\rangle$$
$$+\sqrt{n}\widetilde A_1\langle Z(\zeta,\eta,w,w+y)-v(\zeta,\eta)\rangle/R_0]$$

if $R'\langle\eta\rangle\le\kappa R_0\langle v(\zeta,\eta)\rangle$, $|y|\le\varepsilon_0$, $\langle\xi\rangle\ge R'|\alpha|$, $\langle v(\zeta,\eta)\rangle\ge 2R'(|\beta|+|\gamma|+|\rho|)$, $\langle\eta\rangle\ge\max\{1,3\kappa\}R'(|\lambda|+|\rho|)$ and $\langle\eta\rangle\ge 3R|\rho|$. Therefore, if follows from Lemmas 2.1. 6 and A.1. 13 , (A.26) and (A.27) that

$$\Big|\partial_\xi^{\alpha+\tilde\alpha}D_w^{\beta+\tilde\beta}\partial_\zeta^\gamma D_y^\lambda\partial_\eta^{\rho+\tilde\rho}\Big\{\psi_j^R(\eta)g_\kappa^{R'}(\xi,w,\zeta,y,\eta;$$
$$Z(\zeta,\eta,w,w+y)-v(\zeta,\eta))\Big\}\Big|_{y=0,\,\zeta=0}$$
$$\le C_{|\tilde\alpha|+|\tilde\beta|+|\tilde\rho|,\kappa}(\widetilde A_1/R_0)^{|\alpha|}(\widetilde A_2'/R_0)^{|\beta|}(\widetilde A_0/R_0)^{|\gamma|}(\widetilde A_0'/R_0)^{|\lambda|}$$
$$\times(\widetilde A_3'/R_0)^{|\rho|}\langle\xi\rangle^{m_1-|\tilde\alpha|}\langle\eta\rangle^{m_2+m_3-|\tilde\rho|+|\beta|+|\lambda|+1}$$
$$\times\exp[\delta_1\langle\xi\rangle+(\tilde\delta(T,U,\delta_2\,\delta_3)+\delta'\widetilde B_1/R_0)\langle\eta\rangle] \tag{A.30}$$

if

$$\begin{cases} R'\langle\eta\rangle\le\kappa R_0\langle v(0,\eta)\rangle, \quad \langle\xi\rangle\ge R'|\alpha|,\\ \langle v(0,\eta)\rangle\ge 2R'(|\beta|+|\gamma|+|\rho|),\\ \langle\eta\rangle\ge\max\{1,3\kappa\}R'(|\lambda|+|\rho|), \quad \langle\eta\rangle\ge 3R|\rho|,\end{cases} \tag{A.31}$$

where

$$\widetilde A_2'\equiv\widetilde A_2'(S,T,U,A_1,R_0/R')=\max\{2^6c_2(T)A_0'(T,U)R_0/(c_3(U)R'),$$
$$2^5c_2(T)A_1/c_3(U),2^{10}\sqrt{n}C_0(T,U)\widetilde A_1A_0'(T,U)\},$$
$$\widetilde A_0\equiv\widetilde A_0(S,T,U,A_1,R_0/R')$$
$$=\max\{2^6\sqrt{n}c_2(T)|U_0^{-1}|A_0(T,U)R_0/(c_3(U)R'),$$
$$4\sqrt{n}|U_0^{-1}|\widetilde A_1,2^{10}nC_0(T,U)|U_0^{-1}|\widetilde A_1A_0(T,U)\},$$
$$\widetilde A_0'\equiv\widetilde A_0'(S,T,U,A_1,R_0/R',\kappa)=\max\{2^4A_0'(T,U)R_0/(\max\{1,3\kappa\}R'),$$
$$2A_1,2^{10}\sqrt{n}C_0(T,U)\widetilde A_1A_0'(T,U)\},$$
$$\widetilde A_3'\equiv\widetilde A_3'(S,T,U,A_1,R_0/R',R_0/R,\kappa)=\max\{2^4A_0'(T,U)R_0/R,$$
$$2\widehat C/\kappa+(4\sqrt{n}|U_0^{-1}||T_0|+2)\widetilde A_1+\widehat CR_0/R,$$

$$2^{10}\sqrt{n}C_0(T,U)\tilde{A}_1 A_0'(T,U)\},$$
$$\tilde{B}_1 \equiv \tilde{B}_1(S,T,U,A_1,R_0/R') = 13nc_2(T)C_0(U)\tilde{A}_1/(12c_3(U)).$$

Assume that (A.29) is satisfied. Since κ satisfies (A.11) and $2R(j-1) \le \langle\eta\rangle \le 3Rj$ when $\psi_j^R(\eta) \ne 0$, (A.31) is valid if $|\gamma|, |\lambda| \le j-1$, $\psi_j^R(\eta) g_\kappa^{R'}(\xi, w, v(0,\eta), w, \eta; u) \ne 0$, $\langle\xi\rangle \ge R'|\alpha|$, $\langle\eta\rangle \ge 2\tilde{\kappa}(T,U,\kappa)R'(|\beta| + |\rho|)$ and $\langle\eta\rangle \ge 3R|\rho|$. On the other hand, Lemma A.1. 13 yields

$$\begin{aligned}
&\Big|D_w^{\beta+\beta+\tilde{\beta}} \partial_\zeta^\gamma D_y^\lambda \partial_\eta^{\rho+\tilde{\rho}} \det \frac{\partial Z}{\partial z}(v(\zeta,\eta) - {}^tU_0^{-1}\tilde{\nabla}_y \tilde{T}(w, w+y, \eta); \\
&\qquad\qquad\qquad\qquad w, w+y)\Big|_{y=0,\,\zeta=0} \\
&\le C_{|\tilde{\beta}|+|\tilde{\rho}|}(T,U)(10A_0'(T,U))^{|\beta|+|\lambda|+|\rho|} \\
&\quad \times (10\sqrt{n}|U_0^{-1}|A_0(T,U))^{|\gamma|}|\beta|!|\gamma|!|\lambda|!|\rho|! \\
&\quad \times \langle\eta\rangle^{-|\gamma|-|\rho|-|\tilde{\rho}|}
\end{aligned} \tag{A.32}$$

if $g_\kappa^{R'}(\xi, w, v(0,\eta), w, \eta; u) \ne 0$. By (A.30) and (A.32) we have

$$\begin{aligned}
&\Big|\partial_\xi^{\alpha+\tilde{\alpha}} D_w^{\beta+\tilde{\beta}} \partial_\eta^{\rho+\tilde{\rho}} \Big[(-\partial_\zeta)^\gamma D_y^\gamma \Big\{\psi_j^R(\eta) G_{\kappa,j}^{R',R}(\xi, w, \zeta, y, \eta)\Big\}\Big]\Big|_{y=0,\,\zeta=0} \\
&\le C_{|\tilde{\alpha}|+|\tilde{\beta}|+|\tilde{\rho}|,\kappa}(S,T,U,p)(\tilde{A}_1/R_0)^{|\alpha|}(\tilde{A}_2/R_0)^{|\beta|}(\tilde{A}_3/R_0)^{|\rho|} \\
&\quad \times \langle\xi\rangle^{m_1-|\tilde{\alpha}|}\langle\eta\rangle^{m_2+m_3-|\tilde{\rho}|+|\beta|+1} \\
&\quad \times \exp[\delta_1\langle\xi\rangle + (\tilde{\delta}(T,U,\delta_2\,\delta_3) + \delta'\tilde{B}_1/R_0)\langle\eta\rangle] \\
&\quad \times \chi_{F_j}(\eta)(\tilde{B}_2/(\sqrt{n}R_0^2))^{|\gamma|}\langle\eta\rangle^{|\gamma|},
\end{aligned}$$

if $\langle\xi\rangle \ge R'|\alpha|, \langle\eta\rangle \ge 2\tilde{\kappa}(T,U,\kappa)R'(|\beta| + |\rho|)$, $\langle\eta\rangle \ge 3R|\rho|$ and $|\gamma| \le j-1$, where $F_j = \{\eta \in \boldsymbol{R}^{n''};\ 2R(j-1) \le \langle\eta\rangle \le 3Rj\}$, χ_F denotes the characteristic function of F and

$$\begin{aligned}
\tilde{A}_2 &\equiv \tilde{A}_2(S,T,U,A_1,R_0/R',\kappa) \\
&= \tilde{A}_2' + 5A_0'(T,U)R_0/(\tilde{\kappa}(T,U,\kappa)R'), \\
\tilde{A}_3 &\equiv \tilde{A}_3(S,T,U,A_1,R_0/R',R_0/R,\kappa) \\
&= \tilde{A}_3' + 10A_0'(T,U)R_0/(3R), \\
\tilde{B}_2 &\equiv \tilde{B}_2(S,T,U,A_1,R_0/R',R_0/R,\kappa) \\
&= \sqrt{n}(\tilde{A}_0 + 5\sqrt{n}|U_0^{-1}|A_0(T,U)R_0/R)(\tilde{A}_0' + 5A_0'(T,U)R_0/R).
\end{aligned}$$

Since

$$\begin{aligned}
&\sum_{j=1}^\infty \chi_{F_j}(\eta) \sum_{|\gamma|\le j-1} (\tilde{B}_2/(\sqrt{n}R_0^2))^{|\gamma|}\langle\eta\rangle^{|\gamma|}/\gamma! \\
&\le (\langle\eta\rangle/(4R) + 5/2)\exp[\tilde{B}_2\langle\eta\rangle/R_0^2],
\end{aligned}$$

we have (A.28). (ii) The assertion (ii) can be proved by the same manner. In fact, noting that $1 - g_1^{R'}(\xi,\zeta,\eta) = 1$ if $\tilde{g}^R(\xi,\eta) \neq 0$, we have

$$\begin{aligned} \left|\partial_\xi^{\alpha+\tilde\alpha} D_w^{\beta+\tilde\beta} \partial_\eta^{\rho+\tilde\rho} p_\kappa(\xi,w,\eta)\right| &\le C_{|\tilde\alpha|+|\tilde\beta|+|\tilde\rho|,\kappa}(S,T,U,p) \\ &\times(\widehat{A}_1/R_0)^{|\alpha|}(\widehat{A}_2/R_0)^{|\beta|}(\widehat{A}_3/R_0)^{|\rho|} \\ &\times\langle\xi\rangle^{m_1-|\tilde\alpha|}\langle\eta\rangle^{m_2+m_3-|\tilde\rho|+|\beta|+2} \\ &\times\exp[\delta_1\langle\xi\rangle + (\tilde\delta(T,U,\delta_2,\delta_3)+\delta' B_1/R_0 + B_2/R_0^2)\langle\eta\rangle] \end{aligned}$$

if $\langle\xi\rangle \ge R|\alpha|$, $\langle\eta\rangle \ge 2\tilde\kappa(T,U,\kappa)R_0(|\beta|+|\rho|)$, $\langle\eta\rangle \ge 3R|\rho|$ and $R \ge \tilde\kappa(T,U,\kappa)R_0$, where

$$\begin{aligned} \widehat{A}_1 &= \widehat{C}(S,T)R_0/R + A_1, \\ \widehat{A}_2 &= \max\{2^6 c_2(T)A_0'(T,U)/c_3(U), 2^5 c_2(T)A_1/c_3(U), \\ &\qquad 2^{10}\sqrt{n}C_0(T,U)A_1A_0'(T,U)\} + 5A_0'(T,U)/\tilde\kappa(T,U,\kappa), \\ \widehat{A}_3 &= \max\{2^4 A_0'(T,U)R_0/R, 2\widehat{C}/\kappa + 4\sqrt{n}|U_0^{-1}||T_0|A_1 + 2\widehat{A}_1 + \widehat{C}R_0/R, \\ &\qquad 2^{10}\sqrt{n}C_0(T,U)A_1A_0'(T,U)\} + 10A_0'(T,U)R_0/(3R), \\ B_1 &= 13nc_2(T)C_0(U)A_1/(12c_2(U)), \\ B_2 &= \sqrt{n}(\widehat{A}_0 + 5\sqrt{n}|U_0^{-1}|A_0(T,U)R_0/R) \\ &\quad\times(\widehat{A}_0' + 5A_0'(T,U)R_0/R), \\ \widehat{A}_0 &= \max\{2^6\sqrt{n}c_2(T)|U_0^{-1}|A_0(T,U)/c_3(U), 4\sqrt{n}|U_0^{-1}|A_1, \\ &\qquad 2^{10}n|U_0^{-1}|C_0(T,U)A_1A_0(T,U)\}, \\ \widehat{A}_0' &= \max\{2^4A_0'(T,U)/\max\{1,3\kappa\}, 2A_1, \\ &\qquad 2^{10}\sqrt{n}C_0(T,U)A_1A_0'(T,U)\}. \end{aligned}$$

(iii) By (A.27) we have $(w,\xi) \in \Omega$, $(w,\eta) \in \Omega_1'$ and

$$2c_2(T)|\eta|/c_3(S) \le |\xi| \le 2c_2(T)(8c_2(U)/c_3(U)+3)|\eta|/c_3(S)$$

if $q_\kappa^3(\xi,w,\eta) \neq 0$ and $|\xi|^2+|\eta|^2 \ge 8$. Note that

$$|\nabla_w S(w,\xi) + \nabla_w T(w,\eta)| \ge (1-1/\sqrt{2})c_3(S)|\xi| \ge c_3(S)|\xi|/\sqrt{13}$$

if $q_\kappa^3(\xi,w,\eta) \neq 0$ and $|\xi|^2+|\eta|^2 \ge 8$. Let K_3 be a differential operator defined by

$$^tK_3 = |\nabla_w S(w,\xi) + \nabla_w T(w,\eta)|^{-2}\sum_{j=1}^n (\overline{\partial_{w_j}S(w,\xi)} + \overline{\partial_{w_j}T(w,\eta)})D_{w_j}.$$

Then it follows from Lemma 2.1. 5 that

$$\left|\partial_\xi^\alpha \partial_\eta^\rho K_3^j q_\kappa^3(\xi, w, \eta)\right| \leq C_{|\alpha|+|\rho|,\kappa}(S,T,U,p)$$
$$\times \Gamma(S,T,U,A_1,R_0,R',\kappa)^j \langle\eta\rangle^{m_1+m_2+m_3-|\alpha|-|\rho|+2}$$
$$\times \exp[\delta_1\langle\xi\rangle + (\tilde{\delta}(T,U,\delta_2,\delta_3) + \delta'\widetilde{B}_1/R_0 + \widetilde{B}_2/R_0^2)\langle\eta\rangle]$$

if $\langle\eta\rangle \geq 2\tilde{\kappa}(T,U,\kappa)R'j$, where

$$\Gamma(S,T,U,A_1,R_0,R',\kappa) = 2^6 \cdot 13nA_0(c_3(S)c_2(T))^{-1}C(S,T)$$
$$\times \max\{\widetilde{A}_2/R_0, 2^4 \cdot 13 \cdot 15nA_0^3 c_3(S)^{-2}\tilde{\kappa}(T,U,\kappa)^{-1}C(S,T)^2/R'\},$$
$$C(S,T) = C(S) + c_3(S)C(T)/(2c_2(T)).$$

Applying the same argument as in the proof of Lemma A.1. 1 , we can prove the assertion (iii). □

Let $R_0 \geq 1$, and choose positive constants $\hat{\kappa}'(S,T,U,A_1,\varepsilon_0)$ and $\hat{\kappa}_1(T,U,\varepsilon_0)$ so that $R' \equiv \hat{\kappa}'(S,T,U,A_1,\varepsilon_0)R_0$ and $R_1 \equiv \hat{\kappa}_1(T,U,\varepsilon_0)R_0$ satisfy the conditions on R' and R_1 in this section. Next choose a positive constant $\kappa_0(S,T,U,A_1,\varepsilon_0)$ so that $\kappa \equiv \kappa_0(S,T,U,A_1,\varepsilon_0)$ satisfies the conditions on κ in this section. We can also choose $\kappa(S,T,U,A_1,\varepsilon_0) > 0$ so that $R \equiv \kappa(S,T,U,A_1,\varepsilon_0)R_0$ satisfies the conditions on R in this section. Moreover, there is $\delta(S,T,U,A_1,\varepsilon_0) > 0$ such that δ' (> 0) satisfies the conditions in this section if $0 < \delta' \leq \delta(S,T,U,A_1,\varepsilon_0)$. Then, Lemmas A.1. 1 –A.1. 4 , A.1. 9 , A.1. 12 , A.1. 14 and A.1. 15 prove Theorem 2.4. 4 .

A.2 Proof of Corollary 2.4. 5

We assume that the hypotheses of Section 2.4 are satisfied and use the notations there. Let us first prove the assertion (i) of Corollary 2.4. 5

Repeating the same arguments as in Appendix A.1, we can prove the assertion (i) derectly. However, we shall use Theorem 2.4. 4 to prove the first part of the assertion (i). It is easy to see that

$$p_\kappa(\xi, w, \mathrm{Re}\ Z(\zeta,\eta,w,y), w+y, \eta; i\,\mathrm{Im}\ Z(\zeta,\eta,w,y)))$$
$$-p_\kappa(\xi, w, v(\zeta,\eta), w+y, \eta; Z(\zeta,\eta,w,y) - v(\zeta,\eta))$$
$$= \sum_{j=1}^n \int_0^1 ((\partial_{\zeta_j} - \partial_{u_j})p_\kappa)(\xi, w, (1-\theta)v(\zeta,\eta) + \theta\mathrm{Re}\ Z(\zeta,\eta,w,y), w+y,$$
$$\eta; (1-\theta)(\mathrm{Re}\ Z(\zeta,\eta,w,y) - v(\zeta,\eta)) + i\,\mathrm{Im}\ Z(\zeta,\eta,w,y))\,d\theta$$
$$\times(\mathrm{Re}\ Z_j(\zeta,\eta,w,y) - v_j(\zeta,\eta)).$$

Using Lemmas 2.1. 6 , 2.4. 2 , 2.4. 3 and A.1. 7 , and applying the same argument as in the proof of Lemma A.1. 15 , we can prove that $r^1_{\kappa\, S,T}(D_x, y, D_z)$ is a regularizer like Q in Theorem 2.4. 4 , where $r^1_\kappa(\xi, y, \eta) = p'_\kappa(\xi, y, \eta) - p_\kappa(\xi, y, \eta)$. In doing so, we must show, under suitable assumptions on κ, δ, R/R_0, $c(T)$ and $c(U)$, that

$$\begin{cases} \langle\eta\rangle \le \kappa\langle\hat{\zeta}_\theta\rangle, \quad \langle\hat{\zeta}_\theta\rangle \ge 2R_0(j-1). \\ \langle\eta\rangle \ge \max\{1, 3\kappa\}R_0(j-1), \quad z \equiv -{}^tU_0^{-1}\tilde{\nabla}_y T(w, w, \eta) \in \tilde{\mathcal{C}}_{w,w,\delta} \end{cases} \tag{A.33}$$

if $j \in \boldsymbol{N}$, $\psi_j^R(\eta) \ne 0$, $(w, \eta) \in \Omega_1$, $|\eta| \ge 1$, $0 \le \theta \le 1$, $\hat{\zeta}_\theta \equiv (1-\theta)v(0,\eta) + \theta\mathrm{Re}\, Z(0, \eta, w, 0) \in \Gamma_w$ and $|\hat{\zeta}_\theta| \ge 1$. Let $j \in \boldsymbol{N}$ and $\theta \in [0,1]$, and assume that $(w, \eta) \in \Omega_1$, $|\eta| \ge 1$, $\psi_j^R(\eta) \ne 0$, $\hat{\zeta}_\theta \in \Gamma_w$ and $|\hat{\zeta}_\theta| \ge 1$. We put $\zeta = Z(0, \eta, w, 0)$ and $\tilde{z}_\theta = \hat{\zeta}_\theta + {}^tU_0^{-1}\tilde{\nabla}_y \tilde{U}(w, w, \hat{\zeta}_\theta)$ ($\in \tilde{\Gamma}_{w,w,\delta}$). By (2.37) we have

$$c_3(T)|\eta|/2 \le |{}^tT_0\eta| \le |U_0|\,|v(0,\eta)|.$$

Suppose that $\zeta \notin \mathcal{C}_{w,5\delta/2}$. Then we have $\zeta = z$ and

$$\begin{cases} |\zeta - \hat{\zeta}_\theta| \le 4c(T)|U_0^{-1}|\,|U_0|\,|v(0,\eta)|/c_3(T), \\ |\hat{\zeta}_\theta - v(0,\eta)| \le |\mathrm{Re}\; z - v(0,\eta)| \\ \qquad \le 4c(T)|U_0^{-1}|\,|U_0|\,|v(0,\eta)|/c_3(T). \end{cases} \tag{A.34}$$

Assume that

$$c(T) \le c_3(T)\min\{2^{-3}, 5\delta/16\}/(|U_0^{-1}|\,|U_0|). \tag{A.35}$$

(A.35) gives $|\zeta - \hat{\zeta}_\theta| < 5\delta\langle\hat{\zeta}_\theta\rangle/2$, which is a contradiction. So we have $\zeta \in \mathcal{C}_{w,5\delta/2}$, and there is $\tilde{\zeta} \in \Gamma_w$ such that $|\zeta - \tilde{\zeta}| < 5\delta\langle\tilde{\zeta}\rangle/2$ and $|\zeta - \tilde{\zeta}| \le |\zeta - \hat{\zeta}_\theta|$. From (2.46) it follows that

$$\begin{aligned} &|\zeta - \hat{\zeta}_\theta| + |\hat{\zeta}_\theta - v(0,\eta)| \le 2|U_0^{-1}|(4c(T)|U_0|\,|v(0,\eta)|/c_3(T) \\ &\quad + c(U)\langle\tilde{\zeta}\rangle + (c(U) + 10n^{3/2}C(\tilde{U})A_0^3\delta)|\zeta - \hat{\zeta}_\theta|) \\ &< 2|U_0^{-1}|\{(4c(T)|U_0|/c_3(T) + c(U))\langle\hat{\zeta}_\theta\rangle \\ &\quad + 4c(T)|U_0||\hat{\zeta}_\theta - v(0,\eta)|/c_3(T) + (3c(U) + 10n^{3/2}C(\tilde{U})A_0^3\delta)|\zeta - \hat{\zeta}_\theta|\}. \end{aligned}$$

If

$$\begin{cases} c(T) \le c_3(T)/(2^4|U_0^{-1}|\,|U_0|), \\ 3c(U) + 10n^{3/2}C(\tilde{U})A_0^3\delta \le 1/(4|U_0^{-1}|), \end{cases} \tag{A.36}$$

then we have

$$|\zeta - \hat{\zeta}_\theta| + |\hat{\zeta}_\theta - v(0,\eta)| < 4|U_0^{-1}|(4c(T)|U_0|/c_3(T) + c(U))\langle\hat{\zeta}_\theta\rangle.$$

Moreover, we have

$$\langle\hat{\zeta}_\theta\rangle/2 < \langle\zeta\rangle < 3\langle\hat{\zeta}_\theta\rangle/2,$$
$$\langle\hat{\zeta}_\theta\rangle/2 < \langle v(0,\eta)\rangle < 3\langle\hat{\zeta}_\theta\rangle/2, \quad \langle\tilde{\zeta}\rangle < 2\langle\hat{\zeta}_\theta\rangle$$

if (A.36) is sattisfied and

$$4c(T)|U_0|/c_3(T) + c(U) \le 1/(8|U_0^{-1}|). \tag{A.37}$$

Therefore, (A.33) is satisfied and

$$|Z(0,\eta,w,0) - v(0,\eta)| < 20|U_0^{-1}|^2 c_2(T)(4c(T)|U_0|/c_3(T) + c(U))\langle\eta\rangle,$$

if (A.35)–(A.37) are satisfied, $\kappa \ge 6|U_0|/c_3(T)$, $R/R_0 \ge \max\{1/2, 3\kappa/2\}$ and $22c(T)|U_0|/c_3(T) + 5c(U) \le 11\delta/(96|U_0^{-1}|)$, since

$$|z - \tilde{z}_\theta| < |U_0^{-1}|(22c(T)|U_0|/c_3(T) + 5c(U))\langle\hat{\zeta}_\theta\rangle,$$
$$\langle\hat{\zeta}_\theta\rangle \le \langle\tilde{z}_\theta\rangle + c(U)|U_0^{-1}|\langle\hat{\zeta}_\theta\rangle.$$

If $U(y,\zeta)$ and $T(y,\eta)$ are real-valued, then we replace $f_\kappa^{R'}(\xi,w,\zeta,y,\eta;u)$ by $f^{R'}(\xi,w,\zeta+u,y,\eta)$ in the proof of Theorem 2.4. 4 , where $(\xi,w,\zeta,y,\eta,$ $u) \in \boldsymbol{R}^{n'} \times \boldsymbol{R}^n \times \boldsymbol{R}^n \times \boldsymbol{R}^n \times \boldsymbol{R}^{n''} \times \boldsymbol{R}^n$. This proves the second part of the assertion (i). Next we shall prove the assertion (ii). Assume that $p(\xi,w,\zeta,y,\eta)$ satisfies (2.67), $\langle\eta\rangle \ge 2R$ (≥ 2) and $R \ge 4|U_0|R_0/c_3(T)$. It is obvious that $|\zeta| \ge R_0$ and

$$p(\xi,w,\zeta+u,y,\eta) - p_\kappa(\xi,w,\zeta,y,\eta;u)$$
$$= \sum_{j=1}^\infty \psi_j^{\kappa R_0}(\eta) \sum_{|\mu|\ge j} u^\mu \partial_\zeta^\mu p(\xi,w,\zeta,y,\eta)/\mu!$$

if

$$\begin{cases} \zeta \in \boldsymbol{R}^n, & |\zeta - v(0,\eta)| \le \min\{\varepsilon_0', c_3(T)/(4|U_0|)\}|\eta|, \\ |w-y| \le \varepsilon_0', & u \in \boldsymbol{C}^n, \quad |u| < \langle\zeta\rangle/(\sqrt{n}A_1). \end{cases} \tag{A.38}$$

Moreover, we have

$$\left|\partial_\xi^\alpha \partial_\zeta^\gamma D_y^\beta \partial_\eta^\rho \partial_u^\delta (p(\xi,w,\zeta+u,y,\eta) - p_\kappa(\xi,w,\zeta,y,\eta;u))\right|$$
$$\le C_{|\alpha|+|\rho|}(A_1/R_0)^{|\beta|}(2A_1)^{|\gamma|+|\delta|}(|\gamma|+|\delta|)!\langle\xi\rangle^{m_1-|\alpha|}$$
$$\times\langle\zeta\rangle^{m_2-|\gamma|-|\delta|}\langle\eta\rangle^{m_3-|\rho|+|\beta|}\exp[\delta_1\langle\xi\rangle + \delta_2\langle\zeta\rangle + \delta_3\langle\eta\rangle]$$
$$\times\sum_{j=1}^\infty \chi_{F_j}(\eta) \sum_{k\ge(j-|\delta|)_+} (2\sqrt{n}A_1|u|/\langle\zeta\rangle)^k$$

if (A.38) is satisfied, $R_0|\beta| \le \langle\eta\rangle$ and $|u| < \langle\zeta\rangle/(2\sqrt{n}A_1)$, where $F_j = \{\eta \in \boldsymbol{R}^{n''};\ 2\kappa R_0(j-1) \le \langle\eta\rangle \le 3\kappa R_0 j\}$. Since

$$\begin{aligned}&\sum_{j=1}^{\infty} \chi_{F_j}(\eta) \sum_{k\ge(j-|\delta|)_+} (2\sqrt{n}A_1|u|/\langle\zeta\rangle)^k \\ &\le \sum_{j=1}^{\infty} e^{1-j+|\delta|}\chi_{F_j}(\eta)/(e-1) \\ &\le (\langle\eta\rangle/(4\kappa R_0)+5/2)e^{1+|\delta|}\exp[-\langle\eta\rangle/(3\kappa R_0)]/(e-1)\end{aligned}$$

if $|u| \le \langle\zeta\rangle/(2e\sqrt{n}A_1)$, we have

$$\begin{aligned}&\left|\partial_\xi^\alpha \partial_\zeta^\gamma D_y^\beta \partial_\eta^\rho \partial_u^\delta (p(\xi,w,\zeta+u,y,\eta) - p_\kappa(\xi,w,\zeta,y,\eta;u))\right| \\ &\le C_{|\alpha|+|\rho|,\kappa}(A_1/R_0)^{|\beta|}(2A_1)^{|\gamma|}(2eA_1)^{|\delta|}(|\gamma|+|\delta|)! \\ &\quad \times \langle\xi\rangle^{m_1-|\alpha|}\langle\zeta\rangle^{m_2-|\gamma|-|\delta|}\langle\eta\rangle^{m_3-|\rho|+|\beta|+1} \\ &\quad \times \exp[\delta_1\langle\xi\rangle + \delta_2\langle\zeta\rangle + (\delta_3-(3\kappa R_0)^{-1})\langle\eta\rangle]\end{aligned}$$

if (A.38) is satisfied, $R_0|\beta| \le \langle\eta\rangle$ and $|u| \le \langle\zeta\rangle/(2e\sqrt{n}A_1)$. We note that

$$|\eta| \le 4|U_0|\,|v(\zeta,\eta)|/c_3(T)$$

if $(y,\eta) \in \Omega'$ for some $y \in \boldsymbol{R}^n$ and $|\zeta| \le c_3(T)|\eta|/4$. Assume that

$$\begin{cases} \zeta \in \boldsymbol{R}^n, \quad |\zeta| \le \min\{\varepsilon_0'/|U_0^{-1}|, c_3(T)/4\}|\eta|, \\ |y| \le \varepsilon_0', \quad (w+y,\eta) \in \Omega', \\ c(T) \le c_3(T)\min\{1/(e\sqrt{n}A_1), 5\delta\}/(2^4|U_0^{-1}|\,|U_0|). \end{cases} \tag{A.39}$$

Then, $|v(\zeta,\eta)| \ge R_0$ (≥ 1) and $|v(\zeta,\eta)-v(0,\eta)| \le \min\{\varepsilon_0', c_3(T)|U_0^{-1}|/4\} \times|\eta|$. Put $\zeta^1 = Z(\zeta,\eta,w,y)$, and suppose that $\zeta^1 \notin \mathcal{C}_{w,5\delta/2}$. Then we have

$$\begin{aligned}|\zeta^1 - v(\zeta,\eta)| &\le c(T)|U_0^{-1}|\langle\eta\rangle \le 8c(T)|U_0^{-1}|\,|U_0|\,|v(\zeta,\eta)|/c_3(T) \\ &< \min\{1/(2e\sqrt{n}A_1), 5\delta/2\}\langle v(\zeta,\eta)\rangle.\end{aligned} \tag{A.40}$$

So $p(\xi,w,\zeta^1,w+y,\eta) - p_\kappa(\xi,w,v(\zeta,\eta),w+y,\eta;\zeta^1-v(\zeta,\eta))$ is meaningful, and $v(\zeta,\eta) \in \Gamma_w$ if

$$p(\xi,w,\zeta^1,w+y,\eta) - p_\kappa(\xi,w,v(\zeta,\eta),w+y,\eta;\zeta^1-v(\zeta,\eta)) \ne 0. \tag{A.41}$$

In fact, $p_\kappa(\xi,w,v(\zeta,\eta),w+y,\eta;u) = 0$ and $p(\xi,w,v(\zeta,\eta)+u,w+y,\eta) = 0$ if $v(\zeta,\eta) \notin \Gamma_w$ and $|u| < \langle v(\zeta,\eta)\rangle/(\sqrt{n}A_1)$. Then (A.40) implies that $\zeta^1 \in \mathcal{C}_{w,5\delta/2}$ if (A.41) is valid, which contradicts $\zeta^1 \notin \mathcal{C}_{w,5\delta/2}$. Next assume that $\zeta^1 \in \mathcal{C}_{w,5\delta/2}$, *i.e.*, there is $\tilde{\zeta} \in \Gamma_w$ satisfying $|\tilde{\zeta}| \geq 1$ and $|\zeta^1 - \tilde{\zeta}| < 5\delta\langle\tilde{\zeta}\rangle/2$. From (2.46) and $\langle\tilde{\zeta}\rangle \leq 2\langle\zeta^1\rangle$ it follows that

$$|\zeta^1 - v(\zeta,\eta)| < |U_0^{-1}|\Big((3c(U) + 50n^{3/2}C(\tilde{U})A_0^3\delta^2)\langle\zeta^1\rangle + c(T)\langle\eta\rangle\Big).$$

This gives

$$|\zeta^1 - v(\zeta,\eta)| < \langle v(\zeta,\eta)\rangle/(\sqrt{n}A_1)$$

if

$$\begin{cases} 3c(U) + 50n^{3/2}C(\tilde{U})A_0^3\delta^2 \leq \min\{1/2, 1/(4\sqrt{n}A_1)\}/|U_0^{-1}|, \\ c(T) \leq c_3(T)/(2^5\sqrt{n}A_1|U_0^{-1}|\,|U_0|). \end{cases} \tag{A.42}$$

Therefore, $p(\xi, w, \zeta^1, w+y, \eta) - p_\kappa(\xi, w, v(\zeta,\eta), w+y, \eta; \zeta^1 - v(\zeta,\eta))$ is meaningful, and $v(\zeta,\eta) \in \Gamma_w$ if (A.41) and (A.42) are valid. Similarly, we can show that

$$\begin{cases} |\zeta^1 - v(\zeta,\eta)| < \delta\langle v(\zeta,\eta)\rangle/2, \quad \zeta^1 \in \mathcal{C}_{w,\delta/2}, \\ v(\zeta,\eta) - {}^tU_0^{-1}\tilde{\nabla}_y\tilde{T}(w, w+y, \eta) \in \tilde{\mathcal{C}}_{w,w+y,\delta} \end{cases}$$

if $c(T), c(U), \delta \ll 1$ and (A.41) is satisfied. So, choosing $c(T)$, $c(U)$, and δ sufficiently small, there are positive contants $\tilde{A} \equiv \tilde{A}(T, U, A_1, R_0/R)$ and $\tilde{B} \equiv \tilde{B}(T, U, A_1, R_0/R)$ such that

$$\begin{aligned} &\Big|\partial_\xi^\alpha \partial_\zeta^\gamma D_y^\beta \partial_\eta^\rho \{p(\xi, w, Z(\zeta,\eta,w,y), w+y, \eta) \\ &\qquad - p_\kappa(\xi, w, v(\zeta,\eta), w+y, \eta; Z(\zeta,\eta,w,y) - v(\zeta,\eta))\}\Big|_{y=0,\,\zeta=0} \\ &\leq C_{|\alpha|+|\rho|,\kappa}(\tilde{B}/R_0)^{|\beta|}(\tilde{A}/R_0)^{|\gamma|}\langle\xi\rangle^{m_1-|\alpha|}\langle\eta\rangle^{m_2+m_3-|\rho|+|\beta|+1} \\ &\quad \times \exp[\delta_1\langle\xi\rangle + \delta_2\langle v(0,\eta)\rangle + (\delta_3 - (3\kappa R_0)^{-1})\langle\eta\rangle] \end{aligned}$$

if $j \geq 2$, $|\beta|, |\gamma| \leq j-1$, $\langle\eta\rangle \geq 2R(j-1)$ and $2R \geq R_0$. From the above estimates we can prove that $r^2_{\kappa\,S,T}(D_x, y, D_z)$ is a regularizer like Q in Theorem 2.4. 4 , where $r^2_\kappa(\xi, y, \eta) = \tilde{p}^0(\xi, y, \eta) - p_\kappa(\xi, y, \eta)$. From (2.67) it follows that

$$\begin{aligned} &\Big|\partial_\xi^{\alpha+\tilde{\alpha}} D_w^{\beta+\tilde{\beta}} \partial_z^\gamma D_y^\lambda \partial_\eta^{\rho+\tilde{\rho}} p(\xi, w, z, y, \eta)\Big| \leq C_{|\tilde{\alpha}|+|\tilde{\beta}|+|\tilde{\rho}|} \\ &\quad \times (A_1/R_0)^{|\alpha|+|\beta|+|\lambda|+|\rho|}(2A_1)^{|\gamma|}|\gamma|!\langle\xi\rangle^{m_1-|\tilde{\alpha}|} \\ &\quad \times \langle\zeta\rangle^{m_2-|\gamma|+|\beta|}\langle\eta\rangle^{m_3-|\tilde{\rho}|+|\lambda|}\exp[\delta_1\langle\xi\rangle + \delta_2\langle\zeta\rangle + \delta_3\langle\eta\rangle] \end{aligned}$$

if $z = \zeta + u$, $\zeta \in \boldsymbol{R}^n$, $|\zeta| \geq R_0$, $|\zeta + {}^tU_0^{-1}\,{}^tT_0\eta| < \varepsilon_0'|\eta|$, $|u| < \langle\zeta\rangle/(4\sqrt{n}A_1)$, $\langle\xi\rangle \geq R_0|\alpha|$, $\langle\zeta\rangle \geq R_0|\beta|$, $\langle\eta\rangle \geq R_0(|\lambda| + |\rho|)$. Therefore, by induction on

$|\beta|+|\tilde{\beta}|+|\gamma|+|\lambda|+|\rho|+|\tilde{\rho}|$, we can show that there are positive constants $\tilde{A}_k \equiv \tilde{A}_k(T, U, A_1)$ ($1 \le k \le 4$) such that

$$\Big|\partial_\xi^{\alpha+\tilde{\alpha}} D_w^{\beta+\tilde{\beta}} \partial_\zeta^\gamma D_y^\lambda \partial_\eta^{\rho+\tilde{\rho}} \Big(D_w^{\beta^1+\tilde{\beta}^1} D_y^{\lambda^1} \partial_\eta^{\rho^1+\tilde{\rho}^1} \partial_z^{\delta^1+\delta^2+\tilde{\delta}} p\Big)$$
$$(\xi, w, Z(\zeta, \eta, w, y), y, \eta)\Big|$$
$$\le C_{|\tilde{\beta}|+|\tilde{\rho}|,|\tilde{\alpha}|+|\tilde{\beta}^1|+|\tilde{\rho}^1|+|\tilde{\delta}|}(A_1/R_0)^{|\alpha|+|\lambda^1|+|\rho^1|}(3c_2(T)|U_0^{-1}|A_1/R_0)^{|\beta^1|}$$
$$\times(48|U_0|A_1/c_3(T))^{|\delta^1|}(6A_1/R_0)^{|\delta^2|}(\tilde{A}_1/R_0)^{|\beta|}$$
$$\times\tilde{A}_2^{|\gamma|}(\tilde{A}_3/R_0)^{|\lambda|}(\tilde{A}_4/R_0)^{|\rho|}(|\gamma|+|\delta^1|)!$$
$$\times\langle\xi\rangle^{m_1-|\tilde{\alpha}|}\langle\eta\rangle^{m_2+m_3-|\gamma|-|\delta^1|-|\tilde{\delta}|-|\tilde{\rho}^1|-|\tilde{\rho}|+|\beta^1|+|\beta|+|\lambda^1|+|\lambda|}$$
$$\times \exp[\delta_1\langle\xi\rangle + \delta_2\langle v(\zeta, \eta)\rangle + \delta_3\langle\eta\rangle]$$

if (A.39) is satisfied, $\langle\xi\rangle \ge R_0|\alpha|$, $\langle v(\zeta, \eta)\rangle \ge R_0(|\beta^1| + |\beta| + |\delta^2|)$, $\langle\eta\rangle \ge R_0(|\lambda^1|+|\lambda|+|\rho^1|+|\rho|)$ and $c(T), c(U), \delta \ll 1$. This yields the estimates (2.68). (2.70) easily follows from (2.69). Applying the same argument as in the proof of Lemma A.1. 1 , we can show that $r^3_{\kappa\, S,T}(D_x, y, D_z)$ is a regularizer like Q in Theorem 2.4. 4 , where $r^3_\kappa(\xi, y, \eta) = (1-\tilde{g}^R(\xi, \eta))p^0_\kappa(\xi, y, \eta)$, which proves the assertion (iii). Now the assertion (iv) is obvious by the assertions (ii) and (iii).

A.3 Proof of Theorem 2.4. 6

We assume that the hypothese of Theorem 2.4. 6 are satisfied. Let us first prove the assertion (i). It is easy to see that

$$\mathcal{F}_x[p(D_x, w, D_w, y, D_y)u(x)](\xi)$$
$$= \lim_{\nu,\nu'\downarrow 0}(2\pi)^{-n}\int e^{iw\cdot(\eta-\xi)}\exp[-\nu'|w|^2]a_\nu(\xi, w, \eta)\hat{u}(\eta)\, d\eta dw$$

for $u \in \mathcal{S}_\infty$, where

$$a_\nu(\xi, w, \eta) = (2\pi)^{-n}\int e^{i(w-y)\cdot(\zeta-\eta)}\langle w-y\rangle_{A_1}^{-2M}\langle D_\zeta\rangle_{A_1}^{2M}$$
$$\times\{\exp[-\nu|\zeta|^2]p(\xi, w, \zeta, y, \eta)\}\, dyd\zeta,$$

$\langle\xi\rangle_{A_1} = \langle A_1\xi/(4+4\sqrt{2})\rangle$ and $M = [n/2]+1$. Let $R \ge 2$, and fix $\varepsilon > 0$ so that $0 < \varepsilon \le 1$. We can choose symbols $g_k^R(\zeta, \eta) \in C^\infty(\boldsymbol{R}^n \times \boldsymbol{R}^n)$ ($k = 0, 1, 2$) so that $g_k^R(\zeta, \eta) \ge 0$ ($k = 0, 1, 2$), $\sum_{k=0}^2 g_k^R(\zeta, \eta) \equiv 1$,

$$g_0^R(\zeta, \eta) = \begin{cases} 1 & \text{if } |\zeta| \le \varepsilon\langle\eta\rangle/4 \text{ and } |\eta| \ge R/2, \\ 0 & \text{if } |\zeta| \ge \varepsilon\langle\eta\rangle/2 \text{ or } |\eta| \le R/4, \end{cases}$$

$$g_1^R(\zeta,\eta) = \begin{cases} 1 & \text{if } \varepsilon\langle\eta\rangle/2 \le |\zeta| \le 2\langle\eta\rangle \text{ and } |\eta| \ge R/2, \\ 0 & \text{if } |\zeta| \le \varepsilon\langle\eta\rangle/4 \text{ or } |\zeta| \ge 3\langle\eta\rangle \text{ or } |\eta| \le R/4, \end{cases}$$

$$g_2^R(\zeta,\eta) = \begin{cases} 1 & \text{if } |\zeta| \ge 3\langle\eta\rangle \text{ or } |\eta| \le R/4, \\ 0 & \text{if } |\zeta| \le 2\langle\eta\rangle \text{ and } |\eta| \ge R/2, \end{cases}$$

$$\left|\partial_\zeta^{\gamma+\tilde\gamma}\partial_\eta^{\rho+\tilde\rho} g_k^R(\zeta,\eta)\right|$$
$$\le C_{|\tilde\gamma|+|\tilde\rho|}(\widehat{C}_*/(\varepsilon R))^{|\gamma|+|\rho|}\langle\eta\rangle^{-|\tilde\gamma|-|\tilde\rho|} \qquad (\ k=0,1),$$
$$\left|\partial_\zeta^{\gamma+\tilde\gamma}\partial_\eta^{\rho+\tilde\rho} g_2^R(\zeta,\eta)\right| \le C_{|\tilde\gamma|+|\tilde\rho|}(\widehat{C}_*/R)^{|\gamma|+|\rho|}\langle\eta\rangle^{-|\tilde\gamma|-|\tilde\rho|},$$

if $\langle\zeta\rangle + \langle\eta\rangle \ge R|\gamma|$ and $\langle\eta\rangle \ge R|\rho|$, where $g_k^R(\zeta,\eta)$ ($k=0,1$) also depend on ε and $\widehat{C}_*$ does not depend on ε (see Lemma 2.1. 3 , Proposition 2.2. 3 and the proof of Lemma 2.4. 1). Put

$$a_\nu^k(\xi,w,\eta) = (2\pi)^{-n}\int e^{i(w-y)\cdot\zeta}\langle w-y\rangle_{A_1}^{-2M}\langle D_\zeta\rangle_{A_1}^{2M}$$
$$\times\{\exp[-\nu|\zeta+\eta|^2]g_k^R(\zeta,\eta)p(\xi,w,\zeta+\eta,y,\eta)\}\,dyd\zeta$$

for $\nu > 0$ and $k = 0,1,2$. We note that $a_\nu^k(\xi,w,\eta)$ ($k=0,1$) depend on ε. So we also write $a_\nu^k(\xi,w,\eta) = a_{\nu,\varepsilon}^k(\xi,w,\eta)$ ($k=0,1$). By definitions we have

$$a_\nu(\xi,w,\eta) = \sum_{k=0}^{2} a_\nu^k(\xi,w,\eta) \qquad (\ \nu > 0),$$

$$\lim_{\nu,\nu'\downarrow 0}(2\pi)^{-n}\int e^{iw\cdot(\eta-\xi)}\exp[-\nu'|w|^2]a_{\nu,\varepsilon}^k(\xi,w,\eta)\hat u(\eta)\,d\eta dw$$
$$= (2\pi)^{-n}\int e^{iw\cdot(\eta-\xi)}\langle w\rangle_{A_1}^{-2M}\langle D_\eta\rangle_{A_1}^{2M}\{a_{0,\varepsilon}^k(\xi,w,\eta)\hat u(\eta)\}\,d\eta dw$$
$$\left(\equiv \mathcal{F}[a_{0,\varepsilon}^k(D_x,y,D_y)u](\xi)\right)$$

for $u \in \mathcal{S}_\infty$ and $k = 0,1$. A simple calculation yields

$$\left|\partial_\xi^{\alpha+\tilde\alpha}D_w^{\beta^1+\beta^2+\tilde\beta}\partial_\zeta^\gamma D_y^\lambda\partial_\eta^{\rho+\tilde\rho}\left\{\langle w-y\rangle_{A_1}^{-2M}g_0^R(\zeta,\eta)p(\xi,w,\zeta+\eta,y,\eta)\right\}\right|$$
$$\le C_{|\tilde\alpha|+|\tilde\beta|+|\gamma|+|\lambda|+||\tilde\rho|,\varepsilon}(A_1/R_0)^{|\alpha|+|\beta^1|}(3A_1/(2R_0))^{|\beta^2|}$$
$$\times(\widehat{C}_*/(\varepsilon R)+2A_1/R_0)^{|\rho|}\langle w-y\rangle_{A_1}^{-2M}\langle\xi\rangle^{m_1-|\tilde\alpha|+|\beta^1|}$$
$$\times\langle\eta\rangle^{m_2+m_3-|\gamma|-|\tilde\rho|+|\beta^2|}\exp[\delta_1\langle\xi\rangle+(\delta_2+\varepsilon|\delta_2|/2+\delta_3)\langle\eta\rangle]$$

if $\langle\xi\rangle \ge R_0(|\alpha|+|\beta^1|)$, $\langle\eta\rangle \ge 2R_0(|\beta^2|+|\rho|)$ and $\langle\eta\rangle \ge \max\{R_0,R\}|\rho|$. So we have

$$\left|\partial_\xi^{\alpha+\tilde\alpha}D_w^{\beta^1+\beta^2+\tilde\beta}\partial_\eta^{\rho+\tilde\rho}a_{0,\varepsilon}^0(\xi,w,\eta)\right|$$

$$\leq (2\pi)^{-n} \int \sum_{\beta^{1\prime}\leq\beta^1,\beta^{2\prime}\leq\beta^2,\tilde{\beta}'\leq\tilde{\beta}} \binom{\beta^1}{\beta^{1\prime}}\binom{\beta^2}{\beta^{2\prime}}\binom{\tilde{\beta}}{\tilde{\beta}'}$$
$$\times\Big|\zeta^{\beta^{1\prime}+\beta^{2\prime}+\tilde{\beta}'}\langle\zeta\rangle^{-2N}\partial_\xi^{\alpha+\tilde{\alpha}}D_w^{\beta^1+\beta^2+\tilde{\beta}-\beta^{1\prime}-\beta^{2\prime}-\tilde{\beta}'}$$
$$\times\partial_\eta^{\rho+\tilde{\rho}}\langle D_y\rangle^{2N}\Big[\langle w-y\rangle_{A_1}^{-2M}\langle D_\zeta\rangle_{A_1}^{2M}$$
$$\times\{g_0^R(\zeta,\eta)p(\xi,w,\zeta+\eta,y,\eta)\}\Big]\Big|\,dyd\zeta$$
$$\leq C_{|\tilde{\alpha}|+|\tilde{\beta}|+|\tilde{\rho}|,\varepsilon}(A_1/R_0)^{|\alpha|}((A_1+\varepsilon/\delta)/R_0)^{|\beta^1|}$$
$$\times((3A_1/2+\varepsilon/(2\delta))/R_0)^{|\beta^2|}(\widehat{C}_*/(\varepsilon R)+2A_1/R_0)^{|\rho|}$$
$$\times\langle\xi\rangle^{m_1-|\tilde{\alpha}|+|\beta^1|}\langle\eta\rangle^{m_2+m_3-|\tilde{\rho}|+|\beta^2|}$$
$$\times\exp[\delta_1\langle\xi\rangle+(\delta_2+\varepsilon|\delta_2|/2+\delta_3+\delta)\langle\eta\rangle]$$

if $\delta>0$, $\langle\xi\rangle\geq R_0(|\alpha|+|\beta^1|)$, $\langle\eta\rangle\geq 2R_0(|\beta^2|+|\rho|)$ and $\langle\eta\rangle\geq\max\{R_0,R\}\times|\rho|$, where $M=[n/2]+1$ and $N=[(n+|\tilde{\beta}|)/2]+1$, noting that

$$|\zeta|^{j+k}\leq(\varepsilon/\delta)^{j+k}j!k!e^{\delta\langle\eta\rangle}\quad\text{if } \delta>0 \text{ and } g_0^R(\zeta,\eta)\neq 0.$$

Put

$$p_{0,\varepsilon}(\xi,y,\eta)=a_{0,\varepsilon}^0(\xi,y,\eta).$$

Then $p_{0,\varepsilon}(\xi,y,\eta)$ satisfies (2.71) if we take $R=4R_0$. Let K be a differential operator defined by

$$K=|\zeta|^{-2}\sum_{k=1}^n\zeta_kD_{y_k}.$$

We can easily show that

$$\Big|\partial_\xi^\alpha D_w^{\beta+\tilde{\beta}}\partial_\eta^\rho K^{j+N}\Big[\langle w-y\rangle_{A_1}^{-2M}\langle D_\zeta\rangle_{A_1}^{2M}\{g_1^R(\zeta,\eta)p(\xi,w,\zeta+\eta,y,\eta)\}\Big]\Big|$$
$$\leq C_{|\alpha|+|\tilde{\beta}|+|\rho|+N,\varepsilon}(A_1/R_0)^{|\beta|}(4\sqrt{n}A_1/(\varepsilon R_0))^j\langle\xi\rangle^{m_1-|\alpha|+|\beta|}$$
$$\times\langle\eta\rangle^{(m_2)_++m_3-N}\langle w-y\rangle_{A_1}^{-2M}\exp[\delta_1\langle\xi\rangle+(4(\delta_2)_++\delta_3)\langle\eta\rangle]$$

if $j\in Z_+$, $\langle\xi\rangle\geq R_0|\beta|$ and $R_0j\leq\langle\eta\rangle\,(\leq R_0(j+1))$, where $N=[(n+(m_2)_++|\tilde{\beta}|)/2]+1$. This yields

$$\Big|\partial_\xi^\alpha D_w^{\beta+\tilde{\beta}}\partial_\eta^\rho a_{0,\varepsilon}^1(\xi,w,\eta)\Big|$$
$$\leq\sum_{k=0}^{|\beta|}\binom{|\beta|}{k}(3\langle\eta\rangle)^kC_{|\alpha|+|\tilde{\beta}|+|\rho|,\varepsilon}(A_1/R_0)^{|\beta|-k}$$

$$\times(4\sqrt{n}A_1/(\varepsilon R_0))^j\langle\xi\rangle^{m_1-|\alpha|+|\beta|-k}\langle\eta\rangle^{(m_2)_++m_3}$$
$$\times\exp[\delta_1\langle\xi\rangle+(4(\delta_2)_++\delta_3)\langle\eta\rangle]$$
$$\le C'_{|\alpha|+|\tilde\beta|+|\rho|,\varepsilon}(A_1/R_0+3/(e\kappa))^{|\beta|}\langle\xi\rangle^{m_1-|\alpha|+|\beta|}$$
$$\times\langle\eta\rangle^{m_3}\exp[\delta_1\langle\xi\rangle-\kappa\langle\eta\rangle/R_0]$$

if $\kappa>0$, $R_0\ge 4e^{3\kappa}\sqrt{n}A_1/\varepsilon$, $\langle\xi\rangle\ge R_0|\beta|$, $4(\delta_2)_++\delta_3\le\kappa/R_0$ ($j\in\boldsymbol{Z}_+$ and $R_0j\le\langle\eta\rangle\le R_0(j+1)$), since

$$\sum_{k=0}^{|\beta|}\binom{|\beta|}{k}(3\langle\eta\rangle)^k(A_1/R_0)^{|\beta|-k}\langle\xi\rangle^{-k}$$
$$\le\sum_{k=0}^{|\beta|}\binom{|\beta|}{k}|\beta|^{-k}e^k(\kappa\langle\eta\rangle/R_0)^k(3/(e\kappa))^k(A_1/R_0)^{|\beta|-k}$$
$$\le(3/(e\kappa)+A_1/R_0)^{|\beta|}\exp[\kappa\langle\eta\rangle/R_0],$$
$$1\le e^{3\kappa(j+1)}\exp[-3\kappa\langle\eta\rangle/R_0]$$

if $\langle\xi\rangle\ge R_0|\beta|>0$ and $\langle\eta\rangle\le R_0(j+1)$. We shall next consider $a_\nu^2(\xi,w,\eta)$. Let $\{\chi_j^{\varepsilon_0}(y)\}$ be the family of functions used in Section 2.4, and let $w^0\in\boldsymbol{R}^n$. We can represent for $w\in\boldsymbol{R}^n$ with $|w-w^0|<\varepsilon_0/3$

$$a_\nu^2(\xi,w,\eta)=\sum_{j=1}^{\infty}(a_{\nu,j}^{2,1}(\xi,w,\eta)+a_{\nu,j}^{2,2}(\xi,w,\eta)),$$

where

$$a_{\nu,j}^{2,1}(\xi,w,\eta)=(2\pi)^{-n}\int e^{i(w-y)\cdot\zeta}\psi_j^R(\zeta)\chi_j^{2\varepsilon_0/3}(y-w^0)$$
$$\times\exp[-\nu|\zeta+\eta|^2]g_2^R(\zeta,\eta)p(\xi,w,\zeta+\eta,y,\eta)\,dyd\zeta,$$
$$a_{\nu,j}^{2,2}(\xi,w,\eta)=(2\pi)^{-n}\int e^{i(w-y)\cdot\zeta}\langle w-y\rangle_{A_1}^{-2M}(1-\chi_j^{2\varepsilon_0/3}(y-w^0))$$
$$\times\langle D_\zeta\rangle_{A_1}^{2M}\{\psi_j^R(\zeta)\exp[-\nu|\zeta+\eta|^2]g_2^R(\zeta,\eta)p(\xi,w,\zeta+\eta,y,\eta)\}\,dyd\zeta.$$

A simple calculation gives

$$\Big|\partial_\xi^\alpha D_w^{\beta+\tilde\beta}\partial_\eta^\rho K^j\{\psi_j^R(\zeta)\chi_j^{2\varepsilon_0/3}(y-w^0)$$
$$\times\exp[-\nu|\zeta+\eta|^2]g_2^R(\zeta,\eta)p(\xi,w,\zeta+\eta,y,\eta)\}\Big|$$
$$\le C_{|\alpha|+|\tilde\beta|+|\rho|,R}(A_1/R_0)^{|\beta|}(3e^{12\kappa}\sqrt{n}(C_*/(\varepsilon_0R)+A_1/(2R_0)))^j$$
$$\times j^{m_2}\langle\xi\rangle^{m_1-|\alpha|+|\beta|}\langle\eta\rangle^{m_3-|\rho|}$$
$$\times\exp[\delta_1\langle\xi\rangle+(\delta_2+|\delta_2|/2-4\kappa/R)\langle\zeta\rangle+\delta_3\langle\eta\rangle]$$

if $|w-w^0| < \varepsilon_0/3$, $0 \le \nu \le 1$, $\langle\xi\rangle \ge R_0|\beta|$, $\kappa > 0$ and $R \ge 2R_0$. Therefore, we have

$$\begin{aligned}\left|\partial_\xi^\alpha D_w^{\beta+\tilde\beta}\partial_\eta^\rho a_{\nu,j}^{2,1}(\xi,w,\eta)\right| &\le C_{|\alpha|+|\tilde\beta|+|\rho|,\varepsilon_0,R}((A_1+2/\delta)/R_0)^{|\beta|}\\ &\quad\times\langle\xi\rangle^{m_1-|\alpha|+|\beta|}\langle\eta\rangle^{m_3-|\rho|}j^{-2}\\ &\quad\times\exp[\delta_1\langle\xi\rangle+(2\delta_2+|\delta_2|-8\kappa/R+\delta+\delta_3)\langle\eta\rangle]\end{aligned}$$

if $0 \le \nu \le 1$, $\langle\xi\rangle \ge R_0|\beta|$, $\kappa > 0$, $\delta > 0$, $R \ge 2R_0$, $8\kappa/R \ge 2\delta_2+|\delta_2|+\delta$, $R_0 \ge 6e^{12\kappa}\sqrt{n}A_1$ and $R \ge 12e^{12\kappa}\sqrt{n}C_*/\varepsilon_0$. Similarly, we have

$$\begin{aligned}&\Big|\partial_\xi^\alpha D_w^{\beta+\tilde\beta}\partial_\eta^\rho \widetilde{L}_1^j\Big[\langle w-y\rangle_{A_1}^{-2M}(1-\chi_j^{2\varepsilon_0/3}(y-w^0))\\ &\qquad\times\langle D_\zeta\rangle_{A_1}^{2M}\{\psi_j^R(\zeta)\exp[-\nu|\zeta+\eta|^2]g_2^R(\zeta,\eta)p(\xi,w,\zeta+\eta,y,\eta)\}\Big]\Big|\\ &\le C_{|\alpha|+|\tilde\beta|+|\rho|,\varepsilon_0,R}((A_1+24(1+\sqrt{2})/\varepsilon_0)/R_0)^{|\beta|}\\ &\qquad\times(48e^{12\kappa}((8(\sqrt{2}+\sqrt{3})+\widehat{C}+\widehat{C}_*)/R+A_1/R_0)/\varepsilon_0)^j j^{m_2}\\ &\qquad\times\langle\xi\rangle^{m_1-|\alpha|+|\beta|}\langle\eta\rangle^{m_3-|\rho|}\langle w-y\rangle_{A_1}^{-2M}\\ &\qquad\times\exp[\delta_1\langle\xi\rangle+(\delta_2+|\delta_2|/2-4\kappa/R)\langle\zeta\rangle+\delta_3\langle\eta\rangle]\end{aligned}$$

if $|w-w^0| < \varepsilon_0/6$, $0 \le \nu \le 1$, $\langle\xi\rangle \ge R_0|\beta|$, $\kappa > 0$ and $R \ge 4R_0$, where $\widetilde{L}_1 = |w-y|^{-2}\sum_{k=1}^n (y_k-w_k)D_{\zeta_k}$. Taking $R = 4R_0$, we have

$$\begin{aligned}\left|\partial_\xi^\alpha D_w^{\beta+\tilde\beta}\partial_\eta^\rho a_{\nu,j}^{2,2}(\xi,w,\eta)\right| &\le C_{|\alpha|+|\tilde\beta|+|\rho|,\varepsilon_0,R_0}\\ &\quad\times((A_1+24(1+\sqrt{2})/\varepsilon_0+2/\delta)/R_0)^{|\beta|}\langle\xi\rangle^{m_1-|\alpha|+|\beta|}\langle\eta\rangle^{m_3-|\rho|}j^{-2}\\ &\quad\times\exp[\delta_1\langle\xi\rangle+(2\delta_2+|\delta_2|-2\kappa/R_0+\delta+\delta_3)\langle\eta\rangle]\end{aligned}$$

if $0 \le \nu \le 1$, $\langle\xi\rangle \ge R_0|\beta|$, $\kappa > 0$, $\delta > 0$, $2\kappa/R_0 \ge 2\delta_2+|\delta_2|+\delta$, $R_0 \ge 96e^{12\kappa}(2(\sqrt{2}+\sqrt{3})+(\widehat{C}+\widehat{C}_*)/4+A_1)/\varepsilon_0$. Therefore, we have

$$\begin{aligned}\left|\partial_\xi^\alpha D_w^{\beta+\tilde\beta}\partial_\eta^\rho a_\nu^2(\xi,w,\eta)\right| &\le C_{|\alpha|+|\tilde\beta|+|\rho|,\varepsilon_0,R_0}\\ &\quad\times((A_1+24(1+\sqrt{2})/\varepsilon_0+4R_0/\kappa)/R_0)^{|\beta|}\langle\xi\rangle^{m_1-|\alpha|+|\beta|}\\ &\quad\times\langle\eta\rangle^{m_3-|\rho|}\exp[\delta_1\langle\xi\rangle-\kappa\langle\eta\rangle/R_0]\end{aligned}$$

if $0 \le \nu \le 1$, $\langle\xi\rangle \ge R_0|\beta|$,

$$\begin{cases}\kappa > 0, \quad \kappa/R_0 \ge \max\{2\delta_2, 4\delta_2+2|\delta_2|+2\delta_3\},\\ R_0 \ge 3e^{12\kappa}\max\{2\sqrt{n}A_1, \sqrt{n}C_*/\varepsilon_0,\\ \qquad (2^6(\sqrt{2}+\sqrt{3})+8(\widehat{C}+\widehat{C}_*)+2^5A_1)/\varepsilon_0\},\end{cases} \tag{A.43}$$

where $a_0^2(\xi,w,\eta) \equiv \lim_{\nu\downarrow 0} a_\nu^2(\xi,w,\eta) = \sum_{j=1}^\infty (a_{0,j}^{2,1}(\xi,w,\eta)+(a_{0,j}^{2,2}(\xi,w,\eta))$. Assume that (A.43) is satisfied. Then Lebesgue's convergence theorem gives

$$\begin{aligned}
&\lim_{\nu,\nu'\downarrow 0}(2\pi)^{-n}\int e^{iw\cdot(\eta-\xi)}\exp[-\nu'|w|^2]a_\nu^2(\xi,w,\eta)\hat{u}(\eta)\,d\eta dw\\
&= \lim_{\nu,\nu'\downarrow 0}(2\pi)^{-n}\int e^{iw\cdot(\eta-\xi)}\exp[-\nu'|w|^2]\\
&\qquad\times\langle w\rangle^{-2M}\langle D_\eta\rangle^{2M}(a_\nu^2(\xi,w,\eta)\hat{u}(\eta))\,d\eta dw\\
&= (2\pi)^{-n}\int e^{iw\cdot(\eta-\xi)}\langle w\rangle^{-2M}\langle D_\eta\rangle^{2M}(a_0^2(\xi,w,\eta)\hat{u}(\eta))\,d\eta dw\\
&\left(=\mathcal{F}[a_0^2(D_x,y,D_y)u](\xi)\right)
\end{aligned}$$

for $u \in \mathcal{S}_\infty$, where $M = [n/2]+1$. Putting $q_{0,\varepsilon}(\xi,y,\eta) = a_{0,\varepsilon}^1(\xi,y,\eta) + a_0^2(\xi,y,\eta)$, we have the first part of the assertion (i). It is easy to see that (2.72) is valid. Using Taylor's formula, we can write

$$\begin{aligned}
q_\varepsilon(\xi,w,\eta) &= (2\pi)^{-n}\sum_{j=1}^\infty \psi_j^R(\eta)\int e^{-iy\cdot\zeta}\sum_{|\gamma|=j}\langle y\rangle^{-2M}\langle D_\zeta\rangle^{2M}\\
&\times\left(\int_0^1 j(1-\theta)^{j-1}\partial_\zeta^\gamma\left\{g_0^R(\zeta,\eta)D_z^\gamma p(\xi,w,\zeta+\eta,z,\eta)\right\}_{z=w+\theta y}d\theta\right)\\
&\times\gamma!^{-1}\,d\zeta dy,
\end{aligned}$$

where $M = [n/2]+1$ and $R = 4R_0$. This yields

$$\begin{aligned}
&\left|\partial_\xi^\alpha D_w^{\beta+\tilde\beta}\partial_\eta^\rho q_\varepsilon(\xi,w,\eta)\right| \le C_{|\alpha|+|\tilde\beta|+|\rho|,\varepsilon,R_0}(5A_1/(2R_0))^{|\beta|}\\
&\quad\times\langle\xi\rangle^{m_1-|\alpha|}\langle\eta\rangle^{m_2+m_3-|\rho|+|\beta|}\exp[\delta_1\langle\xi\rangle+(\delta_2+\varepsilon|\delta_2|/2+\delta_3)\langle\eta\rangle]\\
&\quad\times\sum_{j=1}^\infty\exp[-2\kappa\langle\eta\rangle/R_0]j^n(3e^{1+24\kappa}n(\widehat{C}_*/\varepsilon+4A_1)A_1/R_0)^j
\end{aligned}$$

if $\kappa > 0$, $\langle\eta\rangle \ge 4R_0|\beta|$, which proves (2.73) with $R'(A_1,\varepsilon_0,\varepsilon,\kappa) \ge 6e^{1+24\kappa}n\times(\widehat{C}_*/\varepsilon+4A_1)A_1$. Next assume that $p(\xi,w,\zeta,y,\eta)$ satisfies (2.74). We choose $\varepsilon > 0$ so that $0 < \varepsilon \le \min\{1,\varepsilon_0'\}$. A simple calculation gives (2.75). Let $w^0 \in \boldsymbol{R}^n$ be fixed, and represent

$$q_\varepsilon(\xi,w,\eta) \equiv \sum_{j=1}^\infty \psi_j^R(\eta)(q_{\varepsilon,j}^0(\xi,w,\eta)+q_{\varepsilon,j}^1(\xi,w,\eta))+q_\varepsilon^2(\xi,w,\eta)$$

for $(w,\eta) \in \Gamma$ with $|w-w^0| < \varepsilon_0/3$, where $R = 4R_0$,

$$q_{\varepsilon,j}^0(\xi,w,\eta) = (2\pi)^{-n}\sum_{|\gamma|=j}\frac{j}{\gamma!}\int_0^1(1-\theta)^{j-1}\Big[\int_{|y-w^0|<2\varepsilon_0/3} e^{i(w-y)\cdot\zeta}$$

$$\times\langle w-y\rangle_{A_1}^{-2M}\langle D_\zeta\rangle_{A_1}^{2M}\partial_\zeta^\gamma\Big\{g_0^R(\zeta,\eta)D_z^\gamma p(\xi,w,\zeta+\eta,z,\eta)\Big\}_{z=w+\theta(y-w)}$$
$$\times dyd\zeta\Big]\,d\theta,$$

$$q_{\varepsilon,j}^1(\xi,w,\eta)=-(2\pi)^{-n}\int_{|y-w^0|>2\varepsilon_0/3}e^{i(w-y)\cdot\zeta}\langle w-y\rangle_{A_1}^{-2M}$$
$$\times\sum_{|\gamma|<j}\frac{1}{\gamma!}\langle D_\zeta\rangle_{A_1}^{2M}\partial_\zeta^\gamma\Big\{g_0^R(\zeta,\eta)D_z^\gamma p(\xi,w,\zeta+\eta,z,\eta)\Big\}_{z=w}dyd\zeta,$$

$$q_\varepsilon^2(\xi,w,\eta)=(2\pi)^{-n}\int_{|y-w^0|>2\varepsilon_0/3}e^{i(w-y)\cdot\zeta}\langle w-y\rangle_{A_1}^{-2M}\langle D_\zeta\rangle_{A_1}^{2M}$$
$$\times\{g_0^R(\zeta,\eta)p(\xi,w,\zeta+\eta,y,\eta)\}\,dyd\zeta$$

and $M=[n/2]+1$. Then we have

$$\Big|\partial_\xi^\alpha D_w^{\beta^1+\beta^2+\tilde\beta}\partial_\zeta^{\gamma+\tilde\gamma}\partial_\eta^\rho$$
$$\times\Big\{\langle w-y\rangle_{A_1}^{-2M}g_0^R(\zeta,\eta)D_z^\gamma p(\xi,w,\zeta+\eta,z,\eta)\Big\}_{z=w+\theta(y-w)}\Big|$$
$$\le C_{|\alpha|+|\tilde\beta|+|\tilde\gamma|+|\rho|,\varepsilon}(2(A_1+B)/R_0)^{|\beta^1|+|\beta^2|}$$
$$\times((\widehat{C}_*/\varepsilon+4A_1)B/R_0)^{|\gamma|}|\gamma|!\langle\xi\rangle^{m_1-|\alpha|+|\beta^1|}$$
$$\times\langle\eta\rangle^{m_2+m_3-|\tilde\gamma|-|\rho|+|\beta^2|}\langle w-y\rangle_{A_1}^{-2M}$$
$$\times\exp[\delta_1\langle\xi\rangle+(\delta_2+\varepsilon|\delta_2|/2+\delta_3)\langle\eta\rangle]$$

if $(w,\eta)\in\Gamma$, $|w-w^0|<\varepsilon_0/3$, $|y-w^0|<2\varepsilon_0/3$, $0\le\theta\le1$, $\langle\xi\rangle\ge 2R_0|\beta^1|$, $\langle\eta\rangle\ge 4R_0|\gamma|$ and $\langle\eta\rangle\ge 4R_0|\beta^2|$. This yields

$$\Big|\partial_\xi^\alpha D_w^{\beta^1+\beta^2+\tilde\beta}\partial_\eta^\rho q_{\varepsilon,j}^0(\xi,w,\eta)\Big|\le C_{|\alpha|+|\tilde\beta|+|\rho|,\varepsilon_0,\varepsilon,R_0}$$
$$\times((\varepsilon/\delta+2A_1+2B)/R_0)^{|\beta^1|+|\beta^2|}$$
$$\times(n(\widehat{C}_*/\varepsilon+4A_1)B/R_0)^j\langle\xi\rangle^{m_1-|\alpha|+|\beta^1|}\langle\eta\rangle^{m_2+m_3-|\rho|+|\beta^2|}$$
$$\times j^{n+|\tilde\beta|}\exp[\delta_1\langle\xi\rangle+(\delta_2+\varepsilon|\delta_2|/2+\delta_3+\delta)\langle\eta\rangle]\qquad\text{(A.44)}$$

if $(w,\eta)\in\Gamma$, $|w-w^0|<\varepsilon_0/3$, $8R_0(j-1)\le\langle\eta\rangle\le 12R_0j$, $\langle\xi\rangle\ge R_0|\beta^1|$, $\langle\eta\rangle\ge 4R_0|\beta^2|$ and $\delta>0$. We have also

$$\Big|\partial_\xi^\alpha D_w^{\beta^1+\beta^2+\tilde\beta}\partial_\eta^\rho\widetilde{L}_1^{j-|\gamma|}$$
$$\times\Big\{\langle w-y\rangle_{A_1}^{-2M}\partial_\zeta^{\gamma+\tilde\gamma}\Big(g_0^R(\zeta,\eta)D_z^\gamma p(\xi,w,\zeta+\eta,z,\eta)\Big)_{z=w}\Big\}\Big|$$
$$\le C_{|\alpha|+|\tilde\beta|+|\tilde\gamma|+|\rho|,\varepsilon_0,\varepsilon,R_0}((A_1+4B+12(1+\sqrt{2})/\varepsilon_0)/R_0)^{|\beta^1|}$$
$$\times((2A_1+B+3(1+\sqrt{2})/\varepsilon_0)/R_0)^{|\beta^2|}((\widehat{C}_*/(4\varepsilon)+A_1)/R_0)^j$$

$$\times(24n/\varepsilon_0)^{j-|\gamma|}(4B)^{|\gamma|}|\gamma|!\langle\xi\rangle^{m_1-|\alpha|+|\beta^1|}$$
$$\times\langle\eta\rangle^{m_2+m_3-|\tilde{\gamma}|-|\rho|+|\beta^2|}\langle w-y\rangle_{A_1}^{-2M}$$
$$\times\exp[\delta_1\langle\xi\rangle+(\delta_2+\varepsilon|\delta_2|/2+\delta_3)\langle\eta\rangle]$$

if $(w,\eta)\in\Gamma$, $|w-w^0|<\varepsilon_0/3$, $|y-w^0|>2\varepsilon_0/3$, $8R_0(j-1)\le\langle\eta\rangle\le 12R_0 j$, $\langle\xi\rangle\ge R_0|\beta^1|$, $\langle\eta\rangle\ge 4R_0|\beta^2|$ and $|\gamma|\le j-1$. This yields

$$\left|\partial_\xi^\alpha D_w^{\beta^1+\beta^2+\tilde{\beta}}\partial_\eta^\rho q_{\varepsilon,j}^1(\xi,w,\eta)\right|\le C_{|\alpha|+|\tilde{\beta}|+|\rho|,\varepsilon_0,\varepsilon,R_0}$$
$$\times((\varepsilon/\delta+A_1+4B+12(1+\sqrt{2})/\varepsilon_0)/R_0)^{|\beta^1|}$$
$$\times((\varepsilon/(4\delta)+2A_1+B+3(1+\sqrt{2})/\varepsilon_0)/R_0)^{|\beta^2|}$$
$$\times(n(6/\varepsilon_0+B)(\widehat{C}_*/\varepsilon+4A_1)/R_0)^j$$
$$\times\langle\xi\rangle^{m_1-|\alpha|+|\beta^1|}\langle\eta\rangle^{m_2+m_3-|\rho|+|\beta^2|}j^{n+|\tilde{\beta}|}$$
$$\times\exp[\delta_1\langle\xi\rangle+(\delta_2+\varepsilon|\delta_2|/2+\delta_3+\delta)\langle\eta\rangle] \tag{A.45}$$

if $(w,\eta)\in\Gamma$, $|w-w^0|<\varepsilon_0/3$, $8R_0(j-1)\le\langle\eta\rangle\le 12R_0 j$, $\langle\xi\rangle\ge R_0|\beta^1|$, $\langle\eta\rangle\ge 4R_0|\beta^2|$ and $\delta>0$. Similarly, we can see that

$$\Big|\partial_\xi^\alpha D_w^{\beta^1+\beta^2+\tilde{\beta}}\partial_\eta^\rho\widetilde{L}_1^j\Big[\langle w-y\rangle_{A_1}^{-2M}\langle D_\zeta\rangle_{A_1}^{2M}$$
$$\times\{g_0^R(\zeta,\eta)p(\xi,w,\zeta+\eta,y,\eta)\}\Big]\Big|$$
$$\le C_{|\alpha|+|\tilde{\beta}|+|\rho|,\varepsilon_0,\varepsilon}((3A_1/2+18(1+\sqrt{2})/\varepsilon_0)/R_0)^{|\beta^1|}$$
$$\times(3(A_1+3(1+\sqrt{2})/\varepsilon_0)/(2R_0))^{|\beta^2|}$$
$$\times(6n(\widehat{C}_*/\varepsilon+4A_1)/(\varepsilon_0R_0))^j\langle\xi\rangle^{m_1-|\alpha|+|\beta^1|}$$
$$\times\langle\eta\rangle^{m_2+m_3-|\rho|+|\beta^2|}\langle w-y\rangle_{A_1}^{-2M}$$
$$\times\exp[\delta_1\langle\xi\rangle+(\delta_2+\varepsilon|\delta_2|/2+\delta_3)\langle\eta\rangle]$$

if $|w-w^0|<\varepsilon_0/3$, $|y-w^0|>2\varepsilon_0/3$, $j\in \boldsymbol{Z}_+$, $4R_0 j\le\langle\eta\rangle\le 4R_0(j+1)$, $\langle\xi\rangle\ge R_0|\beta^1|$ and $\langle\eta\rangle\ge 4R_0|\beta^2|$, and, therefore,

$$\left|\partial_\xi^\alpha D_w^{\beta^1+\beta^2+\tilde{\beta}}\partial_\eta^\rho q_\varepsilon^2(\xi,w,\eta)\right|\le C_{|\alpha|+|\tilde{\beta}|+|\rho|,\varepsilon_0,\varepsilon,R_0}$$
$$\times((\varepsilon/\delta+3A_1/2+18(1+\sqrt{2})/\varepsilon_0)/R_0)^{|\beta^1|}$$
$$\times((\varepsilon/(4\delta)+3A_1/2+9(1+\sqrt{2})/(2\varepsilon_0))/R_0)^{|\beta^2|}$$
$$\times\langle\xi\rangle^{m_1-|\alpha|+|\beta^1|}\langle\eta\rangle^{m_2+m_3-|\rho|+|\beta^2|}$$
$$\times\exp[\delta_1\langle\xi\rangle+(\delta_2+\varepsilon|\delta_2|/2+\delta_3+\delta-3\kappa/R_0)\langle\eta\rangle] \tag{A.46}$$

if $(w,\eta)\in\Gamma$, $|w-w^0|<\varepsilon_0/3$, $\langle\xi\rangle\ge R_0|\beta^1|$, $\langle\eta\rangle\ge 4R_0|\beta^2|$, $\delta>0$, $\kappa>0$ and

$$R_0\ge 12e^{12\kappa}n(\widehat{C}_*/\varepsilon+4A_1)/\varepsilon_0.$$

Now (2.76) easily follows from (A.44)–(A.46).

A.4 Proof of Corollary 2.4. 7

We assume that the hypothese of Corollary 2.4. 7 are satisfied and that $\delta_1 > 0$. In fact, the proof of the assertion will become simpler in the case where $\delta_1 = 0$. The proof of the corollary is very similar to that of Theorem 2.4. 6 . Let us first prove the assertion (i). Put

$$p_{0,\varepsilon}(x,\xi) = a^0_{0,\varepsilon}(x,\xi),$$
$$q_{0,\varepsilon}(x,\xi) = a^1_{0,\varepsilon}(x,\xi) + a^2_0(x,\xi),$$

where $M = [n/2]+1$, g^R_k ($k = 0,1,2$) are functions in Appendix A.3, $\{\chi^\varepsilon_j\}$ is a family of functions used in Section 2.4 and

$$a^k_{0,\varepsilon}(x,\xi) = (2\pi)^{-n}\int e^{-iy\cdot\zeta}\langle y\rangle^{-2M}_{A_1}\langle D_\zeta\rangle^{2M}_{A_1} \times\{g^{4R_0}_k(\zeta,\xi)p(x,\xi+\zeta,x+y,\xi)\}\,dyd\zeta \quad (k = 0,1),$$

$$a^2_0(x,\xi) = \sum_{j=1}^{\infty}(a^{2,1}_{0,j}(x,\xi) + a^{2,2}_{0,j}(x,\xi)),$$

$$a^{2,1}_{0,j}(x,\xi) = (2\pi)^{-n}\int e^{-iy\cdot\zeta}\psi^{4R_0}_j(\zeta)\chi^{\varepsilon_0}_j(y)g^{4R_0}_2(\zeta,\xi) \times p(x,\xi+\zeta,x+y,\xi)\,dyd\zeta,$$

$$a^{2,2}_{0,j}(x,\xi) = (2\pi)^{-n}\int e^{-iy\cdot\zeta}\langle y\rangle^{-2M}_{A_1}(1-\chi^{\varepsilon_0}_j(y)) \times\langle D_\zeta\rangle^{2M}_{A_1}\{\psi^{4R_0}_j(\zeta)g^{4R_0}_2(\zeta,\xi)p(x,\xi+\zeta,x+y,\xi)\}\,dyd\zeta.$$

Then the first part of the assertion (i) easily follows, applying the same arguments as in Appendix A.3. $q_\varepsilon(x,\xi)$ can be represented as

$$q_\varepsilon(x,\xi) = \sum_{j=1}^{\infty}\psi^{4R_0}_j(\xi)\sum_{|\gamma|=j}\frac{j}{\gamma!}\int_0^1(1-\theta)^{j-1} \times\Big[\int e^{-iy\cdot\zeta}\langle y\rangle^{-2M}\langle D_\zeta\rangle^{2M}\partial^\gamma_\zeta \times\Big\{g^R_0(\zeta,\xi)D^\gamma_z p(x,\xi+\zeta,z,\xi)\Big\}_{z=x+\theta y}dyd\zeta\Big]\,d\theta.$$

So one can easily prove the assertion (i). Next we shall prove the assertion (ii). Assume that (2.77) and (2.78) are valid, and let $(x,\xi)\in\Gamma$. Represent

$$a^k_{0,\varepsilon}(x,\xi) = (2\pi)^{-n}\int e^{i(x-y)\cdot\zeta}\langle x-y\rangle^{-2M}_B\langle D_\zeta\rangle^{2M}_B \times\{g^{4R_0}_k(\zeta,\xi)p(x,\xi+\zeta,y,\xi)\}\,dyd\zeta$$

($k = 0, 1$). Applying the same argument as for $a^k_{0,\varepsilon}(\xi, w, \eta)$ ($k = 0, 1$) in Appendix A.3, we have (2.79), and

$$\left|a^{1(\alpha)}_{0,\varepsilon(\beta)}(x,\xi)\right| \le C_{|\alpha|,\varepsilon}(B+3R_0)^{|\beta|}|\beta|!\langle\xi\rangle^{m_2}\exp[-\langle\xi\rangle/R_0]$$

if $(x,\xi) \in \Gamma$, $R_0 \ge 4e^3\sqrt{n}(3B/4 + A_1)/\varepsilon$ and $4(\delta_1)_+ + \delta_2 \le 1/R_0$. Let $x^0 \in \boldsymbol{R}^n$, and represent

$$a^2_0(x,\xi) = \sum_{j=1}^{\infty}(\tilde{a}^{2,1}_{0,j}(x,\xi) + \tilde{a}^{2,2}_{0,j}(x,\xi))$$

for $|x - x^0| < \varepsilon_0/3$, where

$$\begin{aligned}
\tilde{a}^{2,1}_{0,j}(x,\xi) &= (2\pi)^{-n}\int e^{i(x-y)\cdot\zeta}\psi^{4R_0}_j(\zeta)\chi^{2\varepsilon_0/3}_j(y-x^0) \\
&\quad \times g^{4R_0}_2(\zeta,\xi)p(x,\xi+\zeta,y,\xi)\,dyd\zeta, \\
\tilde{a}^{2,2}_{0,j}(x,\xi) &= (2\pi)^{-n}\int e^{i(x-y)\cdot\zeta}\langle x-y\rangle^{-2M}_B(1-\chi^{2\varepsilon_0/3}_j(y-x^0)) \\
&\quad \times\langle D_\zeta\rangle^{2M}_B\{\psi^{4R_0}_j(\zeta)g^{4R_0}_2(\zeta,\xi)p(x,\xi+\zeta,y,\xi)\}\,dyd\zeta.
\end{aligned}$$

The same arguments as in Appendix A.3 give

$$\begin{aligned}
\left|\tilde{a}^{2,1(\alpha)}_{0,j(\beta)}(x,\xi)\right| &\le C_{|\alpha|,R_0}(B+2/\delta)^{|\beta|}|\beta|!\langle\xi\rangle^{m_2-|\alpha|} \\
&\quad \times j^{-2}\exp[(2\delta_1+|\delta_1|+\delta_2+\delta-2/R_0)\langle\xi\rangle], \\
\left|\tilde{a}^{2,2(\alpha)}_{0,j(\beta)}(x,\xi)\right| &\le C_{|\alpha|,\varepsilon_0,R_0}(B+12(1+\sqrt{2})/\varepsilon_0+2/\delta)^{|\beta|}|\beta|! \\
&\quad \times\langle\xi\rangle^{m_2-|\alpha|}j^{-2}\exp[(2\delta_1+|\delta_1|+\delta_2+\delta-2/R_0)\langle\xi\rangle]
\end{aligned}$$

if $(x,\xi) \in \Gamma$, $|x - x^0| < \varepsilon_0/6$, $\delta > 0$, $R_0 \ge 3e^{12}\max\{\sqrt{n}(C_*/(2\varepsilon_0) + A_1), 8(8(\sqrt{2}+\sqrt{3})+\widehat{C}+\widehat{C}_*+4A_1)/\varepsilon_0\}$ and $2\delta_1+|\delta_1|+\delta \le 2/R_0$. Therefore, we have

$$\begin{aligned}
\left|q^{(\alpha)}_{0,\varepsilon(\beta)}(x,\xi)\right| &\le C_{|\alpha|,\varepsilon_0,\varepsilon,R_0}(B+4R_0+12(1+\sqrt{2})/\varepsilon_0)^{|\beta|} \\
&\quad \times|\beta|!\langle\xi\rangle^{(m_1)_++m_2}\exp[-\langle\xi\rangle/R_0]
\end{aligned}$$

if $(x,\xi) \in \Gamma$, $R_0 \ge \max\{4e^3\sqrt{n}(3B/4+A_1)/\varepsilon, 3e^{12}\sqrt{n}(C_*/(2\varepsilon_0)+A_1), 24 \times e^{12}(8(\sqrt{2}+\sqrt{3})+\widehat{C}+\widehat{C}_*+4A_1)/\varepsilon_0\}$ and $\max\{4(\delta_1)_+ + \delta_2, 2\delta_1, 4\delta_1 + 2|\delta_1|+2\delta_2\} \le 1/R_0$. This proves (2.80). Now assume that (2.81) is valid. (2.82) is obvious. Write, for $|x - x^0| < \varepsilon_0/3$,

$$q_\varepsilon(x,\xi) = \sum_{j=1}^{\infty}\psi^{4R_0}_j(\xi)(q^0_{\varepsilon,j}(x,\xi) + q^1_{\varepsilon,j}(x,\xi)) + q^2_\varepsilon(x,\xi),$$

where

$$q^0_{\varepsilon,j}(x,\xi) = (2\pi)^{-n}\sum_{|\gamma|=j}\frac{j}{\gamma!}\int_0^1(1-\theta)^{j-1}\Big[\int_{|y-x^0|<2\varepsilon_0/3} e^{i(x-y)\cdot\zeta} \times\langle x-y\rangle_B^{-2M}\langle D_\zeta\rangle_B^{2M}\partial_\zeta^\gamma \times\Big\{g_0^{4R_0}(\zeta,\xi)D_z^\gamma p(x,\xi+\zeta,z,\xi)\Big\}_{z=x+\theta(y-x)}dyd\zeta\Big]\,d\theta,$$

$$q^1_{\varepsilon,j}(x,\xi) = -(2\pi)^{-n}\int_{|y-x^0|>2\varepsilon_0/3} e^{i(x-y)\cdot\zeta}\langle x-y\rangle_B^{-2M} \times\sum_{|\gamma|<j}\frac{1}{\gamma!}\langle D_\zeta\rangle_B^{2M}\partial_\zeta^\gamma\Big\{g_0^{4R_0}(\zeta,\xi)D_z^\gamma p(x,\xi+\zeta,z,\xi)\Big\}_{z=x}dyd\zeta,$$

$$q^2_\varepsilon(x,\xi) = (2\pi)^{-n}\int_{|y-x^0|>2\varepsilon_0/3} e^{i(x-y)\cdot\zeta}\langle x-y\rangle_B^{-2M}\langle D_\zeta\rangle_B^{2M} \times\{g_0^{4R_0}(\zeta,\xi)p(x,\xi+\zeta,y,\xi)\}\,dyd\zeta.$$

Then, repetition of the same arguments as in Appendix A.3 yields

$$\Big|q^{(\alpha)}_{\varepsilon(\beta)}(x,\xi)\Big| \le C_{|\alpha|,\varepsilon,R_0}(2B+\varepsilon R_0+6(1+\sqrt{2})/\varepsilon_0)^{|\beta|} \times|\beta|!\langle\xi\rangle^{m_1+m_2-|\alpha|}\exp[-\langle\xi\rangle/R_0]$$

if $(x,\xi)\in\Gamma$, $R_0\ge e^{24}n(12/\varepsilon_0+B)(\widehat{C}_*/\varepsilon+4A_1)$ and $2\delta_1+\varepsilon|\delta_1|+2\delta_2\le 1/R_0$. This proves (2.83).

A.5 Proof of Theorem 2.5. 3

We assume that the hypotheses of Section 2.5 are satisfied and use the notations there. Let $g_1^R(\xi,\zeta,\eta)\in C^\infty(\boldsymbol{R}^{n'}\times\boldsymbol{R}^n\times\boldsymbol{R}^n)$ ($R\ge 3$) be a symbol such that $0\le g_1^R(\xi,\zeta,\eta)\le 1$ and

$$g_1^R(\xi,\zeta,\eta)=\begin{cases}1 & \text{if } |\xi|\ge 2c_2(T)(2|\zeta|+3|\eta|)/c_3(S)\\ & \text{or } |\xi|^2+|\zeta|^2+|\eta|^2\le 1,\\ 0 & \text{if } |\xi|\le c_2(T)(2|\zeta|+3|\eta|)/c_3(S)\\ & \text{and } |\xi|^2+|\zeta|^2+|\eta|^2\ge 4,\end{cases}$$

$$\Big|\partial_\xi^{\alpha+\tilde\alpha}\partial_\zeta^{\gamma+\tilde\gamma}\partial_\eta^{\rho+\tilde\rho}g_1^R(\xi,\zeta,\eta)\Big|\le C_{|\tilde\alpha|+|\tilde\gamma|+|\tilde\rho|}(\widehat{C}(S,T)/R)^{|\alpha|+|\gamma|+|\rho|} \times\langle\xi\rangle^{-|\tilde\alpha|-|\tilde\gamma|-|\tilde\rho|}\quad\text{if } \langle\xi\rangle\ge R|\alpha|,\ \langle\zeta\rangle\ge R|\gamma| \text{ and } \langle\eta\rangle\ge R|\rho|.$$

Similarly, we choose symbols $g_2^{\varepsilon_0,R}(\zeta,\eta), g_3^R(\zeta,\eta)\in C^\infty(\boldsymbol{R}^n\times\boldsymbol{R}^n)$ ($R\ge 3$) so that $0\le g_2^{\varepsilon_0,R}\le 1$, $0\le g_3^R\le 1$ and

$$g_2^{\varepsilon_0,R}(\zeta,\eta)=\begin{cases}1 & \text{if } |\zeta-\eta|\ge\varepsilon_0|\eta| \text{ and } |\eta|\ge 1,\\ 0 & \text{if } |\zeta-\eta|\le\varepsilon_0|\eta|/2 \text{ or } |\eta|\le 1/2,\end{cases}$$

$$g_3^R(\zeta,\eta)=\begin{cases}1 & \text{if } |\zeta|\ge 2|\eta| \text{ and } |\eta|\ge 1,\\ 0 & \text{if } |\zeta|\le |\eta|/2 \text{ or } |\eta|\le 1/2,\end{cases}$$

$$\left|\partial_\zeta^{\gamma+\tilde\gamma}\partial_\eta^{\rho+\tilde\rho} g_2^{\varepsilon_0,R}(\zeta,\eta)\right|\le C_{|\tilde\gamma|+|\tilde\rho|}(\widehat{C}(\varepsilon_0)/R)^{|\gamma|+|\rho|}\langle\eta\rangle^{-|\tilde\gamma|-|\tilde\rho|}$$
$$\text{if } \langle\zeta\rangle\ge R|\gamma| \text{ and } \langle\eta\rangle\ge R|\rho|,$$
$$\left|\partial_\zeta^{\gamma+\tilde\gamma}\partial_\eta^{\rho+\tilde\rho} g_3^{R}(\zeta,\eta)\right|\le C_{|\tilde\gamma|+|\tilde\rho|}(\widehat{C}_*/R)^{|\gamma|+|\rho|}\langle\eta\rangle^{-|\tilde\gamma|-|\tilde\rho|}$$
$$\text{if } \langle\zeta\rangle\ge R|\gamma| \text{ and } \langle\eta\rangle\ge R|\rho|.$$

We assume that $R_0\ge R(U,A_1)$ and $\delta_2+c_1^+(T)+c_1^+(U)\le 1/(3R_0)$, where $R(U,A_1)$ is the quantity in Lemma 2.5. 1 . We may also assume that $\varepsilon_0\le 1/2$. Write, for $u\in\mathcal{S}_\infty(\boldsymbol{R}^n)$,

$$\mathcal{F}_x[p_{S,T,U,-U}(D_x,w,D_z,y,D_y)u(x)](\xi)$$
$$\equiv \widehat{P}u(\xi)+(2\pi)^{-2n}\sum_{j=1}^{3}Q_j\hat{u}(\xi),$$

where

$$Q_1\hat{u}(\xi)=\int E(\xi,w,\zeta,y,\eta)$$
$$\times L_1^M L^M\{g_1^{R_0}(\xi,\zeta,\eta)p(\xi,w,\zeta,y,\eta)\hat{u}(\eta)\}\,d\eta dy d\zeta dw,$$
$$Q_2\hat{u}(\xi)=\int E(\xi,w,\zeta,y,\eta)\sum_{j=1}^{\infty}\psi_j^{R_0}(\zeta)$$
$$\times L_1^M L^M K_1^j\{(1-g_1^{R_0}(\xi,\zeta,\eta))g_2^{\varepsilon_0,R_0}(\zeta,\eta)g_3^{R_0}(\zeta,\eta)$$
$$\times p(\xi,w,\zeta,y,\eta)\hat{u}(\eta)\}\,d\eta dy d\zeta dw,$$
$$Q_3\hat{u}(\xi)=\int E(\xi,w,\zeta,y,\eta)L_1^M L^M\{(1-g_1^{R_0}(\xi,\zeta,\eta))g_2^{\varepsilon_0,R_0}(\zeta,\eta)$$
$$\times(1-g_3^{R_0}(\zeta,\eta))p(\xi,w,\zeta,y,\eta)\hat{u}(\eta)\}\,d\eta dy d\zeta dw,$$
$$\widehat{P}u(\xi)=(2\pi)^{-2n}\int E(\xi,w,\zeta,y,\eta)L_1^M L^M\{(1-g_1^{R_0}(\xi,\zeta,\eta))$$
$$\times(1-g_2^{\varepsilon_0,R_0}(\zeta,\eta))p(\xi,w,\zeta,y,\eta)\hat{u}(\eta)\}\,d\eta dy d\zeta dw,$$
$$E(\xi,w,\zeta,y,\eta)=\exp[iS(w,\xi)+iT(w,\zeta)+iU(y,\zeta)-iU(y,\eta)],$$

$M\ge n+n''+2$ and L, L_1 and K_1 are differential operators defined by

$${}^tL=\left(1+|\nabla_\eta U(y,\eta)|^2\right)^{-1}\left(-\sum_{j=1}^{n}\overline{\partial_{\eta_j}U(y,\eta)}D_{\eta_j}+1\right),$$
$${}^tL_1=\left(1+|\nabla_\zeta T(w,\zeta)+\nabla_\zeta U(y,\zeta)|^2\right)^{-1}$$

$$\times\Big(\sum_{j=1}^{n}\Big(\overline{\partial_{\zeta_j}T(w,\zeta)}+\overline{\partial_{\zeta_j}U(y,\zeta)}\Big)D_{\zeta_j}+1\Big),$$

$${}^tK_1=|\nabla_yU(y,\zeta)-\nabla_yU(y,\eta)|^{-2}\sum_{j=1}^{n}\Big(\overline{\partial_{y_j}U(y,\zeta)}-\overline{\partial_{y_j}U(y,\eta)}\Big)D_{y_j}$$
$$(\zeta\neq\eta).$$

Applying the same arguments as in Appendix A.1, we can prove the following

Lemma A.5.1 *Q_j ($j=1,2,3$) map continuously $\widehat{S}_{\varepsilon_2}(\boldsymbol{R}^n)$ and $\widehat{S}_{-\varepsilon_2}{}'(\boldsymbol{R}^n)$ to $\widehat{S}_{\varepsilon_1}(\boldsymbol{R}^{n'})$ and $\widehat{S}_{-\varepsilon_1}{}'(\boldsymbol{R}^{n'})$, respectively, if $R_0\geq R(S,T,U,A_1,\varepsilon_0)$ and*

$$\max\{\delta_1+c_1(S)+\varepsilon_1,\delta_2+c_1^+(T)+c_1^+(U),\delta_3+c_1^-(U)-\varepsilon_2\}\leq\delta(S,T,U)/R_0,$$

where $R(S,T,U,A_1,\varepsilon_0)$ and $\delta(S,T,U)$ are positive constants.

We can represent

$$\begin{aligned}\widehat{P}u(\xi) &= (2\pi)^{-n}\lim_{\nu\downarrow 0}\int e^{-\nu|w|^2}e^{iS(w,\xi)+iT(w,\eta)}\tilde{p}(\xi,w,\eta)\hat{u}(\eta)\,d\eta dw\\ &= \mathcal{F}_x[\tilde{p}_{S,T}(D_x,w,D_y)u(x)](\xi),\end{aligned}$$

where

$$\begin{aligned}\tilde{p}(\xi,w,\eta) &= (2\pi)^{-n}\lim_{\nu'\downarrow 0}\int e^{-\nu'|y|^2}e^{iT(w,\zeta)-iT(w,\eta)+iU(y,\zeta)-iU(y,\eta)}\\ &\qquad\times f(\xi,w,\zeta,y,\eta)\,dyd\zeta,\\ f(\xi,w,\zeta,y,\eta) &= (1-g_1^{R_0}(\xi,\zeta,\eta))(1-g_2^{\varepsilon_0,R_0}(\zeta,\eta))p(\xi,w,\zeta,y,\eta).\end{aligned}$$

Write

$$\begin{aligned}&T(w,\zeta)-T(w,\eta)+U(y,\zeta)-U(y,\eta)=z(w,\zeta,y,\eta)\cdot(\zeta-\eta),\\ &z(w,\zeta,y,\eta)=\widetilde{\nabla}_\eta U(y,\eta,\zeta)+\widetilde{\nabla}_\eta T(w,\eta,\zeta)\\ &\Big(=U_0y+\widetilde{\nabla}_\eta\widetilde{U}(y,\eta,\zeta)+\widetilde{\nabla}_\eta U_1(\eta,\zeta)+\widetilde{\nabla}_\eta T(w,\eta,\zeta)\Big).\end{aligned}$$

Then $z(w,\zeta,y,\eta)$ can be regarded as an analytic function of y in $\mathcal{U}_{\eta,1/(\sqrt{n}A_0)}$ for $\zeta,\eta\in\boldsymbol{R}^n$ with $|\eta-\zeta|\leq\varepsilon_0|\eta|$, $|\zeta|\geq 1$ and $|\eta|\geq 1$, and $w\in\boldsymbol{R}^{n''}$ with $(w,\zeta)\in\Omega_1'$. Recall that $\delta\equiv\delta(U)$ and $c(U)\leq\min\{\gamma_1\delta,\gamma_2\}/|U_0^{-1}|$, where $\delta(U)$ is the quantity in Lemma 2.5.2 . Let $0<\delta'\leq\delta$, and assume that

$$c(U)\leq\min\{\gamma_1\delta',\gamma_2\}/|U_0^{-1}|.$$

Fix $\zeta, \eta \in \boldsymbol{R}^n$ and $w \in \boldsymbol{R}^{n''}$ so that $|\eta - \zeta| \le \varepsilon_0|\eta|$, $(w, \zeta) \in \Omega_1'$, $|\zeta| \ge 1$ and $|\eta| \ge 1$, and put

$$\Omega_{\eta,\zeta,w,t} = \Big\{\mathrm{Re}\ z(w,\zeta,y,\eta) + i(1-t)\,\mathrm{Im}\ z(w,\zeta,y,\eta) \in \boldsymbol{C}^n;$$
$$y \in \mathcal{U}_{\eta,\delta'} \cap \boldsymbol{R}^n\Big\} \quad (\,t \in [0,1]),$$

$$\Omega_{\eta,\zeta,w} = \bigcup_{t\in[0,1]} \Omega_{\eta,\zeta,w,t}.$$

We define the map $\widetilde{\mathcal{Z}}_\delta$ by

$$\widetilde{\mathcal{Z}}_\delta \,:\, \boldsymbol{R}^n \ni y \mapsto y + U_0^{-1}\Psi_\eta(y;\delta)\,\mathrm{Re}\ \widetilde{\nabla}_\eta \widetilde{U}(y,\eta,\zeta) \in \boldsymbol{R}^n.$$

Then, applying the same argument as in the proof of Lemma 2.4. 2 we can show that $\widetilde{\mathcal{Z}}_\delta$ is a diffeomorphism on $\boldsymbol{R}^n$, $|(\partial\widetilde{Z}_\delta/\partial z)(z;\eta,\zeta)| \le 2$ and

$$\Big\{\widetilde{Z}_\delta(z;\eta,\zeta) \in \boldsymbol{R}^n;\ |z - \tilde{z}| < \delta/2 \text{ for some } \tilde{z} \in \widehat{\mathcal{U}}_{\eta,\zeta,\delta}\Big\}$$
$$\subset \mathcal{U}_{\eta,\delta} \cap \boldsymbol{R}^n,$$

where $\widetilde{Z}_\delta(z;\eta,\zeta)$ is the inverse function (map) of $\widetilde{\mathcal{Z}}_\delta$ and $\widehat{\mathcal{U}}_{\eta,\zeta,\delta} = \widetilde{\mathcal{Z}}_\delta(\mathcal{U}_\eta)$, modifying δ and γ_j ($j = 1, 2$) if necessary.

Lemma A.5. 2 *Assume that $\delta' \le \delta/5$ and $c(T) \le \delta'/(4|U_0^{-1}|)$. Let $z \in \boldsymbol{C}^n$, and put $z^1 = U_0^{-1}(z - \widetilde{\nabla}_\eta U_1(\eta,\zeta) - \widetilde{\nabla}_\eta T(w,\eta,\zeta))$ and $y^1 = Z_\delta(z^1;\eta,\zeta)$.* (i) *If $z \in \Omega_{\eta,\zeta.w}$ and*

$$\gamma_1 \le 1/4, \quad \gamma_2 \le 1/4, \quad \delta' \le 3/(2^4 n^{3/2} C(\widetilde{U}) A_0^3 |U_0^{-1}|), \tag{A.47}$$

then $z^1 \in \widetilde{\mathcal{U}}_{\eta,\zeta,5\delta'} \subset \widetilde{\mathcal{U}}_{\eta,\zeta,\delta}$, $y^1 \in \mathcal{U}_{\eta,\delta}$ and, in particular, $y^1 + U_0^{-1}\widetilde{\nabla}_\eta \widetilde{U}(y^1, \eta,\zeta) = z^1$. (ii) *If $\mathrm{Re}\ y^1 \in \mathcal{U}_\eta$ and*

$$\gamma_1 \le 1/(4(1+2\delta)), \quad \delta' \ge 2^6 n^{3/2} C(\widetilde{U})|U_0^{-1}|A_0^3\delta^2/3, \tag{A.48}$$

then $\mathrm{Re}\ z \in \Omega_{\eta,\zeta,w,1}$.

Proof (i) Assume that $z \in \Omega_{\eta,\zeta,w}$ and (A.47) is satisfied. Then there are $y \in \mathcal{U}_{\eta,\delta'} \cap \boldsymbol{R}^n$, $\tilde{y} \in \mathcal{U}_\eta$ and $t \in [0,1]$ such that $|y - \tilde{y}| < \delta'$ and

$$z = \mathrm{Re}\ z(w,\zeta,y,\eta) + i(1-t)\,\mathrm{Im}\ z(w,\zeta,y,\eta).$$

Therefore, we have

$$|U_0^{-1}(z(w,\zeta,y,\eta) - \widetilde{\nabla}_\eta U_1(\eta,\zeta) - \widetilde{\nabla}_\eta T(w,\eta,\zeta) - U_0\tilde{y} - \widetilde{\nabla}_\eta \widetilde{U}(\tilde{y},\eta,\zeta))|$$
$$\le |y - \tilde{y}| + |U_0^{-1}|(c(U)|y - \tilde{y}| + 4n^{3/2}C(\widetilde{U})A_0^3|y - \tilde{y}|^2/3)$$

since $\delta' \leq 1/(4\sqrt{n}A_0)$. This gives

$$U_0^{-1}(z(w,\zeta,y,\eta) - \tilde{\nabla}_\eta U_1(\eta,\zeta) - \tilde{\nabla}_\eta T(w,\eta,\zeta)) \in \tilde{\mathcal{U}}_{\eta,\zeta,3\delta'}.$$

Similarly, we have

$$\begin{aligned}&|U_0^{-1}\,\mathrm{Im}\; z(w,\zeta,y,\eta)|\\ &\leq |U_0^{-1}|(c(U)(1+|y-\tilde{y}|) + 4n^{3/2}C(\tilde{U})A_0^3|y-\tilde{y}|^2/3 + c(T)) < \delta'.\end{aligned}$$

So we have $z^1 \in \tilde{\mathcal{U}}_{\eta,\zeta,5\delta'} \subset \tilde{\mathcal{U}}_{\eta,\zeta,\delta}$ and, by Lemma 2.5. 2 , $y^1 \in \mathcal{U}_{\eta,\delta}$. (ii) Assume that Re $y^1 \in \mathcal{U}_\eta$ and (A.48) is satisfied. First suppose that $y^1 \notin \mathcal{U}_{\eta,2\delta}$. Then, we have $y^1 = z^1$ and

$$\mathrm{Re}\; z - \mathrm{Re}\; z(w,\zeta,y^2,\eta) = -\mathrm{Re}\; \tilde{\nabla}_\eta \tilde{U}(y^2,\eta,\zeta),$$

where $y^2 = \mathrm{Re}\; y^1$. By assumption,

$$|U_0^{-1}(\mathrm{Re}\; z - \mathrm{Re}\; z(w,\zeta,y^2,\eta))| \leq c(U)|U_0^{-1}|.$$

Since $U_0^{-1}(\mathrm{Re}\; z(w,\zeta,y^2,\eta) - \tilde{\nabla}_\eta U_1(\eta,\zeta) - \mathrm{Re}\; \tilde{\nabla}_\eta T(w,\eta,\zeta)) = y^2 + U_0^{-1} \times$ Re $\tilde{\nabla}_\eta \tilde{U}(y^2,\eta,\zeta) \in \tilde{\mathcal{U}}_{\eta,\zeta,\delta}$ and $c(U)|U_0^{-1}| < \delta'/2$, $y^3 \equiv \tilde{Z}_\delta(U_0^{-1}(\mathrm{Re}\; z - \tilde{\nabla}_\eta U_1(\eta,\zeta) - \mathrm{Re}\; \tilde{\nabla}_\eta T(w,y,\zeta));\eta,\zeta)$ belongs to $\mathcal{U}_{\eta,\delta'} \cap \boldsymbol{R}^n$ and

$$U_0^{-1}(\mathrm{Re}\; z - \tilde{\nabla}_\eta U_1(\eta,\zeta) - \mathrm{Re}\; \tilde{\nabla}_\eta T(w,y,\zeta)) = y^3 + U_0^{-1}\mathrm{Re}\; \tilde{\nabla}_\eta \tilde{U}(y^3,\eta,\zeta).$$

So we have Re z = Re $z(w,\zeta,y^3,\eta)$ and Re $z \in \Omega_{\eta,\zeta,w,1}$. Next suppose that $y^1 \in \mathcal{U}_{\eta,2\delta}$. Then, $|y^1 - y^2| < 2\delta$. By definition we have

$$\begin{aligned}&U_0^{-1}(\mathrm{Re}\; z - \mathrm{Re}\; z(w,\zeta,y^2,\eta))\\ &= U_0^{-1}(\Psi_\eta(y^1;\delta)\mathrm{Re}\; \tilde{\nabla}_\eta \tilde{U}(y^1,\eta,\zeta) - \mathrm{Re}\; \tilde{\nabla}_\eta \tilde{U}(y^2,\eta,\zeta)).\end{aligned}$$

Therefore, we have

$$\begin{aligned}&|U_0^{-1}(\mathrm{Re}\; z - \mathrm{Re}\; z(w,\zeta,y^2,\eta))|\\ &\leq |U_0^{-1}|(c(U)|y^1 - y^2| + 4n^{3/2}C(\tilde{U})A_0^3|y^1 - y^2|^2/3 + c(U)) < \delta'/2,\end{aligned}$$

which gives Re $z \in \Omega_{\eta,\zeta,w,1}$ in the same way. □

We assume that the hypotheses in Lemma A.5. 2 , (A.47) and (A.48) are satisfied. We put

$$f(\xi,w,\zeta,y,\eta;u) = (1 - g_1^{R_0}(\xi,\zeta,\eta))(1 - g_2^{\varepsilon_0,R_0}(\zeta,\eta))p(\xi,w,\zeta,y,\eta;u)$$

for $(\xi,w,\zeta,y,\eta,u) \in \boldsymbol{R}^{n'} \times \boldsymbol{R}^{n''} \times \boldsymbol{R}^n \times \boldsymbol{R}^n \times \boldsymbol{R}^n \times \boldsymbol{C}^n$.

Lemma A.5. 3 *$f(\xi, w, \zeta, y, \eta; u)$ is analytic in u,* supp $f(\cdot; u) \subset$ supp *$f(\cdot)$ for $u \in C^n$ and*

$$\begin{aligned}
&f(\xi, w, \zeta, y, \eta; 0) = f(\xi, w, \zeta, y, \eta),\\
&\left|\partial_\xi^{\alpha+\tilde\alpha} D_w^{\beta^1+\beta^2+\tilde\beta} \partial_\zeta^{\gamma+\tilde\gamma} D_y^{\lambda^1+\lambda^2+\tilde\lambda} \partial_\eta^{\rho+\tilde\rho} \partial_u^{\delta+\tilde\delta} f(\xi, w, \zeta, y, \eta; u)\right|\\
&\le C_{|\tilde\alpha|+|\tilde\beta|+|\tilde\gamma|+|\tilde\lambda|+|\tilde\rho|+|\tilde\delta|}((\widehat{C}(S,T) + A_1)/R_0)^{|\alpha|}\\
&\quad \times (A_1/R_0)^{|\beta^1|+|\beta^2|+|\lambda^1|+|\lambda^2|+|\delta|}((\widehat{C}(S,T) + \widehat{C}(\varepsilon_0) + A_1)/R_0)^{|\gamma|}\\
&\quad \times ((\widehat{C}(S,T) + \widehat{C}(\varepsilon_0) + \widehat{C} + A_1)/R_0)^{|\rho|}\langle\xi\rangle^{m_1-|\tilde\alpha|+|\beta^1|}\\
&\quad \times \langle\zeta\rangle^{m_2-|\tilde\gamma|+|\beta^2|+|\lambda^1|}\langle\eta\rangle^{m_3-|\tilde\rho|+|\lambda^2|+|\delta|+1}\\
&\quad \times \exp[\delta_1\langle\xi\rangle + \delta_2\langle\zeta\rangle + (\delta_3 + \sqrt{n}A_1|u|/R_0)\langle\eta\rangle],\\
&\left|(D_{y_j} - D_{u_j})\partial_\xi^{\alpha+\tilde\alpha} D_w^{\beta^1+\beta^2+\tilde\beta} \partial_\zeta^{\gamma+\tilde\gamma} D_y^{\lambda^1+\lambda^2+\tilde\lambda} \partial_\eta^{\rho+\tilde\rho} \partial_u^{\delta+\tilde\delta} f(\xi, w, \zeta, y, \eta; u)\right|\\
&\le C_{|\tilde\alpha|+|\tilde\beta|+|\tilde\gamma|+|\tilde\lambda|+|\tilde\rho|+|\tilde\delta|}((\widehat{C}(S,T) + A_1)/R_0)^{|\alpha|}(A_1/R_0)^{|\beta^1|+|\beta^2|+|\lambda^1|+|\lambda^2|}\\
&\quad \times (eA_1/R_0)^{|\delta|}((\widehat{C}(S,T) + \widehat{C}(\varepsilon_0) + A_1)/R_0)^{|\gamma|}\\
&\quad \times ((\widehat{C}(S,T) + \widehat{C}(\varepsilon_0) + \widehat{C} + A_1)/R_0)^{|\rho|}\\
&\quad \times \langle\xi\rangle^{m_1-|\tilde\alpha|+|\beta^1|}\langle\zeta\rangle^{m_2-|\tilde\gamma|+|\beta^2|+|\lambda^1|}\langle\eta\rangle^{m_3-|\tilde\rho|+|\lambda^2|+|\delta|}\\
&\quad \times \exp[\delta_1\langle\xi\rangle + \delta_2\langle\zeta\rangle + (\delta_3 + e\sqrt{n}A_1|u|/R_0 - 1/(3R_0))\langle\eta\rangle]
\end{aligned}$$

if $\langle\xi\rangle \ge R_0(|\alpha| + |\beta^1|)$, $\langle\zeta\rangle \ge R_0(|\beta^2| + |\gamma| + |\lambda^1|)$, $\langle\eta\rangle \ge 2R_0(|\lambda^2| + |\rho|)$ and $\langle\eta\rangle \ge 3R_0|\rho|$.

Putting

$$\begin{aligned}
Z(z, \eta, \zeta, w) &= Z_\delta(U_0^{-1}(z - \tilde\nabla_\eta U_1(\eta,\zeta) - \tilde\nabla_\eta T(w,\eta,\zeta)); \eta, \zeta),\\
G(\xi, w, \zeta, \eta; z) &= f(\xi, w, \zeta, \mathrm{Re}\, Z(z,\eta,\zeta,w), \eta; i\,\mathrm{Im}\, Z(z,\eta,\zeta,w))\\
&\qquad \times \det\frac{\partial Z}{\partial z}(z,\eta,\zeta,w)\,\mathrm{sgn}(\det U_0),
\end{aligned}$$

we can write

$$\begin{aligned}
\tilde p(\xi, w, \eta) = (2\pi)^{-n} \lim_{\nu'\downarrow 0} \int \Big(\int_{\Omega_{\eta,\zeta,w,0}} & e^{-\nu' Z(z,\eta,\zeta,w)^2}\\
&\times e^{iz\cdot(\zeta-\eta)} G(\xi, w, \zeta, \eta; z)\, dz_1 \wedge \cdots \wedge dz_n\Big)\, d\zeta.
\end{aligned}$$

Lemma A.5. 2 implies that Re $z \in \Omega_{\eta,\zeta,w,1}$ if $G(\xi, w, \zeta, \eta; z) \ne 0$. Therefore, applying Stokes' formula we have

$$\tilde p(\xi, w, \eta) = (2\pi)^{-n} \lim_{\nu'\downarrow 0} \int \Big(\int_{\Omega_{\eta,\zeta,w,1}} e^{-\nu' Z(y,\eta,\zeta,w)^2} e^{iy\cdot(\zeta-\eta)}$$

$$\times G(\xi, w, \zeta, \eta; y)\, \mathrm{sgn}(\det U_0)\, dy_1 \wedge \cdots \wedge dy_n\Big)\, d\zeta + q(\xi, w, \eta)$$
$$= (2\pi)^{-n} \lim_{\nu' \downarrow 0} \int \Big(\int_{R^n} e^{-\nu' Z(y,\eta,\zeta,w)^2} e^{iy\cdot(\zeta-\eta)} G(\xi, w, \zeta, \eta; y)\, dy \Big)\, d\zeta$$
$$+ q(\xi, w, \eta),$$

where

$$q(\xi, w, \eta) = \pm (2\pi)^{-n} \lim_{\nu' \downarrow 0} \int \Big(\int_{\Omega_{\eta,\zeta,w}} d_z \Big(e^{-\nu' Z(z,\eta,\zeta,w)^2}$$
$$\times e^{iz\cdot(\zeta-\eta)} G(\xi, w, \zeta, \eta; z)\, dz_1 \wedge \cdots \wedge dz_n \Big)\Big)\, d\zeta,$$

d_z denotes the exterior differential with respect to z and "$\pm$" depends on the orientation of the chain $\Omega_{\eta,\zeta,w}$. It follows from Lemma A.5. 2 that $U_0^{-1}(z - \tilde{\nabla}_\eta U_1(\eta, \zeta) - \tilde{\nabla}_\eta T(w, \eta, \zeta)) \in \tilde{\mathcal{U}}_{\eta,\zeta,\delta}$ if $z \in \Omega_{\eta,\zeta,w}$. So $Z(z, \eta, \zeta, w)$ is analytic with respect to z in $\Omega_{\eta,\zeta,w}$. Fix $\zeta, \eta \in \boldsymbol{R}^n$ and $w \in \boldsymbol{R}^{n''}$ so that $|\eta - \zeta| \le \varepsilon_0 |\eta|$, $|\eta| \ge 1$, $|\zeta| \ge 1$ and $(w, \zeta) \in \Omega_1'$, and put

$$\hat{z}(y, t; \eta, \zeta, w) = \mathrm{Re}\ z(w, \zeta, y, \eta) + i(1-t)\mathrm{Im}\ z(w, \zeta, y, \eta),$$
$$y^1(y, t; \eta, \zeta, w) = U_0^{-1}(\hat{z}(y, t; \eta, \zeta, w) - \tilde{\nabla}_\eta U_1(\eta, \zeta) - \tilde{\nabla}_\eta T(w, \eta, \zeta))$$

for $y \in \mathcal{U}_{\eta,\delta'} \cap \boldsymbol{R}^n$ and $t \in [0, 1]$. Then, $\hat{z}(y, t; \eta, \zeta, w) \in \Omega_{\eta,\zeta,w}$ and $y^1(y, t; \eta, \zeta,\ w) \in \tilde{\mathcal{U}}_{\eta,\zeta,\delta}$ if $y \in \mathcal{U}_{\eta,\delta'} \cap \boldsymbol{R}^n$ and $t \in [0, 1]$. We define $F_\ell(y, t; \eta, \zeta, w)$ ($1 \le \ell \le n$) by

$$\sum_{k=1}^n \frac{\partial}{\partial \bar{z}_k} \mathrm{Re}\ Z_\ell(z, \eta, \zeta, w)\, d\bar{z}_k \wedge dz_1 \wedge \cdots \wedge dz_n$$
$$= F_\ell(y, t; \eta, \zeta, w)\, dy_1 \wedge \cdots \wedge dy_n \wedge dt,$$

where $Z(\cdot) = (Z_1(\cdot), \cdots, Z_n(\cdot))$ and $z = \hat{z}(y, t; \eta, \zeta, w)$. Then we can write

$$q(\xi, w, \eta) = (2\pi)^{-n} \lim_{\nu' \downarrow 0} \int \Big(\int_0^1 \Big(\int_{\mathcal{U}_{\eta,\delta'} \cap R^n}$$
$$\times \exp[-\nu' Z_\delta(y^1(y, t; \eta, \zeta, w); \eta, \zeta)^2 + i\hat{z}(y, t; \eta, \zeta, w) \cdot (\zeta - \eta)]$$
$$\times \sum_{\ell=1}^n ((\partial_{y_\ell} - \partial_{u_\ell}) f)(\xi, w, \zeta, \mathrm{Re}\ Z_\delta(y^1(y, t; \eta, \zeta, w); \eta, \zeta), \eta;$$
$$i\,\mathrm{Im}\ Z_\delta(y^1(y, t; \eta, \zeta, w); \eta, \zeta))$$
$$\times |\det U_0|^{-1} \det \frac{\partial Z_\delta}{\partial z}(y^1(y, t, ; \eta, \zeta, w); \eta, \zeta) F_\ell(y, t; \eta, \zeta, w)\, dy \Big)\, dt \Big)\, d\zeta.$$

It is easy to see that

$$\begin{aligned}
&y^1(y,t,;\eta,\zeta,w) = y + U_0^{-1}(\widetilde{\nabla}_\eta \widetilde{U}(y,\eta,\zeta) \\
&\quad -it\,\mathrm{Im}\ \widetilde{\nabla}_\eta \widetilde{U}(y,\eta,\zeta) - it\,\mathrm{Im}\ \widetilde{\nabla}_\eta T(w,\eta,\zeta)) \in \widetilde{\mathcal{U}}_{\eta,\zeta,\delta}, \\
&|y^1(y,t,;\eta,\zeta,w) - y| \le |U_0^{-1}|(c(U)(1+\delta') + 2n^{3/2}C(\widetilde{U})A_0^3\delta'^2 + c(T)), \\
&\quad Z_\delta(y^1(y,t,;\eta,\zeta,w);\eta,\zeta) = y + U_0^{-1}(-\widetilde{\nabla}_\eta \widetilde{U}(Z_\delta(y^1(\cdots);\eta,\zeta),\eta,\zeta) \\
&\quad +\widetilde{\nabla}_\eta \widetilde{U}(y,\eta,\zeta) - it\,\mathrm{Im}\ \widetilde{\nabla}_\eta \widetilde{U}(y,\eta,\zeta) - it\,\mathrm{Im}\ \widetilde{\nabla}_\eta T(w,\eta,\zeta)) \in \mathcal{U}_{\eta,\delta}, \\
&|Z_\delta(y^1(y,t,;\eta,\zeta,w);\eta,\zeta) - y| \\
&\le |U_0^{-1}|(c(U)(2+\delta+\delta') + 2n^{3/2}C(\widetilde{U})A_0^3(\delta^2+\delta'^2) + c(T))
\end{aligned}$$

if $\zeta,\eta \in \boldsymbol{R}^n$, $w \in \boldsymbol{R}^{n''}$, $|\eta-\zeta| \le \varepsilon_0|\eta|$, $|\eta|,|\zeta| \ge 1$, $(w,\zeta) \in \Omega_1'$, $y \in \mathcal{U}_{\eta,\delta'} \cap \boldsymbol{R}^n$ and $0 \le t \le 1$. Put

$$\varphi_{\nu'}(y,t;\eta,\zeta,w) = \hat{z}(y,t;\eta,\eta+\zeta,w)\cdot\zeta + i\nu' Z_\delta(y^1(y,t;\eta,\eta+\zeta,w);\eta,\eta+\zeta)^2$$

Then there are positive constants $C(T,U)$ and $C'(T,U)$ such that

$$|\nabla_\zeta \varphi_{\nu'}(y,t;\eta,\zeta,w)| \ge |U_0 y| - |\nabla_\zeta T(w,\eta+\zeta)| - \nu' C(T,U)|y| - C'(T,U).$$

Let $L_{t,\nu'}$ be a differential operator defined by

$${}^tL_{t,\nu'} = (1+|\nabla_\zeta \varphi_{\nu'}(y,t;\eta,\zeta,w)|^2)^{-2}\Big(\sum_{k=1}^n \overline{\partial_{\zeta_k}\varphi_{\nu'}(y,t;\eta,\zeta,w)}D_{\zeta_k} + 1\Big).$$

Integrating by parts with $L_{t,\nu'}$, we have

$$\begin{aligned}
q(\xi,w,\eta) = (2\pi)^{-n} \int_0^1 \Big(\int_{(\mathcal{U}_{\eta,\delta'}\cap R^n)\times R^n} &\exp[i\hat{z}(y,t;\eta,\eta+\zeta,w)\cdot\zeta)] \\
\times \sum_{\ell=1}^n L_{t,0}^M \Big\{ ((\partial_{y_\ell} - \partial_{u_\ell})f)&(\xi,w,\eta+\zeta,\mathrm{Re}\ Z_\delta(y^1(y,t;\eta,\eta+\zeta,w); \\
&\eta,\eta+\zeta),\eta; i\,\mathrm{Im}\ Z_\delta(y^1(y,t;\eta,\eta+\zeta,w);\eta,\eta+\zeta)) \\
&\times|\det U_0|^{-1}\det\frac{\partial Z_\delta}{\partial z}(y^1(y,t,;\eta,\eta+\zeta,w);\eta,\eta+\zeta) \\
&\times F_\ell(y,t;\eta,\eta+\zeta,w)\Big\}\, dyd\zeta\Big)\, dt,
\end{aligned}$$

where $M \ge [n/2]+1$. By assumptions we have

$$\begin{aligned}
&\Big|\partial_\eta^\rho(\hat{z}(y,t;\eta,\eta+\zeta,w)\cdot\zeta)\Big| \le C_{|\rho|}\langle\eta\rangle^{1-|\rho|} \quad \text{for } |\rho| \ge 1, \\
&\Big|\partial_\zeta^\gamma\partial_\eta^\rho Z_\delta(y^1(y,t;\eta,\eta+\zeta,w);\eta,\eta+\zeta)\Big| \le C_{|\gamma|+|\rho|}\langle\eta\rangle^{-|\gamma|-|\rho|}
\end{aligned}$$

$$\text{for } |\gamma|+|\rho| \geq 1,$$
$$\left|\partial_\zeta^\gamma D_y^\lambda \partial_\eta^\rho \hat{z}(y,t;\eta,\eta+\zeta,w)\right| \leq C_{|\gamma|+|\lambda|+|\rho|}\langle\eta\rangle^{-|\gamma|-|\rho|}$$
$$\text{for } |\gamma|+|\lambda|+|\rho| \geq 1,$$
$$\left|\partial_\zeta^\gamma \partial_\eta^\rho (\partial_z^\beta Z)(\hat{z}(y,t;\eta,\eta+\zeta,w),\eta,\eta+\zeta,w)\right|$$
$$\leq C_{|\gamma|+|\rho|+|\beta|}\langle\eta\rangle^{-|\gamma|-|\rho|} \quad \text{for } |\gamma|+|\rho|+|\beta| \geq 1$$

if $\zeta,\eta \in \boldsymbol{R}^n$, $w \in \boldsymbol{R}^{n''}$, $|\zeta| \leq \varepsilon_0|\eta|$, $|\eta| \geq 1$, $|\eta+\zeta| \geq 1$, $(w,\eta+\zeta) \in \Omega_1'$, $y \in \mathcal{U}_{\eta,\delta'} \cap \boldsymbol{R}^n$ and $0 \leq t \leq 1$. So Lemma A.5. 3 yields

$$\left|\partial_\xi^\alpha \partial_\eta^\rho q(\xi,w,\eta)\right| \leq C_{|\alpha|+|\rho|}\langle\xi\rangle^{m_1}\langle\eta\rangle^{m_2+m_3}$$
$$\times \exp[(\delta_1 - \nu(S,T)/R_0)\langle\xi\rangle + (\delta_3(T,U,\delta_2,\delta_3) - 1/(9R_0))\langle\eta\rangle]$$

if $c(U), c(T), \delta \ll 1$, where $\nu(S,T) = c_3(S)/(4\cdot 3^3 c_2(T))$ and $\delta_3(T,U,\delta_2,\delta_3) = c_1^+(T) + c_1^+(U) + \delta_2 + |c_1^+(T) + c_1^+(U) + \delta_2|/2 + c_1^-(T) + c_1^-(U) + \delta_3$. From Lemma A.1. 7 we have the following

Lemma A.5. 4 *$q_{S,T}(D_x,w,D_y)$ maps continuously $\mathcal{S}_{\varepsilon_2}(\boldsymbol{R}^n)$ and $\mathcal{S}_{-\varepsilon_2}{}'(\boldsymbol{R}^n)$ to $\mathcal{S}_{\varepsilon_1}(\boldsymbol{R}^{n'})$ and $\mathcal{S}_{-\varepsilon_1}{}'(\boldsymbol{R}^{n'})$, respectively, if*

$$\delta_1 + c_1(S) + \varepsilon_1 \leq \nu(S,T)/R_0,$$
$$\delta_3(T,U,\delta_2,\delta_3) + c_1^+(T) - \varepsilon_2 \leq 1/(18R_0)$$

and $c(U), c(T), \delta \ll 1$.

It is easy to see that

$$\tilde{p}(\xi,w,\eta) - q(\xi,w,\eta)$$
$$= (2\pi)^{-n} \lim_{\nu'\downarrow 0} \int \left(\int_{R^n} e^{-\nu'|y|^2} e^{iy\cdot\zeta} G(\xi,w,\eta+\zeta,\eta;y)\,dy\right) d\zeta.$$

Therefore, we can write

$$\tilde{p}(\xi,w,\eta) - q(\xi,w,\eta) = p^1(\xi,w,\eta) + q^1(\xi,w,\eta),$$

where

$$p^1(\xi,w,\eta) = \sum_{j=1}^\infty \psi_j^{8R_0}(\eta) \sum_{|\gamma|<j} \frac{1}{\gamma!}\left((-\partial_\zeta)^\gamma D_z^\gamma G\right)(\xi,w,\eta,\eta;0),$$

$$q^1(\xi,w,\eta) = (2\pi)^{-n} \sum_{j=1}^\infty \psi_j^{8R_0}(\eta) \sum_{|\gamma|=j} \int e^{iy\cdot\zeta}\langle y\rangle^{-2M}$$
$$\times \langle D_\zeta\rangle^{2M}\left(\int_0^1 j(1-\theta)^{j-1}\left((-\partial_\zeta)^\gamma D_z^\gamma G\right)(\xi,w,\eta+\zeta,\eta;\theta y)\,d\theta\right) dy d\zeta/\gamma!.$$

Lemma A.5. 5 (i) *There is* $A_0(T, U, \varepsilon_0) > 0$ *such that*

$$\begin{aligned}&\Big|D_z^\lambda D_w^\beta \partial_\zeta^\gamma \partial_\eta^\rho \Big(D_z^{\lambda'} \partial_\zeta^{\gamma'} \partial_\eta^{\rho'} Z_\delta\Big)(U_0^{-1}(z - \tilde{\nabla}_\eta U_1(\eta, \eta+\zeta) \\ &\qquad\qquad -\tilde{\nabla}_\eta T(w, \eta, \eta+\zeta)); \eta, \eta+\zeta)\Big| \\ &\le C_0(U) A_0(U)^{|\lambda'|+|\gamma'|+|\rho'|}(|U_0^{-1}| A_0(U))^{|\lambda|} A_0(T, U, \varepsilon_0)^{|\beta|+|\gamma|+|\rho|} \\ &\quad \times (|\lambda| + |\beta| + |\beta'| + |\gamma| + |\rho| + |\lambda'| + |\gamma'| + |\rho'|)! \langle\eta\rangle^{-|\gamma|-|\rho|-|\gamma'|-|\rho'|}\end{aligned}$$

if $|\lambda| + |\beta| + |\gamma| + |\rho| + |\lambda'| + |\gamma'| + |\rho'| \ge 1$ $\zeta, \eta \in \boldsymbol{R}^n$, $w \in \boldsymbol{R}^{n''}$, $|\zeta| \le \varepsilon_0 |\eta|$, $|\eta| \ge 1$, $|\eta+\zeta| \ge 1$, $(w, \eta+\zeta) \in \Omega_1'$ *and* $z \in \Omega_{\eta,\zeta,w}$. (ii) *There are positive constants* $A_1(T, U, A_1, \varepsilon_0)$, $A_j(S, T, U, A_1, \varepsilon_0)$ ($j = 2, 3$) *and* $A_4(U, A_1)$ *such that*

$$\begin{aligned}&\Big|\partial_\xi^{\alpha+\tilde{\alpha}} D_w^{\beta+\tilde{\beta}} \partial_\zeta^{\gamma+\tilde{\gamma}} D_y^{\lambda+\tilde{\lambda}} \partial_\eta^{\rho+\tilde{\rho}} f(\xi, w, \eta+\zeta, \mathrm{Re}\; Z(y, \eta, \eta+\zeta, w), \eta; \\ &\qquad\qquad i \,\mathrm{Im}\; Z(y, \eta, \eta+\zeta, w))\Big| \\ &\le C_{|\tilde{\alpha}|+|\tilde{\beta}|+|\tilde{\gamma}|+|\tilde{\lambda}|+|\tilde{\rho}|}((\widehat{C}(S, T) + A_1)/R_0)^{|\alpha|} (A_1(T, U, A_1, \varepsilon_0)/R_0)^{|\beta|} \\ &\quad \times (A_2(S, T, U, A_1, \varepsilon_0)/R_0)^{|\rho|} (A_3(S, T, U, A_1, \varepsilon_0)/R_0)^{|\gamma|} \\ &\quad \times (A_4(U, A_1)/R_0)^{|\lambda|} \langle\xi\rangle^{m_1-|\tilde{\alpha}|} \langle\eta\rangle^{m_2+m_3-|\tilde{\gamma}|-|\tilde{\rho}|+|\beta|+|\lambda|+1} \\ &\quad \times \exp[\delta_1\langle\xi\rangle + (\delta_2 + \varepsilon_0|\delta_2| + \delta_3 + \nu(T, U, A_1, \delta)/R_0)\langle\eta\rangle]\end{aligned}$$

if $\langle\xi\rangle \ge R_0|\alpha|$ *and* $\langle\eta\rangle \ge 4R_0(|\beta| + |\gamma| + |\lambda| + |\rho|)$, *where*

$$\nu(T, U, A_1, \delta) = \sqrt{n} A_1 |U_0^{-1}| (2c(U)(1+\delta) + 4n^{3/2} C(\tilde{U}) A_0^3 \delta^2 + c(T)).$$

(iii) $q^1(\xi, w, \eta)$ *satisfies the estimates*

$$\begin{aligned}&\Big|\partial_\xi^\alpha \partial_\eta^\rho q^1(\xi, w, \eta)\Big| \le C_{|\alpha|+|\rho|, R_0} \\ &\quad \times \exp[(\delta_1 - \nu'(S, T)/R_0)\langle\xi\rangle + (\delta_2 + \varepsilon_0|\delta_2| + \delta_3 - 1/(72R_0))\langle\eta\rangle]\end{aligned}$$

if

$$\begin{cases} R_0 \ge 48e^2 n A_3(S, T, U, A_1, \varepsilon_0) A_4(U, A_1), \\ \nu(T, U, A_1, \delta) \le 1/72, \end{cases} \tag{A.49}$$

where $\nu'(S, T) = c_3(S)/(2^5 \cdot 3^3 c_2(T))$. *Therefore,* $q_{S,T}^1(D_x, w, D_y)$ *maps continuously* $\mathcal{S}_{\varepsilon_2}(\boldsymbol{R}^n)$ *and* $\mathcal{S}_{-\varepsilon_2}'(\boldsymbol{R}^n)$ *to* $\mathcal{S}_{\varepsilon_1}(\boldsymbol{R}^{n'})$ *and* $\mathcal{S}_{-\varepsilon_1}'(\boldsymbol{R}^{n'})$, *respectively, if* (A.49) *is satisfied and*

$$\begin{aligned}&\delta_1 + c_1(S) + \varepsilon_1 \le \nu'(S, T)/R_0, \\ &\delta_2 + \varepsilon_0|\delta_2| + \delta_3 + c_1^+(T) - \varepsilon_2 \le 1/(144R_0).\end{aligned}$$

Proof By induction on $|\lambda|+|\beta|+|\gamma|+|\rho|$ we can prove the assertion (i). Then, applying Lemma 2.1.6 we can prove the assertion (ii). The assertion (iii) follows from the assertions (i) and (ii) and Lemma A.1.7. □

Assume that $c(T) < 2\delta/|U_0^{-1}|$ and $c(U)+4n^{3/2}C(\tilde{U})A_0^3\delta \le 1/(2|U_0^{-1}|)$, and that Re $Z(0,\eta,\eta,w) \in \mathcal{U}_\eta$. Let us estimate $|\text{Im } Z(0,\eta,\eta,w)|$. Put $y = Z(0,\eta,\eta,w)$ and $y^1 = \text{Re } Z(0,\eta,\eta,w)$. Then we have

$$\begin{aligned} &y + \Psi_\eta(y;\delta)U_0^{-1}\nabla_\eta\tilde{U}(y,\eta) = -U_0^{-1}(\nabla_\eta U_1(\eta) + \nabla_\eta T(w,\eta)), \\ &y - y^1 = -U_0^{-1}(\Psi_\eta(y;\delta)\,\text{Im } \nabla_\eta\tilde{U}(y,\eta) + \text{Im } \nabla_\eta T(w,\eta)). \quad \text{(A.50)} \end{aligned}$$

Suppose that $y \notin \mathcal{U}_{\eta,2\delta}$. (A.50) gives $|y-y^1| \le |U_0^{-1}|c(T) < 2\delta$, which is a contradiction. Thus we have $y \in \mathcal{U}_{\eta,2\delta}$, *i.e.*, $|y-y^1| < 2\delta$. (2.84) and (A.50) give

$$|\text{Im } Z(0,\eta,\eta,w)| = |y-y^1| \le 2|U_0^{-1}|(c(U)+c(T)).$$

Define

$$q^2(\xi,w,\eta) = p^1(\xi,w,\eta) - p(\xi,w,\eta).$$

Then we have

$$q^2(\xi,w,\eta) = (1-\tilde{g}^{8R_0}(\xi,\eta))p^1(\xi,w,\eta).$$

Let K_2 be a differential operator defined by

$${}^tK_2 = |\nabla_w S(w,\xi) + \nabla_w T(w,\eta)|^{-2}\sum_{j=1}^{n''}(\overline{\partial_{w_j}S(w,\xi)} + \overline{\partial_{w_j}T(w,\eta)})D_{w_j}.$$

Lemma 2.1.5 yields

$$\begin{aligned} &\left|\partial_\xi^\alpha\partial_\eta^\rho K_2^j q^2(\xi,w,\eta)\right| \\ &\le C_{|\alpha|+|\rho|}(\Gamma(S,T,U,A_1)/R_0)^j\langle\xi\rangle^{m_1-|\alpha|}\langle\eta\rangle^{m_2+m_3-|\rho|+2} \\ &\quad \times \exp[\delta_1\langle\xi\rangle + (\delta_2+\delta_3+\tilde{\nu}(T,U,A_1)/R_0 + B'(S,T,U,A_1)/R_0^2)\langle\eta\rangle] \end{aligned}$$

if $j \in \boldsymbol{Z}_+$ and $\langle\eta\rangle \ge 8R_0 j$, where $\Gamma(S,T,U,A_1)$ is a positive constant and $\tilde{\nu}(T,U,A_1) = 2\sqrt{n}A_1|U_0^{-1}|(c(U)+c(T))$ and $B'(S,T,U,A_1) = nA_3(S,T,U,A_1,0)A_4(U,A_1)$. So we have the following

Lemma A.5.6 *$q^2_{S,T}(D_x,w,D_y)$ maps continuously $\mathcal{S}_{\varepsilon_2}(\boldsymbol{R}^n)$ and $\mathcal{S}_{-\varepsilon_2}{}'(\boldsymbol{R}^n)$ to $\mathcal{S}_{\varepsilon_1}(\boldsymbol{R}^{n'})$ and $\mathcal{S}_{-\varepsilon_1}{}'(\boldsymbol{R}^{n'})$, respectively, if $R_0 \ge R_0(S,T,U,A_1)$, $\max\{\delta_1+c_1(S)+\varepsilon_1, \delta_2+\delta_3+c_1^+(T)-\varepsilon_2\} \le 1/(48R_0)$, $\tilde{\nu}(T,U,A_1) \le 1/96$ and $R_0 \ge 96B'(S,T,U,A_1)$, where $R_0(S,T,U,A_1) > 0$.*

We can also show that

$$\begin{aligned}
&\left|\partial_\xi^{\alpha+\tilde{\alpha}} D_w^{\beta+\tilde{\beta}} \partial_\eta^{\rho+\tilde{\rho}} p(\xi, w, \eta)\right| \\
&\leq C_{|\tilde{\alpha}|+|\tilde{\beta}|+|\tilde{\rho}|}((\widehat{C}(S,T) + A_1)/R_0)^{|\alpha|}(A_1(T,U,A_1,0)/R_0)^{|\beta|} \\
&\quad \times (A_2'(S,T,U,A_1)/R_0)^{|\rho|}\langle\xi\rangle^{m_1-|\tilde{\alpha}|}\langle\eta\rangle^{m_2+m_3-|\tilde{\rho}|+|\beta|+2} \\
&\quad \times \exp[\delta_1\langle\xi\rangle + (\delta_2 + \delta_3 + \tilde{\nu}(T,U,A_1)/R_0 \\
&\qquad\qquad + B(S,T,U,A_1)/R_0^2)\langle\eta\rangle] \qquad\qquad (A.51)
\end{aligned}$$

if $\langle\xi\rangle \geq 8R_0|\alpha|$, $\langle\eta\rangle \geq 24R_0|\rho|$ and $\langle\eta\rangle \geq 8R_0(|\beta| + |\rho|)$, where $A_2'(S,T,U,A_1)$ and $B(S,T,U,A_1)$ are positive constants. (A.51), together with Lemmas A.5. 1 and A.5. 4 –A.5. 6 , proves the first part of Theorem 2.5. 3 . If $U(y,\eta)$ and $T(w,\eta)$ are real-valued, then Im $Z(y,\eta,\zeta,w) = 0$ and the above proof becomes much simpler. So one can easily prove the second part of Theorem 2.5. 3 .

Appendix B

A priori estimates

B.1 Grušin operators

We shall give *a priori* estimates for some special class of Grušin operators, which are used in Section 4.5. About *a priori* estimates for general Grušin operators we refer to [Gr1], [Gr2], [Ta] and [PR]. Let $\mu \geq 0$, $M \in \boldsymbol{Z}_+$ and $n, m \in \boldsymbol{N}$, and let

$$P(x, y, \xi, \eta; t) = \sum_{\substack{|\alpha|+j \leq M \\ |\alpha|+(\mu+1)j-|\beta| \leq M}} \{a_{\alpha,\beta,j}(x, y, \xi, \eta) + t b_{\alpha,\beta,j}(x, y, \xi, \eta)\} x^\beta \xi^\alpha \eta_m^j,$$

where $x = (x_1, \cdots, x_n) \in \boldsymbol{R}^n$, $\xi = (\xi_1, \cdots, \xi_n) \in \boldsymbol{R}^n$, $y = (y_1, \cdots, y_m) \in \boldsymbol{R}^m$, $\eta = (\eta', \eta_m) = (\eta_1, \cdots, \eta_m) \in \boldsymbol{R}^m$, $t \geq 0$, $a_{\alpha,\beta,j}(x, y, \xi, \eta), b_{\alpha,\beta,j}(x, y, \xi, \eta) \in S^0_{1,0}(\boldsymbol{R}^{n+m} \times \boldsymbol{R}^{n+m})$,

$$a_{\alpha,\beta,j}(x, y, \xi, \eta) = a^0_{\alpha,\beta,j}(x, y, \xi, \eta) + a^1_{\alpha,\beta,j}(x, y, \xi, \eta),$$

$a^0_{\alpha,\beta,j}(x, y, \xi, \eta)$ is positively homogeneous of degree 0 in (ξ, η) for $|(\xi, \eta)| \geq 1$, and $a^1_{\alpha,\beta,j}(x, y, \xi, \eta) \in S^{-\nu}_{1,0}$ with some $\nu > 0$. We assume that supp $a_{\alpha,\beta,j} \cup \operatorname{supp} b_{\alpha,\beta,j} \subset \{(x, y, \xi, \eta);\ |(x, y)| \leq 1 \text{ and } \eta_m \geq |(\xi, \eta')|\}$. Put

$$L(x, \xi; \tau) = \sum_{\substack{|\alpha|+j \leq M \\ |\alpha|+(\mu+1)j-|\beta| = M}} a^0_{\alpha,\beta,j}(0, 0, 0, \eta^0) x^\beta \xi^\alpha \tau^j,$$

where $\eta^0 = (0, \cdots, 0, 1) \in \boldsymbol{R}^m$. We impose the following conditions on P (or L):

(G–1) There is $C > 0$ such that

$$C^{-1}\lambda(x,\xi)^M \le |L^0(x,\xi)| \le C\lambda(x,\xi)^M$$

for $(x,\xi) \in \boldsymbol{R}^n \times \boldsymbol{R}^n$ with $|\xi| \ge 1$, where $\lambda(x,\xi) = \langle\xi\rangle + \langle x\rangle^\mu$ and

$$L^0(x,\xi) = \sum_{|\alpha|\le M,\, |\beta|=\mu(M-|\alpha|)} a^0_{\alpha,\beta,M-|\alpha|}(0,0,0,\eta^0)x^\beta\xi^\alpha.$$

(G–2) The equation $L(x,D_x;1)v(x) = 0$ in $\mathcal{S}(\boldsymbol{R}^n)$ has no non-trivial solution.

Let $\varphi(x,y) \in C_0^\infty(\boldsymbol{R}^{n+m})$ and $\psi(\xi,\eta) \in S^0_{1,0}$ be functions such that $\varphi(x,y) = 1$ if $|(x,y)| \le 1$, $\varphi(x,y) = 0$ if $|(x,y)| \ge 2$, $\psi(\xi,\eta) = 1$ if $|(\xi,\eta')| \le \eta_m$ and $\eta_m \ge 1$, and $\psi(\xi,\eta) = 0$ if $|(\xi,\eta')| \ge 2\eta_m$ or $\eta_m \le 1/2$. We put $\Psi_\delta(x,y,\xi,\eta) = \varphi(x/\delta,y/\delta)\psi(\xi,\eta',\delta\eta_m)$ for $\delta > 0$.

Theorem B.1.1 *There are positive constants t_0, δ and C such that*

$$\begin{aligned}\|u\|_{M/(\mu+1)} \;\le\; & C(\|P(x,y,D_x,D_y;t)u\| + \|u\|_{-1} \\ & +\|(1-\Psi_\delta(x,y,D_x,D_y))u\|_M)\end{aligned}$$

if $0 \le t \le t_0$ and $u \in C_0^\infty(\boldsymbol{R}^{n+m})$, where $\|u\|$ denotes the L^2-norm of u and $\|u\|_s = \|\langle D\rangle^s u\|$.

We shall give an outline of the proof here. Define the Riemannian metric on $\boldsymbol{R}^{2n}$ by

$$g_{(x,\xi)} = \langle x\rangle^{-2}|dx|^2 + \lambda(x,\xi)^{-2}|d\xi|^2.$$

Then one can easily show that g is σ temperate in the sense of Hörmander and that $\lambda(x,\xi)$ is σ, g temperate (see §§18.4 and 18.5 of [Hr6] for the terminologies). Moreover, we have

$$\begin{aligned} & g_{(x,\xi)}(\delta x,\delta\xi) = g_{(x,\xi)}(\delta x,-\delta\xi), \\ & g^\sigma_{(x,\xi)}(\delta x,\delta\xi) = \lambda(x,\xi)^2|\delta x|^2 + \langle x\rangle^2|\delta\xi|^2, \\ & g_{(x,\xi)}(X)/g^\sigma_{(x,\xi)}(X) = \langle x\rangle^{-2}\lambda(x,\xi)^{-2} \le 1. \end{aligned}$$

The following lemma can be easily proved.

Lemma B.1.2 (i) $L(x,\xi;1) \in S(\lambda(x,\xi)^M, g)$, *i.e.,*

$$\left|L^{(\alpha)}_{(\beta)}(x,\xi;1)\right| \le C_{\alpha,\beta}\lambda(x,\xi)^{M-|\alpha|}\langle x\rangle^{-|\beta|}.$$

(ii) *There are $c > 0$ and $A > 0$ such that*

$$|L(x,\xi;1)| \ge c\lambda(x,\xi)^M \quad \text{if } \lambda(x,\xi) \ge A.$$

From Theorem 18.5.10 and the remark given at the end of §18.5 of [Hr6] we can use a standard symbol calculus (see, also, [Bl]). Therefore, we can construct a parametrix of $L(x, D; 1)$, *i.e.*, there are $R_N(x,\xi) \in S(\lambda(x,\xi)^{-M}, g)$ ($N \in \boldsymbol{Z}_+$) such that

$$K_N(x,\xi) \in S((\langle x\rangle\lambda(x,\xi))^{-N-1}, g), \tag{B.1}$$

where I denotes the identity operator and

$$K_N(x,D) = R_N(x,D)L(x,D;1) - I.$$

Note that the $K_N(x,D)$ are compact operators in $L^2(\boldsymbol{R}^n)$ (see, *e.g.*, Theorem 18.6.6 of [Hr6]). The condition (G–2), together with (B.1), implies that the equation $L(x,D;1)v(x) = 0$ in $L^2(\boldsymbol{R}^n)$ has no non-trivial solution. Therefore, compactness of $K_N(x,D)$ yields the following

Lemma B.1. 3 *There is $C > 0$ such that*

$$\|v\| \le C\|L(x,D;1)v\| \quad \textit{for } v \in C_0^\infty(\boldsymbol{R}^n).$$

Lemma B.1. 4 (i) *Let $a(x,y,\xi,\eta) \in S^0_{1,0}$, and assume that* supp $a \subset \{(x,y,\xi,\eta);\ |(x,y)| \le 1$ *and* $\eta_m \ge |(\xi,\eta')|\}$. *Then, there is $C > 0$ such that*

$$\begin{aligned}&\left\|Op\left(a(x,y,\xi,\eta)x^\beta\xi^\alpha\eta_m^{(M-|\alpha|+|\beta|)/(\mu+1)}\right)u(x,y)\right\|\\ &\le C(\|L(x,D_x;D_{y_m})u(x,y)\| + \|u\|_{-1})\end{aligned}$$

for $u \in C_0^\infty(\boldsymbol{R}^{n+m})$ if $\mu|\alpha| + |\beta| \le \mu M$, where $Op(b(x,y,\xi,\eta)) = b(x,y,D_x,D_y)$. (ii) *Let $\nu > 0$, and let $a^0(x,y,\xi,\eta) \in S^0_{1,0}$ and $a^1(x,y,\xi,\eta) \in S^{-\nu}_{1,0}$. Assume that $a^0(x,y,\xi,\eta)$ is positively homogeneous of degree 0 in (ξ,η) for $|(\xi,\eta)| \ge 1$ and that $a^0(0,0,0,\eta^0) = 0$. We put $a(x,y,\xi,\eta) = \sum_{j=0}^1 a^j(x,y,\xi,\eta)$ and assume that* supp $a \subset \{(x,y,\xi,\eta);\ |(x,y)| \le 1$ *and* $\eta_m \ge |(\xi,\eta')|\}$. *Then, for any $\varepsilon > 0$ there are $\delta > 0$ and $C_\varepsilon > 0$ such that*

$$\begin{aligned}&\left\|Op\left(a(x,y,\xi,\eta)x^\beta\xi^\alpha\eta_m^{(M-|\alpha|+|\beta|)/(\mu+1)}\right)u(x,y)\right\|\\ &\le \varepsilon\|L(x,D_x;D_{y_m})u(x,y)\|\\ &\qquad + C_\varepsilon(\|u\|_{-1} + \|(1-\Psi_\delta(x,y,D_x,D_y))u(x,y)\|_M)\end{aligned}$$

for $u \in C_0^\infty(\boldsymbol{R}^{n+m})$ if $\mu|\alpha| + |\beta| \le \mu M$.

Proof Assume that $\mu|\alpha| + |\beta| \le \mu M$. It is easy to see that

$$Op\left(a(x,y,\xi,\eta)x^\beta\xi^\alpha\eta_m^\rho\right)u = \sum_{\gamma\le\beta} Op\left(a_{\beta,\gamma}(x,y,\xi,\eta)\eta_m^\rho\right)(x^\gamma D_x^\alpha u),$$

where

$$a_{\beta,\gamma}(x,y,\xi,\eta) = i^{|\beta|-|\gamma|}\beta!/(\gamma!(\beta-\gamma)!)\partial_\xi^{\beta-\gamma}a(x,y,\xi,\eta).$$

Moreover, we have, with $C > 0$,

$$\begin{aligned}&\left\|Op\Big(a_{\beta,\gamma}(x,y,\xi,\eta)\eta_m^\rho\Big)(x^\gamma D_x^\alpha u)\right\|\\ &\le C\Big(\Big(\int_{\eta_m\ge 1}\left\|\eta_m^{\rho+|\gamma|-|\beta|}x^\gamma D_x^\alpha \hat{u}(x,\eta)\right\|^2_{L^2(R_x^n\times R_{\eta'}^{m-1})}d\eta_m\Big)^{1/2}+\|u\|_{-1}\Big)\end{aligned}$$

for $u \in C_0^\infty(\boldsymbol{R}^{n+m})$, where $\hat{u}(x,\eta) = \mathcal{F}_y[u(x,y)](\eta)$. Since

$$x^\beta D_x^\alpha v(x) = x^\beta D_x^\alpha R_M(x,D)L(x,D;1)v(x) - x^\beta D_x^\alpha K_M(x,D)v(x)$$

and $x^\beta D_x^\alpha R_M(x,D)$ and $x^\beta D_x^\alpha K_M(x,D)$ are bounded operators in $L^2(\boldsymbol{R}^n)$, Lemma B.1. 3 yields

$$\|x^\beta D^\alpha v\| \le C\|L(x,D;1)v\|$$

for $v \in C_0^\infty(\boldsymbol{R}^n)$. Note that $L(s^{-1}x, s\xi; s^{\mu+1}) = s^M L(x,\xi;1)$. Therefore, by use of coordinate transformation, we have

$$\eta_m^{(M-|\alpha|+|\beta|)/(\mu+1)}\left\|x^\beta D_x^\alpha v(x)\right\|_{L^2(R^n)} \le C\|L(x,D_x;\eta_m)v(x)\|_{L^2(R^n)}$$

for $v \in C_0^\infty(\boldsymbol{R}^n)$. This gives

$$\begin{aligned}&\Big(\int_{\eta_m\ge 1}\left\|\eta_m^{(M-|\alpha|+|\beta|)/(\mu+1)+|\gamma|-|\beta|}x^\gamma D_x^\alpha \hat{u}(x,\eta)\right\|^2_{L^2(R_x^n\times R_{\eta'}^{m-1})}d\eta_m\Big)^{1/2}\\ &\le C\|L(x,D_x;D_{y_m})u(x,y)\|_{L^2(R^{n+m})}\end{aligned}$$

if $|\gamma| \le |\beta|$, which proves the assertion (i). We can write

$$\begin{aligned}&a^0(x,y,\xi,\eta)\\ &= \sum_{\substack{|\alpha^1|+|\beta^1|+|\gamma^1|+|\lambda^1|=1\\ \gamma_m^1=0}} x^{\beta^1}y^{\lambda^1}(\xi/\eta_m)^{\alpha^1}(\eta/\eta_m)^{\gamma^1}a^0_{\alpha^1,\beta^1,\gamma^1,\lambda^1}(x,y,\xi,\eta),\end{aligned}$$

where $a^0_{\alpha^1,\beta^1,\gamma^1,\lambda^1}(x,y,\xi,\eta) \in S^0_{1,0}$. If $b(x,y,\xi,\eta) \in S^0_{1,0}$, $\nu > 0$, $|\alpha^1| + |\beta^1| + |\gamma^1| + |\lambda^1| = 1$ and $\gamma_m^1 = 0$, then there is $C > 0$ such that

$$\begin{aligned}&\left\|Op\Big(x^{\beta^1}y^{\lambda^1}(\xi/\eta_m)^{\alpha^1}(\eta/\eta_m)^{\gamma^1}b(x,y,\xi,\eta)\Psi_{2\delta}(x,y,\xi,\eta)\Big)w\right\| \le C\delta\|w\|,\\ &\left\|Op\Big(\eta_m^{-\nu}b(x,y,\xi,\eta)\Psi_{2\delta}(x,y,\xi,\eta)\Big)w\right\| \le C\delta^\nu\|w\|\end{aligned}$$

for $w \in L^2(\boldsymbol{R}^{n+m})$ and $\delta > 0$. This, together with the assertion (i), proves the assertion (ii). □

We can write

$$\begin{aligned} P(x,y,\xi,\eta;t) &= L(x,\xi;\eta_m) \\ &+ \sum_{\substack{|\alpha|+j\le M \\ |\alpha|+(\mu+1)j-|\beta|=M}} \Big\{a^0_{\alpha,\beta,j}(x,y,\xi,\eta) - a^0_{\alpha,\beta,j}(0,0,0,\eta^0)\varphi(x,y) \\ &\qquad + a^1_{\alpha,\beta,j}(x,y,\xi,\eta) + tb_{\alpha,\beta,j}(x,y,\xi,\eta)\Big\} x^\beta \xi^\alpha \eta_m^j \\ &+ \sum_{\substack{|\alpha|+j\le M \\ j<(M-|\alpha|+|\beta|)/(\mu+1)}} \{a_{\alpha,\beta,j}(x,y,\xi,\eta) + tb_{\alpha,\beta,j}(x,y,\xi,\eta)\} \\ &\qquad \times \eta_m^{j-(M-|\alpha|+|\beta|)/(\mu+1)} x^\beta \xi^\alpha \eta_m^{(M-|\alpha|+|\beta|)/(\mu+1)} \\ &- \sum_{\substack{|\alpha|+j\le M \\ |\alpha|+(\mu+1)j-|\beta|=M}} a^0_{\alpha,\beta,j}(0,0,0,\eta^0)(1-\varphi(x,y)) x^\beta \xi^\alpha \eta_m^j. \end{aligned}$$

It follows from Lemma B.1. 4 (or its proof) that, with some $C > 0$,

$$\begin{aligned} \|u\|_{M/(\mu+1)} \le\ & C(\|L(x,D_x;D_{y_m})u(x,y)\| + \|u\|_{-1} \\ & + \|(1-\psi(D_x,D_y))u(x,y)\|_M) \end{aligned} \tag{B.2}$$

for $u \in C_0^\infty(\boldsymbol{R}^{n+m})$. By Lemma B.1. 4 there are positive constants t_0, δ and C such that

$$\begin{aligned} &\|L(x,D_x;D_{y_m})u(x,y)\|/2 \le \|P(x,y,D_x,D_y;t)u(x,y)\| \\ &\quad + C(\|u\|_{-1} + \|(1-\Psi_\delta(x,y,D_x,D_y))u(x,y)\|_M) \end{aligned}$$

if $0 \le t \le t_0$, $u \in C_0^\infty(\boldsymbol{R}^{n+m})$ and $\operatorname{supp} u \subset \{|(x,y)| \le 1\}$. Since

$$\|u\|_{M/(\mu+1)} \le \|(1-\Psi_\delta(x,y,D_x,D_y))u\|_M + C_\delta \|u\|_{-1}$$

if $u \in C_0^\infty(\boldsymbol{R}^{n+m})$, $\operatorname{supp} u \subset \{|(x,y)| \ge 1/2\}$ and $0 < \delta < 1/4$, this, together with (B.2), proves Theorem B.1. 1 .

B.2 A class of operators with double characteristics

Let $\xi^0 = (0,\cdots,0,1) \in \boldsymbol{R}^n$, and let $p(x,\xi)$ be a symbol in $S^2_{1,0}$ such that

$$p(x,\xi) = \xi_1^2 + a(x,\xi') + b(x,\xi) + c(x,\xi)$$

in a conic neighborhood of $(0,\xi^0)$, where $\xi' = (\xi_2,\cdots,\xi_n)$, $a(x,\xi')$ is positively homogeneous of degree 2 in ξ' for $|\xi'| \geq 1$, $b(x,\xi)$ is positively homogeneous of degree 1 in ξ for $|\xi| \geq 1$ and $c(x,\xi) \in S^0_{1,0}$. We assume that

(D) there are a conic neighborhood $\mathcal{U}$ of $(0,\xi^0)$ in $\boldsymbol{R}^n \times (\boldsymbol{R}^n \setminus \{0\})$, $\varepsilon > 0$, $C > 0$ and $q_j(x,\xi) \in S^1_{1,0}$ ($1 \leq j \leq 2J$) such that the $q_j(x,\xi)$ are real-valued and

$$(1-\varepsilon)\mathrm{Re}\ p_2(x,\xi) + \mathrm{Re}\ p_1^s(x,\xi) - \sum_{j=1}^{2J} q_j(x,\xi)^2$$
$$+\sum_{j=1}^{J}\{q_{2j-1},q_{2j}\}(x,\xi) \geq -C$$

for $(x,\xi) \in \mathcal{U}$, where $p_2(x,\xi) = \xi_1^2 + a(x,\xi')$ and $p_1^s(x,\xi) = b(x,\xi) + (i/2)\sum_{j=2}^n (\partial^2 a/\partial x_j \partial\xi_j)(x,\xi)$.

Let $\varphi(x) \in C_0^\infty(\boldsymbol{R}^n)$ and $\psi(\xi) \in S^0_{1,0}$ be functions such that $\varphi(x) = 1$ if $|x| \leq 1$, $\varphi(x) = 0$ if $|x| \geq 2$, $\psi(\xi) = 1$ if $|(\xi_1,\cdots,\xi_{n-1})| \leq \xi_n$ and $\xi_n \geq 1$, and $\psi(\xi) = 0$ if $|(\xi_1,\cdots,\xi_{n-1})| \geq 2\xi_n$ or $\xi_n \leq 1/2$. We put $\Psi_\delta(x,\xi) = \varphi(x/\delta)\psi(\xi_1,\cdots,\xi_{n-1},\delta\xi_n)$ for $\delta > 0$.

Theorem B.2.1 *There are positive constants δ and C such that*

$$\mathrm{Re}\ (p(x,D)u,u) \geq \|u\|^2 - C\left(\|u\|_{-1}^2 + \|(1-\Psi_\delta(x,D))u\|_2^2\right)$$

if $u \in C_0^\infty(\boldsymbol{R}^n)$, where $(u,v) = \int u(x)\overline{v(x)}\,dx$.

Proof Choose $\delta > 0$ so that $\{(x,\xi) \in \boldsymbol{R}^n \times (\boldsymbol{R}^n \setminus \{0\});\ |x| \leq 12\delta$ and $|(\xi_1,\cdots,\xi_{n-1})| \leq 12\delta\xi_n\} \subset\subset \mathcal{U}$, where $\mathcal{U}$ is the conic neighborhood of $(0,\xi^0)$ in the condition (D). It is easy to see that

$$\begin{aligned}\mathrm{Re}\ (p(x,D)u,u) \geq\ & ((p_\delta(x,D)+p_\delta(x,D)^*)\Psi_\delta(x,D)u,\Psi_\delta(x,D)u)/2 \\ & -C(\|u\|^2 + \|(1-\Psi_\delta(x,D))u\|_2^2)\end{aligned} \tag{B.3}$$

for $u \in C_0^\infty(\boldsymbol{R}^n)$, where $p_\delta(x,\xi) = \Psi_{6\delta}(x,\xi)p(x,\xi)$. Note that

$$\mathrm{Re}\ \sigma((p_\delta(x,D)+p_\delta(x,D)^*)/2)(x,\xi) \equiv \mathrm{Re}\ p_2(x,\xi) + \mathrm{Re}\ p_1^s(x,\xi)$$

mod $S^0_{1,0}$ in $\{(x,\xi);\ |x| < 6\delta$ and $|(\xi_1,\cdots,\xi_{n-1})| < 6\delta\xi_n\}$, where $\sigma(q(x,D))(x,\xi) = q(x,\xi)$. The condition (D) implies that $\mathrm{Re}\ a(x,\xi) \geq 0$ for

$(x,\xi) \in \mathcal{U}$ and that

$$q_\delta(x,\xi) \equiv \Psi_{3\delta}(x,\xi)\Big((1-\varepsilon)\mathrm{Re}\ p_2(x,\xi) + \mathrm{Re}\ p_1^s(x,\xi) - \sum_{j=1}^{2J} q_j(x,\xi)^2 + \sum_{j=1}^{J}\{q_{2j-1},q_{2j}\}(x,\xi) + C\Big) \geq 0.$$

Therefore, the Fefferman-Phong inequality yields, with some $C' > 0$,

$$\mathrm{Re}\ (q_\delta(x,D)v,v) \geq -C'\|v\|^2 \qquad \text{(B.4)}$$

for $v \in C_0^\infty(\boldsymbol{R}^n)$ (see [FP] and [Hr6]). Moreover, we have

$$\begin{aligned}
&\mathrm{Re}\ (\{(p_\delta(x,D)+p_\delta(x,D)^*)/2 - p_{2,\delta}(x,D)\}\Psi_\delta(x,D)u, \Psi_\delta(x,D)u) \\
&\geq -C_1\|u\|^2, && \text{(B.5)}\\
&\mathrm{Re}\ ((p_{2,\delta}(x,D) - q_\delta(x,D))\Psi_\delta(x,D)u, \Psi_\delta(x,D)u) \\
&\geq \varepsilon\|D_1u\|^2 + \sum_{j=1}^{J}\|(q_{2j-1}(x,D) - iq_{2j}(x,D))\Psi_\delta(x,D)u\|^2/2 \\
&\quad -C_2(\|u\|^2 + \|(1-\Psi_\delta(x,D))u\|_2^2) && \text{(B.6)}
\end{aligned}$$

for $u \in C_0^\infty$, where

$$p_{2,\delta}(x,\xi) = \Psi_{3\delta}(x,\xi)(\mathrm{Re}\ p_2(x,\xi) + \mathrm{Re}\ p_1^s(x,\xi)).$$

Indeed, we have

$$\begin{aligned}
&(q_{2j-1}(x,D) - iq_{2j}(x,D))^*(q_{2j-1}(x,D) - iq_{2j}(x,D)) \\
&= (q_{2j-1}(x,D) + iq_{2j}(x,D) + c_j(x,D))(q_{2j-1}(x,D) - iq_{2j}(x,D)), \\
&\sigma((q_{2j-1}(x,D) + iq_{2j}(x,D))(q_{2j-1}(x,D) - iq_{2j}(x,D)))(x,\xi) \\
&\equiv Q_j(x,\xi) - i\sum_{\ell=0}^{1}\sum_{k=1}^{n}(\partial_{\xi_k}q_{2j-\ell}(x,\xi)\cdot\partial_{x_k}q_{2j-\ell}(x,\xi)) \quad \mathrm{mod}\ S_{1,0}^0,
\end{aligned}$$

where $c_j(x,\xi) \in S_{1,0}^0$ and

$$Q_j(x,\xi) = q_{2j-1}(x,\xi)^2 + q_{2j}(x,\xi)^2 - \{q_{2j-1},q_{2j}\}(x,\xi).$$

So we have

$$\begin{aligned}
&\mathrm{Re}\ (Q_j(x,D)\Psi_\delta(x,D)u, \Psi_\delta(x,D)u) \\
&\geq \|(q_{2j-1}(x,D) - iq_{2j}(x,D))\Psi_\delta(x,D)u\|^2 \\
&\quad -C(\|(q_{2j-1}(x,D) - iq_{2j}(x,D))\Psi_\delta(x,D)u\|\|\Psi_\delta(x,D)u\| + \|u\|^2)
\end{aligned}$$

for $u \in C_0^\infty(\boldsymbol{R}^n)$. This gives (B.6). On the other hand, by Poincaré's inequality, for any $K > 0$ there are $\delta > 0$ and $C > 0$ such that

$$\|D_1 u\|^2 \geq K\|u\|^2 - C\Big(\|(1 - \Psi_\delta(x, D))u\|_1^2 + \|u\|_{-1}^2\Big)$$

for $u \in C_0^\infty(\boldsymbol{R}^n)$. This, together with (B.3)–(B.6), proves the theorem. □

Corollary B.2. 2 *There are positive constants δ and C such that*

$$\|u\| \leq C(\|p(x, D)u\| + \|u\|_{-1} + \|(1 - \Psi_\delta(x, D))u\|_2)$$

for $u \in C_0^\infty(\boldsymbol{R}^n)$.

We may assume that Re $p_2(x, \xi) \geq 0$, and define

$$\begin{aligned} \mathcal{L}_1 &= \{\ell(x,\xi) \in S^1_{1,0};\ \ell(x,\xi) \text{ is real-valued} \\ &\qquad \text{and } \ell(x,\xi)^2 \leq \mathrm{Re}\ p_2(x,\xi)\}, \\ \mathcal{L}_j &= \{\{\ell_1(x,\xi), \ell_2(x,\xi)\};\ \ell_1(x,\xi) \in \mathcal{L}_1 \text{ and} \\ &\qquad \ell_2(x,\xi) \in \mathcal{L}_{j-1}\} \quad (j = 2, 3, \cdots). \end{aligned}$$

We denote by $\widetilde{\mathcal{L}}_j$ the module over $S^0_R \equiv \{h(x,\xi) \in S^0_{1,0};\ h(x,\xi)$ is real-valued$\}$ generated by $\bigcup_{k=1}^j \mathcal{L}_k$, and put $\widetilde{\mathcal{L}} = \bigcup_j \widetilde{\mathcal{L}}_j$ and $\varepsilon_k = 2^{1-k}$ ($k \in \boldsymbol{N}$). $\widetilde{\mathcal{L}}$ is a Lie algebra over S^0_R with the Poisson brackets product. Theorem B.2. 1 yields the following

Theorem B.2. 3 *Assume that $\mu \in \boldsymbol{N}$ and $\ell(x,\xi) \in \widetilde{\mathcal{L}}_\mu$. Then there are positive constants δ and C such that*

$$\|u\| + \|\ell(x, D)u\|_{\varepsilon_\mu - 1} \leq C(\|p(x, D)u\| + \|u\|_{-1} + \|(1 - \Psi_\delta(x, D))u\|_2)$$

for $u \in C_0^\infty(\boldsymbol{R}^n)$.

Remark There are many works concerning the above estimates. The most precise estimates were given by Rothschild and Stein in [RS].

Proof It suffices to prove the theorem when $\ell(x,\xi) \in \mathcal{L}_\mu$. So we may assume that $\ell_k(x,\xi) \in \mathcal{L}_1$ ($1 \leq k \leq \mu$) and $\ell(x,\xi) = H_{\ell_1} \cdots H_{\ell_{\mu-1}} \ell_\mu$. We shall prove by induction that there are a conic neighborhood $\mathcal{U}$ of $(0, \xi^0)$, $c > 0$, $C > 0$ and real-valued symbols $r_k(x,\xi) \in S^1_{1,0}$ ($1 \leq k \leq 2K$) such that

$$\begin{aligned} &\mathrm{Re}\ p_2(x,\xi) - \sum_{k=1}^{2K} r_k(x,\xi)^2 + \sum_{k=1}^{K} \{r_{2k-1}, r_{2k}\}(x,\xi) \\ &\geq c\ell(x,\xi)^2 \langle\xi\rangle^{2\varepsilon_\mu - 2} - C \end{aligned} \tag{B.7}$$

for $(x,\xi) \in \mathcal{U}$. It is obvious that (B.7) is valid for $\mu = 1$ with $K = 0$ and $c = 1$. Now assume that (B.7) is valid for $\mu \le M$, where $M \in N$. Let $\mu = M+1$, and put $\tilde{\ell}(x,\xi) = H_{\ell_2} \cdots H_{\ell_M} \ell_\mu(x,\xi)$ ($\in \mathcal{L}_M$). So there are a conic neighborhood $\mathcal{U}$ of $(0,\xi^0)$, $c > 0$, $C > 0$ and real-valued symbols $r_k(x,\xi) \in S^1_{1,0}$ ($1 \le k \le 2K$) such that (B.7) is valid with $\ell(x,\xi)$ and μ replaced by $\tilde{\ell}(x,\xi)$ and M, respectively. Put

$$r_{2j+1}(x,\xi) = \ell_1(x,\xi)/\sqrt{2},$$
$$r_{2j+2}(x,\xi) = c_1 \ell(x,\xi)\langle\xi\rangle^{\varepsilon_M - 2}\tilde{\ell}(x,\xi),$$

where c_1 is a constant satisfying $c_1^2 \ell(x,\xi)^2\langle\xi\rangle^{-2} \le c/4$. Then we have

$$\begin{aligned}
&\text{Re } p_2(x,\xi) - \sum_{k=1}^{2K}(r_k(x,\xi)/\sqrt{2})^2 - \sum_{k=2j+1}^{2j+2} r_k(x,\xi)^2 \\
&+ \sum_{k=1}^{K}\{r_{2k-1}/\sqrt{2}, r_{2k}/\sqrt{2}\}(x,\xi) + \{r_{2j+1}, r_{2j+2}\}(x,\xi) \\
&\ge c\tilde{\ell}(x,\xi)^2\langle\xi\rangle^{2\varepsilon_M - 2}/4 + e(x,\xi)\tilde{\ell}(x,\xi) - C/2 \\
&\quad +(\text{Re } p_2(x,\xi) - \ell_1(x,\xi)^2)/2 + c_1\ell(x,\xi)^2\langle\xi\rangle^{\varepsilon_M - 2}/\sqrt{2} \\
&\ge c_1\ell(x,\xi)^2\langle\xi\rangle^{\varepsilon_M - 2}/\sqrt{2} - C'
\end{aligned}$$

for $(x,\xi) \in \mathcal{U}$, where $e(x,\xi) = c_1\{\ell_1(x,\xi), \ell(x,\xi)\langle\xi\rangle^{\varepsilon_M - 2}\}/\sqrt{2} \in S^{\varepsilon_M - 1}_{1,0}$. This proves (B.7). Form (B.7) and the condition (D) we have

$$\begin{aligned}
&(1-\varepsilon/2)\text{Re } p_2(x,\xi) + \text{Re } p_1^s(x,\xi) - \sum_{j=1}^{2J} q_j(x,\xi)^2 - \sum_{k=1}^{2K}(\sqrt{\varepsilon/2}\, r_k(x,\xi))^2 \\
&+ \sum_{j=1}^{J}\{q_{2j-1}, q_{2j}\}(x,\xi) + \sum_{k=1}^{K}\{\sqrt{\varepsilon/2}\, r_{2k-1}, \sqrt{\varepsilon/2}\, r_{2k}\}(x,\xi) \\
&\ge \varepsilon c\ell(x,\xi)^2\langle\xi\rangle^{2\varepsilon_\mu - 2}/2 - C
\end{aligned}$$

for $(x,\xi) \in \mathcal{U}$, modifying $\mathcal{U}$ and C if necessary. It follows from Theorem B.2. 1 (or its proof) that there are positive constants δ and C satisfying

$$\begin{aligned}
\text{Re } (p(x,D)u,u) &\ge (\varepsilon c/2)\,\text{Re } (Op(\ell(x,\xi)^2\langle\xi\rangle^{2\varepsilon_\mu - 2})u,u) \\
&\quad + \|u\|^2 - C(\|u\|^2_{-1} + \|(1-\Psi_\delta(x,D))u\|_2^2)
\end{aligned} \tag{B.8}$$

for $u \in C_0^\infty(\boldsymbol{R}^n)$, where $Op(c(x,\xi)) = c(x,D)$. On the other hand, we have, with some $C > 0$,

$$\text{Re } (Op(\ell(x,\xi)^2\langle\xi\rangle^{2\varepsilon_\mu - 2})u,u) \ge \|\ell(x,D)u\|^2_{\varepsilon_\mu - 1} - C\|u\|^2_{\varepsilon_\mu - 1}$$

for $u \in C_0^\infty(\boldsymbol{R}^n)$. This, together with (B.8), proves the theorem. □

As an example, we consider the following case:

$$\xi_1^2 + a(x,\xi') = \sum_{j=1}^{r} \left(\xi_j^2 + \lambda_j^2 x_j^{2k_j} \xi_n^2\right) + \alpha(x,\xi'),$$

$$\text{Re } b(x,\xi) = \sum_{j=1}^{r} b_j(x) x_j^{k_j - 1} \xi_n$$

in a conic neighborhood of $(0,\xi^0)$, where $1 \le r \le n-1$, $k_j \in \boldsymbol{N}$, $\lambda_j > 0$ and $\alpha(x,\xi') \ge 0$. Fix $\varepsilon > 0$ so that $\varepsilon < 1$. If $q_{2j-1}(x,\xi) = \sqrt{1-\varepsilon}\xi_j$ and $q_{2j}(x,\xi) = -(\sqrt{1-\varepsilon}k_j)^{-1} b_j(x) x_j^{k_j} \xi_n$ for $1 \le j \le r$ and $b_j(x)^2 < (1-\varepsilon)^2 k_j^2 \lambda_j^2$ in a neighborhood of 0 in $\boldsymbol{R}^n$ for $1 \le j \le r$, then there a conic neighborhood $\mathcal{U}$ of $(0,\xi^0)$ and $C > 0$ such that

$$\begin{aligned}
&(1-\varepsilon)p_2(x,\xi) + \text{Re } p_1^s(x,\xi) - \sum_{j=1}^{2r} q_j(x,\xi)^2 + \sum_{j=1}^{r} \{q_{2j-1}, q_{2j}\}(x,\xi) \\
&\Big(= \sum_{j=1}^{r} \Big\{ (1-\varepsilon)\left(\lambda_j^2 - (1-\varepsilon)^{-2} k_j^{-2} b_j(x)^2\right) x_j^{2k_j} \xi_n^2 \\
&\qquad\qquad - (\partial b_j / \partial x_j)(x) x_j^{k_j} \xi_n / k_j \Big\} \Big) \\
&\ge -C
\end{aligned}$$

for $(x,\xi) \in \mathcal{U}$. Indeed, we have

$$\left| x_j^{k_j} \xi_n \right| \le \nu \left| x_j^{2k_j} \xi_n^2 \right| + 1/(4\nu)$$

for $\nu > 0$. Thus $p(x,\xi)$ satisfies the condition (D) if $|b_j(0)| < k_j \lambda_j$ ($1 \le j \le r$). Put $\mu = \min_{1 \le j \le r} k_j$. Then we have $\xi_n \in \widetilde{\mathcal{L}}_{\mu+1}$. By Theorem B.2. 3 there are positive constants δ and C such that

$$\|u\|_{\varepsilon_{\mu+1}} \le C(\|p(x,D)u\| + \|u\|_{-1} + \|(1-\Psi_\delta(x,D))u\|_2)$$

for $u \in C_0^\infty(\boldsymbol{R}^n)$. We can easily show that $\varepsilon_{\mu+1}$ can be replaced by $1/(\mu+1)$ in the above estimate, which is the best one.

Bibliography

[ABG] M. F. Atiyah, R. Bott and L. Gårding, Lacunas for hyperbolic differential operators with constant coefficients I, Acta Math. **124** (1970), 109–189.

[Bl] R. Beals, A general calculus of pseudodifferential operators, Duke Math. J. **42** (1975), 1–42.

[BK] L. Boutet de Monvel and P. Kree, Pseudo-differential operators and Gevrey classes, Ann. Inst. Fourier Grenoble **27** (1967), 295–323.

[CH] S.-N. Chow and J. K. Hale, Methods of Bifurcation Theory, Springer-Verlag, New York-Berlin-Heidelberg, 1982.

[FP] C. Fefferman and D. H. Phong, On positivity of pseudodifferential operators, Proc. Nat. Acad. Sci. **75** (1978), 4673–4674.

[Fn] S. Funakoshi, Elementary construction of the sheaf of small 2-microfunctions and estimate of supports, J. Math. Sci. Univ. Tokyo **5** (1998), 221–240.

[GS] A. Grigis and J. Sjöstrand, Front d'onde analytique et sommes de carres de champs de vecteurs, Duke Math. J. **52** (1985), 35–51.

[Gk] A. Grothendieck, Topological vector spaces, Gordon and Breach, London, 1973.

[Gr1] V. V. Grušin, On a class of hypoelliptic operators, Math. USSR Sb. **12** (1970), 458–476.

[Gr2] V. V. Grušin, Hypoelliptic differential equations and pseudodifferential operators with operator-valued symbols, Mat. Sb.

88 (1972), 504–521. (translation: Math. USSR-Sb. **17** (1972), 497–514)

[Hr1] L. Hörmander, Linear Partial Differential Operators, Springer, Berlin-Göttingen-Heidelberg, 1963.

[Hr2] L. Hörmander, On the singularities of solutions of partial differential equations, International Conference of Functional Analysis and Related Topics, Tokyo, 1969.

[Hr3] L. Hörmander, The Cauchy problem for differential equations with double characteristics, J. Analyse Math. **32** (1977), 118–196.

[Hr4] L. Hörmander, Spectral analysis of singularities, Seminar on Sing. of Sol. of Diff. Eq., Princeton University Press, Princeton, 1979, 1–49.

[Hr5] L. Hörmander, The Analysis of Linear Partial Differential Operators I, Springer-Verlag, Berlin-Heidelberg-New York-Tokyo, 1983.

[Hr6] L. Hörmander, The Analysis of Linear Partial Differential Operators III, Springer-Verlag, Berlin-Heidelberg-New York-Tokyo, 1985.

[Hr7] L. Hörmander, The Analysis of Linear Partial Differential Operators IV, Springer-Verlag, Berlin-Heidelberg-New York-Tokyo, 1985.

[Hr8] L. Hörmander, An Introduction to Complex Analysis in Several Variables, North-Holland, Amsterdam, 1990.

[IP] V. Ya. Ivrii and V. M. Petkov, Necessary conditions for the Cauchy problem for non-strictly hyperbolic equations to be well-posed, Uspehi Mat. Nauk. **29** (1974), 3–70.

[Jh] F. John, On linear partial differential equations with analytic coefficients, unique continuation of data, Comm. Pure Appl. Math. **2** (1949), 209–254.

[KW1] K. Kajitani and S. Wakabayashi, Microhyperbolic operators in Gevrey classes, Publ. RIMS, Kyoto Univ. **25** (1989), 169–221.

[KW2] K. Kajitani and S. Wakabayashi, Propagation of singularities for several classes of pseudodifferential operators, Bull. Sc. math., 2e série, **115** (1991), 397–449.

[Kn] A. Kaneko, Introduction to Hyperfunctions, Kluwer Acad. Publ., Dordrecht, 1988.

[KK] M. Kashiwara and T. Kawai, Micro-hyperbolic pseudo-differential operators I, J. Math. Soc. Japan **27** (1975), 359–404.

[KKK] M. Kashiwara, T. Kawai and T. Kimura, Foundations of Algebraic Analysis, Princeton Univ. Press, Princeton, 1986.

[Km] H. Komatsu, Projective and injective limits of weakly compact sequences of locally convex spaces, J. Math. Soc. Japan **19** (1967), 366–383.

[Mt] G. Métivier, Analytic hypoellipticity for operators with multiple characteristics, Comm. in P.D.E. **6** (1981), 1–90.

[Mr] M. Morimoto, Introduction to Sato Hyperfunctions, Kyoritsu, 1976 (in Japanese).

[Ok] T. Okaji, Analytic hypoellipticity for operators with symplectic characteristics, J. Math. Kyoto Univ. **25** (1985), 489–514.

[Ol] F. W. J. Olver, Asymptotics and Special Functions, Academic Press, New York-London, 1974.

[PR] C. Parenti and L. Rodino, Parametrices for a class of pseudo differential operators I, Ann. Mat. Pura Appl. **125** (1980), 221–254.

[RS] L. Rothschild and E. M. Stein, Hypoelliptic differential operators and nilpotent groups, Acta Math. **137** (1976), 247–320.

[SKK] M. Sato, T. Kawai and M. Kashiwara, Hyperfunctions and pseudodifferential equations, Lecture Notes in Math. No. 287, Springer, 1973, 265–529.

[Sr] H. H. Schaefer, Topological vector spaces, Macmillan, New York, 1966.

[Sj] J. Sjöstrand, Singularités analytiques microlocales, Astérisque **95** (1982).

[Ta] K. Taniguchi, On the hypoellipticity and the global analytic-hypoellipticity of pseudo-differential operators, Osaka J. Math. **11** (1974), 221–238.

[Tf] D. Tartakoff, The local real analyticity of solutions of $\Box_b$ and the $\overline{\partial}$-Neumann problem, Acta Math. **145** (1980), 177–204.

[Tp] J. M. Trépreau, Sur la résolubilité microlocale des opérateurs de type principal, Saint Jean de Monts, 1982, Conf. No. 22, Soc. Math France, Paris, 1982.

[Tr1] F. Treves, Analytic hypoellipticity of a class of pseudodifferential operators with double characteristics, Comm. in P.D.E. **3** (1978), 475–642.

[Tr2] F. Treves, Introduction to Pseudodifferential and Fourier Integral Operators I, Plenum Press, New York-London, 1980.

[Wk1] S. Wakabayashi, Singularities of solutions of the Cauchy problem for summetric hyperbolic systems, Comm. in P.D.E. **9** (1984), 1147–1177.

[Wk2] S. Wakabayashi, Generalized Hamilton flows and singularities of solutions of hyperbolic Cauchy problem, Proc. Hyperbolic Equations and Related Topics, Taniguchi Symposium, Kinokuniya, Tokyo, 1984, 415–423.

[Wk3] S. Wakabayashi, Singularities of solutions of the Cauchy problem for hyperbolic systems in Gevrey classes, Japan. J. Math. **11** (1985), 157–201.

[Wk4] S. Wakabayashi, Generalized flows and their applications, Proc NATO ASI on Advances in Microlocal Analysis, Series C, D. Reidel, 1986, 363–384.

[Wk5] S. Wakabayashi, Remarks on hyperbolic polynomials, Tsukuba J. Math. **10** (1986), 17–28.

Index

4. Lecture Notes are printed by photo-offset from the master-copy delivered in camera-ready form by the authors. Springer-Verlag provides technical instructions for the preparation of manuscripts. Macro packages in TEX, LATEX2e, LATEX2.09 are available from Springer's web-pages at

http://www.springer.de/math/authors/b-tex.html.

Careful preparation of the manuscripts will help keep production time short and ensure satisfactory appearance of the finished book.

The actual production of a Lecture Notes volume takes approximately 12 weeks.

5. Authors receive a total of 50 free copies of their volume, but no royalties. They are entitled to a discount of 33.3 % on the price of Springer books purchase for their personal use, if ordering directly from Springer-Verlag.

Commitment to publish is made by letter of intent rather than by signing a formal contract. Springer-Verlag secures the copyright for each volume. Authors are free to reuse material contained in their LNM volumes in later publications: A brief written (or e-mail) request for formal permission is sufficient.

Addresses:

Professor F. Takens, Mathematisch Instituut,
Rijksuniversiteit Groningen, Postbus 800,
9700 AV Groningen, The Netherlands
E-mail: F.Takens@math.rug.nl

Professor B. Teissier
Université Paris 7
UFR de Mathématiques
Equipe Géométrie et Dynamique
Case 7012
2 place Jussieu
75251 Paris Cedex 05
E-mail: Teissier@ens.fr

Springer-Verlag, Mathematics Editorial, Tiergartenstr. 17,
D-69121 Heidelberg, Germany,
Tel.: *49 (6221) 487-701
Fax: *49 (6221) 487-355
E-mail: lnm@Springer.de